W0227395

CAMBRIDGE LIBRARY COLLECTION

Books of enduring scholarly value

Art and Architecture

From the middle of the eighteenth century, with the growth of travel at home and abroad and the increase in leisure for the wealthier classes, the arts became the subject of more widespread appreciation and discussion. The rapid expansion of book and periodical publishing in this area both reflected and encouraged interest in art and art history among the wider reading public. This series throws light on the development of visual culture and aesthetics. It covers topics from the Grand Tour to the great exhibitions of the nineteenth century, and includes art criticism and biography.

Treatise on Architecture

The architect Arthur Ashpitel (1807–69) worked on a wide variety of projects, including churches, houses and schools, and wrote widely on architecture, literature and politics. He became a fellow of the Institute of British Architects in 1841 and the Society of Antiquaries of London in 1847. This 1867 work comprises his compilation and revision of notable tracts on architecture from the *Encyclopædia Britannica*. It includes entries by William Hosking (1800–61) on classical, Gothic and modern architecture, building and construction; Thomas Tredgold (1788–1829) on joinery and stonemasonry; Thomas Young (1773–1829) on carpentry; and John Robison (1739–1805) on roofs, arches and the strength of materials. Intended to be of practical use to architects, craftsmen and 'the building trade', Ashpitel's synthesis remains a valuable resource for scholars interested in nineteenth-century thought on architectural history, practice and technology.

Cambridge University Press has long been a pioneer in the reissuing of out-of-print titles from its own backlist, producing digital reprints of books that are still sought after by scholars and students but could not be reprinted economically using traditional technology. The Cambridge Library Collection extends this activity to a wider range of books which are still of importance to researchers and professionals, either for the source material they contain, or as landmarks in the history of their academic discipline.

Drawing from the world-renowned collections in the Cambridge University Library and other partner libraries, and guided by the advice of experts in each subject area, Cambridge University Press is using state-of-the-art scanning machines in its own Printing House to capture the content of each book selected for inclusion. The files are processed to give a consistently clear, crisp image, and the books finished to the high quality standard for which the Press is recognised around the world. The latest print-on-demand technology ensures that the books will remain available indefinitely, and that orders for single or multiple copies can quickly be supplied.

The Cambridge Library Collection brings back to life books of enduring scholarly value (including out-of-copyright works originally issued by other publishers) across a wide range of disciplines in the humanities and social sciences and in science and technology.

Treatise on Architecture

Including the Arts of Construction,
Building, Stone-Masonry, Arch, Carpentry,
Roof, Joinery, and Strength of Materials

EDITED BY ARTHUR ASHPITEL

CAMBRIDGE
UNIVERSITY PRESS

CAMBRIDGE UNIVERSITY PRESS

Cambridge, New York, Melbourne, Madrid, Cape Town,
Singapore, São Paolo, Delhi, Mexico City

Published in the United States of America by Cambridge University Press, New York

www.cambridge.org
Information on this title: www.cambridge.org/9781108062091

© in this compilation Cambridge University Press 2013

This edition first published 1867
This digitally printed version 2013

ISBN 978-1-108-06209-1 Paperback

This book reproduces the text of the original edition. The content and language reflect
the beliefs, practices and terminology of their time, and have not been updated.

Cambridge University Press wishes to make clear that the book, unless originally published
by Cambridge, is not being republished by, in association or collaboration with, or
with the endorsement or approval of, the original publisher or its successors in title.

TREATISE

ON

ARCHITECTURE.

TREATISE

ON

ARCHITECTURE

INCLUDING

THE ARTS OF CONSTRUCTION,
BUILDING, STONE-MASONRY, ARCH, CARPENTRY, ROOF, JOINERY,
AND STRENGTH OF MATERIALS.

EDITED BY

ARTHUR ASHPITEL, Esq.

F.S.A., F.R.I.B.A., &c.

———

EDINBURGH:
ADAM AND CHARLES BLACK.
MDCCCLXVII.

PREFACE.

THE proprietors of the Encyclopædia Britannica having been advised that the treatises on ARCHITECTURE and the arts connected with it in that work, if published in a separate volume, would be useful to those engaged in the building trade, have selected the following for that purpose :—

ARCHITECTURE, BUILDING, AND CONSTRUCTION, written by William Hosking, F.S.A., a distinguished architect and engineer. When young, he was apprenticed to a surveyor and general builder in Sydney, New South Wales. There the business of the surveyor was of the most general nature, and gave Mr Hosking an opportunity of acquiring a practical knowledge of the rough as well as of the finer operations of the constructor. His preliminary training was thus of the kind to which Telford in his autobiography refers, when he says the young engineer must descend if he would excel.

On his return home, having qualified himself by previous studies in the higher branches of his profession, he spent a year in Italy and Sicily previous to establishing himself in London as an architect. A course of lectures, which he delivered at the Western Literary and Scientific Institution in 1829, and several works on Architecture, which he published about the same time, brought him into very favourable notice.

In 1834, he became Engineer of the West London Railway, for which he designed and executed the curious work near Kensal Green, by which the Paddington Canal is passed over the railway, and a public carriage-way over both the canal and railway. In 1842, he was appointed Professor of Construction in connection with Civil Engineering and Architecture in King's College, London, to which was afterwards added the Professorship of the Principles and Practice of Architecture. He was appointed one of the Official Referees under the Building Act, and held other honorary and official offices, the duties of which he discharged with great benefit to the public and credit to himself.

In 1846, he published his Treatises on Architecture and Building along with those of Masonry, Joinery, and Carpentry, in one volume, which met with such a favourable reception, that it was resolved to revise and republish them with additional articles; and he was engaged in the preparation of these when he was seized with an illness which ended fatally in August 1861.

JOINERY AND STONE-MASONRY were contributed by Thomas Tredgold, one of the ablest engineers of his day; and it is remarkable that his original training greatly resembled that of Mr Hosking. Apprenticed to a carpenter in his native village of Brandon, near Durham, he served for six years, and for five years afterwards he worked as a journeyman in Scotland. He then proceeded to London, where he obtained employment for ten years in the office of Mr Atkinson, an architect. His education in youth was only of that scanty description intended to fit him for a mechanical trade; but instead of wasting his fragments of spare time in listlessness or dissipation, he devoted himself to improvement in his profession. Under circumstances as little favourable for such pursuits as can well be imagined, he applied himself to the study of chemistry, geology, and mathematics, and with such success, that his published works have been received with great approbation by men of science both in his own country and in France, where translations of many of them have been published.

It was not till he was thirty-five years of age that he began to practise as a civil engineer; and it was during the period when he was so engaged that he published the scientific treatises which have entitled him to an honourable place among those who have made important and successful exertions for the advancement of the useful arts.

CARPENTRY was written by Thomas Young, one of the most distinguished men of the present century. He was educated for the medical profession, and was one of the most learned of the physicians of his time; but his attainments and disquisitions on science and literature, on antiquities and the most abstruse inquiries, range over so extensive a field that we can only glance at them, and wonder that a philosopher, distinguished by his works on the medical sciences,—celestial mechanics,—motion of waves and sounds,—on tides,—on Egyptian language and hieroglyphics, and on the most abstruse subjects, should have occupied himself in writing a treatise on Carpentry; but everything that his attention was directed to he followed earnestly and scientifically; and such was the industry and patience with which he pursued every inquiry in which he was engaged, that he could not desist till he had mastered it.

At the time of his death he was preparing an Egyptian Dictionary; and to a friend who expostulated with him on the danger of fatiguing himself while under the pressure of severe illness, he replied it would be a satisfaction to him to have finished the work ; but if it were otherwise, and which seemed most probable, it would still be a great satisfaction to him never to have spent one idle day in his life.

THE TREATISES ON ROOF, ARCH, AND STRENGTH OF MATERIALS were contributed by John Robison, Professor of Natural Philosophy in the University of Edinburgh. It was comparatively late in life that he assumed the character of an author. His first papers, which were on astronomical subjects, were communicated to the Royal Society of Edinburgh ; but his chief writings were contributed to the Encyclopædia Britannica. Of these there were above forty valuable articles on philosophy, science, and mechanical arts, which exhibit a more complete view of the modern improvements in physical science than had been ever given before. Important additions have been made to those included in the present volume by Robert Stephenson, the Editor, and others.

The Editor of the present publication, besides revising the different Treatises and introducing improvements and additions where necessary, has availed himself of the late discoveries in the East to supplement the articles on Egyptian, Jewish, and Assyrian Architecture, and to add a chapter on Indian and Chinese Architecture. He has also written an entirely new Glossary of the terms in Mediæval Architecture, with explanations and sixteen new plates containing nearly 300 subjects, many of which have never been published before, illustrating the Arabic, the Romanesque, and the Pointed styles. He has also given illustrations of Modern French Architecture, and added a chapter on Acoustics, with the view of securing an object of essential importance to Churches, Halls, and Lecture Rooms, where, unless the speaker be distinctly heard, every other accommodation will be of little avail. He has also supplemented the articles, Joinery, Roof, Stone-Masonry, &c., bringing the information down to the present time.

It is hoped that with this volume, embracing the Principles of Architecture and Treatises on the different sections of art connected with Building, written by men of great scientific and practical knowledge, and profusely illustrated with plates and illustrations, those who are engaged in any department of architectural work, from the cottage to the temple, and from the foundation to the roof, will be furnished with such information as will enable them to finish their undertaking with safety and credit.

ARTHUR ASHPITEL.

March 1867.

TREATISE

ON

ARCHITECTURE.

NOTE.—*References are frequently made throughout the book to subjects such as Timber, Harbour, Bridge, &c.; these references are to Articles in the Encyclopædia Britannica, Eighth Edition.*

History.

Architecture, what it is.

ARCHITECTURE (Greek, ὀικοδομικὴ ἀρχιτεκτονία; Lat., *architectura;* Ital., *architettura*), the art of building with convenience, strength, and economy, and also with beauty. It is this last quality which forms the important difference between the art of architecture, the science of construction, and the business of the builder. The term is derived from the name given to its professors by the Greeks, ἀρχιτέκτων, head or chief of the builders. The word τέκτων is probably derived from τέχνη, and signified a builder as far back as the time of Homer, who (*Iliad,* Z, 315) says they are "men who make the bedchamber, the house, and the hall." In *Ib.* O, 411, and *Odyssey,* I, 125, the term is applied to a shipbuilder. Herodotus appears to be the first who uses the phrase ἀρχιτέκτων (Thalia 60), where it is given to the engineer of the great tunnel at Samos; in the other instance (Melpomene, 87) to the architect of the bridge across the Bosphorus. The Romans sometimes used the word *architecton* (Plautus, *Pœnulus* 5, 2, 125), but more generally *architectus*—the word being supposed to have an analogy to *tectum,* the roof of the house.

Architecture, branches of.

Civil architecture

Architecture is generally divided into three distinct branches—military, naval, and civil. It is this last branch alone of which we propose to treat in this place.

Civil architecture teaches to design all buildings, ecclesiastical, palatial, monumental, public, or domestic. The first considerations, of course, are utility, convenience, soundness, and economy; and no beauty of form or decoration can excuse the want of these primary qualifications. But as the minds of almost all men have delight in stately or graceful forms and elegant combinations, the great element of beauty, or that which is pleasing to the eye, is the next consideration.

History.

Architect, qualifications of.

The investigation of the qualifications of a good architect will best exhibit the requirements of good architecture. These have been the same ever since the days of Vitruvius, who tells us that the architect should be a man both of theory and practice; that without study and literary acquirements he will be at fault in all matters which require the weight of authority; while, should he rely on this last without sound practical knowledge, he "seeks a shadow and not a substance."[1] He says the architect should be of a literary turn of mind, a skilled draftsman, thoroughly learned in geometry, and not ignorant of optics (perspective), a good arithmetician and accountant; that he should know much of history, and should have attended lectures on natural philosophy; that he should understand as much of music as teaches acoustics, be not ignorant of medicine (chemistry and sanitary matters), and should know as much of law and of astronomy as applies to his own profession. "I do not mean," Vitruvius says further on, "that an architect should be a philosopher of the highest rank, nor a most eloquent orator, nor a man excelling in the highest branches of literature;" but he goes on to insist that the true architect, besides the absolutely necessary artistic and mathematical acquirements, should be imbued with learning (*literis imbutus*). Just as it is in the present day,

---

[1] "Qui autem ratiocinationibus et literis solis confisi fuerunt umbram, non rem persecuti videntur."

A

History.

the true architect must possess a combination of the qualities of the scholar, the artist, the mathematician, and thorough man of business—qualities the most opposite in their character, not often seen singly, and most rarely to be all found combined in the same individual. However, such is the state of the case, and such is the true reason why in all ages there have been so few architects of the very first rank.

Vitruvius' divisions of.

It is to be regretted that the only author of classic antiquity on the subject of architecture whose works have come down to us is Marcus Vitruvius Pollio. His ten books on the subject are of the highest curiosity and value. Unfortunately, as he himself hints, he was more of the architect than the author. His style is crude and somewhat confused, his technology very difficult to understand, and his definitions sometimes perplex instead of elucidating the subject. Nevertheless, a careful and painstaking perusal of his works show a strong common sense, a love of beauty, a great propriety of taste, and principles which are equally true in every style of architecture. The number[1] and beauty of the MSS. of this author which are extant show the popularity he must have had in the Middle Ages. Indeed his enumeration of the requisites of a good building, *Firmitas, Utilitas, Venustas*—Stability, Utility, Beauty—is as much that of the mediæval as of the classic architect.

Construction.

Of the first of these qualities, firmness and stability, we shall here treat only as regards the bearing such qualities may have in design and effect. The details of stability will be found in the various heads—BUILDING, CONSTRUCTION, CARPENTRY, JOINERY, ROOF, STONE-MASONRY, STRENGTH OF MATERIALS, &c. &c. The others will be treated of as they arise.

It is now proposed to follow our author in his definition of the different branches of his art, and to comment on the same as we go on; first, because he treats so copiously on the matter, and next on account of the interest arising as to his works since the commencement of the Dictionary of the Architectural Publication Society. It has been his misfortune to be over praised sometime ago, and unreasonably neglected lately, and in both cases to have been misunderstood. The dictionary alluded to, and the noble edition of Stratico (4to, Utini, 1826), have, however, thrown much light on the work, and shown how much more valuable it is than has been supposed.

Parts of the Art. Ordination.

Vitruvius begins (lib. i. cap. 2) by stating that architecture consists of *ordination*, which the Greeks call τάξις; of *disposition*, which they call διάθεσις; of *eurythmy*; of *symmetry*; of *propriety* (decor.); and of *distribution*, which the Greeks call οἰκονομία, or stewardship. Commentators have been much puzzled to give a correct explanation of *ordinatio*, but as we are not now editing this author, we shall not enter on the question. The general scope, however, may be shortly stated thus—the first thing to be done is a general consideration of the convenience of the design (*commoditas*), a comparison of the parts, an idea of the size (ποσότης) and dimension, and of the style and general effect of the whole.[2]

Vitruvius, then, defines *disposition* as the getting the work into form, or, as we should say, "upon paper." He subdivides this head into three parts—*Ichnography*, or plan; *Orthography*, or geometrical elevation; *Scenography*, or perspective views, showing the flanks as well as the front. These are to be carried out, he says, by two faculties of the mind,—thought and invention,—by which beauty is arrived at and difficulties overcome.

History.

He then tells us to consider the *eurythmy*, or the general large proportions of height to width and width to length of the members of the composition. In other words, the architect should next study the *masses* of his design. Next comes *symmetry*, or the proportions of the details; as the human body has certain symmetrical proportions, the forearm, the foot, hand, finger, &c., so should the columns, antæ, and other parts of a building also have definite proportions each to the other.

Propriety.

But to effect this well, our author says we must study *propriety* (decor.), and this he divides into three parts— θεματισμός, or site; *custom* (conventionality); and *nature*. Thus, he says, temples to Jove, the Sun, Moon, &c., should be hypæthral, open to the air (*sub dio*); to Minerva, Mars, Hercules, &c., they should be Doric; to Venus, Flora, Proserpine, &c., they should be Corinthian, as more graceful and florid; while to Juno, Diana, Bacchus, &c., they should be Ionic, as a style between the severe Doric and the slender Corinthian. He instances, as a want of propriety, elegant interiors with low and shabby entrances; and, as a solecism against established rules, dentils in Doric and triglyphs in Ionic entablatures. He then treats of propriety, as regards site and aspect, which he calls natural propriety, first as regards the selection of a healthy spot, and one having good air and water; and then tells us, among many other matters, that the cubicula and libraries should face the east; baths and winter rooms the west; and that picture galleries should have a north light, &c. &c.

Economy.

We then come to the division, *Distributio* or economy; and this consists of two branches,—the making the most of the site, and such a saving of funds as is consistent with moderation.[3] "If you cannot get pozzolano except at great price," he says, "you may use river-sand or well-washed seasand. If fir or pine is scarce, use cypress, poplar, &c.; and, last of all, build according to the wealth, dignity, and power of your employer." Then, with a further injunction to regard stability, utility, and beauty, Vitruvius goes on with the details of his art.

True principles.

What principles of architecture can be more sound, real, and true, we are at a loss to know. The architect is first to find out what the requirements of his employer may be, then to get them roughly together; then to set out his plan, study his elevation, and set up a rough perspective. From this he is to proportion his masses, and then study his detail and ornament; all the while to be governed by a sense of general propriety, a deference to religious opinion and the conventionalities of society, and to spare his employer's purse as much as he can, consistently with making the building worthy of his position in society. This seems exactly the mode pursued by Sir Christopher Wren, if we may judge from his drawings preserved at All Soul's College and elsewhere, and is equally applicable to every style of architecture. It is true, in the present day, there are some who try to produce effects by lavish expenditure, especially in ecclesiastical buildings; but our common sense must

---

The Utini editor cites 20 MSS., without including those in the British Museum and in the Library of the Royal Institute of British Architects.

[2] " Universi operis conveniens effectus."

[3] " Locique commoda dispensatio, parcaque in operibus sumptus cum ratione temperatio."

History.

Decorate construction.

tell us that one of the chief virtues of an architect is the economy so strictly enjoined by Vitruvius.

We find, also, that our author held the principle of designing the masses or general construction, and then adding such ornament as might be fitting to the object, use, and expense of the different buildings. It has been lately supposed that this true principle, " decorate your construction," is peculiar to mediæval architecture alone. But this is not so. The ancients having settled their general proportions, proceeded to add such decoration as propriety seemed to dictate. Mouldings were increased in number, and covered with the echinus or other ornament; friezes and architraves were sculptured with rich foliage and figures; capitals and bases enriched; columns fluted instead of being left plain; soffits sunk and coffered; in short, the decoration of the classic architect had as little to do with the broad masses of the construction as that of the mediæval architect. The Doric of the theatre of Marcellus has very much the same proportionate parts as those of the same order in the baths of Diocletian; the Corinthian of the Pantheon is very similar in mass to what is generally called the Jupiter Stator or those in the Forum of Nerva. The same blocks of marble would have constructed the same orders (of course if on the same scale); but the difference is in the enrichments—the labour of carving in the one is threefold that in the other. The construction is the same, but the decoration has been carried much further. The former are plain and simple; the latter, being intended for more luxurious purposes, are naturally much richer.

Utility without beauty.

Let us now consider a mere utilitarian building, where there is no pretension to design. Take the majority of the houses about Bloomsbury and Marylebone. They are roomy, commodious structures, fitted with every convenience for families of the highest respectability. Yet, externally, they have been very truly described as " brick walls with rectangular holes in them." If, instead of a plain, flat coping at the top, we add a handsome cornice; if we put a well-proportioned dressing round each window, with consoles to carry the cills, and rich string courses, or balconies with ornamented balustrades; if we build a handsome portico to the door: if we invest the house with all these architectural features, the tenants will enjoy the interior just as before; they will not eat more, nor sleep sounder; but they will feel a greater pleasure and pride in their dwelling every time they or their friends go in or out; and the house itself will fetch a higher rent. All this results from the pleasure which elevated minds take in works of art, and the appreciation they have of thought, care, and mental ingenuity.

Beauty without utility.

But to affect beauty without utility is the gravest of all mistakes. Houses are made to reside in, for our comfort, our solace, our repose. If the internal arrangements are not suited to all these things, the house, which at first was our pride, becomes at last intolerable, and we are compelled to seek in some plain unpretending abode the qualifications that tasteful decoration will not afford us. When the Earl of Burlington designed the house for General Wade, the front and all the internal decoration was of great beauty; but the rooms were all passage rooms, the staircase being in one corner, and there was not one apartment in the house fitted for its particular domestic use. The wits advised him to take a house opposite, that he might look across and admire the beauty of his own, while he enjoyed the comfort of the latter; and there was much truth and common sense in their advice.

Beauty of design.

Let us now inquire into the principles of designing with beauty. By this is meant designing so as to please the eye and taste; for buildings may be of rude material and form, and yet may please by their size and propriety. Thus, the

jail at Newgate is generally admired, particularly the corner facing Giltspur Street. It is simple, massive, imposing, and tells its own tale. It is a great prison, a building intended by its rugged appearance to deter from crime. So that in speaking of beauty we include those buildings which please from their grandeur, just as in poetry the sublime delights us as well as the beautiful. The chief elements which lead to these feelings of pleasure may be ranked as follows.

Size.

The size of a building has much to do with the pleasure we derive from its contemplation. Pyramids have all the same general form, but those of Egypt astonish and delight us by their vastness. Obelisks have much the same proportion, but we are more pleased with the vast monoliths at Rome than with smaller erections of exactly the same proportion. The great temples in Sicily and the Parthenon please from their size as well as their design; and the temple at Tivoli pleases more than the choragic monument of Lysicrates. The trilithons at Stonhenge, and the cromlech at Aylesford please from their vastness, while the very same forms on a small scale would be unnoticed. The plain, simple early cathedrals in Normandy please more than the smaller and richer churches of a later period. In fact, size, no doubt, is a great element in exciting our admiration. Grandeur can hardly be said to exist without it.

In the same way, the size of the material used has much to do with the effect in some styles of architecture. Large buildings look all the better if the stones are in large blocks, particularly in the level architraves of the Egyptians and Romans. Thus our Lord's disciples (St Mark xiii. 1) admired the vastness of the stones in the Temple at Jerusalem; and the historian of Peterboro', Hugo Candidus, records with much satisfaction the "lapides immanissimos" with which the old cathedral was built. Marble looks better than stone, and stone than brick, in all classic work. But the Gothic architects, whose methods of transport were very imperfect, and who were compelled to cut large stones into small pieces in the quarry for the convenience of carriage, boldly proceeded in a contrary path, and gave artificial grandeur to their buildings by the small size of their component parts. Much of this source of pleasure depends on our next element, costliness.

Costliness.

This also, though not invariably so, is another element which gives us pleasure. Apart from the beauty of colour in marble, comes the consideration of the distance from which it has been brought, and its corresponding price. In contemplating polished granite, we consider the vast labour and expense of reducing so hard a material. We respect the motives which led the founder to what must have been a sacrifice on his part to exalt his object; and we feel also that he meant to please us, the spectators, and are gratified to have beheld such rarities and riches of art as are not to be seen every day, and we feel a sort of reflected pride in the remembrance.

But costliness must be tastefully expended, or it becomes unpleasant. A plain garment which fits well is much more becoming than ill shaped cloth of gold. When we behold lavish expenditure badly carried out, we say, " What a pity it is this rich man had not gone a step further, and paid somebody to find him a little taste." Besides this, it often then assumes the character of ostentation, and nothing is more offensive than this in architecture, as in other matters.

Mass.

Leaving, now, the consideration of size and cost, the first thing no doubt to study is the general disposition of the masses of the composition. Perhaps the best way would be to look at the building at such a distance that the mind may not be disturbed by the contemplation of the details, or to study it by moonlight. Thus, the mass of a Gothic cathedral, the proportion of its parts, the outline of tower,

History. nave, choir, and lady chapel, the deep shadows which show the projection or recess of its various parts—all form one great element of beauty, when there is not light enough to distinguish mouldings, or carvings, or tracery.

Stability. In an analogous way, we are pleased by the appearance of stability in a building. A feeling of security always gives satisfaction ; and to an educated mind the proper disposition of parts, as openings over openings, sufficient support to superincumbent weights, and adequate abutment to arches, all contribute to please. While large weights on slight shafts, huge ponderous pendants in groined roofs, artifices to conceal the means of support, all tend to excite feelings of distrust and dissatisfaction.

On the other hand, the eye is oppressed by heaviness in a design. Any undue waste of material, any unnecessary weights or thicknesses, displease in buildings where such massiveness or heaviness is not an inherent feature. The massive forms and rude rustics of Newgate would be quite out of character in a palace, theatre, or even in private houses.

Another error into which an architect may fall is a want of repose in his composition. Few things are more displeasing in a building than an appearance of flutter, confusion of parts, projections without purpose, ornaments obtruding themselves on our notice, a want of rest and balance,—all this should be avoided if we would please the spectator.

Harmony. There should also be a harmony throughout the design. Parts should balance each other. Those supported should have an adequate ratio to their supports, and so should it be with the solids and voids. If there be columns in one story those in the next may be lighter, but should still retain a definite proportion. If too large they dwarf the lower story, if too small they become poor and petty. So in gothic buildings the clere-stories should neither be too large nor too small for the arcades ; nor the aisles too wide or too narrow in relation to the nave.

Proportion of. This brings us to the most material of all considerations, the proportion of masses. Every one, even the most uneducated, will be struck on entering a building if it be too low, or unreasonably high. A gallery for pictures, or a corridor, may be as long as one pleases without offence to the eye, because length is necessary to its use ; but an apartment for any useful purpose should not be too long for its width, nor too high or low for either. There is no doubt that a system of proportion of dimensions, based on mathematical ratios, gives pleasure. An exact cube ; a double cube, or two cubes placed side by side ; the ratios of aliquot parts each as 1 to 1½,—as 20 feet high, 30 feet wide, and 45 feet long ; or those of the base perpendicular and hypothenuse of a right-angled triangle, 3, 4, and 5, or their multiples, please the eye more than dimensions taken at random. Besides this, they are clearly better for hearing ; but this is treated of under the head ACOUSTICS.

Symmetry. The only remaining branch of the treatment of masses is that of symmetry or uniformity. The unreasonable straining after uniformity by architects of the past age, the sham doors and windows, and the unnatural contrivances to make everything balance, has driven people into a corresponding error in the present day. Many architects now go out of their way to make things irregular and unsymmetrical. Now this is a grave mistake, and, like the former error, arises entirely from the bigotted attachment to the study of one style to the exclusion of that of every other. This betrays men to the belief, that the true principles of architecture are to be found in one style only, and that the only way to arrive at truth is to be ignorant of half your profession. Some years back the architects thought the essence of classicality to be tameness, coldness, and sameness. To avoid these errors the mediæ-

valists ran into the opposite extreme. It is because both have studied under limited views. The truth is, the **History.** ancients placed their buildings irregularly where it was reasonable to do so. The Propylæa at Athens, with the Temple of Victory, was of irregular composition. The Erectheium comprised three several buildings, each differing most materially in design, and each disposed as was most fitting, without any regard to the uniformity of the entire mass. The old Forum at Rome was a collection of buildings, temples, basilicas, atria, triumphal arches, each differing in style, character, and taste. That at Pompeii was the same. Just so it was with the mediæval buildings. The abbey had its church, chapter-house, cloister, dormitory, refectory, guestern-house, abbot's and prior's chambers, all distinct buildings, and all grouped together. But it is utterly forgotten by the advocates for irregularity that each building was symmetrical in itself ; that is to say, if a line were drawn down the middle of each one side would correspond with the other, just as in the case with a leaf, an animal, or with the human figure and face. While such instances of symmetry abound in nature ; while irregularities, a nose askew, a mouth on one side, one eye, or one shoulder higher than another, are accounted deformities, in spite of the dictum of any sect, one great element of beauty will be found in symmetry. This, however, must not be set above utility or convenience. The best possible dictum on the subject is that of the great Lord Bacon, who says: " Houses are built to live in, and not to look on ; therefore let use be preferred before uniformity, except where both may be had."

Another source of pleasure to the eye is the judicious in- **Colour.** troduction of colour. In all ages this has been freely used in interiors. In Jeremiah xxii. 14, we read of buildings " ceiled with cedar and painted with vermilion ;" and the temples of the Egyptians are richly decorated with colour. As long as Nature decks her landscapes with varied tints, as long as flowers and birds are covered with bright hues, so long will colour please in architecture.

It is clearly so in interiors, but it has been doubted whether external colouring is desirable. Some contend for local colour only, that is, the colour of the material. Thus, Mr Ferguson advocates the use of red brick against a green wood. Mr Repton, on the contrary, says, " A red brick house puts a landscape in a fever ;" and demands that such houses should always be white, and gives diagrams, with overlays, to show the advantage of the latter. Others contend that the colour of the material itself should be varied in every possible way. Thus, some short time ago every building had horizontal bands, or stripes of colour on it, without any reason suggested by the construction. This fashion was then carried further, and brickwork was spotted over red, yellow, and black, in patterns so unartistic as to appear childish and *petite*. The former received the name of the " streaky bacon," or " holy zebra " style ; the latter has not unaptly been called the " Tunbridge-ware style." The truth is, that form should always be the first consideration. Colour is but the handmaid of form. Like the graphic art, correct drawing will please in monochrome, while no colour can compensate for bad drawing. Still this is no reason why both may not be used, if employed with discretion.

Next to general beauty or grandeur of form, the eye de- **Ornament.** rives most pleasure from ornament in architectural work. Except in such cases where vases, or other similar ornaments or statues are placed on bases or pedestals, and used partly to make a sort of finish to the work, partly to break the sameness of long lines, and partly to show the elegance of the objects themselves, as we place works of art on our

History. chimney shelves, with these exceptions all ornament should be the decoration of constructive parts of the fabric. Ornament put up without meaning always provokes the remark, " What is this stuck up for ?" Or like the dialogue Canova supposed to have been held between himself and the screen of columns at old Carlton House,—

"Care colonne che fate quà ?
Non sapiamo in verita !"

Ornament. Ornament, no doubt, was originally derived from improvements upon, or finishings to the various parts of construction. The capital of a column has justly been supposed to be derived from the tile placed at the top of the trunk of the tree, or the upright prop which carried the architrave to keep the wet from running down the grain; while the necking is supposed to represent a ring put round to prevent the tree from splitting. In the like manner the triglyphs are supposed to have been channels cut in the ends of the beams carrying the ceilings, and which rest on cross beams which form the architraves. From a passage in the *Odyssey*, A. 128, the flutes in the columns are supposed to have been intended as places in which to rest the points of the spears; for Telemachus, on entering the house, goes up to a column and places his against it. Unless there were some channel to catch the point the weapon must have fallen down. The skulls of oxen, carved in the metopes of classic temples, were no doubt originally the actual skulls of the beasts which had been sacrificed, nailed up as memorials, and afterwards perpetuated in stone. Just so the festoons of wreaths of real leaves and flowers, which to this day are hung up before the fronts of churches abroad on great days, became fitting objects for the carver's skill in later times.

Ornament may be divided into two classes—mouldings and the sculptured representation of natural or fancied Mouldings. objects. Mouldings, no doubt, were derived, first, from the simply taking off the edge of anything that might be in the way, as the edge of a square post, then sinking the chamfer in hollows or various forms, and thence was derived the systems of mouldings we now find in all styles and periods. Each has its own system; and so well are these known, and so clearly is the difference understood, that a skilful architect will tell, not only the period in which any building has been erected, but will even give an idea of its probable size, as the professors of physiology will construct the animal from the examination of a single bone. In fact, mouldings are the comparative anatomy of architectural styles. Of course, like everything else in architecture, their use may be over-done; and, on the other hand, their absence or paucity betokens a poverty which is very unpleasing. They should, however, always be carefully studied. Nothing offends an educated eye like a confusion of mouldings. Roman circular forms in Greek work, or early English in that of the Tudor period, all are disagreeable, not only to the professor, but also to the ordinary spectator. He cannot tell you exactly why, but he feels there is something wrong, something incongruous, and is disappointed accordingly.

Sculptured The same remarks also apply to sculptured ornaments.
ornament. They should not be too numerous nor too few, and, above all, they should be consistent. The carved ox skulls, which are appropriate in the temple of Vesta at Tivoli, or that of Fortuna Virilis at Rome, would be very incongruous on a Christian church; while saints and angels would appear out of place on an arsenal gateway. No rules can be laid down further than general hints what to avoid; the rest must be left to the common sense and good taste of the student. It may be well, however, to say, that ornament should always be architectural in character. That style of carving which indulges in prominent projections, extravagant scrolls, and grotesque work, is very properly called " plate-

resque," or silversmith's work, by the Spaniards, because it History. resembles the magnified designs for jugs, tankards, &c. We must also remember, that when a building is covered with ornament of this kind, it loses all its architectural effect; the architect, in truth, does but make, as it were, a frame for the artist to exhibit his work in.

A third sort of ornament is a mixture of the moulding Enriched and the carved work, and is commonly called enriched mouldings. moulding. Of these, the most usual are the egg and tongue (plate 8, *ovolo*), leaf and tongue (ib. *cyma reversa*), and the reel and bead (ib. *bead*). These are to a great degree conventional. The enrichments in the Gothic mouldings (Plates XXXIII. *et seq.*) are partly imitative of natural objects, as cords, &c., and partly heraldric. Mediæval mouldings are very varied in character, and show great fancy and love of beauty. Having traced the main divisions of the art, and the qualifications necessary to the architect, we proceed now to treat of its origin.

Origin of Architecture.

The necessity for obtaining frequent shelter from the Shelter. great heat, or from the inclemency of the climate, no doubt first suggested the piling up materials in some form to effect this purpose. Shelter was perhaps readily found in some wood, and in rocky countries in some cavern; but as it was necessary, particularly for pastoral tribes, to inhabit plains where there were neither groves nor caves, that which at first was a protection afforded by nature was imitated by a sort of rude art. Branches of trees were no doubt carried into the open country, and there piled up, so that the shepherd might creep under and find shelter from the sun's heat or the chilling storm. On the wild moors, where there are no trees, and where the ground is covered with scattered fragments of rock, the remembrance of the natural caverns no doubt suggested the piling up stones in such form as to be a protection against the elements, just as shepherds do in the present day; and thus, as a distinguished writer has said, " the wigwam became a hut, and the hut a house." Where trees abounded, stone probably was the last material used, as it would entail so much more labour than timber; but of course it was soon found stone had two great advantages—it would neither burn nor rot; so that it soon had the preference for all durable purposes. Where there were many trees, as in Greece and in Lycia, the stone architecture exhibits traces of the original timber construction. As has already been said, the columns were originally posts, and the architraves and triglyphs beams resting on each other. The famous Lycian tomb in the British Museum is also a strong proof that the art of the carpenter there preceded that of the mason, and suggested forms which became conventional, and from which the latter could not venture to depart. On the contrary, in the plains of Egypt, where building timber is scarce, and where there is abundance of large stone in the mountains, the *mason* element seems to prevail. In such plains as those on which Nineveh and Babylon stood, a factitious stone was made, first by lumps of dried, and then, advancing a step, of burnt clay. In the vast sandy deserts, where there are neither trees nor stones, the skins of beasts, sewed together and supported by sticks, was the earliest shelter. This soon grew into the tent, and its form still influences the architecture of the Chinese and Tartars. There has been much time expended on endeavours to prove which of the two materials, timber or stone, first gave birth to the art of architecture; the truth probably is, that the hut, the cairn, and the tent, all contributed their share in their respective countries.

Monumental Architecture must have originated in a de- Monumen-
sire to commemorate important events, such as the death of tal archi-
great men; hence we may suppose that the first considera- tecture

History. tion would be to make such memorial as durable as possible, and this circumstance would lead to the use of stone instead of wood. The piling a few stones on each other to form altars can scarcely be called anything more than preparing a place for fire. Probably the first act which might be called the erection of anything designed to be a lasting memorial, would be the setting up a large stone or pillar as a memorial of any event. In the earliest records of the Scriptures this is frequent. Jacob sets up a stone as a memorial of his agreement with Laban; Joshua, after the covenant, by Shechem; and Samuel, after the battle with the Philistines, at Mizpeh. And though it has lately been pretty clearly proved, that what have been commonly called cromlechs, that is, three or four stones placed on each other like a small chamber or hut, are really sepulchres which have been covered with earth, and are not temples; yet it is clear that such constructions of stone as the circles at Avebury, those in Brittany, and particularly the great monument at Stonehenge, have been used by a rude people for the purposes of assemblage either for civil or religious ceremonial. The existence of an altar, if there be such, would prove it to have been for the latter. The earliest record we have of such a construction is in Exodus xxiv. 4, where Moses builds an altar on Mount Sinai, and sets up twelve pillars according to the twelve tribes of Israel. Joshua (iv. 20) also directed twelve stones to be taken from Jordan in Gilgal, as a memorial of the passage of that river.

Nomenclature of cromlechs. It formerly was the custom to call every construction of this kind a cromlech. But the subject has lately been thoroughly investigated, particularly by Dr Lukis, in a paper read before the Society of Antiquaries, vol. xxxv. 233, and in another, printed in the Journal of the Archæological Association, September 1864, and the following nomenclature is now generally adopted. The single upright stone (see Plate I. fig 1), is called a *Maenhir*. One stone supported on another, or "half table stone," as it has sometimes been called, is (fig. 2) a *demi-dolmen*. A stone supported on two or more such stones, or a "table stone" (fig. 3) is a *dolmen*. One large stone supported on several smaller, so as to form a small chamber, is a *cist-vaen*. Several dolmens in succession form a *cromlech*.

A single *Maenhir* is also called a *monolith.* Several in a straight line, as those in Brittany and Germany, are called *ortholiths*. If in parallel lines, as at Abury, Dartmoor, Carnac, &c., are *paralleliths*. If in circles, as in the Ring of Brogar near Stennis, Stanton Drew, and Arbor Low, they are called *cycloliths*. Dolmen standing in a circle, like Rollrich, L'Ancresse, Stennis Circle, &c., are termed *peristaliths*. Le Couperon, at Jersey, is square, and Abdon Burf is concentric. Stonehenge, from its trilithons arranged in a circle, is called a *cyctotrilith*. It has been pretty clearly proved from excavations, that the cist-vaen and cromlech were sepulchral, all the others were ceremonial, in all probability religious, though Dr Lukis is of opinion, after excavating and otherwise examining about forty of those curious relics of antiquity, that the large, flat, "inclined stones" were not altars, but probably sepulchral memorials. In all instances in the Channel Islands, and in some in England, where a cromlech is surrounded by peristaliths, the circle is exactly sixty feet in diameter.

Sir Gardner Wilkinson, in a very able paper (see the same Journal, March 1862), divides the cromlechs into five classes—1. which he designates the *cromlech proper*, as one large, flat, cap-stone, supported on *three* upright stones; 2. The *cist-cromlech*, on *four* stones; 3. The *many pillared cromlech*, on more than four stones; 4. The *chamber cromlech*, having a roof; and, 5, the *subterranean chambers*. The author supposes the four first divisions never to have been covered with earth. Dr Lukis, however, denies this, and says all stone chambers, whether cist-vaen or cromlech,

were covered by mounds of earth; and he proposes to divide History. the cromlech into only two classes. 1. Simple chamber, *without* passages; and, 2, The like *with* passages, or covered ways, leading into them.

The chief difficulty in treating of these curious erections of stone is to ascertain their age. That they are the work of persons in a rude state of civilisation is clear. Still, the vastness of such stones, as at Stonehenge, would show they Stone- must have been a people of great energy and resources, to henge. effect such an extraordinary labour. All sorts of conjecture have been made as to this latter work. Some have supposed it to be Roman, others have even considered it to be antediluvian. The mortises and tenons, however, show clearly it must belong to the period when iron tools were used; it is impossible to conceive they were worked with flint instruments. The most rational supposition seems that it was erected to commemorate the treacherous murder of the British chiefs at the banquet given by Vortigern to Hengist. It is surrounded by numerous barrows, evidently the graves of men of great importance, a circumstance that adds much probability to the tradition.

Having now discovered the art of quarrying large stones, Places for moving them to different sites, and erecting them in sym- public as- metrical forms—having found out the way to construct sembly. places for civil, military, or religious assemblages, the next step was to cover these large places by roofs. In all probability this was first attempted in the adytum, or cella of temples, and there is every reason to suppose the earliest of these were the Egyptian. The oldest historian, Herodotus, (Cli. 13) tells us the Persians erected neither statues, nor temples, nor altars, and they considered them as foolish who did so. He also says (*Euterpe*, 4) that the Egyptians were the first to give altars, images, and temples to the gods, and to carve the likenesses of animals in stone.

History of the Progress of Architecture.

We now proceed to trace the progress of the science from its earliest regular formations, of which we have sufficient information, down to the present day.

Indian chronology being so vague and undefined, and the Indian connection of the Hindoos with the civilized nations about architec- the Mediterranean Sea having been so much restricted ture. in the earlier ages that we can get little assistance from the Greek historians on the subject, the date of their architectural monuments can be determined only by analogy. That, however, is an uncertain guide, without proper delineations, and, indeed, without any work that gives a competent idea of them. Though we have held India so long, and by a so much more honourable tenure than the French did Egypt, if we were now to be dispossessed we should leave nothing, and we should certainly retain nothing, to show to our credit that we had ever held it. Such an undertaking as the great work of the French Institute on the Architectural Antiquities of Egypt is far beyond the means of individuals; the constitution of our government appears to preclude the application of funds from the public purse to such purposes; and the East India Company, from whom, perhaps, something of the kind on the archæology of India might have been expected, had, it would appear, occupations of more interest to them than the advancement of science and art. It may be generally stated, that, in its leading forms and more obvious features, Hindoo architecture strongly resembles Egyptian, and may be considered as of the same family with it.

No nation that ever existed within the annals of the Egyptian human race has left structures that, in extent, magnifi- architec- cence, and grandeur, can vie with those of ancient Egypt. ture. We have the authority of historians for believing that there were others in the same country which no longer exist, that must have surpassed those which do remain; and they speak also of the cities of Assyria, as unparalleled in the extent and splendour of their edifices, whose sites, even, are not now determinable. The pyramids, however, mausoleums of a nation—and the temples, monuments of human folly—speak more strongly than any historian can, and compel our belief of what they have been by what they are; whereas the others do not exist but in name. Nineveh and Babylon were—but Thebes and Memphis still remain. It is strange, indeed, that a people who displayed such energies in the construction of tombs, pyramids, and temples, should have left no work of any description that could be applied to any really useful purpose. Denon, speaking of Thebes, says, " Still temples—nothing but temples—not a vestige of the hundred gates, so celebrated in history; no walls, quays, bridges, baths, or theatres; not a single edifice of public utility or convenience. Notwithstanding all the pains I took in the research, I could find nothing but temples, walls covered with obscure emblems, and hieroglyphics which attested the ascendency of the priesthood, who still seemed to reign over the mighty ruins, and whose empire constantly haunted my imagination."[1] Champollion, however, in his late researches, speaks of the remains of quays, and calls some of the structures palaces instead of temples; but as the former exist only in connection with the latter, they can hardly be considered as any thing more than mere embankments; and the regal and hierarchical offices having been so closely connected in the economy of ancient Egypt, it is of little or no consequence to our position whether the same edifices be called palaces or temples. Diodorus Siculus says, in one place, that " Busiris," believed to be one of the Pharaohs who persecuted Israel, " built that great city which the Egyptians call Heliopolis and the Greeks Thebes, and adorned it with stately public buildings and magnificent temples, with rich revenues;" and that " he built all the private houses, some four, and others five stories high."[2] Shortly after, speaking of Memphis, to account for the splendour with which the Egyptians built their tombs, and the comparative meanness of their houses, the same author says, " They call the houses of the living inns, because they stay in them but a little while; but the sepulchres of the dead they call everlasting habitations, because they abide in the grave to infinite generations. Therefore they are not very curious in the building of their houses; but in beautifying their sepulchres they leave nothing undone that can be thought of." Strabo also speaks of a splendid dwelling which was erected for the priests at Heliopolis, but that probably was one of the sacred palaces just referred to; for none of the ancient writers describe the domestic structures of the Egyptians, from personal knowledge of them, as being worthy of any notice; and that assertion of Strabo is too loose and unsupported by contemporary authority or analogy to deserve confidence of itself. To the statement of Diodorus, that private houses were built to four and five stories high, we can give no credence whatever; for the construction of edifices in tiers or stories was very imperfectly understood even in his time, which was many centuries after the destruction even of Thebes; and none of the existing remains of that city give the slightest indication of a second story, or indeed of aptitude to construct one, except the rude landings in some of the propylæa. Herodotus says that the Egyptians were the first who erected altars, shrines, and

temples; but of their private houses he says nothing; neither does he describe any of the temples as they existed in his time in Egypt; so that he in fact affords no assistance in determining the comparative antiquity of the various architectural structures which remain to the present time in that country. Indeed the ancient historians and topographers speak for the most part so widely of dates and dimensions, that they are, at the best, most unsatisfactory, if not fallacious, guides; and in the present case, that of Egypt, the style of architecture is so uniform, or so imperfectly understood, that no argument can with safety be drawn from it, as there may in other cases. In Hamilton's *Ægyptiaca*, the author says, with reference to this question: " In Egyptian architecture there is an uniformity of structure, both in the ornaments and in the masses, which, if unassisted by other circumstances, reduces us to mere conjecture; and that not only for the difference of a century or two, but perhaps for a thousand years."[3] Again: " The monuments of antiquity in Upper Egypt present a very uniform appearance; and his first impressions incline the traveller to attribute them to the same or nearly the same epoch. The plans and dispositions of the temples bear throughout a great resemblance to one another. The same character of hieroglyphics, the same forms of the divinity, bearing the same symbols and worshipped in the same manner, are sculptured on their walls from Hermopolis to Philæ. They are built of the same species of stone; very little difference is discernible in the degrees of excellence of workmanship, or the quality of the materials; and where human force has not been evidently employed to destroy the buildings, they are all in the same state of preservation or decay."[4] But we are fortunately now about to be rid of that difficulty by the erudition and industry of those learned men who have given their attention to the hieroglyphic literature of the Egyptians. M. Champollion professes to have determined the date of every monument of antiquity in that country which is inscribed, by the inscriptions, which he has qualified himself to read. As yet, however, we are not in possession of the whole result of his discoveries.

Hypogea, or spea, being caves formed by excavation, are of earlier date than any existing structures. Internally they present square piers, which were left to support the superincumbent mass of mountain or rock when their magnitude rendered it necessary. These were originally tombs; and the cave of Machpelah, of which Abraham made the purchase as a burying-place for his family, was, doubtless, one of that kind. Oratories or chapels were afterwards made in the same manner, but, it would appear, not until columnar architecture had come into use; for their entrances are generally sculptured into the resemblance of the front of a rude portico, or an actual portico or pronaos is constructed before them. Many such are found on the banks of the Nile, in its course through Nubia and Egypt. At Ibrim, which the Greeks call Primis, in the former country, there are several of these cavern temples, the earliest of which, according to M. Champollion, bears date of the reign of one of the Pharaohs, who was contemporaneous with Abraham, or his son Isaac, or about eighteen centuries before Christ; the latest is of the time of Rhameses Sethos, the Sesostris of Greek history. To some of the cavern tombs and temples in Upper Egypt M. Champollion accords even a still higher degree of antiquity. The earliest columnar structures which are found within the same range of country do not appear to bear a higher date than that of the

[1] *Voyage dans la Basse et la Haute Egypte*, p. 176. Par V Denon.
[2] Diod. Sic. lib. 1. cap. iv.
[3] *Ægyptiaca*, by Wm. Hamilton, Esq. F. S. A. Part I. p. 260.
[4] Ibid. p. 18.

ARCHITECTURE.

earliest kings of Pharaohs of the eighteenth dynasty of Manetho, which began about the time of the Jewish patriarch Abraham and ended with the Pharaoh from whom his descendants escaped under the conduct of Moses. The temple at Amada, to which we have already referred, is of the time of Moeris, who was contemporary with the patriarch Jacob, and consists of twelve square piers or pillars, and four columns, which possess the form and character of the Greek Doric, and may it is suggested, be called *protodoric*. The same intention, if it may be so called, is found in others of the early monuments, but in none so perfect as in this, as almost all the structures of ancient Egypt were either destroyed or seriously damaged by the Persians at the time of their invasion under Cambyses; and they are supposed not to have ascended the Nile much above Psalcis or Dakkè, but to have turned off by the way across the desert to Ethiopia, so that the temple at Amada, which is considerably above Dakkè, escaped.

Of all the Pharaohs, Sesostris, the first of the nineteenth dynasty, was the most distinguished for the great and extensive works he executed in architecture. Most of the existing ruins in Egypt, anterior to the Persian invasion, are attributed to that monarch by M. Champollion. The immense ruins at Thebes, which have been called the Memnonium and the tomb of Osymandyas, and are popularly called Medinet Abou, are considered by the same inquirer to be those of the Palatial Temple of Rhameses the Great, or Sesostris, and which he therefore calls the Rhamesseion, the ruins at Luxor being those of the Memnonium; that edifice or series of edifices having been constructed by Amenophis Memnon, of the eighteenth dynasty, one of the good and beneficent princes by whom the children of Israel were protected during their sojourn in Egypt. The magnificent structure at the village of Carnack, within the same city, appears however to excel all the rest in extent and grandeur, and is at least their equal in antiquity. It is generally known as the temple of Carnack, but it has been distinguished as that of Jupiter Ammon. It bears inscribed the name of Thothmosis II., the predecessor of Amenophis Memnon. From the existing remains of Thebes, and the relations of historians combined, that city may be assumed to have attained its highest degree of splendour in the time of Sesostris; few of the ruins it presents being of later date than the time of that monarch. This being admitted, and we believe it can hardly be denied, it must be admitted also that the practice of architecture, and of the allied mechanical arts, were already well understood; for the composition of the monuments displays an exquisite combination of simplicity and harmony, which produce the finest effects of beauty and grandeur; while their construction is the apparent result of perfection in the use of mechanical powers. All the Pharaonic monuments, indeed, throughout Egypt and Nubia, are wonders of science and art. The structures of Ombos, Apollinopolis Magna, and Latopolis, between Thebes and the cataract, M. Champollion determines to be generally of the age of the Ptolemies, and some even of the Roman dominion; those, however, which are of comparatively modern date are evidently restorations; others, probably of the earliest ages, having occupied the same sites. Indeed M. Champollion asserts generally that the Ptolemies, and the Ethiopian Ergamenes himself, only rebuilt temples where they had already stood in the times of the Pharaohs, and to the same divinities that had always been worshipped there; and he remarks, that the religious system of this people was such a complete whole, so connected in all its parts, and fixed from time immemorial in so absolute and precise a manner, that the dominion of the Greeks and of the Romans did not produce any innovation;

that the Ptolemies and the Cæsars only restored in Nubia what the Persians had destroyed, and rebuilt temples where they had formerly stood, and dedicated them to the same gods.

Of the arrangements of an Egyptian temple we shall speak when we come to treat of Egyptian architecture as a style. In construction the Egyptians appear to have used wrought stones at a very early period: this probably was induced by the still earlier habit of excavating rocks to form tombs; for the walls in their oldest structures are composed of rectangularly cut blocks in parallel courses; whereas we shall find that the most ancient specimens of walling in Greece and Italy are not so. In the Pharaonic monuments, besides walls built in parallel courses of wrought stone, we find squared piers also; and frequently, in the same structure with them, the peculiarly formed tumescent column with a bulbous capital or head, covered with an abacus or square tablet, corresponding with the size of the piers, and warranting the supposition that that species of column is a mere refinement on the simple square pillar. What dictated its singular form must remain matter of speculation. The cylindrical column with a bell-shaped capital was the next advance, and that also is found in the same structures, though not in the simplest and earliest of them, in which piers occur. Terminal or Caryatic figures are common in those early works, not absolutely supporting an entablature, but placed before piers which do so, and having the appearance of doing it themselves when seen in front. Bold, massive, rectangular architraves extend from pier to pier and from column to column, and are generally surmounted externally by a deep coved coping, or cornice, with a large corded and torus-formed moulding intervening. This masks the ends of the stones which are placed transversely on the architraves to form the ceiling internally, the whole being flushed square on the top, and forming a flat terrace or floor. The pyramidal form of the moles or propylæa, peculiar to Egyptian temples, may have been suggested by the pyramids, as neither that form nor those adjuncts to a temple appear to have been used before the period at which it is supposed the former were constructed. The grandeur and dignity inherent to that form would indeed hardly be suspected till its appearance in the pyramids themselves; and certainly the impression of its effect must have been strong, to induce men to seek it in a truncated pyramid under a very acute angle, as in the propylæa, relying on the tendency of its outline alone. It was gradually, too, that this tendency was generally applied, for in the earliest Pharaonic structures the vertical outline is most common, except in the propylæa, where they exist; and in the structures of the Ptolemies the inclined outline pervades every thing. The monolithic obelisk is of Egyptian origin also. Its tapering form may be the consequence of the impression the pyramidal tendency had occasioned, though perhaps the object itself is the representative of the single stone by which religious feeling appears first to have expressed itself. Obelisks were set up by the Egyptians, sometimes in the courts or atria of their temples, and sometimes before the entrances to them.

Of all the architectural works of the Egyptians, however, none have excited so much the wonder and curiosity of men as the pyramids themselves; not in consequence of any particular beauty in their composition, or ingenuity in their construction, but simply because of their immense magnitude, and unknown use, and antiquity. Denon makes the following observation on his first visit to the great pyramid of Gizeh, at Memphis. " If we reflect upon these pyramids, we shall be inclined to think the pride that constructed them greater even than these masses themselves, and shall scarcely know whether to reprobate most the insolent tyranny which commanded, or the stupid servility of the people which executed, the undertaking. None but sacerdotal despots would ever have undertaken them, and none but a stupid fanatical people would ever have built them....The most honourable reason that can be assigned for their erection is the emulation of man to excel the works of nature in immensity and duration, and in this project he has not been altogether unsuccessful. The mountains near the pyramids are not so high, and have suffered more from time than the pyramids themselves."[1] But Memphis itself was of late foundation in comparison with other cities on the Nile. According to Professor Heeren,[2] civilization descended by the Nile from Ethiopia with the caste of priests who brought with them the worship of Ammon, Osiris, and Phtha (the Jupiter, Bacchus, and Vulcan of the Greeks), and " the spread of this worship, which was always connected with temples, affords the most evident vestiges of the spread of the caste itself; and those vestiges, combined with the records of the Egyptians, lead us to the conclusion that this caste was a tribe which migrated from the south, above Meroe, in Ethiopia, and, by the establishment of inland colonies around the temples founded by them, gradually extended and made the worship of their gods the dominant religion in Egypt. Proofs of the accuracy of this theory," he asserts, " may be deduced from monuments and express testimonies concerning the origin of Thebes and Ammon from Meroe; that it might indeed have been inferred from the preservation of the worship of Ammon in this last place." The same author goes on to say, that " Thebes was, if not the most, one of the most, ancient cities of Egypt;" and that " Memphis and other cities of the vale of the Nile are known to have been founded from Thebes." Now Thebes exists to the present time in the ruins of her magnificent temples, the works of the Pharaohs, but without the vestige of a pyramid, so that it may be concluded that none was ever built there; and Memphis may be said to exist in the everlasting pyramids of Gizeh and Saccharah, which occupy two of its extremities; but no indication remains of the existence of a temple of any kind: indeed the exact site of the city cannot be determined except by the pyramids. Herodotus, however, speaks of temples at Memphis, particularly of that of Vulcan or Phtha; but certainly no vestige of such has existed for a long period of time within that vicinity. Memphis was a great and ancient capital, and why should it not retain some evidence of the existence of temples in it? But Thebes was a greater and more ancient capital, and indeed the metropolis of all Egypt; and why has it no pyramids? These things are equally unaccountable and inexplicable, affording groundwork for almost any theory, but giving perfect support to none. Mr Hamilton, in his *Ægyptiaca*, before quoted, places Memphis considerably further south, where some ruins have been discovered which may be thought to give a colour to his supposition. But the ruins are of very inconsiderable extent, and are all prostrate, so that nothing can be positively determined by them; and the statement of Pliny as to the relative distances of the Nile and the city from the pyramids of Gizeh being proved to be correct in the one, may be admitted in the other. If Herodotus's account of the building of the pyramids be received, they are of comparatively modern date, the oldest having been constructed several generations after the time of Sesostris, under whom Thebes attained its highest degree of splendour; but this would leave unaccounted for the tendency to pyramidal forms in Egyptian architecture before referred to, unless every example exhibiting that tendency were itself referred to a date posterior to that assigned to Cheops and Cephron, which cannot be done in accordance with the assertions of M. Champollion as to the structures of Thebes, Elephantina, and Nubia generally.

From its immense size, the dimensions of the great pyramid of Gizeh, at Memphis, are variously given by the various persons who have measured it. M. Nouet, who was of the French commission in Egypt, and had perhaps the best means of ascertaining the truth, states its base to be a square whose side is 716 French or 768 English feet in length, which is about the extent of the great square of Lincoln's-Inn-Fields in London; and its height 421 French or 452 English feet, or about one-third as high again as St Paul's Cathedral. It is built in regular courses or layers of stone, which vary in thickness from two to three feet, each receding from the one below it to the number of 202; though even this is variously stated from that number to 260, as indeed the height is given by various modern travellers at from 444 to 625 feet. And the ancient writers differ as widely, both among themselves and from the moderns. On the top course the area is about 10 English feet square, though it is believed to have been originally two courses higher, which would bring it to the smallest that in regular gradation it could be. It is a solid mass of stone, with the exception of a narrow corridor leading to a small chamber in its centre; and a larger ascending corridor or gallery, from about half the distance of the first to another larger chamber at a considerable distance, vertically above the former, in which there is a single granite sarcophagus, not more than large enough for one body, putting the intention of the structure clearly beyond doubt. The other pyramids differ from that of Cheops (as the largest is called) in size, and slightly in form and mode of construction, some having the angles of the steps or courses of stone worked away to an inclined plane, and some not diminishing in a right line. One of the middle-sized pyramids is unlike all the rest, in being neither smooth nor in small steps, but in six large benches or stages, apparently of equal height, and diminishing gradually. But the circumstance which most distinguishes it is, that it is constructed of rude unshapen blocks of stone, cemented together with a very large proportion of mortar. Another is of unburnt brick, and has consequently become ruinous and mis-shapen.

The famous labyrinth, of which Herodotus speaks as having been built by the twelve kings of Egypt, beyond the Lake Mœris, is believed by Denon, after examination of the described site, to be little better than fabulous, and that the historian was imposed on by the priests, from whom he derived most of his information. He says, indeed, that he saw and examined it himself; but his description is so vague, that an architect who should endeavour to make a design from it, would be greatly embarrassed. As we can therefore derive no information from it with regard to architecture, it need not be further discussed here. It has been suggested as probable, and indeed the opinion has been maintained, that the pyramids stand over immense substructures; that their areas are occupied by chambers, in which may be found the arcana of Egyptian lore, of which they are the depositories. If it really be so, may not the labyrinths just referred to have been under the pyramid, which the historian says was constructed at the point where the labyrinth terminates, instead of near it? His expression is so ambiguous, that it leaves room for a suggestion of the kind.

Of the domestic architecture of the Egyptians we have

---

[1] *Voyage dans la Basse et la Haute Egypte*, p. 77. Par V. Denon.   [2] *Manual of Ancient History*, p. 58.

nc knowledge whatever. The statements of the ancient writers on the subject have been already mentioned; but supposing them to be more explicit, and more in conformity with probability, than they really are, without existing remains we could form but a very imperfect idea of what it was. Reasoning from analogy, and the slight information of historians, we should conclude that the habitations of the Egyptians were of a very unpretending description. The already quoted statement of Diodorus Siculus, that " they are not very curious in the building of their houses," even in his time, after their long intercourse with Greece, and their more recent connection with luxurious Rome; added to the fact, that no indications of domestic structures exist in any part of the country, and that the presumed habitations of the priests, in the ancient temples, are small and inconvenient cells; and all these things, taken in conjunction with the mildness of the climate and the salubrity of the atmosphere, we think it must be admitted, warrant the conclusion.

No style of architecture of which we have any knowledge is so well qualified to produce impressive effects on the mind as the Egyptian. The mere assumption of its forms, however, is not sufficient to produce its effects; and drawing is more incompetent to convey an idea of it than perhaps of any thing else in art. To this point the authors of the great work of the French Institute on the antiquities of Egypt bear testimony in strong language. Speaking of the incompetence of drawings to convey just ideas of the grandeur, magnificence, and beauty of the Egyptian temples, and other remains of antiquity, they say, " Despite the care we have given ourselves to describe the Egyptian monuments, we cannot even hope that we have succeeded in giving to others the ideas which we ourselves received from actual views and present contemplation of them; for there are things which drawings and descriptions cannot convey. Geometrical drawings are without doubt quite competent to show the form and proportions of an edifice, its disposition and distribution; but far indeed are they from giving satisfactory ideas of the elegance and effect of structures. Frequently we had to regret how much of the beauty of the original was lost in its geometrical representation on paper; for what in execution was light and graceful, often in the geometrical drawings appeared heavy and inelegant."[1]

The materials used in the construction of the Egyptian architectural monuments are, for the most part, granite, breccia, sandstone, and unburnt brick. The granite was principally supplied by the quarries at Elephantina and Syene, for which the Nile offered a ready mode of conveyance; some species were brought down the river from Ethiopia, but we do not find that the materials were at any time brought from any other foreign country. It may be remarked, too, that in the earliest structures the common *grès* or sandstone is principally employed. Excepting the obelisks and some few of the propylæa, all the temples at Thebes are of that material. In Lower Egypt, on the contrary, and in the works of later date generally, almost every thing is constructed of granite.

Herodotus informs us that the ancient Persians had neither statues, temples, nor altars; and Diodorus Siculus affirms that the palaces of Persepolis and Susa were not built till after the conquest of Egypt by Cambyses, and that they were constructed by architects of that nation. In this case, as in that of India, we are at a great loss for evidence. The Persepolitan remains, though frequently visited and slightly sketched, have not been explored and delineated by such men as Stuart and Revett, or the authors of the great French work we have so often alluded to. That the Persian style, though very different in particulars, does bear a relation to the Egyptian family, however, is very evident. Sir Robert Ker Porter, in his travels in the East, says that the first impression he received in his first walk among the ruins of Persepolis was, that " in mass and in detail they bore a strong resemblance to the architectural taste of Egypt."[2] Nevertheless, there is a strong probability that the Persian is itself an original style, and that the resemblance is merely fortuitous, similar results arising from the same causes, as in Egypt and India; for the eastern parts of that country are believed to have been the earliest seat of the human race. Professor Heeren says of Persia, " It cannot be doubted, that long before the rise of the Persian power, mighty kingdoms existed in these regions, and particularly in the eastern part of Bactria; yet of those kingdoms we have by no means a consistent or chronological history—nothing but a few fragments, probably of dynasties which ruled in Media properly so called, immediately previous to the Persians;"[3] from whom the style of architecture may be derived, though indeed we know of no remains of earlier date than those which are properly called Persian. But we may be said to know nothing of Bactria; it may, and probably does, rival Elora, Salsette, and the banks of the Nile, in primitive specimens of architecture.

We have neither historical nor archæological information that can be depended on to prove what the state or style of architecture was among the ancient Assyrians. Lucian says, however, that their temples were less ancient than those of Egypt. The ruins believed to be those of the great capital of Babylonia present nothing but shapeless masses of brick, from which no idea whatever can be formed as to the style of architecture, or the progress it had made in that country; but some cylindrical and other seals and fragments, in *terra cotta*, found by excavation among those ruins, and now in the British Museum, are sufficiently in accordance with the rest of the eastern antiquities to be received as evidence of the general assimilation of its style of design with that which was common to the neighbouring nations.

The Phœnicians, we are told by Lucian, built in the Egyptian style; but their country retains no memorials of its ancient architecture by which we might confirm or correct his information. Doubtless Carthage and the other colonies of Phœnicia followed their parent country in this particular.

As far as we can judge from the trifling documents we possess of the architecture of the ancient Mexicans and Peruvians, it was of a rude but massive character, and may be thought also to resemble the early architecture of India, Egypt, and Persia more than we can see any reason for, except in the tendency of the mind of man to the same result when he is placed under similar circumstances. An impression to this effect appears to have been made on Humboldt, who, when speaking of a pyramidal mass of ancient Mexico, says, " It is impossible to read the descriptions which Herodotus and Diodorus Siculus have left us of the temple of Jupiter Belus, without being struck with the resemblance of that Babylonian monument to the *teocallis* of Anahuac."[4]

It is an illustration of the fact that the wants and fancies of man lead him to nearly the same results as he becomes civilized, without communication and consequent imitation, that the plans given by Sir William Chambers, of Chinese public and private buildings, might be taken

---

[1] *Description de l'Egypte*, vol i. p. 292.
[2] *Travels in Georgia, Persia, &c.* by Sir R. K. Porter. vol. i. p 579.
[3] *Manual of Ancient History*, p. 26.
[4] Humboldt's *Personal Narrative*. vol. i. p. 82.

History. at the first glance, for either Hindoo, Greek, Roman, or Moresco—of course not considering magnitude of parts, but general forms and arrangements. Indeed, the remark may be extended beyond the mere plans; for all have, to a certain extent, insulated columns placed equidistant, and crowned with an entablature; and the general effect of many Chinese buildings is altogether Moorish.

Jewish architecture. Architecture was not likely to flourish among the shepherd tribes of Israel. It is in agricultural and commercial countries, such as Egypt and Greece, that its noblest works are produced, and not among the nomades of Arabia and Palestine. Saul, the first king of Israel, appears to have had no settled place of abode; and the most sacred ceremonies of the Jewish religion were performed at Gilgal, where was the temple of unhewn stones set up by Joshua on taking possession of the promised land, and making a covenant between God and the people, until the building of the temple at Jerusalem in the place rendered holy by Abraham s great sacrifice. Saul himself was confirmed in the kingdom at Gilgal, and there the nation swore allegiance to him with sacrifices to the Almighty; but as yet nothing existed there in which to perform the rites, except the ancient Celtic structure to which we have alluded. Joshua also set up an altar in Mount Ebal, and in long after times a splendid temple was erected in one of the two neighbouring mounts, Ebal or Gerizim, where the Samaritans worshipped. Like his predecessor on the throne, David appears to have been but indifferently lodged till towards the end of his reign, when he is said to have built himself a house; and until the temple was built in the following reign, the ark of the covenant was never in a fixed place;—it was at one time in a private house, at another in captivity among the Philistines; and, indeed, King David expressed his shame that he had a house of cedar, whilst the ark of the Lord still dwelt in a tent. These things, and the fact that Solomon sent to Tyre for workmen, and indeed for an architect also, are, we think, conclusive evidence, that in whatever state architecture was among the Jews from the building of the temple at Jerusalem, it was very low before that time; and from the descriptions we have of that edifice itself in the Bible, it appears to have exhibited a greater degree of barbaric splendour than of classic elegance. From mere description, however, it is impossible to understand an unknown species of building, as many things we shall have occasion to refer to will clearly prove.

Few things have occasioned controversies more amusing, from the singularity of some assumptions, and the absolute futility of them all, than the style and manner in which Solomon's temple was built. Villalpanda, a Spanish Jesuit, appended to a commentary which he wrote on the prophecies of Ezekiel, a long dissertation on the first and second temples of Jerusalem, in which he insists that the theory and practice of permanent architecture commenced with the building of that temple by Solomon—that with it, " the orders," which, he says, are falsely attributed to the Greeks, came into existence—that indeed the design (from a passage in the first book of the Chronicles), perfect in all its details, was given to David, *drawn by the hand of God!* He moreover pretends to show, that the proportions assigned by Vitruvius to the different orders accord exactly with the descriptions given of the temple of Solomon; and accuses Callimachus of usurping the honour of inventing the Corinthian capital, which could not belong to him, as it was of divine origin, and had been executed in the temple at Jerusalem centuries before he was born. Some learned, and in some respects sensible men, have attempted to support this theory; and others have thought it worth while to controvert it, by proving

that the architect and the principal workmen were all History either Egyptians or Phoenicians, and that consequently the edifice must have been in the Egyptian style. A learned architect of the present day has endeavoured to show that it was in the Greek style, and that its form, proportions, and distribution, were not dissimilar to those of the temple of Ceres at Eleusis. As the Phoenicians, who were principally employed by Solomon, themselves built in the Egyptian manner, we think the probability is great that it was in the Egyptian or Phoenician style, as far as the Jewish ceremonial would permit; and certainly the descriptions of its distribution accord better with that of an Egyptian than of a Grecian temple. The pillars of Jachin and Boaz, which are said to have been set up before the temple, correspond exactly in relative situation with the obelisks in temples at Thebes. Clemens of Alexandria, too, gives a description of an Egyptian temple very much like that of the Jewish : and the palm-leaves, roses, fruits, and flowers, in the latter, are very common in existing specimens of the former, whereas in the Greek remains of early date no such things are to be found. Whether the Jews in after-times possessed a national style of architecture or not, we cannot tell: there is no reason, however, for supposing that they did; for their monotheistic structure at Jerusalem was not repeated in other places, as the temples of the heathen divinities were among the Greeks and Romans, by which they might have acquired a peculiar mode of composition and combination. The non-existence of a national Jewish style of architecture tends also to strengthen our position, that architecture did not originate in the disposition and decoration of buildings for domestic purposes, of which the Jews must, when settled, have made as much use as other nations; and a multiplicity of religious edifices, in the construction of which they might have acquired one, was forbidden by their code.

In various parts of Greece and Italy, specimens of rude Pelasgic architecture. walling are found of such remote antiquity that they are, as by common consent, referred to the fabulous ages, and, for want of a more distinctive term, are called Cyclopæan. Now it appears, from the concurring evidence and opinions of most antiquaries, that a people who have been called Pelasgi, or sailors, migrated from Asia Minor, or the coast of Syria, at a very early period, and possessed themselves of various countries, some of which were unoccupied, and others inhabited by Celtic tribes. Mr Godfrey Higgins says that the Pelasgi were Canaanites, and being a hardy sea-faring race, they soon subdued the Celtic inhabitants of Delphi in Greece, or of Cumæ in Italy, who, from their first quitting the parent hive, never had occasion for an offensive weapon, except against wild beasts; and that they were the people who settled Carthage, Spain, and Ireland. Bishop Marsh has proved the Pelasgi to be Dorians, Dr Clarke has proved the Etrusci to be Phoenicians, and Gallæus has proved the Dorians to be Phoenicians. Thus, says Mr Higgins, the Pelasgi, the Etrusci, and the Phoenicians, are all proved to be the same. According to Professor Heeren, also, who affixes dates to the various migrations, the Pelasgi were of Asiatic origin. " Their first arrival in the Peloponnesus was under Inachus, about 1800 years B. C.; and according to their own traditions," he says, " they made their first appearance in this quarter as uncultivated savages. They must, however, at an early period, have made some progress towards civilization, since the most ancient states, Argos and Sicyon, owed their origin to them; and to them, perhaps with great probability, are attributed the remains of those most ancient monuments generally termed Cyclopic."[1] He adds, that the Hellenes, a people

[1] *Manual of Ancient History*, p. 119.　　　　[2] *The Celtic Druids*, c. vi. § 29, p. 259.

of Asiatic origin also, expelled the Pelasgi from almost every part of Greece, about 300 years after their first occupation of it; the latter keeping their footing only in Arcadia and in the land of Dodona, whilst some of them migrated to Italy, and others to Crete and various islands. The arrival of the Egyptian and Phœnician colonies in Greece, Professor Heeren thinks, was between 1600 and 1400 B. C.

The connection of Greece and Italy with each other, and with Egypt and Phœnicia, is thus made evident. The Cyclopæan structures, however, were the works of the rude Pelasgi before that connection took place, except as far as it existed in their having a common origin. They occupied, either simultaneously or consecutively, both Greece and Italy; and this accounts for the sameness of that peculiar and original mode of structure which, we have said, is found in both countries, though no evidence exists of its ever having been practised elsewhere. If, indeed, the things in question were the work of the earlier Celtic inhabitants, a still more remote date must be assigned them than they could derive from the Pelasgi; and this is the opinion of Mr Higgins, supported, he contends, by the suffrages of Dodwell, Clarke, and others, who say that the doorway called the Gate of the Lions, in the Acropolis of Mycenæ, is built exactly like the remains of Stonehenge. The most ancient specimen of Cyclopic walling is found at Tyrinthus, near My-

Plate I.
Fig. 14.

cenæ. It is composed of huge masses of rock roughly hewn and piled up together, with the interstices at the angles filled up by small stones, but without mortar or cement of any kind. The next species is in stones of various sizes also, shaped polygonally, and fitted with nicety one to another, but not laid in courses. Specimens of this are found at Iulis and Delphi, as well as at the places already mentioned, in Greece, and in various parts of Italy, particularly at Cossa, a town of the Volsci. This also was constructed without mortar. The mode of building walls, which took the place of that, is not called Cyclo-

Fig. 16

pæan; it is in parallel courses of rectangular stones, of unequal size, but of the same height. This is common in the Phocian cities, and in some parts of Bœotia and Argolis. To that succeeded the mode most common in, and which was chiefly confined to, Attica. It consists of horizontal courses of masonry, not always of the same height, but composed of rectangular stones.

The oldest existing structure in Greece of regular form is of far superior construction to the Cyclopæan walling, and must be referred to the Egyptian or Phœnician colonists. It is at Mycenæ, and consists of two subterranean chambers, one of which is much larger than the other. The outer and larger one is of circular form, and is entered by a huge doorway at the end of a long avenue of colossal walls, built in nearly parallel courses of rectangular stones, roughly hewn, however, and laid without mortar. Its external effect is that of an excavation, though the structure of the front is evident; and internally it assumes the form of an immense lime-kiln; its vertical section being of a somewhat conical form, under nearly parabolic curves, like a pointed, or what is vulgarly called a Gothic arch. The construction of this edifice was thought to afford clear evidence that the Greeks were acquainted with the properties of the arch; but in the most material point this was destroyed on finding that it consisted of parallel projecting courses of stone in horizontal layers, in the manner called by our workmen battering, or more correctly perhaps corbelling. It proves, however, that its architect understood the principle of the arch in its horizontal position; for Mr Cockerell has discovered, by excavations above it, that the diminishing rings of which the dome is composed are complete in them-

selves for withstanding outward pressure; the joints of the stones being partly wrought radiating, and partly rendered so by wedges of small stones driven tightly into them behind. The apex is formed, not by a key-stone, for the construction does not admit of such, but by a covering stone, which is merely laid on the course immediately below it. It may be added, that internally the lower projecting angles of the stones are worked off to follow the general outline. Though this is the largest and most perfect, its internal diameter at the base being 48 feet 6 inches, and its height from the floor to the covering stone 45 feet, yet edifices exhibiting similar structure are found in many other places in Greece itself, in Egypt, in Sicily, and in Italy. They all however tend to prove, that the principle of the construction of the vertical arch was unknown at the time of their erection in all those countries; and their erection is as evidently of the most remote antiquity, perhaps of the presumed era of Dædalus, to whom some have assigned many of them, as well as the discovery of so much of the principle of the arch as is exhibited in the arrangement of the horizontal rings or layers in the Mycenæan monument. Neither could the mechanical powers have been unknown to their constructors. In the edifice which we have described, and which is thought by some to be the Treasury of Atreus, or the Tomb of his son Agamemnon, mentioned by Pausanias as existing among the ruins of Mycenæ in his time, the inner lintel of the doorway is 27 feet in length, 16 feet deep, and nearly 4 feet thick, weighing, it is computed, upwards of 130 tons; and the lintel of the Gate of the Lions in the Acropolis of the same city, is, from its immense magnitude, also strongly illustrative of the great mechanical skill of the people of those times. As the Treasury of Atreus at present exists, it exhibits nothing like an attempt at decoration, except that the doorway is, on the outside, sunk in two faces all round, as if to harmonize with some architectural composition; and the interior of the edifice may be supposed to have been lined, probably with plates of metal, like the tower of Acrisius, as bronze nails for attaching them to the vault still remain. Some sculptured fragments of marble which have been found among the ruins of the fallen parts and the rubbish which chokes up the entrance, together with indications on the external front of the edifice that it was cased, have led to an ingenious attempt at restoration, upon the supposition that the fragments were parts of a frontispiece. The fact that such frontispieces were sometimes carved, and sometimes constructed, in connection with the entrances of excavated tombs and other *spea* in Egypt and Nubia, gives a degree of probability to the idea that it would not otherwise have; for the fragments do not resemble the earliest existing specimens of Greek architectural forms; though indeed these latter may be traced to Persepolis, and Ibrim in Nubia, according to several ingenious antiquaries and architects. In curious accordance with this Mycenæan structure is the ancient monument at New Grange, near Drogheda, in Ireland. Ruder in every respect than the former, in form, construction, and mode of access, it bears such a striking similarity to it, that it is almost impossible to be supposed the effect of mere chance. The opinion of Mr Godfrey Higgins, that the Pelasgi, who peopled many of the countries on the shores of the Mediterranean Sea, peopled Ireland also, appears to be supported by this coincidence between the so-called Treasury of Atreus, or Tomb of Agamemnon, in the Peloponnesus, and the monument at New Grange in Ireland.

We know of no columnar edifice in Greece, or elsewhere in the Grecian style, of earlier date than the ruined temple at Corinth, which is in the plainest and sim-

Plate I.
Fig. 10.

History. plest form of what has been called the Doric Order, though it would be more correctly designated the Doric Style; for the term Order is objectionable, because it supposes rules and limitations to what in its best times was subjected to neither. As, however, it is the term best understood, we shall not hesitate to continue it. It is difficult, if not impossible, to ascertain where and in what manner the Doric order originated. The example we have referred to, though the earliest, does not differ in its leading features and characteristics from the more perfect specimens of later date; and it bears no direct and easy analogy to any species of columnar arrangement of other countries and earlier times. The story of Vitruvius, even supposing it rational, does not coincide with the Greek style of Doric at all, but, if with any thing, with the Roman examples of it, which at the best are mean and inelegant deteriorations of the simple and beautiful original. This author says that "Dorus, the son of Hellenus and of the nymph Orseis, king of Achaia and of all the Peloponnesus, having formerly built a temple to Juno in the ancient city of Argos, this temple was found by chance to be in that manner which we call Doric."[1] In another place he deduces the arrangements of this same order from those of a primitive log-hut in the first place, through all the refinements of carpentry, leaving nothing to chance, but settling with the utmost precision what, in the latter, suggested the various parts of the former. Chance in one case, and experience in another, however, are not enough for this author; but he also tells us that the Doric column was modelled by the Grecian colonists in Asia Minor, on the proportions of the male human figure, and was made six diameters in height, because a man was found to be six times the length of his foot; and that eventual improvements occasioned the column to be made one diameter more, or seven instead of six. "Thus the Doric column was first adapted to edifices, having the proportions, strength, and beauty of the body of a man!" The earliest examples of this order, however, are those which least agree with the primitive forms and proportions of Vitruvius; the columns at Corinth hardly exceed four diameters in height, while in later examples they gradually extend, till, in the temple of Minerva on the promontory of Sunium, the columns are nearly six diameters, being one of the tallest specimens of pure Greek origin ever executed. If the trunks of trees used in the structure of tents suggested the first idea of columns, and of the Doric in particular, as many contend, how is it that the earliest specimens discovered are the most massive? For the merest saplings would have formed the wooden proto-columns, and necessarily, when imitated in stone, they would not have been made more bulky than the less tenacious nature of the material required; much less would the slender wooden architrave have been magnified into the ponderous entablature of the primitive permanent architectural structures of all nations. In the construction of edifices with the trunks of trees, and timber generally, then, we do not find the origin of Doric architecture. If we have recourse to Egypt, the mother of the arts and sciences, we shall indeed find many things even in the more ancient structures which *may* have furnished an idea of the Doric arrangements to the fertile imagination of a Greek. The later works of that country cannot be trusted for originality, as they may themselves have been influenced by Greek examples; but we hardly dare assert that the Doric order was suggested by any thing in Egyptian architecture, though in making such assertion we should be supported by the opinions of many competent

judges. The temple at Amada in Nubia can hardly be positively assumed as an example of the *proto-Doric*, though it may of the *proto-columnar*. Nevertheless, the example is striking, as it certainly possesses the Doric character. The broad square abacus, and the cylindrical or even conoidal tendency of the shaft, marked as it is, as if for fluting, with the plain, simple, and massive epistyle or architrave superimposed, are all in accordance with the Hellenic columnar ordinance; still there is nothing to connect that rude model with the positive and somewhat formally arranged example at Corinth with which we began. It must be remembered, however, that two connecting links between Egyptian and Greek architecture are lost; Lower Egypt, with its splendid capital Memphis, and Phœnicia; through which latter the learning and taste of the inhabitants of the former country appear to have taken their course; but of neither of these do we possess architectural remains that bear on the subject in question. In the Pharaonic structures of Thebes we find both the tumescent and the cylindrical columns; and an amalgamation and modification of the two would easily produce the Doric column, or something very much like it, which may have been executed in those places, and so transferred to Greece. Of the triglyphs, the most distinguishing part of the Doric entablature, there are many indications in the early works of Upper Egypt; and in the structures of the Ptolemies they are still more evident though it may be objected that, in these, those indications were borrowed from the Greeks after the Macedonian conquest. But it must be borne in mind that the Egyptian nation did not change its character, religion, or usages by the change of its governors; and the Egyptians were, through the whole period of their existence as a nation, an originating and not an imitative people; whereas the Greeks seized on a beauty wherever they found one, and made it their own by improving it. The forms and arrangement, too, of many of the Greek mouldings, and the manner of carving to enrich them, are common in the earliest ornate works of the Egyptians; and such things are as strong evidence of community of origin, as the existence of similar words having the same meaning in different languages is of theirs. We may be asked, why the Greeks cannot be allowed to have originated that beautiful style of architecture which they brought to the perfection it displays in their works? To which we think it a sufficient answer, that it would be against the common course of events if it were so. In Egypt we can trace a progress from the ruder to the more advanced, and, with trifling discrepancies, to the most perfect; but in Greece, the earliest specimen of columnar architecture that presents itself displays almost all the qualities and perfections which are found in works of periods when learning and civility were at their acme in that country. We cannot find in Greece a stepping-stone from the Celtic or Pelasgic Gate of the Lions of Mycenæ, to the Doric columns at Corinth, and hardly to the Fane of Minerva in the Acropolis of Athens; and have therefore to seek the gradations among the people with whom we have seen they were connected, and whose country furnishes them in a great measure, if not entirely. Differences in climate and in political constitution, as well as in forms of religion, account sufficiently for the differences between the arrangements of the religious structures of the Greeks and those of Egypt. At the present day we find, that though they may be built in the same style, and for the worship of the same divinity, there is a wide difference between a church in Italy and a church in England, and a still greater between a church in the former country and

---

[1] Vitruvius, lib. i. cap. 1

History. one in Scotland. The model, however, of the Greek temple is found in many places in Egypt, generally placed as a chapel or ædicula, subsidiary to, and in connection with, the larger structures, as well as in the earlier Nubian temples themselves.

None other than the Doric style or order was used in Greece till after the Macedonian conquest, about which period that beautiful and graceful variety called the Ionic was brought into use. It is as difficult to determine its origin as that of the Doric. Vitruvius says that the Ionian colonists, on building a temple to Diana, wished to find some new manner that was beautiful; and by the method which they had pursued with the Doric, proportioning the column after a man, they gave to this the delicacy of the female figure; in the first place by making the diameter of the column one eighth of its height, then by putting a base to it in twisted cords, like the sandals of a woman, and forming the capital with volutes, like the hair which hangs on both sides of her face. To crown all, he says that they channelled or fluted the column, to resemble the folds of female garments, by which it would appear that Vitruvius did not know that the Greeks never executed the Doric order without fluting the columns. "Thus," he goes on to say, "they invented these two species of columns, imitating in the one the naked simplicity and dignity of a man, and in the other the delicacy and the ornaments of a woman." It can hardly be doubted that the voluted or Ionic order did originate in Ionia, at least we know of no earlier examples of it than those which exist there; and it does not appear to have been known to the European Greeks, and certainly was not practised by them, till after the period we have indicated. It probably took its rise from some peculiarities in Persian architecture; though many believe that the Ionic order had a much earlier origin, deriving it from Egypt, where, it is true, many indications are found of its volutes in the spiral enrichments of capitals; but it must be observed that they are in edifices now ascertained to be of the age of the Ptolemies, and consequently later than the structures which exhibit the voluted order in Ionia and its islands. We think, too, that many persons are influenced in assigning a higher degree of antiquity to this style than facts will bear out, by their respect for the authority of Vitruvius; though Mr Gwilt (his latest translator into English) confesses that "upon his authority in matters of historical research not much reliance is to be placed."[1] We are willing to admit that much may be adduced in support of the opinion, that this style was known and used in Greece even before the age of Pericles; specimens of it having been found in connection with sculpture, certainly less perfect, and therefore presumed to be of earlier date, than the works of Phidias and his pupils and compeers.

It is no less difficult to determine the origin of what is called the Corinthian order. The not inelegant traditionary tale by Vitruvius of the invention of its capital, is the only *reason* of the name it bears. His account of the origin of this third species of columnar composition is more summary, and not less absurd, than that of the preceding. He says that it was arranged "to represent the delicacy of a young girl whose age renders her figure more pleasing and more susceptible of ornaments which may enhance her natural beauty." With much more reason might the Doric be called the Corinthian order; for, as we have stated, at Corinth there exists the oldest example of that style; whereas there is nothing, either in ruins or authentic record, to prove that the latter was ever known in that city. Columns with foliated capitals are not of very early date in Greece; earlier examples exist in Asia Minor, and foliage adorns the capitals of columns in some of the Pharaonic monuments of Egypt; not arranged, indeed, as in the later Corinthian capital, which by possibility may have been the result of some such accident as Vitruvius relates of Callimachus and the basket on the grave of the Corinthian virgin. The interior of the temple of Apollo Didymæus at Miletus in Ionia exhibits the earliest example of the acanthus leaf arranged round the drum of a capital in a single row, surmounted by the favourite honeysuckle; but that edifice was constructed about a century before Callimachus is understood to have lived. The only perfect columnar example in Greece itself of this species of foliated capital is of later date than, and is a great improvement on, that of Miletus; it is the beautiful little structure called the Choragic monument of Lysicrates at Athens. Specimens are less uncommon in Greece of square or antæ capitals, enriched with foliage, than of circular or columnar capitals; but they are almost invariably found to have belonged to the interior of buildings, and not to have been used externally. In considering Greek architecture, it is necessary to bear in mind that it ceases almost immediately after the subjection of Greece to the Roman power; for there are many edifices in that country in the style of columnar arrangement of which we are now speaking besides those referred to, but they belong to Roman and not to Greek architecture. The earliest of them perhaps, and certainly the least influenced by Roman taste, is the structure called the Tower of the Winds, or of Andronicus Cyrrhestes, at Athens. A spurious example of Greek Doric, evidently executed under the Roman domination, may be referred to here; it is that of the Agora, or Doric portico, as it *is* sometimes distinguished, in the same city.

Besides the three species of columnar arrangement we have enumerated, the Greeks employed another in which statues of women occupied the place of columns. The *reason* of this too Vitruvius furnishes in a story which is as usual, totally unsupported by history or analogy; but the consequence of it is, that such figures are called Caryatides; and the arrangement has been called by some the Caryatic order. The use of representations of the human and other figures with or instead of columns is, however, common in the structures of Egypt and India; and to the former the Greeks were doubtless indebted for the idea, though they appear to have restricted its application to human female figures. Mr Gwilt infers, from various facts connected with the worship of Diana Caryatis, "that the statues called Caryatides were originally applied to or used about the temples of Diana; and instead of representing captives or persons in a state of ignominy (as the Vitruvian story goes), were in fact nothing more than the figures of the virgins who celebrated the worship of that goddess."[2]

The only architectural works of the Greeks that remain to us of any consequence, besides temples, propylæa, and Choragic monuments, are theatres; but these latter do not retain any thing connected with architectural decoration to make them interesting, except to the architect and antiquary. They are generally situated on the side of a hill, and were rather excavated or carved out in the earth or rock, than built; except the proscenium and parascenium, which being at the lower part, in front, and requiring elevation, must of necessity be built; but very little of the constructed portions in any case exists. It does not appear that the theatres afforded any provision for sheltering the spectators, or indeed the ac-

History

Plate VII. Figs. 1, 2, & 3.

Gwilt's *Chambers's Civil Architecture*, p. 30.   [2] *Ibid.* p. 57.

tors, from rain, except perhaps a covered cyrtostylar colonnade within the upper boundary wall, which, even when it existed, was of necessity very narrow and small for so large a number of people as were generally assembled; for the theatres were calculated to hold from five to fifteen and even twenty thousand persons. This, to say the least of it, must have subjected the public to great inconveniences, even in so fine a climate as that of Greece; for they were unsheltered from the sun at all times, and effectually debarred from a favourite amusement in wet weather.

No remains exist of the domestic structures of the Greeks; and we are too well aware, from the example of others, of the futility of following mere descriptions on the subject, to attempt it here; especially as the most explicit are those of Vitruvius, whom we know to have been ignorant of the arrangement of Greek temples, by those of them which exist, and may therefore reasonably suspect of ignorance with regard to things of which we have no remains. It may be taken for granted that the houses of the Greeks were less extensive than those of the Romans, as they were a poorer and less luxurious people; and we shall be able to determine those of the latter nation with great exactitude, from the actual remains of Roman towns and country mansions. The exquisite beauty of form and decoration which pervades every article of Greek origin, whether coins, medallions, vases, implements of war or husbandry, or even the meanest article of domestic or personal use of which we have specimens or representations, is evidence of the fine taste with which the mansions of the Greeks were furnished. However, ignorance of the use of the arch, inferior carpentry, the absence of glass, and ignorance of the use of chimneys, were disadvantages which the Greeks laboured under in the construction and convenient arrangement of their houses, that no degree of taste and elegance could completely countervail.

In the construction of their edifices, the Greeks seldom, if ever, had recourse to foreign materials; the stone used in their temples being almost invariably from the nearest convenient quarries, which supplied it of sufficiently good quality. The structures of Athens are built of marble from the quarries of Pentelicus, and those of Agrigentum of a fossil conglomerate which the place itself furnishes.

We have taken it for granted that the Greeks were ignorant of the properties of the arch, having too high an opinion of their good sense to think that they could be acquainted with so useful and admirable an expedient, and never use it; and no instance of its adaptation occurs in the construction of Greek edifices before the connection of Greece with Rome took place. Whether its invention should be referred to Italy or not is another question. If the great sewer at Rome called the Cloaca Maxima was constructed in the time of Tarquinius Priscus, it must be conceded that the properties of the arch were known and the arch applied in that country at an earlier period than we know the principle to have been understood and applied elsewhere; for neither Egypt nor Greece, nor any of the Grecian colonies, can furnish evidence that it was known to either Egyptians or Grecians till a long time after the period referred to, and when it may have been communicated from Italy. But it is contended that the Cloaca Maxima, as it now exists, is a work of much more recent date, and that it may have succeeded the sewer constructed by the first Tarquinius, who was moreover himself a Greek. If the first part of the objection be correct, the evidence in favour of Italy is destroyed, as far as that

work is concerned; the second is fallacious, because it is not necessary that the monarch should have brought the knowledge with him; though indeed he might have acquired it in Etruria, or it might have existed in Rome before his arrival there. Most writers on the subject are of opinion that the principle of the arch was not known, in Europe at least, nor to the nations of Western Asia and Africa, till after the Macedonian conquest, about which time it may have been invented, or acquired from some of the eastern nations who were visited by the conquerors. To these suggestions the objections hold that the arch was not applied in Egypt in the architectural works which remain of the Ptolemies, nor is it found in the Persian and Indian monuments which date beyond that period. The author of the *Munimenta Antiqua*, after a comprehensive review of all the authorities and examples on the subject, gives it as his own opinion that " Sicily was the country where this noble kind of ornament first appeared, and that Archimedes was the inventor of it."[1] The evidence appears, we think, generally stronger in favour of its Italian origin; but to whomsoever the invention may be attributed, and whensoever it was made, the Romans were the first to make extensive practical use of it; and by its means they succeeded in doing what their predecessors in civilization had never effected. It enabled them to carry secure and permanent roads across wide and rapid rivers, and to make a comparatively frail and fragile material, such as brick, more extensively useful than the finest marbles were in the hands of the Greeks without that principle. To the Greeks, however, the Romans were indebted for their knowledge of the more polished forms of columnar architecture; for, before the conquest of Greece, the structures of Rome appear to have been rude and inelegant. The few specimens of architecture which exist of date anterior to that period evidently resemble the works of the ancient Etrurians, who, though they had made considerable advances on the architecture of their Pelasgic ancestors, were far inferior in taste and refinement to the Greeks; yet it is to that people we are inclined to attribute the invention of the arch, from whom the Romans acquired their knowledge of its use, and that degree of civilization which they possessed before the epoch referred to. It may be presumed that the Etrurians had also originated the style of columnar architecture which Vitruvius describes and calls Tuscan; but as no example of it exists, at least nothing that answers his description of it, we cannot tell positively what it really was; for, as we have before remarked, descriptions without a model, of architecture particularly, are quite unintelligible, as far as understanding a new style goes. Whatever then was the style of architecture in Rome before the conquest of Greece, it was either exploded by the superior merit and beauty of what the Romans found in that country, or combined with it, though frequently the combination tended to destroy the beauty of both. In the porticoes of the temple of Antoninus and Faustina, and of the Pantheon, at Rome, the chaste simplicity of a Greek columnar composition is preserved, and in the magnificent dome of the latter edifice, and in the long extended aqueduct, it is fully equalled. But the triumphal arch of the Romans, a hybrid composed of columns and arches, is devoid alike of simplicity and harmony, indeed of every quality which constitutes beauty in architecture.

In the transference of Greek columnar architecture to Rome, a great change was effected, independently of those combinations. The less refined taste of the Romans

[1] *Munim. Antiq.* by Edward King, Esq. F. R. S. and S A. vol ii. p. 268.

could not appreciate the simple grandeur and dignified beauty of the Doric, as it existed in Greece. They appear to have moulded it on what we suppose their own Tuscan to have been; and the result was the mean and characterless ordinance exemplified in the lowest story of the theatre of Marcellus at Rome, and in the temple at Cora, between 30 and 40 miles south of that city. Not less inferior to the Athenian examples of the Ionic order, than the Doric of Cora is to the Doric of Athens, are the mean and taste-less deteriorations of them in the Roman temples of Manly Fortune and Concord. It was different, however, with the foliated Corinthian, which became to the Romans what the Doric had been to the Greeks—their national style. But though they borrowed the style, they did not copy the Greek examples. In Rome the Corinthian order assumed a new and not less beautiful form and character, and was varied to a wonderful extent, but without losing its original and distinctive features. The ex-ample of the temple of Vesta at Tivoli differs from that of the temple of Jupiter Stator in Rome, as much as the latter does from the ordinance of the choragic monu-ment of Lysicrates at Athens; and all three are among the most beautiful examples of the Corinthian order in exist-ence—if indeed they are not pre-eminently so—and yet they do not possess a single proportion in common. It must be confessed, moreover, that if the Romans had not good taste enough to admire the Doric and Ionic models of Greece, they had too much to be fond of their own, for they seldom used them. Both at home and abroad, in all their conquests and colonies, wherever they built, they employed the Corinthian order. Corinthian edifices were raised in Iberia and in Gaul, in Istria and in Greece, in Syria and in Egypt; and to the present day Nismes,[1] Pola, Athens, Palmyra, and the banks of the Nile, alike attest the fondness of the Romans for that peculiar style. We cannot agree with the generally received opinion, that Greek architects were employed by the Romans after the connection between the two countries took place; for the difference between the Greek and Roman styles of archi-tecture is not merely in the preference given to one over another peculiar mode of columnar arrangement and com-position, but a different taste pervades even the details: though the mouldings are the same, they differ more in spirit and character than do those of Greece and Egypt, which certainly would not have been the case if Roman architecture had been the work of Greek architects. In-deed, were it not for historical evidence, which cannot absolutely be refuted, an examination and comparison of the architectural monuments of the two countries would lead an architect to the conclusion, that the Corinthian order had its origin in Italy, and that the almost solitary perfect example of it in Greece was the result of an acci-dental communication with that country, modified by Greek taste; or that the foliated style was common to both, with-out either being indebted to the other for it. The Ro-mans conquered Egypt as well as Greece, but we do not find that they adopted any of the peculiarities of Egyptian architecture. They carried away indeed the obelisks and many of the sculptures of Egypt, as trophies of their con-quests or as ornaments of their city; but they neither made obelisks nor constructed temples to Egyptian divinities in

the Egyptian style. If, however, Greek architects were employed by the Romans, they must have made their taste and mode of design conform to those of their con-querors much more readily than we can imagine they would as the civilized slaves of barbarian masters; and it is too clear to be disputed, that the Roman architecture is a style essentially distinct from the Greek. This is elucidated by the fact that many of the minor works of sculpture in connection with architecture, such as candela-bra, vases, and various articles of household furniture dis-covered at the villa of Adrian, near Tivoli, and at Her-culaneum and Pompeii, are fashioned and ornamented in the Greek style, while others are as decidedly Roman in those particulars; rendering it evident that such things were either imported from Greece, or that Greek artists and artisans were employed in Italy, who retained their own national taste and modes of design. It is probable, nevertheless, that both the architects and the artists, natives of Rome, qualified their own less elegant productions by reference to Greek models; but that the Romans derived their architecture entirely from the Greeks, may certainly be disputed.

Half the extent and magnificence of the architectural works of the Romans is attributable to their knowledge and use of the arch, which enabled them, as we have already intimated, to make small parallelopipedons of burnt earth more extensively applicable to useful purposes than any other material could be, from the greater cost of provid-ing and preparing it; whereas brick can, in almost every place, be made on the spot in which it is wanted. There is a very false notion abroad as to the richness of the ma-terials used for building in Rome, induced by the inflated accounts of travellers and poets, who attempt to disguise their ignorance, while they display it by filling Rome with what it never contained—marble temples, palaces, and baths. The truth is, that Rome was built, not of marble, nor even of stone, but of brick; for in compari-son to the quantity of brick, it may be safely asserted that there is more stone in London than there was in imperial Rome. Almost all the structures of the Romans indeed were of brick—their aqueducts, their palaces, their villas, their baths, and their temples. Of the present remains, it is only a few columns and their entablatures that are of marble or granite, and two or three buildings of Traver-tine stone;—all the rest are brick. The Colosseum, the Mausoleum of Adrian, the Tunnel Sewer, the Temple of Manly Fortune, and the ancient bridges on the Tiber, are of Travertine stone; the remaining columns of the more splendid temples, the internal columns and their accessories of the Pantheon, the exterior of the imperial arches, and the cenotaphial columns of Trajan and of Antonine, are of marble: but the Imperial Mount of the Palatine, which holds the ruins of the Palace of the Cæsars, is but one mass of brick; the Pantheon, except its portico and in-ternal columns, &c., is of brick; the Temples of Peace, of Venus and Rome, and of Minerva Medica, are of brick; and so, for the most part, were the walls of others, though they may have been faced with marble or freestone. The Baths of Titus, of Caracalla, and of Diocletian, are of brick; the city walls are of brick; so are the extensive remains of the splendid villa of Adrian, and those of

---

[1] Bordeaux *did.* A century and a half ago there existed at Bordeaux very considerable remains of a most interesting Roman edifice, of which no authentic record is preserved but a slight sketch by Perrault, the architect of the great front of the Louvre, who delineated it a few years before its destruction by the government, and who termed it one of the most magnificent and most entire of the Roman monuments then remaining in France. The editor of the new edition of Stuart's Athens, speaking of this, says, " on this occasion the reflection presents itself, that while the Turks are reprobated for appropriating the columns of ancient Athens, in their haste to raise a wall to defend their town from the predatory Albanians, here, in the vaunted age of Louis XIV. (in his kingdom and under his government, may be added) the finest production of ancient architecture in France was more recklessly demolished to make place for the fortifications of Bordeaux, deliberately constructed by Vauban; and no architect, either of the city or government, has preserved for posterity the details of so noble a monument." (*Antiquities of Athens,* new edition, vol. iii. p. 120.)

the villa of Mecænas at Tivoli; the palaces of the Roman emperors and patricians at Baiæ and in other parts of Italy; and so, it may be said, are the remains of Herculaneum and Pompeii, for the houses in those cities are generally built of alternate double courses of brick, and courses of stone or lava. In most cases, at Rome and in the provinces, stucco formed the surface which received the decorations. From the above enumeration, it will appear how much more variously the Romans built than any of their predecessors in civilization did. In Egypt we find no indications of edifices of real utility or convenience, nothing but temples and tombs,—and in Greece there is but a small addition to this list; but in Rome are found specimens of almost every variety of structure that men in civilized communities require. Much of this also may be attributed to the knowledge they possessed of the properties of the arch, which may be considered among the most admirable and useful discoveries ever made in the practical applications of mechanical science. It entered into the composition of every structure, and made the rudest and cheapest material of more real value than the most costly. It not only superseded the use of long stone beams, but was constantly used in places where indeed joists of wood would have been much more convenient, giving support to the opinion that even the Romans were not skilled in the application of timber to their edifices; though, on the contrary, it is difficult to understand how Rome could become subject to such a dreadful conflagration as that which occurred in the reign of Nero, if timber had not been employed in the ordinary houses of the city to a much greater extent than would appear from existing remains. The domestic structures of Herculaneum and Pompeii were evidently never very susceptible of fire, from the small quantity of timber required in their construction; and discoveries which are made from time to time, of portions of the ordinary houses of ancient Rome, under the pavements of the modern city, evince that they were very similar to them in almost every particular. The infrequency of stairs, and the meanness of those which exist leading to upper apartments in the houses of those cities, induce the belief that the Romans seldom built above the ground story, and that their skill in carpentry was not very great; otherwise they would more frequently have had recourse to so easy and convenient a mode of extending room as upper stories offer. There are, however, other things which tend to prove that carpentry was well understood by the Romans; and the most remarkable is the bridge that Trajan built over the Danube, the piers of which are said by Dion Cassius to have been 150 feet high and 170 feet apart. Now, whether the bridge itself consisted of a wooden platform, as there is much reason to believe, or was of stone arches, as the historian intimates, the skill which constructed centring for the latter, or laid the platform from pier to pier in the former case, of that immense extent, was amazing; nevertheless, such skill in carpentry is not evinced by the remains of the civic and domestic structures of the Romans, in which arching in all its varieties was used where carpentry would have been better. Of their joinery we know nothing; but it does not appear, from the last-quoted mode of ascertaining such things, to have been much practised by them—mosaic pavements supplying the place of flooring, and stucco that of wainscoting: the luxury of windows being unknown, their fittings were not required; and doors, it would appear, were uncommon, except externally—the internal doorways being most probably covered with something equivalent to the quilted leather mats suspended from the lintel, which are used instead of swinging doors at the entrances of the churches in Italy at the present time. Although the Romans did not use marble to the extent

that has been supposed, yet they were extremely luxurious in the use of costly stones. Marbles of every variety, and from all parts, were used in Rome; and columns were made of Egyptian and other granites, and of porphyry. In Greece, and the Grecian colonies which were conquered by Rome, the edifices of the Romans might be distinguished by the foreign marbles used in them, if the style of their execution were not sufficient otherwise to determine them.

The mingling of columnar and arcaded arrangements in the same composition appears to have been the grand cause of the deterioration of Roman architecture. It occasioned unequal and inordinately distended intercolumniations and broken entablatures: these a vitiated taste repeated, where the necessity that had first occasioned them did not exist; and harmony and simplicity being thus destroyed, the practice went on deteriorating until it was made to produce such monstrous combinations as the Palace of Diocletian at Spalatro, and the Temple of Pallas, or ruins of the Forum of Nerva in Rome, present. It was indeed a fall from the grandeur, harmony, and noble simplicity of the interior of the Pantheon in its pristine state, to the hall or xystum of the baths of Diocletian, which now exists as the church of Santa Maria degli Angeli, with its straggling columns and broken and imperfect entablature; or from the temple of Jupiter Stator to that of Concord or the arch of Septimius Severus.

Architecture as a fine art was already extinct among the Romans when the seat of empire was transferred to Constantinople; so that, however great were the extent and splendour of the edifices, we cannot suppose them to have possessed any of those qualities which give to the Parthenon at Athens, and to the interior of the Pantheon at Rome, the charm they possess; unless the Greeks had recourse to the monuments of their own country, and used them as founts from which to draw matter for the composition of the edifices of their new capital. This, indeed, is possible, for there appears to have been, even in Rome, at and after the time of Constantine, a recurrence to the ancient simplicity, though, truly, without any of that beauty and elegance of form in the details, and of proportion in the general arrangement, which constitute half the merit of works of architecture. The change of religion which took place under Constantine led to the destruction or destitution of many of the noblest structures in Rome. The ancient Christian basilicas are for the most part constructed of the ruins of the more ancient Pagan temples, baths, and mausoleums; and in them a much greater degree of simplicity, and consequent beauty, pervades the columnar arrangements than existed perhaps in some of the previous combinations of the same materials. Frequently, however, the collocation of various parts was most unapt; and gross inconsistencies were recurred to, to get rid of the difficulty of combining discordant fragments. Sometimes it was necessary to make up with new, what was wanting of old materials, whose forms were rudely imitated.

In those countries which received the Christian religion from Rome, but which did not contain mines of architectural material in temples, amphitheatres, and palaces, as Italy did, and indeed in the other parts of Italy itself which did not contain them as Rome did, churches were constructed in imitation of those of the metropolis of the Christian world. These, being the work of a semibarbarous and unpolished people, were of necessity rude and clumsy. Hence arose the Gothic architecture of the middle ages, and not from any previously existing style of architecture among the northern nations who overran Italy and subverted the Roman power. The rude Celtic

Gothic or Pointed architecture.

ARCHITECTURE.

monuments were the only specimens of architecture they possessed, and the performance of their unhallowed rites appears to have been long transferred even from them to the groves, or it may be that the stone circles and temples themselves were called groves. This, however, is of but little consequence to our purpose. The fact is indisputable, that nothing existed among those nations that could have given rise to the rude style of architecture referred to, which was indeed introduced to them by the Christian religion in the manner we have stated. It will be found in what are called the Saxon and Norman styles of this country, and to a greater or less extent in all the countries of Europe in which the Romans had been masters, and particularly in those which adhered to the Roman communion in the great division of the churches. The general forms and modes of arrangement peculiar to Roman architecture may be traced throughout; in some specimens they are more, and in others less obvious, but the leading features are the same. This is more evident in Italy than elsewhere. In the early Roman basilicas and churches, some of which are of the Constantinian age, and which were constructed with the matter and in the manner related, the first divergencies occur; in those which are later they are still greater, and distance of time and place appears still to have increased them, till what may be called a new style was formed, having peculiarities of its own, but yet more clearly deducible from its origin than Roman is from Greek or Greek from Egyptian. As might be expected, this style was not the same in all the countries which practised it; it was derived, in them all, from the same source as we have shown, but was materially influenced by the habits, manners, and state of civilization in which the various nations were, and much too by their means of communication with Rome. This, with strict propriety, may be called Gothic architecture, as it was partly induced by the Gothic invasions of Italy, and was most generally practised by the nations to whom that term may with equal propriety be applied. It arose in the fourth century, and was subverted in the twelfth by the invention or introduction of the pointed arch, which marks a new era, and was destined to give birth to a new style in architecture. Where, when, and with whom the pointed form originated, has been more discussed and disputed than the discovery of the properties of the arch itself. Some have contended that it was suggested by the intersections of semicircular arches, as they were employed in ornamenting the fronts of edifices in the preceding style; some, that groined arches of the same form gave the idea; others have referred it to the interlacing of the branches of trees when planted in parallel rows,—to an imitation of wicker-work,—to a figure used on conventual seals,—to the principle of the pyramid,—to Noah's Ark, —to chance. Its invention has been accorded to almost every nation, civilized and uncivilized It has been claimed by Germans for Germany, by Frenchmen for France, by Scotsmen for Scotland, and by Englishmen for England. Italians have not directly laid claim to the honour for themselves, but it has been given them by others. Such a mass of conflicting opinions, almost all supported by some show of reason, and more or less by evidence, may be called a proof of the impossibility of determining the question, and therefore we shall not attempt it. There is one striking fact, however, which has been too much overlooked by many of the theorists in the discussion of the question; it is, that the pointed arch made its appearance almost at the same moment of time in all the civilized countries of Europe. This is proved by the controversies of those who, more patriotically than philosophically, claim its invention for their respective nations; for none of them can produce genuine specimens of it before a certain period, to which they can all reach. Now, if it had been invented in any of the European nations, that one would certainly have been able to show specimens of it of a date considerably anterior to *some* of the others; for though it might by chance have been soon communicated to any one of them, the improbability is great that it would immediately have reached them all, and have been at once adopted by all, to the subversion of their previously practised forms of construction. The infrequent and imperfect modes of communication between the different countries of Europe at the period referred to, furnish another reason why it is not probable that a discovery of the kind should travel rapidly from one to another. Considering these things, and particularly the fact of the almost simultaneous introduction of the pointed arch to the various nations of Europe, as it appears by their monuments immediately after the first crusade, in which they all bore a part, connected with existing evidence that it was commonly used in the East at and anterior to that period, it seems to be the most rational theory, that a knowledge of it was acquired by the crusaders in the Holy Land, and brought home to their respective countries by them. This, indeed, is the opinion of many of those who have written on the subject; and without contending that the evidence in its favour is quite conclusive, we think it more satisfactory than any other. In Europe there are found rude approaches to the pointed form in some of the earlier Gothic structures; but we believe it may be safely asserted, that nothing can be indicated of a date beyond that of the first crusade, approaching the simple but perfect lancet arch, which, it is not denied, came into use immediately after that period; whereas tolerably well authenticated examples of it are found in the East, of sufficient antiquity to induce the opinion that it was at that time imported from thence. It is, moreover, indisputable that the Saracenic or Mahometan nations do use, and have used, the pointed arch; but they were never known to adopt any European custom or invention of any kind till very lately How then can they be supposed to have availed themselves so readily as they must have done, if it be of European origin, of so unlikely a thing to attract a Moslem's attention, as the peculiar form and structure of an arch? and when and where in Europe had they an opportunity of contemplating it until long after it is known to have been in common use among them? With what nation of the East, and in what manner, the pointed arch originated, are points equally difficult to solve. We have not been able to discover that the properties of the arch were known to the Egyptians or to the Greeks, and there is no evidence to show that they were known to the Persians or to the Indians of ancient times; but structures are found in the countries of those nations in which chambers are domed, and apertures headed in the form of a pointed arch, produced, however, by gathering or corbelling over, and not by arched structure. It is not improbable, therefore, that such things being before the eyes of men, when the properties of the arch became known that form would be repeated upon it, and the result would be the lancet arch,— the prototype, the germ of the style. The pointed arch, on its introduction into Europe, does not appear to have been accompanied by its ordinary accessories in after-time; its light clustered pillars—its mullions, foliations or featherings, and graceful tracery—these resulted from its adoption: so that whether the arch itself was invented in Europe, or imported from the East, to the European nations must be assigned the credit of educing the beautiful style of architecture whose distinguishing feature it is.

It may be doubted whether Venice was not the parent of the style, for very early specimens of the pointed arch are certainly found there, in private houses as well as in the basilica of St Mark. In the former they are generally of the ogee or contrasted form, in windows formed by columns or mullions, with, in certain places, approaches to foliations and tracery; and in the basilica the lancet arch is not uncommon. The commercial connection of that city with the eastern nations may easily account for its presence there, even before the first crusade; and Venice is known to have been one of the thoroughfares from the other parts of Europe to the Holy Land. But the peculiar mode of arrangement in the Venetian style does not appear to have been adopted north of the Alps; so that, however original it may be, it can hardly be considered the progenitrix of the school, or the model on which it was formed.

Before proceeding further with this subject, it is necessary to determine by what name to call the style whose progress we have yet to contemplate. There would be no greater propriety in calling it Saracenic because its distinctive feature originated in the East, even if that point were conceded, than in calling all architectural combinations which derive their character from the use of columns in them by the name of the nation in whose works we find columns first used, and from whom the idea of them may have been acquired. Neither can it with any degree of fitness be called Gothic: that term, we have seen, applies to the style that preceded it, and was first given to the pointed-arch style opprobriously, during the offuscation of good taste that succeeded its subversion. In Italy it had never taken root, as in the countries north of the Alps—the ancient Roman monuments having continued to influence the national architecture, it would appear, throughout the middle ages; for the ecclesiastical structures of that country, though rude, were never so rude as they were in other places, and a better style had so far formed itself before the introduction of the pointed arch, that it was hardly received there. Indeed, whatever edifices of merit Italy possesses in its manner, are, with hardly an exception, by German architects, few Italians having ever qualified themselves to practise it. When, therefore, what has been called " the revival of architecture" took place in the fifteenth century, under Brunelleschi and his successors, the rude structures of their own country, the precursors and contemporaries of our Saxon and Norman edifices, were called Gothic; but the style of which the pointed arch is the characteristic feature, was always distinguished as the German manner, *Maniera Tedesca*. The disgrace of applying the opprobrious term Gothic to it attaches itself to an Englishman, Sir Henry Wotton. It was continued by Evelyn, who applied it more directly; and the authority of Sir Christopher Wren finally settled its application. Its injustice is, however, Plate XXI. rendered very obvious, by comparing the front of Pisa Cathedral, the best example, perhaps, of Gothic, or merely deteriorated Roman architecture, with that of York Minster, which holds an almost equal rank in the style of which it is an exemplar. The presence of the pointed arch, on the singular oriental-looking cupola of the former, shows it to be one of the latest edifices in its style, overtaken by that before it was completely finished. Within the present century a better taste has been formed, in this country particularly, and has led to the appreciation of that, which is, indeed, our national style; and within that period many attempts have been made to explode the universally-decried, unjust, and totally irrelevant appellation, but without effect. Sir Christopher Wren himself attempted to change it to Saracenic, believing that not merely the arch, but the style generally, was borrowed from the Saracens It was, however. too late—he had already used the other. Dr Stukely wished to call it Arabian. Some writers called

it Italian, others German, others Norman or French, others British, and many have contended for the exclusive term English; and to this last the Society of Antiquaries lent its influence, but with equal inefficiency, for the term Gothic still prevails. Mr Britton, than whom perhaps no man possesses a greater right to affix an appellation to the pointed-arch style, from the splendid services he has done it in the publication of his Cathedral and Architectural Antiquities, wishes to introduce a term which is not at all unlikely to succeed, as it is equally appropriate and independent of national feeling and hypothetic origin. He calls it Christian architecture. This, as a generic term, would admit each nation possessing specimens of it to distinguish its own species or style; and as the varieties of Hellenic architecture are known by the names of the tribes or nations who are presumed to have originated them—Dorian, Ionian, and Corinthian—so might Christian architecture be English or British, German, French, &c for each has its peculiarities. These species would again individually admit of classification, according to the changes each underwent in the course of its career. One strong objection, however, in our view of the case, lies to Mr Britton's distinctive appellation. It is, that " Christian" applies as well, if not better, to the real Gothic style—that which arose on the extinction of Roman architecture, and was subverted by the introduction of the pointed arch, and which, indeed, owed its diffusion and progress, if not its origin, to the Christian religion. We are therefore still left to seek an appellation; and, in the absence of a better, will use the term Pointed, which is not only distinctive, but descriptive; and it has, too, the merit of being general, so that it may mark the genus,[1] while the national species and their varieties may be distinguished by their peculiarities as before.

The Pointed Arch was a graft on the Gothic architecture of northern Europe, as the circular arch of the Romans had been on the columnar ordinances of the Greeks; but with a widely different result. The amalgamation in the latter case destroyed the beauty of both the stock and the scion; while in the former the stock lent itself to the modifying influence of its parasitical nursling, gradually gave up its heavy, dull, and cheerless forms, and was eventually lost in its beautiful offspring, as the unlovely caterpillar is in the gay and graceful butterfly.

We have seen that architecture had its origin in religious feelings and observances—that its noblest monuments among the pagan nations of antiquity were temples to the divinity—that the rude nations of the north in the middle ages devoted their energies, after their conversion to Christianity, to the construction of edifices for the worship of the Almighty; and we find, again, that the most extensive and most splendid structures raised by the same people, when the light of learning had begun to shine upon them, and a new and more beautiful style of architecture was introduced, were dedicated to the same purpose. In addition, however, many, hardly less magnificent, and not less beautiful, were raised for the purposes of education, and became the nurseries of science and literature. Kings and nobles also employed architecture in the composition, arrangement, and decoration of their palaces and castles; but still, for domestic purposes, its aid was hardly required beyond the carving grotesque ornaments on the wooden fronts of houses in towns.

When the practice of building houses in stories commenced cannot be correctly ascertained; but it appears to have arisen during the middle ages. We frequently, indeed, find an apparent equivalent for the term *story* used by the ancient writers, both sacred and profane; but it must have reference to something else—some peculiarity of which we are not aware; for none of the ancient re-

---

[1] See also " Styles," *Glossary*.

mains, whether of public or private structures, give reason to believe that it was a common practice even among the Romans; much less was it likely to be so among the eastern nations, with whom the practice is not very general, nor is it carried to great extent even at the present day. Indeed, without considerable proficiency in the art of construction, it is hardly practicable to build in stories with such slight materials as were used by the Romans in their domestic edifices; and their remains do not evince the requisite degree of proficiency. We find, however, in the oldest existing works of the middle ages, and particularly in some of the secular structures of Venice which are among them, a degree of intelligence evinced in that respect far surpassing any thing in those of the ancients. Possibly the skill was principally acquired in that city from the necessity of making artificial foundations, which consequently required a superstructure not unnecessarily cumbrous; and again, to make slight walls sufficiently strong, they must be skilfully bonded in themselves, and bound together, which could only be done by means of a material possessing considerable length and great fibrous tenacity—whence framed floors of timber. These, by their strength, their obvious utility and convenience, added to the want of space which existed in a thriving and populous community on a very restricted spot of dry land, superinduced, in the second place, the building of additional stories, which would soon be imitated in other places. But in what manner soever the improvement took place, the fact is certain that the acquisition was made; and we find it applied in all the works of the European nations, both ecclesiastical and civil, from the ninth and tenth centuries downwards. The combination of masonry and carpentry in building tended greatly to the advancement of both; for, it being required at times to make them act independently of each other, additional science and art were necessary, as the proportions must be retained that were given to similar works in which they co-operated. Hence the wondrous skill evinced in the vaulted roofs and ceilings, in the towers and lofty spires, of some of our Pointed cathedrals for the one, and the splendid piece of construction in the roof of Westminster Hall for the other. To this point Sir William Chambers, who was no depreciator of the merits of the Romans in architecture, says, " In the constructive part of architecture the ancients do not seem to have been great proficients :"[1] then having referred many of what he calls the " deformities observable in Grecian buildings" to want of skill in construction, he continues, " neither were the Romans much more skilful ; the precepts of Vitruvius and Pliny on that subject are imperfect, sometimes erroneous, and the strength or duration of their structures is more owing to the quantity and goodness of their materials than to any great art in putting them together. It is not, therefore, from any of the ancient works that much information can be obtained in that branch of the art. To those usually called Gothic architects we are indebted for the first considerable improvement in construction. There is a lightness in their works, an art and boldness of execution, to which the ancients never arrived, and which the moderns comprehend and imitate with difficulty. England contains many magnificent specimens of this species of architecture, equally admirable for the art with which they are built, the taste and ingenuity with which they are composed." To this Mr Gwilt, in his new edition of Sir William's work, adds in a note, " there is more constructive skill shown in Salisbury, and others of our cathedrals, than in all the works of the ancients put together."

Pointed architecture took root and grew with the greatest vigour in Germany and Great Britain, and in those provinces, principally, of France which were connected with England; but in this country its course is the most marked, and its advances are the most easily traceable. We find in various portions of the same edifice, according to the period of its construction, exemplifications of the style, from the ingrafting of the simple lancet arch on the Norman or Gothic piers in the time of Henry II. to the highly enriched groinings and ramified traceries of the age of Henry VII.; but the changes are so gradual, and are so finely blended, that the one in advance appears naturally to result from that which comes before it. Whether the nations of the Continent, then, borrowed from us, or were themselves originators, it is very clear that we did not borrow; for our structures bear the strongest possible marks of originality, as the advances can be traced from one thing to another on them; and such is not so completely the case with theirs. Moreover, the latest manner, and certainly not the least beautiful, the Corinthian order of Pointed architecture, is almost peculiar to this country. Neither Germany nor France can produce edifices in the style of St George's Chapel at Windsor, King's College Chapel at Cambridge, and Henry VII.'s Chapel at Westminster. The structures of Scotland in the Pointed style so much resemble those of England, that they must be considered of the same school; Roslin Chapel is one of the few specimens which indicate a connection with the Continent. Ireland contains but few examples in any degree of perfection, and they are, of course, of the English school. The German school was next in merit to the English in the practice of Pointed architecture. In the extent and magnificence of its attempts, perhaps, that country excelled; but few of the great structures in Germany were ever completed. In regularity, however, they have generally an advantage over those of England, being mostly in the same manner throughout, as far as they were carried; whereas few of the greater edifices of this country were begun and completed without considerable variations in the style. But the Germans were never so successful in the splendour and beauty of their interiors as the English; indeed in that particular our Pointed structures are strikingly superior to every other: nor is their ornament generally so effective as ours. The Flemish style of Pointed architecture is hardly a variety of the German, but may be classed with it through the whole course of its history. Italy, we have said, possesses but few structures in the Pointed style, and they are for the most part the work of German architects, which their appearance indeed bespeaks. Milan cathedral, or " the Duomo," as it is called, is the most renowned edifice in the style that Italy contains; but it has few beauties in the eyes of those who are accustomed to the models of Great Britain, France, and Germany. The Patriarchal Church of St Mark at Venice is a genus *per se*. It was constructed by a Constantinopolitan architect in the ninth or tenth century, and may be a specimen of the architecture of the Byzantine capital at that time. The few examples in Sicily of the pointed arch may be attributed to the Norman conquerors of that island; and so indeed may most of those which are found in the continental part of the same kingdom. Although France contains many fine specimens of Pointed architecture, it can hardly be considered indigenous to that country. On the German frontier they resemble the German style; and in the provinces which were formerly connected with England they are different, and more like the English styles: cer-

---

[1] Gwilt's *Chambers's Civ. Arch.* p 128.

tain it is, that after their connection with this country was broken off, its practice fell into disuse, and nothing of consequence in it was posteriorly produced in any of them. This fact is an argument against the presumption that England was indebted to Normandy and Norman architects for its improvements in Pointed architecture; it tends rather to prove the opposite. It must be confessed, moreover, that there is an air of cumbrous massiveness in the Pointed style of Normandy which renders it peculiar, and perhaps marks its more close relationship with the earlier Norman Gothic. Like the German examples, too, those of France are generally inferior in internal richness and beauty to those of England. Pointed architecture in Spain never acquired that degree of consistency and elegance which might justify us in speaking of the Spanish specimens of it as forming a style. The edifices in Spain of the ages of Pointed architecture are more in accordance with the Moorish than with the European manner, and may perhaps be more correctly considered as an off-shoot of the former than of the latter. Though not in the Pointed style, the Moorish or Saracenic structures in Spain may be referred to here. They are in a very peculiar manner, which, it would appear, their authors brought with them from the East. Probably it grew out of some of the earlier styles of architecture, as the Gothic and Pointed did out of the Roman, and was not the result of design. The really distinctive feature of the Saracenic style is the horse-shoe arch, which is the greater segment of an ellipsis, nearly, on a conjugate chord. The columns from which the arches are sprung are slender, and the superincumbent masses are broad and heavy, giving an air of the intermingling of Chinese and Egyptian, both of which this style may be said to assimilate. The enrichments of Saracenic architecture are very much confined to flat surfaces, the walls being sculptured all over with monotonous ornaments, which produce an effect very similar to that produced by the hieroglyphics on the flat surfaces of the Egyptian temples, and possibly were derived from them. The most distinguished monument of this style in Spain is the Alhamrā at Granada. The most distinguished specimen of Pointed architecture in Portugal is the church of the convent of Batalha, which was constructed by an Irish architect, who appears to have modified the style of his own country (the English), by the manner of the country itself, which is nearly that of Spain.

What the expansive dome is to Roman architecture, the graceful spire is to Pointed. Bell-towers appear to have been added to Christian churches at a very early period; but it is much to be doubted whether the pyramidal pinnacle or spire was ever used before the introduction of the pointed arch, though one or two doubtful examples exist. These, certainly the earliest specimens of it, are simple cones, whose vertical bisection would be nearly an equilateral triangle: the angle at the apex was gradually made less, and as it diminished the altitude was increased, till at length resulted an object even more beautiful than an Egyptian obelisk, which would of itself indeed be a sufficient warranty for the appellation we have given to the style that it crowns. The spire was at first round, solid, and unornamented; it then became polygonal, and finally octagonal, though there are examples of square spires. They are sometimes plainly ribbed, sometimes crocketed, and in some instances pierced; and in the finer examples are almost invariably surmounted by a rich finial in the style of ornament peculiar to the time of its execution. In some cases the whole structure is a pyramis or spire, but most commonly the spire rests on a rectangular and upright tower. The Rev. W L. Bowles has suggested that the spire was at first built on the bell-tower as a beacon or land-mark

for the guidance of the traveller and the distant parishioner; and adduces as evidence, the fact, that in the hilly parts of England spires are hardly to be found except in modern churches. The old village church on a hill has a plain square tower, merely consisting of about two cubes, which can be seen at the greatest distance the nature of the country will allow any object to be distinguished; whereas in the level parts of the country, where a low tower would be lost amidst the foliage of its own church-yard, and be completely indistinguishable at a very short distance, spires are their almost invariable accompaniment. It may be added, that the tapering spire is almost unknown in Italy, and in France it is frequent only in Normandy; but in no part of the Continent is it so common as it is in this country.

We have already given our reasons for thinking that the pointed arch originated in the East; but whether it did or did not, it has been very extensively used in various parts of Asia, and nowhere in more sumptuous edifices, or to such effect, as by the Mohammedan conquerors of India in various parts of that country.

The opening of the Italian school of architecture on the resuscitated dogmas of Vitruvius was the signal for the extinction of that of the beautiful Pointed style. Fortunately, however, its effects were a full century in reaching this country, and during that period many of our most elegant structures came into being; and many of those of earlier date which had been commenced before, or during the wars of the Roses, and left unfinished, were completed. The first indication we have of the presence of the *Cinquecentist*, the real Goth, is in the tomb of Henry VII., which was executed by Torregiano, an Italian artist, who, it would appear, was obliged to have some respect to the style of the edifice in which his work was to rest; but his preconceived ideas of propriety and beauty were too strong to allow him to omit the characteristics of his school, and the result is a strange mixture of both. From that time the Pointed style was rapidly deteriorated, being overborne by the devices of Italy. On the Continent the latter were already predominant, for during the whole of the fifteenth century the current had been setting from Italy over every part of Europe which received its religion from Rome; and this country was only the last to be overwhelmed by it. Before quitting the part of the subject having reference to our national style of architecture, it may be well to controvert the absurd but too prevalent idea, that we are indebted to foreigners, and particularly Italians, for the excellence of our ancient works. After what has been already said, perhaps it may be unnecessary to do so here, seeing that we have described our specimens of the Pointed style as being not only fully equal in composition, construction, and execution to those of any other country, but as being absolutely in a different manner, having peculiarities which no other nation has ever equalled in beauty and elegance. But, to put the case in a clear point of view: If foreigners were employed to design and execute for us, it is not less strange that they should surpass their own works at home, than that they should make inventions and improvements for us (or let them be called mere variations) which were not in turn executed in their own countries. We know very well that works of architecture and sculpture which have been executed by foreigners in this country, since the explosion of Pointed architecture, are in the style of Italy or France, and not according to the manner prevalent in this country at the time of their execution. Moreover, for one whole century this nation alone adhered to the Pointed style, during which works were produced, that, for originality, exuberance of fancy, and beauty, spirit, and excellence of execution, have been seldom equalled and never surpassed; while all the architects, artists, and workmen of the Continent were rendered totally unable to assist us by the change which had taken place in their practice. If this required proof, it is proved in the case of Torregiano just referred to, who sculptured Henry VII.'s tomb, and in that of Hans Holbein, who designed architectural works for us in the *classical* manner; and if the Torregianos and Holbeins had been employed during the 15th century, would they not have done the same?

If the architecture of Italy never fell away so much from the more classic style of Imperial Rome as that of the northern nations did, neither did the Italians ever possess that more than equivalent, whose splendid course we have last noticed. Whilst the Pointed style was almost exclusively known and practised in Germany, France, and the British Islands, the Italians were gradually improving on their Gothic style; yet the improvement was more evinced in their secular than in their ecclesiastical structures. Florence, Bologna, Ferrara, Venice, and many other cities of Italy, contain palaces and mansions of the twelfth, thirteenth, and fourteenth centuries, which for simplicity and classical beauty far excel most of those in the same and other places of the three subsecutive centuries. The contemporary churches, however, do not exhibit the same degree of improvement, forming, as it were, an anomaly in the history of architecture; a change in it being first developed in secular structures, and then applied to those devoted to the worship of the divinity; for many of the churches of the fifteenth century are in this, which may be called the early Italian style, or *Trecento*, as that which followed it is known as the *Cinquecento*.[1] Circular arches, and plain continuous horizontal cornices, and pilasters but slightly projected, with simple but generally tasteful and elegant enrichments of foliage and carved mouldings, are the most striking characteristics of the *Trecento ;* but columns, and the arrangements depending on them, except as collocated with pilasters, are very infrequent in it. In various parts of Italy, and particularly in Venice and some of the Venetian cities, this style produced many of its best works, both secular and ecclesiastical, even during the fifteenth century; but it gradually gave way to, though in some instances its influence may be traced even when it had been overborne by, the new style.

The first step taken towards the reformation of architecture was by Filippo Brunelleschi, a Florentine architect, who was employed to finish the cathedral or duomo of his native city early in the 15th century; a work which had been commenced more than a century before on the design of Arnolpho, a Florentine also, but which still required the cupola when its completion was intrusted to Brunelleschi. The edifice is in the Italian Gothic style, slightly modified by what we have termed the *Trecento*, which his superior taste and talent induced him to attempt to supersede, and bring the world back to the classic style of ancient Rome. The construction of the cupola gained him great reputation and the confidence of the public, which he employed to advance his favourite scheme. To use the words of an Italian writer on the subject, " On the example of so wise and skilled a man, other architects afterwards devoted themselves to free architecture of the monstrosities introduced by barbarism and excessive license, and to restore it to its primitive simplicity and dignity."[2] But to what did they have recourse to effect this? Did they examine and study the remains of antiquity in Greece and Rome, in Italy and elsewhere? No! they referred to the writings of an obscure Latin author, who professed to give the principles and practice of architecture among the Greeks and Romans, but paid no more attention to the existing architectural works of those nations than if they had never been, although one could hardly walk the streets of any of the old cities in the south of Italy without seeing Roman edifices, whilst Rome and its vicinity was, as it still is, full of them. All the use, however, that these self-called " restorers" of architecture made of the works of the ancients, was to use them as lay-figures, or frame-work, to model on, according to the proportions and directions given by Vitruvius; and the effect was formality and mannerism in those who adhered to the dogmas of the school, and wild grotesqueness in those who allowed themselves to wander from them, whilst simplicity, and its consequence good taste, were effectually banished from the works of them all.

It will be necessary here, perhaps, before we advance further in our remarks on the Italian school, to disabuse the public mind as to the merit of the works of Vitruvius, whose anilities have so long passed for authorities, that a writer would be suspected of prejudice who spoke of them slightingly without adducing reason and evidence to prove them valueless; except, indeed, as records of the architectural practice, and the opinions and acquirements of an architect of a distant age. It is of very little consequence that Vitruvius is only known by his own writings, but that mention of him by a contemporary or other ancient author would probably determine the age in which he lived. From several things he mentions, and his inscription or dedication to the " Imperator Cæsar," it has been concluded that he lived in the time of Augustus; but certainly without sufficient reason; for if the man he speaks of as the son of Masinissa had been the son of the celebrated Numidian of that name, in the course of nature neither he nor Vitruvius could have lived to the time of Augustus. But he addresses an emperor who succeeded his father an emperor, and speaks of a temple of Augustus; so that he must have been a contemporary of some period of the empire. If that period had been the Augustan, he would doubtless, have made some reference to some of the many distinguished men of that age, or have been referred to by some of them if he had himself been at all known or distinguished as his admirers insist he was; neither of which is the case; and, moreover, his language is not that of an educated man of the Augustan age. This, however, does not affect the merit of his work as a treatise on architecture; but his fables about the origin of building, the invention of the orders, and the arrangements which grew out of certain modes of construction, do so; by proving his total ignorance not only of the architectural works of the more ancient eastern nations, but of those of Greece itself, which he professes to describe. Now his classical taste, in consequence of his knowledge of antiquity, is vaunted by Perrault, one of his commentators, and given by him as a reason why Vitruvius was not much employed by the whimsical Romans in their love of variety, to which he would not administer. How far his knowledge of antiquity, that is, according to himself, of the works of the Greeks, extended, may be readily determined by comparing the designs of Greek structures, made by Perrault and others, according to the directions of Vitruvius, with the Greek structures

---

[1] *Cinquecento* means literally *five hundred*, but it is used as a contraction for *fifteen* hundred, or rather for *one thousand five hundred*, by the omission of *mille*, the century in which the revival of architecture, of which we are about to speak, took place ; and the manner consequent is so designated. *Trecento* would be *three* hundred, or the *third*, for the thirteenth century.

[2] *Le Fab. e i Disegni di A. Palladio da O. B. Scamozzi,* tomo i. p. 4.

History.

History themselves as they exist at the present time, and are faithfully delineated in various modern works, but especially in Stuart and Revett's *Antiquities of Athens*. It is indeed not less strange than true, that not a single example of Greek architecture will bear out a single rule which Vitruvius prescribes, professedly on its authority; and not an existing edifice, or fragment of an edifice, in form or proportion, is in perfect accordance with any law of that author, nor indeed are they generally referable to the principles he lays down. Examples might be cited almost to infinity in support of this statement, and to prove the inutility of a work consisting of mere descriptions without delineations, even if it were otherwise correct. The latter may certainly be supplied from the ancient remains when they exist; but to a man in possession of the specimens, descriptions and directions for their composition are quite unnecessary. Even Sir William Chambers, a distinguished disciple of the Italian or Vitruvian school, speaks very lightly of the advantage to be gained from the study of the Vitruvian principles of construction; and Mr Gwilt, in the introductory treatises to and notes on Sir William Chambers's work, has done much to undermine the authority, by exposing the absurdities and fallacies of the Magnus Apollo of pseudo-classical architecture. A student would acquire as correct a knowledge of history and geography from the Seven Champions of Christendom and Gulliver's Travels, as of architecture from the text of Vitruvius!

The adoption of the Vitruvian laws by the Italian architects of the 15th century led to the formation of the "Five Orders." It will have been observed that, in speaking of the course of Greek and Roman architecture, the Doric, Ionic, and Corinthian styles were mentioned. Vitruvius describes, in addition to these, another, which he calls Tuscan—possibly a style of columnar arrangement peculiar to Italy, and most likely of Etrurian origin; but, in the absence of delineations, the *Cinquecentists* could only apply the proportions he laid down for it, to what appeared to approximate them in the ancient remains; and hence arose a fourth, or "the Tuscan Order." It is, however, a mere modification of the Roman debasement of the Doric, and may be considered, in its present form, as of purely modern Italian origin. The same "Revivers," on looking among the ruins of ancient Rome for the forms of their Vitruvian orders, found specimens of a foliated ordinance, which the bad taste of the Romans had compounded of the foliated and voluted styles of the Greeks. This was seized upon as a fifth style, subjected to certain rules and proportions, and called "the Composite Order." The very poor Roman specimens of Doric and Ionic fitted themselves without much difficulty to the Vitruvian laws; but the examples Rome afforded of the Corinthian were less tractable, and being as various as they are generally beautiful, they were all passed over, and their places supplied by a mere changeling—an epitome of the Vitruvian theory. Thus we have the "Five Orders" of the Italo-Vitruvian school, and in this manner they are arranged: *First*, the Tuscan, of which there is no recognised example of antiquity, but which owes its form to the descriptions of Vitruvius and the fancies of the revivers; *second*, the Doric, a poor and tasteless arrangement of the general features of the style on a Roman model; *third*, the Ionic, which is almost as great a debasement of the Grecian originals, and was produced in the same manner as the last-mentioned; *fourth*, the Corinthian, a something totally unlike the ancient examples of both Greece and Rome in beauty and spirit; and, *fifth*, the Composite, an inelegant variety of the Corinthian, or a hybrid mixture of the horned or angular-Ionic volutes, with a deep necking of the foliage of the preceding order.

The first to publish this system was Leon Battista Alberti, a pupil of Brunelleschi. He has been followed by many others, the most distinguished of whom are, Palladio, Vignola, Scamozzi, Serlio, and De Lorme, architects; and Barbaro, a Venetian prelate, and an esteemed translator of, and commentator on, Vitruvius. None of these, it must be understood, agreed with any other of them, but each took his own view of the meaning of their common preceptor; and yet none of their productions evince the slightest approach to the elegance of form and beauty of proportion which distinguish the classic models of the columnar architecture of antiquity. Palladio and Serlio were the first to publish delineations and admeasurements of the Roman architectural remains in Italy; but the total absence of verisimilitude to the originals, and, in many cases, the absolute misrepresentations, in both works, prove how incompetent the authors were to appreciate their merits; and the exaggeration of their defects proves with equal clearness the general bad taste of the school in which they are masters. The worst qualities of the Roman school of architecture were embraced and perpetuated by the *Cinquecento*. The inharmonious and unpleasing combinations which arose out of the collocation of arches with columnar ordinances became the characteristics of the Italian: unequal intercolumniations, broken entablatures, and stylobates, enter alike into the productions of the best and of the worst of the *Cinquecento* architects. The style of this school is marked, too, by the constant attachment of columns and their accessories to the fronts or elevations of buildings; by the infrequency of their use in insulated (their natural) positions to form porticoes and colonnades; by the thinness or want of breadth in the smaller members of their entablatures, and the bad proportion of the larger parts, into which they are divided, to one another; by the general want of that degree of enrichment which fluting imparts to columns; by the too great projection of pilasters, and the inconsistent practice of diminishing, and sometimes fluting them; by the use of circular and twisted pediments, and the habit of making breaks in them to suit the broken ordinance they may crown; and by various other inconsistencies and deformities, which will be rendered more evident when we come to treat of the style in detail. The merit of the Italian school consists in the adaptation and collocation of the prolate hemispheroidal cupola, which appears to have grown out of its opposite in the Roman works during the Gothic ages, as we find it in the early cathedrals; though it is highly probable that the idea was brought from the East, in the forms exhibited by the cupolas of St Marks at Venice, and of Pisa Cathedral. A very noble style of Palatial architecture also was practised by many of the Italian architects. It consists of the use of a grand crowning cornice, running in one unbroken line, unsurmounted by an attic, or any thing of the kind, superimposing a broad, lofty, and generally well-proportioned front, made into graceful compartments, but not storied, by massive blocking courses and other things, which are at the disposal of the judicious architect. Not unfrequently, however, the faults of the school interfere to injure a composition of this kind; for, to produce variety in the decorations of the windows, some of them have been made like doors, with distyle arrangements of columns, surmounted by alternations of circular and angular pediments, and sometimes with all the vagaries which deform the front of an Italian church. It is indeed the ecclesiastical architecture of the school in which its faults are most rife and its merits most rare. An Italian church possesses nothing of the stern simplicity and imposing grandeur of an Egyptian sacred structure—nothing of the harmonious beauty and classic dignity of a Grecian fane—nothing of the ornate and attractive elegance of a Roman temple —and nothing truly of the glittering grace and captivating harmony of a Pointed cathedral. No other style of architecture presents so great a contrast, in any two species of its productions, as the Italian does, in one of its ordinary church fronts, with the front of a nobleman's mansion or palazzo, in the manner already referred to; and in no city of Italy is the contrast so strong, by the egregiousness of the examples it contains of both, as Rome. The stately portico is hardly known in Italian architecture; and in the rare cases in which insulated columns are found, they are for the most part so meagre in themselves, and so thinly set, according to the Vitruvian laws, that the effect produced by them is poor and wretched in the extreme. This applies most particularly to Italy itself: in some other countries, and especially in this, those architects who have been of the Italian school have generally preferred the proportions and arrangements which they found in the Roman examples of antiquity, to those laid down by their Italian masters. Still, Italian church architecture boasts the cupola,—certainly its redeeming feature; and the architects of Italy must have full credit for the use they have made of it, both internally and externally. Perhaps no two edifices display more, and in a greater degree, both the merits and defects of the school which produced them, than the Farnese palace and the basilica of St Peter in Rome. The principal front of the former edifice is exquisite in its proportions, but frittered in its details. It has an immense crowning cornice, whose general effect is surpassingly grand; but the mouldings are too much projected, and its vertical parts want the breadth which the blocking courses possess. The lowest of its three tiers of windows is characterized by the most charming simplicity and good taste in almost every particular; but the other two are crowded with sins against both those qualities, in the dressings of the windows. The cortile and back front, though both very differently arranged from the front, and from each other, are not less filled with contrarieties; and so of the structure throughout. The front of St Peter's is not more distinguished by its magnitude than by its littleness and deformity. It contains the materials of a noble octaprostyle, and consists of an attached tetrastyle. It is divided into three unequal stories, within the height of the columns, whose entablature is surmounted by a windowed attic. In length it is frittered into a multitude of compartments, between which not the slightest harmony is maintained; while tawdriness and poverty are the distinguishing characteristics of its detail. A total absence of every thing which produces grandeur and beauty in architecture, marks, indeed, the whole of the exterior of the edifice, except the glorious cupola, than which architecture never produced a more noble and magnificent object. Internally, the structure is open to similar praise and similar dispraise. Gorgeousness in matter and meanness in manner characterize the interior of St Peter's, except the sublime concave which is formed by its redeeming feature without.

The *Cinquecento* architects of Italy were exceeding mannerists: but besides the manner of the school, each had his own peculiarities; so that there exists in their works what may almost be called monotonous variety. Brunelleschi's designs are distinguished by a degree of simplicity and comparative good taste, which causes regret that he had not referred more to the remains of antiquity in Italy, and sought out those of Greece, and less to the dogmas of Vitruvius; for then his works would have been more elegant than they are, and the school he founded would have done him much more honour than it does.

The works of Bramante possess a more classical character than those of any other architect of the school. Bramante's design for St Peter's was preferred by Pope Julius II. to a great many others by the most esteemed men of the time. He it was who suggested the cupola; but, unfortunately, after his death men of less taste and talent were allowed to alter the design, and the edifice has resulted very differently from what it would have done had Bramante been adhered to. This we judge from his works generally, and not from any positive knowledge of the design, which indeed does not exist. The elder Sangallo was far inferior to his contemporary and rival Bramante, and his works are full of the faults of the school. Michel Angelo Buonaroti was a man of great genius, but of coarse taste in architecture; and to him may be attributed many of the coarser qualities of the Italian style. His principal works are the buildings of the Capitol, and the College della Sapienza in Rome, and the Laurentian Library at Florence; and these are all distinguished for their singular want of architectural beauty and propriety in every particular. Michel Angelo was the Dante of Italian painting, but the Berni of its architecture. Raffaelle, too, had a very bad style in architecture, and so indeed had almost all the painters who professed to be architects also. They generally carried to extremes all the faults of the school. Sansovino and Sanmichele were men of considerable talent: their works display more originality and less servility than those of most of their contemporaries. Peruzzi was less employed than many who had not half his merit: his productions are with reason considered among the most classical of the Italian school. Vignola had a more correct taste than perhaps any other Italian architect of the 16th century: his works are indeed distinguishable by their superiority in harmony of composition and in general beauty of detail. Palladio very much affected the study of the antique, but his works do not indicate any appreciation of its beauties. He appears to have been very well qualified by nature for an architect, but spoiled by education. He did not look at the remains of antiquity with his own eyes, but with those of Vitruvius and Alberti, and he seems to have been too much influenced by the admired works of some of his predecessors. Palladio made greater use of insulated columns than the Italian architects generally, but his ordinances are deficient in every quality that produces beauty; his porticoes may be Vitruvian, but they certainly are not classic; and all his works evince that he studied the Colosseum, the Theatre of Marcellus, and the Triumphal Arches, more than the columns of Jupiter Stator and Mars Ultor, the Temple of Antoninus and Faustina, the Pantheon, the Portico at Assisi, and the other classic models, which he drew, but clearly did not appreciate. His columns upon columns, his attached and clustered columns, his stilted post-like columns, his broken entablatures, his numberless pilasters, straggling and unequal intercolumniations, inappropriate and inelegant ornaments, circular pediments and the like, are blemishes too numerous and too great to be passed over because of occasional elegance of proportion and beauty of detail. Scamozzi did not improve on the style of his master, which, however, he very much affected. Indeed the term *Palladian* has long been in general use throughout the civilized world for beautiful and excellent in architecture, so that it cannot be wondered at that Palladio's pupils and successors should imitate him; nor is it surprising that they did not surpass, or even equal him, for they were taught to look to his works as the *ne plus ultra* of excellence. Giacomo della Porta, a contemporary of Palladio, followed Michel Angelo in several of his works, and imbibed much of his manner, on which

Plates.
XV.
XVI. and
XVII.

he certainly improved; but still his own is far from being good; Della Porta was much employed in Rome; and it fell to him, in conjunction with Domenico Fontana, to put the cupola on St Peter's. Fontana's style of architecture is not particularly distinguished for its good or bad qualities; he obtained more reputation as an engineer than as an architect, having been engaged in removing and setting up most of the obelisks which give so much interest to the architectural scenery of Rome. The Lunghi, father, son, and grandson, the Rainaldi, Maderno, Borromini, Bernini, Carlo Fontana, Fuga, Vanvitelli, and many others in the course of the seventeenth and eighteenth centuries, carried the peculiarities of the Italian school to the greatest extremes. Of those enumerated, Bernini was perhaps the least offensive, and Borromini the most extravagant; but throughout that period, except in extreme cases, individual manner is less distinguishable, and that of the school more strongly marked.

It may be gathered from the preceding remarks, that the secular architecture of the Italian school is generally preferable to the ecclesiastical, and that the architects of the fifteenth and sixteenth centuries were generally superior to those who followed them. In Italy the school has not yet ceased to exist, nor indeed has its style ceased to be studied. Designs are still made by the students of the various academies in the manner of the *Cinquecento*, and on the models with which the country abounds. The precepts of Vitruvius are yet inculcated, and the works of the men whose names we have mentioned are looked up to as master works of architecture in the country which contains the Roman Pantheon and the Greek Neptunium, besides the power of referring to the more exquisite works of Greece herself.

In the fifteenth century such was the reverence of men for the revived works of ancient literature and science, that the profession of the Italians, that they had restored ancient classical architecture on the precepts of an architect of the Augustan age, was sufficient to open the way for them all over civilized Europe. In the course of that and the following century Italian architecture was adopted and Italian architects employed in France, Spain, Germany, Great Britain, and their respective dependencies; and now, in the nineteenth century, Vitruvius and Palladio are as predominant on the shores of the Baltic as on those of the Mediterranean Sea; though in this country and in some parts of the Continent their influence is considerably diminished since the time of Inigo Jones and Claude Perrault. It has been already remarked, too, that the *Cinquecento* was later in gaining a footing here than on the Continent, in consequence of the existence of a beautiful national style of architecture, which our ancestors do not appear to have been induced to resign to the barbarian innovators of the South, as readily as the interjacent nations were to give up theirs; for which indeed the reason exists in the greater attractions of ours, and the consequent greater difficulty of inducing the nation to part with it. The French, though they received the Vitruvian architecture from the Italians, were patriotic enough, as soon as they had acquired its principles, to confine the practice of it almost entirely to native architects, in whose hands it assumed a different character from that which it possessed in Italy, and became what may be called the French style of *Cinquecento*. Its ecclesiastical structures are less faulty than are those of the corresponding period in Italy, but its secular edifices are as far inferior to those of that country. The grand palatial style, which is exemplified in the Farnese palace in Rome, never found its way into France; but instead, there arose that monstrous and peculiarly French manner, of which the well-known

palaces of the Tuileries and Luxembourg are egregious examples. In the age of Louis XIV. the French appear to have reverted to the Italian manner in a certain degree; for the palace of Versailles includes almost all the extravagancies of that school in its worst period, and contains moreover architectural deformities which Italy never equalled till it imitated them. They consist in the style of enrichment which is distinguished by the name of that monarch in whose reign it had its origin, and of whose gross taste and vulgar mind it is an apt emblem. The same period produced one of the most classical architects of the French school—its Palladio or Inigo Jones, Perrault, whose design for the buildings of the Louvre was preferred to that of Bernini, though indeed the preference was no compliment to the one nor discredit to the other, considering to whom the decision was of necessity referred. The Hôtel des Invalides is of the same age: it exhibits the graces of the Italian cupola, surmounting a composition which includes more than all the faults of St Peter's in Rome. The church of Sainte Genevieve, or the Pantheon, a work of the following reign, was intended to be in the ancient Roman style, and of Roman magnificence; but it is rather papally than imperially so. Ancient Rome was regarded in the columnar ordinance, but modern Rome in the architectural composition. In it the ecclesiastical style of the *Cinquecento* is commingled with the simple beauties of Roman architecture, almost indeed to the destruction of the latter: to this structure also there is a handsome cupola. Of late years the works of the ancients have been studied by the architects of France, greatly to the amelioration of their style; as yet, however, they are but imperfectly acquainted with the peculiarities of the Greek, and many of them still appear to retain their devotion to Vitruvius and the fifteenth century. Spain servilely received the Italo-Vitruvian architecture, and to the present day knows no other. Less patriotic than the French, the Spaniards have for their greatest works employed the architects of France and Italy. The Escurial, though the work of a native architect,[1] does little credit to Spanish art. The Italian Revival was the means of extinguishing the Pointed style of architecture in Germany, and certainly without affording it an equivalent. Italian architects were employed in Germany, and Germans acquired their manner; but they did not improve it, nor did they make it productive of so many good effects as the Italians themselves did. The change in religion which supervened the change in architecture in so large a part of Germany, may have tended to prevent the latter from acquiring that degree of exuberance there which it did in Italy; but even in Catholic Germany the splendid Pointed cathedrals have never given way to modifications of the pseudo-classic St Peter's. In the use of *Cinquecento* architecture for secular structures, it may be truly said that the Germans have not excelled the Italians; nor, on the other hand, have they equalled them in the absurdities and extravagancies which are so frequently observable in the works of some of the latter. The Germans also have lately turned their attention to the works of the ancients, and the fruit of it is already evident in many parts of the country, and most particularly in Prussia: still, however, they appear to have yet to learn the right use of the Greek models, and a proper sense of the exquisite perfection of their detail; as well as to emancipate themselves from many of the trammels of the Vitruvian school. The northern continental nations have been dependent on Germany, France, or Italy for their architecture, and can produce nothing that gives them a claim to our consideration in such a review as the

---

[1] It has been sometimes erroneously attributed to a Frenchman. The work was begun by Juan Bautista de Toledo in 1563, and completed by his pupil Juan de Herrera in 1584.—*Descripcion del Escorial* por Francisco de los Santos; *Viage de Fspana* por Antonio Ponz: *Varia Commensuracion* de Juan de Arfe; Laborde, *Itin. Descrip. de L'Espagne*, t. iii.; Townsend's *Spain.*—ED.

present. St Petersburg is exclusively the work of architects of the nations just enumerated, and presents a mass of the merest common-places of Italian architecture, in structures calculated by their extent, like Versailles, the Escurial and St Peter's, to impose on the vulgar eye.

We have already more than once had occasion to refer incidentally to the introduction of *Cinquecento* architecture into Britain; and in noticing it more particularly, and tracing its course, we are saved the trouble of keeping up a distinction between the different parts of our triple nation, because, at the time it actually crossed the channel, the amalgamation of the kingdoms had taken place by the union of their crowns on the head of the Scotish sovereign.

When the Pointed style received its deathblow in England, in the reign of Henry VIII., it did not immediately cease to exist; nor was it immediately succeeded by the Italian when it became extinct. It was gradually declining through all the 16th century, during the latter part of which period, what has been called the Elizabethan style became somewhat permanent. It consists of a singular admixture of the Italian orders, with many peculiarities of the Pointed style, and in many examples the latter appears predominant. With such difficulty, indeed, did that fascinating manner give up its hold on the minds of men in this country, that the *cinquecentists* appear to have relinquished the hope of effecting its destruction,—unfortunately, however, not until the injury was done; and for some time we were left without a style of any kind, unless that may be called by the name which marks the edifices of the reign of James I., and of which the oldest parts of St James's palace are a specimen.

The destruction of the Pointed style has been referred by some to the change in religion which took place under the Tudor line of English monarchs; but such was certainly not the case. It was the "Reformation" of architecture in Italy, and not that of religion in Great Britain, that effected it; and it may be doubted whether the change would not have taken place sooner in this country, if its connection with Italy had not been so materially affected by the moral change here, and so delayed that of architecture; for it was Germany and France that supplied us with architectural reformers during the reigns of Henry VIII. and his children, and not Italy, whose professors might possibly have obtained more credit than their disciples did.

So dilatory were we, indeed, in the cultivation of the Italian style, that the first professor of it who was actually employed on edifices in this country came hither from Denmark! It is true, he was an Englishman; but so little hope did he appear to have of success at home, that he accepted an invitation from the king of that country, when he was at Venice, whither he had gone to study painting; but becoming enamoured of architecture, as he saw it in the works of Palladio, he had made that his study instead, and had already acquired considerable reputation in that city, when Christian IV. of Denmark invited him to his court to occupy the post of his first architect. A train of circumstances, to which we need not here advert, brought him to England a few years after James I. came to the English crown, and he was appointed architect at first to the queen, and subsequently to Henry prince of Wales. But he does not appear in consequence to have then obtained employment; for after the death of the prince, he went again to Italy, where he remained till the office of surveyor general, which had been promised him in reversion, fell vacant. This was the celebrated Inigo Jones, who has been called the English Palladio; and indeed he succeeded so well in acquiring the peculiar manner of that architect, that he richly deserves whatever credit the appellation conveys. It is unfortunate, however, for his own re-

putation, that he had not looked beyond Palladio and their common preceptor Vitruvius, to the models the latter pretends to describe; in which case he might have been the means of restoring, or at least of introducing, to his own country, the truly classical architecture of the ancients. But instead of that he brought nothing home but Italian rules and Italian prejudices. Jones commenced the truly Gothic custom of thrusting *Cinquecento* fittings into our Pointed cathedrals, by putting up an Italian screen in that of Winchester; and he barbarized the ancient cathedral of St Paul in London, by repairing it according to his notions of Pointed architecture, for it was in that style, and affixed to it an Italian front. Fortunately the great fire supervened, and made room for the present magnificent structure, by clearing away that early specimen of pseudo-classic taste. Of the Palladian style, however, it must be confessed Jones was a complete master. He designed a royal palace which was to have been built at Whitehall, in a manner as far superior to those of Versailles and the Escurial, as the works of Palladio are to those of Borromini. The only part of Jones's design ever executed is the structure called the Banqueting-House, whose exterior is an epitome of many of the faults, and most of the beauties, of the Palladian school. It rises boldly from the ground with a broad, simple, and nearly continuous basement or stereobate, and the various compartments of its principal front are beautifully proportioned; but the circular pediments to the windows, the attached unfluted columns, with broken entablatures and stylobates, the attic and balustrade, though they be the materials of Palladian, it may be confidently denied that they are consistent with classical architecture. Another well-known work of this architect is the Italo-Vitruvian Tuscan church of St Paul, Covent-Garden, whose eastern portico is well-proportioned in general, but grossly deformed in detail.

Architecture was in abeyance in this country, again, from the troublous times of Charles I. till the restoration of the monarchy in the person of his son, whose French taste would have completely Gallicized the architecture, as well as the manners and morals, of the nation, if the resplendent genius of Sir Christopher Wren had not been present to avert the infliction, or rather to modify it; for it cannot be denied that the influence of the French manner had an effect on the architecture of this country from that period down to the middle of the last century. Indeed Wren himself only knew the style he practised from books and the structures of France, except the few that existed of Inigo Jones in this country; and, in consequence of his visit to France, the peculiarities of the French style are obvious in many of his less esteemed works. Fortunately, however, he was proof against the grosser peculiarities of the *Cinquecento*, whether in the books of the Italians or in the edifices of the French; and his own productions evince that he had imbibed much of the spirit of the antique monuments of Italy, which he could have known only from engravings, and those very imperfect ones. The field that was opened to his genius by the great fire of London in 1666, and its result, are equally well known. It is true that the general offuscation of taste and feeling with regard to the Pointed style extended even to him. Wren was guilty of many offences in that respect, besides giving authority to the opprobrious term Gothic; and in no case more so than in the construction of the towers to Westminster Abbey, which are a lasting proof of his ignorance of its most obvious principles. Nevertheless, to the influence of our beautiful native style on his mind, architecture is indebted for some of its most charming works. If Wren had not been accustomed to contemplate the graceful and elegant pyramids or spires of his native country, he would never

ARCHITECTURE.

History. have originated the tapering steeple, in the composition of which with the materials of Italian architecture he still stands as unrivalled as he was original. Witness the steeples of Bow Church and St Bride's in London, the former of which is hardly surpassed in grace and elegance by the Pointed spires themselves. It must remain a constant subject of regret that this great head of the English school of *Cinquecento* architecture did not know the remains of ancient Greece and Rome from personal observation. With his splendid genius and fine taste, if he had not been imposed on by the specious pretence of the Italo-Vitruvian school, his works would have been models for imitation and study, as they are objects of admiration: as it was, he avoided many of the faults of that school, and improved on many of its beauties. Without knowing the Greek style at all, and knowing the Roman only through imperfect mediums; without, indeed, ever having seen an example of either; whenever he has varied from the Italian practice, it has been towards the proportions and peculiarities of the Greek! The great west front of St Paul's, though it is said to be imitated from that of St Peter's in Rome, or rather from what it was proposed to be, with the two towers to form its wings, is a much finer, a more imposing, and more classical specimen of architecture than its prototype; for the advantage the latter should have in being of columns in one height is lost entirely in their poverty and in the miserable arrangement of the whole front; whereas that of St Paul's is in two noble pseudo-prostyle and recessed porticoes, with the columns fluted, and generally conceived and executed in much better taste than those of St Peter's. The entablatures, though massive, are finely proportioned, and sufficiently ornate to be elegant; they are, too, quite continuous, and the upper one is surmounted by a noble pediment, whose pyramidal form gives at the same time dignity and a finished appearance to the whole front. The coupling of the columns, however, and the putting of one columnar ordinance over another, can only be defended by the practice of the Italian school; though, in the present case, both are rendered less offensive by the judicious management of the architect. Nothing shows more strikingly the superiority of St Paul's to St Peter's, as an architectural composition, than a parallel of their flanks. The great magnitude of the latter may strike the vulgar eye with admiration in the contrast; but the rudest taste must appreciate the surpassing merit of the former in the form and arrangement of the cupola, and the noble peristyle with its unbroken entablature and stylobate, out of which it rises, when compared with the sharper form and depressed substructure of that of St Peter's. The superiority of St Paul's in the composition of the main body of the edifice is not less in degree, though perhaps less obvious, than in the superstructure. In the one it is broken and frittered, and in the other almost perfectly continuous, in broad, bold, and effective masses.

The history of the works of Sir Christopher Wren is the history of the architecture of the period in this country; and as it must be admitted that he was not so successful in the composition of the architecture of secular structures as of ecclesiastical, it will follow that our secular edifices of that time are of inferior merit. If it were not indeed an historical fact, it would hardly be credited, that Chelsea College, the old College of Physicians in London, and the halls of some of the city companies, are by the architect of Bow Church and St Paul's.

The style introduced by Sir John Vanbrugh, who may be said to have succeeded Sir Christopher Wren in the direction of architecture in England, was distinguished by massiveness unsuited to the style in which he built, which was of course Italian. It was, however, free from the va-

garies and extravagancies which characterize that style generally in other countries at the same period, but was certainly more suited to the soberer character of ecclesiastical than of secular structures, whereas his principal works were noblemen's mansions. Vanbrugh's faults were generally those of Michel Angelo: he was a painter's architect, and did not understand beauty of proportion and detail so well as the pictorial arrangement of lights and shadows; to produce which in the *Cinquecento* it is almost necessary to part with all the higher beauties of architecture. Hawksmoor added to the style of his master that noble ornament in which Italian works are so very deficient—a prostyle portico. His compositions are marked by severe simplicity, and only want to be absolved from a few faults and enriched with a few elegancies to be among the best of modern times. Not the least distinguished architect of the same age (the first half of the 18th century) was the Earl of Burlington, who was a passionate admirer of the style of Palladio and Inigo Jones. Many of the edifices erected by Kent are believed to be from the designs of that amiable nobleman, who, with considerable talent, was, however, a somewhat bigoted devotee to Vitruvius and the *Cinquecento* generally, as well as to Palladio in particular; for he has frequently used columns representing half-barked trees, in conformity with the silly tales of Vitruvius, and the sillier whims of his disciples. The portal of his own house in Piccadilly, and that of the King's Mews, are special examples of this bad taste, and of other faults of the school besides. Lord Burlington built for himself at Chiswick a villa on the model of the Villa Capra, or Rotonda, near Vicenza—a structure which has been called the masterpiece of Palladio. In form and proportion it is certainly elegant, but its details strongly exhibit the poverty of Italian columnar architecture, when unaided by the frittering which is its bane, and almost its sole dependence for effect. Gibbs was a contemporary of the same period. He too, like Hawksmoor, had imbibed a taste for the classic prostyle portico, which he evinced in St Martin's church in London; but that he also was in the trammels of the Italian school is no less evident, in the same structure, to a considerable extent, and still more so in the church of St Mary in the Strand, which is a bad specimen of architecture, and a favourable one of its style. During the following half-century (the latter half of the 18th) Sir William Chambers and Sir Robert Taylor were the most distinguished architects of this country. They were both men of talent and genius, who had availed themselves of the remains of Roman antiquity to good purpose; for as yet those of Greece were either unknown or unappreciated; and the former of them has left us, in the Strand front of Somerset House in London, perhaps the best specimen of its style in existence. Other parts of the same edifice, however, are far from deserving the same degree of praise: indeed, as an architectural composition, the river front is altogether inferior in merit to the other, though of much greater pretence. The inner fronts to the great quadrangle, though exhibiting good parts, are, as a whole, not above mediocrity. An air of littleness pervades them; and the general effect of the fronts themselves is made still worse by the little clock towers and cupolas by which they are surmounted; and to this may be added the infinity of ill-arranged chimneys, which impart an air of meanness and confusion that nothing can excuse. While Sir William Chambers and a few others were applying the best qualities of Italian architecture, indeed improving its general character, and, it may be said, making an English style of it, there were many structures raised in various parts of the country in a manner hardly superior to that of the time of James I; structures in which all the meanness and poverty of the

Cinquecento are put forth, without any of its elegance of proportion. or that degree of effectiveness which men of talent contrived to give it. During the same period, too, the seeds of a revolution were sown, which has almost succeeded in ejecting the Italian style and its derivative from this country, without perhaps having as yet furnished a complete equivalent.

In the year 1748 James Stuart and Nicholas Revett, two painters pursuing their studies in Rome, having moreover paid some attention to architecture, issued " Proposals for publishing an accurate description of the Antiquities of Athens, &c." These proposals met with general approbation, and in consequence they determined on prosecuting their plan; but various hinderances prevented their arrival in Athens till March 1751, when they commenced measuring and delineating the architectural monuments of that city and its environs. In this work Messrs Stuart and Revett were unremittingly employed (as far as their own exertions went, for they were frequently interrupted by the Turks) for several years, so that they did not reach England with the result of their labours until 1755; and, by a series of almost unaccountable delays, the first volume of their work did not appear until the year 1762. Sixteen years more expired before the second issued from the press; and the third was not published until 1794, being nearly fifty years from the time the work was first announced! Avarice and envy had induced a Frenchman of the name of Le Roy, who was at Rome when our countrymen issued their proposals, to forestall them with the public, and rob them of the profit and reputation they were so hardly labouring to earn. This man went to Athens, and in a very short time collected some loose materials, with which he published at Paris, in 1758, a work which he called *Les Ruines des plus beaux Monumens de la Grèce, &c.*, in which he makes not the slightest mention of Stuart and Revett, nor of their labours or intentions, with all of which he was well acquainted. This work is moreover notoriously and grossly incorrect; so incorrect indeed, as to make it difficult of belief that its author ever saw the objects of which he professes to give the representations. Such as it is, however, it was from M. le Roy's work that the public had to judge of the merits and beauties of Greek architecture; for we have said that the first volume of Stuart and Revett's did not appear for several years after it, and that does not contain any pure specimen of the national or Doric style: the second, which does, was not published for twenty years after Le Roy's. Considering, therefore, the source from which the public had to derive information on the subject, it can hardly be wondered at that Greek architecture was vituperated on all sides; and by none with greater acrimony than by Sir William Chambers, whose apology must be, ignorance and the prejudices of education. He really did not know the style he carped at; and his education in the Italo-Vitruvian school had unfitted him for appreciating its grand, chaste, and simple beauties, even if he had known it. Notwithstanding the misrepresentations of Le Roy, the vituperations of Chambers, the established reputation of Italian architecture, and the trammels which Vitruvius and his disciples had fixed on the public mind, when Stuart and Revett's work actually appeared, the Greek style gradually advanced in esteem, by dint of superior merit alone—for it has had no factitious aids; and since that period Greece and all her colonies which possess remains of her unrivalled architecture have been explored, and we now possess correct delineations of almost every Greek structure which has survived, though in ruins, the wreck of time and the desolation of barbarism. To our country and nation, then,

is due the honour of opening the temple of Greek architectural art,—of drawing away the veil of ignorance which obscured the beauties it contains,—and of snatching from perdition, and consequent oblivion, the noble relics of ancient architecture which bear the impress of the Grecian mind. Not only indeed were we the first to open the mine, but by us it has been principally worked; for among the numerous publications which now exist on the Hellenic remains, by far the greatest number, and indisputably the most correct, are by our countrymen, and were brought out in this country. It required, however, a generation for the effects of ignorance and prejudice in some, and imperfect knowledge in others, to wear away, before the beneficial effects of the Greek style could be obvious in our structures. The works of the Adams, who were contemporaries of and immediate successors to Sir William Chambers, evince a taste for the beauties of Greek architecture, but a very imperfect knowledge indeed of the means of reproducing them. The architects who have had the direction of our principal works during the first quarter of this century had the disadvantage of being pupils of those who were themselves, as we have shown, incompetent to appreciate the Greek style; and at a time too when the state of Europe shut up all access to the remains of Greece and Rome; so that no great improvement could perhaps be expected from them. When they shall have passed away, it is to be hoped that we shall find a new class, some of whom, indeed, are already before the world, who, having received their education since peace has opened the Continent, are prepared by the actual contemplation and study of the works of Egypt, Greece, Rome, and Italy, in all their varieties, to form new and pleasing combinations of their beauties, adapted to our wants,—to produce what may equal, if not surpass them all. The structures of Egypt may show how to arrange large masses harmoniously and effectively; those of Greece and Rome how to impart grace and dignity; and the structures of Italy how the materials of ancient architecture may be moulded to modern uses, while at the same time they give practical warning of what may result from the abuse of the most obvious principles of the art.

The difference between the representations of the Athenian antiquities by Stuart and his colleague, and the misrepresentations of them by Le Roy, appears to have opened the eyes of the world to those of ancient Rome, to see if they too had not been dealt with unjustly; for of late years much more correct delineations of them have appeared than those of Palladio and Desgodetz,—delineations of them as they exist, exhibiting the spirit of the originals, and not warped to the Vitruvian precepts, and thereby stripped of their best quality, truth. The excavation of the ancient cities of Herculaneum and Pompeii has opened to us much interesting matter, and some that is instructive: their ruins too have the advantage of being correctly delineated; so that we are at this time in possession of more knowledge of the architecture of the ancients, acquired in a few years by the actual examination of its relics, than our predecessors of the last generation were, after talking, and writing, and reading Vitruvius about it, for nearly four centuries.

It is an argument in proof of the classical beauty of the Pointed style, that when the eyes of men were opened to the perfections of Greek architecture, they began to discover its merits also. Pointed architecture, under the opprobrious name Gothic, had long been a subject of discussion among antiquaries; that is, essays were written by them to prove how the Pointed arch originated, but none appreciated its beauties. Our Pointed cathedrals and churches were, after the example of Inigo Jones, ruthlessly barbarized in repairing and fitting up. If an architect were employed to do anything in or to one of them, he appears to have thought it incumbent on him to convert it to the doctrines of his own faith—to Italicize it. Deans and Chapters for the most part entrusted their commissions to country masons and plasterers, who also operated according to the laws of the "five orders." About the middle of the 18th century one Batty Langley endeavoured to draw the attention of the world to Pointed architecture, by reducing it to rules, and dividing it into orders. Fortunately he was only laughed at, and both he and the book he published on the subject were soon forgotten. One of the first men in rank and influence of his time, in matters of taste particularly, Horace Walpole, *patronized* Pointed architecture, but ineffectually. He had himself neither taste nor feeling to appreciate its beauties, as his Strawberry Hill clearly evinces ; so that his patronage of it must have been the effect of mere whim, or a wish to lead a fashion. Delineations were indeed put forth from time to time, but generally so rude and imperfect, that, like M. le Roy to Greek architecture, they did more harm than good. The Society of Antiquaries, however, at length took up the subject, engaged Mr John Carter, an ardent and judicious admirer of our national architecture, and commenced the publication of a series of splendid volumes, containing engravings of its best specimens, from drawings and admeasurements by him. The "Antiquities of Athens" had already done much to dispossess men of their prejudices, by showing that Greek architecture, though neither Vitruvian nor Palladian, was nevertheless beautiful ; and the great work of the Society of Antiquaries did the same for Pointed architecture. Since the death of Mr Carter, our native style has been beautifully illustrated, in a series of valuable works by Mr Britton, and elucidated in detailed "specimens" by Mr Pugin, a French gentleman but an English artist, and by a great variety of other useful and excellent publications ; so that, at the present time, the Pointed style, too, is studied and understood, and not a few of our architects are now competent, not only to be entrusted with the repairs and restorations of the ancient structures, but also to originate new ones, which may rival all but their prototypes in beauty.

Egyptian Architecture.

This is in all probability the oldest architecture in the world, and now that those mysterious stumbling-blocks to the former antiquary, the hieroglyphics, have been deciphered and clearly understood, a flood of light has been thrown on what a few years ago was all vague conjecture. We are indebted to the labours of the late eminent architect and antiquary, Luigi Canina, and to those of Sir Gardner Wilkinson, for an immense mass of information on the subject. In fact, these learned men seem to have been the first to have digested the matter into tangible shape, and their friendship and mutual association can never be overvalued. The drawings for the chief works of the latter were made out in the studio of the former.

Oldest
Egyptian
works.
The oldest works of the Egyptians (Herodotus, *Euterpe*, 4, 10, 99) seem to have been the embankment of the Nile by Menes, the foundation of the city of Memphis, and the commencement of a temple to Vulcan. Before his time, our author says, Egypt was a morass, and that for seven days' journey from the sea. This embankment must have been a gigantic work, especially as it was executed between 400 and 500 years before the time of Abraham, or about 4000 years ago.

We then hear from Manetho, as cited by Eusebius in his Chronicle, part i., that Venephes, the fourth king of the 1st dynasty, built some pyramids at a place called Cochomen, but this record is all we know of them. The Chronicle shortly after records that Tosorthus, or Sosorthus, the 2d king of the 3d dynasty, found out how to build with polished or smooth stone, καὶ τὴν διὰ ξεστῶν λίθων οἰκοδομὴν εὕρατο.

The great
pyramid.
The next of which we have notice is the most gigantic work in the world ; one which never has, and perhaps which never will be, surpassed. It is the great pyramid. At this time the Egyptians must have advanced in the mechanical arts beyond anything of which we have an idea. They seem to have been able to quarry rocks of the hardest stone, even granite; to transport them to any distance they pleased; to raise huge blocks, vast monolith obelisks that would puzzle our engineers with their best tackle ; and what is the most wonderful of all, not only to polish granite as well as we can, but they did more—they seem to have had the power of carving on that most stubborn and difficult of all materials with the utmost facility, large surfaces and even huge statues being covered with hieroglyphics of the most minute kind, and of the highest finish. The most extraordinary thing is to know how this was done, for though Herodotus (*Euterpe*, 124, 125) tells us they had iron tools, yet it was long before the conversion of that metal into steel had been found out; and with all our best tools that the best steel can afford us, it is very difficult and very costly even to carve plain letters in granite.

Cheops or
Suphis.
Herodotus tells us the occasion of the great work was the caprice of a king, whom he calls Cheops, and who is supposed to be the Suphis of Syncellus, and the Chembes of Diodorus, about 2100 years before the Christian era. The account of Herodotus is, that this king was a tyrant of the very worst kind; that he closed all the temples throughout Egypt, forbade every sort of religious observance, and forced all his subjects to labour for him as he pleased. Among other whims, he determined to build this pyramid as a tomb for himself. Our author tells us the stones were quarried in the Arabian mountains, none less than 30 feet long. They were then conveyed by the Nile to where a new road had to be made, three-quarters of a mile long, 60 feet broad, and in a cutting of 48 feet. This road, our author says, is of polished stone, and carved with figures, and that it took ten years to complete. Twenty years were spent in building the pyramid itself.

Dimension
of the great
Pyramid.
This, Herodotus tells us, was 8 plethra, or about 808 feet square, and as many high; but this is not the case. The old Greek author probably measured up the slant side. The dimensions, as given by Colonel Vyse, are 764 feet length of base, 720 feet slant side, and 480 feet high. The theories as to the law on which they were designed will be considered hereafter. This huge mass of stone covers about 13¼ acres. It is said that Lincoln's Inn Fields was set out the exact size of its base, and it is about 150 feet higher than St Paul's, so that our English readers may judge of its comparative size. As compared with the largest building in the world, St Peter's, the great pyramid covers an area as 58 : 22, or nearly three times as much, and it is 50 feet higher, of course far the loftiest structure in the world ever piled up by human hands. Like almost all the pyramids, it faces the cardinal points, and is entered from the north by a descending passage, which leads to a few small chambers or cells, the largest of which is but 17 feet wide. In one of these a solitary sarcophagus was found. So this vast mass, on which the labours of 100,000 men were bestowed for twenty years, which contains 100 millions of cubic feet of stone, and which, reckoning quarrying, transport—twice by land and once by river—squaring, hoisting, and setting at 2s. per foot of our money, would have cost as much as has been laid out on the London, Chatham, and Dover Railway : this more than gigantic

work is the tomb of one person only—a wondrous monument of human pride !

A second pyramid, close to the first, was built by his successor, whom Herodotus calls Chephren; the inscriptions on the stones, however, give the name Sursuphis. The base of this measures about 60 feet less than the former. Its date is about 2083. About forty years after this, his successor, whom Herodotus calls Mycerinus, and the inscription Mencheres, built a third; but the dimensions of the base are only about 350, less than half the great pyramid; but this last was entirely faced with polished granite, while the others were of limestone.

Canina (*Archittetura Antica*, part I.) has described altogether twenty large and twenty-seven small pyramids, some not more than 30 feet square. He imagines that they all had a propylon or entrance doorway, which was closed by huge stones fitting close. They all seem to have been places of sepulture, with cells and winding passages like those described before. Some were built of unburnt bricks, but these are now only shapeless masses of clay.

Such as exist are almost solid masses of masonry, whose bases are squares, and whose inclined sides are nearly equilateral triangles : some of them are truncated, and some run up to a point. They are generally much injured on the surface by long exposure, so that it is impossible to say whether any of them were considered finished while in steps or receding courses, or if the angles were either filled up or worked off, to make smooth surfaces on the exterior. Some of them not only were made plain by working off, but remain so still; whilst others bear no indication of ever having been finished in that manner. In one existing example, that of the great southern pyramid of Dashour, the angles of the receding courses have been wrought off; and it is singular that the blocks of stone are not laid in horizontal courses, but at an angle inclined to the base; nor are its sides carried up to the top in one continued plane, but at about two thirds from the base they incline towards each other under a more obtuse angle.

It has been imagined, but not determined, that most of them have natural hills, either of earth or stone, for cores, or rather that hills have been cut to the shape, and built over with large courses of stone to give them the appearance of being solid masonry. If this be the case, the chambers and the passages to them, which have been discovered in some of the pyramids, have been carefully built around to have the appearance of being left in the construction, which is not very probable.

A great deal of trouble has been given to discover the principles on which the Egyptians set out these erections; the most reasonable theory is that each side was meant for an equilateral triangle, four of which laid sloping, and brought to a point, would compose the pyramid; but neither the dimensions nor the angles agree. It is true that the sides of the three great pyramids have an angle with the horizon of from $51\frac{1}{4}°$ to $52\frac{1}{4}°$, or thereabouts; but those at Abusir and at Saccara, as given by Canina, measure 55°, at Assou 68°, while at Barkal, near Meroe, the angle is no less than 72°. At Dashoor the pyramid has a slope about half-way up of 53°, which afterwards is flattened to 44°. At Meydoum is a pyramid in three great steps.

The great reverence paid by the Egyptians to the bodies of their ancestors, and their careful preservation of them by embalmment, necessitated a great number and vast extent of tombs. Some of these are built up like small houses, others are caves cut in the sides of rocks, others are long passages tunnelled under ground to very great extent. The pyramids, however, seem particularly to have been the places of sepulture of distinguished men, at least in Egypt. In Thebes there are no pyramids, and all the tombs are excavated in the solid rock. There is no great

pretence at architectural decoration in these, except the propylon or façade; but they are filled with the most interesting paintings, delineative of every possible subject, down to the minutest incident of private life, which give more information as to their manners and customs than volumes could convey. A model of one of these sets of chambers was exhibited some years ago in London by Belzoni; and at present there is a valuable series built up, and painted in fac-simile, in the Vatican at Rome. From a careful study, it appears, as soon as a king succeeded to the throne, the excavation of his tomb commenced, and this proceeded year by year—first a series of corridors and then square halls, some supported by columns, till, at the demise of the monarch, the work comes to an abrupt termination. Canina has given plans and sections of several of the royal tombs, varying from about 80 to 120 metres in length, or extending from 250 to 400 feet direct into the solid rock.

Next in order the temples must be noticed. Sir Gardner Wilkinson, in that excellent work, *The Architecture of Ancient Egypt* (8vo, London, 1850), gives a very full account of these edifices. From this work, and those of Canina, we find that the earliest temples were small, consisting of a simple chamber to hold the statue of the deity, with one opening or doorway in front, through which the votary might look, and with an altar for sacrifice. This was a sort of sanctuary into which only priests might enter. It is curious that, though, with very few exceptions, all pyramids face the cardinal points, the temples stand in any direction which convenience may dictate. The building was surrounded with a wall of brick forming a court or temenos, which was entered by a tall stone gate-way or propylon, not unlike the first temple of the Israelites, with its sanctuary and court. This enclosure was often planted with trees, which were, no doubt, the origin of the groves of Ashtaroth, Chemosh, and Milcom, which were the haunt of every sort of abomination, and which the good kings of Judah and Israel so zealously destroyed.

In process of time these temples grew larger, and chambers for the priests were added, and large door-ways, flanked with a sort of towers with sloping sides (Plate II. fig. 5), and sometimes a sort of portico or pronaos (fig. 2), supported by columns. The vestibule, or court-yard, was surrounded by a colonnade (see fig. 4), and the propylon became of gigantic proportions, and was full of chambers (see fig. 1, 4), all which are taken from the temple at Apollinopolis Magna or Edfoo, which will be shortly described. The sanctuary or σήκος (marked adytum, fig. 4) still contained the idol and its altar. Across the court, and, in fact, sometimes for an immense distance outside, there was a δρόμος, or avenue of sphinxes, through which the procession defiled. At the commencement of this avenue there was frequently an open or hypæthral building, or peristyle of columns, where it is supposed the processions assembled and were marshalled. This building is called a canopy by Sir Gardner Wilkinson.

The last-named author classifies Egyptian temples thus: —1. *Sanctuary Temples*, or those with only one single chamber. 2. *Peripteral Temples*, or the like, but surrounded with columns. 3. *Temples in Antis*, with a portico of two or four columns in front; 4. With porticos of *many* columns, as Esné, Dendera, &c., and many inner chambers. 5. The like, with large *courts*, and with *pyramidal towers* or propylons in front.

These images form so conspicuous a figure in Egyptian architecture as to merit remark. They all have the body of a lion, sometimes winged. Those with the head of a ram, or crio-sphinxes, are supposed to be dedicated to Amun (Jupiter Ammon); those with the head of a hawk are called Hierco-sphinxes, and are sacred to Re, or the

Egyptian Architecture.

sun; while those with the human heads are called andro-sphinxes, and are supposed to be intended to compliment the reigning king or queen, the sphinx being the emblem of mingled force, swiftness, and wisdom.

The largest of these sculptures deserves a few lines, as it shows what gigantic conceptions the Egyptian artists must have had. The body is that of a lion with a woman's head, and is crouched close to the ground. The height from the floor or platform on which it lies to the top of the head is 100 feet. The total length is 146 feet; across the shoulders it measures 34 feet. The head, from the top to the chin, is 28 feet 6 inches, and is calculated to be 40,000 times the bulk of an ordinary human head. A small temple or sanctuary was built between its paws. With the exception of this, and the paws themselves, which are of masonry, the whole is carved out of the solid rock. Let us imagine one stone a little longer than the church of St Martin's-in-the-Fields, as high, and about half as wide. Such a block would represent the body; while the tower, up to the height of the clock, would represent the head and neck.

Having passed through the avenue of sphinxes, we come to the propylon, or gate, at the main entrance. The one at Edfoo (Plate II. fig. 1; see also plan, fig. 4) is about 50 feet high, and is flanked by two massive towers 110 feet high. The whole façade measures, according to Canina, 69 metres, or 230 feet, or about 10 feet higher than that of St Paul's, and 50 feet more in length. Before the propylon two obelisks are often found, and sometimes two gigantic seated figures; and, we gather from pictorial representations, there were also tall flagstaffs with long streamers. The propylon at Edfoo is covered with a multitude of figures, all of colossal proportions, and some as high as 40 feet. It is full of small chambers, story after story, supposed to have been intended for the priests or attendants. We then enter a magnificent court, 160 feet by 140 feet,[1] surrounded on three sides by columns 32 feet high, which leads to the pronaos or vestibule of the great hall. This measures 110 feet by 44, and has twelve columns, 34 feet high. From this we pass to the inner vestibule, or hypostyle hall, which is 70 feet by 44 feet.

These halls are so called by Diodorus Siculus, because the middle ranges of columns, with the roof, &c., are higher than the side parts, and admit light by a range of windows opening over the side roofs, something like the clere-stories of our cathedrals.

Adytum.

After passing two small courts, we come to the most sacred part of the temple, the adytum, or shrine of the deity or deities to whom the temple was dedicated. This is always a small building, without any light, except that which enters by the doorway. In this instance it measures only about 33 feet by 17 feet, or the size of a moderate drawing-room, while the whole edifice within the walls covers about as much ground as our St Paul's.

But this temple is small, indeed, when compared with that at Karnak. This is about 350 metres long, by 110 wide, and therefore covers 420,000 feet, or five times as much as our St Paul's, and more than double the superficies of St Peter's at Rome. The propylon here is 240 feet long, or more than half as much again as St Paul's. The hypostyle hall is the most wondrous apartment in the world, being a parallelogram of about 342 feet long and 170 feet wide. It has fourteen rows of columns, nine in each row, and 43 feet high; and two rows, six in each, of the enormous height of 72 feet, 11 feet 6 inches in diameter, and carrying capitals which measure 22 feet across. This hall (with the two gigantic pylones) is said to cover 4000

superficial feet more than St Paul's. Beyond this is the adytum or shrine, a small apartment, measuring only 26 feet by 16 feet. So inconsiderable is the place for the deity compared with the vast structure which surrounds it—a puny chamber, enclosed by buildings which cover ten acres.

Egyptian Architecture.

The Egyptian temples, without possessing that entire uniformity of plan which those of the Greeks do, are very similar in arrangement and manner. The larger and more perfect structures do not externally present the appearance of being columned, a boundary wall or peribolus girding the whole, and preventing the view of any part of the interior, except perhaps the towering magnificence of some inner pylones; of the lofty tops of an extraordinary avenue of columns, with their superimposed terrace; of the tapering obelisks which occupy, at times, some of the courts; or of a dense mass of structure, which is the body of the temple itself, inclosing the thickly columned halls. The immense magnitude of these edifices may perhaps have made them independent, in their perfect state, of considerations which have weight in architectural composition at the present time, and on which indeed its harmony depends. The various portions of the same temple differ in size and proportion; and being intermingled, the cornices of the lower abut indefinitely against the walls of the higher parts, while the latter are not at all in accordance among themselves.

The structure we produce to exemplify Egyptian archi- Pl. II. tecture, though not, according to M. Champollion, one of the Pharaonic monuments, is perfectly characteristic of the style and arrangement of Egyptian temples, and is a more regular specimen than any other possessing the national peculiarities. It is known as the temple of Apollinopolis Magna, or of Edfoo, in Upper Egypt, on the banks of the Nile, between Thebes and the first cataracts.

The plan of the inclosure behind the propylæa is a long parallelogram, the moles or propylæa themselves forming another across one of its ends. The grand entrance to the great court of the temple is by a doorway between the moles, to which there may have been folding gates, as the notches for their hinges are still to be seen. Small chambers, right and left of the entrance, and in the core of the propylæa, were probably for the porters or guards of the temple: a staircase remains on each side, which leads to other chambers at different heights. To furnish these with light and air, loop-holes have been cut through the external walls, which disfigure the front of the structure. The court-yard, cloister, or vestibule, has on three of its sides a colonnade, against the wall of the peribolus, forming a covered gallery. This, and the gradual ascent by corded steps to the great portico or pronaos, will be better understood by reference to the plan and section. The pronaos, Figs. 3 & 4. or covered portico, consists of three rows of six columns each, parallel and equidistant, except in the middle, where the intercolumniation is greater, because of the passage through. The front row of columns is closed by a sort of breast-work or dado, extending to nearly half their height, in which moreover they are half-imbedded; and in the central opening a peculiar doorway is formed, consisting of piers, with the lintel and cornice over them cut through, as exhibited in the elevation of the portico. From the Fig. 2. pronaos another doorway leads to an atrium or inner vestibule, consisting of three rows of smaller columns, with four in each, distributed as those of the pronaos are. Beyond this vestibule there are sundry close rooms and cells, with passages and staircases, whose intention is not obvious. The insulated chamber within the sixth door was most probably the adytum or inmost sanctuary, which

---

[1] All these measurements are reduced from Canina.

may be supposed, in an Egyptian temple, to have contained a presentment of the divinity : the rest is inexplicable.

In many cases the temples are without the peribolus and propylæa, the edifice consisting of no more than the pronaos and the parts beyond it ; and in others, particularly in those of Thebes, this arrangement is doubled, and there are two pairs of the colossal moles, the second being placed where the pronaos is in this, and another open court or second vestibule intervening between them and the portico. In these the central line across the courts is formed by a covered avenue of columns, of much larger size than ordinary ; and the galleries around are of double rows of columns instead of one row with the walls, as in this case. The obelisks marked in the plan, and indicated in the section, before the propylæa, occupy the situation in which they are generally found, though they do not exist with this example. Colossal sedent figures are sometimes found before the piers of the gateway ; and from them, as a base, a long avenue of sphinxes is frequently found ranged like an alley or avenue of trees from a mansion to the park-gate, straight or winding, as the case may require.

Fig. 3.   The longitudinal section of the edifice shows the relative heights of the various parts, and the mode of constructing the soffits or ceilings, which are of the same material of which the walls and columnar ordinances are composed : that is in some cases granite, and in others freestone.

Fig. 2.   The elevation of the pronaos shows also a transverse section of the colonnades and peribolus. It displays most of the general features of Egyptian columnar architecture ; the unbroken continuity of outline, the pyramidal tendency of the composition, and the boldness and breadth of every part. The good taste with which the interspaces of the columns are covered may be remarked. Panels standing between the columns would have had a very ill effect, both internally and externally ; and if a continued screen had been made, the effect would be still worse, as the columns must then have appeared from the outside absurdly short ; but as it is, their height is perfectly obvious, and their form is rendered clear by the contrast of light and shade occasioned by the projection of the panels, which would not exist if they had been detailed between
Figs. 6 & 7. the columns. The lotus ornament at the foot of the panels is particularly simple and elegant ; and nothing can be more graceful and effective than the cyma above their cornice, which is singularly enriched with ibis mummy-cases. The jambs forming a false doorway in the central interspace, are a blemish in the composition ; and they injure it very much by the abruptness of their form, and their want of harmony with anything else in it. It may be remarked, that the effect of the front generally is that of an excavation rather than of a structure, the end piers and entablature having a unity of purpose, which leads to the idea that the rest was similar, or the whole at first a plain wall afterwards pierced and carved into its present form. This view of it would support the supposition that the excavations or hypogea are the antitypes of columnar architecture.

Fig. 1.   The front elevation of the moles or propylæa, with the grand entrance between them, is peculiarly Egyptian ; and very little variety is discoverable between the earliest and latest specimens of this species of structure. It is an object that must be seen to be appreciated ; simplicity and an inherent impressiveness in the pyramidal tendency are all on which it has to depend for effect, except magnitude, which alone would certainly make no agreeable impression on the mind. The projecting fillet and coving which form a cornice to the structures, though large and bold, appear small and inefficient when compared with the bulk they crown ; and there is nothing particularly striking in

the torus which marks the lateral outline and separates the straight line of the front from the circular of the cornice. Neither are they dependent for their effect on the sculpture, for their appearance is as impressive at such a distance as to make the latter indistinct, as when they are seen near at hand. The effect of the sculptures and hieroglyphics generally on Egyptian architecture is to enrich the surfaces, but not to interfere with the general form of a structure, or even with the forms of its minor parts.

A portion of the portico is given on a larger scale, to show more clearly the forms and arrangement of Egyptian columnar composition. The shaft of the column in this example is perfectly cylindrical. It rests on a square Fig. 5. step, or continued stylobate, without the intervention of a plinth or base of any kind ; and has no regular vertical channelling or enrichment, such as fluting, but is marked horizontally with series of grooves, and inscribed with hieroglyphics. The capitals are of different sizes and forms in the same ordinance. In this example the capital, exclusive of its receding abacus, is about one diameter of the column in height. Its outline is that of the cyma, with a reversed ovalo fillet above, and its enrichment consists principally of lotus flowers. The capital of the column next to Fig. 2. this, in the front line, is much taller, differently formed, and ornamented with palm leaves ; the third is of the same size and outline as the first, but differently ornamented ; and the corresponding columns on the other side of the centre have capitals corresponding with these, each to its fellow, in the arrangement. Above the capital there is a square block or receding abacus, which has the effect of a deepening of the entablature, instead of a covering of the columns when the capitals spread, as in this case. In the earlier Egyptian examples, however, in which the columns Pl. I. are swollen, and diminished in two unequal lengths, the Fig. 7. result is different, and the form and size of the abacus appear perfectly consistent. The height of this column Pl. II. and its capital, without the abacus, is six diameters. Fig. 5. The entablature consists of an architrave and cornice, there being no equivalent for the frieze of a Greek entablature, unless the coving be so considered, in which case the cornice becomes a mere shelf. The architrave, including the torus, is about three-quarters of a diameter in height, which is half that of the whole entablature. The architrave itself is in this example sculptured in low relief, but otherwise plain. The torus, which returns and runs down the angles of the building, is gracefully banded, something like the manner in which the fasces are represented in Roman works. The coving is divided into compartments by vertical flutes, which have been thought to be the origin of triglyphs in a Doric frieze ; but these are arranged without reference to the columns, and are in other respects so totally different from them as to give but little weight to the suggestion. The compartments are beautifully enriched with hieroglyphics, except in the centre, where a winged globe is sculptured, surmounting another on the architrave, as shown in the elevation of the pronaos. The crowning tablet or fillet is quite plain and unornamented. Angular roofs are unknown in ancient Egyptian buildings, and consequently pediments are unknown in its architecture.

These are common in Egypt, and many very large. Rock-cut Some writers have supposed the cave or quarry was, in Temples. fact, the original of all temples. Sir Gardner Wilkinson has, however, proved the contrary, from the simple fact that there are architraves over the columns, which support the rock above, evident imitations of beams in constructed buildings, but which are of course unnecessary in this method of construction. No rock temple, our author states, dates so early as those which are built. Our author classifies these temples thus :—1. *Simple sanctu-*

Egyptian Architecture.

aries, or small square chambers, with a plain door. 2. *In antis*, with pilaster and two columns in front. 3. *Hall Temples*, with one or more halls supported by columns or by caryatides, and with or without a portico. At Aboo Simbel there are two large halls, three adyta or shrines, and six other chambers, besides corridors. In front are four gigantic colossi in a sitting position. In some instances, as at Sabora, the sanctuaries are excavated in the rock; while the courts, porticoes, &c., are built out in front of them.

Palaces.

It is believed the kings inhabited parts of these temples, and that the huge halls above described were used for the purposes of state. An immense mass of buildings,

Labyrinth.

however, called the Labyrinth, has been mentioned by Herodotus as the palaces built for the twelve kings, who ruled in the time of King Josiah, or nearly 700 B.C. According to the description of this author, this was as great a work and wonder as the pyramids themselves. It was close to Lake Mœris, and contained, in the time he wrote, 3000 chambers, half above and half below ground, besides immense halls, corridors, courts, gardens, &c. The roofs were wholly of stone, and the walls covered with sculpture. On one side stood a pyramid 40 orgyœ, or about 243 feet high.[1] Denon doubted whether this description was at all correct; but subsequent investigations have so confirmed the truth of the representations of the old Greek traveller that there can be but little doubt of the accuracy of his description of what he saw, and he always has the candour to tell us whenever he relates anything from hearsay. Sir Gardner Wilkinson, however, supposes he mistook the lake of Mœris for the canal, and is inclined to fix the locality of the Labyrinth near the pyramid of Howara. Subsequent investigations have proved this to be correct. It appears from the ruins that huge masses of buildings once occupied three sides of an open quadrangle about 200 yards square in the inside, the two wings being about 300 yards long, and the third side about 400, measured on the outside. The pyramid, as stated by the various authorities, occupied the greater part of the fourth side, and measured about 300 feet square. There are a multitude of small chambers in two stories, as described by Herodotus; and Canina supposes there was a third story above these, supported on columns—a sort of open gallery.

A small palace, however, at Medinet Abou, deserves our notice. It is said to have been erected by Rhamses IV. (Sethos), about 1200 years before the Christian era, and supposed to be a summer-house or pavilion.

The plan is very peculiar. In front are two pylons, each about 28 feet in front, and 50 feet high. These contain two ground chambers, which are only 20 feet by 8 feet. Two passages turning at right angles, probably for better defence, lead to the main building, which contains a saloon 64 feet by 68, and 48 feet high: this is gorgeously painted. The pylons in front, are covered with sculpture. The building is three stories in height, with three rooms on each story.

Private dwellings.

Of these little is known, except from paintings found in the tombs. They seem, like the houses in the Labyrinth, to have been in two stories, with an open gallery at the top, supported by columns probably of wood. The larger houses consisted of rooms ranged round three sides, and sometimes four, of a large court-yard planted with trees, and with a tank, and perhaps a fountain in the middle. There was an entrance porch, on which are hieroglyphics, being, as Sir G. Wilkinson supposes, the name of the inhabitant. Still larger houses are supposed to have had two courts—the outer, in which to receive visitors, the inner for the females of the family. Smaller houses, particularly in the country, had a similar court, with granaries and store-

rooms below, and living apartments above, like those of the modern Fellah in Egypt, or the small *vigna* houses in Italy. The roofs seem to have been flat, like those of the modern Egyptians; and the houses appear, from a painting found at Thebes, to have been ventilated in the same way as at present, by the contrivance called a *mulkuf*, or windshaft, over which are two screens, like large square fans back to back, bending forward each way to catch any air that may chance to be stirring, and direct it down the shaft into the house.

Egyptian Architecture.

Statues.

The statues also bear the same enormous proportions. Canina makes the sitting figure of Memnon 16 metres, or 53 feet; and that in the sepulchre of Osymandyas 18 metres, or nearly 60 feet high. The face of Memnon is 7 feet high, and its ear 3 feet 6 inches long, and it is nearly 26 feet across the shoulders. If a figure of these proportions were supposed to be standing, and its arm to be stretched upwards, it would touch the underside of the gallery of the Duke of York's column.

OBELISKS.—These astonishing objects are peculiarly Egyptian. They are generally huge monoliths of red granite or syenite. Their use originated no doubt in the custom of setting up a stone to commemorate some particular event. The Egyptians seem to have improved in this idea, first, by working these stones to a fine face, and afterwards by covering them with carvings. They stand sometimes singly, as the famous Cleopatra's needle, but more frequently in pairs before the propylon of the temples, as at Karnak, Philae, and Edfoo (see Plate II. fig. 3). After the conquest of Egypt, the emperors seem to have delighted in transporting these monuments to Rome. At the time the celebrated Regionaries (accounts of each of the wards or *Regiones* of Rome) were written, there were six great obelisks, and forty-two small ones at Rome. Of these, twelve only are left, and they vary in height from over 100 feet to 8½ feet. It may be sufficient to describe the first, which now stands close to the church of St John Lateran. It is 148 Roman palms in height, or a little over 108 feet English, about 8 feet square at the base, and weighs, as is estimated, nearly 450 tons. It is covered with hieroglyphics, from which we gather it was erected in honour of Tothmes IV. the king who reigned over Egypt shortly after the Exodus. It stood originally before the temple of Ammun Re, and was brought over by Caligula. This wonderful work must therefore have been in existence between 3000 and 4000 years, and unless it should again fall into the hands of savage invaders, is likely to endure quite as long as it has lasted. Each obelisk diminishes equally towards the top—that near the Lateran, ·253 part of the base; at St Peter's, ·261; two at Thebes, ·3; and the one near S. Maria Maggiore, ·307, or nearly one-third. We may therefore assume the diminution from the base roughly at from one-quarter to one-third.

The Barberini obelisk is about 7¾ times as high as the diameter of the base; Cleopatra's needle, 8½; the one at St Peter's 9 times; at Luxor 10 times; at the Lateran 11 times; while two at Thebes, and one in the Piazza Navona at Rome, have an altitude of no less than 12 times the diameter of the base.

The obelisks have no entasis or swell from top to bottom like a column, but in almost all cases there is a slight convexity on the horizontal section of each face. The one in the Place de la Concorde at Paris has the peculiarity of being convex on one side, and slightly concave on the other.

In all ancient examples, the small pyramid which covers the obelisk, is at least 1½ times as high as the diameter of the top of the obelisk in which it is placed. In modern

---

[1] Accounts of this Labyrinth are given by Herodotus, ii. 148; Diodorus, i. 61; Strabo, xvii. 1; and Pliny, xxvi. 13.

Egyptian Architecture.

Orders of Egyptian Architecture.

examples, the pyramidion is almost always too flat, which gives a very bad effect.

The excellent author we have so often quoted has classed the columns of this style into eight orders. *First*, The *square pillar*, or post of stone. This is sometimes painted, or otherwise ornamented in devices (see Plate I. fig. 5.) *Second*, The *polygonal* column, plain or fluted (see the same fig. left column). This often has a line of hieroglyphics running down it vertically. *Third*, The *bud* capital, or one formed like the bud of the papyrus. In this there are three varieties, the oldest, from Beni Hassan, is composed of four plants bound together by a sort of necking of fine bands under the buds, the columns coming down straight to the plinth. Then there are columns of eight similar shafts, and these generally turn in at the bottom. After the reign of Amunoph III., simple round shafts came into use. The next variety is composed of similar shafts, capitals, and neckings; but there are similar bands or necking on the bud itself, and a sort of short rods or reeds, descending vertically from the neckings on the sides of the column. The third variety has a single circular shaft, without any indication of the united water-plants, but still with bands round the necking, and the capital itself. In these two last varieties the lower part of the shaft is generally ornamented with a sort of sheath or spathe, resembling the lower part of the water plant.

In the *fourth* order the capital is like an *inverted bell*. It formerly was called the lotus capital, but in reality it has no resemblance to that flower. The capital is much undercut, so much so that the ornaments on its edge are not visible, except a person is nearly under them.

The *fifth* order of Sir Gardner Wilkinson is the *palm-tree column*, and resembles the head of that tree, supposing the lower or drooping boughs to be cut off. The neckings, like those of the preceding styles, are composed of five bands, but have the peculiarity of a piece hanging down like a knot at the end. These columns are found as early as the time of Amunoph III. In the times of the Ptolemies, the shafts came straight down to the plinths, and were not drawn in at the bottom, as in the earlier periods.

The *sixth* order is called the *Isis-headed order*, the capital being formed of one or more heads of that deity, surrounded by a representation of a doorway, or small shrine with an image, and sometimes a votary worshipping placed over it. At Denderah the faces, exclusive of the head-dress, are five feet across. Sometimes the Isis head is formed on a square or polygonal column. Sometimes the head is that of Athor, the Venus of the Egyptians. In this order also our author includes the capitals at the tomb of Rhamses III. at Thebes. These are the heads of cows painted blue and red, and with long reverted horns.

The *seventh* order is called the *composite order*. The shafts are generally round, and the capitals are, as its name imports, a mixture of styles. The bell and palms are frequently found in combination. Another favourite mixture is that of the palm over which is the Isis head. A most curious instance is ranked under this style, of columns of the third order with inverted shafts, and also inverted capitals, taken from Karnak, the work of Tothmes III.

The *eighth* order is called the *Osiride*, from the statue of that deity. (Plate I. fig. 6). This order is something like the Persian, or Caryatides of the Greeks and Romans; but it differs inasmuch as the figure does not support the entablature, but stands in front of a square pier which discharges that duty. This order is sometimes used in the courts and sometimes in the halls. Grotesque figures of Typhon are found in a building called the Typhonium at Barkal. These, however, partly support the entablature with odd-shaped caps.

Heights, Intercolumniations, &c.

The following list of heights, diameters, intercolumnation, &c., selected from those given by Canina and Sir G.

Jewish Architecture.

Wilkinson, is given as showing at one view the peculiarities of the various styles ; they are in feet and inches :—

	Height.	Diameter.	Inter-columniation.
Beni Hassan, square, .	18·4	3·8	6·1
Do., polygonal,	16·8	3·7½	10·5
Do., four reeds,	15	2·4	7·2
Karnak, bud capital,	18·11	3·8	6·6
Do., side colonnades, do.,	39	9·6	9·4
Do., central, fourth order	60·3	11·8	12·6
Memnonium, do., do.,	29·4	7·2	7·2
Do., side do., do.,	22·10	5·8½	8 7
Denderah, sixth order,	43·2	8	8·5
Esné, seventh order, .	29·11	5·9	8
Edfoo (see Plate III), do.,	43·2	8	8.5

Pilaster

PILASTERS.—These are employed in all ages, though they do not always accord with the order of the columns. For instance, at Thebes the pilasters of the Temple in Antis of Dayr el Medeeneh are of the sixth or Isis-headed order, while the columns between are of the seventh or composite order. They are generally square, and without diminution.

Entablatures.

ENTABLATURES.—These are nearly alike in all orders, and may be described as a cornice and architrave without a frieze. The former consists of a fillet or regula, beneath which is what is generally called a large hollow or cavetto; but in reality the upper half is a quarter round, and the lower nearly straight. Under this moulding is a bold torus which separates the cornice from the architraves, and runs down the sloping sides of the angle of the building to the ground. The cornice is generally ornamented with divisional vertical lines, like trygliphs. (See Plate II. fig. 5). Between these the cartouches or ovals of the reigning monarch, or other devices, are carved. The centre is generally occupied by the winged globe, or emblem of the Good Dæmon. The torus is often ornamented as if strings were bound round a bundle of sticks, like the *fasces* of a Roman lictor. (See fig. 6.) On the upper part of the smaller cornices there is often a row like the antifixes of a Greek temple, but their form is peculiar (see fig. 7), and is supposed to represent the pots in which the mummies of the sacred cats and ibis were preserved. The architraves are plain, without being broken into faciæ, and are generally covered with hieroglyphics. The plinths are often beautifully carved. (See *ib.* fig. 6.)

Roofs.

ROOFS.—These were always flat, formed of long stones reaching from column to column, or from wall to wall, as the case may be. The joints were grooved, and strips of stone carefully cemented therein to keep out the rain ; for our author, to whom we are so much indebted, tells us that though showers only fell five or six times in the year, yet sometimes there is heavy rain for hours, and now and then violent storms occur. As the ceilings were almost always painted in distemper, of course it was of great importance to keep out the wet.

Materials.

MATERIALS—These were mostly those of the country, either sandstone or limestone. Granite seems to have been kept for statues, sarcophagi, &c. In fact, it is said there is but one temple in all Egypt of that material. The bricks seem almost always to have been merely sun-dried.

As will be shown hereafter, the Egyptians were acquainted with the use of the arch, but, like the Greeks and Hindus, preferred the level architrave, whether from the same principle as the latter, who say, "The arch never sleeps," or whether it was matter of taste, is doubted.

Jewish Architecture.

Jewish architecture.

The long sojourn of the Jews in Egypt, and the fact that their chief employment there seems to have been the manufacture of bricks, must have made them acquainted with the

Jewish Architec- ture.

architecture of that country, probably the noblest in the world. If we accept the date of the Exodus as B.C. 1491, all the great pyramids must have been finished long before the " going out of Egypt." The works at Beni Hassan and Heliopolis, the temples to Phtah at Memphis, and of Ammon at Thebes, that of Thothmes III. at Eilethyia, and of Amenses, his wife, at El-Assasif, must have been completed; and the vast edifice at Karnak, which was begun forty years previously to the Exodus, must have been in a very advanced state. Of course, the long wanderings in the wilderness, the nomad life, and the dwelling in tents, would do much to dissipate these remembrances, especially as the holy Temple itself was a tabernacle of boards and rails, and the sacred courts enclosed by pillars and curtains, the gates being hangings, and the altar itself of wood (except the necessary ash-pans and other works relating to the fire), and capable of being carried on staves of shittim wood overlaid with brass.[1]

On the conquest of Canaan, the Israelites seem to have taken possession of the dwellings of the vanquished people; and we have no record of any important building constructed by the Jews till the days of Solomon. There must, however, have been some large edifices in the country, for we read that, at the destruction of the house of Dagon by Samson, he slew more Philistines at his death than he had done in his active and warlike life; and he had judged Israel more than twenty years.[2]

It is true that Joshua had erected an altar in Israel, but this was only of twelve stones, each no larger than one man could carry.[3] He also built another altar of whole stones on Mount Ebal, on which the law was inscribed;[4] and a third, an altar of testimony;[5] but this, we are expressly told, was not for burnt-offering, nor meat-offering, nor even for peace-offering, " unless the Lord require it." It was simply a monumental testimony. Some have thought, where there was an altar, there must have been residences for priests; but as the Tabernacle at that time was pitched at Shiloh, and afterwards was moved from place to place, this does not seem to follow. At any rate, we have no account of any such buildings.

At the establishment of the kingly government it is plain that Saul had a house.[6] That David also had a house, with a flat top, on which people might walk, is also clear;[7] and that this house was only of wood (cedar), we learn from the same book.[8] From the same passage we also learn the strong wish of this king to build a proper house for the ark of God, then dwelling " within curtains." Whether these houses had been built by the Jews, or whether they were those of the expelled Canaanites, does not appear.

The piety of Solomon seems to have induced him to carry out his father's wishes, and we find[9] that at so low an ebb was the art of building, that the Jews did not even know how to hew timber properly.[10] The king therefore sent to Hiram, king of Tyre, with whom it appears he was on very friendly terms, and that monarch appears to have sent him an architect and staff of skilled workmen. Materials were collected for the building, and careful accounts of the whole work are given, both in the Books of Kings and of Chronicles.

Opinion of various authors.

An immense deal of erudition has been bestowed on the subject, and, in particular, by the Spanish Jesuit Villalpandus, who wrote a work filling three large folios, to prove the design was the work of the fingers of the Almighty himself, and that the order was a species of Corinthian, palm leaves forming the foliage on the capitals instead of

the acanthus leaf, but having on the frieze the very anomalous ornament of Doric triglyphs. There is a very curious book in that Talmudical collection the *Mishna*, called the *Middoth*, and which is supposed to have been written about the time of the Antoninus, and which treats of the Temple and its dimensions. On this book Dr Lightfoot and other Hebraists have expended a vast amount of learning. Later than this, is the work of Wilkins, who wishes to prove that the first temple was of Greek Doric architecture, of massive design, much like those at Pæstum. The most rational theory, however, seems to be that of the late Luigi Canina, of which we shall give a short epitome.

Jewish Architec- ture.

Scriptural descrip- tion.

The Scriptures describe the early temple as a building of stone, roofed and floored with cedar, the work all being prepared at a distance and fixed in its place, so that no noise of hewing timber or cutting stone was heard. It appears to have been a simple rectangular building under one roof, and divided into two parts by a wall, in which was an open door, generally covered by a curtain or "veil." The dimensions (which appear to have been given in the clear or exclusive of walls), being 60 cubits[11] long, 20 wide, and 30 in height, or about 110 feet by 36 feet, and 55 feet high. Compared with the new church in Euston Square, the temple itself would have been about 7 feet shorter and 20 feet narrower, but 15 feet higher. It must, however, be remembered that none but the priests entered its walls. In front of this was a porch the same width as the temple (20 cubits), but only 10 cubits in depth. The description goes on to say, that in the house were " windows of narrow lights," as our version translates the words. The Hebrew[12] seems rather to infer windows which were closed or secret. Rabbi Kimchi interprets the words, " windows of a sloping form—wide within and narrow without." The Greek of the LXX.[13] would seem to imply sloping secret windows, while the Vulgate[14] calls them simply oblique windows. But to this we propose to allude further on.

Priests' chambers.

Round the house—which, of course, must mean on three sides only, as the porch occupied the front—were the priests' chambers, in three stories, one over the other, the lowest 5 cubits broad, the middle 6, and the upper 7,—a passage which has puzzled most commentators, but which will be considered presently. On the right side was a winding-stair leading to the upper stories of chambers. The walls of the house, as well as the ceiling, were lined with boards of cedar. The joists of the floor seem also to have been of cedar; but the floor itself was of planks of fir. The cedar was carved with "knops" and open flowers.

Outer tem- ple and oracle.

The house was, as has been said before, divided crossways into two parts—the outer temple and the oracle, or Holy of Holies. The one was 40 cubits long, by 20, as before stated; the other 20 cubits square. The oracle had doors and door-posts of olive-wood. The temple door-frames were of olive, and doors of fir, all being hung folding. Both doors were carved with cherubim, palm trees, and open flowers. The entire fabric, even the floors, were gilt "overlaid with gold." The account in the Book of Chronicles[15] is substantially the same, except (a difference easily to be accounted for) that it says the greater house, *i.e.*, the outer temple, was ceiled with fir tree; and we gather also from the description, that the whole was roofed with tiles of gold: the nails were also of gold, and weighed 50 shekels. In the oracle, of course, were the ark, overshadowed by the cherubim, the tables of the law, the candlestick, and other objects. There was also the altar of brass, the molten sea, about 18 feet across, supported on

---

[1] Exodus xxv., &c. [2] Judges xv., xvi. [3] Joshua, chap. iv. [4] *Ib.* chap. viii. [5] *Ib.* chap. xxii. 23.
[6] 1 Sam. xix. 9. [7] 2 Sam. xi. 2; xvi. 22. [8] *Ib.* vii. 2. [9] 1 Kings v. *et seq.* [10] *Ib.* v. 6.
[11] Canina makes the sacred cubit = ·554 of a French metre, or 21·81 English inches, or not quite 1 foot 10 inches.
[12] חלוני שקפים אטמים. [13] Θυρίδας παρακυπτομένας κρυπτάς. [14] Fenestras obliquas. [15] 2 Chron. iii. iv., &c.

twelve oxen; lavers, tables, and candlesticks, and many other things, which it is not our province to describe.

But there is one thing which deserves special notice. There were two columns of bronze, or "pillars of brass," at the door of the porch, 18 cubits each, or 33 feet high, about the height of the columns in front of St Martin's-in-the-Fields. They were 12 cubits round, or about 7 feet in diameter. They had capitals (chapiters) also of molten brass, five cubits high, decorated with lily work, chain work, and pomegranates.

The description of Flavius Josephus,[1] the Jewish historian, is substantially the same. He differs a little as to the dimensions, and says the material of the walls was white marble. He also says the floor was paved with plates of gold, and that throughout the whole house, within and without, there was not one part which was not covered with gold, the lustre of which was such, that it completely dazzled the eyes of the spectators. He also describes the columns of brass, and says the capitals were lilies, covered with a sort of tracery, from whence hung, in two rows, 200 pomegranates.

The gorgeous effect of the temple must have exceeded all description, particularly as it is believed that in those days gold was always used in its purity, and without alloy—a belief which is confirmed by an examination of the Jewish shekels, the darics of Alexander's time, and the coins and goldsmiths' work dug up at Pompeii and other places. The Scripture tells us the gold used, exclusive of the gold nails, was worth 600 talents.[2] If we take the value of the talent at L.5475[3] of our money, the gilding alone must have been worth more than three millions and a quarter of pounds sterling; and when we reflect that this was placed on masses of rich carving, we must confess that we can never form an adequate idea of the glory of the old temple. Well might the ancient men, who remembered the former, weep over the diminished glories of the second temple.

In front of the porch was the altar, surrounded by a low wall three courses of stones in height. The whole building was enclosed by a walled court, called the inner court, or that of the priests. In front of this was another, called the lower court; and the whole of this area was enclosed by a circumscribed court going round the whole of the other courts and buildings; and this was called the outer court, or that of the Gentiles.

Canina[4] conceives the style of the building to have been Egyptian; that the temple was lighted like the hypostyle, halls, by a range of windows over the roofs of the cells or priests' chambers; that these windows were like those of the clerestory of a church splayed at the bottom and sides, which would explain the meaning of the different versions of the Hebrew, the Greek, and Vulgate before cited; that the walls of the temple itself sloped towards the top on the outside, or, what is technically termed, were built "battering," while the walls of the priests' chambers were built perpendicular, and for this reason each story measured a cubit more than the room below. He also supposed the capitals of the columns, which are described as of lily work, were in fact the lotus capital of Egypt—that plant being, in other words, the water-lily of the Nile. The porch itself he considers to have been like the propylon, containing other chambers like those of Egypt. Two other circumstances add weight to this opinion—the known connection between Tyre and Egypt, and the fact that Solomon[5] had married the daughter of the Pharaoh of that day, probably one of those who are called the doubtful kings of the 21st dynasty.[6]

Jewish
Architec-
ture.

Temple in
time of
Ezekiel.

In the prophecy of Ezekiel[7] we have a very full and interesting account of what the temple was in his time. The house itself and the oracle do not seem to have been altered, but the old courts seem to have been swept away and succeeded by some vast atria, and a mass of halls and chambers. The external dimensions, as measured by the angel, were, a square of 500 reeds of six cubits each on all sides, nearly 5500 feet English, or more than a mile square—covering a space as large as Hyde Park and Kensington Gardens together. This appears so vast an area that Canina is disposed to consider it a misreading of reeds for cubits. This would then give an area of nearly double that of Lincoln's Inn Fields, and falls in so well with the detailed dimensions that it may be accepted, unless the meaning of the text be that the temple and its courts stood within a large enclosure of vacant ground like a cathedral close, which may have been the case. As nearly as may be judged from the prophet's description, there must have been more than two hundred apartments on the ground floor, without calculating the upper stories. There were also large chambers, which are supposed[8] to have been treasuries, storehouses, refectories, places for sacrifice, chambers for priests in attendance on the altar, and for the singers; oratories, or places for prayer, for those who were not allowed to enter the temple; and, in short, every requisite for so large and important a community as must have been collected about the sacred edifice.

There were, however, but two courts—the outer, or court of the Gentiles, and the inner, or court of the priests. The former was entered on the east, north, and south sides by three gateways or propylons, each containing on the ground story eight chambers, a vestibule, and a hall. On the east side of this court was a sort of cloister composed of three ranges of columns, behind which, and against the outer wall, were thirty cells or chambers. On the north and south sides were colonnades of one row of columns only to each. At every angle of the court were buildings like towers, containing sixteen chambers on the ground floor, surrounding a small atrium each.

The inner court was entered on the north, south, and east sides by gateways or propylons exactly like those of the outer court. It had a colonnade of a single range of columns round the north, east, and south sides, and, as far as we can understand the description of the prophet, forty-six cells or chambers for the priests. On the west side of the inner court was the temple looking eastward, and, as has been said before, in the same state as it stood in the days of Solomon. To the right and left stood a refectory—one for the priests of the altar, the other for the singers; behind these, separated from the temple by two small courts, were two oratories or houses of prayer for those who were forbidden to enter the temple itself. In front of this, and in the middle of the court, the great altar approached by a flight of steps, and surrounded by a low wall, is supposed to have stood as in the time of Solomon. This altar is said by Ezekiel to have been called Ariel, and was 12 cubits square. Behind the temple was an area, probably having a building thereon, called in our version, "the separate place,"[9] in each side of which, it is supposed, was a colonnade and two ranges of chambers, nine in each. Besides these chambers on the ground floor ranges of galleries are mentioned, so that it is difficult to form an adequate idea of the extent and accommodation of this vast edifice.

In the year 588 Nebuzaradan, one of Nebuchadnezzar's generals or "captains" besieged the city, and took it. They plundered and burnt every building, including the temple,

[1] Jewish Antiquities," viii. 2.   [2] 2 Chron. iii. 6.   [3] See Encyclop. Britan. vol. viii. p. 66, art. Coinage.
[4] Tempio di Gerusallemme, Rome fo.   [5] 1 Kings iii. 1.   [6] Sir Gardner Wilkinson's Egypt, vol. i. page 135.
[7] Chap. 40, et seq.   [8] Partly from the LXX. and partly from the Vulgate.   [9] Ezekiel xli. 12.

Jewish
Architec-
ture.

Second
temple.

first stripping it of all its utensils and ornaments, and then carrying off the bronze pillars and the gold covering. The walls of the city were demolished, and the whole people carried into captivity by the victorious Chaldeans.

In the time of Cyrus the temple was rebuilt; and though the erection of any such edifice was matter of joy to the young men who had never beheld the former building, the old men, who remembered the ancient temple, wept when they saw how inferior it was to the first: in their eyes "it was as nothing."[1]

Herod's
temple.

From Flavius Josephus[2] we learn that Herod demolished this building, and began a third temple, hoping to rival the glories of the first. His edifice must have been very magnificent. The ground was first levelled and the foundation laid, immense substructures being made by the side of the rock to obtain sufficient room. The temple itself seems to have been of exactly the same dimensions as that of Solomon. The outer house, 40 cubits long—and the holy of holies 20 cubits; but it was nearly double the height of the first temple. The porch, also, instead of being the width of the house, 20 cubits, was 100 cubits long and 100 high, crossing the temple in the form of a T, and forming a magnificent façade much longer than that of St Paul's in London. Round the house were three stories of priests' chambers also like those about the old temple. It stood in a court in which was the altar of burnt-offering, surrounding which were chambers for various purposes. The court had three entrances on each side, which were called respectively the water-gates, the fire-gates, and the oblation-gates. But the peculiar difference between Herod's temple and the earlier building was that, in front of the court last described, was another of about the same size, surrounded by a colonnade and chambers, which was the place set apart for the women. These courts were surrounded by an outer court, having a species of cloister on the north, east, and west sides, composed of a double row of columns. On the south side was a similar construction a furlong in length. It consisted of three rows of columns, forming, with the outer wall, three aisles—the two outer being 30 feet wide and 50 feet high; the centre being 45 feet wide and 100 feet high—no doubt with a species of clerestory. These columns are described to have been so large that it took three men with extended arms to span their circumference. The outer court was elevated six steps above the surrounding country; the inner courts stood on a sort of plateau, with retaining walls and parapets round, which was ascended by fourteen steps; this was on the level of the women's court. Between this and the inner court was a gate called that of Nicanor, in front of which was a semicircular flight of fifteen steps. The temple itself was entered by a flight of twelve steps, so that its floor must have been between 20 and 30 feet above the ground of the surrounding city. The whole was built of the most beautiful white marble. Some of the blocks were of the gigantic size of 47 feet by 22 feet, and 14 feet thick, and must have weighed 900 tons, while the largest at Stonehenge is estimated only at 42 tons. The front of the temple was entirely gilt. The sight from a distance, our author says, was dazzling to the eyes like looking at snow. On the roofs were pointed ornaments, like crest-ridging, to prevent the birds settling, which were also gilt. The candlestick, altar, and other ornaments and utensils, were exactly those of the old temple, and most of them are sculptured on the arch of Titus at Rome, as borne in triumph by that Emperor, and they exist to this day.

Solomon's
palace.

We read in the First Book of Kings, that Solomon built the "house of the forest of Lebanon"[3]—"his house where

he dwelt,[4] and a house for Pharaoh's daughter."[5] Some have supposed these to have been three distinct palaces, but Canina considers them as all connected and as three parts of one large structure. The house of the forest of Lebanon —so called, in all probability, from the cedar of which it was constructed—is described as being 100 cubits long, 50 wide, and 30 high.[6] It had four rows of cedar pillars, or, as the Vulgate more correctly renders it, four "ambulatories,"[7] formed by three rows of cedar columns, fifteen in each, or forty-five in all, with cedar architraves, and covered with cedar. This Canina restores as an Egyptian hall, lighted, as described by Vitruvius, with a portico in front of 50 cubits by 30. The great hall, he supposes, led on one side to the palace occupied by the king, in which was the hall of judgment and the throne, and on the other side, to the queen's palace and the women's apartments, or, as it has been called in later times, the harem. The porticoes and hall seem to have been of cedar, but the houses were of "costly stones, even great stones."[8]

Jewish
Architec-
ture.

Other
Jewish re-
mains.

With the exception, perhaps, of some relics of ancient walls, it is supposed no part of old Jerusalem is extant, so complete was the destruction. There are, however, some objects generally considered to belong to Jewish times, that should be mentioned. There are, first, what are called the sepulchres of the kings of Judah. These are a number of sepulchral chambers hewn out of the solid rock, and containing sarcophagi. They vary from 10 to 20 feet square, and are entered by a portico in antis, exactly like the tombs at Beni-Hassan, about 40 feet wide. There are two columns and two pilasters in front, of Greek Doric character, about 13 feet high. The most curious feature is, that on each side, running down 4 or 5 feet, and over the columns horizontally, is a broad band, about 3 feet wide, richly sculptured with foliages. Above this last, quite independent of the lower construction, is a regular Doric architrave and frieze, of a character between Grecian and Roman; this is ornamented with triglyphs, pateræ, and foliage. In front of the portico is a large court-yard, about 100 feet square.

In the valley of Jehoshaphat, near Jerusalem, are three extremely curious relics; two stand alone, on platforms excavated from the rock, and the third is scarped into the rock itself. The first is called the tomb of Absalom. It is a square building with a solid wall, in which are engauged Ionic columns, about 13 feet high; over this is a Doric entablature with triglyphs, and an altar, surmounted by a very curious sort of hollow-sided cupola of trumpet-mouth section, and a terminal. It is no doubt intended to represent the Arab tent. The whole, including the flight of steps, is about 60 feet high.

Another similar building, of about the same size, is commonly called the tomb of the prophet Zachariah. This is surmounted by a simple pyramidal roof. Beneath it is a handsome doorway leading to a sepulchral chamber. Over the ordinary classic entablature is the regular Egyptian cornice or torus, surmounted by a bold quarter hollow and fillet, exactly like those on the Propylons of Egypt, which have been described before.

The third building is entirely rock cut, and consists of a large façade, about 90 feet wide and 100 feet high, and is reported to be the place to which the apostles retired before the siege of Jerusalem. Below is a plain face, about 45 feet high, on each side of which are wings with two pilasters, both running up to the top of the building. Between these is a sort of portico, about 40 feet wide, with columns and pilasters, of nearly pure Grecian Doric. There are also several other rock-cut tombs or sepulchres scattered

---

[1] Haggai ii. 3. [2] Josephus' Jewish Antiquities, xv. 11. [3] 1 Kings vii. 2. [4] *Ibid.* v. 8. [5] *Ibid.*
[6] About 180 feet by 90 by 54 feet. About the same superficial contents as Westminster Hall. [7] "Deambulacra." [8] *Ibid.* v. 9.

Indian
Architec-
ture.

about in the neighbourhood of Jerusalem, but none of which possess much architectural interest. One is called the sepulchre of Jehoshaphat.

From the character of the architecture, it is impossible that these buildings should have anything like the age ascribed to them. The Ionic capitals are evidently Roman, and therefore cannot date earlier than the conquests by that people; probably they are of the time of Herod; while the Egyptian cornices show that the traditional ornaments of that people had not been entirely forgotten. Besides this, the general plan of a temple, *in antis*, scarped into a rock, so entirely resembles the work at Beni-Hassan, that it is impossible to deny these very interesting remains strongly corroborate the views of Canina, that the architecture of the early temples, if not purely so, was at any rate *based* on the architecture of the Egyptians.

Indian Architecture.

Buddhist
Architec-
ture.

The late researches of the Royal Asiatic Society, and their published works, particularly that of Rám Ráz, who was a learned native of the Carnatic, and afterwards one of our native Judges at Bangalore, in the Mysore; and still more so the works of Mr Fergusson, have thrown a great and unexpected light upon the history of architecture in the East. The traditions as to the extremely remote antiquity of the rock-cut temples, the caves of Ellora, and the wonderful pagodas, have disappeared before the searching eye of critical investigation. In the time of Herodotus, as has been before stated, the Persians had no temples; and even as late as that of Tacitus, the great Indo-Germanic races "would not confine their gods within walls." In the emphatic words of Mr Fergusson, "throughout the Vedas there is no allusion to temples, nor to images, nor indeed to any public form of worship. Every man stood forth in the presence of his God, and without intercessors offered up his prayers with the prescribed form." This religion is generally known as Brahminism. About the period of the foundation of Rome, the first of the Buddhas seem to have commenced a perversion of the ancient faith. The struggle seems to have gone on for years, till three quarters of a century after the time of Alexander the Great, about 250 years before the Christian era, when a powerful ruler named Asoka, a grandson of Chandragupta, who is supposed to be the Sandracottus of the Greek writers, abjured Brahminism, and made Buddhism the religion of the state. About this period the building topes, or *sthupas*, commenced. These at first, we are told, were *sthambas*—that is, simply monoliths—or were *Lats*, pillars, which were built up in several pieces. Before this we are told no Indian building or inscription exists.

These topes are at first supposed to have been set up to commemorate some event, or to show the spot was sacred; but after a time they seem devoted for the purpose of containing relics, such as the tooth or collar-bone of one of the Buddhas. After a time a number of these topes took the form of large towers. The relics seem in some cases to have been preserved in a sort of box or case at the top of the tope, called a *tee*; in others, in regular relic chambers. Where there were relics, the place was called *dagoba*, or relic shrine, from whence, perhaps, our phrase pagoda is corrupted.

Other topes took the form of hemispheres. One of these, the Sanchi tope, is given by Mr Fergusson. The diameter is 106 feet. It has a curious species of gate entrance, and is surrounded by a stone fence. Others are partly cylindrical, and finished with either a flat circle, or pointed terminals like a dome at the top.

Our author tells us that the next two classes of Bud-

Indian
Architec-
ture.

Rock-cut
Temples.

dhist architecture are the temples (Chaityas) and monasteries (Viharas), but that no *built* examples exist in India. They are, in fact, rock-cut caves. Of these at least one thousand are known—one-tenth probably Brahminical or Jain, the rest Buddhist. Of the monasteries, twenty or thirty only are believed to exist. The series is said to be uninterrupted from the first by the grandson of Asoka, 200 years before Christ, to those at Ellora, which, strange as it may seem to our ears, instead of being of an almost Diluvian antiquity, as has generally been reported, have proved to have been the work of Indra-dyumna, as late as about the time of our Norman Conquest. One of these cave-temples at Karli, presents exactly the features of a Roman basilica, or early Christian church. It has a circular end or apex, and is divided into three aisles by two rows of columns. Others are simple square buildings, with a circular or oval chamber at the end entered by a small door. The viharas or monasteries seem generally to have been square caves supported by pillars of the natural rock left in their places, and surrounded by a number of small sleeping places or cells.

Caves at
Ellora.

But the most wonderful excavations are those at Ellora. These are a series of hypogea or caves sunk in the solid rock, and extending a distance of between three and four miles. Canina has given plans and interior views of six of these caves. These, called Parasova Rama and Diajannata, are simply halls supported on massive piers with level architraves. The piers are richly carved with figures and friezes, and have a sort of cushion capitals, and square abaci, and stand round, forming a sort of atrium. That called Indra has a court open to the sky, in which is a small shrine or temple. In the solid rock are two halls similar to those above described, a larger and smaller. By some this has been supposed to be Jainiti. At Tin Tali, the piers are quite plain. At Visovakarmalla is a quadrangle, open to the sky and surrounded by pillars. This leads into an atrium with three aisles and an apsis, and is exactly like a basilican church. But the greatest wonder of all is the collection of chambers and halls called the Kylas, or Kailasa. These are sunk into the rock, and occupy a space of 270 feet deep and 150 feet wide. Of course, the roofs are solid rock, supported by pillars, or rest on the walls, or other division of the assemblage of chambers. There is a porch, on each side of which are two columns. This conducts into a hall, supported on 16 such columns, and leading into a sort of adytum. Round this is passage space and five chambers. The whole forms a temple, with its usual appendages, just such a one as would be built on the ground, and round this a wide open space with a sort of colonnade or cloister round the whole. Great part is open to the sky, for the sake of light and air, but the work is entirely cut out of the solid. Two of the columns are given, Plate I. fig. 8.

The Kylas.

We shall not follow our author through the details of Buddhist architecture in Ceylon, Birmah, Java, Nepaul, and Thibet, further than to state that examples are cited as early as 161 years before our era. In the latter country, the famous traveller, M. Huc, mentions a monastery which has accommodation for 15,000 lamas. In one of the monasteries outside the city of Lassa it is said there is a vast building, the residence of the chief Lama, who is there worshipped as a deity, consisting of five stories, surmounted by a dome covered with plates of gold.

Under this head, Mr Fergusson describes the rock-cut Raths at Mahavellipore. They are five stories in height, and are copies of the Buddhist monasteries, but used as temples. The Raths, in fact, our author thinks, link the two styles, the *Chaityas* and *Viharas*. The wonderful domical topes deserve, however, a passing notice. A diameter of 200 and 300 feet occurs not unusual; while at

Anuradhapoora the external diameter of the domes is 360 feet, or double that of St Peter's or the Pantheon at Rome.

After giving an account of Buddhist architecture in Ceylon, Burmah, Thibet, and Nepaul, as before stated, Mr Fergusson enters on the subject of that of the Jainas, or Jains. This was a sect which arose in the endeavour to re-establish Brahminism. It probably had been in existence for many years, but first seems to have acquired importance about A.D. 450. This sect rejects the doctrines of Asoka, as also the practice of monasticism. The famous temple at Somnauth belongs to them. Mr Fergusson has given as an illustration of that built by Vimala Sah, on Mount Abu, as a type of the ordinary Jain temple. In the centre is a cell in which is a cross-legged figure of one of the twenty-four saints worshipped by this sect; in this case it is that of Parswanath. The cell is always terminated by a pyramidal roof. In front of this is a portico of 48 pillars, disposed much like a cross church with a dome at the intersection of the transepts. The whole is surrounded by a sort of cloister formed by double rows of columns, and a series of small chambers like the cells of a *vihara*. But as the sect refuse monasticism, each cell is used not as a dwelling, but as a sort of small chapel, and contains one of their cross-legged deities. One of the peculiarities of this style is that richly-carved brackets spring from the pillars from about two-thirds of their height, and extend to the level architraves, forming a sort of diagonal strut to strengthen and support them.

Our author goes at great length into the subject of domes, their invention and construction. It is his opinion that the Jains adopted the dome at a very early period, and he doubts whether the Buddhists ever used this species of construction. "No tope," he observes, "has the smallest trace of such a structure, though of domical shape outside, and the design of the rock-cut temples, with the upright supports, the raking struts, and the level architraves, have manifestly been deduced from timber construction." The Indian dome has no voussoirs radiating from the centre, as in European architecture. The courses are all horizontal, so that they are necessarily pointed in section, or they would not stand if circular. The Indian dome, however, has this merit, it requires no abutments, and has no lateral thrust. The pressure is entirely vertical; and if the foundation be sound, and the pillars stout enough, there can be no failure.

The leading idea of the plan of the Jain Temple is that of a number of columns arranged in squares. Wherever it was intended to have a dome, pillars were omitted, so as to leave spaces in the form of octagons. By corbelling over the pendentives in level courses the dome was gradually formed. Our author gives the plan, and a view of the temple at Sadree, a building as large as most cathedrals. This has the great number of 20 domes, varying from 36 feet to 24 feet in diameter, and supported by 420 columns. The effect, we are assured, is highly pleasing and picturesque.

Like most architectural people, the Jains were also fond of tower-building. Mr Fergusson has given a perspective view of the *Jaya Sthumba*, a tower of victory erected by Khumbo Rana, to commemorate the conquest over Mohammed of Malwa, in 1439. It is nine stories high, the two topmost of which are open. The general outline is not unlike that of an Italian campanile, with pilasters at the angles, and an overhanging corbelled out top. It is richly ornamented from bottom to apex, and affords a very favourable idea of Indian art. The Jains are said to have constructed the Indian group of rock-cut caves at Ellora before described, but the point seems undetermined.

None of these can be older than the 9th century of our era, when the Jainite religion was first revived. The

oldest existing are said to be in Gugerat. Those on Mount Abu are said to date from A.D. 1030 to 1230. The Jains seem to have declined under the rule of the Mohammedans. No Jainite building is mentioned after A.D. 1500.

The Hindoo architecture, no doubt, originally rose from that of the Buddhist's, and has been divided with those of the Arian or Sanscrit races of North India, that of the South or of the Tamul races, and that prevalent in the Punjab and Cashmere. Of the first and last we have comparatively but little knowledge, but South Hindoo work is treated of at great length by Rám Ráz, the native author before mentioned. The remains of the buildings are numerous, as the Tamul races were perhaps the greatest temple builders in the world; and the whole subject has been so well elucidated by the author last referred, that its principles may be considered to be clearly ascertained and settled.

This last branch has been chiefly illustrated by the drawings in Vigne "Travels in Cashmere," and those of Major Cunningham in his Memoir to the Asiatic Society of Bengal. The temple of Martund, reduced from the last-named author by Mr Fergusson, shows a cloistered court surrounded by pillars and cells, and entered by a porch. In the middle of this is a temple with a species of *naos* and *pronaos*. But the most curious feature is a series of doors with acute pediments over them shaped very much like Gothic gablets, and containing trefoil arches. A similar feature occurs at Pandrethan, the latter built about A.D. 1000, and the former A.D 752. It seems by no means improbable that the pointed domes, these gablets, and trefoiled arches may have strongly affected the architecture of the Saracens, and have given rise to that which we call Gothic.

Of this style Mr Fergusson gives some extraordinary, and by no means elegant examples, as the black Pagoda at Kannaruc, and temples at Barolli, and Benares. The chief features are a sort of entrance porch, sometimes walled, and sometimes carried on pillars called the nuptial hall, leading into a great pagoda, square in plan, and finishing with a sort of tub-shaped dome. The ornamentation is profuse, so much so as to detract from the greatness of the design. The rock-cut caves of North India have been described before. There are no buildings in this style anterior to the Mohammedan conquest. The temple of Juggernath is dated A.D. 1174, about the time of the Ellora caves, and the black Pagoda, A.D. 1236.

We are told by Rám Ráz that many treatises on architecture, some say sixty-four, existed in India. The collection he calls the *Silpa Sástra*. The most perfect of them he calls *Mánasára.* This is imperfect, but our author says he had forty-one chapters of it in his possession. He also cites several others, one of the chief of which he calls Casyápa. He gives an epitome of the *former* work, which is worth a short notice. In the first chapter it treats of the various measures in use in the country in the second it describes the *sthapati*, or architect—the *sátragrúhi*, or measurer, probably the surveyor or clerk of works, and then the various builders. Of the architect it says, "He should be conversant in all sciences; ever attentive to his avocations; of an unblemished character; generous, sincere, and devoid of enmity or jealousy." And further on he says, "Woe to them who dwell in a house not built according to the proportion of symmetry. In building an edifice, therefore, let all its parts, from the basement to the roof, be duly considered." Subsequent chapters treat of the religious rites necessary in commencing a building, one of which is finding the orientation, or rather the laying down a true meridian line, minute directions for the finding which are given, and afford a high opinion of the mathematical knowledge of the Hindoos. The author of the *Manasara* then treats of pedestals, bases, pillars, and entablatures, the proportions of which, with sections of the mouldings, &c., are as carefully given as in any treatise on classic architecture. In fact, it would almost seem as if the one were derived from the other. The Hindoo architecture seems to have no affinity with that of Egypt. He treats of seven sorts of columns with their dimensions, of which the second resembles the Tuscan, the third the Doric, the fourth the Ionic, and the fifth the Corinthian. It would be too long for our pages to follow our author through his various directions for planning villages, towns, and cities; we therefore proceed to his description of temples, of one of which, the Pagoda of Tirùvalúr, a plan is given by Rám Ráz, with an isometrical perspective view. They generally consist of a large quadrangular court; that before us measures over 700 by 945 feet, with a lofty wall surmounted by a blocking course. The enclosure is always entered by gateways, surmounted by several storys, diminishing somewhat in a pyramidal manner. These are called *gopuras*. Our author gives designs for various species of these, varying from one to twelve stories in height, and very imposing structures they are. The next great feature of the south Hindoo temple is the *choultries*, pillared halls, or covered spaces supported on columns. In the instance before us there are several, varying from small peristyles of four columns square, up to the largest, which has nearly 700 columns, 30 feet high; only about one half, however, is roofed. At Tinnevelly the *choultrie* is on the east side of the entrance, and consists of nearly a thousand such columns. The same feature occurs at Chilianbrum. Within the front enclosure there is generally a second, and in this instance a third, also entered by *gopuras*. Within these enclosures are one, and sometimes two porches, called *Mantapas*, leading to the temple itself, usually square buildings. Mr Fergusson remarks, that while in other buildings the Hindoos used a profusion of columns, they were very sparing of them in the *mantapas*, adopting the use of struts and brackets, and even of iron, that they should be as open as possible. The next great feature is the *Vimana*, or temple itself. Of these Rám Ráz has given designs, varying from one to fifteen stories in height. These all partake of the pyramidal form of the *gopuras*, but are square in plan, while the latter are generally oblong. Mr Fergusson describes that at Tanjore as having a base 82 feet each way, and as containing fourteen stories in all, nearly 200 feet high. Our limits will not permit us to proceed further with the subject, we therefore refer the reader with pleasure and confidence to the works above cited, and to those of Kittoe and of Daniell.

Chinese Architecture.

The buildings of the Chinese are very inferior in character to those last described; in fact, Mr Fergusson goes so far as to say, " China possesses scarcely anything worthy of the name of architecture." Sir W. Chambers has described one of the Buddhist temples, that at Ho-nang, which is not unlike those of India in arrangement. There is an extensive court, with avenues of trees, leading to a flight of steps and portico of four columns. In a second vestibule behind this are four colossal figures bearing various emblems. Beyond this is a very large second court, entirely surrounded by colonnades and small sleeping cells for the priests or *bonzes ;* in other words, a huge cloister, much like the viharas before described. In the same ranges are four pavilions filled with idols, and large rooms for refectories, behind which are the kitchen, courts, &c. At the extreme corners of the grand court are four other pavilions, which are the dwellings of the higher order of priests.

At equal distances behind each other, down the centre of the court, are three larger pavilions, called *tings*, entered on each side by a flight of steps, and a fourth engaged in the cloister itself, and having a front portico and one flight of steps only. The first three are square, two stories in height; the lowest surrounded by fourteen columns, each face or front showing six. They have rude caps, composed of eight brackets, projecting various ways. Sir William Chambers says there are four species of *tings*,—three used for temples and the fourth for gardens ; some having a gallery and fretted railing round the first floor on the outside, the upper story being set back. The roofs all have the peculiar hollow dip, which leads one to suppose their prototype was the tent, the *sag* of the cloth of which would suggest the form. They are frequently surmounted with a sort of cresting and finial, and each angle is turned up sharply, and ornamented with a dragon. Sometimes the columns have a frieze perforated in the form of frets ; sometimes the same is also under the eaves of the upper roof. The author before cited also gives examples of smaller octagonal *tings*, intended to cover the large vessels in which the Chinese burn gilt paper to their idols.

The most striking buildings in China are, however, the tapering towers which they call *taas*, and our old writers pagodas. These are of brick covered with marble, or most generally with glazed tiles ; and are built in stories, one over the other, from three, four, or five, to as many as nine in number. Each story is reduced in width, and has a gallery round to walk in. The roofs are hollow or sagging, like those formerly described. They project a great deal, the corners being turned up sharply. On these light bells are suspended, which make a constant ringing when the wind blows. The roofs are covered with glazed tiles of various colours, and the summit ornamented with a species of spire and finial. The most celebrated of those was at Nankin, and is better known as the porcelain tower. It had nine storys, and was about 200 feet high, exclusive of the iron spire before mentioned. At each angle was a bell, making, of course, seventy-two in number ; and there were eight chains hanging from the top of the finial to the angles of the spire, and carrying nine bells each, or seventy-two more. This celebrated building, it is reported, has just been destroyed by the rebels. The *taa* is not a pagoda or temple, as has been supposed, but is a memorial of some event,—as a victory, or of some great personage. At Pekin is one used as an observatory. At Nanganfoo one was erected simply to bring good luck ; and Mr Davis relates one was decreed to be built at Macao, with a similar idea.

Buildings called Toov Tang, or halls of ancestors, are found in all considerable towns. These much resemble temples, but instead of idols, memorial tablets are placed in the niches to record the transactions and deeds of the " worthies" or celebrated inhabitants of the neighbourhood.

The *Pai Loo*, or *Pai Fang*, is another common object in China. They are monumental memorials, though they have been mistaken for triumphal arches. M. Quatremere says, the Chinese annals reckon 3636 of these erected in honour of literary men, philosophers, princes, generals, &c. The smaller are of wood, forming a sort of doorway. The larger have three openings side by side, and over these are several broad panelled facias for inscriptions and carving, which is often very bold and in high relief, and over all is a projecting cornice carrying a tiled roof. Chambers has given one, the side gateways of which have semi-circular arches with festoons of drapery.

There is not much variety of design about their houses, as every one must be on a scale according with the rank of the inhabitant. Le Comte mentions a case where a Mandarin was obliged to pull down one that he had constructed of a somewhat better quality than those of the others. Chambers has given a plan of a house which he says is of very common design. It is about 260 feet from front to back, and about 65 feet wide. It is entered at the front by a passage nearly 20 feet wide, which goes nearly through the entire building. On each side of this, fronting the street, is a shop, with its back shop. It should be stated, that the divisions on each side of the central passage exactly correspond with each other. The room on the right exactly resembles that on the left, till we arrive at the great dining-hall, which runs right across the house. First we have two studies, and two small bed-rooms, then two saloons or reception rooms about 24 feet by 18. These look into open courts or gardens, with fish ponds, fountains, flowers, &c., divided by walls; then two more saloons with bed-rooms, and then the great dining-hall before mentioned. This is about 60 feet by 30, and is carried on eight columns. Behind this is the kitchen and other offices.

The first floor is called the *Leou*, and, beginning at the end fronting the street, has two bed-rooms, one on each side of a passage, for the shopkeeper; then on each side is a saloon, and the bed-rooms for the family. Between these last, and also carried on columns, is the hall where the family idol is worshipped. This looks on to the open gardens before-mentioned. At the further end of these courts are two more saloons and bed-rooms, and then a hall, said to be devoted to the use of strangers or visitors, which is over the ground floor dining-hall, and of the same size. Chambers tells us, every house has a number of movable partitions kept ready, to be put up to subdivide the larger rooms.

The Chinese never use square timber when they can get round trees of a suitable size, probably on account of the lightness, strength, and convenience of the bamboo. The roofs are of very peculiar construction, which our space will not permit us to give, and all timbers are left visible. The windows are filled in with the lining of the oyster shell, which looks like talc, is quite as transparent, and the main door is frequently a perfectly round aperture.

Ancient American Architecture.

It was not long before the exhumation by Mr Layard in Central Asia, of the wonderful remains of fine art in monstrous combinations of brute and human forms, entombed in earthen mounds, that Mr J. L. Stephens, when engaged on a mission from his government—that of the United States of North America—to some of the mutable states of Central America, heard of and tracked out in the forests of Yucatan[1] the remains of a bygone time, exhibited in sculptural and architectural monuments of a coarse character, and revolting as regards the representations of human and bestial forms in sculpture, but not more revolting than the more elegantly designed and beautifully wrought works of the Central Asians; whilst the architectural remains afford a strange counterpart to those which Mr Layard describes, as he imagines them to have existed in and about the valleys of the Tigris and Euphrates. Huge mounds, constructed pyramid-wise and of stone, form in Yucatan the bases of the temples and shrines upon which some former inhabitants practised the horrid rites by which they thought the Almighty was to be worshipped; as in Assyria, there were mounds of the material the country afforded upon which the temples and shrines of an equally unholy worship was performed by the more graceful but most abominable sovereigns, subjects, and slaves of the Eastern tyrannies.

Among the objects represented in Mr Catherwood's *Views of Ancient Monuments in Central America, Chiapas, and Yucatan*, are several examples of vaults having the arch form but not being arched vaults; that is to say, of vaults presenting the appearance internally, or upon the soffit, of arches, but formed by the gathering over of horizontally-coursed masonry, with the inner and lower angles worked away—or cleaned off, as it is technically expressed—to the appearance on the inside which an arched vault would present. See fig. 5.

Fig 5.

The circumstance that the arch form presented in the American monuments is produced by the gathering over of horizontally-ranged masonry, and not by means of arch structure, would seem to show clearly that if their builders ever had intercourse with the Old World, it was before the properties of the arch were known and practised in it. These remains show an advance on the Pelasgic and Celtic monuments of the cis-Atlantic world, and take the general character of the stone-works of Egypt and India; but like those works they exhibit the vaulted form by gathering over and not by arching.

Mr Catherwood states, that he and Mr Stephens concur in the opinion of Mr Prescott, author of the *History of the Conquest of Mexico*,—"that though the coincidences are sufficiently strong to authorize a belief that the civilization of Anahuac (Ancient Mexico) was in some degree influenced by Eastern Asia, yet the discrepancies are so great as to carry back the communication to a very remote period, so remote, that this foreign influence has been too feeble to interfere materially with the growth of what may be regarded, in its essential features, as a peculiar and indigenous civilization;" and this opinion the monuments, as presented by Mr Catherwood, would seem fully to justify. But Mr Catherwood adds to this, as the ground, it would appear, for coinciding with Mr Prescott's opinion, that the results arrived at by Mr Stephens and himself "are briefly, that they (the American monuments) are not of immemorial antiquity, the work of unknown men; but that, as we now see them, they were occupied and probably erected by the Indian tribes in possession of the country at the time of the Spanish conquest, that they are the production of an indigenous school

---

[1] Lord Kingsborough's great work on the *Antiquities of Mexico* contains, in some of the later volumes, representations of monuments which would almost appear to be the same as some of those subsequently explored by Mr Stephens.

of art, adapted to the natural circumstances of the country, and to the civil and religious polity then prevailing; and that they present but very slight and accidental analogies with the works of any people or country in the old world."

Grecian Architecture.

As no nation has ever equalled the Egyptians in the extent and magnitude of their architectural monuments, neither have the Greeks been surpassed in the exquisite beauty of form and proportion which theirs possess. Extreme simplicity and perfect harmony pervade every part of a Greek structure; and to the evanescence of the finer spirit of these qualities may be referred the difficulty—for great difficulty certainly exists—of applying Grecian architecture to modern practice. The national style, or Doric Order, is in every respect the most distinguished and the most intractable. The voluted Ionic being more complicated, is more plastic; and the foliated Corinthian, from its still greater divergence from Doric simplicity and harmony, is the most easily moulded to various purposes. Unfortunately nothing remains from which we might acquire a knowledge of the practice of the Greeks themselves in the architecture of domestic and general structures; but it may be inferred from some existing edifices, particularly the Choragic monuments, that the Doric columnar style was not used by them except for the temples of the gods and some of their accessories. But whether this arose—if the feeling really did exist—from the sanctity of its character, in consequence of that appropriation, or from the difficulty of moulding it to general purposes, cannot be determined. It is very certain, however, that the few structures which do exist of Greek origin, not of a religious character, are either Ionic or Corinthian, or a mixture of one of them with some of the features of the Doric; and in all Greece and the Grecian colonies, except Ionia, there are very few examples of religious edifices not of the Doric order, and none which are of the Corinthian.

We have already given our reasons for mistrusting the descriptions of ancient writers on architectural subjects; and when they merely make reference to different parts of a structure, without pretending to describe, in the absence of examples or models they must be unintelligible, and therefore no more valuable to the architectural antiquary than those of the writers, whom existing specimens of what they profess to describe prove to have been totally ignorant of their subject. We shall therefore not attempt to develope what does not exist, either from inferences to be drawn from Homer and others, from the professional dicta of Vitruvius, or from the description of Pausanias; but confine ourselves to the remains of the structures themselves of the Greeks, which are actually before our eyes, for the elucidation and exemplification of the Grecian style.

Like the architecture of Egypt, that of Greece is known to us principally by means of its sacred monuments, and from them is deduced almost all we know of its principles. The Doric temples of the Greeks are uniform in plan, and differ only in arrangement and proportion, as they are of greater or less size; for every part depends on the same modulus. If the dimensions of a single column, and the proportion the entablature shall bear to it, were given to two individuals acquainted with the style, with directions to compose a hexastyle peripteral temple, or one of any other description, they would produce designs exactly similar in size, arrangement, features, and general proportions, differing only, if at all, in the relative proportions of minor parts, and slightly perhaps in the contour of some of the mouldings. This can only be the case with the Do-

ric, and it arises from the intercolumniation being determined by the arrangement of the frieze with triglyphs and metopes; the frieze bearing a certain proportion in the entablature to the diameter of the column, and so on, in such a manner that the most perfect harmony is preserved between every part. Thus, in the example, the column is so many of its diameters in height; it diminishes gradually from the base upwards, with a slightly convexed tendency or swelling downwards; and is superimposed by a capital proportioned to it, and coming within its height. The entablature is so many diameters high also, and is divided, according to slightly varying proportions, into three parts—architrave, frieze, and cornice. A triglyph bearing a certain proportion to the diameter of the column is drawn immediately over its centre; the metope is then set off equal to the height of the frieze; another triglyph is drawn, which hangs over the void; then a metope as before; and a second triglyph, the centre of which is the central line for another column; and so on to the number required, which, in a front, will be four, six, eight, or ten columns, as the case may be, the temple being tetrastyle, hexastyle, octastyle, or decastyle; and on the flanks twice the number of those on the front and one more, counting the columns on the angles both ways. Thus a hexastyle temple will have thirteen columns on each flank, an octastyle seventeen, and so on. It must be observed, however, that to ease the columns at the angles, they are not placed so that the triglyph over them shall impend their centre as the others, but are set in towards the next columns so far that a line let fall from the outer edge of the triglyph will touch the circumferential line of the column at the base, or at its greatest diameter. It has been generally thought that the object in this disposition was to bring the triglyph to the extreme angle, to obviate the necessity of a half-metope there; and many imitators have puzzled themselves to no avail to effect it without contracting the intercolumniation or elongating the first metope; though it is perfectly obvious that the intention of the Greek architects was to ease the columns in those important situations of a part of their burden, and for no such purpose as Vitruvius and his disciples have thought. Indeed, this has been a problem to the whole school, which their master proposed, and which they have settled only by putting a half-metope beyond the outer triglyph; thus preserving the intercolumniation equal, but rendering the quoins more infirm, or perhaps less stable, than the Greek architects judiciously thought they should be. Besides contracting the intercolumniation, the Greeks also made the corner columns a little larger than the rest, thus counteracting in every way the danger that might accrue to them, or to the structure through them, from their exposed and partly unconnected situation. The graduated pyramidal stylobate on which the structure rests also bears a certain proportion to the standard which is the measure of all the rest; and so every part is determined by the capacity of the sustaining power. Though the Doric order thus possesses, as it were, a self-proportioning power, which will secure harmony in its composition under any circumstances, yet skill and taste in the architect are necessary to determine, in every instance, the number of diameters the column shall have in height, and according to that assign the height of the entablature. For these two points in proportioning, and for appropriate detail and enrichment, he may, without servility, refer to the ancient examples; with the confidence, moreover, that in availing himself of their beauties he acquires the power of producing an object that shall be itself beautiful, while he can avoid being a mere copyist in the adaptation and arrangement of the materials of his composition, as well as in the selection

Grecian
Structures.

Plate IV.
Fig 1 & 4.

Plate V.
Fig 3.
Plate IV.
Fig. 3.
Fig. 1 & 4.

of them. We cannot discover that the elevation of the pediment depended so immediately on the common standard, though in the best examples the tympanum will be found to be about one diameter and a half in height.

The Ionic and Corinthian, or Voluted and Foliated orders, do not possess that innate principle of harmony which pervades the Doric, and therefore they are, as styles, less perfect, and depend more on factitious combinations. The Greek compositions of Ionic and Corinthian are of such consummate beauty in every particular, that their examples appear perfect, and may therefore be taken as models for study, in preference to the rules which have been laid down for those orders, without a knowledge of these exemplifications. With a consciousness of their inferior capacity to produce grand and harmonious effects in such arrangements as their temples require, the Greeks never applied either the Ionic or the Corinthian peripterally, and, as far as we have certain knowledge, only the latter in prostyles. Whether the Ionians did or did not, cannot be satisfactorily ascertained, as their temples are in every case so much destroyed, that it is impossible, at least without more care and attention than they have yet received, to make out satisfactorily what their plans were. In the Ionic and Corinthian orders, the proportions of the various parts are generally made dependent on the diameter of the column, as in the Doric; but the intercolumniations, and consequently the general proportions, of a composition, are not determined by the column and its accessories according to their capacity, but must be left to the taste and skill of the architect, as well as the columnar proportions themselves. This gave rise to the rules referred to, which are laid down by Vitruvius, for what he calls the " Five Sorts of Edifices," or, more correctly, species of intercolumniation. They are pycnostyle, systyle, diastyle, aræostyle, and eustyle, to each of which a fixed space is assigned. Architects will, however, act more wisely in judging for themselves, by reference to the best models of antiquity, what proportion constitutes an *Eustyle* intercolumniation, according to the application of his ordinance, than by attending to such irrational dogmas as are contained in that classification.

The temples of the Greeks are described, according to their external arrangement, as being either in antis, prostyle, amphiprostyle, peripteral, pseudo-peripteral, dipteral, or pseudo-dipteral; and internally, as cleithral or hypæthral. The columnar arrangement in antis is not common in Greek architecture, though there are examples Plate IV. Fig. 2, and Plate V. Fig. 2. of it, generally of the Doric order. The inner porticoes or pronaoi of peripteral temples are for the most part placed in antis, as may be seen by reference to the examples, in which columns stand between the antæ. The Ionic temples of Athens are the principal examples of the simple prostyle. They may be called apteral, if it be necessary to distinguish them from peripteral, as the latter are prostylar; but the former term alone is sufficient. Neither does Greek architecture present more than one example, and that is at Athens also, of an amphiprostyle, except in the same peripteral structures, which are also amphiprostylar. Almost all the Doric temples are peripteral, and being peripteral, they are, as a matter of course, amphiprostylar, as we have just remarked; so that the former term alone is used in describing an edifice of that kind, with the numeral which expresses the number of columns in each of its prostyles. There is but one known example of Greek antiquity of a pseudo-peripteral temple, and that is the gigantic fane of Jupiter Olympius at Agri-

gentum in Sicily. It is not even prostylar, for the columns on its fronts are attached, as well as those on its flanks. The dipteral arrangement is found at Selinus, in an octastyle temple; and in some cases the porticoes of peripteral temples have a pseudo-dipteral projection, though no perfect example of the pseudo-dipteros exists.

Most of the temples of the Greeks were cleithral; those to the inferior and demi-gods were invariably so. The fanes of the supreme divinity were almost as invariably hypæthral, and frequently those of other superior gods were of the latter description also. The Doric order was never used by the Greeks in mere prostyles; consequently there is no Doric temple of the tetrastyle arrangement, for it is incompatible with the peripteral, the tetrastyle examples which do exist being all Ionic.[1] With very few exceptions, all the Doric temples of the Greeks are hexastyle. Their queen, however, the unmatched Parthenon, Plate III. is octastyle; and the pseudo-peripteral fane of Jupiter Olympius at Agrigentum, just referred to, presents the singular arrangement, heptastyle. No example exists in Greek architecture of a portico of more than eight columns, except the mis-shapen monument called the Basilica at Pæstum, the Thersites of its style, be so considered, and that has a front of nine columns, or an enneastyle arrangement.

It may be here remarked, in support of the opinion we have given as to the authority of Vitruvius, that, according to him, peripteral temples have on each flank twice the number of intercolumniations they have in front; thus giving to a hexastyle eleven, to an octastyle fifteen columns, and so on, whereas in the Greek temples this is never the case, for they always have more. The best examples have two, some have only one, but many have three, and in one instance there are four intercolumniations more in flank than in front. Again, he limits the internal hypæthral arrangement to those structures which are externally decastyle and dipteral, though an example, he says, existed in Greece of an octastyle hypæthros, and that was a Roman structure. Now, the Parthenon is an octastyle hypæthros, but all the other hypæthral temples, both in Greece and her colonies, are hexastyles, except perhaps the octastyle dipteral at Selinus; and there is no evidence in existence that the Greeks ever constructed a decastyle dipteral temple.

A Greek temple, whose columnar arrangement is simply in antis, whether distyle or tetrastyle, consists of pronaos and naos or cella. A tetraprostyle may have behind it a pronaos and naos. An amphiprostyle has, in addition to the preceding, a posticum, but is not understood to have a second entrance. The porticoes of a peripteral temple are distinguished as the porticus and posticum, and the lateral ambulatories are incorrectly called peristyles. It may indeed be here suggested, that as the admixture of Latin with Greek terms in the description of a Grecian edifice cannot be approved of, it would perhaps be better to apply the term *stoa* to the colonnaded platform or ambitus altogether, and distinguish the various parts of it by the addition of English adjectives: or the common term portico would be quite as well with front, back, and side or lateral, prefixed, as the case may be. Within the back and front stoas or porticoes, then, a pe- Plate V. Fig. 3. ripteral temple has similar arrangements in antis, which are relatively termed the pronaos and opisthodomus, with an entrance only from the former; unless there should exist, as there does in the Parthenon, a room or chamber Plate IV. Fig. 3 within the opisthodomus, supposed to be the treasury, when a door opens into it from the latter. Besides these,

---

[1] Athens itself containing a Doric tetraprostyle, may seem to contradict this; but it must be recollected that we have already said (page 442), that in speaking of Greek architecture, we exclude all the examples, even in Greece itself, which were executed under the Roman dominion, for they bear the Roman impress; and that is one of them.

a Greek temple of the most ramified description consists only of a cell, in those which are cleithral; and of a naos, which is divided into nave and aisles, to use modern ecclesiastical terms, in an hypæthral temple.

The only pure Greek architectural works that remain to us, and of which we have certain information, besides temples, are, it has been already stated, propylæa, choragic monuments, and theatres. The Propylæum, or Propylæa, as applied to the Acropolis of Athens, is the entrance or gateway through the wall of the peribolus into it. It consists of a Doric hexaprostyle portico internally, with a very singular arrangement of its columns, the central intercolumniation being ditriglyph. This was done probably to allow a certain procession to pass, which would have been incommoded by a narrower space. Within the portico there is a deep recess, similar to the pronaos in a temple, but without columns in antis; a wall pierced with five doorways corresponding to the intercolumniations of the portico, close the entrance; and beyond it is a vestibule, divided into three parts by two rows of three Ionic columns, and forming an outer portico, fronted externally by a hexaprostyle exactly similar to that on the outside. Right and left of it, and setting out about one intercolumniation of the portico from its end columns, at right angles, are two small triastyle porticoes in antis, with chambers behind them. These have been called temples, but most probably they were nothing more than porters' lodges or guard-houses. The whole structure, though extremely elegant, and possessing many beauties, is not a good architectural composition: the unequal intercolumniation detracts from its simplicity and harmony. The use of Ionic columns in a Doric ordinance is equally objectionable; and their elevation from the floor of the portico on insulated pedestals is even worse, though their intention is obvious; and without raising them, the ceiling might have been too low, or they must have been made taller.[1] The uneven style of the small temples or lodges is not pleasing, even though they be taken as flank and not as front compositions; and, moreover, their entablature abuts indefinitely against the walls of the larger structure, both internally and externally, to the total destruction of the harmony of the general composition. Indeed the unequal heights of the entablature of the greater ordinance involves a fault, if there were not something to prevent them from being seen in the same view, which it requires more than all the beauties of detail and harmony of proportion to countervail.

Plate VII.
Fig. 1 & 2.
The choragic monument of Lysicrates, vulgarly called the Lantern of Demosthenes, at Athens, is a small structure, consisting of an elegant rusticated quadrangular basement or podium, which is more than two fifths of the whole height, surmounted by a cyclostyle of six Corinthian columns, attached to, and projecting rather more than one half from, a wall which perfects the cylinder up to the top of their shafts, where it forms a stand for tripods the height of the capitals. A characteristic entablature rests on the columns, and receives a tholus or dome, which is richly ornamented, and terminates in a foliated and heliced acroterium. To this Stuart has added dolphins as supporters, and has placed on the summit a tripod, which was the prize in the choragic festival; thus completing perhaps the most beautiful composition in its style ever executed. In Vitruvian language the arrangement of this edifice would be called monopteral; but it is more correctly cyclostylar, or, perhaps, because of the wall or core, it may be termed a pseudo or attached

cyclostyle. The basement of this monument is eminently bold and simple, admirably proportioned to the rest of the structure, and harmonizing perfectly with it. The columnar ordinance is the only perfect specimen of the style in existence of pure Greek origin, and it has never been surpassed, perhaps not equalled, in beauty elsewhere. The most exquisite harmony reigns throughout its composition: it is simple without being poor, and rich without being meretricious; and the same applies to the superimposed tripod and its supports.

Grecian
Structures.
Fig. 3.

Totally different in style and arrangement, and far inferior in merit, is the choragic monument of Thrasyllus. It bears, however, the impress of the Grecian mind. This composition is merely a front to a cave, consisting of three pilasters proportioned and moulded like Doric antæ, and supporting an entablature similar in style, but too shallow to harmonize with them. Above the entablature there is an attic or parapet, divided into three compartments horizontally. The two external form tablets with a cornice or impost on them, and the central is composed of three receding courses, on the summit of which is seated a draped human figure, whether male or female, in its mutilated state is not determinable. The entablature, instead of triglyphs in the frieze, has laurel wreaths; and it would appear as if the absence of the triglyph had deranged the whole composition. The two outer pilasters are of good proportion, and the architrave is well proportioned to them; but the frieze and cornice are both too narrow, and the spaces between the pilasters, equivalent to intercolumniations, are too wide. The third pilaster, itself inharmonious, is absurdly narrow, being narrower than the others; and, standing immediately under the statue, evidently to support it, its meagreness is the more obvious and striking. In spite of all this, the general outline of the structure is simple and pleasing, the detail is elegant, and the execution spirited and effective. This little monument is, however, a proof that the Greeks were not so excellent in architectural compositions at all times, as in the self-composing Doric temples, and in the choragic monument of Lysicrates; and to this evidence may be added that of the triple temple in the Acropolis of Athens. It consists of an Ionic hexaprostyle in front, resting on a bold, continuous, and well-proportioned stylobate, and forming the entrance to a parallelogramic cella, but, from all that has yet been discovered, without a pronaos in antis. The back-front consists of four columns like those of the portico, attached in antis; and the flanks are broad and bold, crowned by the well-proportioned and chaste entablature, with the enriched congeries of mouldings and running ornament of the antæ under it. In the absence of a pronaos to give depth to the portico, the composition was defective, but otherwise simple and harmonious. It was, however, completely spoiled by the attachment of a tetraprostyle to one of its sides, Ionic certainly, like that in front, but not only in a different manner, but of a different size; beautiful in itself, but a blot on the main building, with which it harmonizes in no one particular, being altogether lower; for the apex of its pediment only reaches to the cornice of the former. This and the Caryatidean portico are omitted in the example. In a similar situation, against the other side is attached a similar arrangement of Caryatides, a tetraprostyle of female figures raised on a lofty basement, and yet not reaching to the entablature of the main building,—according in no one particular either with it or with the portico on the other side, and altogether forming the most heterogeneous

Plate VI.

---

[1] An editorial note in the new edition of *The Antiquities of Athens* says that " they are incorrectly mounted on pedestals" in Stuart and Revett's *Restoration.* This structure cannot perhaps fairly be judged of, until its site and remains shall have been examined without the jealous supervision of a Turkish governor.

and inharmonious combination imaginable. Yet the two Ionic porticoes are the most beautiful examples of their Order in existence, and perhaps, it might be added, that were ever executed,—arranged in the finest proportion, and with the most exquisite details and enrichments. The Caryatidean frontispiece, also, for more it cannot be called, is full of architectural beauties, though it is most injudiciously collocated.

The theatres of the Greeks, it has been already intimated, present but little to interest in the view we are taking of architecture. They were not structures, but excavations; and whatever decoration they may have received to make them objects of interest externally, is, in every known example, entirely gone; and attempts to restore them from their remains as now existing, aided by all the information to be derived from ancient writers, would be futile, in the absence of a knowledge of the Greek style of art obtained from some other source. And as no existing example of the Greek theatre furnishes matter for architectural illustration, we should gain no information in furtherance of our present subject by treating of it here.

The division of the columnar architecture of the Greeks and Romans into orders by the Italian architects of the fifteenth century, according to the laws of Vitruvius, and the universal reception of that mode of arranging it, almost impose upon us the necessity of adopting the same course, and laying down a standard or model for each. But instead of so doing, we think it better to give each school separately, and describe the general features of the orders as they occur in the works of each,—pointing out, moreover, the varieties that exist, and prevent the monotony consequent on restricted forms and proportions. We retain, too, the term " Order," and the names in general use, without consenting to the propriety of either the one or the other; for if it be judicious to divide Greek and Roman columnar architecture into orders, there can be no reason why Egyptian, Hindoo, Persian, or any other style, should not be classed in a similar manner. Moreover, there is nothing in any one " order" that, were it not for custom, would not be thought as fitting in any other as in that to which it may belong. The Greeks did not hesitate to put triglyphs in the frieze of an entablature whose columns were fillet-fluted and had foliated capitals, as some ruins at Pæstum attest. As to names, the Doric might, as we have said, be called Corinthian with more propriety; the Ionic, Samian; and the Corinthian, Athenian; referring to the oldest known examples of each. The term Style would be more correct than Order, as it would indicate the column as the feature referred to, without conveying the idea of fixed rules; and architectural works into which columns do not enter need not be constrained to admit the arrangement of some Order in the composition, proportion, and detail of its various parts. In naming, too, the Doric might be called the Greek sacred or triglyphed style; the Ionic, the Voluted style; and the Corinthian, the Foliated; thus admitting any varieties of combination which could be expressed as composites of the voluted and foliated, or of the foliated and triglyphed, as the case might be.

An Order, according to Mr Gwilt, is " an assemblage of parts, consisting of a base, shaft, capital, architrave, frieze, and cornice, whose several services requiring some distinction in strength, have been contrived or designed in five several species,...each of which has its ornaments as well as general fabric proportioned to its strength and character." Perrault says that an order may be defined " a rule for the proportion of columns, and for the form of certain parts which belong to them, according to the different proportions which they have." We would have

it understood to be a species of columnar arrangement, differing in its forms and general proportions, and in some leading features, from any other. Greek columnar architecture may thus be divided into the three arrangements or orders, Doric, Ionic, and Corinthian, which form its classes or styles. In considering them, however, it is necessary to discharge the mind of all the absurdities of the Italo-Vitruvian school about the proportions of the human figure being applied to columns, whether virile, matronal, or virginal; about the trunks of trees and rafters' feet; whether Doric columns should not have bases because men have feet, or that Ionic columns should have them because women wore sandals; that the guttæ in a Doric entablature should be conical, and not pyramidal, because they are to look like drops of water; that sculls, furies, thunderbolts, and daggers may be used to enrich a Doric frieze, but that spears, and swords, and stars, and garters may not;—these, with the thousand other puerilities of the *Cinquecentists*, whether Italian, French, or English,—whether acquired from the writings of Palladio and Scamozzi, of Perrault and Leclerc, or of Wotton and Chambers,—must be forgotten, and the greater or less degree of beauty resulting from this or that mode of arrangement and detail alone attended to.

Not to induce the idea that the quoted examples of the antique should be imitated to the line and letter, but rather in spirit, we shall speak of the proportions of their various parts generally; though it must at the same time be understood that much of the beauty of a columnar composition depends on its minutiæ: still it is not necessary that these minutiæ should be mere repetitions of an original; it is in the spirit of the antique models that excellence is to be sought, and not in crude rules for their reproduction.

Of the Grecian Doric.

This order may be divided into three parts, Stylobate, Column, and Entablature. The stylobate is from two thirds to a whole diameter of the column in height, in three equal courses, which recede gradually the one above from the one below it, and on the floor or upper step the column rests. That graduation, it may be remarked, does not appear to have been made by the ancients to facilitate the access to the floor of the stoa or portico, but on the principle of the spreading footings of a wall, to give both real and apparent firmness to the structure, both of which it does in an eminent degree.

The column varies in different examples from four to six diameters in height, of which the capital, including the necking, is rather less than half a diameter: in those cases in which a necking does not exist, the capital itself occupies nearly the same proportion. The shaft diminishes in a slightly curved line, called entasis, from its base or inferior diameter upwards to the hypotrachelium, leaving it at that place, or at the superior diameter, from two thirds to four fifths of the lower or inferior, which latter is the diameter always intended when the term is used as a measure of proportion. The capital consists of a necking, an echinus or ovalo, and an abacus; the necking is about one fifth of the height of the capital, and the other two members equally divide the remaining four fifths: when there is no necking, the ovalo occupies the greater proportion of the whole height. The abacus is a square tablet, whose sides are rather more than the inferior diameter of the column. The corbelling of the ovalo adapts it to both the diminished head of the shaft and the extended abacus, flowing into the one, and forming a bed for the other by means of a graceful cyma-reversa; but its lower part is encircled by three or four rings or annulets, which are variously formed in different examples.

Plate I.
Fig. 10.
Plate IV.
Fig. 4 & 9.
Plate V.
Fig. 4.

Fig 6 & 7.

Plate IV.
Fig. 12.
Plate V.
Fig. 8 & 14.

and which are the means of giving the echinus form to the great moulding, although it is, as we have said, part of a cyma-reversa. The shaft is divided generally into twenty flutes; but there are several examples with sixteen, and there is one with twenty-four. The flutes are sometimes segments of circles, sometimes semiellipses, and sometimes eccentric curves. They always meet in an arris or edge, and follow the entasis and diminution of the column up through the hypotrachelium to the annulets, under which they finish, sometimes with a straight and sometimes with a curved head. At the base they detail on the pavement or floor of the stylobate.

Fig. 4.

The third part of the order, the entablature, ranges in various examples from one diameter and three quarters to rather more than two diameters in height, of which about four fifths is nearly equally divided between the architrave and frieze, and the cornice occupies the remaining one fifth : this is in some cases exactly the distribution of the entablature. The architrave is in one broad face, four fifths, and sometimes five sixths of its whole height; and the remaining fifth or sixth is given to a projecting continuous fillet called the tænia, which occupies one half the space, and a regula or small lintel attached to it, in lengths equal to the breadth of the triglyphs above in the frieze. From the regulæ six small cylindrical drops called guttæ depend. There are examples to the contrary, but it may be taken as a general rule, that the architrave is not in the same vertical line with the upper face of the shaft, or its circumferential line, at the superior diameter, but is projected nearly so much as to impend the line or face of the column at the base. The frieze, vertically, is plain about six sevenths of its whole height, and is bounded above by a fascia, slightly projecting from it, which occupies the remaining seventh. Horizontally, however, it is divided into triglyphs and metopes, which regulate the intercolumniations in the manner that has been already described; the former being nearly a semidiameter in width, and the latter the space interposed between two triglyphs, generally an exact square, its breadth being equal to the whole height of the frieze, including the fascia. This latter breaks round the triglyphs horizontally, and is a little increased in depth on them. Each glyph, of which there are two whole ones and two halves to every tablet, is one fifth of the width of the whole, and the interglyphs are each one seventh of the whole tablet or triglyph. The glyphs detail on the tænia of the architrave, but are variously finished above. In some examples they are nearly square-headed, with the angles rounded off; in others the heads are regular curves, from a flat segment to a semiellipsis. The semiglyphs are finished above in a manner peculiar to themselves, with a turn or drop; but hardly two examples correspond in that particular. The tablets in which the glyphs are cut are vertical to the face of the architrave, the metopes recede from them like sunk panels; these are often charged with sculptures, and indeed almost appear contrived to receive them. The third and crowning part of the entablature, the cornice, in what may be considered the best examples, projects from the face of the triglyphs and architrave about its own height. Vertically, it is divided into four equal parts, one of which is given to a square projecting fillet at the top, with a small congeries of mouldings, different, and differently proportioned to each other, in various examples. Two other parts are given to the corona, and the remaining fourth to a narrow sunk face below it, with the mutules and their guttæ. These latter form the soffit or planceer of the cornice, which is not horizontal or at right angles to the vertical face of the entablature generally, but is cut up inwards at an angle of about 80°. The width of the mutules themselves is regulated by that

of the triglyphs over which they are placed, to which it is exactly equal. They are ornamented each with three rows of six small cylinders, similar to those which depend from the regulæ under the triglyphs and on the architrave. There is twice the number of mutules that there is of triglyphs, one of the former being placed over every metope also in the manner the examples indicate.

This completes the Greek Doric Order according to the generally received sense of the term; but there are other parts necessary to it. In the front or on the ends of a temple, or over a portico, a pediment is placed. Its intention is obviously to inclose the ends of the roof, but it forms no less a part of the architectural composition. In reason, it should be raised as much as the roof required; but when the span is great that would be unsightly; and reference appears to have been made to the common standard of proportion, as the pediments of most Doric temples are found to be about one diameter and a half in height at the apex of the tympanum, which in a hexastyle arrangement makes an angle at the base of about 14°, and in an octastyle about 12½°. The pediment is covered by the cornice, without its mutules, rising from the point of its crowning fillet, so that no part of it is repeated in profile. Another moulding, however, is superimposed: sometimes this is an ovalo with a fillet over it, and sometimes a cymatium. It varies much in its proportion to the cornice, but in the best examples it is about one half the depth of the latter without its mutules. Ornaments of various kinds, statues or foliage, are believed to have been placed on the apices and at the feet of pediments as acroteria. Of these, however, we have no actual remains; but indications of the plinths or blocks which may have received them exist, and such things appear represented in ancient coins and medallions. The tympana of pediments are well known as receptacles of ornamental sculpture. On the flank of a Doric temple, the cornice supported a row of ornamented tiles called antefixæ. These formed a rich and appropriate ornament, but they rather belonged to the roof than to the columnar arrangement or order. The antefixæ covered the ends of the joint-tiles as the pediments did those of the roofs; and corresponding ornaments called stelai rose out of the apices of the joint-tiles, forming a highly enriched ridge.

Plate V.
Fig. 1.

Plate IV.
Fig. 1.

A secondary Doric order arises in the disposition of a Grecian temple, from the columns of the pronaos and the inner part of the external entablature continued and repeated. Of this the frieze is generally without triglyphs, though there may be regulæ and guttæ on the architrave. The fascia of the frieze is either moulded or enriched on the face; and, instead of a cornice, the beams of the ceiling are laid at equal intervals to support sunk panels or coffers, in which there may be flowers or other enrichments.

Fig. 9.

Propriety in the composition and arrangement of antæ is as necessary to the perfection of the Doric order as to that of the columnar ordinance itself, especially if the latter be in antis. Slight projections are made on the end and side faces of a wall, so as to form a species of pilaster, whose front shall be nearly equal to the diameter of the columns to which it is attached, exactly equal indeed to the soffit of the entablature, whose faces have been described as impending the circumferential line of the column at a little above its base. This rests on the stylobate in the same manner as the columns do, with sometimes a small continuous moulding as a base; and its capital is a congeries of mouldings, about the depth of the abacus, with a plain fascia corresponding to the ovalo of the columnar capital. The entablature of the order to which it is attached rests on it, and, continuing along the flank of the building, is received by a similar combination at the other end. These, it may be remarked, were never diminished or fluted, being projections from and upon the ends and faces of walls, they could not be diminished without involving an absurdity; and fluting on a straight surface must be productive of monotony, as the flutes can only project a series of equal and parallel shadows. Not so, however, with columns, on whose rotund surface fluting produces a beautiful variety of light and shade in all their gradations, which it could not possess without that enrichment; for on a plain column neither are the lights so bright nor the shadows so dark as in the former case, nor are they so finely diffused over the whole surface in the one as in the other.

In the only example which occurs in the ancient architectural remains of attached Doric columns,—that of the pseudo-peripteral temple of Jupiter Olympius at Agrigentum—the stylobate is peculiarly arranged. The upper gradus is grooved, and detailed round the columns and along the walls between them; and a congeries of vertically arranged mouldings and fillets rests on it, and receives the base of the column.

Such are the materials of which the Greeks composed their beautiful temples, the manner of whose composition has been already described. Of their effect, however, it is impossible to form a competent idea without seeing one. And whence, it may be asked, does their interest arise? Plate III. From their simplicity and harmony;—simplicity, in the long unbroken lines which bound their forms, and the breadth and boldness of every part; such as the lines of the entablature and stylobate, the breadth of the corona, of the architrave, of the abaci, of the capitals, and of their ovalos also; in the defined form of the columns, and the breadth of the members of the stylobate;—harmony, in the evident fitness of every part to all the rest. The entablature, though massive, is fully upborne by the columns, whose spreading abaci receive it, and transmit the weight downwards by the shafts, which rest on a horizontal and spreading basement; the magnitude of every part, as we have before had occasion to remark, being determined by the capacity of the sustaining power. Besides graceful and elegant outline, and simple and harmonious forms, these structures exhibit a bewitching variety of light and shade, arising from the judicious contour and arrangement of mouldings, every one of which is rendered effective,—by the fluting of the columns and the peculiar form of the columnar capital, whose broad, square abacus projects a deep shadow on the bold ovalo, which mingles it with reflections, and produces on itself almost every variety. The play of light and shade, again, about the insulated columns, is strongly relieved and corrected by the deep shadows on the walls behind them; and in the fronts, where the inner columns appear, the effect is enchanting. For all the highest effects which architecture is capable of producing, a Greek peripteral temple of the Doric order is perhaps unrivalled.

Of the Grecian Ionic

Not less Hellenic in its detail than the national Doric is the graceful and elegant style called the Ionic, whose proportions and peculiarities we take from the perfect examples of the Athenian Acropolis. Plate VI. This order may also be considered in three similar parts, Stylobate, Column, and Entablature. The stylobate is in three receding equal courses or steps, whose total height is from four fifths of to a whole diameter. The column, consisting of base, shaft, and capital, is rather more than nine diameters in height, of which the base is two fifths of a diameter; and the capital, including the hypotrachelium, is in one case three fourths, and in the other seven eighths of a diameter high. The base consists of a congeries of mouldings, extending gradually from a diameter and a third to a diameter and a half, and its height is in three

nearly equal parts, two equal fillets separating them. The Grecian lowest, a torus, rests on the top of the stylobate or floor of Ionic. the portico, a fillet divides that from a scotia, a second fillet intervenes the scotia and a second torus, and a third fillet bases the apophyge or escape of the shaft. The upper torus of the base is, in one example, fillet-fluted horizontally; and, in the other, the same member is enriched with the guilochos. The shaft diminishes with entasis from its lower or whole diameter, to above five sixths of it immediately under the hypotrachelium. It is fluted with twenty-four flutes and alternating fillets, which follow the diminution and entasis of the column. The flutes in plan are nearly semiellipses, and they finish at both ends with the same curve: a fillet is in thickness nearly one fourth the width of a flute. The difference in the height of the capital is in the length of the necking, which in one case is separated from the head of the shaft by a carved bead, and in the other by a plain fillet. Above the necking, a height of about one third of a diameter is occupied by a congeries of three spreading or corbelling mouldings, a bead, an ovalo, and a torus, which are all appropriately carved. On these rests the parallelogramic block, on whose faces are the volutes, and whose ends are concaved into what is technically termed a bolster, to connect them. This part is about one third of a diameter in height, and includes a rectilinear abacus, whose edges are moulded to an ovalo, and carved with the egg and tongue ornament. The volutes are three fifths of a diameter in depth, and extend in front to one diameter and a half; and they are nearly a semidiameter apart. The flowing lines which connect the volutes can only be understood by reference to the example. The Fig. 5. bolsters are fluted vertically, with alternate fillets, on Fig. 12 which are carved beads. An ornament composed of the honeysuckle with tendrils encircles the necking of the column. It must be remarked, that as the capitals are parallelogramic, and present but two similar fronts, to preserve the appearance of volutes externally on all sides, the capitals of those columns which occupy the external angles of porticoes are differently arranged. The outer Fig. 16. volute is bent out at an angle of 45°, and volutes are put on the end or side-front of the capital also, the outer one being the other side of the angular volute of the front. To suit the angle internally, the two volutes of the inner face are placed at right angles to each other: this is, however, at best but an awkward expedient, and need not be employed when a portico projects only one intercolumniation.

The entablature, which is rather more than two diame- Fig. 5. ters in height, is also divided into three parts—architrave, frieze, and cornice—which may be proportioned by dividing the whole height into five parts, four of which, as in the Doric, may be again equally divided between the architrave and frieze. The cornice, however, in the examples referred to, does not occupy one fifth of the entablature; but if it had a fillet over the upper moulding, which it appears to want, that would be just its proportion. If the architrave be divided into nine parts, seven of them may be given to three equal fascias, which slightly project the one before the other; the first or lowest, which is vertical to the circumferential line of the inferior diameter, being covered by the second, and the second by the third. The remaining two ninths form a band of mouldings corbelling a broad fillet, which separates the architrave from the frieze: these mouldings are enriched. The frieze, which does not project quite so much as the lowest fascia of the architrave, is, in the Athenian examples, quite plain; but it may be enriched with foliage, or made the receptacle of sculpture in low relief. The cornice projects from the face of the frieze rather more than

Grecian Ionic.

as much as its whole height, and is composed of bed mouldings, a corona, and crown mouldings. The first are a carved bead and carved cyma-reversa, the former of which only occupies a portion of the height of the cornice, as the planceer is cut up inwards in the manner represented by dotted lines in the example, to a sufficient depth for it; the crown mouldings, which consist of a carved ovalo above a carved bead, are rather more than one fourth of the whole cornice; and the corona occupies the rest of its height, except that small portion given to the bead of the bed mould. A fillet above the crown mouldings, as already intimated, is certainly necessary to complete the order and receive the antefixæ, as described in the Doric, for the flank of a temple.

Plate VI. Fig 1 & 2.

The pediments in the examples of Ionic are rather flatter than in those of the Doric, the angle made by the covering cornice with the base being, in a hexastyle, less than 14°. A vertical fillet, with a small moulding, equal in depth to the two crown mouldings of the cornice, covers them in the pediment, in the place of the cyma-recta or ovalo used in the Doric order. The intercolumniation differs in these examples; in the one it is two diameters, and in the other three diameters and one-sixth.

A much greater variety is found in the composition of the Ionic than of the Doric order. Indeed the examples of the Athenian Acropolis alone have neckings; in all the others the shaft runs up to the corbelled mouldings, which bed the block of the volutes, and the flutes finish under them. Neither have they a torus in that congeries, but a bead and ovalo alone, which latter makes an inconvenient projection under the pendent lines that connect the volutes, and thus the capital is not more than half a diameter in height.

Fig. 10.

The Asiatic or the truly Ionian examples of this order are far inferior to those referred to. Their bases are differently, and certainly less elegantly composed. They are without hypotrachelia, as may have been inferred; they want the torus in the capital; and, in most cases, instead of flowing, pendent lines, they have straight lines connecting the volutes. Their entablatures are not so finely proportioned, nor so delicately executed. The coronas want breadth, and the bed moulds of the cornice are as much too heavy as those of Athens are perhaps too light. Indeed, upon the whole, they have more of the grossness of Roman architecture than of the delicacy and elegance of Grecian, though the Ionian examples are supposed to be the models of those of Athens.

The width of the antæ of the Ionic order is determined, as in the Doric, by the soffit of the entablature; and it will, of course, be exactly the same as, or rather less than, the inferior diameter of the column. It is slightly raised, too, from the face of the wall at the ends of which it stands. The base of the antæ is, in one of the two examples of the Acropolis, a little deeper than that of the column, having a small projecting moulding between the lower torus and the floor; and the lower torus itself is reeded. In the other example there is no difference in the form and proportion of the antæ and columnar bases, but both the tori are fluted horizontally, with beaded fillets between the flutes. The antæ cap consists of a congeries of corbelling mouldings, nearly one third of a diameter in height. It is divided into three nearly equal parts, the lowest of which is composed of a bead and an ovalo; the second of another bead and a cyma-reversa, all carved; and the third of a plain flat cavetto, with a narrow fillet and small crowning cyma-reversa, forming an abacus. The necking is like that of the capital, and is enriched in the

Fig. 3.

same manner. The cap or cornice thus formed breaks round the projection of the antæ, and is continued along the wall under the entablature the whole length of the

building, or till it is impeded by some other construction, and the base is continued in like manner.

Grecian Corinthian.

Fig. 2.

Attached columns have the voluted capital, but their base is that of the antæ; and it is detailed round them and along the wall to which they belong, as with the antæ. It must be remembered, however, that the attached columns in the triple temple are about one ninth less in diameter than those which are insulated, though they are similar in other respects, and have the same entablature.

The back of the triple temple, between the attached columns, presents the only example in Greek architecture of windows. These are rather more than twice their width in height, and are narrower at the top than at the bottom. They rest on a broad, bold sill, which is equal in depth to two sixths of the opening, and are surrounded externally by a congeries of mouldings, which, with a plain fascia, constitute an architrave. This architrave is one fourth the opening in width; it diminishes with the window, and in the same proportion, and is returned above in two knees, which are made vertical to its extreme point at the base.

Of the Grecian Corinthian.

Plate VII. Fig. 1.

The importance which the Greeks attached to a graduated stylobate, and the necessity of giving it a relevant proportion in a columnar ordinance, are evinced in the only example of this order which remains to us of Grecian origin. Unlike the Doric and Ionic in its application, which is in temples of rectangular form, whose whole height they occupy, this is attached to a small circular structure, resting on a lofty square basement; and yet, like those orders, it has a stylobate in receding courses, and in plan, too, corresponding with the arrangement of the

Fig. 2.

columns, and not with that of the substructure; in this too offering further proof that the stylobate was considered a part of the columnar ordinance. Thus the Corinthian

Fig. 1 & 3.

order also consists of stylobate, column, and entablature. The stylobate is rather more than a diameter in height, and is divided into three parts, but not equally, in consequence, it is probable, of the peculiar position of the ordinance. The two lower grades have vertical faces, are of equal depth, and occupy three-fourths of the whole height; whilst the third step occupies the remaining fourth and is moulded on the edge, in exquisite harmony with the more ornate style, of which it forms a part. Like the column of the Ionic order, that of the Corinthian consists of base, shaft, and capital: it is ten diameters in height. The base is rather more than one third of a diameter high, and is composed of a torus and fillet, which are nearly two fifths of its whole height; a scotia and another similar fillet, rather less than the former; and a second torus or reversed ovalo, one fifth the height of the base, on which rests a third fillet basing the apophyge of the shaft. The extent or diameter of the base, at the lower torus, is rather more than one diameter and a half. The shaft diminishes with entasis to five sixths of its diameter at the hypotrachelium, and, like that of the Ionic order, has twenty-four flutes and fillets. The flutes are semiellipses, so deep as nearly to approach semicircles: they finish in the apophyge at the foot of the shaft, in the same manner and form; and at the head they terminate in leaves, to which the fillets are stalks. The fillets are rather more than one fourth the width of the flutes. The hypotrachelium is a simple channel or groove immediately under the capital. The capital itself is a diameter and rather more than one third in height: its core is a perfect cylinder, in bulk rather less than the superior diameter of the shaft. This is banded by a row of water leaves, whose profile is a flat cavetto, one sixth of the whole height, and another of leaves of the acanthus, with flowered buttons attaching them to the cylinder. These latter have the contour of a cyma-recta, and are twice the height of the last, or one third of the whole capital. Rather more than another third is occupied by helices and tendrils, which latter support a honeysuckle against the middle of the abacus; and the abacus itself, resting on and covering the whole mass, is but little more than one seventh of the whole height. This member in plan can only be described as a square whose angles are cut off at 45°, and whose sides are deeply concaved. In profile it consists of a narrow fillet, an elliptical cavetto or reversed scotia, and another fillet surmounted by a small ovalo, or rather a moulding whose profile is the quadrant of an ellipsis.

The entablature of this order is two diameters and two sevenths in height. It also consists of architrave, frieze, and cornice, of which the first occupies one tenth more than a third, the second rather more than as much less than that proportion, and the cornice is so much more again above one third. The architrave is divided, like that of the Ionic order, into three equal fascias, which occupy all but one sixth of its whole height, and that is given to a corbelled band, consisting of a bead, cyma-reversa, and fillet, separating the two members of the entablature. The fascias of the architrave, it must be remarked, are not perpendicular, but incline inwards, so that their lower angles are all in the same vertical line, which impends the surface of the shaft about one third of its height from the base. The frieze is one plain band, slightly inclining inwards, like the fascias of the architrave, and slightly projected beyond them: in this example it is enriched with sculptures. The cornice consists of a deep congeries of bed mouldings, and a corona, with the accustomed small crown mouldings and fillet. Its extreme projection is nearly equal to its whole height: of this the bed-mouldings project about two fifths. As in the Ionic cornice, additional height is given to the bed-moulds, by undercutting the planceer. In this case, the undercutting extends to nearly one-fourth the height of the corona. One-sixth the height of the cornice is given to a flat bead and an ovalo, which are immediately above the frieze, and which base a broad dentilled member that occupies more than one-fourth of the whole cornice. This is surmounted by a listel or broad fillet, above which is a cymarecta, whose narrow fillet nearly reaches the horizontal plane of the planceer, and separates it from a cyma-reversa that beds the superimposed projecting corona. This latter is only three eighths of the whole cornice, and nearly one of the three is given to the ovalo and fillet, the bed-moulds alone occupying five eighths. The cornice is surmounted by a cut fascia supporting honeysuckle antefixæ, which may indeed be taken as a part of the order, as the solitary example in question presents it. This, however, we know, from the Doric and Ionic structures, to be a modification of the flank ornament of temples; and we may suppose from analogy, that if used in a portico, the cornice of this order would have a cyma-recta to crown it on the inclined side of the pediment. The intercolumniation of this example is two diameters and one third.

Of Corinthian antæ we have no examples, nor indeed have we of insulated columns; but as we find in the Ionic examples quoted, that the attached columns are less in proportion to the entablature than those which are insulated, we may conclude that it would be the same with this; thus reducing the entablature to two diameters, the ordinary average of that part in Greek columnar architecture.

Of the Caryatides, or Caryatic Order.

Plate VII.
fig. 4, 5,
& 6 The solecism in architecture of which we have now to speak has but the one existing example in the works of the Greeks. to which we have already referred. It is the

third portion of the triple temple in the Athenian Acropolis, and is a projection from the flank of the principal Ionic structure; formed by a stereobatic dado raised on the stylobate and antæ-base mouldings of it, with a surbase consisting of a carved bead and carved ovalo covered by a broad listel, with a narrow projecting fillet above it. On this rests a square plinth, which bases a draped female figure, on the head of which there is imposed a circular moulded block, with a deep rectangular abacus, two thirds of whose face is vertical, and the other third is a cavetto, fillet, and small cyma-reversa. The stereobate, including the moulded base of the temple, is about three fourths the height of the statue-pillar with its base and capital. The entablature is rather less than two fifths of the same, but it consists of architrave and cornice alone, between which parts the height is nearly equally divided. Rather more than one fifth of the former is given to a carved bead and carved cyma-reversa, with the flat, plain cavetto and fillet which they support; the other four-fifths is divided nearly equally into three fascias, of which the third or upper one has a fraction more than the other two, and is studded with plain circular tablets, whose diameter is five sixths of its depth. The cornice consists of bed-mouldings, corona, and crown mouldings. Two fifths of its whole height is given to the bed-mould, to which one seventh of that may be added for the portion cut up in the planceer. Half that increased height is occupied by a dentilled member, and the other half by a broad plain fillet, a carved bead, carved cyma-reversa, and a narrow fillet above it. The remaining three fifths of the whole cornice being again divided into five parts, rather less than two of them is given to the corona; a little more than one to a plain cyma-reversa and fillet, of which the latter is the wider; and of the rest a carved ovalo occupies five sevenths, and a listel or crowning fillet, with a carved bead on it, the other two. A pier, pilaster, or antæ, projects from the wall of the greater temple, and receives the end of the entablature behind the inner figure; for the projection is of two statues and their interspaces. It does not, however, rest on the stereobate, but runs down to the base-mouldings of the temple, the dado and surbase abutting against it. The antæ is capped by a congeries of carved mouldings, which support a narrow cavetto and fillet: the height of the cap is half the diameter of the antæ. There is also a hypotrachelium, consisting of a carved bead and the honeysuckle ornament, occupying about one third of a diameter in height. This Caryatidean portico displays very clearly the arrangement of the ceiling, with its coffers or cassoons. Internally the architrave is plain two thirds of its height; of the remaining third rather more than one half is a plain, slightly projected, fascia; the other half is occupied by a carved bead and ovalo. In the absence of a frieze, the ceiling rests on this, and is divided by carved beads into panels, which are deeply coffered, and diminished by three horizontal moulded fascias.

Of Grecian Mouldings and Ornament.

Greek architecture is distinguished for nothing more Pl. VIII. than for the grace and beauty of its mouldings; and it may be remarked of them generally, that they are eccentric, and not regular curves. They must be drawn, for they cannot be described or struck; so that though they be called circular, or elliptical, it is seldom that they are really so: not but that they may be, but, if they are, it is evidently the result of chance, and not of design. Hence all attempts to give rules for striking mouldings are worse than useless, for they are injurious: the hand alone, directed by good taste, can adapt them to their purpose, and give them the spirit and feeling which render them effective and pleasing.

G

The leading outline of Greek moulding is the gracefully flowing cyma. This will indeed be found to enter into the composition of almost every thing that diverges from a right line; and even combinations of mouldings are frequently made with this tendency. It is concave above and convex below, or the reverse; and though a long and but slightly flected line appears to connect two quickly-curving ends, it will always be found that the convexity and the concavity are in exactly the same curve, so that if the moulded surface were reversed, and the one made to assume the place, it would also have the appearance, of the other, and the effect would be the same. It is, in fact, the Hogarthian line of beauty; and it is not a little singular that Hogarth, in his well-known *Analysis of Beauty*, although he did not know, and indeed could not have known, the contours of Greek architectural mouldings, has given the principle of them, and, under his line of beauty, has described many of the finest Greek forms. The Roman and Italian mouldings were called Greek in his day, and he assumed them to be so; but they evidently do not agree with his theory, whereas, in principle, the now well-known Greek forms do most completely.

The cyma-recta is generally found to be more upright and less deeply flected than the cyma-reversa; it is almost always the profile of enrichments on flat surfaces, of foliage, of the covering moulding of pediments, of the undercut or hooked mouldings in antæ-caps, the overhanging not affecting the general principle; and it pervades, as we have said, flected architectural lines generally, whether horizontal or vertical. The cyma-reversa has all the variety of inflection that its opposite possesses, but the line connecting its two ends is, for the most part, more horizontal, and its curves are deeper. It pervades many architectural combinations, but is most singularly evinced in the composition of the Greek Doric capital,

Plate IV.
& V.
Fig. 6 & 7.

which is a perfect cyma-reversa, with the ends slightly but sharply flected, as it flows out of the shaft below, and turns in under the abacus above. The obviousness of the former is prevented by the annulets which divide the cyma into an ovalo and a cavetto, but the principle is

Plate VI.
Fig. 5.
Plate VII.
Fig. 3.

clear.[1] The cyma is the governing outline in the congeries of mouldings in bases also, as may be noticed in the Ionic and Corinthian examples quoted and referred to.

An ovalo is but the upper half of a cyma-reversa, even when it is used as a distinct moulding, and unconnected

Plate VIII.

with the waving form. The name expresses its apparent rather than its real tendency; for its contour is not that of an egg in any section, though the ornament which is carved on it, when used as a running moulding, is formed like an egg; and from that the moulding was named.

The upper torus of a base forms, with the escape or apophyge of the shaft, a perfect cyma, and the scotia and lower torus do the same; so that the torus and scotia are referable to the same principle when in composition, and they are not found together except in the combination referred to.

The bead is an independent moulding, varying in contour; but it is generally the larger segment of a circle. It is used, however, sometimes to mask the waving form, and sometimes to separate it.

The cavetto, or simple hollow, is part of a cyma also, as we have shown; but it is also applied independently, to obviate a sharp angle, or to take from the formality of a vertical line, as in the abaci of Ionic antæ-caps. Its form,

nevertheless, is not the segment of a circle, for the upper part of a cavetto is the most flected, and it falls below almost into a straight line.

Grecian
Mouldings.

There is a hooked moulding common in Greek architecture, particularly in the Doric antæ-caps, which is technically called the hawk's-beak. It is a combination of curves which cannot be described in words; but it has been already referred to in speaking of the cyma-recta, which is brought into its composition.

Plate IV.
Fig. 5 & 11.

Plate V.
Fig. 5 & 10.

The cyma-recta is never found carved, or sunk within itself; but it sometimes has the honeysuckle, or other ornament of the kind, wrought on it in relief, particularly when used as the covering moulding—the cymatium—or a pediment. The enrichment of the cyma-reversa consists of a contrasted repetition of its own contour meeting in a broad point below, and joining by a circular line above, and making a sort of tongued or leafed ornament, whose surface is inflected horizontally also. Between the leaves a dart-formed tongue is wrought, extending from the circular flexure above to the bottom of the moulding, whose contour it takes in front alone. As this would not mitre or join well on the angles of the cyma, a honeysuckle is gracefully introduced in the manner shown in the example. This enrichment is not wrought in relief on the moulding, but is carved into it, so that the surfaces of the parts of the ornament alone retain the full outline of the cyma. The ovalo is enriched with what is called the egg and dart ornament. This will be best understood by reference to the example. Its angles also are made with a honeysuckle, and the inflections are made in the moulding itself. The torus is sometimes enriched with the interlaced ornament called the guilochos: this too is cut into the moulding itself. We have no Greek example of an enriched scotia, and from its form and position, which, to be effective, must be below the eye, it hardly seems susceptible of ornament which could operate beneficially. The bead is carved in spheres or slightly prolate spheroids, with two thin rings or buttons, dilated at their axes, placed vertically between them. A cavetto is not enriched at all, nor is the hawk's beak, except by painting, which does not appear to have been an uncommon mode of enriching mouldings among the Greeks; that is, the ornament was painted on the moulded surface instead of being carved into it. Fascias are also found enriched by painted running ornaments, such as the fret or meander, the honeysuckle, and the lotus. Sometimes plain colour was given to a member, to heighten the effect it was intended to produce. Ornaments were painted and gilt on the coffered panels of ceilings too.

The few examples which exist of sculptured ornament on straight surfaces exhibit varieties of nearly the same combinations as those last mentioned,—the honeysuckle with the lotus, and sometimes a variety of itself on scrolls, either throwing out tendrils, or plain. This is found on the necking of the Ionic columns of the Athenian Acropolis, and on those of their antæ, and continuing along under the congeries of mouldings, as previously described. The varieties of foliage used in the enrichments of Greek architecture are few, and will be found generally exemplified in the Corinthian capital of the choragic monument of Lysicrates, and in the rich acroteral pedestal or stem of the same edifice, than which we possess no more elaborate specimen of foliated enrichment of the Greek school. There exist many specimens of architectural ornament on vases and fragments, in marble and terracotta,

Plate VII.

Fig. 3

Fig. 1.

---

[1] The presence of the cyma in the Doric capital was, we believe, first pointed out by Mr T. L. Donaldson, in the supplementary volume to the new edition of Stuart's *Athens*, though the true contour of the cyma itself appears to have escaped that gentleman's attention.

in which human figures, both male and female, are composed, with a greater variety of foliage than is generally found in Greek architectural works; and many of the beautiful marble and bronze utensils discovered in Herculaneum and Pompeii have enrichments obviously of Greek origin, from which, as well as from the specimens of ornament on positive architectural monuments, we may judge of their productions generally, as well as acquire or imbibe somewhat of the fine taste which originated them.

It would be puerile to speculate on the domestic edifices of the Greeks any further than we have done, as we possess no genuine data on which to proceed. Their sacred structures have taught us their style of architecture; but for its application to general purposes we have no resource but to consult the Roman remains of the exhumated Campanian cities and other places, and gather from analogy what Greek domestic architecture was.

The instances which present themselves on all sides of endeavours to adapt the columnar ordinances derived from the noble structures of ancient Greece are numerous; but it is to be added, that so far as regards the most noble and the most distinguished of the Grecian styles, every instance must be regarded as a failure. Even the Walhalla in Bavaria, which is understood to be professedly a reproduction, as to exterior form and distribution, of the Parthenon, is without the inner portico at one of its fronts—the opisthodomus is wanting. In what manner, or with what effect, the work is executed, the present writer has no personal knowledge; but at best the Walhalla is an exceptional case, and cannot be taken to be an adaptation of the Greek style to a modern purpose, but a copy, and an imperfect one, of a Greek temple. The interior of the Walhalla cannot be accepted as an embodiment of either Greek taste or structural propriety, whilst the rapid constructed approach from a low level to the elevated site of the building cannot fail to be injurious to the effect of its exterior, if it be only by taking the roof out of the view; but the graduated stylobate must also be intercepted and thereby lost to the view throughout much of the circuit.

The Grecian Ionic ordinance may be thought to have yielded in some degree to the requirements of modern uses; but having less of dignity to lose than the more purely Greek style—the Doric—and being exhibited in no ancient remains in such effective combinations as the Doric, modern compositions, of which the Ionic ordinance forms the basis, have the advantage over those which exhibit the Doric features of being tried by a less severe test. Nevertheless, the results, in the more prominent instances which London presents of the application of the Grecian Ionic in modern structures, do not tend to encourage the endeavour to mould this more plastic ordinance to modern purposes.

In compositions of less pretence, nearer approach may have been made to success in the application of the Greek Ionic ordinance. But the great defect of one in particular of the Athenian examples—that of the portico of Minerva Polias—of an inordinately wide intercolumniation, whereby structural weakness is induced, is too commonly taken as an authority for such a disposition of the columns as that example presents, to the great detriment of most reproductions in modern instances of the Greek Ionic.

Corinthian, as an ordinance, is more Roman than Greek, and is much more plastic than the more purely Grecian ordinances; and being more plastic, it has been applied in modern practice and to modern purposes more extensively, and with greater success, than the less exacting, as it regards enrichment, but more exacting as it regards the circumstances under which they may be applied in practice, Doric and Ionic of the Greeks. It may be that the Corinthian column, being a

more finished and acceptable object in itself than either the Doric or the Ionic column of whatever style, the eye exacts less, in respect of accessories and concomitants, from the Corinthian column than it does from the Doric and Ionic, which as columns are nothing—that is to say, they are unmeaning, and to some extent ungainly, objects, unless it be in appropriate composition, and with all the requisite accessories to complete the respective ordinances; whilst the Corinthian column, with its richly foliated and otherwise enriched capital, coped with a moulded abacus, and bound below by a moulded hypotrachelium, its fluted shaft and its moulded and enriched base, is itself an object to satisfy the eye, while the mind inquires less urgently for what purpose it is intended, and how it is to be disposed in composition.

However this may be, there can be no doubt, that whether the peculiar form in which the Corinthian ordinance presents itself be Greek or Roman—that is to say, whether the characteristics and details be those of the Greek example, or of a Roman example, to which the designation Corinthian is conventionally applied, or to that form of the foliated style which is by the Italo-Vitruvian school distinguished as the Composite order—it can be applied, whether in its full proportions, or shorn of some of its parts and details, with a success that has not attended the application of Doric and Ionic, in whatever guise.

But even the classical Corinthian ordinance is seldom found applied in modern works either at home or abroad, without sacrifice being made, in a greater or less degree, of the purposes of the work to the demands of the classical exemplar employed, or the more common sacrifice of the proportions found in the exemplar to the necessity of the case. Now the proportions found in the purer columnar architecture of Greece and Rome, of whatever style, are necessary results of the materials employed and of the mode of applying the materials in composition. The exemplars are all in stone of some kind or other, and the proportions and dispositions are such as the use of stone in legitimate structure of the sort applied would impose. Columns are placed as far apart as stone of the kind used in the case will bear over with certain safety in an architrave or lintel from column to column; such due proportion being observed between the horizontally disposed stone beam bearing over, and the vertically disposed stone column that has to support the superstructure, as the weight may require. Hence intercolumniation, or the space to be borne over, is not a matter of fancy, but of structural necessity; and the applicability of a classical columnar ordinance is structurally limited by the quality and character of the stone employed. Stone beams cannot be pieced out in length to bear over a void by any legitimate, or, what is commonly but expressively termed *workmanlike* manner; and arched structure is inadmissible in a pure columnar composition, as arches of whatever form require to be buttressed; and the essence of a columnar ordinance as a structure is in the weights being applied to act upon the column in that direction only in which the column is qualified to bear pressure— that is to say, vertically, or in the direction of its vertical axis. Marble will bear both transverse strain and vertical pressure better than freestone, and granite better than marble; and hence it is that the several parts of columnar ordinances in the works of the Greeks are more massive in their proportions when freestone is used, and lighter in the examples in which marble is employed; and the same relation will be found to exist between heavy and light proportions in the Egyptian temples. It is not according to their respective dates, but according to the material employed in the structure, that the proportions of the parts to each other are heavier or lighter,—the structure is more condensed or more drawn out as sandstone or porphyry is the constituent material of the building. It is the same with the better

ARCHITECTURE.

works of the Roman period; that is to say, in the composition of the columnar ordinances of their sacred structures; and the Romans indulged in the use of almost every kind of stone available for the purpose in the columnar ordinances applied to their temples.

Thus out of truth in structure, having reference to the material employed, grew the forms and proportions found in the classical exemplars of columnar architecture, stone being the material of the structures, and the capability of the stone to bear and to bear over giving the limit to the columns, and to the parts required to bear over, and thereby prescribing the proportions of the parts to each other, and to the spaces between the bearing parts. To falsehood in structure are attributable the failures to reproduce in modern compositions the admirable effects found in the columnar ordinances of the Greeks and Romans in which truth prevailed. Proportions permitted by marble are repeated in tender freestone, and this failing, foreign substances and unworkmanlike tricks of handicraft are applied to cheat the sense; but the eye refuses to be deceived, even if the structure endure. There may be sufficient strength in reality, for the effective material employed may be timber or iron, or other substances capable of yielding the required result in strength with far slighter proportions than stone of any kind truthfully applied will permit; but a tutored eye will detect the weakness in effect; and not unfrequently the trick is exposed to the common observer by some failure in the means adopted in the case.

The entablature of the central bay of the lower pseudo-prostyle of the west front of St Paul's in London, reaching from column to column, over a disproportionately wide intercolumniation, exhibits a well-marked instance of the failure of the false structure whereby it was endeavoured to cheat the eye. The architrave is in three stones; that is to say, the architrave stones bearing on the adjacent columns are pieced out by a third stone, which having no true structural connection with them, the middle length has torn itself away from its illegitimate attachments, and the stone above it in the frieze is broken across; or it may be, indeed, that the suspended block in the architrave was hung up to the frieze course; but be that as it may, there it is a striking example of undue proportions exhibiting themselves in the failure of the means by which they were obtained.

Structural untruth is not to be justified by authority. Neither Sir Christopher Wren, nor the Athenian exemplars of Doric and Ionic in the Propylæum, and in the Minerva Polias, with their irregular and inordinately wide intercolumniation, can persuade even the untutored eye to accept weakness for strength, or what is false for truth. Probably no eye ever regarded the middle bay of the Propylæum with satisfaction after having looked upon the Parthenon; and certainly none could turn with pleasure from the well and truly proportioned disposition of the columns of the Erechtheum prostyle to the straggling columns of the adjacent portico of Minerva Polias.

This doctrine may be supposed to preclude the use as models of the classical exemplars in architecture of the most tasteful nations of antiquity, unless reproductions be made of the materials as well as in the manner of the antique

exemplar. But this no more follows than that a man who cannot afford to possess himself of sculptures in marble shall not accept casts; the architect is not precluded from applying the truth and beauty of form and proportion found in the classical examples in consistent composition, in any fitting material proper to the purpose of the work. Mr Welby Pugin did not hesitate to employ brick-work in the face as well as in the heart of the work, in building the Romish church known as St George's, Southwark, though the great exemplars of the style he employed are of stone; but he modified the proportions to the material, and produced a truthful and in many respects an admirable work.

To do this or that, however, merely because this or that is found in canonized examples of a period or of periods in which works of high merit and great beauty were produced, is not less preposterous than it is to endeavour to apply the classical columnar ordinance, and to pay no regard, in doing so, to the principles upon which their beauties are founded. And the same observation applies alike to every style of architectural composition.

Out of truth in structure, and that structure of a very inartificial sort, grew the beautiful forms of the admirable proportions found in the works of the Greeks, and out of truth in structure, with the strictest regard to the necessities of the composition and to the materials employed—and that structure of a sort as full of artifice as the artifice employed is of truth and simplicity—grew the classical works vulgarly called Gothic, but more characteristically designated Pointed, from the form of the arch which is the basis of the style. But the classical columnar degenerated into the Italian of the *cinquecento*, in which columns and their accessories are constructed as ornaments, and are no part of the structure as such with which they are mixed up, or to which they are affixed; and the classical Pointed fell into desuetude, or was found only in a debased connection with the Italian—picturesque at times, but more commonly grotesque, and never with any claims to decent bearing.

Truly, however, the purposes of modern life are not fully answered by either of the grand and ruling styles—the classical columnar of the Greeks and Romans, and the classical Gothic or Pointed. They both had their birth in those frailties of man which lead him through devotion to superstition, and in structures intended for the worship of the unknown God, the noblest results of both styles were and still are exhibited;[1] and neither the classical columnar, nor the classical pointed bends itself, or allows itself to be bent to secular purposes without loss of both truth and beauty.

Doubtless there is much to admire in many of the productions of the Italian school of architecture, from the *cinquecento* period down to the present time; and many recent works in the pointed style, and some in the mongrel manner that arose out of its debasement in connection with the Italian, exhibit, if not originality, at the least careful copying.[2] Copying, with an affectation of originality, indeed, is carried to such an excess in modern English practice, that mere monstrosities are sought out to be copied. Wearied with building spires upon the towers of churches of such reasonable proportions that they may at once please the eye and be safe as structures, recent practice has set up

---

[1] It is a remarkable fact, indeed, that the greatest excellence in architecture has been obtained in connection with the grossest errors in theology. Witness Egypt, Greece, and Rome.

[2] Mere copying, and the most servile copying, is the ruling vice of the day, and it cannot be denounced in more eloquent language, or in more just terms, than it has been done in the following passage by an eminent person, a distinguished member of his sect in England:—"We have almost canonized defects, and sanctified monstrosities. What was the result of ignorance or unskilfulness, we attribute to some mysterious influence or deep design. A few terms give sanction and authority to any outrageousness in form, anatomy, or position; to stiffness, hardness, meagreness, unexpressiveness—nay, to impossibilities in the present structure of the human frame. Feet twisted round, fingers in wrong order on the hand, heads inverted on their shoulders, distorted features, squinting eyes, grotesque postures, bodies stretched out as if taken from the rack, enormously elongated extremities, grimness of features, fierceness of expression, and an atrocious contradiction to the anatomical structure of man,—where this is displayed,—are not only allowed to pass current, but are published in the transactions of societies, are copied into stained glass, images, and prints, and are called 'mystical,' or 'symbolical.'"

Roman Structures — in London and its vicinity several offensively acute, and dangerously poised erections, as spires. Monstrous dwarf towers, having the appearance of wind-mills without arms, almost alternate with the feeble spires, and consistently enough, the natural order is commonly reversed in placing these things; towers are set up in the plain, and spires erected on the hills.

London is indebted to Sir Charles Barry for an admirable example of the better kind of Italian street architecture in the Travellers' Club House, which shows how much of beauty may be obtained by just proportions, tasteful disposition, and unpretending decoration. The north or Pall Mall front would seem to be the result of a study of the Pandolfini Palace at Florence, to which it is fully equal in merit; but the south or Carlton Gardens front, which is in the manner of the Venetian school of design, is really beautiful, and as far excels anything else hitherto produced by the same architect as his works in Pall Mall excel all others in that street of great pretence.

OF ROMAN STRUCTURES.

Temples and other public buildings. — With a treatise on Roman architecture by a Roman architect, in our hands, mere transcription would appear to be all that is necessary in writing on the subject. But finding the writer and existing specimens of the art at variance, we cannot help determining in favour of the superior authority of the latter; to which, therefore, we shall refer to elucidate Roman edifices and the Roman style, as we did to the Greek remains to elucidate the Grecian.

Though far inferior in simplicity and harmony to the columnar architecture of the Greeks, that of the Romans, whether derived from it or not, is evidently of the same family, and is distinguished by boldness of execution and elaborate profusion of ornament. The tastes of the two nations are exemplified in the Doric of the former Plate IV. & V. and the Corinthian of the latter; the one a model of simple grandeur, perfect in its peculiar adaptation, but al-Plate IX. most inapplicable to any other purpose; and the other, less refined, but more ornate, making up in extrinsic what it wants in intrinsic beauty—imperfect in every combination, but almost equally applicable to every purpose. As in Greece, so also in Rome, the noblest specimens of columnar architecture are in the temples of the divinity; but it does not appear that the Romans were in the habit of constructing them peripterally, as the Greeks so constantly did. There are indeed ruins which induce the belief that they at times built dipteral temples; but their common practice (as far as existing examples are authorities) was to make them pseudo-peripteral, or apteral and prostylar: of an amphiprostyle, even, we have not an example. It certainly is the custom to restore the ruined temples, whose remains are a few columns only, as if they had been peripteral; but it is done not only without sufficient authority, but against that which the more perfect structures present. The great projection, too, that the Romans gave their porticoes, is evidence that they were dependent entirely on themselves for effect; for they are generally projected three columns and their interspaces before the cella, which, however, has no pronaos with columns in antis; nor does it appear from existing remains that the Romans were accustomed to use that arrangement. Circular or peristylar temples are not uncommon in Roman architecture; and there are temples to which it can hardly be supposed that columns were ever attached: these are for the most part polygonal. Neither do the Romans appear ever to have constructed hypæthral temples with columns internally, as the Greeks did. Indeed it is a question whether all their temples were not cleithral; for it is not generally admitted that the Pantheon, which is hypæthral by the open eye of the dome, was originally a temple; and where the structures remain tolerably per-

fect, the ceilings and roofs appear to have been formed Roman Structures. by arching from flank to flank, and thereby quite inclosing them.

Plate XI. Fig. 4 & 5. The application of columns internally is most strikingly effective in the Pantheon, where they are arranged in front of niches, or deep recesses, composed with antæ to carry a crowning entablature round under an attic on which the cupola rests. No representation can convey even the most incompetent idea of the effect of this arrangement, to those who cannot gather it from the plan. A section presents only one compartment correctly; all the rest must of necessity be foreshortened. It is far otherwise with the temple of peace and the hall of the baths of Diocletian, in which columns stand before the piers to have the entablature broken over them. This, indeed, was the result, as we have before intimated, of the combination of columns with arches; and it is most clearly exemplified in those works which most probably originated the practice, and which are next in pretence to the temples:—these are the triumphal arches.

Fig. 6. 7. & 8. The Romans had not adopted the simple graduated stylobate of Greek columnar architecture in their temples, but made the access to their porticoes in front with thin steps, and built vertical stereobates along the flanks for the walls of the cella, or as stylobates, if there were attached columns. In applying a columnar arrangement to Fig. 10 & 11. the triumphal arch, this lofty stylobate was taken also. The breadth of the opening prevented the columns from being placed equidistant; they were, therefore, coupled, the entablature was broken over them, and necessarily the stylobate was cut through, leaving mere attached pedestals to stilt the columns, so that the whole ordinance was deprived of every thing that could render it as a composition beautiful: its simplicity and harmony were entirely gone; and instead of giving a graceful character to the structure, it became a mere attached frontispiece, that could only deform it. As if conscious that the Corinthian was too beautiful to maltreat in such a manner, the Roman architects produced the hybrid, which has since been called the Composite order, to use in these compositions: Plate X. in them, indeed, it is chiefly found; and if it were not Ex. 2. evidently a mere deterioration of the Corinthian, it might with truth and propriety be called the Roman order.

Coupled columns, broken and recessed entablatures and pedestals, and the Composite order, are among the greatest blemishes in Roman architecture; for the misformed and inappropriate abortions which have obtained the name of Ionic and Doric, in Roman works, are hardly Ex. 3 & 4. to be attributed to the school, but to individuals of it, as they are of very infrequent occurrence, and generally appear only in works which are otherwise ungainly. Such are the amphitheatres, whose elliptical forms can never be graceful, and whose architecture was invariably the worst the time produced. The immense structure in Plate XI. Rome, which, from its magnitude, has been called the Fig. 1. Colosseum, bears in relief the gross architectural solecism of columns in stories, which, moreover, have recessed stylobates and immense intercolumniations, with large arches in them, which again reduce the effect of the column still more, making the continuity of the entablatures themselves a fault, by their consequent infirmity. The architectural details of this structure are coarse and inelegant, plain without simplicity, and laboured without elegance. But internally these blemishes disappear, columns and arches piled upon columns and arches give way to the long continuous lines which graduated from the arena to the gallery, and must have produced as grand an effect as almost any object in architecture: its magnitude and ruined state produce the imposing effect so striking at the present time; but the mind can easily restore it, or it may be contemplated in miniature in the amphitheatre at Verona.

The most perfect specimens of the Roman theatre remaining are those of Pompeii and Herculaneum. Like those of the Greeks, they rest on the side of a hill; but instead of being hewn out of, they are built on it, as there is no rock out of which they might have been excavated. Their general form, however, is very similar to that of the Greek theatre, but they received a greater degree of architectural decoration than the latter was susceptible of. Of this the theatre of Marcellus in Rome is an example; for though otherwise destroyed, its external wall remains, and presents columnar ordinances, with intervening arches in stories, according to the vicious and inelegant practice of the Roman school. This, however, is on a plain, and presents external walls, which other examples do not so completely.

The baths of the Romans were structures of immense extent, and of splendid appearance internally. What their exteriors were we have no competent means of determining; fragments, however, give us reason to believe, that in architectural merit they did not surpass the exteriors of the amphitheatres. The walls internally were covered with stucco, and painted with foliage, figures of animals, and compositions, architectural landscape or history: the floors were of mosaic, laid in compartments, and variously ornamented: the ceilings were vaulted and stuccoed like the walls; sometimes they were enriched with coffered panels containing sculptured flowers or other architectural ornament, and sometimes they were merely painted with what are termed arabesques. Columnar ordinances do not appear to have been much used, and when they were, it was not always with good taste, as we have had occasion already to remark; though the structure called the Pantheon, which, with a great show of probability, is believed to have been a saloon in the baths, perhaps of Agrippa, on the contrary presents a beautiful adaptation of one. Fig. 2 & 3. That the Pantheon was part of a more extended edifice, is very clear from its external form and appearance, which are unsightly in the extreme, presenting a mis-shapen and unfinished mass. Now, domed chambers are very common in the baths and palaces of the Romans. They are not only more effective than rectangular apartments, but were much more convenient in the absence of glass; for a small opening left in the apex lights and ventilates the domed saloon most completely, whilst the rain that could pass through it was necessarily small in quantity, and could be easily avoided by those walking on the floor. In rectangular vaults this could not be effected, so that rooms of that form depended on lateral openings for light and air, and were thereby exposed and uncomfortable. Again, there is a rectangular portico attached to the Pantheon, having no single feature in common with it, the former being a noble Corinthian octaprostyle of three intercolumniations projecting, with two others of antæ and pilasters behind; and the other a polygonal, bulbous mass of brickwork, much loftier than the portico, having cornices and blocking courses, too, none of which range with the entablature or any part of it. A conclusive argument, moreover, against the commonly received opinion, that the portico and the circular temple are an original composition, proving indeed that the former was an adaptation of what most likely had previously existed elsewhere in a different situation, and of course could not be intended for its present adjunct, is, that it now fronts north, and consequently the sun never shines full on it, so that it is in fact always in shadow; and that was never permitted even by the Romans in their original compositions; for two-thirds of the beauty of a portico, consisting in the beautiful play and contrast of light and shade it affords, are thus

sacrificed. But in an after-appropriation, as we imagine in this case, it might have been, and clearly was done; probably through the ignorance or inattention of those who did it. If, then, the Pantheon, whose diameter is nearly 150 feet, was but an apartment—suppose the grand saloon or xystum—in the baths, the whole structure must have been immense; and if its proportions and internal architecture be taken, as they certainly may be, as a specimen of the style and manner of the interior of the edifice generally, we shall obtain a very high opinion of the magnificence of the Roman baths or thermæ. That they were adorned with admirable works of sculpture, too, is proved by the fact, that some of the noblest specimens of that art have been discovered in, and among the ruins of, the baths in Rome. It may be further remarked of the Pantheon, that its effect has been seriously injured since its original construction, by the removal of the columns from the re- Fig. 4 & 5. cess opposite to the entrance, making the opening greater, fixing the columns before the antæ with the entablature broken round them, and turning an arch over the whole; thus destroying, as far as that part could affect it, the simplicity and perfect harmony of the primitive composition. The same bad taste which dictated that alteration affixed little pedimented excrescences, now used as altars, against the piers which alternate with the compartments of columns.

The still extensive remains of the villa of Adrian, near Palaces and villas Tivoli, bespeak its original magnificence; and the architectural fragments with which the site even now abounds, though it has furnished specimens to almost every country in Europe, after having suffered the spoliation and destruction attending the incursions of barbarians and the lapse of so many centuries, attest its pristine beauty, and the fine taste of its imperial builder. This, however, furnishes no evidence that its exterior was attractive. Everything appears to have been directed to internal splendour and effect alone; and indeed all collateral evidence tends to the conclusion, that the exterior of Roman palaces and mansions was not heeded; being merely plain brick walls. This is the case at Pompeii, as we shall see; and the ruins of mansions in various parts of Italy, from that of Sallust on the Benacus or Lago di Garda, to those of other Roman nobles on the shores of the Bay of Baiæ, present no indications whatever that their exteriors were subjected to architectural decoration. The palace of Diocletian at Spalatro, and the splendid remains of Balbec and Palmyra, some of which perhaps belonged to secular structures, offer evidence to the contrary of this, if they are correctly restored in the works which treat of them; for they present in their elevations so many of the worst features of the Italian school, that there would be room for doubt, if views of the ruins did not help to justify the restorers. But this does not appear to have been the case in the earlier ages of the empire, when architecture among the Romans was in its best state. Notwithstanding the extent of the structure, and its general magnificence, however, the mouldings and ornaments in the interior of the villa of Adrian, though in themselves classical and elegant, are small, and have a general air of littleness, especially when compared with the apartments to which they belong;—not that the apartments are generally large, but they are for the most part lofty. The ceilings appear to have been formed by vaulting; there are no indications of windows, and none of stairs of any magnitude—so that the rooms must have been nearly if not quite open at one end, to admit light and air; and the probability is that there were no apartments above the ground floor, though it is likely enough that terraces formed on the vaulted roofs were used for the purposes of recreation and pleasure. The floors were of mosaic, several of

ARCHITECTURE.

Roman Structures. which are preserved entire in the Museum of the Vatican; and many fine specimens of ornamental sculpture in vases and candelabra, besides busts, statues, and groups in bronze, marble, porphyry, and granite, of various styles, the remains of the noble collection Adrian made during his progress through his extensive dominions, found among the ruins of the villa, are conserved in the same place.

The ruins of the palace of the Cæsars present no forms or arrangements from which it would be possible to form a rational notion of its original plan, still less of its general elevation, or indeed of the elevation of any part of it. Large vaulted apartments, with here and there a little stucco, sometimes moulded and sometimes plain or merely painted, and a few small unconnected chambers scattered up and down in a mountain of brick rubble, convey an exceedingly vague idea of a palace, or rather they are incompetent to convey an idea of a palace at all.

Roman streets. If evidence were required to prove the futility of written descriptions of buildings when their general model is unknown, it would be enough to compare the house of a Roman gentleman in Pompeii with the various designs which have been made of the same thing from the descriptions and directions of Vitruvius, before the exhumation of that city. Their authors could only work upon the notion they had of laying out houses; and therefore the plans produced are those of ill-contrived modern residences, so arranged that they may present a uniform and architectural external elevation, which the Roman houses have not; with windows properly lighting every apartment, which are totally wanting in the latter; with staircases to upper stories that did not exist; with corridors and doors uniformly disposed, which was unheeded in laying out a Roman house. The Vitruvian restorers put columns wherever they could, whereas the Roman architects appear only to have put them where they could not avoid it. In dimensions, too, the former erred no less than in distribution; they thought none too extensive for a Roman domicile: but the apartments in Roman houses, wherever they are, are generally small, and in ordinary cases their whole site is exceedingly restricted. In proportioning the various parts, they adhered to rules the Romans never heeded; and applied the details of the architecture of temples and triumphal arches to domestic edifices, in whose composition the plasterer, painter, and mason, almost appear to have been the only architects!

Far inferior as Pompeii was to Rome in magnitude and splendour, there is no more reason for supposing that houses in the latter were so very differently arranged from those in the former that the same general description of them should not apply to both, than there would be for a future antiquary to hesitate in applying the plan of a Brighton or Bath house, which may be preserved, to a London mansion; for we know that in ordinary cases they nearly coincide. It is, too, a recorded fact, that wealthy Roman citizens had mansions at Pompeii and Herculaneum; and we have already stated that discoveries made of ordinary houses under the present level of Rome show them to be exactly like the more perfect ones of the Campanian city, except in their state of preservation; so that, " *parvis componere magna,*" in Pompeii we may see the domestic as well as public architecture of ancient Rome.

The streets of Pompeii are very narrow, their average width being not more than twelve or fifteen feet; frequently they are not more than eight feet wide, and very few in any part exceed twenty. The principal excavated street in the city, that leading from the Forum to the gate towards Herculaneum and the street of the tombs, is, at the widest, twenty-three feet six inches, including two footways, each five feet wide The streets are all paved

Roman Structures. with lava, and almost all have side pavements or footways, which, however, are for the most part so narrow, that, with few exceptions, two persons cannot pass on any of them. That the cars or carriages of the inhabitants could not pass each other in most of the streets, is proved by the wheel-ruts which have been worn on the stones, and the recesses made here and there for the purpose of passing. Their narrowness and inconvenience are aptly exemplified in London by the narrow lanes which come between St Paul's and Thames Street, about Doctors' Commons. The Forums, on the contrary, though not very spacious, are of regular forms, and have wide and convenient footways, completely colonnaded. In immediate connection with them are the theatres, the principal temples, the basilica, the courts of justice, and other public edifices: the amphitheatre is by itself, in an extreme angle of the city. The use of some of the buildings on one flank of the great Forum is not obvious: they are not arranged like temples, and indeed possess no peculiar character by which they may be distinguished: it is tolerably clear, however, from circumstances, that they were for the use of the public. The temples differ very little from the ordinary Roman structures of that description, but are generally inferior to them in the quality of the materials of which they are constructed, in the style of their architecture, and in the manner of its execution. The basilica is not unlike a modern church, it being a long rectangular edifice, having an arcaded porch at one end, being divided internally by rows of columns into nave and aisles, and having a columned recess at the west end of the nave for a tribunal. There are, however, no indications of windows, so that it was probably hypæthral, though that arrangement would have made the place very inconvenient for its purpose.

The streets of Pompeii are lined on either side with small cells, which served for shops of various kinds; and **Plate XII. Fig. 1.** they are strikingly like the ordinary shops in towns in the south of Italy and in Sicily at the present time. Like these, too, there appear in very few cases to be accommodations in connection with the shops for the occupiers and their families, who must have lived elsewhere, as modern Italian shopkeepers very commonly do. They present no architectural decoration whatever; the fronts are **Fig. 3.** merely plain stuccoed brick walls, with a large square opening in each, part of which is the door, and part the window, for lighting the place and showing the goods.

Whenever a private house or gentleman's mansion occurs in a good place for business, like the ground floor of **Houses and domestic accommodations** many modern Italian noblemen's palaces, the street-front, or fronts, was entirely occupied with shops, a comparatively narrow entrance being preserved to the house in a convenient part between some two of them. The door to this is sometimes quite plain, but at times it is decorated with pilasters. When the site permitted such an arrangement, the entrance door being open, a passer by could look completely through the house to the garden, or, in the absence of a garden, to the extreme boundary wall, on which was painted a landscape or other picture. An arrangement, it may be observed, not unlike this, is common in some of the Italian cities at the present day; but the mansions being now built in stories, and the upper stories alone being occupied by the families, a merely pleasing effect is produced; whilst in the former, persons crossing from one apartment to another were exposed, and domestic privacy thus completely invaded to produce a pretty picture. Within the entrance passage, which may be from ten to twelve feet in depth, there is a vestibule or atrium, generally square, or nearly so, on which various rooms open, that vary in size from ten feet square to ten feet by twelve, or even twelve feet square: they

55

Roman Structures. have doors only, and were probably used as sleeping-chambers by the male servants of the family. In the centre of this court there is a sunk basin or reservoir for receiving the rain, called the compluvium, rendering it likely that this was roofed over, with a well-hole to admit light and air, and allow the rain to drop from the roof into the reservoir. Connected with this outer court was the kitchen and its accessories. If the site allowed the second court to be placed beyond the first in the same direction from the entrance, the communication was by a wide opening not unlike folding doors between rooms in modern houses, generally with a space intervening, which was variously occupied; if the site did not allow of that arrangement, a mere passage led from one to the other. The second or inner vestibule, atrium, or court, is generally much larger than the first, is for the most part parallelogramic, but variously proportioned. It forms a tetrastoön, being open in the middle and arranged with a peristyle of columns, colonnading a covered walk all round. On this the best and most finished apartments open; but they are of such various sizes, and are so variously arranged, that it is not easy to determine more than that they included the refectory, the library, and sleeping-rooms. Some of them, indeed, are such as must have been useless except for the last purpose: these, perhaps, were the apartments of the female branches of a family, at least in most cases. Some houses, however, have a nest of small cells in an inner corner or secluded recess, which may have been the gynæceum; but that is far from being common. Exhedræ or recesses, open in front to the atrium, are common, and are often painted with more care and elegance than any other part of the house; but generally the walls are everywhere painted—in the more common places flat, with a slight degree of ornament perhaps, and in the best rooms with arabesques and pictures in compartments. The architectural decorations are mostly painted: the ornaments are not unfrequently elegant, but the architecture itself of the mansions is bad in almost every sense. The rooms being windowless, would, when covered, be necessarily dark; the doors are arranged without any regard to uniformity, either in size or situation. The street-fronts of those houses which, not being in a good business situation, were not occupied with shops, were not merely unadorned, but were actually deformed by loop-holes to light some passage or inner closet which had no door on one of the courts. The columns of the second courts are generally in the worst style possible: those which have foliated capitals, and may be considered compositions of the Corinthian order, are the best; but the imitations of Doric and Ionic are both mean and ugly. From the duty they had to perform, and the wideness of their intercolumniations, together with the fact that none of them remain, it is probable that the entablatures were of wood, and were consequently burnt at the time of the destruction of the city, and broken up by the inhabitants, almost all of whom certainly escaped, and who, it is very evident, returned, when the fiery shower and the conflagration had ceased, to remove whatever they could find of their property undestroyed; for it must be remembered that the roofs and ceilings all over the city are entirely gone, and the uncovered and broken walls remain, from eight to ten feet only in height. Every thing, indeed, clearly demonstrates that great exertions were used to recover whatever was valuable; and it is very probable, moreover, that the place was constantly resorted to by treasure-seekers for perhaps centuries after the calamity occurred. It may also be remarked that the loftier edifices, which would have been unburied by the ashes, had been thrown down by a terrible earthquake about sixteen years before the volcanic shower fell, and

Fig. 8.

therefore were the more easily covered. Other showers must have fallen since that which destroyed the city, to produce the complete filling up of every part and the general level throughout; as the one must have been prevented by those roofs and ceilings which were fire-proof in the first instance, and the other would be the result of the same, if it were not deranged by the subsequent excavations. It is indeed the fact, that the superstrata of ashes are evident and unbroken, while the substratum is mingled with ruins. Hence we are still uninformed as to the structure and disposition of the roofs and ceilings of the houses of the ancients. The doors, too, of whatever materials they were composed, are entirely gone: there remain, however, here and there, indications of wooden door-posts—in some cases, indeed, charred fragments of them—but they are to outer or street-doors, leaving it probable, as we have before suggested, that a matting of some kind, suspended from the lintel, formed the usual doors to rooms,—or perhaps they were closed by curtains only. In these particulars, unfortunately, Herculaneum affords but little assistance, as the mode of its destruction was similar to that of Pompeii, and it, too, was doubtlessly exposed in nearly the same manner; its subterranean situation, moreover, at present, renders it difficult to examine; but, upon the whole, Herculaneum is more likely to furnish information on these particulars than its sister in misfortune. Although it has been ascertained that the Romans understood the manufacture of glass, or at least that they possessed some utensils of that material, it must not be supposed that they were accustomed to apply it to exclude the weather and transmit light; for in no case has a glass window of any kind been discovered in any ancient structure; and, without contemplating the houses of Pompeii, it is impossible to appreciate the advantages we derive in our habitations from the application of that beautiful production of the useful arts, and how much superior it alone renders them to those of the ancients. The floors of the houses of Pompeii and Herculaneum are all of mosaic work, coarser and simpler in the less esteemed parts, and finer and more ornate in the more finished apartments: the ornaments are borders, dots, frets, labyrinths, flowers, and sometimes figures. In this, too, the superior advantages the moderns enjoy are evident. The ancients did not understand how to construct wooden floors, at least the application of timber to that use was not made by them; for, though it were admitted, which, however, it cannot be with justice, that, in the warmer climate of the south of Italy, paved floors would be more grateful, that would not be the case in this country; and we find the remains of Roman houses, baths, &c. in England, with floors of mosaic, as in Naples and Sicily. All the indications which are found in Pompeii of an upper story consist in a few rude and narrow staircases, which, it is very probable, were to afford access to the terraces or flat roofs, for they are not common, and no portion of an upper story remains in any part, though it is most likely that the lower or ground-floor rooms were arched over. In one part of the city the houses on one side of a street are on a declivity: there a commodious flight of stairs is found to lead from the atrium in front, to another atrium and rooms below, not under the houses, but behind them; for neither do we find an under-ground or cellar story in the Pompeian houses. On the shores of the Bay of Baiæ, and of the Gulf of Gaeta, at Cicero's Formian Villa, however, there are crypts or arched chambers under the level of the mansions; for the sites require substructions; but it may be questioned whether even these were used as parts of the house, and as we use cellars; for they present no indications of stairs, and have no regular means of intercommu-

Roman Structures.

nication. Neither had the houses of the Romans chimneys of any kind; their only mode of warming their apartments was by means of braziers, many specimens of which have been taken out of both Herculaneum and Pompeii; and their cooking fires were on fixed gratings over a sort of stove, but without flues; so that most probably charcoal alone was burnt for domestic purposes. In this respect the modern Italians are not far beyond their predecessors; and the mode used by them of applying fire in warming and cooking appears very similar to that used by the Romans. Indeed many of the peculiarities we have noticed in the Pompeian houses are still found in various parts of Italy and Sicily; the cortili, courts, or cloisters of palaces, mansions, monasteries, and inns, are representatives of the cavædia, vestibula, atria or courts, of Pompeian or Roman mansions. It is common, too, in the former, for bed-rooms to open on open galleries, as on the colonnaded courts of the latter. There are instances also in the countries referred to, of rooms which have no aperture but the doorway. Shops, we have said, are frequently mere cells, having an opening towards the street, part of which is a door, and the other part, with a low dado, a window. It was only in the forums and public places, then, that architectural beauty and magnificence were displayed in a Roman city. Street architecture was unknown, and the decoration of houses was the work of the plasterer and painter rather than of the architect.

If such as we have described were the imperfections and inferiority of the domestic architecture of the Romans, who knew that of the Greeks and Egyptians, and had, moreover, knowledge of the use of the arch, and were, we have reason to believe, better carpenters than either, besides possessing greater wealth, and a greater taste for luxury than they, with a less mild and serene climate than Greece and Egypt,—what must the domestic edifices of these nations have been! A person accustomed to the comforts and conveniences of houses in this country finds much to complain of in a modern Italian mansion, but not so much as an Italian would in the house of an ancient Roman; and from analogy we may believe that a Roman of the empire would have had reason to complain of a Grecian domicile, even of the Periclean age; and a Greek, again, might have been abridged of the comforts of his house in the palace of an Egyptian.

Superior as the habitations of civilized men in modern times may be to those of the ancients, a degree of classic beauty and elegance pervaded the decorations and furniture, and even the domestic utensils, in the houses of Pompeii and Herculaneum, which we do not equal, though we imitate them; and from the Hellenic taste which reigns in their forms and enrichments, their origin may probably be attributed to Greek artists; so that, it may be supposed, in these particulars the Greeks even excelled the Romans. It is indeed not a little singular, that though the architecture of these cities is completely Roman, the painted ornaments and ornamental sculpture generally are in style and manner perfectly Greek. There are certainly modifications found of Greek Doric columns in Pompeii; but they bear so slight a degree of relationship to their original, that its existence may almost be denied,—they have the form without the feeling.

As works of architecture, the sepulchral monuments of the Romans were of more importance than their domestic structures. There is more architectural display in the street of the tombs at Pompeii, than in any street of the city itself; and the mausoleum of Adrian on the right bank of the Tiber at Rome was a much more important object in its perfect state than his villa near Tivoli could ever have been. It was perhaps the most splendid struc-

ture of the kind ever executed; excelling the Memphian pyramids as much in architectural pretence as they surpass it in magnitude. There was, too, a degree of harmony and simplicity in its composition, which can only be accounted for by supposing that the imperial builder, who was himself an adept in architecture, had acquired better taste than the architects of Rome generally possessed, by the contemplation of the monuments of Greece and Egypt. It consisted of a deep quadrangular basement, each of whose sides was about 250 feet in length. This was surmounted by a lofty circular mass, on which was a graduated stylobate, supporting a noble peristyle of Corinthian columns, with their entablature; forming, with its circular cone, a species of peristylar temple, something like that below the cupola of St Paul's Cathedral in London. Above this there was, most probably, a species of dome, whose acroterium is said to have been a metal pine-cone, which was the receptacle of the ashes of the emperor. The mausoleum of Augustus was, we may believe, from the descriptions which exist of it when in a more perfect state than it is at present, inferior in size, in splendour, and in good taste, to that of Adrian; but it was nevertheless a magnificent monument. Its form was conical; it diminished in stories and terraces, probably columned round, and terminated at the apex in a bronze figure of its founder. The sepulchral monuments of Cecilia Metella, of the Plautian family, and others, are evidences of the same fact. The sarcophagus from the first-mentioned of these is simple and elegant in the extreme; and indeed it exhibits a greater degree of simple beauty than almost anything of the same kind that remains to us of the Romans.

Of the Roman Corinthian.

Like the Greek orders, the Roman Corinthian may be said to consist of three parts, stylobate, column, and entablature; but, unlike them, the stylobate is much loftier, and is not graduated, except for the purposes of access before a portico. Its usual height is not exactly determinable, in consequence of the ruined state of most of the best examples; but it may be taken at from two and a half to three diameters. In the triumphal arches the height of the stylobate sometimes amounts to four, and even to five diameters. It is variously arranged, moreover, having, in the shallower examples, simply a congeries of mouldings to form its base, with perhaps a narrow square member under it, a plain dado, and a covering cornice or coping, on the back of which the columns rest. In the loftier examples a single, and sometimes a double plinth, comes under the base mouldings; and a blocking course superimposes the coping, to receive the bases of the columns. This last is only necessary when the height of the stylobate is such as to take the columnar base above the human eye, when the coping cornice would intercept it if a blocking course did not intervene.
Plate XI.
Fig. 6, 8,
9, & 10.

The column consists of base, shaft, and capital, and varies in height from nine and a half to ten diameters. The base has, ordinarily, in addition to the diminishing congeries of mouldings which follows the circular form of the shafts, a square member or plinth, whose edges are vertical: with this the whole height of the base is about half a diameter. The rest of this part of the column is variously composed, but it generally consists of two plain tori and a scotia, with fillets intervening, as in the Greek examples of this order, but differently proportioned and projected, as the examples indicate. Sometimes the scotia is divided into two parts, by two beads with fillets, as in the Jupiter Stator example, in which also a bead is placed between the upper torus and the fillet of the apophyge. The spread of the base varies from a diameter and one third to a dia-
Plate IX.
E v. 1.

H

Roman
Corinthian.

meter and four ninths. In the best Roman examples, as well as in the Greek, the shaft diminishes with entasis: the average diminution is one eighth of a diameter. The shaft was always fluted when the material of which it was composed did not oppose itself; for the Romans often used granites, and sometimes an onion-like marble called therefore *cipollino*, for the shafts of columns; the former of which could not be easily wrought and polished in flutes, and the latter would scale away if it were cut into narrow fillets. Like the Greek Corinthian and Ionic orders, the Roman Corinthian has twenty-four fillets and flutes. The flutes are generally semicircles, and they terminate at both ends, for the most part, with that contour. Dividing the space for a fillet and a flute into five parts, four are given to the latter, and one to the former. The hypotrachelium is a plain torus, about half the size of the upper torus of the base, or half the width of a flute, as these nearly correspond: it rests on a fillet above the cavetto at the head of the shaft.

The ordinary height of the capital is a diameter and one-eighth; but there is a very fine example in which it hardly exceeds a diameter, and another in which it is not quite so much. It is composed of two rows or bands of acanthus leaves, each row consisting of eight leaves ranged side by side, but not in contact; of helices and tendrils trussed with foliage; and an abacus, whose faces are moulded and variously enriched. The lower row of acanthus leaves is two-sevenths the whole height of the capital; the upper row is two-thirds the height of the lower above it, and its leaves rest on the hypotrachelium below, in the spaces left between the others. They are placed regularly, too, under the helices and tendrils above, which support the angles, and are under the middle of each side of the abacus. The construction and arrangement of the next compartment above must be gathered from the examples; for a competent idea cannot be conveyed in words. The abacus is one-seventh of the height of the capital; in plan it is a square whose angles are cut off, and whose sides are concaved in segments of a circle, under an angle at the centre of from 55° to 60°. Its vertical face is generally a flat cavetto, with a fillet and carved ovalo corbelling over at an angle of about 125°. The cavetto is sometimes enriched with trailing foliage, and a rosette or flower of some kind overhangs the tendrils from the middle of each side of the abacus.

Every example of this order differs so much in the form, proportion, and distribution of the various parts, of its capital particularly, that it cannot be described in general terms, like the Greek Doric and Ionic: the example

Ex. 1.

we have referred to in this definition is that of the Jupiter Stator, the most elegant, perhaps, of all the Roman specimens.

The entablature varies in different examples from one diameter and seven eighths to more than two diameters and a half in height. Perhaps the best proportioned are

Ex. 4.
Ex. 3.

those of the portico of the Pantheon, and of the temple of Antoninus and Faustina; the former being rather more than two diameters and a quarter, and the latter rather less than that ratio. The entablature of the temple of Jupiter Stator is more than two diameters and a half in height, of which the cornice alone occupies one sixth more than a full diameter, leaving to the frieze and architrave somewhat less than one diameter and a half between them. In this latter particular it nearly agrees with the other two quoted examples, so that the great difference in the general height is in the cornice almost alone, the cornices of the others being about a sixth less, instead of as much more, than a diameter in height. The Roman Corinthian entablature may be taken, then, at two diameters and a quarter in height. Rather more than

three fifths of this is nearly equally divided between the architrave and frieze, the advantage, if any, being given to the former; the cornice, of course, takes the remaining two fifths, or thereabouts. The architrave is divided into three unequal fascias and a small congeries of mouldings, separating it from the frieze. The first fascia is one fifth the whole height; one third of what remains is given to the second, and the remainder is divided between the third fascia and the band of mouldings,—two thirds to the former, and one to the latter. A bead, sometimes plain and sometimes carved, taken from the second fascia, which is itself enriched in the Jupiter Stator example, marks its projection over the first; and a small cyma-reversa, carved or plain as the bead may be, taken from the third fascia, marks its projection over the second. The band consists of a bead, a cyma-reversa, carved or plain according to the general character of the ordinance, and a fillet. In non-accordance with the practice of the Greeks, the face of the lowest or first fascia of the architrave, in the Roman Corinthian, impends the face of the column at the top of the shaft, or at its smallest diameter; and every face inclines inwards from its lowest face up. The whole projection of the architrave, that of the covering fillet of the band, is nearly equal to the height of the first fascia. The frieze impends the lowest angle of the architrave. Its face is either perpendicular, or it slightly inclines inwards, like the fascias of that part of the entablature: in some cases it is quite plain, and in others is enriched with a foliated composition, or with sculptures in low or half relief. The cornice consists of a deep bed-mould, variously proportioned to the corona; but it may be taken generally, when it has modillions, at three fifths, and when it has none, at one half the whole height. It is composed of a bead, an ovalo or cyma-reversa and fillet, a plain vertical member, sometimes dentilled, another bead, and a cyma-reversa and fillet or ovalo, as the lower may *not* be: this is surmounted, when modillions are used, by another plain member, with a small carved cyma-reversa above it. On this the modillions are placed, and the cyma breaks round them. They are about as wide as the member from which they project, and are about two thicknesses apart. In form they are horizontal trusses or consoles, with a wavy profile, finishing at one end in a large, and at the other in a small, volute; and under each there is generally placed a raffled or acanthus leaf. In proportioning the parts of this bed-mould in itself, one third of its height may be given to the modillion member, and the other two thirds divided nearly equally, but increasing upwards into three parts, one for the lowest mouldings, one for the plain or dentil member, and the third and rather largest portion for the mouldings under the modillion member. The mouldings of this part of the cornice are carved or left plain, according to the character of the ordinance; and its greatest projection, except the modillions themselves, that of the modillion member, is about equal to half its height. The upper part of the cornice,—the corona, with its crown mouldings,—consists of the vertical member called the corona, which is two fifths the whole height;—this, in the examples of the temples of Jupiter Stator and Antoninus and Faustina, is enriched with vertical flutes;—a narrow fillet, an ovalo, and a wider fillet, occupy one third of the rest, the other two thirds being given to cyma-recta, with a covering fillet which crowns the whole. Its extreme projection is nearly equal to the whole height of the cornice.

The ordinance of the temple of Vesta, or of the sibyl

Ex. 2

at Tivoli, whose entablature is the very low one mentioned, is not generally in accordance with the scale we have given, and it must be referred to for its own peculiar proportions.

Roman Corinthian. Pediments with the Roman Corinthian order are found to be steeper than they were made by the Greeks, varying in inclination from eighteen to twenty-five degrees; but they are formed by the cornice of the entablature in the same manner Antefixæ do not appear to have been used on flank cornices as in Greek ordinances, in which the cymatium is confined to pediments; but in Roman works it is continued over the horizontal or flank cornice, as we have described; and frequently it is enriched with lions' heads, which were at the first introduced as water-spouts. The planceer or soffit of the corona is, in the Jupiter Stator example, coffered between the modillions, and in every coffer there is a flower. The soffit of the entablature in this order is generally panelled and enriched with foliated or other ornament. The intercolumniation is not the same in any two examples. In the temple of Vesta in Rome it hardly exceeds a diameter and a quarter; in the Jupiter Stator example it is a fraction less than one diameter and a half, in that of Antoninus and Faustina, nearly a diameter and three quarters; in the portico at Assisi, rather more than that ratio; in the portico of the Pantheon, almost two diameters; and in the Tivoli example, a fraction more than that proportion.

The antæ of the Roman Corinthian order are generally parallel; but pilasters are mostly diminished and fluted as the columns. Of two of the existing examples of antæ, in one—that of the temple of Mars Ultor—they are plain, to fluted columns; and in the other—that of the Pantheon portico—they are fluted, to plain columns. The capitals and bases are transcripts of those of the columns, fitted to the square forms.

Ceilings of porticoes are formed, as in the Greek style, by the frieze returning in beams from the internal architrave to the wall or front of the structure, supporting coffers more or less enriched with foliage or flowers. This, however, could only have been effected when the projection was not more than one, or at the most two intercolumniations, if stone was used; and it is only in such that examples exist. Porticoes ordinarily must have had arched ceilings, as that of the Pantheon has, or the beams must have been of wood; in which latter case probably the compartments of the ceiling would be larger. How, in the former, it was arranged we cannot tell, as the arches only remain; and they may not be of the date of the rest of the portico.

Plate X. Ex. 2. The ancient examples of what is called the Composite order do not differ so much from the ordinary examples of the Corinthian as the latter do among themselves, except in the peculiar conformation of the capital of the column. In other respects, indeed, its arrangement and general proportions are exactly those of the Corinthian. The Composite was used, we have said, in triumphal arches, and, in the best ages of Roman architecture, in them alone. The difference in the capital consists in the enlargement of the volutes to nearly one fourth the whole height of the capital, and in connecting their stems horizontally under the abacus, giving the appearance of a distorted Ionic capital. The central tendrils of the Corinthian are omitted, and the drum of the capital is girded under the stem of the volutes by an ovalo and bead, as in the Ionic. Acanthus leaves, in two rows, fill up the whole height from the hypotrachelium to the bottom of the volutes, and are consequently higher than in the Corinthian capital: this difference is given to the upper row. Besides this Composite, however, the Romans made many others, the arrangements and proportions of the ordinances being generally those of the Corinthian order, and the capitals corresponding also in general form, though in themselves differently composed. In these, animals of different species, the human figure, armour, a variety of foliage, and other peculiarities, are found. Shafts of columns also are

sometimes corded or cabled instead of being fluted: those of the internal ordinance of the Pantheon are cabled to one third their height, and the flutes of the antæ of that ordinance are flat, eccentric curves. There are fragments of others existing, in which the fillets between the flutes are beaded; some in which they are wider than usual, and grooved; others, again, whose whole surface is wrought with foliage in various ways; and it would be no less absurd to arrange all these in different orders, than it is to make a distorted and hybrid capital the ground-work of an order.

Of the Roman Ionic.

Roman Ionic. The only existing example of this in Rome, in which the columns are insulated, is in the Temple of Manly Fortune, except that of the Temple of Concord, which is too barbarous to deserve consideration. Its stylobate, like that of the Roman Corinthian, is lofty and not graduated, having a moulded base and cornice or surbase. The column is nearly nine diameters in height; its base is half a diameter in height, and consists of a plinth, two tori, a scotia, and two fillets; the shaft has twenty fillets and flutes, and diminishes one tenth of a diameter; the capital is two fifths of a diameter in height; the volutes, however, dip a little lower, being themselves about that depth without the abacus; the corbelling for the volutes is formed by a bead and large ovalo, half the height of the capital; the latter of these is carved: a straight band connects the generating lines of the volutes, whose ends are bolstered and enriched with foliage; and a square abacus, moulded on the edges, covers the whole. The entablature is rather less than two diameters high; three tenths of this are given to the architrave, the same to the frieze, and the cornice occupies the remaining two fifths. The architrave is unequally divided into three fascias, and a band consisting of a cyma-reversa and fillet; the lowest angle impends the upper face of the shaft of the column. The frieze is in the same vertical line, and is covered with a fillet which receives the cornice; it is also enriched with a composition of figures and foliage. The cornice consists of a bed-mould, two fifths of its height, and a corona with crown mouldings. The bed-mould is divided nearly equally between a cyma-reversa and fillet, a square dentilled member and fillet, and another fillet and ovalo. The corona is two fifths the height of the rest of the cornice; another fifth is occupied by two fillets and a cyma-reversa, and the rest is given to a cyma-recta and crowning fillet. The whole projection is nearly equal to the height of the cornice. The cymatium is enriched with acanthus leaves and lions' heads, and the mouldings of the bed-mould and architrave band are carved. The soffit of the corona is hollowed out in a wide groove, whose internal angles are rounded off in a cavetto, but without ornament of any kind, forming indeed a mere throating. Like the angular capitals of the Greek Ionic, the external volute of this is turned out and repeated on the flank: either that or the abuse of it in the Composite capital gave rise to distortions of this order, in which all the volutes of the capital are angular, and consequently all its four faces are alike. In other respects, however, it does not differ generally from the ordinary Roman examples of Ionic. The Temple of Manly Fortune is pseudo-peripteral, and consequently has neither antæ nor pilasters, nor do ancient examples exist of either.

Of the Roman Doric.

This is even a ruder imitation of the Grecian original than the mean and tasteless deterioration of the voluted Ionic is of the graceful Athenian examples. The specimen of it which is considered preferable to the others is

Margin notes (right): Plate XI. Fig. 12. — Plate X. Ex. 2. — Plate XI. Fig. 12. — Plate X. Ex. 1.

Roman
Doric.

that of the theatre of Marcellus in Rome. The column is nearly eight diameters in height: it consists of shaft and capital only. The shaft is quite plain, except fillets above and below, with escape and cavetto; and it diminishes one fifth of its diameter. The capital is four sevenths of a diameter high, and is composed of a torus which forms the hypotrachelium, and, with the necking, occupies one third of the whole height. Three deep fillets, with a semitorus cr quarter-round moulding, are intended to represent the ovalo and its annulets of the Greek capital. They occupy three sevenths of the rest; the other four sevenths are given to the abacus, three fifths of whose depth is plain and vertical; and the other two is divided between a cyma-reversa and a fillet.

The corona and crown mouldings of the cornice being destroyed, the whole height of the entablature cannot be correctly ascertained; but from analogy it may be taken, with the bed-mould, part of which exists, at about two thirds of a diameter, making, with the architrave and frieze, an entablature nearly two diameters high. Of this the architrave is rather more than one fourth, indeed exactly half a diameter. Three tenths of its depth are unequally occupied by the tænia, regula, and guttæ, the first being rather the widest, projecting more than its own depth, and the second the narrowest. The guttæ are six in number, and are truncated semicones in form. The rest of the surface of the architrave is plain and vertical, impending a point rather within the superior diameter of the column. The frieze is two fifths the whole height of the entablature. A fascia, one eighth of its own height, bands it above the triglyphs, and projects about one third of its depth; the rest of its surface is plain vertically, but horizontally it is divided into triglyphs, which are half a diameter in width, and are placed over the centres of the columns. These are channelled with two full and two hemiglyphs, whose heads are cut square on the outer edge, but inclined downwards at the angle of the glyphs. The space between the triglyphs is equal to the height of the frieze without its plat-band or fascia, making in effect perfectly square metopes. All that can be traced of the cornice is a small cyma-reversa, immediately over the frieze, and a square member with dentils on it. In the example the cornice is completed from that of the Doric of the Colosseum.

The temple at Cora presents a singular specimen of the Doric order, evidently the result of an examination of some Greek examples, but moulded to the Roman proportions and to Roman taste. The columns are enormously tall, but the shafts are partly fluted and partly chamfered for fluting, like the Greek. The capital is ridiculously shallow, but the abacus is plain, and the echinus of a somewhat Hellenic form. The entablature is very little more than a diameter and one third in height, and the architrave of it is shallower even than the capital; but the frieze and cornice are tolerably well proportioned, though the triglyphs in the former are meagre, narrow slips, and the latter is covered by a deep widely projecting cavetto, that would be injurious to even a better composition. Instead of regular mutules with guttæ, the whole of the planceer of the cornice is studded with the latter; but, like the Greek, the triglyph over the angular column extends to the angle of the architrave, which does not appear to have been the practice of the Romans; yet the reason for so doing does not appear to have been understood, for the external intercolumniations are the same as the others.

As far as we have the means of judging, the Romans made the antæ of their Doric similar to the columns, only that they were of course square instead of round; though indeed an attached column appears to have been generally preferred.

It may, however, be here again intimated, that these two orders, the Ionic and Doric of the Roman school, ought hardly to be considered as belonging to the architecture of the Romans. They are merely coarse and vulgar adaptations of the Greek originals, of which we now possess records of the finest examples. If it were not, therefore, that custom required it, we should have omitted all mention of them, or at least have left them to the Italo-Vitruvian school, to which they properly belong. Yet their meanness and tastelessness, when compared with the Grecian models, will more strikingly show the superiority of the latter, and show, moreover, how the architects of the Italian school must have been blinded by their system, when they fancied such wretched exemplars as those of which we have been speaking to be beautiful.

Roman
Mouldings.

Of Roman Mouldings and Ornament.

The mouldings used in Roman architectural works are the same as the Grecian in general form, but they vary materially from them in contour. The Roman cyma-recta is projected much more than the Greek, with a deeper flexure; and the two parts or ends seldom correspond, the one being generally larger than the other. On the contrary, the Roman cyma-reversa does not project so much, or at so large an angle with its base, as the Grecian, nor is it so deeply flected as the Greeks made it. The upper or convex part of this moulding is almost always larger than the lower or concave; and it is frequently allowed to finish below in a sharp arris projecting from whatever is below it, and above it abuts upon the horizontal soffit of its covering fillet in a similarly harsh manner. The ovalo of Greek architecture is represented in the Roman style by a moulding whose outline is nearly the convex quadrant of a circle, or a quarter round, and sometimes it is nearly that of the quadrant of an ellipsis. The Roman torus is either a semicircle or a semiellipsis; and the bead is a torus, except in its application, and in being smaller, and generally projected rather more than half the figure whose form it bears. The cavetto, in Roman architecture, is nearly a regular curve, being sometimes the concave quadrant of a circle, or indeed the reverse of an ovalo, and sometimes a smaller segment. A Roman scotia is more deeply cut, and is consequently less delicate than the same member in a Greek congeries: its form frequently approaches that of a concave semiellipsis.

This correspondence in general form, and disagreement in spirit, of Greek and Roman mouldings, appear to have arisen entirely from the ignorance or inattention of the Romans to the governing principle of Greek combinations; as we have seen that in these the individual mouldings are not independent, as the Romans made them, but that they take their contour and direction from each other, under a certain pervading outline.

The enrichments of Roman mouldings are for the most part similar to those of the Greek, but less delicate and graceful both in design and drawing. Those of the cyma and ovalo are particularly referred to, but the Romans used others besides. Raffled leaves form a favourite enrichment in the architecture of the Romans: indeed these are hardly less frequent in their works than the honeysuckle is in those of the Greeks. Mouldings were enriched with them; and a raffled leaf masks the angles of carved cymas and ovalos in the former, as a honeysuckle does in the latter. Nevertheless, the honeysuckle and lotus are both found in Roman enrichments, particularly the latter, and perhaps even more than in Greek. It is not uncommon to find examples of Roman architecture completely overdone with ornament,—every moulding carved, and every straight surface, whether vertical or

Pl. VIII.

horizontal, sculptured with foliage, or with historical or characteristic subjects in relief. This fault is most obvious in those works which exhibit similar bad taste in the general composition. The triumphal arch of Septimius Severus, the little arch of the goldsmiths, and the half-buried ruin called the temple of Pallas, in the forum of Nerva at Rome, are egregious specimens. The entablature of the arch of Titus, too, is overloaded with ornament.

Frieze enrichments, consisting of foliage composed with animals, and a variety of other things, are very common in Roman architecture. Many specimens indeed are not found in existing structures, but there are numerous fragments of entablatures of destroyed edifices which exhibit them in great variety. Their general character is exuberance, and a tendency to frittering, from the variety and incoherence of form in their composition; but their effect can only fairly be judged of when seen in appropriate situations. One existing example of an enriched frieze of the kind referred to, that of the temple of Antoninus and Faustina, speaks strongly in its favour, for nothing can surpass its efficiency and simple beauty; but it must, moreover, be confessed that, when examined in detail, the enrichment is less exuberant, and is composed of fewer parts, than most others of the species to which that example belongs. Architectural ornament, however, is not confined to purely architectural works. We find many beautiful specimens of it on the vases and candelabra which decorated the baths and mansions of the ancient Romans, and whose elegance of form rivals even the beauty and delicacy of their enrichments. Whether these should be referred to the Romans or not, is doubtful; for it has been already intimated, from the style of many of them, both in outline and ornament, which appertain more to the Greek, that they are the productions of Grecian artists; but indeed they belong exactly to neither, for they frequently possess the beauties, and sometimes exhibit the defects, of both. There are existing works, clearly of Roman origin, far inferior in every respect to the objects last mentioned. These are for the most part cenotaphial monuments, sarcophagi, and altars, whose composition, details, and enrichments, are gross and inelegant when compared with the objects alluded to. The difference may arise merely from the inferiority of the artists of the one to those of the other, and not from the difference of their schools; but the prevalence of Greek taste in the superior productions is not the less striking because it was acquired by education, while it is wanting in the inferior, whose authors had not been imbued with the spirit and fine feeling of the Greek style.

OF ITALIAN ARCHITECTURE.

Gothic architecture,—that is, the style which preceded the Pointed,—being for the most part a mere deterioration of Roman, and possessing no peculiar character which can recommend it as a subject for study and imitation that may not be deduced from the Roman style, and Pointed architecture being a genus *per se*, we have thought it better to allow the Italian, or revived Roman style, to usurp its chronological place; as the latter more naturally follows what it pretends to be derived from, than it would follow the Pointed, or than the Pointed would the Roman.

We have already stated that Italian architecture, though professedly a revival of the classical styles of Greece and Rome, was formed without reference to the existing specimens of either, but on the dogmas of an obscure Roman author, and the glosses of the " revivers" on his text. Vitruvius described four classes or orders of columnar composition; and, on the principles which go-

verned him in subjecting to fixed laws all the varieties with which he appears to have been acquainted, they formed a fifth, of a medley of two of his, thus completing the Italian orders of architecture. The school which was founded on the Vitruvian theories has systematized every thing, and laid down laws for collocating and proportioning all the matter it furnishes for architectural composition and decoration. It teaches that columns are modelled from the human figure; that the Tuscan column is like a sturdy labourer—a rustic; the Doric is somewhat trimmer, though equally masculine—a gentleman, perhaps; the Ionic is a sedate matron; the Corinthian a lascivious courtesan; and the Composite an amalgam of the last two! In a composition which admits any two or more of them, the rustic must take the lowest place; on his head stands the stately Doric, who in his turn bears the comely matron, on whose head is placed the wanton, and the wanton again is made to support the lady of doubtful character! But as we in this place are neither apologists for nor impugners of any particular doctrines, we proceed at once to point out the general features of the Italian style; premising only, that, according to the practice of the school, every thing is confined to an exclusive use and appropriation; such columns may be fluted, and such must not; such a moulding may be used here, but not there; and so on. The proportions and arrangements of an order, of any part of one, or of any thing that may come within an architectural composition, are fixed and unchangeable, whatever may be the purpose or situation for which it is required; whether, for instance, an order be attached or insulated, the column must have exactly the same number of modules and minutes in height. It is true that the masters of the school are not agreed among themselves as to those things in which they are not bound by Vitruvius; but every one not the less contends for the principle, each, of course, prescribing his own doctrine as orthodox on the unsettled points.

Mouldings are considered as constituent parts of an order, and are limited to eight in number, strangely enough including the fillet. They are the cyma-recta, cyma-reversa, commonly called the ogive or ogee, the ovalo, the torus, the astragal or bead, the cavetto, the scotia, and the fillet. They are gathered from the Roman remains, but reduced to regular lines or curves, which may be drawn with a rule or struck with a pair of compasses. Arranged according to certain proportions, with flat surfaces, modillions, and dentils, a profile is formed; no two conjoined mouldings may be enriched, but their ornaments, as well as the modillions and dentils, must be disposed so as to fall regularly under one another, and, when columns occur, above the middle of them.

An order is said to be composed of two principal parts, the column and the entablature; these are divided into base, shaft, and capital, in the one, and architrave, frieze, and cornice in the other, and are variously subdivided in the different orders. The Tuscan column must be made seven diameters in height, the Doric eight, the Ionic nine, and the Corinthian and Composite ten. The height of the entablature, according to some authorities, should be one fourth the height of the column, and, according to others, two of its diameters. The parts of the entablature of all but the Doric may be divided into ten equal parts, four of which are given to the cornice, three to the frieze, and three to the architrave; and in the Doric, the entablature being divided into eight parts, three must be given to the cornice, three to the frieze, and the remaining two to the architrave. For the minor divisions a diameter of the column is made into a scale of sixty minutes, by which they are arranged; but this is obviously irrelevant if the whole height of the entablature is determined by

Plate XIII.

the height of the column, and not by its diameter; in this case, therefore, they must be proportioned from the general divisions already ascertained. Columns must be diminished, according to Vitruvius, more or less as their altitude is less or greater; those of fifteen feet high, or thereabout, being made one sixth less at their superior than at their inferior diameter, and that proportion is lessened gradually, so that columns fifty feet high shall be diminished one eighth only. On this subject, however, many of his disciples controvert the authority of their master; and some of them have fixed the diminution at one sixth of a diameter for columns of all sizes in all the orders. The entasis of columns is disputed also, some authorities making it consist in preserving the cylinder perfect one quarter or one third the height of the shaft from below, diminishing from thence in a right line to the top; while others, following Vitruvius, make the column increase in bulk in a curved line from the base to three sevenths of its height, and then diminish in the same manner for the remaining four sevenths, thus making the greatest diameter near the middle.

It being difficult to determine among the masters of the Italo-Vitruvian school whose designs of the various orders are to be preferred, we have selected those of Palladio, certainly not for any superior merit they possess, but because he is more generally esteemed than any other, and because he the most strictly adhered, as far as he could understand them, to the precepts of Vitruvius. It should be remarked, however, that although Palladio has fluted all but the shaft of the Tuscan column, he very seldom fluted columns in his own practice; and indeed it may be called the custom of the Italian school not to flute, how much soever their doctrine may be to the contrary; for fluted columns in Italian architecture are exceptions to the general practice. Swelled or pillowed friezes are not peculiar to Palladio; they are more ɯ less common to the works of most of the masters of the same school. Prostyles being almost unknown in Italian architecture, antæ are not often required; but when they are, the meanest succedaneum imagin-

Pl. XVII.
Fig. 3.

able is recurred to. Of this, Palladio's Villa Capra near Vicenza, and Lord Burlington's Palladian Villa at Chiswick, afford striking examples. Pilasters, however, are very common, so common, indeed, that they may be called pro-columns, as they are often used as an apology for applying an entablature. They are described as differing from columns in their plan only, the latter being round, and the former square; for they are composed with bases and capitals, they are made to support entablatures according to the order to which they belong, and are fluted and diminished with or without entasis, just as columns of the same style would be. When they are fluted, the flutes are limited to seven in number on the face, which, it is said, makes them nearly correspond with the flutes of columns; and their projection must be one eighth of their diameter or width when the returns are not fluted; but if they are, a fillet must come against the wall. Pedestals are not considered by the Italo-Vitruvian school as belonging to the orders, but they may be employed with them all, and have bases and surbases or cornices to correspond with the order with which they may be associated. The dado of a pedestal must be a square whose side shall be equal to that of the plinth of the column or pilaster which rests on it, or a parallelogram a sixth or even a fourth of a diameter taller. The intercolumniations of columns are called pycnostyle, systyle, eustyle, diastyle, and aræostyle, and are strictly adhered to in Italian architecture when columns are insulated, and that is not very often; when they are attached, the interspaces are not limited, except when a peculiar arrangement called aræosystyle is adopt-

ed. This consists of two systyle intercolumniations, the column that should stand in the mid-distance between two others being placed within half a diameter of one of them, making in fact coupled columns or pilasters. It is applied to insulated columns as well as to those which are attached. Following Vitruvius, the Italian school makes the central intercolumniation of a portico wider than any of the others. Arched openings, in arcades or otherwise, are generally about twice their width in height; if, however, they are arranged with a columnar ordinance, having columns against the piers, they are made to partake of the order to which the columns belong, being lower in proportion to their width with the Tuscan than with the Doric, and so on: and the piers are allowed to vary in the same manner, from two fifths to one half of the opening. With columnar arrangements, moulded imposts and archivolts are used; the former being made rather more than a semidiameter of the engaged columns in height, and the latter exactly that proportion. Variously moulded key-stones are used, too, projecting so that they give an appearance of support to the superimposed entablature. Smaller columns with their entablature are sometimes made to do the duty of imposts, and sometimes single columns are similarly applied; at others, columns in couples are allowed to stand for piers to carry arches. In plain arcades the masonry is generally rusticated, without any other projection than a plain blocking course for an impost, and a blocking course or cornice crowning the ordinance. Niches and other recesses are at times introduced in the plain piers, which are in that case considerably wider than usual, or in the spandrels over wide piers. Very considerable variety is allowed in these combinations, which will be best understood by reference to the examples. Doors and windows, whether arched or square, follow nearly the same proportions, being made, in rustic stories, generally rather less than twice their width in height, and in others either exactly of that proportion, or an eighth or a tenth more. If they have columned or pilastered frontispieces, these are sometimes pedimented; and, except in rustic stories, whether with or without columns, a plain or moulded lining called an architrave is applied to the head and sides of a door or window. This architrave is made from one sixth to one eighth the width of the opening it bounds, and it rests on a blocking course or other sill, as the case may be. In the absence of columns or pilasters in the frontispiece, their place is frequently supplied by consols or trusses of various form and arrangement, backed out by a narrow pilaster, which may be considered as the return of the frieze of the entablature, and supporting the cornice. It is not uncommon for the architrave lining to project knees at the upper angles, and this is sometimes done even with consols and their pilasters. With columned frontispieces to gateways, doors, and windows, arose the custom, so frequent in Italian architecture, of rusticating columns, by making them alternately square and cylindrical, according to the heights of the courses of rustic masonry to which they are generally attached, and with which they are less offensive than in other collocations. The practice of the *Cinquecento* school of piling columns on columns, with their accessories. is warranted by the doctrine of its master; but his precepts not being practicable, recourse has been had to the inferior works of the Romans, which present examples of it. The difficulty of preserving any thing like a rational arrangement is acknowledged on all hands to be great, if not insurmountable; for if the first or lowest order be at an intercolumniation fitting its proportions, the second or next above it, though diminished ever so little, is already deranged, for it has the same distance from column to column that the inferior order has, whilst the columns themselves are smaller in diameter, and their entablature

Plate XIV.

Fig. 11 &
12.

Fig. 12

Fig 10.
Fig. 13.

Fig. 2 & 4
and Plate
VII.
Fig. 2.

Fig. 1.
Plate XI.

consequently shallower. This derangement must of course increase with every succeeding ordinance, rendering it indeed impossible to make such a composition consistent. The most approved practice in arranging order above order appears to be, that the upper column shall take for its diameter the superior diameter of the one below it; that when the columns are detached their axes shall be in the same perpendicular line; but when attached or engaged, the plinth of the pedestal of the upper shall impend the top of the shaft of the lower column. The most rational mode, however, for diminishing, if reason can be applied to such compositions, is to carry the diminution through, the outlines of the columns of the lowest order being drawn up in the same direction, and so the columns of every story would take up their place and be diminished in regular gradation.

Pl. XVII. Fig. 2. Pl. XVI.

When columns are attached, or pilasters used, in Italian architecture, the almost invariable custom is to break the entablature over every column or pilaster, or over every two when they are in couples. Because of the great length of the intercolumniation, it would appear to have been done at first; but it has frequently been done by some of the most esteemed practitioners of the school, even without that excuse, so that it may be held as approved by them. A basement is either a low stereobate or a lofty story, as it may be intended to support a single ordinance the whole height of the main body of the structure, or indeed the lowest of two or more orders; or as it may occupy the ground story of a building, and support an ordinance, or the appearance of one, above. In either case, much is necessarily left to the discretion of the architect; but in the latter the height of the order it is to support is the generally prescribed height of the basement. A basement may be rusticated or plain; if it be low, and is not arranged like a continued pedestal, it must have neither cornice nor blocking course; but if lofty, a deep, bold, blocking course is indispensable. An attic may vary in height from one quarter to one third the height of the order it surmounts; attics are arranged with a base, dado, and coping cornice, like pedestals, and generally have pilasters broken over the columns below. The rule for the form, composition, and application of pediments in Italian architecture, if it may be gathered from the practice of the school, appears to be to set good taste at defiance in them all. We find pediments of every shape, composed of cornices, busts, scrolls, festoons, and what not, and applied in every situation, and even one within another, to the number of three or four, and each of these of different form and various composition. The proportion laid down for the height of a pediment is from one fourth to one fifth the length of its base, or the cornice on which it is to rest. Balustrades are used in various situations, but their most common application is in attics or as parapets, on the summits of buildings, before windows, in otherwise close continued stereobates, to flank flights of steps, to front terraces, or flank bridges. Their shapes and proportions are even more diversified than their application: that of most frequent use is shaped like an Italian Doric column, compressed to a dwarfish stature, and consequently swollen in the shaft to an inordinate bulk in the lower part, and having its capital, to the hypotrachelium, reversed to form a base to receive its grotesque form. The base and coping cornice of a balustrade are those of an ordinary attic, or of a pedestal whose dado may be pierced into balusters. The general external proportions of an edifice, when they are not determined by single columnar ordinances, appear to be unsettled.

Pl. XVII. Fig. 1.

The grand front of the Farnese Palace in Rome is in two squares, its length being twice its height; the length of each front of Vignola's celebrated pentagonal palace of Caprarola is twice and a quarter its height above the bastions. In Palladio's works we find the proportions of fronts to vary so considerably, as to make it evident that he did not consider himself bound by any rule on that point. In some cases we find the length to be once and one sixth the height, in others once and a fourth, once and a half, twice, twice and a sixth, and even thrice and a sixth; and elevations by other masters of the school are found to vary to the same extent. The proportions of rooms, again, range from one to two cubes inclusive, though it is preferred that the height should be a sixth, or even a fifth less than a side when the plan is a square; but the sesquialteral form, with the height equal to the breadth, and the length one half more, is considered the most perfect proportion for a room. There is considerable variety and beauty in the foliate and other enrichments of an architectural character in many structures in Italy, but very little ornament enters into the columnar composition of Italian architecture. Friezes, instead of being sculptured, are swollen; the shafts of columns, it has been already remarked, are very seldom fluted, and their capitals are generally poor in the extreme; mouldings are indeed sometimes carved, but not often; rustic masonry, ill-formed festoons, and gouty balustrades, for the most part supply the place of chaste and classic enrichments. This refers more particularly to the more *classic* works of the school; in many of the earlier structures of Italy, especially those of the Trecento period, and on monuments of various kinds, we find what may be called a graceful profusion of ornament, of the most tasteful and elegant kind; few carved mouldings, however, and very few well-profiled cornices, are to be met with in Italian compositions of any kind. In many of the later architectural works of that country we find again a profusion of ornament of the most tasteless and inelegant description, chiefly in the gross and vulgar style, which is distinguished as that of Louis XIV. of France.

Of Pointed Architecture.

There are so many varieties of this beautiful style, and the variations are at the same time so considerable and so minute, that it is impossible to describe it generally. Every country in which it was practised had some peculiarities in its composition, and, to develope it perfectly, all of them should be pointed out. This, however, would far exceed our limits; and as the specimens of our own are not excelled, if indeed they are equalled, by those of any other country, a consideration of the style as exhibited by them will afford us a better opportunity of developing it than could be obtained by making our observations more general.

Various classifications of Pointed architecture have been made, and in almost all of them the arch is considered the index, as the column is in columnar architecture; for, like that, it is more expressive of variety than any other feature in the composition to which each belongs. These, too, form the grand distinctions between the Greek and its derivative styles, and the Pointed; but, independently of the column in the one and the arch in the other, the two species of architecture may be said each to have certain governing principles, which sufficiently distinguish and make it impossible to mould them together in one composition, and almost to apply any of the leading forms of the one to the other. They may be thus generally laid down. In Greek and Roman architecture the general running lines are horizontal, as in entablatures and single cornices. In Pointed, the general running lines are vertical. In the former, arches are not necessary to a composition; in the latter, arches are a really fundamental principle. In Greek and Roman, again, columns require an entablature; in the Pointed style no such thing as an entablature composed of parts is applicable to the pillars, columns, or shafts. (*Vide* Rickman's *Attempt,* &c. p. 110.)

These, however, only determine the generic differences which exist; the varieties in the former styles we have found to be marked by such and such distinctive features in the columns and their accessories, which allowed them to be divided into orders. In the latter the varieties arise chronologically, and, consisting for the most part in the forms and arrangements of details, are not incoherent; nor are certain proportions either fixed or determinable, and consequently they cannot be rendered into orders.

It has been customary, in treating of Pointed architecture, to class with it the Saxon and Norman Gothic styles. This is at least unnecessary, as they have no direct relation to it, except that of immediate precedence in point of time, and that the one was the stock on which the other was grafted. The peculiarities of Pointed architecture are indeed totally independent of those of its predecessor the Gothic; nevertheless we should hardly be excused for passing over the latter in total silence.

According to the best authorities, there are very few specimens of architecture now in existence in this country which can properly be called Saxon, that is, of a date anterior to the Conquest, and not of Roman origin; and those few are of the rudest and most inferior description. Saxon, therefore, as far as the architecture of this country is concerned, is an improper term. All the ancient structures which are distinguished by the semicircular arch may be called Anglo or Anglo-Norman Gothic. It consists principally of massive columnar piers supporting semicircular arches, similarly arched doors and windows, and arches on small columns in relief, against a dead wall, to ornament it. The pier when round has a rudely foliated or a rounded capital, and generally a moulded base, and is variously ornamented on the surface, being altogether a rude resemblance of the columns of Roman architecture: it is at times polygonal, and sometimes piers consist of clusters of small round shafts. Frequently in doors and windows thin columns with rude capitals and bases receive the mouldings of the arched head; and, when the opening is divided, such columns are placed like mullions, to support the inner arches. There are examples of this style which are quite plain in every particular; but it is generally enriched by deep congeries of mouldings on the arches, which, when there are no columns, run down the jams of doors. These are again frequently carved, and mostly with the zigzag or chevron ornament: grotesque masks, and rude representations of animals, foliage, and flowers, form also common enrichments in Anglo-Gothic architecture.

This style prevailed down to the reign of Henry II. of England, when the pointed arch made its appearance. A degree of impressive grandeur pervades its productions, notwithstanding their clumsiness, arising from the great simplicity of manner and massiveness of proportion by which it is distinguished. The best existing specimens in London are the vestibule of the Inner Temple church, which, moreover, exemplifies the transition; many parts of the church of St Bartholomew in Smithfield, and the chapel of the Tower of London. Exemplifications of the style are also to be found in the interiors of Norwich, Chichester, Gloucester, Canterbury, Worcester, Rochester, Winchester, Durham, Peterborough, Oxford, and Hereford Cathedrals. According to Mr Rickman, the naves of Peterborough and Rochester are the most unmixed specimens. Parts, which are easily distinguished, of the exteriors of many of the same edifices, portions of Lincoln, and the towers of Exeter Cathedrals, Bigod's Tower at Norwich, and the White Tower in the Tower of London, afford characteristic external examples of the Anglo-Gothic style. The most striking castellated remains are

those of Rochester in Kent, Hedingham in Essex, Conisbrough in Yorkshire, and Guildford in Surrey. Many minor edifices, principally ecclesiastical, exist in almost every county in England. Mr Rickman remarks two specimens of this style as peculiarly deserving of attention; the one in the vestibule or entrance of the chapter-house at Bristol, and the other in the staircase leading to the registry of Canterbury Cathedral; the former for its simplicity and beauty of composition, and the latter for its singularity, and as exhibiting a very fine specimen of enrichment. The roofs, or ceilings rather, of the Anglo-Gothic edifices, were mostly of wood; but there are various examples of stone-groined ceilings to be found in crypts, which appertain to this style. Spires were unknown: there are, however, turrets crowned with large pinnacles of a date anterior to the introduction of the pointed arch, as in Rochester Cathedral, and the Church of St Peter in the East at Oxford. Towers were not uncommon; they are square massive structures, rising to no great height above the roof of the buildings to which they are attached. It may be remarked in addition, that many of our ancient structures retain the circular-headed or Anglo-Gothic door, when all the rest has been removed, and replaced by work of a later date.

Architects and antiquaries have generally agreed in dividing Pointed architecture into three styles of three succeeding periods. The first commences with the establishment of the pointed arch, and the formation of the style or manner which accompanies it, in the latter part of the twelfth century, the time of Henry II. of England; the second arose in the beginning of the fourteenth century, in the latter part of the reign of Edward I., and was itself superseded before that century closed, about the time of Richard II., by the third style, which is the latest, for with it, on the introduction of the *Cinquecento*, Pointed architecture ceased to exist. A difficulty arises in appropriately naming these three styles, for on that point there is no degree of accordance among those who are best qualified to be considered as authorities Mr Rickman calls the first the "Early English" style the second the "Decorated English," and the third the "Perpendicular English;" to all of which terms Mr Britton objects, and, without giving appellations, except to the first, which he calls the "Lancet Order of Pointed Architecture," suggests that the second might be named with propriety the "Triangular Arched," and the third the "Obtuse Arched." Objecting strongly to the term "Order," used by Mr Britton, we think with him that the first might be appropriately called the "Lancet Arch" style; but his other distinctions are certainly not more defensible than Mr Rickman's. In the absence, therefore, of unobjectionable distinctive terms, as the varieties arise chronologically, we will speak of them as Periods.

Of the First Period of Pointed Architecture.

Mr Rickman describes this style as being distinguished by pointed arches, and long narrow windows without mullions, and a peculiar ornament, which, from its resemblance to the teeth of a shark, he calls the toothed ornament. There is very considerable variety in the forms and proportions of its different examples, as they retain the massive character of the Anglo-Gothic, or tend to the more florid style of the next period. In the former the sharp lancet arch is found at times, in a series of its narrow windows, with rude piers between them, occupying the place of the precedent large circular-headed opening; and in other places springing from the round columnar piers of the former period. In its more advanced works we find the same long narrow window systematically arranged, singly or triply, with light clustered columns, against the

Pl. XVIII.
Fig. 1, 2,
3, 4, 5, 6.
7, 8, & 9.

Plate XIX.
Fig. 1.

Pl. XVIII.
Fig. 12.

Fig. 11 &
13.

Plate XIX.
Fig. 1.

Pointed
Architec-
ture.

Pl. XVIII.
Fig. 14.
Plate XX.
Fig. 1.

piers which divide them, receiving the deep congeries of mouldings which forms the archivolt. Its columned pier, too, consists of clustered shafts, generally on a round core, and always forming cylindrical masses, girded at different heights with slight rings or belts of mouldings. Their capitals consist for the most part of congeries of mouldings following the form of the shafts, though rich and flowering capitals are not uncommon. Moulded bases, too, are generally used, not dissimilar in form to what is called the attic base of Italian architecture.

The lancet arch is described from two centres about an acute-angled isosceles triangle in the line of its base, with a radius equal to twice and one third (in some cases more, and in some less) the length of that base, or of the span the arch is to embrace. This, though the ordinary, is not, however, the universal form of the arch in the first period; but the absence of mullions, and in general of tracery, may almost be considered a criterion: yet foliations or featherings are not uncommon, especially in doors, and as enrichments to flat surfaces, though every thing of the kind certainly indicates an approach to the style of the succeeding period. Ribs on the angles formed by the intersections of arches in groined ceilings, not in ramified tracery, but with bosses at their apices alone, appertain to works of the first period. These ribs sometimes spring from corbels, and sometimes from the heads of slight shafts, which may run uninterruptedly from the floor to the springing of the arched ceiling, against the walls or against the columnar piers; and a small cornice or tablet continuing round them, runs along horizontally to separate the vertical from the vaulted surface. Buttresses in general, of various forms, sometimes in diminishing stages and sometimes upright, with acutely gabled heads without crockets, but having finials—and flying buttresses in particular—belong to this style. The tablets or cornices, mouldings, ornaments, and the variety and arrangement of niches, must be gathered from examples. The parapet or battlement is straight and uninterrupted, and is either plain or ornamented with series of arches or panels with foliations. Turrets are in some cases square, in others octagonal; but the pinnacles which surmount them are almost always of the latter form, and plain or crocketed, as the work may be more or less ornate. Towers, in the style of this period, were generally made to receive that beautiful characteristic of Pointed architecture, the spire. This, in the best examples, is octagonal in its plan, and of pyramidal elevation, running to a point, or nearly so, under an angle of about 12°, the angle at the base being consequently 84°. In some cases the spire is richly crocketed like the pinnacles; but whether plain or crocketed, it is surmounted by a bold finial.

The most perfect structure in this style throughout is Salisbury Cathedral, which, unlike any other Pointed cathedral in England, except perhaps that of Bath, was begun and finished in the same manner; and so excellent an example is it, that it has been proposed to call the style of the period the Salisbury style. Not inferior in merit, and hardly less perfect a model of the same, is Beverley Minster. That which is of later date in it is easily distinguishable; and being confined to particular parts, it hardly interferes with the unity of the composition. The transepts of York Minster are also of the first period, and so is a great part of Westminster Abbey. The fronts of Ely, Lincoln, and Peterborough Cathedrals exhibit good specimens of it. Indeed there is hardly one of all our Pointed cathedrals which does not partake of this style in a greater or less degree. It will be most generally found interwoven with and superimposing the Anglo-Gothic where that exists, and inferior to, when in connection with, works of a later period. Many of the mo-

Plate
XVIII.
Fig. 1.

nastic structures with which the country abounds present very beautiful specimens of this style also. Among other excellent examples of it may be particularized the chapter-houses of Lincoln and Lichfield. Those beautiful monuments which the affection of Edward I. induced him to raise to the memory of his wife, called the Crosses of Queen Eleanor, are in the style of the first period, though they verge on that of the second, and indeed mark the transition which took place in the latter part of that king's reign.

Pointed
Architec-
ture.

Of the Second Period of Pointed Architecture.

The style of this period, which is thought by many to be the classic age of Pointed Architecture, is described by Mr Rickman as being distinguished " by its large windows, which have pointed arches divided by mullions, and the tracery in flowing lines forming circles, arches, and other figures, not running perpendicularly; its ornaments numerous and very delicately carved." Mr Britton says that " during this period the Pointed style received its greatest improvements;" and that, limiting it to the time of Edward III., " the form of the arch then principally in vogue admitted of an equilateral triangle being precisely inscribed between the crowning point of the arch and its points of springing at the imposts." The mullions of this style clearly result from the slender shafts which were used in that of the first period against the piers dividing a number of windows. The piers being removed, it became necessary that an arch should be turned from side to side, leaving a space to be filled up in the head above the smaller arches. This was done by repeating and continuing their contours, and connecting them by gracefully flowing lines and foliations. It is indeed but an extension of the former; for in some of the early examples the mullions are thin columnar shafts having capitals and bases, and the head of the arch is generally filled up with regular figures, such as foliated circles, leaving spandrels or triangular circular-sided spaces in various parts. It is in the more advanced works of this period that the tracery becomes what may be truly called flowing. The mullion is angular and moulded, and the mouldings run all through the composition; the jamb or architrave mouldings also run through, and for the most part without the intervention of any horizontal mouldings at the impost or springing of the arch. Besides the ordinary covering cornice or drip-stone following the form of the arch, we find a moulded cornice, generally arranged pediment-wise, embracing a window or door, having crockets and finials, and resting on corbels, which are almost always masks. This may be called an attached canopy. The columnar piers of this period are nearly square in plan, and are placed diagonally. They are sometimes composed of clustered shafts, and sometimes of shafts separated by deep hollows. Their capitals are either moulded simply in rather a deep congeries, or with woven foliage under a moulded abacus. Their bases are a diminishing series of bold mouldings, supported generally by a vertical-faced octagonal plinth. The shafts which support the ribs of the roof or ceiling tracery, in the finest examples of this style, spring from rich and bold corbels in the angles of the arches, or the spandrels, immediately above the piers. The groining ribs do not adhere to the angles of the groins merely, but are set more profusely to form tracery; and rich bosses are put at every intersection. Buttresses of the second period are exceedingly various: on quoins they are mostly set diagonally. They either diminish gradually in heights or stories, and finish under the cornice, or they run through and are surmounted by pinnacles. In some cases the sets-off in diminishing are made simply with an inclined shelf; in others every set-off is formed with a pediment properly

Plate XIX.
Fig. 2 & 3

Fig. 4 & 5.
and Plate
XXI.
Fig. 1.

Pl. XX.
Fig. 2.

Pl. XXI.
Fig. 1

enriched, and the face of the buttress is commonly ornamented with blank tracery in panels or in niches. Flying buttresses of this period are also more ornate than those of the preceding; indeed in this they became ornaments, whereas in the former they appear to have been kept out of sight as much as possible. Parapets are either pierced or embattled, and a similar variety is maintained in pediments. Pinnacles are generally square, but they stand diagonally with regard to the turret or buttress on which they are placed, their angles resting on the apices of the pediments which surmount the faces of the substructure. These pinnacles are richly ornamented with crockets and finials. Spires are less common in the more extensive works of this period than in the precedent; but in those of minor importance they are frequent, differing little, however, from the same object in works of the first period, except in being more highly enriched. Towers are richly pinnacled; but the pinnacles rest for the most part on small turrets rising from the angles of the tower itself, and seldom from projecting turrets or from the heads of buttresses, which latter are generally found to die away below the cornice. The details and enrichments of this style are too curious and complicated for verbal description, but they may be gathered from the examples.

We possess no one complete cathedral of the second period, but almost all our larger Pointed structures present specimens of it in a greater or less degree. Excepting perhaps the upper story or belfries of the towers of York Minster, which are of the third period, its west front is a model for this style, and it presents specimens of almost all its external peculiarities. The nave of the same edifice, and the interior of Exeter Cathedral, are perhaps the finest examples of the second period. The latter edifice, indeed, has the reputation of presenting a greater and more pleasing variety of tracery than any other of the same style. To these may be added the cathedrals of Lincoln and Ely, both of which contain much that is valuable. Next to these cathedrals may be placed Beverley Minster, which is not only a mine of beauty of the first, but it presents many exquisite specimens of this period also. The steeple of St Mary's Church, Oxford, is a fine example in this style of the combination of tower and spire. Many minor works in England, and several in Scotland, are excellent; particularly much of what remains of the High Church, Edinburgh, much of the remains of Elgin Cathedral, and the largest portion of those of Melrose Abbey, which, it would appear, was not excelled, when perfect, by any thing in the kingdom.

Plate
XXI.
Fig. 1.

Of the Third Period of Pointed Architecture.

This is that period of the style commonly known as florid Gothic. The first authority quoted with regard to the styles of the two preceding periods calls it the Perpendicular English, and says that this name clearly designates it; " for the mullions of the windows and the ornamental panellings run in perpendicular lines, and form a complete distinction from the last style." Mr Britton, however, insists that the term *perpendicular*, though perhaps proper enough, if the style could be sufficiently distinguished by the mullions of the windows and the upright forms and continuity of the panelling over entire surfaces, " gives no idea of the increased expansion of the windows, nor of the gorgeous fan-like tracery of the vaultings, nor of the heraldic description of the enrichments which peculiarly distinguished this period; neither does it convey any information of the horizontal lines of the door-ways, nor of the embattled transoms of the windows, nor of the vast pendents that constitute such important features in the third division." Although windows with tracery in them may be determined as belonging to this

period, by the perpendicular and parallel lines found in the head or arch, and by the use of transoms to divide the bays into heights, yet the presence of a window of this kind does nothing towards fixing the style of the edifice generally to which it may belong; for in hundreds of cases this sort of window will be found where it is the only specimen of its age or style in the structure. Other points must therefore be attended to.

Plate
XIX.
Fig. 6.

The simpler arches of this style are, like those of the preceding periods, struck from two centres only; the two sides or halves of the arch are similar segments of a circle whose radius in this case is about three fourths the width of the opening. Others are segments of ellipses, and are of course struck from four centres; but some are eccentric curves, which may be drawn, but cannot be described. Many of both the latter descriptions are extremely flat or depressed, the angle at their apex being very obtuse. The ogee or contrasted arch is also found in works of this period, but this is more common in internal tracery than in external form. The modes of arranging tracery must be gathered from examples, for they possess no degree of regularity to render it possible to describe them generally in words. Mullions are richly moulded, and so are the architraves of both doors and windows; the deep congeries of mouldings forming architraves are not intercepted by horizontal or impost mouldings, but run through from the head down the sides or jambs. The angular or pedimented canopy to an arched opening in the style of the second period assumes in this the form of a contrasted arch; it is corbelled and enriched with crockets and finials as in that. Doors, however, in this style are peculiar, because whatever the form of the arched head may be, it is inscribed in a square frame or canopy, the spandrels being variously enriched. Columnar piers of this period are of almost parallelogramic form, thinner in the direction of the arches, and generally plain on the longer sides, but deeply moulded and running to a thin shaft on the outer edges. These mouldings are those which enrich the arch, there being no capital of any kind to intercept them, so that they run, as in windows and doors, all round the opening. To this, however, there are exceptions. The thin shaft which is formed on the outer edge of the pier continues through from floor to ceiling, to receive the groining ribs; and it has a thin congeries of mouldings at either end to form base and capital. The tracery of the ribs of groined ceilings of this period is most profuse, and beyond description intricate. To this also belongs the absurdity called basket groining, in which the arches are made to spring on one of their sides from a pendent mass, which, though rich and gorgeous in appearance, threatens constant ruin. Quoin buttresses standing diagonally are not so common in this as in the preceding style; in form, however, they are not dissimilar, excepting that the sets-off are plain moulded slopes for the most part, instead of having pedimented or triangular vertical heads, as in that. Flying buttresses are, like the style generally, very much enriched, and are very commonly used. Parapets are variously arranged; indeed they embrace almost every peculiarity, being either plain, panelled, pierced, or embattled; each decorated arrangement is effected by different means. Pinnacles in this style are generally square, but there are examples of pinnacles which have a greater number than four sides; in the former and most usual case they are sometimes placed with their sides parallel to those of their pedestals, and sometimes diagonally: they are of course in every case highly enriched with crockets and finials. Spires of this are hardly distinguishable from spires of the preceding period; and towers, of which there are innumerable specimens, may be known by the construction of their buttresses, and by the arrangement of the tracery in the heads

Plate
XXII.
Fig. 1.

Fig. 8 & 11

Plate
XXI.
Fig. 1.
Windows
of towers
Plate
XXII.
Fig. 1.
Fig. 4.

Fig. 8 & 11

Fig. 1.

Pointed
Architec-
ture.

of their windows, as the windows of towers are generally contemporaneous with that story, or stage of it at least, to which they belong. Octagonal or otherwise polygonal turrets at the angles of buildings are not uncommon, and they generally finish with an embattled parapet. The pedestals which support the pinnacles on the angles of towers, and at the heads of buttresses, seldom have pedimented faces, as in the preceding period, but finish with a corbelled battlement, and not unfrequently send up minor turrets and pinnacles from its angles.

In the more ornate works of this style the enrichment of flat surfaces is carried to great excess, and it is generally effected by means of panelling. Niches with their canopies, tabernacles, screens, and stalls, exhibit the most exuberant profusion of ornament, for the most part effected in this manner; but we find, besides, a considerable variety of ornaments, foliate and heraldic; of the former the Tudor flower, which is a combination of the roses, is pleasingly predominant.

Plate
XXII.
Fig. 1.

The only one of the cathedrals entirely of this period is that of Bath; but being generally inferior in merit to many other examples, it need not be cited. Many of the cathedrals, however, have large portions in this style, which can hardly be mistaken if the form of the arches, the arrangement of the tracery, and the mode of enrichment, be attended to. The finest west fronts of the third period to cathedrals are those of Gloucester, Winchester, and Chester; but that of Beverley Minster is by far the most perfect and most classic specimen in existence, not excepting the front of Westminster Hall, which is also of great merit, and presents a classic exemplification of most of the peculiarities of the style. Taken as separate edifices, the chapels of St George at Windsor, of Henry VII. at Westminster, and of King's College at Cambridge, are the most complete, as they are entirely and peculiarly of the third period. The central towers of the archiepiscopal fanes of Canterbury and York, the tower of Gloucester Cathedral, that of Magdalene College, Oxford, Boston Tower, and the tower of St Mary Magdalene at Taunton, are singularly excellent examples of the style. To smaller edifices, those of Wrexham and Gresford in Wales, and of St Neot's in Huntingdonshire, are particularly beautiful. Of steeples, that is, towers having spires superimposed, there are many fine specimens; but the most perfect, perhaps, in composition are those of Bloxham in Oxfordshire, and Louth in Lincolnshire: the former is most admirable rather in general than in detail. Many of the monastic ruins throughout the country present excellent specimens of this style also; indeed it is to ecclesiastical structures we must look for architectural display in Pointed architecture, as in that of the Egyptians, Greeks, and Romans. We have just specimens enough existing of the architecture of the secular structures of our ancestors to show how inferior it was in merit to that of the ecclesiastical; and if the castellated mansions of the nobility, and the palaces of the sovereigns, cannot vie in excellence with the cloistered cells of the monks, we may be well assured that ordinary domestic architecture was still more of an inferior cast.

Elements of Beauty in Architecture.

Simplicity and harmony are the elements of beauty in architecture; simplicity in the general form and arrangement of a subject, and harmony in the collocation and combination of its various parts. Without these qualities a structure can never possess either dignity or grace, and with them it will certainly possess the attractions of both. The outline, then, most conducive to beauty in architecture, is that which bounds the most simple forms. These are the parallelogramic and pyramidal, in which the lines

are straight and uninterrupted throughout their whole length. The ancient monuments of Egypt, of Greece, and of Rome, offer the most complete exemplifications of this. No other than the long, unbroken line which bounds the temples of Egypt could produce an effect so grand; and no other than the simple, square, and pyramidal forms, could be productive of so much dignity as they possess. In the pyramids and obelisks of the same country the effect of this simplicity is even more obvious. In the temples of Greece, again, the same dignified simplicity is still predominant; for although in them the parallelogram and pyramid are combined, they are not confused; their mass consisting of a parallelopipedon whose ends are surmounted by vertically faced pyramids, connected by an unbroken line of ridge running parallel to the horizontal boundaries of the sides. Those of the Roman monuments which are deficient in simplicity are also deficient in beauty. Such are the triumphal arches, whose general form is broken by columns and arches which subject themselves to no commanding outline, but are all at the same time prominent features of and excrescences from the general composition. In the temples which are on the Greek model it is not so; nor is it so in the long series of arches in the Roman aqueducts, which are crowned and connected by commanding lines, unimpeded by projections or protuberances of any kind. The crucial form of the Pointed cathedral may be thought to detract somewhat from its simplicity, and so much from its beauty; but it is an aggregation of simple forms, perfectly coherent with the tendency of the leading lines in the style, which, we have seen, is vertical; and the lines are therefore not broken by the projected masses of the transepts, as they would be in the Egyptian and other styles, the tendency of whose commanding lines is horizontal. Otherwise the Pointed cathedral is a modification merely of the form of a Greek temple, with other parallelogramic forms added to it, as towers, or pyramidal, as spires. The same principle will be found to pervade the best works of the Italian school, more or less modified according to its application.

Next to the straight line is the circular; but the greater complexity of this latter, and the variety of which it is capable, render it more subtile, and for the most part less competent to produce grand and impressive effects, except under peculiar circumstances of situation and combination. A cupola such as the cupolas of St Peter's at Rome and St Paul's in London, if placed on its base on the ground, or even on a low structure, like a large beehive, would be not merely ineffective, but absolutely ugly; and if, in the situations they occupy, the cupolas referred to were without the diminishing pinnacles above them, to bring their general outlines within that of the pyramid, it is a question whether they would possess the attractive beauty they now do. If St Paul's be looked at in the gray twilight of morning or evening, or when a mist renders its form indistinct, the impression conveyed by the mass is that of a lofty pyramid or cone, rising out of the substruction which the cathedral forms, and running off to a point in the sky. The superstructure of St Peter's is, as we have seen, more depressed, and less perfectly formed in this particular; yet nevertheless it may be submitted to the same test, and the same or nearly the same result will follow. Furthermore, let a hemisphere or an oblate hemispheroid be supposed in the place of the prolate hemispheroid, as at present, and this reasoning will be rendered more clear; for neither of those forms, even with the accessories these possess, would be as beautiful; and without them they would be ungainly deformities, as is proved by that example on the new palace in London, on the site of Buckingham House.

Beauty in
Architec-
ture.

Plate II.

Pl. XXVII.

Plate XI.
Fig. 10 &
11.

Plate XVI.

The cupola of University College, London, exemplifies this point also; for though its profile is elegant, and its accessories are generally good, the composition does not resolve itself into a simple form, and the result is far from being beautiful.

When the circular form is employed cylindrically, the utmost simplicity is required to be preserved in its horizontal, as well as in its vertical lines, or the result will be totally devoid of all architectural beauty. In proof of this, let the broken and dentilled columnar ordinance which surrounds the tholobate of St Peter's be compared with the noble, unbroken peristyle in the corresponding part of St Paul's. In the former the cylindrical mass is studded with a series of minute excrescences of coupled columns; and in the latter it forms a grand, beautiful, and effective compartment of the composition.

The preceding remarks do not of course apply to the interior of a structure in the same manner; for although as high a degree of simplicity is required internally as externally, similar combinations are not necessary, nor are they indeed always available. A spacious concave, of whatever form its profile may be, so that its plan be a perfect circle, is one of the grandest works of architecture, and at the same time one of the most simple, whether it occupy a compartment of the structure to which it belongs, as in St Peter's and St Paul's, or cover the complete edifice to which it appertains, as in the Pantheon at Rome. In such situations it is indeed almost impossible to destroy its inherent simplicity; and being unconnected with external circumstances, it requires no coherence with any thing else, being as independent of its substructure as of its external contour for effect. Irregular and intricate forms, however, in works of architecture, whether internally or externally, will be found unpleasing. Few can admire the external effect of the Pantheon, or of the structure in London called the Colosseum, which has been subjected to the same arrangement, though certain features in both may be indisputably good. To these may be added the church in Langham Place, London, and indeed many others; but that is an egregious example in point. The complication of straight and circular in their composition, and the consequent irregular forms and undefined outlines, totally destroy both simplicity and harmony. The comparison of an Egyptian obelisk with a monumental column of the same relative size will afford the strongest proof of the superiority the more simple form possesses over the more complicate. None, however, but those who have visited Rome, in which city alone the comparison can properly be made, can duly appreciate this evidence; but London furnishes a contrast almost as much to the purpose, in the monument on Fish Street Hill, and the lofty shot tower by the south-west abutment of Waterloo Bridge. They are both of nearly cylindrical form; but the one is crowned by a square abacus, and the other by a bold cornice, which follows its own outline. The greater simplicity and consequent beauty of the latter, having regard to general form alone, are strikingly obvious.

Not only in general form and outline is simplicity necessary to beauty in architecture, but also in all its details, and even in its enrichments. In exemplification of this, a plain Greek entablature may be compared with one in the Roman style, in which every thing is sacrificed to profuse ornament; and the style of ornament in the latter may again with equal advantage be compared with that of the age of Louis XIV. of France. In the arrangement of the parts of a composition, as well as in the composition itself, simplicity is essentially necessary to the beauty of the whole; every style will afford exemplifications of this also, in the comparison of the more simple with the more complicate specimens of the same. Compare the

few simple and well-defined parts of a Grecian Ionic entablature with a Roman or Italian example of that order in the latter will be found a complexity and straining at effect not at all consistent with beauty and dignity, determining the comparison much in favour of the former; and so in many other cases which might be cited. That the more simple arrangement of columns at equal distances is superior to that in which they are coupled or placed only alternately equidistant, is clear from the fact that the latter mode was first proposed, and is only used to obviate difficulties, and not from choice, except in the works of the merest pretenders.

Harmony, concord, or fitness—of proportion, of form, of one part of a composition to another, and in the collocation of the various enrichments which architecture requires,—is as necessary to its beauty as simplicity. We do not speak of the agreement which should exist between the manner or character of a structure and its application, for that is purely conventional, and totally independent of any architectural consideration. The merit or demerit of a composition is not at all affected by the use to which the edifice is applied; neither would its front be more tolerable, nor its cupola less beautiful, if St Peter's in Rome were, by the course of events, to become a democratic forum instead of a papal basilica; nor is the monument of London a more or less elegant object, whether it be understood to record a triumph or a defeat—the burning of the city, or its re-edification. Harmony in architecture is that agreement which exists between its various parts, as in the relation of a column to its entablature and stylobate, in the accordance of a cornice with the elevation it crowns, and in the coherence of one part of a composition with another. It is that which exists in the common tendency of the leading lines of a structure; and it is that which blends the straight and circular in enrichment or decoration, as in the capital of an Ionic column whose square and horizontal form is harmoniously adapted to the vertical lines and cylindrical form of the shaft, by the intervention of the volutes. An inharmonious combination arises out of the collocation of the same voluted capital with a pilaster or squared pier. This quality requires a judicious arrangement of ornament. That a certain degree of enrichment should pervade the whole of a composition, and not be confined to one part of it—for instance a Corinthian ordinance, in which the columns are unfluted and the entablature is quite plain—is inharmonious; for the capitals being masses of rich foliage, are spots, having nothing to connect them with the rest. A degree of harmony must exist, too, between the solids and vacuities of an edifice. An Italian portico, with its thin and straggling columns, is an inharmonious object, for it conveys an idea of infirmity and poverty, which is not the case with one proportioned like the best Greek and Roman examples. In the front of a house, windows and the piers between them being too wide or too narrow will affect its character in this respect. The comparative size of various portions of the same composition, though they be in themselves simple and harmonious, may be such that they shall not be so in combination. The portico of University College, London, is of great extent and considerable beauty, and the cupola behind and above it is of elegant form, though deficient in another particular, as we have already stated; yet they do not harmonize—the one is much too large for the other, and their forms are incoherent.

Thus harmony has reference to comparative magnitude, strength, decoration, disposition, and proportion. To acquire a knowledge of all these sufficient to produce a worthy result, a long course of study and careful observation are necessary: but such can only be necessary to the architect; it is enough for the general student to be

Beauty in
Architec-
ture.

Plate X.
Ex. 3.
Plate XIII
Fig. 4.

Plate XI.
Fig. 4.

Figs. 1 and
2.

Plate XVI.
Fig. 5.

Composi-
tion.

able to appreciate them when present, and to detect their absence.

PRINCIPLES OF ARCHITECTURAL COMPOSITION.

These must be different in the widely differing species of architecture, whose tendencies in the one are to horizontal or depressed, and in the other to vertical or upright lines and forms; the former including all those varieties which proceed from the Greek and Roman modes of design, or columnar and circular-arched architecture; and the latter embracing those which arise out of the pointed arch, and which we have distinguished by the term Pointed. Except in the elements of architectural beauty, which must be the same in all architectural works, there is no similarity whatever between the principles which govern composition in the two species. Simplicity of form, and harmony between the parts, are as essentially necessary to the one as to the other; but instead of the leading horizontal lines required by the former, the latter is distinguished by the absence of commanding lines having that tendency, and by the presence of strongly marked lateral projections and vertically inclined lines. The rectangular figure formed by the front of a Greek temple, below the pediment, rests on one of its longer sides as a base. In a Pointed composition that order is reversed, and one of the shorter sides becomes the base; and the pediment, instead of being a depressed obtuse-angled triangle, becomes upright and acute-angled; the whole mass, moreover, follows the change thus described, so that the same figure, a parallelopiped, is set for horizontal or vertical composition, as a larger or smaller side is made the base. This being the case, it will be necessary to treat of them separately; for rules which apply to the one are totally inapplicable to the other, and the former, being of most common application, may be taken first. We shall quote the principles which appear to have actuated the Greek and Roman architects in the production of their best works, or rather the principles which those works develope, instead of citing all existing ancient works as authorities; and determine on those principles how to produce similar results in cases of which examples do not appear in ancient practice. In the same manner, we must deduce the principles for general composition in the Pointed style, from those which appear to enter into its best existing works.

Of Horizontal Composition.

Pl. XVII.
Fig. 1.
Pl. XVI.
Fig. 2.

Every thing tending to break the continuity of the leading horizontal lines in a composition should be avoided. The advantage of adhering to this, and the disadvantage resulting from the breach of it, are clearly exemplified in the front of the Farnese Palace, and in the flank of St Peter's at Rome. In London, too, the fronts of the Banqueting House at Whitehall, and of Somerset House to the Strand, offer similar exemplifications of the principle; the former having both the entablatures and the stylobate of the upper ordinance broken round every column, which makes the ordinances mere excrescences, and the latter preserving the leading lines continuous and unbroken throughout, to the manifest advantage of the whole composition. This applies equally to columned and arcaded ordinances, and to compositions in which neither is used; and it is as much opposed to the projection of masses to form wings and centres, whether shallow or deep, as to the breaking of an entablature or stylobate round one or two columns. Sufficient variety of light and shadow is attainable without the use of columnar ordinances at all, as the Farnese Palace evinces. But if, however, it be required to give a greater degree of importance to an elevation than can

Composi-
tion.

be attained in that manner, it may be produced without either attaching or insulating columns the whole extent, by means of antæ and recessed compartments with columns in them, as on either side of the gates in the north or Lothbury front of the Bank of England, and on the flanks of the churches of St Pancras and St Martin in London; but a mere pilastraded ordinance, or pilasters with an entablature and without columns, is bald, tasteless, and unmeaning, as the front of Crockford's Club-house, in London also, very clearly shows. In speaking of the Italian style, we have shown the injudiciousness of putting order above order, because of the impossibility of maintaining a rational arrangement with regard to diminution and intercolumniation. We made that, too, an objection to the elevation of the Roman Colosseum; but the practice is more over objectionable, because of the repetition of the similar parallel lines of the entablatures, similarly projected too, which destroy the breadth a composition should possess; and because the upper and crowning cornice, if in proportion to its own ordinance, must be disproportioned to the whole elevation which takes from that member a character of grandeur or meanness, as it may or may not be fitted to its whole height. This is made very evident by the opposed fronts of the United Service and Athenæum Club-houses in London, the former of which is finished by the thin shelf-like cornice of a second order, and the latter by a crowning cornice which has some relation to the whole height of the building. In a similar manner, and for the same reason, the practice of raising lofty basements to support columnar ordinances is injudicious; and this detracts much from the merit of the front of Somerset House just referred to, by making the crowning cornice of less importance than it should be. In St Paul's this fault is partially relieved by the somewhat exaggerated size of the cornice of the upper order, and by the insertion of cut blocks, in the manner of upright modillions, under it the whole depth of the frieze. Nothing, again, should be allowed to superimpose a crowning cornice, except what may form a part of itself, as antefixæ; where, however, something is absolutely necessary, as on a bridge, a close simple parapet, as low as it may be conveniently, should be resorted to. On the principle first developed, porticoes should not be projected from the front of a building, unless they occupy the whole extent of it, as in a Greek or Roman temple, and so carry the horizontal lines unbroken to the flanks; or they should be made distinct and independent objects, to which the rest of the composition may be subservient, as in University College. A portico should moreover be considerably projected, or the surface behind it recessed, that the columns may have a background of shadow, otherwise it will be poor and inefficient. Of this the Greek temples offer a favourable exemplification; and most of our churches, and other modern edifices which have porticoes to them, prove the correctness of the principle in the breach of it. Exceptions more or less favourable certainly exist, whose superior merit is sufficient to indicate them. A pediment should never be used unless it is made to embrace the whole of the end or front to which it is attached. Numberless absurdities have arisen in Italian architecture from the injudicious application of this form; so general, indeed, is it, that the fact of a pediment existing under any circumstance in a work of that style is almost a sufficient reason for avoiding a similar use of it. Nothing is more difficult than to combine straight and circular, or otherwise bending lines, with propriety and good taste, and therefore their collocation, in general composition particularly, should be seldom attempted. It is when they harshly contrast, as in circular pediments, and in mixed compositions of columns or pilasters, with their accessories, and arches and

Plate XI.
Fig. 1.

their piers, that the combination is bad; but not so in the connection of the arch with its pier, so that the former be semicircular or semielliptical, and not smaller segments, in which cases they fall naturally and gracefully together. The incoherence and inelegance of contrasted straight and circular forms are very evident in the New Exchange at Paris, where two tiers of circular-headed windows are seen within a Corinthian peristyle. Circular prostyles or cyrtoprostyles should be avoided, as their horizontal lines cannot be made to harmonize perfectly with any form to which they may be attached. This however does not apply to peristyles; and both the one and the other are exemplified by the transept porticoes and columned tholobate of St Paul's. The use of coupled columns is so absurd, and they are confessedly so inelegant, that it seems almost unnecessary to denounce them. Suppose apertures, such as windows, arranged in couples throughout an elevation, with very narrow and very wide piers alternating, and both the absurdity and the inelegance become manifest: now, neither the one nor the other can be either lessened or changed by reversing the case, and putting alternately wide and narrow openings, as in coupled columnar ordinances. Columns may with propriety be put further apart when they are attached than when they are insulated, because the entablature, resting in part on the wall, is neither in fact nor in appearance made infirm by the distension, as it would be if it rested on the columns alone. All the parts of the same edifice which come into view, under any circumstances at the same time, should correspond; but insulated and attached columns of the same ordinance and in the same elevation may, under certain circumstances, without impropriety be arranged with a different intercolumniation.

An arcaded ordinance should be considered as only more massive than, and differently shaped from, a columnar, and may therefore be governed by nearly the same principles. A pier is but a differently shaped and more massive column, and the archivolt but a succedaneum for the architrave; while a bold blocking course, or a commensurate cornice and frieze, as the composition may be more or less ornate, will complete the ordinance. Under this view nothing can be more absurd than to affix columns or pilasters to the piers of an arcade to support an entablature, and certainly nothing can be more inharmonious, from the contrast which arises, as we have just remarked, between the rectangular lines of the latter, and the inscribed circular lines of the arch, as well as the incongruity necessarily attending the interspaces of the columns.

In speaking of Greek and Roman architecture, we have shown why columns should, and why antæ and pilasters should not, be fluted; and have shown also, that a certain degree of richness or plainness of surface should pervade a composition, and not be confined to particular parts of it. It will now be enough to add, that in composing, lights and shadows should not be scattered on a surface as they are on the front of the Banqueting House, by broken ordinances; nor should either be too much narrowed, as the light on the corona of a Roman cornice too frequently is, by the too great projection of the cymatium. It will be found, moreover, that shadows projected horizontally are more in coherence with the horizontal style of composition, than those which fall laterally, or from a vertically projecting object.

Columns, &c.—The proportions of the columnar orders will be best sought in the existing examples of the ancients; and those we give of them afford sufficient variety. What is deficient in one may be made up from another; and what appears superfluous in one example may be omitted, as its omission may appear beneficially to affect another. The Doric may be adopted from the

Parthenon or the temple of Theseus, as the best existing models of the order. If an ungraduated stylobate be used, which should be avoided if possible, it should not exceed one diameter in height. The intercolumniation should not exceed one triglyph, as in the Greek temples, though for compositions of a generally less dignified character it may, perhaps, be extended to two. A good modern example of the Doric order, in a work of the latter description, may be seen in the small entrance portico to the University Club-house in London. Plate IV. & V The Ionic example Plate VI. from the Erechtheum, which we have given, may be used as a model for that order, with the same restriction with regard to the stylobate which is made to the Doric. Additional depth may with advantage be allowed to the bed-mould of the cornice, and it may be effected by the insertion of a dentilled member, which indeed some of the ancient Greek (though not Athenian) examples possess. The intercolumniation should not be less than one diameter and a half, nor should it exceed two diameters. In London this order is admirably applied in the front of an Episcopal chapel on the east side of North Audley Street; and this particular example is very correctly copied on the exterior of the church of St Pancras. The great inferiority of the Roman examples of the Doric and Plate X. Ex. 3 & 4. Ionic orders is too evident to require that what it consists in should be pointed out, and they are the models of the Italian. The Greek example of the Corinthian order Plate VII. Fig. 3. might perhaps be improved by making the dentil member of the cornice a little shallower, by projecting the corona rather less, and by correcting the form of some of the mouldings of the entablature generally. If the columns be used in a prostyle or other position to insulate them, they may with advantage be made half a diameter less in height; and the intercolumniation should be made less than it appears in the original, where the columns are attached. This example has been well executed in the entrance to the Philadelpheion or Exeter Hall, in the Strand; but the pedestals and the attic are blemishes in the composition. Of the Roman examples of this order, that of the temple Plate IX. Ex. 11. of Jupiter Stator is certainly the best. Its greatest fault is the too great magnitude of the cornice, of which every member, except the corona, might advantageously be restricted one tenth of its height; that which is dentilled might indeed be reduced one fifth. The projections might also be diminished in the same proportion, removing the greater diminution of one fifth in this particular from the dentilled member to the cymatium, and the ovalo under it, both of which project by far too much. The three fascias of the architrave are too unequally divided. The lowest might be made as wide as the middle one, by deducting their difference from the third or upper one. In Ex. 2. the Tivoli example the architrave is too shallow, and so are the dentil band and corona of the cornice; and the cymatium is both too deep and too much projected. The cornice would moreover be improved by denticulating the dentil band, and by enriching the frieze with an ornament less coarse and less massive. If this example be used in a generally ornate composition, some of the mouldings of the entablature should be enriched. The parts of the Ex. 3. entablature of the temple of Antoninus and Faustina are well proportioned to each other. The cornice of this example would be improved by giving additional height to the dentil band at the expense of the moulding above it, and by denticulating it also. The cymatium is rather too shallow, and might be widened out of the moulding under it; and both should be restricted in their projection at least one fifth. The capital of this example is poor, and its abacus is too shallow. The shaft requires fluting, and one half the depth of the upper fillet of the base might be added with advantage to the tori and scotia. The cited example from the portico of the Pantheon has, like the last mentioned, the parts of its entablature well proportioned to each other. As in the Jupiter Stator example, the architrave should be more equally divided. The mouldings, too, separating the fascias, should be made less; and the superior moulding, at least of the architrave, carved, unless the frieze were enriched, and then it would not be necessary. In the cornice a fifth or sixth should be taken from every member of the bed-mould and added to the corona. In the presence of modillions, however, the dentil band is judiciously kept plain, though the moulding below it would be better if enriched. The capital of this example is as faulty as that of Antoninus and Faustina, and in the same particulars. The shaft also requires fluting, and the base might with advantage be made to spread more.

Plate X.
Ex. 1.

The ordinance of the temple of Mars Ultor, though the most masculine, is, from its good proportions, and the bold character of its foliage, one of the most excellent of the Roman Corinthian examples. Most of the entablature being supplied from a not well authenticated source, may not be original; but that is of no consequence, if it be beautiful. The corona, like that member in most Roman entablatures, wants greater depth; and the cymatium perhaps less, and certainly less projection. In this, as in the first-mentioned Roman example, with modillions there are dentils. This is injudicious; the member would be better plain, as in the Pantheon ordinance. The architrave, which is authentic, is exceedingly well proportioned, and the column is fine in all its parts. These examples all vary in their intercolumniation, from rather less than one diameter and a half to a fraction more than two diameters, beyond which proportions, either less or more, it would not be well to go. A stylobate to the order might judiciously be adapted from the Greek; for the stilted effects produced by insulated pedestals, and even by continuous vertical stylobates, are injurious to the general appearance of a columnar composition; and the thin steps in common use detract exceedingly from its beauty under any circumstances.

There are many varieties of the foliated capital which may be used with advantage; one of the least elegant, however, is that which assumes the distinction of being called the Composite order. The example of it from the

arch of Titus is one of the best, if not the best; but it will be seen, on comparison, to be strikingly inferior to the Corinthian examples, or those in which the volutes of the capital are made subservient to the foliage, instead of being distended into huge mis-shapen knobs. The entablature, too, is only an exaggerated Corinthian. If it be wished to use foliated capitals differently composed from the ordinary, it may be well to preserve the character and proportions of the entablature the same, or nearly so. Under any circumstances, however, care should be taken in composing an entablature, that it have sufficient height, and yet not be too heavy; that it be sufficiently divided, and yet not frittered; that the parts have sufficient breadth, and be not so much projected as to bury all that is below them in shadow; and that ornament be properly distributed, and in sufficient quantity, without overloading the composition with it, as in the ordinance of the arch of Titus.

If again it be wished, under any circumstances (though

Plate VII.
Fig. 4, 5,
& 6.

the practice cannot be recommended), to use human figures as columns, there appears to be no reason why the entablature should be executed without a frieze, as it is in the example of the Pandroseum; and if a frieze be inserted, it should be by lessening the other parts, and not by increasing the whole, as that entablature (taking it as a model) is quite deep enough in proportion to the height of the ordinance.

Entasis in columns need not be regarded, unless they exceed eighteen or twenty feet in height; but it adds much to their beauty, and should not be neglected when they are above that magnitude. No rule can be given for its production, but it may be thus described. The shaft, instead of being the frustum of a regular cone, is the frustum of a cone whose outline is not straight, but slightly convex; so that if it were perfect, its vertical section would have the form of a very acutely pointed arch. This convexity should, however, be so slight as in the finished shaft to be hardly distinguishable. Its abuse is evident in the columns of the east front of the church of St Paul, Covent Garden, and indeed in some of the less esteemed works of the Greeks themselves. The modes of fluting in the different orders may be gathered from the examples. The flutes should be deeper or shallower, as the collocation of the ordinance may require a greater or less depth of shadow on the surface of the columns. The elliptical or nearly elliptical contour seems to be the most generally pleasing. The flutes meet in an arris on columns of the Doric order, and are separated from each other by alternating fillets in the Ionic and Corinthian.

Antæ and Pilasters.—These should seldom be used, externally at least, unless with columns, for their real use is to connect a columnar ordinance with the walls to which it is attached; and being, as they are, but slight projections from walls for that purpose, nothing can be more absurd than to give them the features of columns, either by the application to them of similar capitals and bases, by diminishing, or by fluting. The use of antæ was rightly understood by the Greeks, but not by the Romans; and their proper use may be seen in the works of the former. The examples in London of their judicious application, most worthy of remark, are in those edifices already mentioned as exhibiting good specimens of the Greek orders, in the Bank of England, and in the portico only of University College. The adaptation in these of other than the bold foliage and branching cauliculi to the columnar capitals in the Corinthian ordinances to the antæ caps is particularly worthy of notice (though they are not all of equal merit as compositions), as the Greek remains are without a regular example of Corinthian antæ, and the Roman practice is inelegant.

Pediments.—As there is no mode by which the pitch of a pediment can be determined, it must be left to the taste of the designer to be governed or not by the examples of Greek and Roman antiquity: it may, however, be premised of them generally, that those of the former school are too flat, and those of the latter too steep. The pediment of the portico last above referred to is admirably proportioned to the rest of the composition, but its pitch would be absurdly flat if applied to a tetrastyle portico. The inclined sides of a pediment are covered by a cornice similar to that which forms its base, except that all blocks, modillions, and dentils are omitted, even if the bed-mould itself be retained, and a cymatium superadded.

Cornices, &c.—Although a perfect entablature should not be applied to crown an edifice, except it be in connection with columns of some sort, or their legitimate representatives, piers, yet a single cornice, or a cornice and frieze, is not so; and it forms the most pleasing termination to an elevation in which columns are not used. The proportion of one or the other may be best found by setting out a columnar ordinance of the style preferred at the height of the elevation; and the size of the cornice or cornice and frieze thus given will aptly become it. The Vignolan or block cornice, in which the frieze is occupied by cut blocks, is exceedingly effective: it is this which Sir Christopher Wren has employed in the upper entablature of St Paul's, and Vignola himself in the front

Pl. XVII.
Fig. 2.

Pl. XIII.

of the Villa Giulia. With these cornices rustic quoins consort very pleasingly, and so they do indeed with all single cornices which are of a bold character, and all such should be so.

Arcades, &c.—The most graceful average proportion for these is, that the opening be twice the width of the pier, and twice its own width in height to the crown of the arch. The practice of the Italian school in the composition of arcaded ordinances may be generally followed with advantage, except in mingling and confusing them with columnar. The pier is based by a deep square plinth, and surmounted by a square or moulded cap or impost, the upper surface of which is the base line of the arch. In rusticated work the radiating stones of the arch show their joints, and are cut to a uniform appearance with the ordinary surface of the wall. In other cases there is a moulded archivolt, whose width varies from an eighth to a tenth of the opening of the arch. A dropping keystone is generally used; but this very much injures the simplicity, and consequently the beauty of the arch, and should be avoided.

Doors and Windows, &c.—The most approved proportion for these apertures, also, is twice their width in height. In an elevation which comprises several tiers or stories, it is customary to make those of the lowest or ground story rather less than that proportion in height; those of the first or principal story rather more; those of the second somewhat less again; and those of the third (if there be so many) square or even lower. If, however, the elevation consist of but two, the ground story should be the principal, and its windows of the most importance (if any difference be made between them at all), those of the upper story being then less than the stated proportion in height. The modes of ornamenting doors and windows are so various, and they depend so much on the coherent parts of the composition, that it is impossible here to go into their varieties, or to give particular instructions for their adaptation. The practice of the Italian school may in this case also be generally followed, avoiding those things in it which are injurious, and referring to the Greek for the details of mouldings and ornament. The application of a columnar ordinance to every door or window, giving it the effect of a little edifice in relief, exemplified by the windows of the principal story of the Farnese Palace, must be censured as injudicious; and so must pediments of all kinds, but particularly those formed with circular lines, or lines twisted in any way, or, though right lined, not meeting in a point at the apex. In basements or ground stories windows or doors may be lined with rustic courses with good effect, though the face of the wall be not rusticated; and if it be so, no other lining is thought necessary. The windows of a principal story may be lined with an architrave, either quite straight or returning in knees at the head, and resting on a continuous blocking course below. This architrave may be surmounted by an enriched frieze and cornice, the former bounded at the ends, and the latter upborne by trusses or consoles, which may rest on or be affixed to a species of pilaster, outside the architrave, and parallel to it; if detached sills are preferred, a shorter and bolder truss may be judiciously applied below the sill, under the foot of each pilaster, to complete the composition: the architrave is generally a sixth or a seventh of the opening in width, and the console and its pilaster about a ninth or tenth. Upon no account should rustics be run through the architrave lining of a window, as on the flanks of St Martin's Church in London. A series of circular-headed windows conjoined, as in the earlier works of the Venetian school, is productive of a pleasing effect; but the large circular-headed, with two conjoined smaller rectangular windows, found in

the later works of the Italian school, and called Venetian, is radically inelegant; and there is such a one in the east end of the structure last mentioned. Blank windows should be recurred to as seldom as possible; and when they cannot be avoided, they should have sash-frames and sashes as if they were real windows, otherwise they give a maimed effect to an elevation.

Niches.—There are very few cases in which these do not act injuriously on a composition, from the difficulty of making them cohere with the other parts: the usual mode in Italian practice is to give them the effect of windows, which cannot be approved of. Internally they may be used with much better effect than on exteriors. If a niche is intended to receive a statue, it should have a circular head; if a vase, it will perhaps be better straight: the plan of a niche is semicircular.

Parapets.—The pierced parapet or balustrade is not inelegant when the forms of which it is composed are simple and chaste, as piers; but the close continuous parapet is generally preferable, because of its greater simplicity, and its accordance with the principles developed in the most classic works of architecture. The parapet of a projected balcony, to give an appearance of lightness, may perhaps be better pierced; but if a stereobate continue straight through a window without projection, it should remain close and uncut, unless there exist some special reason for wishing to make the window appear so much higher.

Balconies.—These, whether continuous or broken to every window, act for the most part injuriously in a composition. In the former case they cannot be kept sufficiently under not to appear of too much importance; and in the latter they have the effect of a broken cornice or entablature. In both cases, when a balcony is above the eye, it destroys the proportion of the windows opening on it, by intercepting more or less of their height.

Proportion and Arrangement of Rooms.—Whatever the length of a room may be, it will not be disagreeably proportioned if its height and breadth are the same; and if the length may be limited, once and a half the breadth is the most pleasing. Galleries, of course, will be much longer than that proportion; and corridors will necessarily be narrower than they are high. Entrance-halls should be cubical, regularly polygonal, or circular. Access should be given to a room by the end; it should be lighted on one side, and the fire-place may be at the other end, or on the other side: if the former, there should be two doors, or one and the appearance of another, that the fire-place may not be immediately opposite to a door. Many things, however, from locality and otherwise, frequently occur to make it practically impossible to attend to such suggestions as these. In halls and saloons not commanding a pleasing view, the windows may be advantageously placed above the usual level, for agreeable effect, for light, and for ventilation. In rooms lighted from above, as the Pantheon in Rome is, a columnar ordinance may be judiciously applied; but otherwise columns and their accessories can seldom be well disposed internally.

Chimneys.—If a chimney be in the end of a room, it should be similarly proportioned, the height and breadth of its opening corresponding with the height and breadth of the room; if it be on a side, it should be somewhat wider than it is high; if the room be longer than the sesqialteral proportion, it should have two fire-places, either at the two ends or equidistant from the centre of one of the two sides. The chimney-piece should be bold and massive, not frittered into small parts and much moulded; it may, however, have its vertical faces enriched with great advantage.

Ceilings.—The ceiling of a room should be nearly plain,

but it may rest on a bold and enriched cornice, not composed like an external cornice, as it is differently lighted, but with deep covings instead of broad flat surfaces. Such cornices are highly susceptible of ornament, and they may have additional effect given to them by means of colour. In large rooms the area of the ceiling may be pleasingly contracted, and so made to appear lighter, by coving the angles altogether, and thus bringing the cornice on which it rests lower down on the walls. This mode of arrangement is used, too, in the small rooms of a lofty story, to take off from their too great height. The horizontal surface of a ceiling may be treated like a large panel, with broad borders and slight sinkings; or, if it be very large and lofty, coffering or panelling all over, with moulded or painted ornaments, will produce an agreeable effect. Domed ceilings should be coffered, especially when they are lighted from above; but if the light be from below, as in St Paul's and St Peter's Cathedrals, ribbing is far better. Heavy cumbrous masses of foliage in a ceiling should be avoided; frets, guiloches, and arabesque ornaments, are the best suited enrichments for a ceiling on which ornament is necessary.

Stairs.—In a structure whose principal apartments are on the ground floor, the staircase is a secondary consideration, and should be secluded; but where they are above the level of the entrance door, it becomes an important part of the interior, and should be of immediate and easy access. The rise of a step should not be more than six inches, and the tread not less than twelve. In a square staircase winders should not be used; and in no case should there be more than ten or twelve flyers without a quarter or half space, both to prevent fatigue in ascending, and to avoid even the appearance of danger in the descent. Winding staircases are less convenient and less pleasing in effect than those which are square and without winders. Much room may be saved, however, where it is of consequence, by using the former. Handrails should follow the character of the staircases to which they are attached; but a somewhat square form, with the sides or edges moulded, should be given to them under all circumstances, because of its simplicity, as well as the greater degree of firmness or solidity which the whole composition derives from it, both in effect and in appearance, than can be acquired for it otherwise. The handrail and balusters of an in-door staircase are indeed but the parapet of an external flight of steps, or of a terrace, executed with more lightness and a greater degree of delicacy because of their location. The balustrading, also, should therefore be characterized by boldness and simplicity, though it is indeed a difficult thing to compose with propriety, because of its inclination, and the want of parallelism between the graduating base formed by the ends of the steps and the hanging level of the coping or handrail. The first step of a staircase has a voluted or curtail end (or ends if it be insulated, as in a staircase with a double returning flight) supporting a column or newel, on which the voluted or scrolled end of the handrail rests. The steps of a staircase are wrought with moulded nosings, which are returned at the exposed ends; the under surface is either cut straight and parallel to the inclination of the flight, or moulded to form a sightly object when seen from below.

Mouldings and Ornament.—The Greek examples offer the most beautiful forms for mouldings, and the Grecian mode of enriching them is unsurpassed for beauty and efficiency. By adhering to them, and observing the manner in which they are produced and combined, it will not be difficult to produce and combine mouldings in sufficient variety for every purpose.

For ornament the Roman examples may vie with the

Greek; but in composing or adapting, it is necessary to avoid alike the tendency to too great luxuriance in the one, and to poverty in the other. The remains of Herculaneum and Pompeii have furnished us with a great deal of ornament that is new and beautiful; and much that is excellent may be found on the earlier architectural and sculptural monuments of Italy of the middle ages.

It should nevertheless be always borne in mind that the object in architectural enrichment is not to show the ornament, but to enrich the surface, by producing an effective and pleasing variety of light and shade; but still, although the ornament should be a secondary consideration, it will develope itself, and should therefore be of elegant form and composition, as well as the means of producing a good effect on the surfaces to which it is applied.

Of Vertical or Pointed Composition.

The towers of Westminster Abbey are an excellent practical illustration of the essential difference which exists between the horizontal and vertical styles of architectural composition. In general form they belong to the Pointed style, and in so far cohere with the structure generally; but the running lines of the buttresses, if their quoin piers may be so called, are constantly intercepted by transverse cornices; and all the details are strangely in discordance with the character derived from the pointed arch.

Buttresses in a Pointed composition must not be considered simply as abutments to the arches and aids to the walls of a structure, any more than a cornice in horizontal composition may be thought only necessary to cover or protect the wall on which it rests. That these were the uses for which they were severally applied originally, cannot be doubted; but although such may be their purposes, we must now consider them as aids to architectural effect. Buttresses, then, are of the same use in the vertical style that cornices are in the horizontal—to give character to an elevation, by throwing a mass of shadow, to relieve it of the monotony necessarily attendant on a flat surface, however it may be pierced or enriched. The sides of the buttresses should be either quite perpendicular the whole height they have to run or be slightly diminished, if the wall behind them diminishes, in lengths and not by inclined lines. Their faces also must run up vertically to the sets-off, and these should be in the same inclined line, and that line pointing to the apex of their pinnacles, when pinnacles surmount them. Indeed it cannot be too strongly enforced that there should be a constant tendency in the outlines of compositions in this style to meet, although the surfaces be themselves so generally perpendicular; and the more acute the angle under which they incline, the more graceful and becoming the style the result will be. The commanding lines of every part of a composition should lead through from its summit to the base. Thus, a spire or pinnacle should rest on a tower or turret whose angles are not interrupted, but never on a merely flat wall, however it may be faced with buttresses to give an apparent projection. Neither should low porches be projected from the face of a structure, for such can only have the effect of excrescences, and tend to injure a composition; nor should external doors be made but in places where the harmony of the composition is not injured by them as irregular apertures. Internally, square forms are seldom used; but piers consist of clustered cylindrical shafts, and thin shafts of the same form, lofty, and uninterrupted by crossing lines, act as pilasters. On these, capped with deeply inflected congeries of mouldings or foliage for the former, and lighter ones made continuous and breaking round them for the latter, rest the arches and arched ceilings. Flat surfaces are susceptible of high enrichment by means of tracery and panelling;

K

mouldings are enriched, not by carving on them, but by rounding out foliage and other ornament in covings and other deep inflections. Corbels should not be substituted for shafts to support arches when it can be avoided; but they have a pleasing effect as supports to the dripstone or canopy of a door or window; and indeed there are many other situations in which they are almost necessary, but they should always be considered as succedaneous, and not as necessary to a composition.

To avoid glaring inconsistencies in composing, it will be well to adhere generally to the style of some particular period, and to employ the proportions and enrichments, as well as the forms, peculiar to it; but, nevertheless, a more ornate may superimpose a plainer part, so that the difference be not violent. Windows of the second period may be placed over an arched composition of the first, and appear naturally to result from it; but the transition would be so great from the first to the third, as to make the result inharmonious. It need not, however, be denied, to those who feel themselves competent to use the materials with good taste and propriety, to select matter from examples of the various periods, and make compositions not exactly in the style of any of them. With a clear perception of the principles of the style generally, which we have endeavoured to point out, and a practical acquaintance with the classic exemplars of it, such may certainly be produced; and they may as certainly be adapted to all the purposes to which any style of architecture can be applied.

Rules for practice might be made to infinity, but they are unnecessary in this case, there being no authorized modern practice, like that of the Italian school in horizontal composition, to counteract. It is but to use the forms, proportions, decorations, and enrichments, and follow the mode of combination, which appear in the examples: these, with constant reference to the principles we have attempted to develope, will be the surest and safest guides in composing and arranging any subject. They are, too, so rife with materials for general purposes, that few cases can occur in which there need be any difficulty in finding parallels. Buttresses, piers, shafts, arches, pediments, parapets, turrets, pinnacles, windows, doors, niches, ceilings, tablets, with mouldings and ornaments in great profusion,—indeed almost everything that can be required in practice,—appear in existing works of the pointed style; preventing the necessity of determining from the mode of procedure in one case how we should act in another, as the comparative paucity of materials in the Greek and Roman remains rendered it necessary to do in developing the horizontal style.

A true architect said, with equal truth and point, in commencing a lecture on architecture, which he published in 1841, and entitled *The True Principles of Pointed or Christian Architecture,*—though the doctrine is alike true in respect of every style of architecture,—" The two great rules for design are these: 1st, *That there should be no features about a building which are not necessary for convenience, construction, or propriety;* 2d, *That all ornament should consist of enrichment of the essential construction of the building.* The neglect of these two rules is the cause of all the bad architecture of the present time. Architectural features are continually tacked on buildings with which they have no connection, merely for the sake of what is termed effect; and ornaments are *actually constructed*, instead of forming the decoration of construction, to which, in good taste, they should be always subservient.

The object in architectural enrichment is not to show the ornament but to enrich the surface. It is in propriety or fitness that the condition of delight which Sir Henry Wotton claims as one of the three conditions of building well, that architecture consists, and not in inconsistent devices.

The readiness of the world to accept the shadow, outward presentment, or even mere colour, for the substance in architecture, of which the true embodiment is construction, has made it the aim and pretence of the student in architecture to be an artist rather than a constructor, instead of grafting the artist upon the constructor. In the practice of the law the graces of oratory are taken to be of value only when supported by the sound learning of the lawyer; and in the practice of architecture the taste of the artist ought to be held merely ancillary to truthful disposition for structure and service.

But the soundest constructor is the most apt in the production, or the reproduction it may be, of real art. If Welby Pugin had been the mere artist-draughtsman his father was, which alone he might have been but for the superadded skill of the carpenter, and thereby of the constructor, he would not have arrived at the truth in respect of building well, nor have built so well and with such excellent effect as, in his brief lifetime, he did. The mechanician Smeaton, who was not a mere artist-architect, was employed to build the Eddystone lighthouse, and the means placed at his disposal being truly directed to the end aimed at, the result obtained presents a compendium of commodity, firmness, and delight. The Eddystone lighthouse is well adapted to its uses—that is to say, it is commodious, it is firm and stable almost to a miracle, and its form is as beautiful in outline to the delight of the eye, as it is well adapted to break and thereby to mitigate the force of the sea in defence of its own structure. Smeaton was in that work a true architect, the artist engrafted upon the constructor; whereas in dropping uncombined stone rubble into the Thames about the piers of old London Bridge to protect the bed of the river, and thereby the foundations of the bridge, from the scour of the already pent-up stream; and in building Hexham Bridge after the manner of old London Bridge, as to the effect of the piers upon the stream, and after the manner of Westminster Bridge as to the character of its foundations, Smeaton might have been taken for nothing more than a mere artist-architect. In like manner the botanist and landscape gardener, Paxton, having learnt to help himself in the contrivance and construction of works of architecture of a kind adapted to the purposes of his art, had become qualified to suggest the employment of a material unthought of by professional architects—though as ready to their hands as to his, and often misused by them—as the main constituent of a structure, and to devise a plan by which the material employed might be applied alike with constructive propriety, with marvellous rapidity, and so, moreover, as to produce a magical effect, in a Conservatory designed to fulfil, and amply fulfilling, a noble object, which must probably have failed for the time, or have been carried out in an ignoble manner, if the horticulturist-architect, Paxton, had not already himself learnt to be a constructor. The Great Exhibition building in Hyde Park, originated by Sir Joseph Paxton, possessed in an eminent degree the qualities arising from the conditions of commodity, firmness, and delight; it was most commodious for the purposes of an exhibition; it was firm enough for the temporary service required of it; and there was delight in its luminousness and in the simplicity and truth of its combinations; and all this may be said to have grown out of propriety of construction as applied in and to the material, cast iron.

Cast iron in its proper character, and in the forms which the duty to be imposed requires it to assume, would not be more inappropriate as the material of a superstructure upon piles of timber, than a tower of masonry is; and an open structure, well composed of cast iron to form trestle-like framework, would, in the case of the Brighton Jetty, have left the bridged way wholly unobstructed and convenient for use.

PRACTICE OF ARCHITECTURE.

The practice of architecture in whatever form, and under whatever alternative designation it presents itself—whether as civil architecture in works upon and above ground to be covered from the weather, or as civil engineering in works upon and in the ground and exposed to the weather—consists in originating, designing, composing, and arranging in detail, specifying, estimating, directing, and supervising the formation, construction, and fitting for use of works, and particularly works of the nature of buildings for the use and convenience of man, and mainly of social man in civilized communion.

Originating involves the devising a practicable scheme for effecting any desired object in the manner the best adapted to the end.—A canal, a road, or a railway, is required between any two places.   By the aid of hydrographical maps and by personal exploration of the country, the most fitting line the country affords may be struck out and laid down ; and by investigation of the levels, examination of the soil, computation of the available quantity of water at the summit in the case of a canal, and inquiry into the cost of labour, and of the various materials necessary for the work contemplated at the place where they are to be applied—the cost of land and the contingencies connected with its tenure and occupation—the line struck out as naturally the most fit, is modified to the most economical in execution, consistently with efficiency and economy in its use.—A town, or a new quarter of a town, a new street or a single house or other building, is required ; the town, or part of town, the street, the house, warehouse, or workshop, is to be considered first with reference to the essentials, *commodity, firmness,* and *delight ;*—and it will be convenient and salubrious, or unfit and probably the seat of disease—it will be cheerful and agreeable, or dull and cheerless—and, although the commercial estimation may depend on circumstances beyond the control and not wholly within the modifying power of architecture, even wealth or poverty may result from the propriety or impropriety with which the work has been originated.

Designing includes the bringing together and combining in the most efficient manner permitted by the circumstances of the case for convenience in use, all the parts of which a work may consist, shaping the parts within themselves to the duties they are respectively to perform, or to the uses they are to fulfil ; adapting the work to the materials which necessity, prudential economy, or luxury may dictate ; giving, indeed, a congruous body to the originated idea.   The origination of a bridge has determined its position, the manner in which it shall lie across the river or whatever else it is to be built over—the manner in which it is to be approached—the waterway, in the case of a river, that may be occupied by its substructions, or rather the waterway that the bridge must leave unobstructed—the headway that must be preserved under it, whether for navigation or for flood water, and with reference to the country above it or to its own security—the level or other inclination of the roadway, and the materials of which the bridge shall be composed, together with the general outline of its form, determining thereby the class of composition to be adopted.   The design gives form and consistence to the outline—it arranges the spans, bays, or openings —proportions the solids to the voids—determines, upon close and minute investigation of the site, the depth, mass, and composition of the foundations, and how they shall be executed—settles the form and rise of the arches, and prescribes the requisite appliances in erecting them, as centering and scaffolding—shows the mode of composing and constructing the roadway, and provides all the requisite accessories in completing the work for use.

Composing and arranging in detail are further preliminary considerations, and fall strictly, perhaps, within the province

of design.   These comprise the artistical and constructive adaptation and adjustment of the materials in the forms and combinations in and by which they are to be placed and held together to fulfil the ends of the design—the degree of finish that shall be given to a work, and the means by which it is to be brought about, as well as the modes of combining the parts and protecting them from injury from whatever source, whether from influence within the substance of the materials to be employed, from those which may be brought to bear upon them in the use of the finished work, or from external influences in and through the atmosphere. The composed and arranged design is exhibited in clearly expressed and accurately executed drawings made to scale— the scale being in ordinary English practice some aliquot part of a foot, or of an inch as an understood division of a foot— and the drawings being plans, sections, and elevations in general and in detail, made, by colouring, or by hatching, and by written descriptions, to express the main component materials severally and distinctly.

Specifying is a describing in writing everything required to be used and applied, and the modes of using and applying all the matters proposed to be incorporated in the work, to give effect to the design, the specification having reference to the drawings, of which it is, indeed, a further expression in words ; with the addition of a description of the kinds and qualities of the materials to be employed, the modes of operating upon them, the sequence of the operations, the time to be expended upon them, or within which they are required to be effected (for time has always an important influence upon the cost of a work), and the result to be produced ; and this must be done in such manner as to render misapprehension or misunderstanding as nearly impossible as knowledge and perspicuity can make them.

By estimating is understood the computing the quantities of materials of their various kinds required to be used or applied in the execution of a work ; the power, whether it be that derived from inorganic matter through the aid of machinery, or from the labour of beasts of draught and burden, or of men, and the labour of skilled workmen as artizans necessary in procuring the materials, and in preparing, shaping, and applying them to effect their combination in the manner and to the end proposed ; and, furthermore, the cost of such materials and labour, together with all the charges and expenses proper and incidental to the execution of the work according to the drawings and specification.

Directing and supervising the execution of a work imply the instructional explanations the workmen may require while the works are in preparation and progress, and the overlooking them to see that they do properly what the drawings and specification require to be done.   In the case of a contract for the execution of a work, the effective directions must be supposed indeed to appear in the drawings and specification, and so that supervision shall be only necessary to secure adherence to such directions.

" In architecture, as in all other operative arts, the end is to build well."   But, with whatever limitation in his own mind Sir Henry Wotton may have made that declaration, the conditions of well building declared by himself, give a much wider meaning to it than the words are ordinarily taken to convey.   Architecture in its broad and catholic sense admits of no limitation—it is subterrene as well as superterrene, formative as well as structural, and includes the general disposition of a site as well as the particular arrangement of the parts into which it may be divided.   It is thus applicable to towns, or aggregations of buildings, as such, as well as to the individual buildings which go to make up the town, and ought to be applied in such manner that the general result, in building a town, shall be in accordance with the conditions, and that they shall be fulfilled in the aggregate as completely as they are required to be fulfilled in each particular building which may go to compose it.

In the choice or selection of a site for a town, intrinsic circumstances,—some peculiar advantages in a commercial, or a merely pleasurable, or it may be indeed a military point of view,—commonly determine the general question, and impose upon the architect to fit it in all particulars for the contemplated purpose. Whether the site have the advantage of railway communication, or not, easy access by carriage-roads must be provided; which roads must be laid out, and set out with such inclinations proper for use as are attainable; bridges and culverts must be built to carry the roads over streams and gullies, and they must be cut, formed, drained, bottomed and metalled, or otherwise paved; quays to the river or the harbour must be arranged and formed, or embanked and constructed,—encroaching upon the tideway, or widening the waterway as the case may require; or, in the absence of natural facilities for navigation, it is a question for consideration whether artificial navigation can be obtained, and how and in what manner the arrangement of the site may be modified to render it available, or whether a railway may not be preferable to any other mode of giving commercial access to the town; the site of the town must be drained, both as regards water in the ground, and surface water, and the soilage must be taken off in a fluid state; water must be led to, or be raised by artifice within the site of a town, and accumulated in reservoirs at such levels that it may pass freely from them to all the parts of the town, and to the highest contemplated building in it. In the distribution of the parts of a town, not only are the requirements of the community as such to be considered, but the particular requirements of every class of a community ought to be provided for. There should be public markets and bazaars, and there should be also provision made for shops and warehouses, from which those who may choose to take their supplies may obtain them without recourse to the public markets. There may be within every town exclusive gardens, as in squares for the use of those who, dwelling around, may undertake to maintain them for their own use; but there should be certainly places as gardens, maintained at the public expense, open to general use for the resort and recreation of men, women, and children, of whatever class, as well as parks or pleasure grounds in the environs for general pedestrian, equestrian, and carriage exercise and open-air enjoyment. Streets, squares, and public places, should be disposed upon the site with regard to light, aspect, and ventilation; and sites for all requisite public buildings, sites for the requisite varieties of private dwellings; and places for storing, manufacturing, and commercial buildings and establishments, according to their kinds and requirements, are to be provided and arranged, and so provided and arranged as to produce the greatest possible general convenience, as well as to be capable of producing private benefit. Towards effecting these objects everything likely to be offensive should be placed to leeward, having regard to the prevailing winds. The relative levels of the lowest floor of buildings in every case to the drainage level should be settled, and provision made for the ventilation by legitimate means, of the drains of every building, so that it be not left to chance, or to the ignorance or caprice of individuals in building in detail, whether the town may be pleasant and wholesome, or be the seat of discomfort and disease.

There are certain essentials to every building which it is within the province of architecture to provide, and within the duty of the architect to secure. There must be convenient and easy access to the building itself; protection from water from whatever source and in whatever form water can present itself,—from the earth or from the skies,—by the foundation or by the roofs,—as water supplied for use, and as foul water or water the vehicle of the foulest matters;—air is to be supplied to the interior by legitimate channels from the best accessible

source, and the ejection of vitiated air provided for and secured, and so that there be ventilation without injurious draughts; and light is to be admitted and diffused, so that every part of an interior may be appropriately and sufficiently lighted, and lighted from the general source directly, or with as little recourse as possible to borrowed and artificially created light; equal temperature throughout the varying seasons is to be maintained by making provision to check excess of heat at one season, and by the promotion and diffusion of heat to warm at another. Next to the general essentials—which are as much so to a penitentiary or to an hospital as to a palace—are the particular arrangement and distribution of buildings according to their respective uses, with reference to the demand which the use may make or impose. But the uses of buildings are so various, and the requirements of each particular use so different from those of every other in all the classes of buildings, while every individual building of a class may be forced into peculiarities of disposition, dictated by site and other circumstances, that no general rules can be laid down for the design of buildings in classes, or severally, without running too much into detail for the pages of an encyclopædia. These are, or they ought to be, the special study of the architect, the merit and value of whose services consist mainly, indeed, in giving special effect to the general requirements in a congruous manner, and at the smallest cost consistent with efficiency.

It will be necessary, however, to add in this place some observations upon matters which, though general in their nature, are of particular interest to every community. Whether it be in laying out a new township, or in adding to a town, provision should be made for securing that it be done in such manner as may best tend to secure to it commodity, firmness, and delight, or, in other words, the health, comfort, and safety of the future inhabitants. And first, as to the carriage roads leading into and out of a town. No such road should be permitted to have a greater inclination in any part than one in thirty, such being the maximum at which ordinary four-wheeled carriages will remain at rest upon a well-made road surface. The width of the carriage roadway, clear of any side drains, should be such as to permit three carriages abreast, taking one to be standing on one side or the other, and two in motion, meeting or passing. Eight feet should be allowed to the carriage at rest, and ten feet to each of the carriages in motion, making twenty-eight feet as the minimum width between the water channels, or at the least thirty feet clear between the footways; for there ought to be to every such road a footway on each side, and each of sufficient width for four persons, two and two meeting, to pass abreast; and this requires eight feet at the least, so that there should be in the whole little short of fifty feet of public way wholly devoted to the public in what may be termed the country roads about a town where there are no houses, or but few, before which carts or carriages of any kind can be required to stand.

In laying out a town, there need be no mere lounging places provided, such as the paved area of Trafalgar Square in London, or as the *Place* so common in the towns and cities of the Continent. Markets should be provided for in sheltered and inclosed buildings, and not be held in a *place*; and a more than equivalent for the *place* or even the village green ought to be provided for outdoor recreation, and, it may be, rest; but wholly irrespective of business. To this effect plots of from five to ten acres each, making in all not less than one-tenth of the whole area, should be reserved in laying out a town, or in adding to a town; such plots being so disposed as not be more than a short half-mile apart, or so that there shall be one such plot within less than a quarter of a mile (a sufficiently long walk at once for a child, or for a woman or a girl carrying a baby) of every domicile; and every such plot inclosed, but accessible on every side, and laid out in the best manner to make it a pleasant resort at all times for men and women seeking rest or healthful recreation, and as a play-ground for children. With such a provision in a town, idle men and boys may be reasonably required by the police to "move on;" and with almost equal advantage to children and to the community at large, the trundling of hoops in the streets may be prohibited; whilst river or sea-side quays for business, or terraces for pleasure, need not be the permitted haunts of thieves and beggars. Out of a ten-acre plot (the size of the whole area of Russell Square in London,—and about that of Lincoln's Inn Fields, up to the inclosures before the houses), one acre disposed in four distinct quarters of an acre may be assigned to the four essential requisites of every hundred-acre area in a town—a church, a school, a library with reading-rooms, and a building to contain baths and washhouses; one at each of the four corners of the Town-Garden, in its own quarter-acre plot, and each communicating directly with the garden as well as with the streets by which, if houses or other buildings front towards it, the garden should be belted, without taking them out of the ten-acre area. The town-garden need not supersede the square and its garden, which may be formed, and the garden maintained in all its exclusiveness, with great public benefit, wherever the prospective demands of a future population may seem to require squares in connexion with the streets, by which and by the buildings fronting to them, the greater part of the whole area will certainly be covered. Nor need the town-garden vie with the square-garden in the relative extent of its plantations, or in the picturesque disposition of its paths. It should be laid out with broad walks, and hardily-turfed lawns; it should have a fountain, and trees should not be wanting; upon the whole, more like Hyde Park, the Green Park, and the public grounds of the Regent's Park, than like the too elaborately beautiful grounds of St James's Park. The town-garden should not be too delicate for cricket and quoits; nor should it be supposed to render the suburban park a superfluity: the town-garden for children in the day, and for work-day evenings for men and women, and the park for holidays.

The courses or directions to be given to streets must be greatly influenced by the circumstance of the locality, whether it be plain or hilly, and greatly by the climate, whether sunshine or shade is more to be desired. No general rule in this respect can be laid down; but having reference to the high latitude in which the British Islands are placed, and the consequent obliquity of the sun's rays at all seasons, without great intensity at any season, the light and heat of the sun are to be sought and not to be evaded; and these are more equably diffused, and the roadway exposed to the sun's rays more certainly, the more nearly the direction is that of the meridian. Perhaps, however, it is of more importance that buildings of whatever kind in which human beings are to live or to labour, shall be so disposed, that every room in every such building shall be capable of receiving the rays of the sun at some time in the course of every day that the sun shines.

The width of a street—and by the term street is intended any line of communication, whether adapted for carriages or not, in and through a town having houses or other buildings fronting to it on both sides—ought not in any case to be less from front to front of the buildings, taking an even course, than the height of the buildings, the dimensions of which—the street being once formed with such limit to its width—it ought not to be permitted at any time to exceed, without setting the higher buildings back accordingly. To this effect it is necessary to determine in what manner the height of a building is to be ascertained, the object in view being the due ventilation of the streets as a means of assuring a sufficient flow of air to the buildings, whereby they in their turn may be duly ventilated.

Take a minimum width for a street—say 30 feet—as a horizontal base line transversely of the street, and at a level, say one foot lower than the level assigned, in manner hereafter described, to the floor of the lowest story of the contemplated buildings. Erect a perpendicular line upon each end of such base, and bisecting the inner angles, draw lines which—the angles being right angles—will of course intersect the opposite perpendiculars at the height of 30 feet, the limit in height of any buildings standing up to such lines at the limited distance of 30 feet apart; the diagonals drawn out will limit the heights of the opposite buildings as to the roofs or any erections upon the roofs, beyond the perpendiculars. This rule is applicable in like manner to any greater width than 30 feet; but it will be seen that the rule imposes no limit to the height of any building standing in the narrowest street, if the front be set back at the level of the street, or be in any manner arranged by building terrace-wise; or in any other manner to the same effect of falling wholly under the diagonal line drawn from the opposite extremity of the base. Lofty chimneys, towers, and such like erections, of small extent in plan, should be admitted exceptions to such general rule.

The drainage of a town must be settled, and the levels determined throughout the course of every street, square, or other place, before the levels of the road surfaces, and of the floors of the lowest stories of buildings, can be fixed, so that they may be prescribed; but it must be obvious, that no street should be formed, in the technical sense of the term, by making the road at such a level that it cannot be efficiently drained. In like manner, as regards any building; every building should be placed at such a level that the floor of the intended lowest story can be efficiently drained, and in all respects relieved of fluid matters by the means provided; *and the means of efficient drainage should be made, and be in a condition to operate, before a building to be relieved is begun to be built.*

The widths of streets for the purpose last above referred to are to be considered as minimum widths, and irrespective altogether of the widths which ought to be imposed and required in respect of the roadways and footways, as means of communication and of passage in, through, and about a town.

A minimum width of 30 feet from front to front of buildings, ought to allow of no more than a single-line carriage-way in short lengths, or with turning and passing places; such single-line ways being necessary to give access for carts in the supply of fuel and other heavy matters, and for the removal of solid refuse, as cinders, ashes, bones, and other hard rubbish. Shops requiring frequent supplies by carts, and making their deliveries by carts, require greater width of carriage-way, and consequently greater width of street. As the demands of the carriage road for width increase, so do those of the footway; for shops making a show of their goods attract the attention of passers-by, and thereby tend to cause obstruction. Houses built as private dwelling-houses, in streets which are leading thoroughfares, or which become such, may be turned into shops; and when this is done, the space usually retained in towns between the footway and the house should be, for the reason stated, given up to the footway. In respect of the carriage-way, however, no street being, or being in a position to become, a much-used thoroughfare, ought to be of less width from footway to footway than sufficient for four carriages abreast, reckoning one to be standing on each side, and three to be in motion; for if only two be provided for, as in motion, the fast will be checked by the slow, and the whole traffic of a street be interrupted. Space for two standing carriages at 8 feet each, and for three in motion at 10 feet each, make together 56 feet as the least width that should be given to the carriage-way of a leading thoroughfare in any town. Any addition to this should be by 9 feet at the least, or it will add nothing to the convenience of the road. The footways to such

Site for a town.

a street, where the houses are private dwellings, may be taken at three couples abreast, or 12 feet at the least, and where they are shops, at 16 feet, allowing 4 feet for loungers. Any bridge occurring in a town ought not to require the carriage-way or road to be as wide, by the two stopping carriages, as the roadways in the streets leading up to it.

The relief of a town from water in the ground, if there be any—of the water that will fall upon it in the form of rain, hail, or snow—of the waste and refuse of water raised or supplied for the use of the inhabitants, in whatever condition the refuse may present itself, must, so far as regards the mode of relief, depend, in the first instance, upon the circumstances of the locality. If the choice of a site for a town, or for an addition to a town, lay between one from which all fluids could be passed away by gravitation, and another the general level of which was so low with reference to the means of eventual discharge that nothing could pass off in a natural course, the inducements to prefer the latter to the former must be great; but cases do nevertheless occur, and when they do occur, means must be applied to remedy the defect. London, in its great extent, contains many varieties, or rather instances of many varieties, of soil and surface; and it seems desirable to show somewhat in detail to what an extent natural deficiencies have been exaggerated in that notorious example of folly and neglect, and so to teach by warning. Suffice it for the present purpose to suggest, that the main considerations in providing for the relief of a town from water in all its forms and with all its concomitants, whatever the means of eventual discharge may be, are,—first, whether there is water in the ground so near the surface as to require to be tapped and kept down; secondly, whether the water falling from the clouds upon the surfaces of the roads and streets, and carrying gravel, sand, silt, and other heavy matters, either not soluble in water, or incapable of being held in suspension sufficiently to be carried along by the obtainable current in a reasonably well-formed drain, but inoffensive, shall be passed into special drains forming a separate system; and, thirdly, whether the rain-fall and other surface-water shall be passed into the system of drains necessary to relieve the buildings of the fluids which, coming through and from the habitations of human beings, are, though foul and offensive, for the most part soluble in the water with which they pass away. It is in favour of the separate system, that surface-water (and water falling upon the roofs of buildings in towns may be to a great extent treated and passed away as surface-water), being taken off at a higher level, may often be discharged by gravitation, that is to say, by natural fall, when house drainage, which is commonly at a lower level, cannot be so. Moreover, surface-water may, without offence, be discharged into rivers and other water-courses, whilst the fluids from houses, being foul, ought not to be turned into any far inland waters, and certainly not into any tidal river far from the sea, if they can by any economical possibility be otherwise disposed of. There can indeed be no reasonable doubt, that if the circumstances of any case be such as to permit the discharge of surface-water, or any considerable part of it, by gravitation, it ought to be so discharged, though it may involve the necessity of a separate system of drains; cloacal waters being permitted to pass down to a low level, though it be to be lifted again by artifice, rather than allow it to contaminate the surface-water, and involve the necessity of lifting this also. But there is the further and important reason for keeping the two kinds of drainage-water separate, even if they must pass at or to the same level, and that a low one. The surface-water being necessarily charged with heavy matters, and commonly taking up lime and other cementitious matters with silt and sand from the roadways, a concrete is deposited in slow-coursed drains, which requires to be removed by manual labour.

Thus, when cloacal drainage and surface drainage go to-

gether, the drains must have a greater fall than the former alone requires, to hurry on the surface-waters with their heavy concomitants, and to lead both down to a greater depth than either would require if separate; and this last is the case with London.

Until within the present century, the common sewers of London received surface-water alone. Cloacal matters were not admitted into them, but human ordure and other dejecta fell or was thrown into cess-pools dug deep in the ground; and as London in the eighteenth century hardly extended beyond the limits of the dry gravel bed, the fluids were in a great degree absorbed, and the more solid matters were removed under the name of night-soil by hand-labour, and carried away in carts and applied as manure. But as the clay land became building land, and as the water-closet came into general use, the gravelly subsoil could not take up the increased quantity of fluid matter, and the clays of course refused to receive them at all; so that it became necessary to provide for the discharge in a fluid state of the matters dejected all together. This is done, so far as it is done, by drains into the common sewers, which have thus become cloacal vents, as well as conduits for waste and surface water. As more capacious sewers are found necessary for this double duty, such have been built, and are constantly in course of building, and at such depth that houses may be relieved by their drains of waste-water, and of refuse soluble in water, while the roads are kept drained by the same conduits of their surface-water. The result of such deepening, however, is, that to a large extent the outfalls of the sewers are below the level of high tide in the Thames and its confluent creeks; so that the commingled waters in the sewers are penned back to await the ebb, when the sand, silt, and other heavy and insoluble matters which the surface-water brings from the roadways deposit, and form a concrete which no force of any backwater will wholly remove and carry out at the reflux; and this it is which constitutes the great difficulty in relieving London, and in providing for the discharge of its cloacal refuse—whether for use as manure, or into the river as waste—so much below London, seaward, that it shall not return with the tidal flood to or within the inhabited area. This enormous difficulty is now happily in course of being overcome by a system of intercepting sewers on both sides of the Thames, making a total length of seventy-two miles. On the north side those independent arterial lines converge at the river Lea, thence the sewage is carried to fourteen miles below London Bridge; the discharge of the south London sewage will be between Woolwich and Erith; near the sea the tunnels are 9½ feet in height internally, and 12 feet wide.

The considerations which lead to the choice of sites for new towns are seldom those which have regard to the personal accommodation of future inhabitants. Neither pleasantness of position, nor salubrity of air and dryness of soil, will at any time prevail against commodiousness for the trade or manufacture whereby the community to be established may obtain the means of living and of becoming wealthy. Even facilities for defence are in modern times made secondary to the objects stated, with the certain conviction that wealth can always provide the means of defending itself. The province of architecture is, so to dispose and so to form, that the advantages which dictated the site shall be rendered available and enduring, and that any defects in site, soil, or position shall be remedied—that is to say, so far as artful disposition and skilful operation can bring relief or lead to cure. In like manner as regards the increment of ancient towns. Towns outgrow the sites upon which they were first established, and widen upon soils and at levels which possess none of the advantages which may happen to attach themselves to the ancient site. Whatever were the inducements which led to the first establishment of a town on the site of the city of London—

whether the facilities which it presented for defence, or for commerce, or both; or the beauty of the situation, and of the country around and within view from it; or the excellence of its soil—dry, but covering water of the finest quality—there can be no doubt that its facilities for commerce have prevailed, and that they have made it the centre of a province rather than the heart of a metropolis. But in widening as a town, London has outgrown the beautifully environed high ground, with its deep bed of dry gravel, and brought within its area cold clays and low marsh lands; so that it has long become a subject for the application of all the resources of architecture in its most extended sense, and with all the applications that it can bring in aid.

Sir Christopher Wren's scheme for rebuilding the city of London after the great fire, included every consideration that in his time was recognised as essential to "commodity, firmness, and delight;" and it may be believed that Sir Christopher's plan for re-distributing the site, with a view to fulfil the conditions which had been already laid down by Sir Henry Wotton, was not carried out, only because the case was one of *re*-building, not of building; the site had been cleared, but the rights to the ancient though ill-disposed seats or sites remained. The city was refilled with buildings, and many of them were by the great master in architecture himself; but it was not rebuilt architecturally—the buildings may have fulfilled, and certainly many of them did fulfil, the conditions of well building, so far as they could be fulfilled in a town which had not been, as a whole, submitted to them; and the city of London remains to this day, and will remain—notwithstanding the millions devoted at the expense of the metropolis to "city improvements"—a monument of bad architecture.

An urgent reason prevailed against the application of the great architect's plan to the site of the devastated city in the latter part of the seventeenth century; but there is no good reason whatever why the London that has grown up since that time around and about the city, upon the suburban gardens, fields, and marshes, should have been absolved from all wholesome rule in respect of the great conditions of catholical architecture.

Reverting, however, to the city itself, it may be remarked, that under a crushing despotism, commodiousness and delight were firmly established in the capital of France, when its ancient limits were to be extended, by a broad belt of almost park-like avenue being formed upon the site of the demolished fortifications; whilst the civic body that built an obstruction—which is by the same body still retained—upon the most crowded thoroughfare in Europe, wherewith to exercise their useless and now unmeaning and absurd power of shutting a door in the sovereign's face—whilst the corporation of the city of London, in the plenitude of their freedom, when they demolished their city walls and filled up the moat with the rubbish, covered the new site so obtained,—not with broad avenues for commodiousness and delight,—but with lanes, courts, and alleys—unless, indeed, Houndsditch is to be called a street. Smithfield might have been a park, or at least a garden connected with Tower-Hill by an equivalent for the Boulevards of Paris, at no cost but that of a little self-restraint on the part of the corporation. Fleet-Ditch might have been converted into a dock with broad quays locking into the river at the south, as the Grosvenor Canal further west now does, and capable of being extended inland, as necessity might dictate or convenience require, and so as to deliver fuel and other heavy goods east and west, to the protection of the heart of the town from the heavier traffic north and south,

to the present time and for ever. Elevated viaducts thrown over the valley of the Fleet would have connected the main lines of street running east and west; so that the means of personal communication would have been uninterrupted by, and have left uninterrupted, the trade of the lower parts of the sides of the valley and the commerce upon the quays of the dock.

It may seem useless to speculate in this place upon what might have been done in and for London a century and a half or two centuries ago; but there are many old towns and cities which hardly yet extend beyond their ancient boundaries, but which are likely to be subjected to the same process of extension, and are susceptible of improvement in like manner as they become extended. It is, indeed, to the extension of towns and cities by the bringing within the inhabited area of lands not heretofore built upon, and to the formation of new townships, that such observations are more immediately applicable. It may be said with no exaggeration, that the city of London has been twice rebuilt since the great fire, and with no improvement in its condition as it regards commodiousness, that has not been more than counterbalanced by the being hedged in on almost all sides by suburbs in which all the ends of architecture are coarsely and vilely set at nought; and now large portions of the city are in a course of demolition to provide routes and stations for railways within the heart of the city, to the great distress of thousands of the humbler ranks, whose houses are removed to make way for the intruders; but this grievance will be more than compensated if the railways furnish the means of procuring more commodious dwellings in more salubrious quarters, and at the same time supply an easy, quick, and cheap conveyance from their rural dwellings to their place of business.

Within the last quarter of a century architecture has proceeded by wide strides throughout the world, particularly in London. The great increase of population and the extent of business has created a demand for new buildings to an almost fabulous extent. The railway system has worked in a very different way to what was expected, and instead of the facilities for travelling having attracted London merchants to run down to the great manufacturing towns to make purchases, it has induced the great manufacturers to build large establishments in the metropolis, where their goods are stored for the convenience of customers, so that London has now become the warehouse as well as the counting-house of the world. The railways have also been a cause for a great demand for houses. The metropolis being intersected in all directions by them, every one of these, except a small length of underground line up the New Road, has destroyed thousands of houses, particularly those of the poorer classes.

The enormous increase in the system of banking, and of discounting, of fire and life insurance, and, in fact, of every branch of what is now popularly called financial schemes, and the active rivalry of the various companies, have led to a system of competition which has necessitated the erection of really palatial structures, more as a means of showy advertisement than as necessitated by their business requirements. As these undertakings must of necessity be close to the Bank and Stock Exchange, the value of property in that locality has increased in an unprecedented, and, it may be, in a feverish degree. Land which formerly was valued by the foot frontage is now estimated by the foot superficial, and has sold for as much as L.1 per foot per annum, and at twenty-five years purchase.[1] It seems almost impossible such prices should be maintained, especially as

---

[1] A curious calculation has been gone into as to this value. Ten half-crowns placed side by side will measure 1 foot very nearly, therefore 10 × 10 = 100 will be the number which would cover a foot superficial; but 100 half-crowns are only worth £12, 10s. So that to pay for the ground would be equal very nearly to covering it over with half-crowns two layers deep

Composi-
tion.

the system of speculation has already received a severe check.

Another circumstance has caused very extensive building in London. Six-sevenths of the city was burned down exactly two hundred years ago, and rebuilt all at once with any accessible material, and, for the most part, in the most hurried and cheapest way. The consequence is whole districts of the city have grown old and decrepid, and it has been necessary to rebuild them all at about the same time.

It would almost be impossible, and quite beyond our limits, to record the vast piles of buildings erected for the purposes of business in London—the stately shops, the piles of chambers and counting-houses, the warehouses, quays, and other works erected by merchants, shipping-owners, &c. It would be equally difficult to give any adequate idea of the multitudes of houses erected by private enterprise, of all rents and values, from those of the working-classes to aristocratic mansions of £500 a-year and upwards. Many millions must have been expended by private hands in addition to the public works before named, and this in architectural work, and not including those of engineering character, such as the railways, above and below ground, with their bridges and tunnels, and the vast works of the Thames embankment, and the main drainage.

A few words should be said as to the endeavours to assist the poorer classes, and to ameliorate their condition, both physically and morally; and this, as well as some other remarks we now propose to offer, will apply to the country at large as well as to London. One of the most important was the introduction of public baths and wash-houses; these have been erected in most towns where there is a working population. The architects to these have been chiefly Mr Baly, and Messrs Ashpitel and Whichcord. The swimming baths by these architects at Lambeth are said to be the largest in Europe.

Model Dwellings.

The great demand for dwellings for the working-classes—consequent on the destruction in the poorer districts by the railways, which has been alluded to before, has led to the erection of what are often called model dwellings. These at first were erected much after the fashion of barracks, where a great number of chambers all led out of one common corridor or gallery; but this was found detrimental both to the comfort and morals of the inhabitants. The last-named architects erected a large set at Lambeth, where every set of chambers has not only its own conveniences, but is approached by outside balconies, so that every family has its own front door and separate dwelling within. This plan has since been extensively followed in various parts of both London and the country, and is found to answer extremely well.

Churches.

The number of churches erected, or enlarged and restored, is also beyond calculation. The greater part of the works have been executed exactly in the style of our mediæval ancestors. In many later instances Gothic of foreign character has been introduced; and not only so, but it has been attempted to make a modern or Neo-Gothic.

Styles lately prevailing. Pure English Gothic.

This consideration leads us to a short review of the changes of taste and variations of fashion in architecture. Thanks chiefly to the labours of the various archæological societies, and the publications of Professor Willis, Mr Parker, and others, a correctness and purity of detail has been arrived at, such as only required that an architect should enter into the *spirit* of the mediæval design to rival his illustrious predecessors. That the architecture of the churches of our forefathers did or should appeal most strongly to the reminiscences and sentiments of their descendants seems to be recognised; and it cannot be denied that many churches worthy of the thirteenth and fourteenth centuries were erected. A cognate taste prevailed in building private houses of importance; and

palatial mansions, in the style of the Tudors, almost rivalling Knole, Holland House, Crewe, and Woollaton, were built. But in England, as soon as anything arrives near to perfection there is, first of all, a desire to apply it to incongruous purposes, and next to supersede it by some partially understood novelty. People were startled to see workhouses decked out in the ornaments such as Cecil and Sidney delighted in; and hospitals for the idiot, the insane, and the pauper, take the form of aristocratic palaces. Others wished for change for the sake of change; and architects were obliged to rack their brains for something which was to be at once old and new. In the meantime, others, who sought for fame, or rather fortune, without possessing that intuitive perception which is another word for genius, or the quiet industry which often arrives at most satisfactory results, —these gentlemen found it a cheap and easy road to choose the odd and eccentric. If they could not make the world admire, they could make people stare, and so attract notice somehow. The consequences have been absurd, and in some instances painful.

Foreign Gothic.

The desire for new forms of Gothic architecture in building churches had induced the adoption of some foreign arrangements, such as gables over the windows of every bay of the aisles, a different character of towers and spires, tall apsidal east ends, and many other features; and though these things did not commend themselves to English archæologists, nor indeed to any one acquainted with the churches of our fathers, still there was a degree of keeping and elegance about them, when in the hands of "deacons of their craft," that was pleasing. About the time that the architecture of France and Germany had been brought into notice, a very clever book was published by Mr Street on that of the north of Italy, particularly the style in which brick is the material of the fabric, and marble that of the decoration. It has always been the misfortune that a true artist has first his followers and then his incompetent imitators. So it has been with both these branches. Those of this latter class, who took France and Germany as models, chose the heaviest, the barest, and baldest examples they could find. A flat piece of stone with a few holes cut through supplied the place of tracery, and if a tower had pointed windows, with exaggerated louvre boards, it was thought to possess quite sufficient artistic features to make a design.

The "brick and marble" was exaggerated into equally bad results. No one seems for a moment to have reflected that in the plains of Lombardy there is plenty of clay to make bricks with, and in the mountains both marble and granite, but that in most places good building stone is rare. They forgot also that in England we have most excellent stone, and no marble fit for general building purposes. What was the consequence? Brick was canonized. "Honest brickwork," as was the phrase, forgetting by the way that brick after all is only factitious stone. Then we fell back into the errors of the first attempts at building Gothic churches, and we had a revival of what was then called "brick and a half Gothic"—thin walls, no tracery, and especially no buttresses. Then there was a cry for colour, and horizontal courses of red bricks were used to stripe the walls. This first obtained the name of the "streaky bacon style," and then the sobriquet of "holy zebra." To this succeeded a mania for dotting work over with headers of black, red, yellow, and white brick. This was called the "Tunbridge ware style."[1] In fact, an affectation of the novel and the striking, without the slightest regard to beauty, proportion, or propriety prevailed, just as a century ago everybody went mad after Chambers' Chinese style. Nay, it was worse, for this last *was* an endeavour to design in a style used by a civilized people; but our late mania has been to be modern and to be ancient at the same time. That acute observer, the late Mr Thackeray, has

Composi-
tion.

---

[1] Since the above was written, late events have given it the unenviable title of the "London, Chatham, and Dover style."

Composition.

characterised it as the "bran new and intensely old." To illustrate a modern design, let any man select a plain brick house in any of the streets about Bloomsbury; take out the cambered arches, and replace them with similar cut and guaged arches, but of pointed form (no splays, mind); let him draw some stripes with red paint horizontally every 18 inches or 2 feet, and dot the upper courses with red, black, and white paint, and he will have the model of a house such as may be counted by hundreds round the metropolis. It has been urged in excuse that this is better after all than the plain Bloomsbury house without any attempt at ornament. But it is not so. The Bloomsbury house has no pretensions. It stands in its plain reality, while the other professes to be what it is not.[1] It is sham-mediæval, and, like all shams, it is sinking fast into oblivion.

Modern requirements.

Even where houses have been fully carried out, both as to interior and exterior, in the mediæval style, they are totally incongruous with our modern habits and customs, and at variance with modern sentiment. It has been asked, What has the architecture of the temples of Theseus and Minerva to do with a Christian church? and it may also well be asked, What has the architecture of feudal times, of the days of serfs and retainers, of archers and bill-men, of warders and jesters, of portcullises and battlements, of helmets, gauntlets, and mail shirts, to do with the days of butlers, boudoirs, grand pianofortes, and kid-gloves? However, a style has been gradually springing up on the Continent, and, in fact, has appeared on this side of the Channel, which bids fair to be the architecture of the future.

Institutions to encourage architecture have increased and flourished in a remarkable degree. The Royal Institute of British Architects has removed from the old and inconvenient apartments in Grosvenor Street to the spacious mansion belonging to the late Earl of Macclesfield in Conduit Street. Its transactions, which formerly filled two thin quarto volumes, are now published regularly every fortnight, with copious illustrations, and extend to many volumes. The library has trebled in number, and is now of the greatest value. It is open to any student, properly recommended, on certain evenings of the week, besides which its contents may be consulted at any time of the day on proper application. The Institute, or rather a committee of its body, have been appointed examiners of candidates for the office of District Surveyor under the Metropolitan Buildings Act. A most important step has also been

Voluntary examinations.

taken in the establishment of a voluntary examination of the younger men about to enter the profession, as to their knowledge of the scientific, artistic, and practical branches of the profession, as well as of letters. It comprehends a mathematical and algebraic course, as well pure as applied, including trigonometry and mechanics; a course of languages, ancient and modern—Latin and Greek, French, Italian, &c.; construction, professional practice, and the history and literature of all periods. At present the students are divided into two classes only, but it is intended to extend these classes as the examination progresses. Similar Institutes exist at Edinburgh, Dublin, Liverpool, Manchester, &c. The younger members of the profession in

The Association.

London have instituted the Architectural Association, the objects of which are chiefly for self-instruction, and to pave the way for joining the older body. They also meet at Conduit Street, where classes for the study of scientific and practical subjects, and for drawing, are held very frequently. This body numbers 247 members, which are increasing yearly.

The Architecture Exhibition.

A room has been devoted to an annual exhibition of architectural drawings, &c., and that of any new material,

inventions, or other objects useful in building. This is also held in the galleries in Conduit Street, where there is annual conversazione, and a series of lectures on cognate subjects by some of the first men in the profession, as well as by distinguished amateurs and literary men.

Modern French Architecture.

Another important institution bears the name given in the margin. Its primary object was the republication of some works on architecture. But their attention lately has been turned to the composition of a Dictionary, which, it is supposed, when finished, will be the most complete and comprehensive ever published on any science or art. The subjects take the widest range. Every style of architecture is treated of as well as its minutest details. The biography of every architect of note is given, and an account of every city which may contain any remarkable building. A very copious dictionary of all the terms used in ancient, mediæval, and modern architecture, particularly the difficult words in Vitruvius—of all the materials, with their chemical properties—of all manufactured materials—of all methods of construction—of all new inventions—in short, on any point about which information can possibly be required—from the highest scientific question down to the enumeration and description of the use of the tools in a workman's basket—complete articles may be found.

Architectural Publication Society.

Where there are disputed points, the theories on all sides are fully and fairly stated, and the reader is left then to judge for himself. There is therefore not the slightest bias or partisanship in any way. It is believed no better method could be devised to ensure correctness. The copiousness of the work may be guessed by the fact, that, including letter H, more than a thousand pages of folio letterpress in double columns have been published, besides many thousand illustrations in lithography, chromo-lithography, and wood-cuts. It is supposed about three-fifths of the work is now completed.

Little more than twenty years ago there was no periodical devoted to this subject. There are now four in London alone. *The Builder* established in 1843, goes on with undiminished popularity, and with increasing circulation, under the able editorship of George Godwin, Esq., F.R.S., F.S.A., &c., and has maintained its rank as the earliest and most valuable of all of this school of literature. Next to this ranks the *Building News*, an able journal, conducted by Passmore Edwards, Esq., which has been established twelve years. Besides, there are the *Building Reporter*, and *Building Journal*, both of which are the rather connected with the business than the theory of architecture.

Architectural Periodicals.

Modern French Architecture.

The gigantic improvements in Paris, the construction of miles of streets and boulevards, of a number of parks, squares, public buildings, and fountains, and, above all, the completion of the palace of the Tuilleries, has gradually introduced a new style, which may be termed " modern French architecture."

Improvements at Paris.

This may be defined as classic rénaissance, a style without the meretriciousness of the Louis Quinze period, and with more fancy, spirit, and play than was given to the classic art some fifty years back. In fact, at this last period the world fell into the great mistake of supposing that nothing could be classic except it was regular, cold, and severe. As every art degenerates into extremes, the classic architecture of a few years back became mannered, pedantic, and constrained. No doubt the revival of me-

Definition.

---

[1] The idea that bits of Gothic stuck on to a regular nineteenth century house, will make it mediæval, has been happily illustrated by his question, " If you put a pair of epaulettes on a ploughboy's smock-frock, will that make him a soldier?"

Modern French Architecture.

diæval art has done much to break our trammels, and it is now found that picturesqueness and fancy may be found as well in classic architecture as in that of any other period. The old mistake was much the same as to suppose because there was a Homer and Æschylus, there could be no beauties in Pindar or Theocritus, or as classical Latinity boasted of a Lucretius and a Virgil, there could be no excellencies in Horace or Terence.

Characteristics.

The modern French, as employed at the Tuilleries and other important public buildings, is purely classic. The character of the detail is rather that of Vignola than of Palladio, or Scamozzi. The mouldings are in general carefully studied and pure, and, as a general rule, the whole arrangement is remarkable for lightness and grace. Proportion seems to have been carefully studied. To those whose eyes have been accustomed to the cumbrous, ponderous style of some of our later works, that of the modern French appears to be too light. But it should be remembered that weight or lightness are qualities to be employed as the use, requirement, or sentiment of the building may dictate. Our early churches are all massive, with very small windows, having broad splays inside, where archers or spearmen might stand. The church was often then by necessity the fortress of the place. As these necessities grew less, the style of building became lighter and more elegant. A large window filled with slight tracery would have been easily dashed in, and the place taken, while the little narrow Norman arrow-slit was a formidable means of defence. We admire the massiveness of Newgate, which seems to suggest the means of safely securing the malefactor inside, and of awing the would-be malefactor outside. Surely a nineteenth century house ought to have an air of lightsomeness and cheerfulness.

Lightness and grace.

Sentiment in Design.

We must take leave here to say a few words as to what is called *sentiment* in design—a feeling which depends much on the history or antecedents of similar structures. Thus it was in consonance with the existence of old Westminster Hall, that the new Houses of Parliament were designed in the same style. Under its roof our laws, based on those of our Anglo-Saxon ancestors, had grown to their present state, and had been administered by Littleton, Gascoyne, More, Hale, Somers, and a crowd of other great names. In chambers of similar architecture the feudal system had gradually given way to free soccage, and the petition of right and act of habeas corpus had passed. But however excellent in other respects, it seems very incongruous to design the palace where English law sits in her majesty in the Venetian style. What is there in common between the recollections of the cold cruel tyranny of the ten, their spies, their secret accusations, their mock trials, their tortures, the secret death of the victim, his grave in the sea with a sack for his shroud, and the open noontide, even-handed justice of our British courts? Let us always study congruity. And in this way we should consider our dwellings; and we think the French have chosen rightly. Whatever faults they may have had, the Greeks and Romans were a highly intellectual, polished, and literary people. Everything they touched in the way of art, whether oratory, poetry, music, painting, sculpture, or architecture, had that refinement and elevation of thought and finish of execution that we call classicality. And classic architecture seems a necessary adjunct to refined manners and customs.

Details.

But to proceed to more detailed points. Modern French architecture still adheres to some of the features of the Louis Quatorze style. Thus the high roofs, with their lucarne lights and spikels, prevailing from the time of Francis I. till that of Louis XVI., are still in vogue (see Plate XL. figs. 1, 2, 3, 4; Plate XLI. fig. 2). The modern roofs are very light, being chiefly constructed of zinc, as

are the decorations of the windows themselves. Plate XL. figs. 1 and 5, gives examples made by the Vieille Montagne Company. Another feature is the frequent use of a marquise, or sun-shade over a door, also of the same material (Plate XL. fig. 2). Another is the constant use of balconies with extremely rich cast-iron railings (Plate XLI. figs. 1 and 2). Similar castings, but, of course, much lighter, are also used to fill in the panels of doors. The windows are all hung like folding doors, and, as will be seen, whether decorated or not, are all of graceful proportions. The French adopt Lord Bacon's motto, and build with symmetry, where both this qualification and convenience may be had (see Plate XL. figs. 2 and 3); where it is difficult or impossible, on account of the plan, as in fig. 1, they disregard this quality, and place the door where it is best for the use of the inhabitant. We give fig. 4, a little pavilion or summer-house built near St Cloud, as an example of lightness, grace, and simplicity combined.

Modern French Architecture.

High Roofs.

In street architecture we give three examples (Plate XLI.) of different degrees of richness and decorations, according to the taste or fortune of the owner. Fig. 3 is a plain shop in the Rue de Rivoli, with apartments over, and with only one balcony on the upper story. Fig. 2, from the Boulevard de Sebastopol, shows a shop with a gallery running round to enlarge the opportunities of displaying goods, and which may be made a mezzanine by the introduction of a floor. Here are balconies to the *étage principal* as well as to the top stage. The stories in street architecture are seldom more than 10 feet high. In fact, in matters of business, it seems a great mistake to build such lofty stories as we now often see. A merchant or lawyer can surely transact business as well in a room 10 feet high as in one of 15 feet. But the one will cost nearly half as much again as the other, and the additional 5 feet will entail the climbing up eight steps at least more in the staircase than the necessity of the case requires. Suppose each story (Plate XLI. fig. 3) were 5 feet higher than as drawn, the staircase to the top floor, already a pretty good climb of 72 steps, must be increased to 104—no small consideration both in point of labour and expense. Fig. 3 is a private house in the Rue de Turin. In this, as will be seen, the reception rooms on the ground floor have an extra height, owing to the *caves* or basement being a little above ground; the other stories range. Here again we have greater richness of style.

Street Architecture.

As will be seen by Plate XLII., this is sometimes very simple, and often very rich. Figs. 1, 3, and 4 are examples of the former, and 7 to 11 of the latter quality. There is one peculiarity as to the simple decoration, the flowing lines round the porticos (figs. 1 and 2) are incised, and not in relief. This, of course, is an immense saving of labour, and at a short distance looks equally well. In all these examples there is a lightness and grace as well as originality. Nothing can be simpler than the window head (fig. 3), which is that of the fourth floor of Plate XLI. fig. 1, to a larger, yet still we see at a glance it is not a copy. Instead of concealing their chimneys, as we have been too much in the habit of doing, the French generally make a feature of them, and carry them up so high that they need neither pots nor cowls. Plate XLII. fig. 6, is an example from the Boulevard Monçeau. The pilaster caps (figs. 9 and 11), the one quasi-Corinthian and the other Ionic, are also examples of originality. As to consoles, terminals, escutcheons, agraffes, vases, &c., their number is endless in Paris. Figs. 4, 2, and 7, 5, 8, and 10, are favourable examples.

Decoration.

We would strongly advise our readers to study the subject of modern French architecture as particularly adapted to modern times and modern cities; as flexible and adaptable, and capable of being carried out either in a very simple or highly florid manner.

ARCHITECTURAL ACOUSTICS.

Architectural Acoustics.

Definition.

This branch of science teaches how to construct large public rooms, either for oratory or music, in such a manner that the speaker may be easily and distinctly heard, or the music produce a clear and pleasing effect. As there are so many unavoidable circumstances which necessarily modify the arrangements we would desire to make in all our buildings, and so many adventitious causes which disturb them when made, this branch has been considered by many architects as too vague and unsettled to be treated of scientifically; and it is a curious circumstance, that in the best modern treatise on architecture, the word acoustics is wholly omitted.

It is a fact, however that some buildings are "bad for hearing," and some the reverse; some are "bad for speaking," and yet good as concert rooms. It is a fact that some architects scarcely ever build a public room in which persons can hear well; and some, like Sir Christopher Wren, seldom have erected a church or other public building which is not eminently suited to its purpose in this respect. Clearly, then, so important a subject is worthy of our investigation.

For information on the *abstract* science of acoustics, we must refer to the volumes of the "Encyclopædia Britannica;" but it is our intention to give some short, plain remarks on practical architectural acoustics, which, we hope, will be useful to our readers, particularly to any architect employed on public buildings.

We must, however, premise that *sound* is air set in motion in certain undulations or vibrations. These undulations travel from the source or cause of the sound, spreading in all directions in a way somewhat analogous to the undulations on the surface of still water when a stone is thrown in, or when it is disturbed in any other way. Sound is *communicable* through various substances, particularly air; in fact, without this fluid it is not transmitted. If a musical snuff-box, or any other self-acting mechanical contrivance to produce sounds, is placed under the exhausting glass of an air-pump, and if the air be pumped out, the sounds gradually become fainter and fainter, till they are almost entirely lost. There is always more or less actual *force* accom-

Accompanied with force.

panying the transmission of sound. A clap of thunder will shake a house to its foundation. A salvo of artillery will break windows. Large double drums will affect substances in a less degree. The tones of a violoncello will cause a tumbler perceptibly to vibrate; and in lesser or greater proportion all sounds are accompanied by forces which affect matter. If the time of the modulations or vibrations of two or more sounds are in certain ratios, or regular proportions to each other, they are then *harmonious* or musical;

Harmonious.

if not in such ratios, the sounds are discordant. The laws for this are given in the treatises on Music, and are beside our present purpose, though the facts are all-important in that science. In the former case, the sounds are pleasing; in the latter they are very unpleasant to the ear. Harmonious sounds, however, may be unpleasant if too loud, but even then they can scarcely be said to degenerate into *noise*. *Discordant* noises always partake of the latter character, and if too powerful are doubly painful to the ear. Sound *travels* at a rate varying, according to the condition of the atmosphere, from 1090 to 1120 feet in a second; and it is said that, even when diminishing in force, it still continues to travel at the same speed as long as it is perceptible. Sound always travels in straight lines, but is easily *deflected*, or carried into various directions by the wind or other cur-

rents of air, but is generally rendered much less distinct by such causes. Thus, the tones of a peal of bells that might be heard clearly a mile off in a calm day, may perhaps be carried by a strong wind to perhaps double that distance, and that so as to appear nearly as loud to the ear as in the former case; but though nearly as powerful they lose their distinctness, and the sounds become very much confused. Sounds are not only capable of being deflected or carried aside by currents, but they are capable of being *reflected*, or sent off from any solid surfaces; and not only so, they are capable of rebounding or coming back, a quality of sound which is commonly called *reverberation*, and which is of the greatest consequence in considering the form of public buildings. The laws of the *rebounding*, or reflection of sound, are very analogous to those of light, or of any elastic body, and may be familiarly illustrated by the rebounding of a billiard ball from cushion to cushion. If the times of the refraction, or repetition of sounds, are regular and distinct, it is called *echo*,—a phenomenon so well known as to need no further description. If, however, they be irregular and confused, it is called simply *reverberation*,—a condition the most annoying in public rooms, and the most difficult of correction. Sometimes large rooms will give a reverberating sound of harmonic character; this is called *sonorousness* or *resonance*, and is a valuable quality in concert rooms, and sometimes even for oratory, as it gives a pleasing vibration to the voice, and sometimes seems to increase its intensity; if, however, too powerful in degree, it is apt to degenerate into reverberation.

Architectural Acoustics.

Echo.

Resonance.

There are many other most important conditions regarding sound. Indeed, if we consider it, the science which affects one of the most important of those five senses which are everything to humanity, it is not to be wondered at that the philosophy of acoustics is both profound and extensive. It is believed, however, the outline given above will be sufficient for the purposes of our investigation, and reference has already been made to the sources for further information.

Buildings Acoustically bad, or in which it is difficult to Hear.

A frequent cause of difficulty of hearing is that the rooms are so *large* that a man having only average powers of voice is not able to fill them.

Several investigations have been made as to the distance at which a moderate voice could be heard, the most important of which are those of Sir Christopher Wren, as regards the interior of buildings, and of Saunders as to the power of hearing in the open air. The experiments of the latter were made when he was about to write his work on theatres (about 1790). Sir Christopher says (*Parentalia*, 318), "A remarkable voice may be heard 50 feet distant before the preacher, 30 feet on each side, and 20 feet behind the pulpit" (see fig. 1). Saunders (see fig. 2) tells us that a person in an open plain, and on a still day, read from a book, and was heard 92 feet in front, 75 feet on each side, and 31 feet behind him. The marginal diagrams (P being the place of the speaker) show the extraordinary difference between the two theories, which is still greater if we take into consideration that Wren calculated for a building roofed in, where the preacher or speaker would have the advantage of so much of the voice as might be reflected downwards, and not lost upwards, as it must be in the open air. The pro-

How far a voice may be heard.

Architectural Acoustics.

bability is, Wren meant that a rectangular apartment, about 70 feet by 60, was a *moderate* size for a moderate voice when preaching, and that Saunders gave his as a maximum

Fig. 1.

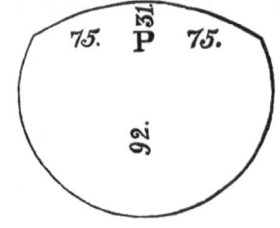

Fig. 2.

size for a theatre. One of Wren's largest churches—St James,' Westminster—is nearly 90 feet by 70, about half as much again as the size he gives. This, he says, holds 2000 persons; and he remarks, "I can hardly think it practicable to make a single room so capacious, with pews and galleries, as to hold above 2000 persons, and all to hear the service, and both to hear distinctly and see the preacher." Concert rooms, halls for public oratory, and theatres *must*, however, be so contrived that they *will* hold large numbers; and in the two former cases in particular, the problem becomes to construct a room of the maximum size, in which a moderate voice may be best heard. Of course, there are some stentorian voices which may be heard almost anywhere, but these are exceptional cases. The best way of treating such rooms will be found in the last section. Some-

Bad proportion.

times rooms are bad for hearing, because they are of *bad proportions*. If *too long*, the voice decays, or is lost before it reaches the end; or it may sometimes be heard at the end by reflection, and lost between that point and the middle. If *too wide*, the voice is lost at the sides, unless the speaker turns first to one and then to the other, in which case he is alternately inaudible to both. In the like manner, a recess coming out of one side of a room is a great impediment; and a dome rising suddenly out of a flat ceiling will deflect the voice upwards in such a way that the speaker will be quite inaudible beyond its margin, though perfectly heard within a few yards nearer to his place. One of the worst forms of churches for preaching is a *transeptal* church. The preacher, in effect, is addressing three congregations; if he looks towards the nave he is audible there, but not so in either of the transepts; if towards either of the transepts, he is not heard in the other, nor in the nave. One of the finest modern churches near London is liable to this defect in an eminent degree.

If a room is *too high*, the voice disperses too much and is wasted, not from reverberation, but because the voice seems as it were *diluted*, when the volumes of air are to large for its powers. If *too low*, the voice is beaten down, as is felt, in a very unpleasant way, both to the hearers and speakers. Probably there is not room for the undulations to develope themselves. In general, it may however be said, that a room intended for singing, should be more lofty than

Arched ceilings.

one intended for oratory. An arched ceiling is also better than one that is flat, as will be treated of hereafter; and a groined vaulting, or open timber roof, better still. But the principal impediments to distinct hearing, as has been said, is *reverberation* or *echo*, and this is principally perceptible in rooms which have plain, unbroken, or bare surfaces, as to their walls and ceiling, the rebounding of the voice from side to side is then very unpleasant and perplexing. For it must be understood, that reverberation does not lessen or absorb sounds: it rather increases their intensity, but

Obstructions.

mixes and confuses them. *Obstruction* is another cause of difficulty of hearing, this however, seldom occurs in public rooms; for though persons immediately behind a

column may not hear so well as others, yet in general, a room so divided, is even better for hearing than a large void space. The obstruction, however, of one person being behind another, is now common, and can only be remedied by raising the speaker, or the more distant seats of the audience.

Architectural Acoustics.

Remedies to Rooms already Erected.

If the room be *too large* for the voice of the person, and yet not too large for the congregation or audience who frequent it, there is no remedy, except the room has also other faults which *may* be lessened or removed. An architect is often blamed, when the fault is only in the speaker. If the latter has not sufficient physical power, he resembles a man too weak to carry a heavy weight. It however often happens that the room is too large both for the voice and for the audience. Such is the case with many country churches, which have been originally built for the ceremonies of great monastic bodies, and are now used by small parochial congregations. In these cases, it is best at once to reduce them to smaller dimensions, by screens framed of wood, or some sonorous material. In Canterbury Cathedral, the elegant stone screens behind the altar are glazed with plate glass, which reflects the voice forward in a remarkable degree. In very large buildings, like the Crystal Palace, the concert rooms are better if screened off altogether; when this is difficult, as at the Handel Festival, the orchestra should be so, and that in such a way as to send the sound directly forward. This should be done with thin boarding, or some sonorous material as far as possible. Before the opening of the Great International Exhibition, the sound of the music at the rehearsal went up, and was lost in the great dome. This was very materially remedied by hanging up a large canvass or "velum" across the same, though, if it could have been made of resonant material, it would have been far better. Side recesses and transepts may also be cut off from the main building. It is a fact, that in almost all large mediæval buildings, the transepts were screened off from the choirs, and used as side chapels, or other ecclesiastical purposes.

Screens.

The remedies in the more common cases against *reverberation* or *echo*, must depend on the *causes* which create it. If we enter an empty room and bang the door, or clap our hands, we are surprised at the confused continuous echo we hear. This is frequently lost when the room is filled with people. If not, it is often remedied by putting up flags or large curtains. In the Law Courts at Guildhall, where the reverberation at one time was positively painful, the hanging up some large pictures was found materially to assist the hearing distinctly. Even laying down felt carpets, has been found of service. It is remarkable, indeed, how non-elastic substances, or those from which sounds cannot rebound, seem to *absorb* those sounds; in fact, they may do so to such a degree, as to be positively detrimental. An eminent Italian vocalist informs us, that the worst possible place to sing in is a large library, even though a well-proportioned room, and that nothing deadens the voice so much as a number of books. If, therefore, a room is wanting in *resonance* or sonority, all superfluous curtains, carpets, &c., should be removed. In fact, as it is in almost every branch of art or science, it is the just medium between too many reflecting and too many absorbing surfaces we must seek.

Reverberation.

Resonance wanted.

The *construction* of rooms can frequently be much altered for the better. Thus an arched *ceiling* may be substituted for a flat one, or its surface may be broken by ribs instead of being quite plane. The surfaces of *walls* may also be broken up by pilasters, or other decoration. It is, however, said, that deep coffers in ceilings absorb or destroy too much sound. The square *corners* of rooms seem to add very much to unpleasant reverberation. Sounds seem to

Architectural Acoustics.

End galleries.

bound across the angles, and thence to the opposite corners, and back again, as a ball will often do on a billiard-table. Such corners can be cut off, or rounded, often without doing harm to the design, but never without assisting the power of hearing. An *end gallery*, however, has often been found of great service, as it checks the rebounding of the sounds backwards and forwards from end to end. As a general rule, anything that breaks up large reflecting surfaces, so that it is not carried so far as to become an obstruction, is very valuable against the effects of too much reverberation. We must not, however, forget what has been said before as to *resonance*, and in all alterations should make use of resonant materials, as will be hereafter treated of.

As we have now spoken of the methods of *checking too much* reverberation, we are now naturally led to the consideration of those by which the distinct transmission of sounds may be *assisted*. These have been called "reflectors," but not very correctly, as reflection implies the *sending back* that which has before been *sent out* or transmitted. They may more properly be designated "directors."

Sounding-boards.

The most common of these are the *sounding-boards* of pulpits. These are frames of wood sometimes quite flat and sometimes slightly arched, suspended over the head of the speaker to prevent the voice ascending to the ceiling, and to deflect it downwards to the congregation. Complaints are however made, that, where there are galleries, the occupants of seats below are benefited at the expense of those above.

Parabolic sounding-boards.

A sort of sounding-boards, of *parabolic* shape, have been made to throw the voice forward, and have succeeded to some degree; but unfortunately every sound in the body of the church is conveyed to this sounding-board, and reflected to the focus of the parabola where the preacher is placed. He not only hears the echo of his own voice repeating every word he says in a most confusing manner, but every whisper of the congregation is reproduced behind him, and he is often compelled to listen to uncomplimentary remarks on the length and dulness of his sermon. These circumstances have caused such annoyances, that the parabolic sounding-boards have been discontinued for some time past. A large frame, made of thin boards nearly of hemispherical form, was hung up over the orchestra at the Surrey Music Hall, slightly tilted upwards, and was found to be very favourable in conducting the sounds forward. In the Handel Orchestra, at the Crystal Palace, something of the kind has also been fitted up, as has been said before. It is not at all improbable that the "concha," or shell-shaped covering over the "bema" of the basilican churches abroad has suggested these auxiliaries.

Resonance.

In concert rooms the resonance, or power of sound, may sometimes be much increased by forming the orchestra of resonant material, and especially contriving it shall be hollow below. The alteration of benches, so as to form what is called an isacoustic curve, will also materially assist the hearing, as will be shown in the next section. The substitution of thin boarding for plastered ceiling is often found materially to assist the voice.

Some Suggestions as to Rooms about to be Erected.

Proportion.

There can be no doubt that a proper proportion of dimensions has much to do with the acoustic properties of a room. If the length, breadth, and height are in a regular ratio to each other, the waves of sound come back, or flow round in regular intervals, or vibrate at isochronal periods. This, as has before been explained, is the very essence of harmonious sounds. If the proportions have no relation to each other, the vibrations are irregular, and at last become broken and confused; and, as has been shown, this forms the difference between harmony and discord, and between resonance and noise. The author has in his mind a room

Architectural Acoustics.

in most respects very unfavourable for sound. It has a flat ceiling, bare unbroken walls, and an awkward recess at one side; yet it is by no means a bad concert room. But it happens to be exactly 60 feet long, 30 feet wide, and 20 feet high, or in the ratios of 1, $1\frac{1}{2}$, and 3. A double cube, that is, a room of equal height and width, and twice as long as wide, is said to be a very good proportion for a public room. The Surrey Music Hall was as high as it was wide (68 feet), and $2\frac{1}{2}$ times as long (170 feet nearly). The Free Trade Hall at Manchester, also remarkable as a good room, both for oratory or music, is $1\frac{1}{2}$ times as wide, and $2\frac{1}{2}$ times as long as it is high; or on an average 52 feet high, 78 feet wide, and 130 feet long. This is, perhaps, a better proportion, as the same height as width is too high for oratory. It must, however, be remembered, that the former building had three galleries, while the latter had but one. The music room lately erected at Edinburgh by Professor Donaldson is also eminently successful. Its proportions are 36 feet in width, 48 feet in height, and 90 feet in length; or (dividing each by 12, as a common multiple), 3, 4, and $7\frac{1}{2}$.

General form.

The question of general form of public rooms may now conveniently be treated of. They may be classed—1st, As simple parallelograms; 2d, As the same, divided into nave and aisles, like churches; 3d, As portions of circles or ellipses, like lecture-rooms, or the legislative chambers abroad; or 4th, As of horse-shoe forms, like our theatres.

Simple parallelograms.

In the *first* of these instances little can be added to what has been already said, viz., that proper proportions should, if possible, be carefully preserved; that all square corners be rounded, or in some way cut off; that recesses be avoided; that all plain surfaces of walls and ceilings be broken up; that the latter be either portions of circles, or, better still, of elliptic sections, or, at any rate, if flat in the centre, should be deeply coved at the sides and ends; and that, in case of music halls, some contrivance be placed over the orchestra to throw the voice forward.

Divided parallelograms.

In the *second* of these instances the same remarks will also very generally apply. The best form for hearing is undoubtedly that of the early Christian basilicæ, where the end of the nave is terminated with a semi-circular bema or chancel, surmounted with a hemispherical concha, or semi-dome. This tends to throw the voice forward into the church in a remarkable degree. The ceiling of the nave should be of much higher pitch than in the former case, seldom less than semi-circular in section; those of the galleries may be flat, or, better still, sloping downwards towards the outer walls. Pointed Gothic vaulting, covered with ribs, seems materially to assist the music of the service. In Canterbury Cathedral the tones travel with remarkable sonorousness; but it is not a good building for preaching. These circumstances, no doubt, were the original cause of what is called "intoning" the service.

Semi-circular.

The *third* instance comprehends those buildings of semi-circular or segmental form, with a straight side, like the theatres of the ancients, the legislative chambers at Paris and Washington, and the lecture theatres at our various institutions. These seem very well suited to the latter purpose, as more persons can see the performance of the experiments, which are generally the most important part of a lecture. They also suit the political arrangements abroad, as opinions there shade off in many ways, and are not treated, as with us, as ministerial and opposition. The speeches, too, are generally prepared, and read from a sort of pulpit or tribune. Acoustically, the objections are that no speaker can be heard well from any other place, and though some few rooms of this form, like the lecture theatre at the Royal Institution, are successful, the greater part are not generally favourable for speaking; and it seems to be for the reason mentioned both by Wren and

Architectural Acoustics.

Saunders, that the voice travels forward far more easily than it does sideways, and consequently those on the right and left of the speaker do not hear like those in front. It is thus at Paris, at the old theatre the Lyrique,[1] which was half an ellipse, on the major axis of which was the proscenium. When, however, circumstances compel the adoption of these forms, the ceiling should be domed if possible, or, at least, have a very deep cove, and if lighted from above the lantern, should be glazed beneath with horizontal sashes or glass, curved so as to follow the general line of the ceiling, or an echo, or at least reverberation, is inevitable. Some have recommended the speaker to be placed in a sort of niche; this would assist the hearing of those in front of him, who are already in the best place for that purpose, but would take away both sight and hearing, to some considerable degree, from those to the right and left.

Horse-shoe form.

As regards the *fourth* of these forms, that of our modern theatres, it is so seldom that a piece of ground can be got in cities of the proportions the architect requires, that, practically, it is little use to go at length into the question of exact plan. If, however, the sides of the horse-shoe be drawn in too much near the stage, and the house be so long that the plan resembles an ellipse, with one end cut off, the place of the speaker will be, as it were, in one focus of the ellipse, and the probability is he will be heard well in the other, or two-thirds up the pit; but there will be reveberation or indistinctness everywhere else. The ceiling should be curved, but not too much so. If it takes too much the form of a dome, the voice is lost upwards, as has before been described. The curve also should so finish over the proscenium as to catch the sounds and convey them forward. For this purpose also it is desirable that the stage itself should advance well into the house, that the speaker or singer may be more among the audience, as it were, and more under the curve intended to conduct the voice.

Effect of crowds.

It has been already explained that the presence of a crowd of people deadens or absorbs sounds to a very great degree. In churches where there are wide aisles, no galleries, and large unbroken masses of air, this is frequently an advantage; but in theatres and concert-rooms, where people are pushed closer together, it acts so disadvantageously as to necessitate some artificial mode of increasing

their sonorousness. This can be done, first, by the use of sonorous materials, and next, by disposing of them to the best effect. If a watch be held in the hand, or placed on a soft cushion, its ticking is scarcely perceived. If laid on a table, the sound is plainly heard; and if on the sounding-board of a pianoforte, or upon the back of a violoncello, the sound is increased to a remarkable degree. Thus it is desirable that everything possible should be made of elastic material, like wood, and that walls and ceilings should be lined with thin boarding rather than plastered; the latter material absorbing sound to a great degree. Where there is too much reverberation, it has already been explained, it is well to break up all flat surfaces; but in theatres and other similar buildings this should be avoided. It is said the difference between plain, thin panelled box fronts and those ornamented with heavy composition ornaments is much greater than would ordinarily be believed.

Architectural Acoustics.

Sonorous materials.

If these materials can be disposed so as to form hollow spaces or chambers, particularly beneath the room, it is also of great advantage. If a pianoforte be placed on a carpet, its sound is neither so great nor so distinct as if it stands on the bare floor, and both these qualities are much augmented if it is placed on a sort of large hollow box. For this reason, if it be not possible to build the entire room with a hollow chamber under it, at any rate, the stage or orchestra should be so treated, and all the framing supports and boarding should be of wood.

Disposition of materials.

Disposition of Seats.

Although the arrangement of seats does not increase the sonorousness of a room, much of the comfort of the auditor and his ability to hear, depends on this. As sound, like light, always travels in a straight line, it is clear where there is an obstruction to sight there must also be to hearing. Where the seats are on a level floor, like those of a church, the preacher must be elevated to be heard, or the heads of the people in the front seats will block up the transmission both of sight and sound to those behind. The diagram (fig. 3) will shew the difference of hearing between the pulpit and reading-desk. But inasmuch as the hearer is assisted by his previous knowledge of the service as well as

Seats.

in Churches,

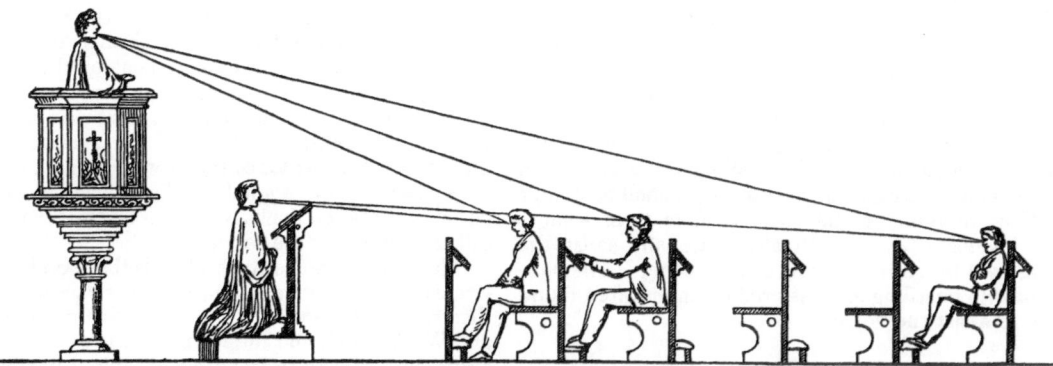

Fig. 3.

by his prayer-book, it is usual to keep the latter lower than the other. The higher, however, the pulpit is raised, the more the voice ascends, and is lost in the roof.

There are many reasons why the seats in churches must be kept on the same level, and there is but one remedy, to place the preacher higher or lower according to the length of the church, which is generally best determined by actual

inspection; but in lecture, concert-rooms, and theatres, the case is different. Here the appearance of the room is improved, and the purposes of the speaker or artist better effected by raising the seats. The question is, How is this best to be done? The old way was to draw a straight line showing the proposed incline or hanging level from back to front, and to subdivide that equally into breaks or

In concert-rooms, &c.

[1] This now pulled down for the public improvements, and rebuilt in the horse-shoe plan.

steps; but an inspection of the diagram (fig. 4) will show, supposing the lines are drawn to the Coping of the respective seats, that the angle of view, and of course of hearing of the last seat, is not half that of the foremost. To

obviate this difficulty, and to avoid the struggle for front seats, Mr Scott Russell, in an admirable paper published in the *Edinburgh New Philosophical Journal*, 1839, proposed a system which, from its affording an equal chance

Archi-
tectural
Acoustics.

Mr Scott
Russell's
isacoustic
curve.

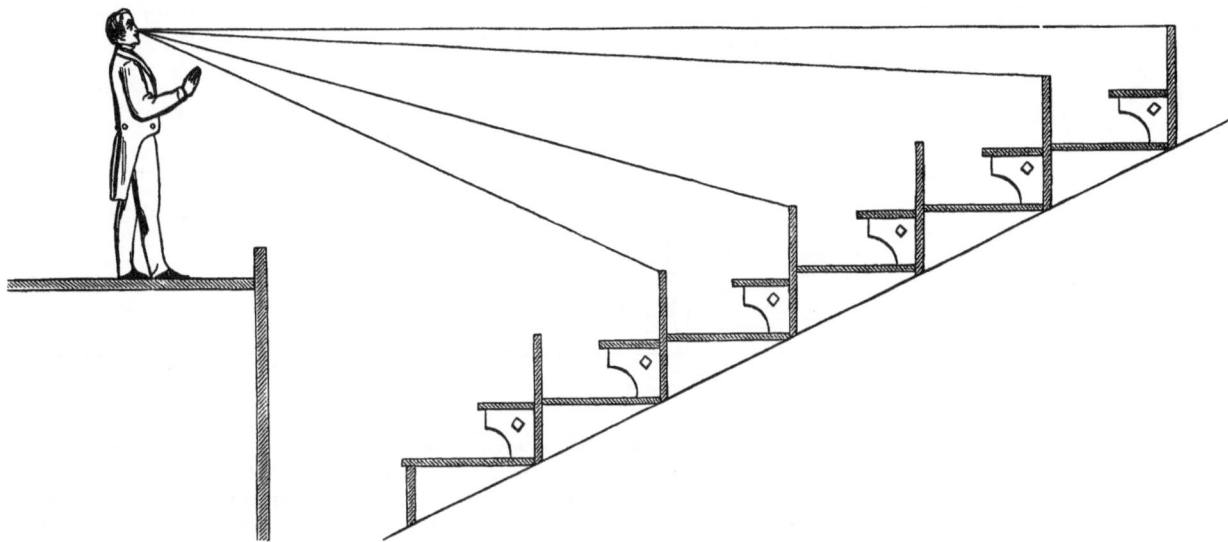

Fig. 4.

of hearing to all, he entitled "The isacoustic curve." (Fig. 5.)

He commences by setting out the position of the first seat in front of the speaker. He then sets out the position of the other seats at any convenient distance—say 2 ft. 6 in. apart, and draws perpendiculars to show the places of the backs. He then returns to the front bench, and sets up the seat 1 ft. 6 in. above the floor, above which he

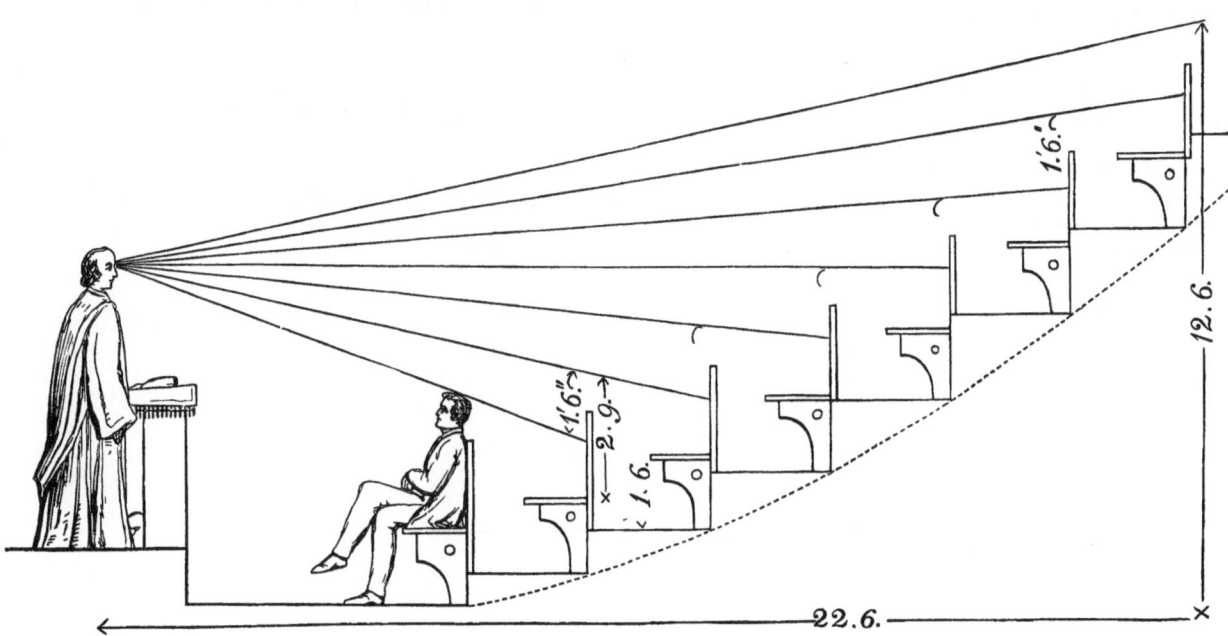

Fig. 5.

sets up a point, 2 ft. 9 in. in height, to represent the top of the head of the foremost auditor. He then draws a line from the head of the speaker over the last found point, and continues it to the perpendicular of the back of the bench No. 2, where he sets up a height of 1 ft. 6 in., which he calls face height, or clear view over the head of the auditor seated on bench No. 1. This line is continued to the

plumb of the back of bench No. 3, where 1 ft. 6 in. again is set up, and the operation continued as long as is necessary. From their respective points thus found he sets down, first, 2 ft. 9 in. for the top of the seats of the bench, and, again, 1 ft. 6 in. for the line of the floor. A line joining any set of these points will describe an isacoustic curve.

This forms an excellent arrangement for a small room or

a lecture theatre; but there is this difficulty with regard to iconcert rooms or other large apartments—the face height s too great, and the rise too sharp. The top of the head of

the auditor when seated on the seventh bench, at a distance of only 22 ft. 6 in. from the speaker, will be 12 ft. 6 in. above the floor of the apartment. The principle, however,

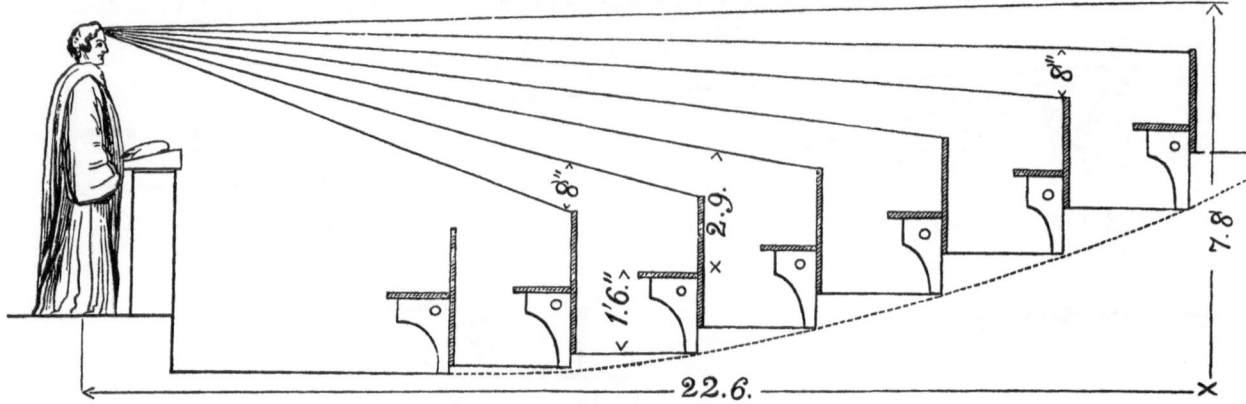

Fig. 6.

may easily be carried out for larger rooms by taking a less face height. From the chin to the level of the eye is scarcely 8 inches. If this be taken as face heights, and a similar diagram set up (fig. 6), it will be seen that the seventh bench will be nearly 5 feet less in height than by the former system.

We have now treated of those points which principally concern the architect; the general principles of acoustics; the causes of defects in existing buildings, and their remedies; suggestions as to buildings about to be erected, and the best way of disposing the seats for the auditory. It only remains to give a short list of the best works of reference for those who desire to go more deeply into the subject.

Contributions to Standard Works, Periodicals, &c.

Sir John Leslie.—Articles in the " Encyclopædia Britannica."
Sir John Herschel.—The like in the " Encyclopædia Metropolitana."
Savart.—Papers in the " Journal des Académies." Paris, 1839–1840.

Wheatstone.—Various Papers in the " Quarterly Journal of Science," &c. &c., from 1827 to 1831.
Scott Russell.—Various papers printed by the British Association, 1844; and Royal Institute of British Architects, 1847.

Separate Treatises :—

Pouillet.—" Cours. de Physique." Paris, 1856.
Brewer.—On Sound. London, 1854.
Chladni.—Traité d'Acoustique. Paris.
Weber.—Wellenlehre. Leipzig, 1825; and the various works on Physics by Somerville, Peschel, Arnott, &c., more especially applied to public buildings.
Saunders.—On Theatres. London, 1790.
Wyatt.—Do. do., 1811.
Inman.—Abstract of Evidence before Parliament, 1835.
Lachez—Acoustique et Optique des Salles de Réunion. Paris, 1848.
T. Roger Smith.—Treatise on the Acoustics of Public Buildings. Weale's Rudimentaries. London, 1861—which last treatise is probably the best yet produced on the subject, and contains an epitome of most of the foregoing works.

GLOSSARY OF NAMES AND TERMS USED IN CLASSIC AND MEDIÆVAL ARCHITECTURE.

ABACISCUS (diminutive of Abacus, *q. v.*) This term is applied to the chequers or squares of a tessellated pavement.

ABACUS, from the Gr. *ἄβαξ*, a tray, or flat board, It. *abaco*, Fr. *tailloir*, the upper part of the capital of a column, pier, &c. The early form of an abacus is simply a square flat stone, probably derived from the Tuscan order. In Saxon work it is frequently simply chamfered, but in the crypt at Repton, it has also two horizontal grooves (Plate XXVI. fig. 1); and in the arcade of the refectory at Westminster, which is known to be the work of Edward the Confessor, there is a deep notched groove between the square and the chamfer (Plate XXIII. fig. 1.) The abacus in Norman work is square where the columns are small; but on larger piers it is sometimes octagonal, as at Waltham Abbey. (Plate XVIII. fig. 4.) Sometimes the square is notched at the corners so as to make it somewhat like a cross, as at Ely (Plate XVIII. fig. 3), and sometimes round, as at Gloucester. The square of the abacus is often sculptured, as at the White Tower (Plate XXVI. fig. 6), and at Alton (ib. fig. 7), and Sutton (ib. fig. 9). In early English work the abacus is generally circular, as at Stone (Plate XXVII. fig. 3), and in larger work, a continuation of circles, as at Salisbury (ib. fig. 1), sometimes octagonal, as at Warmington (Plate XXVII. fig. 4), and occasionally square. The mouldings are generally rounds, which overhang deep hollows, as at Salisbury (ib. fig. 1; see also *Capital*). The abacus in early French work is generally square, as at Blois. (Plate XXVI. fig. 14.) They are also generally quite flat at the top, while in English work (except in Norman), all capitals, with scarce an exception, are more or less chamfered at the top. The Decorated abacus is much like the early English, but the hollows and rounds are not so bold. The upper moulding is frequently the overlapping roll moulding which resembles a scroll. (See *Capital*, Plate XXVII. fig. 12.) In the Perpendicular period, they seem to have gone back to the octagonal form in the majority of instances, particularly on the continent,

for the various mouldings of all these styles. (See *Capital*, Plate XXVII.)

ABATED, a term for such work in mediæval masonry as is worked down or sunk.

ABBEY (Fr. *abbaye*, It. *abbadia*, or contractedly, *badia*, Ger. *abtei*, *kloster*), a term for the church and other buildings used by those conventual bodies presided over by an Abbot or Abbess, in contradistinction to *cathedral*, which is presided over by a Bishop, and a *priory*, the head of which was a Prior or Prioress. The word is supposed to be derived from the Hebrew *Abba*, Father, by which name the Abbot was formerly generally designated. The buildings in all these instances are much the same, and generally consisted of a *church*, with *vestries*, *sacristies*, &c., either within the same or detached, but close to the church; a *cloister* with its usual *alleys*, and a *cemetery*, either in the centre of or attached to the same, a passage from these close to the church called a *slype*, the *chapter-house*, generally with a *vestibule ;* a *refectory* with *kitchens*, and various *stores*, *brew-houses* and *bake-houses ;* a *hospitium* or *guestern hall ; locutory* or parlour, a *library*, a *scriptorium* or writing-room, a *treasury ;* these were generally all on the ground-floor, and over them or part of them, the *dormitories, infirmaries*, &c. Some large abbeys had prisons, and even a mint, and some separate buildings, as an *almonry, sanctuary*, &c., and most were inclosed by a wall for defence, with *gate-houses*, &c. In larger establishments, as at Glastonbury, the Abbot and Prior had separate houses, with large halls, and every appliance necessary for a feudal lord; and all mitred Abbeys, or those presided over by an Abbot who had a seat among the Lords as a spiritual peer, had a house in London generally called the Abbot's Inn, where he resided during the sitting of Parliament. They had also *cells* or smaller houses in the country for change of air, and *granges* or farm-houses, all of which see under their respective heads.

ACROTERIUM (Gr. ακρωτηριον, the summit or vertex), a statue or ornament of any kind placed on the apex of a pedi-

M

ment. The term is often incorrectly restricted to the plinth, which forms the podium merely for the acroterium. The statue of the saint on the apex of the pediment of the western front of St Paul's is an acroterium; the other statues may be called acroterial figures.

AISLE, sometimes written Isle, Yle and Alley (Lat. and Ital. *ala*, Fr. *aile, bas-coté*, Ger. *seitenschiff, seitenchor.*) In its primary sense the wing of a house, but generally used to describe the alleys or passages at the sides of the naves and choirs of churches. When reckoning their number, the nave is usually counted. Thus a nave with an aisle on each side, is generally called a three-aisled church. If with two aisles on each side, a five-aisled church. In England there are many churches with one side aisle only; but there is only one cathedral with five aisles, that at Chichester. There are, however, very many such in the continent, the most celebrated of which are at Milan and Amiens. Others have three aisles of a side, or seven aisles in all, as the cathedrals at Antwerp and Paris. The most extraordinary, however, is that at Cordova, originally erected for a mosque. It was first built with a nave and five aisles on each side, eight others afterwards were added, making nineteen aisles in all. Old English writers frequently call the transepts, " the cross isle, or yle," and the nave the " middle ile."

ALIEN HOUSES, religious houses in England belonging to, or under the control of, foreign ecclesiastics. They generally were built where property had been left by the donors to foreign orders to pray for their souls. They were frequently regular *priories*, and sometimes only *cells*, and even *granges*, with small chapels attached. Some, particularly in cities, seem to have been a sort of mission houses. There were more than 100 in England. Many alien houses were suppressed by Henry the Fifth, and the rest by Henry the Eighth.

ALLEY, also called Ambulatory (Lat. *deambulatorium*), the covered passages round a cloister. (See also ALURE.)

ALMERY, also Aumery, Aumbrie, and Ambry (Fr. *armoire*; Ital. *almario*), a recess in a wall of a church, sometimes square headed, and sometimes arched over, and closed with a door like a cupboard, and used to contain the chalices, basins, cruets, &c., for the use of the priest; many of them have stone shelves. They are sometimes near the piscina, but more often on the opposite side. The word also seems in mediæval times to be used commonly for any closed cupboard, and even book-case. In fact the word *armoire* is in use abroad, to signify such objects to the present day.

ALMONRY (Lat. *eleemosinarium*, Fr. *aumonerie*, Ger. *almosenhaus*), the place or chamber where alms were distributed to the poor in churches, or other ecclesiastical buildings. At Bishopstone Church, County Wilts, it is a sort of covered porch attached to the south transept, but not communicating with the interior of the church. At Worcester Cathedral the alms are said to have been distributed on stone tables, on each side, within the great porch. In large monastic establishments, as at Westminster, it seems to have been a separate building of some importance, either joining or near the gatehouse, that the establishment might be disturbed as little as possible.

ALMSHOUSES, small buildings for the residence of the aged poor, generally endowed with some yearly stipend. The greater portion were built after the Reformation. Two early examples may be cited in England, as affording a study for the architect; that at St Cross, near Winchester, and that near the Preaching Cross of the Black Friars at Hereford.

ALTAR, anciently written Auter, or Awter (Lat. and Ital. *altare*, Fr. *autel*), the elevated table devoted to the celebration of the Eucharist. It is foreign to the purpose of this work to treat of the history of the word altar, and the controversies thereon. It is sufficient to say, in an architectural sense, that altars appear originally to have been plain wooden tables; then to have been of stone, probably because the martyrs were interred in or under them; some at Rome are actually ancient sarcophagi. In mediæval times, it is supposed all were of stone; some were plain slabs, supported by legs or brackets of the same material, jutting out of the wall, and are generally marked with five small crosses, one at each corner, and one in the middle. Viollet-le-Duc gives an example (*autel*) where the slab has only one leg.

ALURE (Lat. *alura—allorium*, probably from *alatorium*), an alley, passage, the water way, or flat gutter behind a parapet, the galleries of a clerestory, sometimes even the aisle itself of a church. (See also VALURING, VAMURE.)

AMBO (Gr. ἄμβων, Fr. *ambon*, Ger. *lesepult*), a raised desk, used in the early Christian Church for preaching or reading from. In some places there were two; one for the epistle, and one for the gospel. They frequently were ascended on each side by steps. The finest, and probably oldest, existing examples are at San Clemente, at Rome; these are of various marbles, enriched with mosaics, and are placed one on each side of the choir. The ambo is frequently called *anagogium, analogium*, and *lectorium*, and sometimes *jubé*.

AMBRY. (See ALMERY.)

AMPHIPROSTYLE (Gr. αμφι, around or about, and *prostyle*, *q. v.*) A temple with a portico at each end is said to be amphiprostylar. This term would be more correctly applied to a structure having projecting porticoes on all its sides, especially if it be equilateral like the *Bourse* or Exchange at Paris, allowing no distinction of flanks or wings to make it peripteral. See Plate VI. fig. 3 and 4.

ANGEL-LIGHTS, the outer upper lights in a perpendicular window, next to the springing. (See Plate XXXVI. fig. 13 C.) It is probably only a corruption of the word angle-lights, as they are nearly triangular.

ANGLO-SAXON ARCHITECTURE. (See SAXON.)

ANGLO-NORMAN. (See NORMAN-TRANSITION; also DATES OF STYLES.)

ANNULET (Lat. *annulus*, a ring). This term is applied to the small fillets or bands which encircle the lower part of the Doric capital immediately above the neck or trachelium.

ANTÆ (probably from the Gr. αντιος, or some other derivative of the preposition αντι, for, or opposite to; it has no singular), the pier-formed ends of the walls of a building as in the portico of a Greek temple. A portico is said to be *in antis* when columns stand between antæ, as in the temple of Theseus, supposing the peristyle or surrounding columns removed. Plate V. fig. 1, 2, and 3.

ANTE-CHAPEL, a small chapel forming the entrance to another. There are examples at the Cathedral and at Merton College, Oxford, and at King's College, Cambridge, and several others. The ante-chapel to the Lady chapel in cathedrals, is generally called the Presbytery.

ANTE-CHOIR, the part under the rood loft, between the doors of the choir, and the outer entrance of the screen, forming a sort of lobby. It is also called the FORE-CHOIR.

ANTEFIXÆ (Lat. *ante*, before, and *fixus*, fixed), upright blocks with an ornamented face placed at regular intervals on a cornice. Antefixæ were originally adapted to close and hide the lower ends of the joints of the covering tiles on the roof of a temple as they appear in the examples. Plate IV. fig. 1, 2, and 4; and Plate VI. fig. 3.

APOPHYGE (Gr. αποφυγη, a flying off), the lowest part of the shaft of an Ionic or Corinthian column, or the highest member of its base if the column be considered as a whole. The apophyge is the inverted cavetto or concave sweep, on the upper edge of which the diminishing shaft rests. Plate XIII. fig. 1.

APSE (Gr. ἀψις, Lat. *absis, tribuna, concha,* Fr. *abside, chevet, rond-point,* Ital. *apside, tribuna,* Ger. *ablauf*), the semicircular or polygonal termination to a church. These forms were no doubt derived from the *concha* or *bema,* in the classic and early Christian basilica. Sometimes the apsis is a simple semicircle; sometimes in large churches out of this a smaller semicircle springs, as Becket's Crown at Canterbury, and at Sens, Langres, and many others abroad. Sometimes the choir finishes with three apses—one to the central aisle, and one to each side aisle, as at Autun. Sometimes the plan is a semicircle, each bay of which has a projecting semicircular apse, as at Beauvais, Troyes, Tours, &c., forming a sort of cluster of apses. The later choir at Mans is encircled by no less than thirteen apses, the centre one being double the depth of the others, and forming the Lady-chapel. In some small churches of the Norman period, there is a sort of *double chancel,* one square, and the other an apse projecting eastward, each of which has its own arch, as at *Sutton, East Ham, Darent,* &c. Large circular and polygonal apses generally have radiating chapels within, one of the finest of these is at Westminster Abbey. The earliest cathedral at Canterbury had an apse at each end, if we trust the old plan; and there are several others in France and Germany. Apses project from the north and south ends of the transepts in some few examples abroad, and from the east sides of the same, the only example of which in England, is at Norwich.

APTERAL (Gr. α priv. and πτερον, a wing), a temple without columns on the flanks or sides. The Greek Ionic temple, Plate VI., is apteral.

ARÆOSTYLE (Gr. αραιος, rare or weak, and στυλος, a column), a wide intercolumniation. (See EUSTYLE.) The space assigned to this term is four diameters.

ARÆOSYSTYLE (compounded of *aræostyle* and *systyle, q. v.*) This term is used to express the arrangement attendant on coupled columns, as in the western front of St Paul's Cathedral. Plate XV. fig. 1.

ARCADE (Fr. *arcade, arcature,* Ital. *arcata,* Ger. *bogen gang*), a range of arches, supported either on columns or piers, and detached or attached to the wall. The word is used in contradistinction to colonnade, which is a range of columns carrying level entablatures. The oldest known in England is probably that of the Old Refectory in Westminster Abbey, the work of the Confessor. (Plate XXIII. fig. 1.) Examples are given, on this Plate, of attached arcades, at St John's, Devizes, fig. 2; Canterbury, fig. 3; Amiens, fig. 6; Notre Dame at Paris, fig. 7; Lincoln, Plate XX. figs. 1 and 2 (below the lower tier of windows); and at Romsey, in Hampshire, Plate XVIII. fig. 11. Of detached arcades, at Wells, Plate XXIII. fig. 4; at Lucca, fig. 8; at Pisa, Plate XXI. fig. 2: of arcades over piers and arches, at Lincoln, Plate XX. figs. 1 and 2: of arcades, the openings of which form seats or canopied stalls (which is very common in chapter-houses), from Lichfield, Plate XXIII. fig. 5. Arcades are found both inside and outside of mediæval buildings. If under the windows inside a building, their columns generally rest on a stone seat or bench table. Sometimes the arcades are pierced here and there to form windows. Sometimes the triforium is a series of arcades, as in parts of Canterbury, Exeter, Beverley, St John's Chester, and many other xeamples. In detached arcades, a very pleasing effect is often produced by placing the columns which are attached to the

wall opposite the centre of the opening instead of immediately behind the front column, so as to alternate and enhance the effect of the perspective, as at Wells (Plate XXIII. fig. 4). Arcades of interlaced circular arches, the intersections of which were at one time supposed to have been the origin of the Gothic or Pointed arch, are given from St John's, Devizes, Plate XXIII. fig. 2, and Canterbury, fig. 3. Many of the cathedrals abroad have arcaded fronts, as at Pisa, Plate XXI. fig. 2. That at Lucca has a range of double arcades, one behind the other (see Plate XXIII. fig. 8), the columns alternating, as at Wells and Lincoln.

ARCH (Lat. *arcus,* It. *arco,* Fr. *arc, arçean,* Ger. *bogen*). The different kinds of arches used in mediæval architecture, and the methods of finding the centres, &c., are given in the art. STONE-MASONRY, sec. V. Their mouldings will be found under the various heads. Saxon arches, and early and late Norman, are invariably circular. (See Plate XVIII. figs. 1 to 8; Plate XXIII. fig. 1, &c.) In the Transition style, they began to be pointed. (Plate XVIII. figs. 10 and 11.) In the early English style, they are sometimes lancet arches, and sometimes equilateral (see STONE-MASONRY, MEDIÆVAL, sec. V. fig. 63, 64); and in some cases depressed, ib., fig. 65. See also the various examples, Plates XVIII., XIX., XX., XXIII., XXX., &c. In the Decorated, they are generally depressed, but are sometimes ogee; see Plate XXIII. fig. 5, and Plate XXX. fig. 7. In the early Perpendicular, they are still flatter; and in the late Perpendicular, or Tudor, they become four-centred. See Plate XXX. fig. 9. (See also DATES OF STYLES.) In some cases where the springings of arches are above the levels of the imposts, and the mouldings are continued perpendicularly downwards, they are called STILTED ARCHES.

ARCH BUTTRESS. (See BUTTRESS, FLYING.)

ARCH JOINTS, to find. (See STONE MASONRY, MEDIÆVAL, figs. 67, 68.)

ARCHITRAVE (Gr. αρχη, chief, and Lat. *trabs,* a beam), the chief beam,—that part of the entablature which rests immediately on the heads of the columns, and is surmounted by the frieze; it is also called the epistylium or epistyle. Plate XIII. fig. 1. The moulded enrichment on the sides and head of a door or window is called an architrave.

ARCHIVOLT. This term is a contraction of the Italian *architrave voltato.* It is applied to the architrave moulding on the face of an arch, and following its contour.

ARRIS, the sharp edge or angle in which two sides or surfaces meet.

ARMATURE, the French term for the iron stays by which the lead lights are secured in windows. The uprights, or STANCHEONS, are called *montans,* and the horizontal or SADDLE BARS, *traverses.* (See their Articles.)

ASHLAR, spelt Ashelere, and various other ways (probably from the Lat. *assella*), squared stones, generally applied to those used for facing walls. The Durham Historian, 180, says—"Murus exterius de puro lapide, vocato *achiler,* plane incisso, interius vero de lapide fracto vocato *roghwall.*" "Clene hewen" ashler often occurs in mediæval documents; this no doubt means tooled or finely scappled, in contradistinction to rough axed faces.

ASHLAR PIECES, upright pieces of wood going from the common rafters to the inner wall plate. (See ROOFS, MEDIÆVAL, figs. 35, 36, and 38, at B.

ASTRAGAL (Gr. αστραγαλος, a vertebral joint), a convex moulding. This term is generally applied to small mouldings, and torus to large ones of the same form. See TORUS.

ATTIC, a low story above an entablature, or above a cornice which limits the height of the main part of an elevation. The etymology of this term is unsettled; probably the upper range of columns in a Greek hypæthral temple (see Plate IV. fig. 1.; and see also *Archæologia*, vol. xxiii. p. 412), was called ατειχον or ατοιχον, from having no co-herent wall; whence the Latin *atticum*, and its application to a story superimposing the general ordinance. Otherwise such a thing is unknown in Greek architecture; but it is very common in both Roman and Italian practice. What is here termed the tholobate in St Peter's and St Paul's Cathedrals are generally termed attics.

BACK CHOIR, a place behind the altar in the principal choir, in which there is a small altar standing back to back with the former. They are said to be chiefly used by the Capuchin Monks, but are very common on the Continent—also called RETRO CHOIR-PRESBYTERY.

BAHUT, the French term for a wall of plain masonry on which there is some superstructure. Thus the stone-work forming the lower part of a porch which carries the wood-framing; the plain part of a parapet between a projecting cornice or gutter, and the coping are both called *bahut*.

BAILEY, said to be a corruption of *Ballium* by some, and by others from the French "baille," a corruption of "bataille," because there the soldiers were drilled in battle array; the open space between the inner and outer lines of a fortification. Sometimes there were more than one, as the "inner and outer" Bailey; the "old" Bailey as at London and York, or the "upper and nether" Baileys as at Colchester.

BALISTRARIA. (See LOOPHOLE.)

BALL-FLOWER, a very pretty ornament which came into use in the latter part of the 13th century, and was in great vogue in the early part of the 14th. It resembles, as its name imports, a ball inserted in the cup of a flower. It is generally placed in rows at equal distances in the hollow of a moulding, frequently by the sides of mullions. The earliest known is said to be in the west part of Salisbury (Plate 33, fig. 10), where it is mixed with the tooth ornament. A long enumeration of the places where it occurs is in the Oxford Glossary. It seems to have been used more and more frequently, till at Gloucester Cathedral, in the south side, it is in profusion; in fact, there it is almost overloaded, and it is by no means improbable that, from this fact, people got tired of it, and so this very elegant little ornament went out of fashion. A very curious example with striated sides is found at Rouen, and given in Plate XXXIII. fig. 13.

BALUSTER, a small column or pier supporting the coping in a pierced parapet: the parapet itself when pierced is hence called a balustrade.

BALUSTER SHAFT, the shaft dividing a window in Saxon architecture, which see. See also Earl's Barton tower. (Plate XXXV. fig. 2.) At St Alban's are some of these shafts, evidently out of the old Saxon church, which have been fixed up with Norman capitals. These shafts in the Norman period become regular columns, and were, no doubt, the origin of mullions.

BAND, a sort of flat frieze or fascia running horizontally round a tower, or other parts of a building, particularly the base tables in perpendicular work. It generally has a bold, projecting moulding above and below, and is carved sometimes with foliages, but in general with cusped circles, or quaterfoils, in which frequently are shields of arms.

BAND OF A COLUMN (Fr. *bague*), a series of annulets and hollows going round the middle of the shafts of columns, and sometimes of the entire pier. (See Plate XVIII. fig. 14, Plate XX. fig. 1 and 2.) They are often beautifully carved with foliages, &c., see as from Amiens, Plate XXXIII. fig. 15. In several cathedrals there are

rings of bronze apparently covering the junction of the frustra of the columns. At Worcester and Westminster they appear to have been gilt; they are there more properly called SHAFT RINGS.

BAPTISTERY, a separate building to contain the font, for the purpose of the rite of baptism. The only examples of such are at Cranbrook and at Canterbury—the latter, however, though lately used for the purpose, is supposed to have been originally part of the treasury. They are more frequent abroad—that at Rome near St John Lateran, and those at Florence, Pisa, Pavia, &c., are all well-known examples.

BARBICAN, an outwork for the defence of a gate or draw-bridge; also a sort of pent-house or construction of timber to shelter warders or sentries from arrows or other missiles.

BARGE BOARD. (See VERGE BOARD.)

BARTIZAN, supposed to be derived from the Ger. *bartizene*, Fr. *échaugette*, a small turret, corbelled out at the angle of a wall or tower to protect a warder and enable him to see around him. They generally are furnished with oylets or arrow-slits.

BASE (G. Σπειρα, Lat. *spira basis*, Fr. and It. *base*, Ger. *fuss*), that part of a column on which the shaft stands. Early bases resemble the classic Atticurges base, *i.e.* two rounds or toruses separated by a hollow or scotia. Plate IX. example 3, fig. 1.; see also Plate XXXIV. *base*, fig. 1 and 2, which are examples from the Norman work at Peterborough. At this period also they frequently are decorated with a leaf at each angle, as Plate XXIII. fig. 9, which is from Rochester. After this the Norman bases generally consist of two or more rounds touching each other, each smaller than that beneath it. They are often richly carved, as Plate XXIII. fig. 13, which is from Cresset. The early English are also much like these; the hollow, however, is generally more deeply cut, and the lower round projects more, see Plate XXIII. fig. 10, which is from Lincoln. See also Plate XXXIV. figs. 5 and 6, bases given by Viollet-le-Duc in a rough state previously to the torus being rounded, and fig. 8, which is from Rheims. In the decorated period they often assume the form as Plate XXIII. fig. 11, which is from Beverley Minster, and sometimes they have a considerable number of rounds, as fig. 12 from Tintern Abbey. The peculiar curve, fig. 11, then began to overhang, as Plate XXXIV. figs. 9 from Monreale and fig. 10 from Dijon, and the bases generally are round or plain. In the Perpendicular style they are generally higher, and get more of the character shown, fig. 11. They are frequently composed of two reversed ogees, one over the other, and the plinths are generally octagonal on plan, with deep chamfers. In piers which have several shafts or clustered piers, each shaft has its own base, and they generally stand on a sub-base, which some authors call a patin and some a footstall. These sometimes have mouldings of their own, but are more generally chamfered, and have deep perpendicular notches dividing the several bases from each other.

BASE COURT (Fr. *basse cour*, i.e., the lower court, the first open space within the gates of a castle.) It was used for exercising cavalry and keeping live stock during a siege. (See ENCEINTE.)

BASE OF A WALL OR GROUND TABLE, mouldings round a building just above ground; they mostly consist of similar members to those above described (BASE), and run round the buttresses, as at Plate XIX. fig. 1, Plate XXV. figs. 11 and 12. The flat band between the plinth and upper mouldings is frequently pannelled and carved with shields, as Henry VII. Chapel at Westminster.

BASEMENT. A basement story is a story in any building placed below the level of the ground on the outside of and about the building. Basement applied specially, as architects apply it, means the compartment in the elevation of a building upon which any columnar pilastraded or arcaded ordinance may rest; as in the Strand front of Somerset House, of which the basement begins at the level of the floor of the vestibule, being about that of the street pavement, and extends upwards to half the height of the adjoining building east and west.

BASILICA (Gr. βασιλικὴ, i.e., the royal house), a term given by the Greeks and Romans to the public buildings devoted to judicial purposes. In later times, almost every wealthy man of note had one of these buildings attached to his house, for the assembly of clients, something after the manner of the Justice Hall in a country mansion of the olden-time. The buildings generally consisted of a nave, ναος, or navis, from a fancied resemblance to a ship, and from its symbolism with the ark. Two aisles, ἔμβολος, ala, or circuitus (see AISLE). A circular recess at the end, βημα, tribunal, where the seat of the judge, and afterwards the altar stood. Two raised pulpits, ambones, one for reading the gospel from, the other the epistle. A sort of inner lobby, εσοναρθηξ, an outer portico, ναρθηξ. In front there was often a sort of cloister, atrium, in which, in hot countries, there was frequently a fountain. The aisles of the basilica were generally separated from the nave by a range of columns supporting an arcade, over which are windows, and forming a clerestory (which see). The roofs appear to have been open timbered, with the exception of the basilica of Constantine and Maxentius, which was vaulted and coffered. Some are said to have had galleries, υπερωα. St Agnese, without the walls at Rome, is a well known example. In the times of the persecution of the early Christians, we are told they assembled in these halls by the connivance of their owners for the purpose of worship, and this form or plan being found very convenient for such purpose, was continued in the churches built by Constantine and others after the general conversion.

BATEMENT LIGHTS, the lights in the upper part of a perpendicular window, abated, or only half the width of those below. (See Plate XXXVI. fig. 13, B.)

BATTER (Fr. battre, to beat). Building over on projecting courses, like inverted steps, is termed battering, gathering, or corbelling over. The term is often applied to the converse operation of throwing back, as in a revetement or retaining wall.

BATTLEMENT (Fr. bretesse, Ital. merlo, Ger. zinne), a parapet with a series of notches in it, from which arrows might be shot, or other instruments of defence hurled on besiegers. The raised portions are called merlons, and the notches embrasures, or crenels. The former were intended to cover the soldier while discharging his weapon through the latter. Their use is of great antiquity; they are found in the sculptures of Nineveh, in the tombs of Egypt, and on the famous François vase, where there is a delineation of the Siege of Troy. In ecclesiastical architecture, the early battlements have small shallow embrasures at some distance apart. In the Decorated period they are closer together, and deeper, and the mouldings on the top of the merlon and bottom of the embrasure, richer. During this period, and the earlier part of the Perpendicular, the sides or cheeks of the embrasures are perfectly square and plain. (See those Plate XXII. fig. 1.) In later times the mouldings were continued round the sides, as well as at top and bottom, mitring at the angles, as over the doorway of Magdalene College, Oxford. (Plate XXX. fig. 9.) The battlements of the Decorated and later periods are often richly ornamented by panelling, as in the last example. (See also the the west front of York Cathedral, Plate XXI.) In castellated work, the merlons are often pierced by narrow arrow slits. (See OYLET.) In South Italy some battlements are found strongly resembling those of old Rome and Pompeii; Plate XXIV. fig. 4, is from a campanile at Brescia. It is a curious fact, that in foreign ecclesiastical architecture the parapets are very rarely embattled.

BAY (Fr. travée, Ital. compartimento, Ger. abtheilung), any division or compartment of an arcade roof, &c. Thus each space from pillar to pillar, as in Lincoln Cathedral (Plate XX. figs. 1 and 2), is called a bay, or severey.

BAY WINDOW, any window projecting outwards from the wall of a building, either square or polygonal, on plan, and commencing from the ground. If they are carried on projecting corbels, they are called ORIEL windows. Their use seems to have been confined to the later periods. In the Tudor and Elizabethan styles, they are often semicircular on plan, in which case some think it more correct to call them BOW WINDOWS. For those in Mediæval Halls, see DAIS, HALL.

BEAD, a small cylindrical moulding of frequent use. Plate VI.

BED-MOULD, the congeries of mouldings which is under the projecting part of almost every cornice, and of which indeed it is a part. Plate XIII. fig. 1.

BELFRY (Fr. clocher, if applied to a church, beffroi, if to the tower of a hotel de ville, Ital. campanile, Ger. glockenthurm), properly speaking, a detached tower or campanile containing bells, as at Evesham, but more generally applied to the ringing room or loft of the tower of a church. (See TOWER.)

BELL of a capital. In early English and Decorated work, immediately above the necking is a deep, hollow curve. (See Plate XXXIV. figs. 4 to 11.) This is called the bell of a capital, and is often enriched with foliages. (See Plates XXVI. XXVII.)

BELL-COT, BELL-GABLE, or BELL-TURRET. The place where one or more bells are hung in chapels, or small churches which have no towers. The simplest form is (Plate XXIV. fig. 1), from Froissy, and of the Norman period. Double bell-cots are given (figs. 14 and 15), from Northboro, and Coxwell; a very common form in France and Switzerland is given (Plate XXIV. fig. 2), which is for three bells. In these countries also they are frequently of wood, as ib. fig. 3, and attached to the ridge. Those which stand on the gable, dividing the nave from the chancel, are generally called SANCTUS BELLS. A very curious and it is believed unique example from Cleve Abbey, juts out from the wall. (Plate XXIII. fig. 16.) In later times bell turrets were much ornamented (Plate XXIV. figs. 8 and 11), the latter of which is from the Sainte Chapelle at Paris; these are often called FLÉCHE.

BEMA (Gr. βῆμα; Lat. tribunal), the semicircular recess or hexedra where the judges sat, and in after times the altar was placed. (See APSIS, BASILICA, CHANCEL.) It generally is roofed with a half dome or concha. The seats, θρονοι, of the priests were against the wall looking into the body of the church, that of the bishop being in the centre. The bema is generally ascended by steps, and railed off by CANCELLI.

BENCH TABLE, the stone seat which runs round the walls of large churches, and sometimes round the piers, and very generally in the porches.

BILLET (Fr. billette), a species of ornamented moulding much used in Norman, and sometimes in early English work, like short pieces of stick cut off and arranged alternately. (See Plate XXXIII. fig. 6.)

BEZANTÉE, a name given to an ornamented moulding much used in the Norman period, resembling pieces of money (bezants, coins struck in Byzantium.) (See Plate XXXIII. fig. 1, from Canterbury.)

BLACK AND WHITE HOUSES. (See POST AND PAN WORK, TIMBER HOUSES.

BLOCKING-COURSE, a deep but slightly projecting course in an elevation, to act as cornice to an arcade, or to separate a basement from a superior story. (See STRING-COURSE.)

BOND STONES. (See PERPENT STONES. THROUGH STONES.)

BOSS (Fr. *agraffe, clef de voute*, Ital. *bozzo*, Ger. *buckel*), an ornament generally carved, forming the key-stone at the intersection of the ribs of a groined vault. Early Norman vaults have no bosses. A later example is given (Plate XXV. fig. 1.) from Saint Sepulchre's, Cambridge. The carving is generally foliage, and resembles that of the period in capitals, &c. Fig. 2 is early English, from Stone Church in Kent; 3, is Decorated, from the Sainte Chapelle; 5, is Perpendicular, from York; and 6, Tudor, from Windsor. (See FOLIAGE.) Sometimes they have human heads, as fig. 4, from Notre Dame at Paris, and sometimes grotesque figures. In later vaulting there are bosses at every intersection.

BOWTELL, supposed to be derived from *Bottle*, the mediæval term for a round moulding or torus. When it follows a curve, as round a bench end, it is called a ROVING BOWTELL.

BRACE MOULD, two ressaunts or ogees united together like a brace in printing, sometimes with a small bead between them.

BRACKET (Fr. *corbeau, cul-de-lampe*, Ital. *mensola*, Germ. *tragstein*), a projecting ornament carrying a cornice, as Plate XXV. fig. 7, from Pietraperzia, and fig. 8 from Mas d'Agen, or at the foot of a niche to support a statue, as on the front of Westminister Hall. (Plate XXII. figs. 1 and 9). Those which support vaulting shafts or cross springers of a roof are more generally called CORBELS.

BRATTISHING or BRANDISHING, variously spelled; no doubt from the French *breteche*, a sort of crest ridge on a parapet, or species of embattlement. The term, however, is generally employed to describe the range of flowers which form the crest of so many parapets in the Tudor period. (See Plate XXXIII. fig. 20, which is from Hull.)

BROACH, (from *broche*, a spit), now used to designate a particular form of spire, the sides of which, with the angles of the tower, finish with a sort of haunching, as that in Plate XXXV. figs. 4 and 5. (See SPIRE.)

BUTTRESS, anciently written Botrasse, or Boterasse (Ital. *puntello*, Fr. *contrefort, éperon*), masonry projecting from a wall and intended to strengthen the same against the thrust of a roof or vault. Buttresses are no doubt derived from the classic pilasters which serve to strengthen walls where there is a pressure of a girder or roof timber. In very early work they have little projection, and in fact are "strip-pilasters." In Norman work they are wider, with very little projection, and generally stop under a cornice or corbel table. Early English buttresses project considerably, sometimes with deep sloping weatherings in several stages, and sometimes with gabled heads, as at Beverley. (Plate XIX. fig 1.) Sometimes they are chamfered, and sometimes the angles have jamb shafts, as in the last example. In the Decorated period they became richly panelled in stages, and often finish with niches and statues, and elegantly carved and crocketted gablets, as at York. (Plate XXI.) Sometimes they have pinnacles as at the Hotel de la Tremouille. (Plate XXV. fig. 13.) A curious example where a buttress finishes with two little shafts from Beauvais is given

in fig. 9. In the Perpendicular period the weatherings became waved, and they frequently terminate with niches and pinnacles (Plate XXV. figs. 11 and 12), the last of which is from Evesham. Buttresses at an angle of 45° with the corner of wall are often found in the later periods; it is supposed they are what in some old MSS. are called French botrasses. A very extraordinary buttress against the west wall of the Tower at Brasted is given in. fig. 10.

BUTTRESS, FLYING (Fr. *arc-boutant*, Ital. *puntello arcuato*, Germ. *strebe-pfeiber*), a detached buttress, or pier of masonry at some distance from a wall, and connected therewith by an arch, or portion of an arch, so as to discharge the thrust of a roof or vault on some strong point. Plate XXII. fig. 3, shows one from Westminster Hall. They are generally crowned by massive pinnacles, the weight of which tends to render them more stable and less likely to be overthrown, as in Plate XXXIX. fig. 4, from Amiens. They are sometimes in stages one over the other, as ib. fig. 5, from Clermont Ferrand, and on the Continent frequently have piers one before the other, to which a second set of arches spring, as at Narbonne and at Amiens, as above, where they form a sort of double flying buttress, one springing out of the other, and which by the French architects is called *arc-boutant á double volée*. In those cases, as at fig. 5, where there are no pinnacles, the external buttress, or abutment of the flying buttress, has usually more projection and general solidity, the more effectually to counteract the thrust. In the latter styles, the flying buttress is frequently ornamented with piered tracery, as at Amiens, as above, and at Henry the Seventh's Chapel at Westminster, which has a very pleasing effect, and when judiciously proportioned is equally strong.

BUTTRESS SHAFTS, slender columns at the angle of buttresses, chiefly used in the early English period. (See those in Plate XIX. fig. 1.)

CABLING. The flutes of columns are said to be cabled when they are partly occupied by solid convex masses, or appear to be refilled with cylinders after they had been formed.

CAMPANILE, a name given in Italy to the bell tower of a town hall or church. In that country this is almost always detached from the latter. (See BELFRY.) They are generally tall square towers, finishing with a battlement, as Plate XXIV. fig. 4, which is from Brescia. When of more importance the battlement is generally carried on large projecting corbels, from which a short second stage, or upper tower, rises as at Sienna (ib. fig. 5), and sometimes they are finished with a sort of spire. as at S. Marco, at Venice. The lower part is usually a plain, square tower like a shaft.

CANOPY, the upper part or cover of a niche, or the projecting ornament over an altar or seat. (See Plate XXII. figs. 8 and 11.) Early English canopies are generally simple, with trefoiled or cinquefoiled heads; but in the later styles they are very rich, and divided its compartments with pendants, knots, pinnacles, &c. See as above. The word is supposed to be derived from *conopæum*, the gauze covering over a bed to keep off the gnats (κώνωψ), a mosquito curtain. The triangular arrangement over an early English and Decorated doorway is often called a canopy. (Plate XIX. fig. 1, Plate XXX. figs. 5 and 8.) These triangular canopies in the north of Italy are peculiar. Those in England are generally part of the arrangement of the arch mouldings of the door, and form, as it were, the hood moulds to them, as at York. (Plate XXI. fig. 1—given on a larger scale, Plate XXX. fig. 5.) The former are above and independent of the door mouldings, and frequently support an arch with a tympanum, above which is a triangular canopy, as in the Duomo at Florence (ib. fig. 8). Sometimes the canopy and arch project from the wall, and are carried on small jamb shafts, as at San Pietro Martire at Verona. An extremely curious canopy, being a sort of horse-shoe arch, surmounting and breaking into a circular arch, from Tourny is given (Plate XXX. fig. 6). (See DOORWAY, LABEL MOULD, &c.) Similar canopies are often over windows, as at York Minster, over the great west window, and lower tiers in the towers. These are triangular—while the upper windows in the towers have ogee canopies. (See Plate XXI. fig. 1.) (See also TABERNACLE.)

CANT. When the corner of a square is cut off octagonally, it is said to be *canted*. Thus a bay window, with octagonal corners, is called a canted bay. (See BAY WINDOW.)

CAPITAL (Gr. Κεφάλαιον, Lat. *capitulum*, Ital. *capitello*, Sp. *chapitel*, Ger. *knauff, kapitel*, Fr. *chapiteau*), the upper part of a column, pilaster, pier, &c. The earliest known examples in England are probably those at Repton (Plate XXVI. fig. 1), which are formed of a plain abacus, with some small notches on it, and a deep, wide chamfer below. Plate XXXIV. fig. 1, also shows a very common early section. The more usual form of what is generally considered to be Saxon is that from the refectory built by Edward the Confessor at Westminister. (Plate XXIII. fig. 1.) This form is also very much used in early Norman work, and looks as if a capital of the Tuscan order (Plate XIII. fig. 2) had been taken from some old Roman building; and, being found too large, a piece had been cut off perpendicularly from each of the four sides. These are commonly called *cushion capitals*. This form prevailed for a long time, and by degrees began to be decorated with carving. A curious instance is given from Canterbury (Plate XXIII. fig. 3), where two of these capitals are placed side by side—one plain the other ornamented. (See also Plate XXVI. fig. 8, from Durham, and fig. 9, from Sutton, in Kent.) By degrees the Norman capital became subdivided into reversed truncated cones, also with the sides cut off, as at fig. 10, from Buildwas, and fig. 11, from Sutton. These last have been compared to strawberry pottles set side by side, and afterwards cut off at the lips. During this time, and through the transition, there was always some attempt to imitate the Roman Corinthian capital. Some of the early attempts are rather rude, as from St Nicholas, Caen, fig. 2, from Buildwas, fig. 4. Some, however, show much talent in carving, fig. 13, from Mantes, and fig. 14, from Blois. It is said when animals are used in the ornamentation the work is late, but they are common at Rochester, and an example is given from Alton (fig. 7).

In the Transition we sometimes find the round abacus. (See fig. 3, from Folkestone.) This is very usual in later styles in England, but was rarely used abroad, as is described under ABACUS. The plain early English capitals generally consist of an overhanging moulding called a lip mould, which forms the abacus; under these are a series of deep under-cut hollows, evidently intended to throw sharp shadows, which alternate with various projecting mouldings, below which there is generally a deep, long hollow called the BELL, under which is the necking. (See Plate XXXIV. figs. 4 to 8.) Where the capitals are ornamented the foliages are carved on the hollow of the bell. The early ones, as at Salisbury (Plate XXVII. fig. 1), seem to be suggested by the ancient Roman Corinthian; later, the foliage flows more freely round the bell, as figs. 2 and 3, from Stone, Kent. The early English foliage is of a very stiff, crisp character, and is supposed by M. Viollet-le-Duc to

be derived from the young shoots of the large fern. (See FOLIAGE.) On the Continent the capitals are much the same, except the bell is generally much longer. (See Plate XXVII. fig. 5, from Freiburg, and fig. 6, from Paris, the Sainte Chapelle.) In fact, the nearer central Italy, the more the capitals resemble the Corinthian, as might be expected, the architects continually seeing so many ruined ancient examples.

In the Decorated style, the mouldings are chiefly an alternation of hollows and rounds, but not so deeply undercut as in the preceding style. The foliages are more natural, and appear as if wrapped round the bells instead of springing out of them. They are in many instances absolute copies of our own plants, as the oak, vine, ivy, white thorn, and are carved so well as to prove the original plants to have served for models. Of course, the stiff conventional forms of the early English were abandoned.

Perpendicular caps are mostly plain—the mouldings are ogees, squares, and hollows, but not deeply undercut. The bells are shorter. The foliage often consists of single leaves in each division of the bell. Where it goes round the cap is generally of a stiff character, as at Stoke in Dorsetshire. (Plate XXVII. fig. 11.) See also the foliage in the Corbel (Plate XXII. fig. 12), and that on the Crocket. (Plate XXVIII. fig. 18.) Sometimes a strong, main stalk passes up and down, carrying vine leaves and grapes, as Plate XXXIII. fig. 22, from Westminster. These are called VINETTES. (See also ABACUS, FOLIAGE.)

CARRELS (Lat. *carola*), small chapels or oratories inclosed by screens; also sometimes the rails of the screens themselves. It was supposed, some little time back, that the scrolls, on which inscriptions of texts, &c., are formed, were called carrels, but this seems a mistake.

CARYATIDES. Human female figures used as piers, columns, or supports, are called *Caryatides;* and, adjectively, *Caryatic* is applied to the human figure generally, when used in the manner of Caryatides. Plate VII. figs. 4 and 6.

CASEMENT, a deep, hollow moulding, as Plate XXXII. figs. 21, 24, &c. They are sometimes filled with foliage (as Plate XXXIII. fig. 22), where they are called VINETTES. Also the frame which holds the lead lights of a quarry glazed window.

CASSOON (Ital.), a deep panel or coffer in a soffit or ceiling. This term is often written, after the French, *caisson*, whereas we derive it directly from the Italian *cassone*, the augmentative of *cassa*, a chest or coffer.

CASTLE. The mediæval castle consisted of walls with flanking TOWERS or TURRETS, in which were one or more GATE HOUSES, a BASE COURT, an ENCEINTE—buildings for residence of the family retainers and soldiers, sometimes an inner court—a KEEP, DONJON, or citadel, chapel, places for state prisoners, &c. These are all treated of under their separate heads.

CATHEDRAL (Ital. *duomo, cattedrale*, Fr. *cathédrale*, Ger. *domkirche*), the principal church where the bishop has his seat (cathedra) as diocesan. It is so called in contradistinction to abbey or priory, which may be churches of equal or greater size or importance, but are not presided over by a bishop. It is said no town can correctly be called a city unless there be a cathedral therein.

CATHETUS (Gr. καθετος, a perpendicular line). The eye of the volute is so termed because its position is determined, in an Ionic or voluted capital, by a line let down from the point in which the volute generates.

CAULICULUS (Lat. a stalk or stem), the inner scrolls or tendrils of the Corinthian capital are called *Cauliculi*. It is not uncommon, however, to apply this term to the larger scrolls or volutes of the same also. Plate XIII. fig. 1.

CAVETTO (Ital. *cavare*, to dig out), a moulding whose form is a simple concave, and impending. Plate VIII.

CEILING (Ital. *soffito, soppalco,* Fr. *plafond, lambris*), that covering of a room which hides the joists of the floor above, or the rafters of the roof. Most churches have either open roofs, or are groined in stone. At Peterboro' and St Albans, there are very old flat ceilings of boards curiously painted. In later times, the boarded ceilings, and in fact some of those of plaster, have moulded ribs locked with bosses at the intersection, and are sometimes elaborately carved. In the cloisters at Lincoln, the side aisles of the choir at Winchester, the church at Warmington, and several other places, there are ceilings formed of oak ribs, filled in at the spandrels with narrow, thin pieces of board, in exact imitation of stone groining. In the Elizabethan and subsequent periods, the ceilings are enriched with most elaborate ornaments in stucco. (See GROIN, VAULTING.)

CELL, small monastic houses, generally in the country, belonging to large conventual buildings, and intended for change of air for the monks, as well as places to reside in to look after the lands, vassals, &c. Thus Tynemouth was a cell to St Albans; Ashwell, Herts, to Westminster Abbey. (See GRANGE.) Also the small sleeping apartments of the monks; also a small apartment used by the anchorite or hermit.

CHAMFER, CHAMPFER, or CHAUMFER. When the edge or arris of any work is cut off at an angle of 45°, in a small degree it is said to be chamfered; if to a large scale, it is said to be a canted corner. (See CANT.) The chamfer is much used in mediæval work, and is sometimes plain (as Plate XXVIII. fig. 4), sometimes hollowed out (as figs. 1, 2), and sometimes moulded (as fig. 5).

CHAMFER STOP. Chamfers sometimes simply die on to the arris by a plane face; more commonly they are first stopped by some ornament, as by a bead (Plate XXVIII. fig. 1, fig. 2); they are sometimes terminated by trefoils or cinquefoils, double or single (as figs. 2, 3, and 4), and in general form very pleasing features in mediæval architecture.

CHANCEL, a place separated from the rest of a church by a screen (cancellus). The word is now generally used to signify the choir of a small church. (See CHOIR.)

CHANCEL DOUBLE. (See art. APSE in fine.)

CHANTRY (Lat. *cantuaria,* Fr. *chantrerie*), small chapels generally built out from churches. They mostly contain a founder's tomb, and are often endowed places where masses might be said for his soul. The officiator, or mass priest, being often unconnected with the parochial clergy, the chantry has generally an entrance from the outside.

CHAPEL (Lat. and Ital. *capella,* Fr. *chapelle,* Ger. *kapelle*), a small, detached building, used as a substitute for a church in a large parish; an apartment in any large building, a palace, a nobleman's house, an hospital or prison, used for public worship; or an attached building running out of, and forming part of, a large church. The first may be of any design, preserving ecclesiastical features; the second is generally made to harmonise with the main building. The third are generally dedicated to different saints, having their own altar, piscina, &c., and are screened off from the body of the building. (See LADY CHAPEL.)

CHAPITER, the old English name for a CAPITAL. The word chapitrell seems a diminution of this.

CHAPTER HOUSE (Lat. *capitolium,* Ital. *capitolo,* Fr. *chapitre,* Germ. *kapitel-haus*), the chamber in which the

chapter or heads of the monastic bodies assembled to transact business. They are of various forms; some are oblong apartments, as Canterbury, Exeter, Chester, Gloucester, &c.; some octagonal, as Salisbury, Westminster, Wells, York, &c. That at Lincoln has ten sides, and that at Worcester is circular; most are groined over, and some, as Salisbury, Wells, Lincoln, Worcester, &c., depend on a single slight vaulting shaft for the support of the massive vaulting.

CHARNEL HOUSE (Med. Lat. *carnarium,* Fr. *ossuaire*), a place for depositing the bones which might be thrown up in digging graves. Sometimes, as at Gloucester, Hythe, and Ripon, it was a portion of the crypt; sometimes, as at Old St Paul's and Worcester, it was a separate building in the churchyard (both these last are now destroyed); sometimes chantry chapels were attached to these buildings. M. Viollet-le-Duc has given two very curious examples of *ossuaires*, one from Fleurance, the other from Faouët.

CHOIR (Lat. *chorus,* Ital. *coro,* Fr. *chœur,* Ger. *chor,* Old English *quire, quere*), that part of a church or monastery where the breviary services or "horæ" are chanted. In the early Christian church, it was simply part of the nave in front of the altar inclosed by a low wall or *podium,* and forming in effect the seating, or a sort of open pew, where the singers sat. In later times, it was inclosed by rails or *cancelli,* which were shut against the laity as soon as these services commenced. Hence it is frequently called the CHANCEL, which see. The choirs abroad are often in side chapels, and most large churches have a summer and a winter choir, used alternately according to the time of the year. In England, the eastern limb of a small church is called the Chancel, and that of a large church is called a Choir or Quire; the proper name, however, for this is the Tribune or *Bema.* The semicircular end of the early Christian church became more and more elongated till it grew to be a long limb of the building, terminated by a circular apse or chevet; by degrees this became octagonal, and then square, the building being made larger as the number of monks increased. It was afterwards still further lengthened by the Lady Chapel, and its vestibule the Presbytery. Choirs of cathedrals generally are inclosed by screen work, and have a throne for the bishop, stalls at the west end for the dean and precentor, and on each side for as many of the prebends, canons, &c., as the chapter is composed of. There are also seats on each side for the singers, who are often, but improperly, called the choir, a pulpit, litany desk, &c. The organ at one time stood over the entrance screen, but is now generally placed on one side. (See APSE, BACK CHOIR, BASILICA, BEMA, LADY CHAPEL, TRIBUNE.)

CHOIR SCREEN. (See ROOD.)

CHRONOLOGY of Mediæval Architecture. (See the various Styles separately, and see DATES OF STYLE).

CIBORIUM (Fr. *Baldaquin,* Ital. *Baldacchino*), a tabernacle or vaulted canopy supported on shafts standing over the high altar. Gervase of Canterbury calls every bay of the quire there a ciborium, probably because the groining rose and formed a sort of canopy over each bay.

CINQUEFOIL, a sinking or perforation like a flower of five points or leaves, as a quaterfoil is of four. They sometimes are in a circle, as in the lower windows at Lincoln (Plate XX. fig. 2), and sometimes form the cusping of a head, as Plate XXVIII. fig. 3. It is said that in foreign achitecture the cinquefoil is rarely found. (See QUATRE-FOIL.)

CLEARSTORY, CLERESTORY (It. *chiaro piano,* Fr. *clairevoie, claire étage,* Ger. *lichtgaden*). When the middle of the nave of a church rises above the aisles and is pierced with windows, the upper story is thus called. Sometimes these windows are very small, being mere quatrefoils, or spherical triangles. In large buildings, however, they are important objects, both for beauty and utility. (See the Upper Ranges of Windows, Plate XX. figs. 1 and 2.) The window of the clerestories of Norman work, even in large churches, are of less importance than in the later styles. In early English they became larger; and in the Decorated are more important still, being lengthened as the triforium diminishes. In Perpendicular work the latter often disappears altogether, and in many later churches, as at Taunton, the clerestories are close ranges of windows.

CLEITHRAL (*vide* CLEITHROS). This is used of a covered Greek temple, in contradistinction to *Hypæthral*, which designates one that is uncovered.

CLEITHROS (Gr. κλειθρος, an inclosed or shut-up place). A temple whose roof completely covers or incloses it is a Cleithros. (Plate V. figs. 1, 2, 3; and Plate VI. figs. 1, 2, 3, and 4.)

CLOISTER (Lat. *claustrum*, Ital. *chiostro*, Fr. *cloître*, Ger. *kloster*), an inclosed square, like the atrium of a Roman house, with a walk or ambulatory round, sheltered by a roof generally groined, and by tracery windows, which were more or less glazed. Cloisters are intended for exercise in bad weather, and for reading and meditation in the shade when the season is hot. There are generally stone benches round, and a lavatory. The tracery is also particularly beautiful. At Gloucester the entire cloister is glazed. In other places the upper part of the tracery seems to have been thus proctected, if we may judge by the glass grooves; in others only that side most exposed to the weather has been glazed. At Wells there have been shutters to keep out the weather. The usual place for the cloister is the south side of the nave. At Canterbury, Gloucester, and Chester, however, they are on the north. In many cathedrals, where there was an establishment of monks as well as the secular clergy, there have been two cloisters, as at Canterbury, Westminster, &c. At Old St Paul's, London; St Stephen's, Westminster; and at the Chetham College, Manchester, the cloisters are in two stories, one above the other.

CLOSE, the precinct of a cathedral or abbey. Sometimes the walls are traceable, but now generally the boundary is uninclosed, and only known by tradition.

COFFER, a deep panel in a ceiling.

COLUMN (Lat. *columna*), a tapering cylindrical mass, placed vertically on a level stylobate, in some cases with a spreading congeries of mouldings called a base, and having always at its upper and smaller end a dilating mass called a capital. Columns are either insulated or attached. They are said to be attached or engaged when they form part of a wall, projecting one half or more, but not the whole of their substance. Plate VI. fig. 1 exhibits insulated, and fig. 2 attached columns. (See also Plate XIII. fig. 1.)

CONSOL or CONSOLE, a bracket or truss, generally with scrolls or volutes, at the two ends, of unequal size and contrasted, but connected by a flowing line from the back of the upper one to the inner convolving face of the lower.

COPING (Ital. *coperto, corona;* Fr. *chaperon*), the capping (from whence the name is probably derived) or covering of a wall. This is of stone weathered to throw off the wet. In Norman times, as far as can be judged from the little there is left, it was generally plain and flat, and projected over the wall with a throating to form a drip. Afterwards it assumed a torus or bowtell at the top, and became deeper, and in the Decorated period there were generally several sets-off. The copings in the late

Perpendicular period assumed something of the wavy section of the buttress caps (Plate XXV. fig. 12), and mitred round the sides of the embrasure, as well as the top and bottom. (See EMBRASURE, MERLON, BATTLEMENT.)

COPING, the covering course or cornice of a wall or parapet. The term *coping* is generally applied to a plain, slightly projected, covering course, and *cornice* to a larger moulded coping.

CORBEL (from the low Latin *corbeyus*, a basket, It. *mensola*, Fr. *corbeau, cul-de-lampe*, Ger. *kragstein*), the name in mediæval architecture for a piece of stone jutting out of a wall to carry any superincumbent weight. A piece of timber projecting in the same way was anciently called a tassel or a bragger. Thus the carved ornaments from which the vaulting shafts spring at Lincoln (Plate XX. figs. 1 and 2) are corbels. Norman corbels are generally plain, as at Buildwas. (Plate XXVIII fig. 6.) In the early English period they are sometimes elaborately carved, as at Lincoln above cited, and sometimes more simply so, as at Stone. (Plate XXVIII. fig. 7.) They sometimes assume a fantastic form, ending with a point apparently growing into the wall, or forming a knot, as at Winchester (fig. 8), and often are supported by angels and other figures. In the later periods the foliage or ornaments resemble those in the capitals, and are more natural or more conventional, as the case may be. (See CAPITAL, FOLIAGE.)

CORBEL TABLE, a projecting cornice or parapet, supported by a range of corbels a short distance apart, which carry a moulding, above which is a plain piece of projecting wall forming a parapet, and is covered by a coping. Sometimes small arches are thrown across from corbel to corbel to carry the projection. (See Plate XXXIII. fig. 11.)

CORBIE STEPS, a Scottish term for the steps formed up the sides of a gable by breaking the coping into short horizontal pieces.

CORNICE, the projection at the top of a wall finished by a blocking course, so common in classic architecture, is never found in mediæval work, though the upper moulding is often called so. In Norman times, the wall finished with a CORBEL TABLE, which carried a portion of plain projecting work, which was finished by a coping, and the whole formed a PARAPET. In early English times the parapet was much the same, but the work was executed in a much better way, especially the small arches connecting the corbels. A curious example is given (Plate XXXIII. fig. 11) from Oadby in Leicestershire. In the Decorated period the corbel table was nearly abandoned, and a large hollow, with one or two subordinate mouldings substituted; this was sometimes filled with the ball flowers, and sometimes with running foliages. In the Perpendicular style, the parapet frequently did not project beyond the wall-line below; the moulding then became a string (though often improperly called a cornice), and was ornamented by quatrefoil or small rosettes set at equal intervals (see Plate XXII. fig. 1) immediately under the battlements.

CORONA (*vide* CORNICE). This term is applied to the deep vertical face of the projected part of the cornice between the bed-mould and the covering mouldings. (Plate XIII. fig. 1.)

COVE—COVING. The moulding called the cavetto,—or the Scotia inverted,—on a large scale, and not as one of a mere moulding in the composition of a cornice, is called a cove or a coving.

CRENELLE, a word generally considered to mean the embrasures of a battlement, but latterly proved to apply to the whole system of defence by battlements. In mediæval times no one could "crenellate" a building without special license from his supreme lord. (See BATTLE-MENT, EMBRASURE, MERLON.)

CRESTING, an ornamental finish in the wall or ridge of a building, which is common on the Continent, but only one early example of the latter is known in England; this is at Exeter cathedral, the ridge of which is ornamented with a range of small fleur-de-lys in lead. Walls are often finished here by perforated parapets, but no example of a regular crest is known. At Venice, however, they are common; Plate XXIV. fig. 17, shows that at the Ducal Palace. Of crest ridges, our space only permits us to give these examples, 12 and 13, which stand up from the ridge, and 16, which lies on the slope of the roof under the ridge.

CROCKET (It. *uncinetto*, Fr. *crochet, crosse*, Ger. *haklein knollen*, anciently called crockytts, and sometimes called creepers), an ornament running up the sides of gablets, hood-moulds, pinnacles, spires; generally a winding stem like a creeping plant, with flowers or leaves projecting at intervals, and terminating in a finial. The idea is probably taken from the ornaments running up the sloping sides of the Roman pediments. They were not in use in England in the Norman period; at any rate no example remains. In the corresponding period abroad they are a sort of knots, as at Pisa. (Plate XXI. fig. 2.) The earliest in England are probably those at Lincoln, which are simple knobs turned back like that at Plate XXVIII. fig. 10, which is from Amiens. Shortly after these became foliated, as at Salisbury, fig. 14. In the Decorated period they became richer, as at Strasbourg, fig. 11, and sometimes had heads, figures of animals, birds, &c.; and sometimes developed themselves as a very rich leaf, as fig. 12, which is from the Palace of the Signoria at Sienna. Latterly they became more like part of a budding-leaf, as fig. 16, from Canterbury; and fig. 13, from Milan; fig. 17, from St Albans; and Plate XXII. fig. 5, from Westminster, is common throughout the Perpendicular period, though the late crockets sometimes assume the form of the stiff, strap-like leaf, as fig. 18, from Litcham.

CROP (Ang.-Sax. *crop*), the top of anything. A word anciently used for a FINIAL, which see.

CROSS. This religious symbol is almost always placed on the ends of gables, the summit of spires, and other conspicuous places of old churches. In early times they were generally very plain, often a simple cross in a circle, as at Beverly. (Plate XIX. fig 1.) Sometimes they take the form of a light cross crosslet, or a cross in a square, or in a square of metal, as at the Sainte Chapelle (Plate XXIV. fig. 7), or wide, as at Freiburg (fig. 9); by degrees they became of less simple design, as Plate XXVIII. figs. 19, 20, the first of which is from Freshton, the second from Doddington. In the Decorated and later styles they became richly floriated, and assumed an endless variety of forms.

CROSS-AISLE, an old name for a transept.

CROSS-SPRINGER, the transverse or lierne ribs of a vault. (See GROINING, VAULTING.)

CROW-STEPS. (See CORBIE STEPS.)

CROW-STONE, the upper stone of a gable, see also as last.

CRYPT (Lat. *crypta*, It. *confessione, sotteraneo*, Fr. *crypte*, Ger. *gruft*), a vaulted apartment of greater or less size, usually under the choir. They are, no doubt, enlargements of the small place under the altar in the Christian basilicas, where the relics of the saints were placed, and which were gradually built of larger and larger dimensions till they became the "undercrofte, laigh church," or church below a church; the oldest in England is that at Repton, the capital of the column of which is given (Plate XXVI. fig. 1.) The next that of

York, which is probably Saxon, as are also those at Hexham and Ripon. After the Conquest, the oldest are at Winchester and Worcester; the largest and most important are those at Canterbury and Rochester, which last is partly early English; there is also a very fine one at Gloucester. Norman crypts are also found at Oxford, Lastingham, St Mary-le-Bow, London, and others of less note. A perspective view and section of this last is given, Plate XXXVIII. figs. 1 and 2. This crypt is extremely curious, but is now unfortunately walled up. The finest crypts, however, were at Old St Paul's, as large as the whole choir, and which formed the church of St Faith, and was destroyed by the fire; and that at Glasgow, which likewise extended under the choir and chapter-house, and is now happily entire, and the most beautiful in the kingdom. Both these were of the transition between the early English and the Decorated periods. It is a curious fact, that just after this time the use of crypts seems to have been abandoned, not one of any importance having been constructed since. They seem to have reached their most palmy period, and then at once to have entirely gone out of fashion.

CUPOLA (Ital. *cupo*, concave, profound), a spherical or spheroidal covering to a building, or to any part of it. (Plate XI. figs. 2, 3, and 4; Plate XVI.; and Plate XXI. fig. 2.)

CURSTABLE, probably *course-table*, because they run horizontal as courses do. (See STRINGS.)

CUSHION CAPITAL, the name given to the early Norman capital. See Plate XXIII. figs. 1, 3; Plate XXVI. figs. 6, 8, 9, 10, and 11; and Plate XXXVIII. fig. 1. (See also CAPITAL.)

CUSP (Fr. *feuille*, Ital. *cuspide*, Ger. *knopf*), the points where the foliations of tracery intersect. A great variety are shown in Plates XXXVI. XXXVII. The earliest example of a plain cusp is probably that at Pythagoras School, at Cambridge (Plate XVIII. fig. 9); of an ornamented cusp at Ely Cathedral, where a small roll, with a rosette at the end, is formed at the termination of a cusp. In the later styles the termination of the cusps were more richly decorated, as Plate XXVIII. figs. 21, 22, and 23; they also sometimes terminate not only in leaves or foliages, but in rosettes, heads, and other fanciful ornaments.

CYCLOSTYLE (Gr. κυκλος, a circle, and στυλος, a column). A structure composed of a circular range of columns without a core is cyclostylar; for with a core the range would be a peristyle. This is the species of edifice falsely called by Vitruvius Monopteral. (See MONOPTEROS.)

CYMA (Gr. κυμα, a wave), the name of a moulding of very frequent use. It is a simple, waved line, concave at one end and convex at the other, like an Italic *f*. In that manner it is called a cyma-recta; but if the convexity appear above, and the concavity below on the right hand, it is then a cyma-reversa. (Plate VIII.)

CYMATIUM. When the crowning moulding of an entablature is of the cyma form, it is termed the Cymatium.

CYRTO-PROSTYLE. An alternation of Cyrtostyle (*q. v.*), but indicating more clearly than Cyrtostyle does an external projection.

CYRTOSTYLE (Gr. κυρτος, convex, and στυλος, a column), a circular projecting portico. Such are those of the transept entrances to St Paul's Cathedral. (Plate XVI. fig. 1.)

DADO or DIE, the vertical face of an insulated pedestal between the base and cornice or surbase. It is extended also to the similar part of all stereobates which are arranged like pedestals in Roman and Italian architecture.

DAIS (Fr. *estrade, haut pas*, Ital. *Predella*), a part of the floor at the end of a mediæval hall, raised a step above the rest of the building, and called in France "estrade." On this the lord of the mansion dined with his friends at the great table, apart from the retainers and servants. In mediæval halls there was generally a deep recessed bay window at each end of the dais, supposed to be for retirement, or greater privacy than the open hall could afford. In France, the word is understood as a canopy or hanging over a seat : probably the name was given from the fact that the seats of great men were then surmounted by such an ornament. (See ESTRADE, FOOTPACE.)

DATES OF STYLES. It is not to be supposed that the styles of architecture became changed all at once, or without opposition from those who, from old custom, loved and adhered to that of their forefathers ; nor is it to be supposed that those new methods of design adopted in the metropolis, or the larger towns, spread directly to the provinces or little country places. We therefore find that works were going on in a new style at one place, when at another the old forms were preferred, and it took some time to eradicate them. Styles, therefore, in a very great degree "overlap each other," as it has been very properly expressed. The following short approximate list may, however, be useful to the student, though nothing but the study of *dated* examples, and the history of the bodies or individuals by whom they were erected, can give the faculty of correct discrimination.

Early, or plain Norman, from.........1066	to about	1100
Richer, do., ...	,,	1160
Transition from do., to Anglo Norman,...	,,	1190
Pure early English,...........................	,,	1260
Transition to Decorated, ...	,,	1300
Pure Decorated, ...	,,	1370
Transition from Decorated, to Perpendicular,....................................	,,	1400
Early Perpendicular, ...	,,	1490
Later do., or Tudor, ...	,,	1550

Or it may be given more roughly thus, in centuries, or parts of centuries.

From the Conquest to about the beginning of the 12th century,........	Early Norman.
To about the middle of the 12th, ...	Richer do.
Do. do. beginning of the 13th,	Transition.
Do. do. middle of the 13th,.....	Early English or first pointed.
Do do. beginning of the 14th,	Transition.
Do. do. third quarter of the 14th,..................................	Decorated or second pointed.
Do. do. beginning of the 15th,	Transition.
Do. do. beginning of the 16th,	Perpendicular or third pointed.
Do. do. middle of the 16th,......	Tudor or four-centred.

Or thus, in the reign of the various monarchs :—

William I. to Stephen, from 1066 to 1154, ...	Norman.
Henry II. .... to 1189,	Anglo-Norman.
Richard I. to Henry III., to 1272,	Early English.
Edward I.,................... to 1307,	Transition.
Edward III.,................. to 1377,	Decorated.
Richard II.,................. to 1399,	Transition.
Henry IV., Henry VIII., } .... to 1546,	Perpendicular and Tudor.

DECASTYLE (Gr. δεκα, ten, and στυλος, a column), a portico of ten columns in front. (See note to the term HEXASTYLE.) The portico to University College, London, is of this description ; more particularly described, it is decaprostyle and recessed.

DECORATED STYLE, also called second pointed (Fr. *ogival secondaire, gottique rayonnant,* Ger. *spitzbogen stiil*),

the style which developed itself from the early English. Its chief characteristics are the tracery of the *windows,* the greater delicacy of the *mouldings of piers, arches, &c.,* the *foliage of capitals,* &c., which is more natural, the lessening of the importance of the *triforium,* and the increasing that of the *clerestory;* the increased richness of the *groining* and introduction of more ribs ; the lowering the unreasonable steepness of the *roofs;* the panelling the *buttresses,* and the introduction of *niches* therein ; the peculiarly *ornamented mouldings,* as the *ball-flower;* the triangular *gablets* and *hood-moulds.* The description of all these, and that of the *doorways, piers, arches, towers, windows, groinings,* &c., will be found under the respective heads. (See also DATES OF STYLES.) One of the finest specimens of geometrical Decorated is supposed to be the choir at Lincoln, given at Plate XX. fig. 2. Another is the lower part of the west front at York Minster (the tops of the towers are Perpendicular); this is given, Plate XXI. fig. 1. There are but few fine specimens of this style in France and Belgium ; but the constant wars that devastated these countries during that period may account for this circumstance.

DECORATED TRANSITION. This style shades off, as it were, into the early Perpendicular. Its chief features are that the arches become more depressed, and the tracery has more or less a tendency to perpendicular lines. It lasted so short a time, that little more can be said of it.

DENTIL (Lat. *dens,* a tooth). The cogged or toothed member, so common in the bed-mould of a Corinthian entablature, is said to be dentilled ; and each cog or tooth is called a dentil. (Plate XIII. fig. 1.)

DEPRESSED ARCHES or DROP ARCHES, those of less pitch than the equilateral. (See STONE-MASONRY, MEDLÆVAL, Part V. fig. 64.)

DESIGN. Architects apply this term to what is vulgarly called a plan, intending by it the scheme or design of a building in all its parts, the term *plan* having a distinct application to a technical portion of the design. (See PLAN.) The plans, elevations, sections, and whatever other drawings may be necessary for an edifice, exhibit the design.

DETAIL. As used by architects, detail means the smaller parts into which a composition may be divided. It is applied generally to mouldings and other enrichments, and again to their minutiæ.

DIAMETER (superior and inferior). The greater diameter of the shaft of a column is technically termed its inferior, because it is that of the lower end ; and the lesser, that of the upper end, its superior diameter.

DIAPER (It. *diaspro,* Fr. *diaspré,* Ger. *gelumterwerk*), a method of decorating a wall panel, stained glass, or any plain surface, by covering it with a continuous design of flowers, rosettes, &c., either in squares or lozenges, or some geometrical form resembling the pattern of a diapered table-cloth, from which, in fact (drap d'Ypres), the name is supposed by some to have been derived. Several examples are given, Plate XXIX.; the most curious is one, fig. 7, but little known from Auxerre.

DIASTYLE (Gr. δια, through, and στυλος, a column), a spacious intercolumniation, to which three diameters are assigned. (See EUSTYLE.)

DIPTERAL. (See DIPTEROS.)

DIPTEROS (Gr. δις, twice, and πτερον, a wing), a double-winged temple. The Greeks are said to have constructed temples with two ranges of columns all round, which were called dipteroi. A portico projecting two columns and their interspaces is of dipteral or pseudo-dipteral arrangement. See description of fig. 3, Plate V.

DISCHARGING ARCH, an arch generally of hard stone over the opening of a door or window, to discharge or relieve the superincumbent weight from pressing on the freestone. (See STONE-MASONRY, Part V. art. 136, fig. 67, often called Relieving Arch.)

DISTEMPER. (See FRESCO, TEMPERA.)

DISTYLE (Gr. δις, twice, and στυλος, a column), a portico of two columns. This term is not generally applied to the mere porch with two columns, but to describe a portico with two columns *in antis*. The elevation of the pronaos of the hexastyle peripteral temple, Plate V. fig. 2, exhibits an example of distyle *in antis*.

DITRIGLYPH (Gr. δις, twice, and triglyph, *q. v.*), an intercolumniation in the Doric order, of two triglyphs. (See MONOTRIGLYPH.)

DODECASTYLE (Gr. δωδεκα, twelve, and στυλος, a column), a portico of twelve columns in front. (See note to HEXASTYLE.) There is no portico of this description in London at present. The lower one of the west front of St Paul's Cathedral (Plate XV.) is of twelve columns, but they are coupled, making the arrangement pseudo-dodecastyle. (See PSEUDO-PROSTYLE.) The Chamber of Deputies in Paris has a true dodecastyle.

DOG-TOOTH, a favourite enrichment in use from the latter part of the Norman period to the early part of the Decorated. It is in the form of a four-leaved flower, the centre of which projects, and probably was named from its resemblance to the dog-toothed violet. (See Plate XXXIII. figs. 8, 9, and 12.)

DOME (Gr. δωμα, a structure of any kind; whence the Latin *domus*, a house or temple), a cupola or inverted cup on a building. The application of this term to its generally received purpose is from the Italian custom of calling an archiepiscopal church, by way of eminence, *Il duomo*, the temple; for to one of that rank, the cathedral of Florence, the cupola was first applied in modern practice. The Italians themselves never call a cupola a dome: it is on this side of the Alps the mistake has arisen, from the circumstance, it would appear, that the Italians use the term with reference to those structures whose most distinguishing feature is the cupola, tholus, or (as we now call it) dome. (See CUPOLA.)

DONJON, the principal tower of a castle. (See KEEP.)

DOOR, DOOR-WAY (Lat. *porta*, Ital. *porta, portone*, Fr. *porte*, Ger. *thur*), the entrance into a garden-court or inclosed space, also to a building or chamber. In the few remains which are considered Saxon, the doorways are very plain, being sometimes only a semicircular arch over square jambs, without even a necking; the voussoirs being sometimes of long, thin rough stones set endways, and the character of the work like common Roman arches. At Barton on Humber, Barnack, and at Deerhurst, instead of arches, two large stones cover the opening, joined together at the top like a triangle, very much resembling the doorway into which the cup-bearer is entering, taken from the Saxon MS. generally called the Ælfric Pentateuch. (See ROOF, MEDIÆVAL, fig. 34.) In other examples there are rude imposts, sometimes square pieces of stone, sometimes roughly moulded. At Earl's Barton (Plate XXXV. fig. 2) there is a narrow pilaster, which is continued round the circular head of the door in pieces of various lengths and rough imposts. At Corhampton and at Stanton Lacy there is the same arrangement, but narrower, and exactly like what is called the strip pilasters (see Plate XXXV. figs. 1 and 2), which are from Sompting and Earl's Barton. In later Saxon times, and in the early Norman, the doorways, as well as the windows, partook of the character of the work of the Confessor (Plate XXIII. fig. 1); later they became more deeply recessed, the mouldings arranged so as to compose a series of recessed squares (see MOULDINGS); they generally have jamb-

shafts, as at Ifley (Plate XVIII. fig. 8), but sometimes the mouldings are continued round the whole opening without even an impost, as fig. 7. The label moulding sometimes is flush with the wall, and sometimes projects slightly. Its termination is frequently on a grotesque head, and often upon the impost (see LABEL TERMINATIONS). In later Norman work the mouldings are very richly decorated (see as above, and with the detail at greater size Plate XXXIII. figs. 1 to 7). Some Norman doorways fill the whole face of a tower or porch between buttress to buttress, and sometimes the whole west front of the nave of a cathedral, as at Lincoln and Rochester. One very curious feature of the Norman doorway is, that it frequently has a level impost running through from jamb to jamb, above which is a flat space, generally called a tympanum, which is often enriched by sculptured figures or other carving. Near Newark is a Norman doorway with such a tympanum, the impost carrying which is part of the stem of a Runic cross, showing that these curious monuments are older than the Norman period. A few of these examples have a triangular canopy over them. Early English doorways are also deeply recessed, generally with pointed heads, but the smaller examples are often square-headed, with small haunchings. The arrangement of the mouldings will be described (MOULDINGS), as also the jamb-shafts, which generally are more slender, and are banded. The capitals, also, are generally foliated, as will be shown under the various heads. The label mouldings in all cases project, and have foliaged terminations or heads; there are frequently triangular canopies over those doorways. One remarkable feature in this style is, that many of the doors are double, that is, with two arches side by side, separated only by a shaft under a series of recessed arches spanning both. Perhaps the earliest example in this style is at Beverley, in the south transept, Plate XIX. fig. 1; a very fine example is also given from Higham Ferrers, Plate XXX. fig. 3. Decorated doorways possess very much the same features as these last, but, as is general in this style, the mouldings, jamb-shafts, &c., are more slender, less undercut, and generally of finer proportions; like the early English, the hollows are often ornamented, particularly with the ball flower, and sometimes, as at the Chapter House at Rochester, with small niches and canopies with figures in them running round the whole jamb and arch. Sometimes the doors are surmounted by ogee arches, as at Crick. The double doors in this style are very fine; one in the cloister at Westminster, lately laid open, is given, Plate XXX. fig. 2; this leads into the vestibule to the Chapter House, and is enriched in every possible way. Another equally beautiful is in the West front at York Minster, Plate XXI. fig. 1; and at larger scale, Plate XXX. fig. 5. Early Perpendicular doors differ but little from these last described, but by degrees the labels were carried up so as to form a square, under which was the arch, the spandrels between which were decorated in various ways, sometimes with circles containing cinque-foils, sometimes with shields. (See DOORWAY, Plate XXII. fig. 1.) At this period the four centred arches came gradually into use; the method of delineating which is given in STONE-MASONRY, MEDIÆVAL (Part V. fig. 66); the doorways then became very richly ornamented, see Plate XXX. fig. 9, which is from Magdalene College, Oxford. *Doors* themselves are of all characters. The early ones have mostly perished; in fact, it has been said that no Norman door exists. For the other periods, we have sometimes plain ledged doors, sometimes panelled in the inside, and ledged on the out. The ledges sometimes are plain; sometimes fixed lapping over each other like a park paling; sometimes they are raised to an arris in the middle ; very often plain ledged, with champhered fillets nailed over them to cover the joints. The better class of doors are panelled outside, as Plate XXX. fig. 3, from Higham Ferrers, and in the Decorated and Perpendicular periods, have rich tracery with cusping, &c., see Plate XXII. fig. 1; the door to Westminster Hall, said to be a reproduction of the original, and Plate XXX. fig. 9, that at Magdalene College, Oxford. A very great number of mediæval doors are simply studded with large projecting squares, or polygonal nails (*ib.* fig. 5). Some have rich iron-work in the way of hinges, &c. (See IRON WORK.) In general, the internal side of mediæval doors is less ornamented than the outside.

DORMER BEAM or DORMANT BEAM, is said by Nicholson to be a tie-beam, but more probably, as its name imports, is a sleeper.

DORMER WINDOW (Fr. *lucarne*, Ital. *abbaino*, Ger. *dachfenster*), a window belonging to a room in a roof, which consequently projects from it with a valley gutter on each side. They are said not to be earlier than the 14th century. In Germany there are often several rows of dormers, one above the other. In Italian Gothic they are very rare ; in fact, the former have an unusual steep roof, while in the latter country, where the Italian tile is used, the roofs are rather flat.

DORMITORY (Fr. *dortoir*, It. *dormitorio*, Ger. *schlafgemach*), the place where the monks slept at night. It was sometimes one long room like a barrack, and sometimes divided into a succession of small chambers or cells. The dormitory was generally on the first floor, and connected with the church, so that it was not necessary to go out of doors to attend the nocturnal services. In the large houses of the late Perpendicular period, and also of some of the Elizabethan, the entire upper story in the roof formed one large apartment, said to have been a place for exercise in wet weather, and also for a dormitory for the retainers of the household, or those of visitors, on occasions of great hunting parties, &c., as at Knole, Eastbury, &c.

DOSSEL or DORSAL. (See REREDOS.)

DOUBLE DOORWAY. These are in pairs, and supposed to be intended for the convenience of processions ; they are generally found at the entrances of important buildings, as Beverly Minster, Plate XIX. fig. 1 ; York Minster, Plate XXI. fig. 1 ; and almost invariably at the entrances of Chapter Houses. (See DOOR.)

DRIPSTONE, the moulding or cornice which acts as a canopy to doors and windows. Horizontal running mouldings are sometimes called tablets and sometimes dripstones.

DROPS. (See GUTTÆ.)

DROP ARCHES. (See DEPRESSED ARCHES.)

DUNGEON, the prison in a castle keep, so called because the Norman name for the latter is donjon, and the dungeons or prisons are generally in its lowest story. (See KEEP.)

EARLY ENGLISH, also called first pointed, *ogival primaire;* the style which developed itself after the Transition or Anglo-Norman; which style, though having pointed arches, preserved the massiveness, not to say heaviness, of the pure Norman. The early English, on the contrary, assumed the lightest appearance, the *piers* were surrounded by slender *shafts,* frequently detached, and often of Purbeck or other marble; the *windows* became very narrow and elongated, sometimes ten times as high as wide ; the *arches* were often lancet, very frequently equilateral, and sometimes depressed. (See STONE-MASONRY, MEDIÆVAL, Part V. figs. 63, 64, and 65.) Now and then the *windows,* and sometimes the smaller *doors,* had trefoil heads; the *groins* had a greater number of ribs; the *roof* became of much higher pitch; the *towers* were

loftier and lighter, and surmounted by spires; the *buttresses,* from the flat, pilaster-like shape of the former styles, became very massive, and of great projection, and finished with rising gablets; and in this period the *flying buttresses* were developed; the *mouldings* became more numerous, bolder, and deeper undercut; the *foliages* took a very peculiar form, being of crisp stiff character, executed with extreme boldness, and deeply undercut; the ornaments in the mouldings were much less frequent than in the Norman style, that chiefly used being the *dog-tooth.* All these are described under their respective heads.

EARLY ENGLISH TRANSITION. Later in the style, the various things enumerated gradually changed their character, till they were ultimately developed as the *Decorated* style. The great and marked change being the rudiments of *tracery,* which at first consisted of mere perforations in the flat stone, these passed in this period into *Geometrical Tracery.* At this time, too, we get lighter *pinnacles,* which were developed from the heavy gablet heads of the buttresses, and the *crockets,* the earliest perhaps of which are given, Plate XXVIII. figs. 10–14.

ECHAUGETTE. (See BARTIZAN.)

ECHINUS (Gr. εχινος, an egg), a moulding of eccentric curve, which (when it is carved) being generally cut into the forms of eggs and anchors alternating, the moulding is called by the name of the more conspicuous. It is the same as Ovalo, *q. v.*

ELEMENT, the outline of the design of a Decorated window, on which the centres for the tracery are formed. (See STONE-MASONRY, MEDIÆVAL, Part V. fig. 67.) These centres will all be found to fall on points which, in some way or other, will be equimultiples of parts of the openings. Before any one can draw tracery well, or understand even the principles of its composition, he must give much attention to the study of the element. (See TRACERY, &c.)

ELEVATION, the front, or *façade,* as the French term it, of a structure. A geometrical drawing of the external upright parts of a building. Architects speak of front, back-front, and side or end elevations.

EMBRASURE, the opening in a battlement between the two raised solid portions or merlons, sometimes called a crenelle. (See BATTLEMENT, CRENELLE.)

ENCIENTE, a French term for the close or precinct of a cathedral, abbey, castle, &c.

ENTABLATURE, or INTABLATURE (Lat. *in,* upon, and *tabula,* a tablet). The superimposed horizontal mass in a columnar ordinance, which rests upon the tablet or abacus of a column, is so called. It is conventionally composed of three parts, architrave, frieze, and cornice, *q. v.* (Plate XIII. fig. 1.)

ENTAIL, ENTAYLE, sculptured ornaments, generally of rich design, most probably derived from the Italian *intaglio.*

ENTASIS, the swelling of a column, &c. (See former glossary.) In mediæval architecture, some spires, particularly those called "broach spires," have a slight swelling in the sides, but no more than to make them look straight, for, from a particular " deceptio visus," that which is quite straight, when viewed at a height, looks hollow. (See SPIRE.)

ÉPI, the French term for a light finial, generally of metal, but sometimes of terra cotta, forming the termination of a pointed roof or spire, see Plate XXXI. fig. 8, from Bourges. (See SPIKEL.)

EPISTYLIUM, or EPISTYLE (Gr. επι, upon, and στυλος, a column). This term may with propriety be applied to the whole entablature, with which it is synonymous; but it is restricted in use to the architrave or lowest member of the entablature.

ESCAPE, a familiar English equivalent for the term Apophyge, *q. v.*

ESCONSON, stated to be a French term for the arris of the internal splay of a window, and Esconson Shaft to be the phrase for the slender column with which such jambs often finish. (Plate XX. fig. 2.) The word, however, is not to be found in M. Viollet-le-Duc. (See SPLAY, JAMB-SHAFT, HOOD-MOULDING.)

ESCUTCHEON (Lat. *scutum*, Ital. *scudo*, Fr. *écusson*, Ger. *wappenschild*), a term for the shields used on tombs in the spandrels of doors or in string courses. (See Plate XXII. figs. 4 and 7.) Also the ornamented plates from the centre of which door-rings, knockers, &c., are suspended, or which protect the wood of the key-hole from the wear of the key. In later times these are sometimes decorated in a very grotesque way.

ESTRADE, a French term for a raised platform. (See DAIS, FOOT-PACE.)

EUSTYLE (Gr. ευ, well, and στυλος, a column), a species of intercolumniation, to which a proportion of two diameters and a quarter is assigned. This term, together with the others of similar import,—pycnostyle, systyle, diastyle, and aræostyle,—referring to the distances of columns from one another in composition, is from Vitruvius, who assigns to each the space it is to express. It will be seen, however, by reference to them individually, that the words themselves, though perhaps sufficiently applicable, convey no idea of an exactly defined space, and by reference to the columnar structures of the ancients, that no attention was paid by them to such limitations. It follows, then, that the proportions assigned to each are purely conventional, and may or may not be attended to without vitiating the power of applying the terms. Eustyle means the best or most beautiful arrangement; but as the effect of a columnar composition depends on many things besides the diameter of the columns, the same proportioned intercolumniation would look well or ill according to those other circumstances, so that the limitation of eustyle to two diameters and a quarter is absurd, and so it is in the case of the other similar terms. With Doric intercolumniation it is different, as may be seen by reference to the word MONOTRIGLYPH.

FAÇADE. (See ELEVATION.)

FAITAGE, the French term for the wooden ridges.

FAITE, the like for the ridge of a roof.

FAITIÉRE, the ornaments running along the ridge of a building. (See CRESTING.)

FAN VAULTING. (See GROINING.)

FASCIA (Lat. a band). The narrow vertical bands or broad fillets into which the architraves of Corinthian and Ionic entablatures are divided, are called fasciæ or fascias; and the term is generally applied to any similar member in architecture.

FEATHERING. (See CUSPS.)

FEMERELL, properly FUMERELL, a sort of lantern in the ridge of a hall (when the fire was in the middle of the floor and not in a chimney), for the purpose of letting out the smoke.

FENESTRALL, a frame or " chassis," on which oiled paper or thin cloth was strained to keep out wind and rain when the windows were not glazed.

FERETORY, a sort of parclose which inclosed the feretrum, shrine, or tomb, as in Henry the Seventh's Chapel.

FILLET (Fr. *filet, liste*, Ital. *listello*, Ger. *binde*), a narrow vertical band or listel, of frequent use in congeries of mouldings, to separate and combine them, and also to give breadth and firmness to the upper edge of a crowning cyma or cavetto, as in an external cornice. The narrow slips or breadths between the flutes of Corinthian and Ionic columns are also called fillets. In

mediæval, a small, flat, projecting square, chiefly used to separate hollows and round. In early English and Decorated work a fillet is often found in the outer parts of shafts and bowtells. (See Plate XXXII. fig. 16.) In this situation the centre fillet has been termed a keel, and the two side ones wings, but apparently this is not an ancient term.

FINIAL (Fr. *fleuron*), the flower or bunch of flowers with which a spire, pinnacle, gablet, canopy, &c., generally terminates. Where there are crockets, the finial generally bears as close a resemblance as possible to them in point of design. They are found in early work where there are no crockets. The simplest form seems to be as at Paris (Plate XXXI. fig. 1), which is more like a bud about to burst than an open flower. They soon became more elaborate, as at Lincoln (fig. 2), and still more, as at Westminster (fig. 3), and the Hotel Cluny at Paris (fig. 4); fig. 5 shows one at Winchester; fig. 6 shows a bud of the lily just bursting into flower, and is cited by M. Viollet-le-Duc as the probable origin of the later form of this object of decoration; fig. 7 is from York, and strongly resembles two crockets placed opposite each other. Many Perpendicular finials are like four crockets bound together; fig. 8 is from Bourges, a form not to be found in England, and which in France is generally called Épi. Almost every known example of a finial has a sort of necking separating it from the parts below. (See ÉPI, CROP.)

FLAMBOYANT, a name applied to the third pointed style in France (*ogive tertiale*), which seems to have been developed from the second as our Perpendicular was from the Decorated. The great characteristic is, that the element of the tracery flows upwards in long wavy divisions like flames of fire. In most cases, also, every division has only one cusp on each side, however long the division may be. The mouldings seem to be as much inferior to those of the preceding period, as our Perpendicular mouldings were to the Decorated, a fact which seems to show that the decadence of Gothic architecture was not due to one country more than another.

FLÉCHE (Ital. *guglio*), a general term in foreign architecture for a spire, but more particularly used for the small slender erection rising from the intersection of the nave and transepts in cathedrals and large churches (see Plate XXIV. figs. 8 and 11), and carrying the sanctus bell.

FLUTE, a concave channel. Columns whose shafts are channelled are said to be fluted, and the flutes are collectively called flutings.

FOIL, FOIL ARCH, FOLIATION. (See CUSP, FEATHERING.)

FOLIAGE, carvings of leaves, buds, flowers, tendrils, &c., used as decorations to columns, bases, cornices, &c. Of Saxon foliage there is scarce any remains. Norman seems to have been either a coarse imitation of the Roman Corinthian capital, as Plate XXVI. fig. 2, from Caen, fig. 6 from the White Tower, figs. 9 and 11 from Sutton, or of a species of interlaced, twisted design, as Plate XXIII. fig. 3, from Canterbury, fig. 13 from Cresset, Plate XXVI. fig. 5, also from Canterbury. In the early English period the foliages became more artistic, but are composed of stiff leaves and stems with crisp-looking tendrils, which some have supposed to be the stalks and leaves of a species of celery, but which M. Viollet-le-Duc supposes, with much greater probability, to have been derived from the shoots of the fern as they first emerge from the earth; those of the Transition period from the Norman are very simple, as Plate XXVI. fig. 3, from Folkestone, figs. 4 and 12 from Buildwas. In the complete early English style, Plate XXVII. fig. 1 is from Salisbury, figs. 2 and 3 from

Stone, Plate XXIII. fig. 10 from Lincoln, Plate XXVIII. fig. 8 from Winchester. In this style, also, the dog-tooth is very common. In the Decorated period the foliages are more natural; in fact, there can be no doubt that they were strictly copied from nature, as may be seen at the celebrated Chapter-House at Southwell. They generally, however, are wrapped round the bell of the capital instead of standing erect, as the examples before cited. The leaves of oak, maple, vine, ivy, strawberry, in fact, of every picturesque plant, are beautifully rendered. Plate XXVII. shows, in twelve examples, the early English foliage and its gradual transition to the pure Decorated. See also Plate XXVIII. fig. 9, and Plate XXV. figs. 2 and 3, which show the difference of the foliages in two bosses. The ball-flower is very common at this period. In the Perpendicular style the foliages are less natural; see the various examples in Plate XXII. A very common form is the Tudor flower (Plate XXXIII. fig. 20), and a sort of vine leaf, with flat bunches of grapes running in a hollow (see *ib.* fig. 22); this is often called *vignette.* (See the various heads—CAPITAL, BASE, BOSS, MOULDINGS, ORNAMENTED do., DOG-TOOTH, BALL-FLOWER.)

FONT, the vessel used in the rite of baptism. The earliest extant is supposed to be that in which Constantine is said to have been baptised; this is a porphyry *labrum* from a Roman bath. Those in the Baptisteries in Italy are all large, and were intended for immersion; as time went on, they seem to have become smaller. What they were in Saxon times is uncertain, though it is not improbable that some of the plain examples, called Norman, may have been of earlier date. Norman fonts are sometimes mere plain, hollow cylinders, generally a little smaller below than above; others are massive squares, supported on a thick stem, round which sometimes there are smaller shafts. In the early English this form is still pursued, and the shafts are detached; sometimes, however, they are hexagonal and octagonal, and in this and the later styles, assume the form of a vessel on a stem. Norman fonts have frequently curious carvings on them, approaching the grotesque; in later times the foliages, &c., partook absolutely of the character of those used in other architectural details of their respective periods. The font is usually placed close to a pillar near the entrance, generally that nearest but one to the tower in the south arcade, or, in large buildings, in the middle of the nave, opposite the entrance porch, and sometimes in a separate building. (See BAPTISTERY.)

FOOT-PACE. (See DAIS, ESTRADE.)

FOOT-STALL, a word supposed to have been a literal translation of piedestalle, or pedestal, the lower part of a pier. (See BASE, PATIN.)

FOOT-TABLE. (See BASE OF A WALL.)

FORMERET, the half ribs against the walls in a groined-ceiling. (See VAULTING, GROINING.)

FRATERY, *Frater House.* Supposed to be the hall where the friars met for dinner, or other purposes; the same as *refectory* among the monks.

FREE-MASON, the worker in *freestone*, as distinguished from the scappler or rough stone mason. These men, like most other trades in the middle ages, formed guilds or societies for the regulation of their craft, and also for religious and charitable purposes; but they could not resemble the modern freemasons, as the Church of Rome always has prohibited (to this day) any secret societies.

FREESTONE (Fr. *pierre de taille*, Ital. *pietra molle*). Stone used for mouldings, tracery, and other work required to be executed with the chisel. The oolitic stones are generally so called, although in some countries the soft sandstones are so used, and in some churches an indurated chalk called clunch is employed for internal lining and for carving.

FRESCO, the method of painting on a wall while the plastering is wet. The colour penetrates through the material, which therefore will bear rubbing or cleaning to almost any extent. The transparency, the chiaroscuro, and lucidity, as well as force, which can be obtained by this method, cannot be conceived unless the frescoes of Fra Angelico or Raphael are studied. The word, however, is often employed improperly to mediæval delineations in ancient churches, which are only painted on the surface in distemper or body colour, mixed with size or white of egg, which gives them an opaque effect. (See DISTEMPER, TEMPERA.)

FRIARY, the set of buildings intended for the accommodation of the various orders of friars; and, in contradistinction to monastery, which was intended for the monks. (See ABBEY, MONASTERY.)

FRIEZE (Ital. *fregio*, from the Lat. *Phrygionius*, enriched or embroidered), that portion of an entablature between the cornice above and the architrave below. (Plate XIII. fig. 1.) It derives its name from being the recipient of the sculptured enrichments either of foliage or figures which may be relevant to the object of the structure. The frieze is also called the zoöphorus, *q. v.*

FRONTISPIECE, the front or principal elevation of a structure. This term, however, is generally restricted in application to a Decorated entrance.

GABLE, sometimes *Gavil* (Fr. *pignon*, Ital. *colma*, Ger. *giebel*). When a roof is not hipped or returned on itself at the ends, its ends are stopped by carrying up the walls under them in the triangular form of the roof itself. This is called the gable, or, indeed, the pediment. The latter term, however, is restricted to the ornamental and ornamented gable; and gable itself is applied to a plain triangular end. The triangular end of the wall of a building where the roof comes to the front, and is not hipped. It answers to the pediment of classic architecture. Of course gables follow the angles of the slope of the roof, and differ in the various styles. In Norman work they are generally about half pitch. In early English, seldom less than equilateral, and often more (see Plate XIX. fig. 1). In Decorated work they become lower, and still more so in the Perpendicular style. In all important buildings they are finished with copings or parapets. In early times the copings were nearly flat. In the later styles gables are often surmounted with battlements (Plate XXI. fig. 1), or enriched with crockets (Plate XXII. fig. 1); they are also often panelled or perforated, and sometimes very richly so. The gables in ecclesiastical buildings mostly are terminated with a cross; in others, by a finial or pinnacle. In later times the parapets or copings were broken into a sort of steps, called *corbie steps* (which see). In buildings of less pretension, the tiles or other roof covering passed over the front of the wall, which then of course had no coping. In this case the outer pair of rafters were concealed by moulded or carved verge boards. (See BATTLEMENT, COPING, CORBIE STEPS, PARAPET, VERGE BOARDS, &c.)

GABLE BOARD. (See VERGE BOARD.)

GABLED TOWERS, those which are finished with gables instead of parapets, as at Sompting (Plate XXXV. fig. 1).

GABLE WINDOW, a term sometimes applied to the large window under a gable (Plate XXI. fig. 1, Plate XXII. fig. 1), but more properly to the windows in the gable itself. See the upper window, Plate XIX. fig. 1. Also LUCARNE.

GABLETS, triangular terminations to buttresses, much in

use in the early English and Decorated periods, after which they generally terminate in pinnacles. The early English are generally plain, and very sharp in pitch (see those, Plate XIX. fig. 1, on the buttresses to the aisles), or with small trefoils cut in between the slopes (see *ib.*) on the nave. In the Decorated period they are often enriched with panelling and crockets. They sometimes are finished with small crosses, but oftener with finials.

GALILEE, a species of porch where, it has been said, the female relatives of the monks went to confer with them, they not being permitted to enter the conventual buildings. The name is supposed to be derived from the scriptural expression, " He shall go before you into Galilee." At Durham the galilee is a sort of chapel at the main entrance into the nave. For a description of similar buildings on the Continent, see PORCH, PORTAL.

GALLERY, any long passage looking down into another part of a building, or into the court outside. The triforium and clerestory passages (Plate XX. figs. 1 and 2) are galleries; as also are the arcades at the cathedral at Pisa (Plate XXI. fig. 2). In like manner, any stage erected to carry a rood, or an organ, or to receive spectators, was latterly called a gallery, though originally a loft. In later times any very long room was called a gallery, particularly those intended for purposes of state, (See LOFT, TRIFORIUM, ARCADE, &c.)

GARGOYLE or GURGOYLE (Fr. *gargouille, canon, lanceur,* Ital. *doccia di gronda,* Ger. *ausguss*). Carved terminations to the spouts which conveyed away the water from the gutters, and are supposed to be called so from the gurgling noise made by the water passing through them. They are, with few exceptions, mostly grotesque figures; Plate XXXI. fig. 9 is from Petherton, fig. 10 from Ruislip, figs. 11 and 12 from Amiens, fig. 13 from York, and 14, which is an elegant figure pouring water out of a sort of leather vessel, is from Troyes. In Plate XXXIX. fig. 3, is a curious example of a gargogle from Meaux Cathedral, which exhibits the use of the channel in a clear manner. Many early gargoyles in France are constructed in this way.

GARRETTING, properly GALLETTING; from gallet, a small piece of stone chipped off by the chisel. A method of protecting the mortar joints in rough walls by sticking in chips of stone while the mortar is wet.

GATE HOUSE, a building forming the entrance to a town, the door of an abbey, the enceinte of a castle or other important edifice. They generally had a large gateway, protected by a gate, and also a portcullis, over which were battlemented parapets with holes (machicolations) for pouring down darts, melted lead, or hot sand, on the besiegers. Gatehouses always had a lodge, with apartments for the porter, and guard-rooms for the soldiers; and generally rooms over for the officers, and often places for prisoners beneath. They are sometimes open in the rear, as at Cooling Castle, and often have doors with portcullises, &c., on both sides, in case the enemy should scale the walls, and attack them both in front and rear. In this case, the space between, on the ground floor, is generally groined over, with holes for missile weapons.

GEOMETRICAL STYLE. The early part of the Decorated period is often so called, because the element of the tracery is composed of circles, or other regular geometrical figures, see Plate XXXVI. figs. 5 to 8; see also the lower windows, Plate XXI. fig. 1, the contrast between which and the main window, which is of flowing tracery, is very remarkable. (See DECORATED STYLE, TRACERY, &c.)

GLASS, GLAZING (Fr. *verrerie, vitrail,* Ital. *vetro, invetrata,* Ger. *glas*). The use of glass to protect apartments from the weather is at least as old as the time

of the Romans, this material being found in windows at Herculaneum and Pompeii. There are also many numerous specimens of coloured glass, forming various utensils, but it does not appear they used this material for windows. In England, glass for windows is mentioned as early as the time of Benedict Biscop. The earliest stained glass known in this country is at Canterbury; this is of the Norman period. These windows generally are filled with circles, lozenges, or other figures, sometimes nearly touching each other, in which scriptural subjects are painted; they have small borders, and the rest of the lights are filled up with regular patterns. Norman windows are remarkable for the extreme brilliancy of their colours, particularly the deep blue and the bright ruby, which have never been surpassed. In the early English period these panels are smaller, and are often in the form of the visica piscis. The borders are narrow, and the filling in is sometimes of scroll foliage, and sometimes of diaper patterns, sometimes the windows are entirely filled with stalks and leaves, running in scrolls over the whole surface, the designs being somewhat stiff, like the foliage of the period. About this period windows of less importance were often filled in with *quarry* glass, that is, lozenge-shaped squares on which a flower, leaf, or monogram was painted in black, heightened sparingly with colour. This method continued in use till the latest time. In the Decorated period the pictured subject was gradually abandoned, and single figures substituted, standing on a pedestal, and with a small canopy over them. The filling-in foliage was more flowing and natural. At the conclusion of this period and commencement of the Perpendicular, the figures were larger, and the canopies and pedestals extremely rich and elaborate. The tracery was also filled in with angels bearing scrolls and heraldic devices. In the Tudor style the windows gradually became large, and regular pictures, covering the whole surface, as at King's College Chapel, St Margaret's, Westminster, &c. The use of stained glass continued up to the time of the great Rebellion, when the iconoclastic spirit which prevailed induced a destruction which was almost universal, so that original glass is now a rarity. (See LEAD WORK, QUARRY.)

GLYPH. (See TRIGLYPH.)

GRADINO (Ital. dim. of *gradus,* a step). Architects frequently use the plural of this term, gradini, and to gradinate, instead of the English, steps, and to graduate, perhaps without sufficient reason, though they find them useful to distinguish what they intend from the meaning of the latter words in their ordinary acceptation.

GRANGE, a word derived from the French, signifying a large barn or granary. They were usually long buildings with high wooden roofs, sometimes divided by posts or columns into a sort of nave and aisles, and with walls strongly buttressed. In England the term is applied not only to the barns, but to the whole of the buildings which formed the detached farms belonging to the monasteries; in most cases there was a chapel either included among these or standing apart as a separate edifice.

GREES or GREECES, a corruption of the French *dégrés,* steps or stairs.

GRIFFE, a French term for an ornament at the angles of the base of early pillars, for which we have no proper term, see Plate XXIII. fig. 9, from Rochester. It first consisted of a single leaf, which became more elaborate, and was, no doubt, the origin of the foliated bases (*ib.* figs. 10 and 13).

GRILLE, the iron work forming the inclosure-screen to a chapel, or the protecting railing to a tomb or shrine; they are more common in France than in England. Our best example, perhaps, is that round the tomb of Queen

Eleanor in Westminster Abbey—of course they are all of wrought iron, ornamented by the swage and punch, and put together either by rivets or clips. (See IRON WORK).

GROIN, by some described as the line of intersection of two vaults where they cross each other, which others call *the groin point;* by others the curved section or spandrel of such vaulting is called a groin, and by others the whole system of vaulting is so called.

GROIN ARCH (Fr. *arc doubleaux*), the cross rib in the later styles of groining, passing at right angles from wall to wall, and which divides the vault into bays (travées) or *severies.*

GROIN CEILING, a ceiling to a building composed of oak ribs, the spandrels of which are filled in with narrow thin slips of wood. There are several in England; one at the early English church at Warmington, and one at Winchester Cathedral, exactly resembling those of stone.

GROIN CENTERING. In groining without ribs, the whole surface is supported by centering during the erection of the vaulting. In later mediæval work the stone ribs only are supported by timber ribs during the progress of the work, any light stuff being used while filling in the spandrels.

GROIN POINT, the name given by workmen to the arris or line of intersection of one vault with another where there are no ribs.

GROIN RIB (Fr. *nerf d'arrête*, Ital. *costola*, Ger. *rippe*), the rib which conceals the groin point or joints where the spandrels intersect. (See GROINED VAULTING. For their mouldings, see RIB.)

GROINED VAULTING (Lat. *fornix, testudo*, Fr. *voute d'arête*, Ital. *fornice*), the system of covering a building with stone vaults which cross and intersect each other, as opposed to the barrel vaulting (*voute de berceau*) or series of arches placed side by side. The earliest groins are plain without any ribs, except occasionally a sort of wide band from wall to wall to strengthen the construction. The Saxon vaulting at Repton and the early Norman had the arches intersecting each other without any cross ribs, as Plate XXXVIII. figs. 1 and 2, which are a perspective view and plan of the crypt under part of Bow Church—a very curious specimen, which is now unfortunately walled up. In later Norman times ribs were added on the line of intersection of the spandrels, crossing each other, and having a boss as a key common to both; these ribs the French authors call *nerfs en ogive.* Their introduction, however, caused an entire change in the system of vaulting; instead of arches of uniform thickness and great weight, these ribs were first put up as the main construction, and spandrels (*remplissage*) of the lightest and thinnest possible material placed upon them, the haunches only being loaded sufficiently to counterbalance the pressure from the crown (see fig. 3). Shortly after, half ribs against the walls (*formerets*) were introduced to carry the spandrels without cutting into the walling, and to add to the appearance. Thus, in fig. 3, *a c, b d,* which cross the church, would be *arcs doubleaux, a d* and *b c* the *ogives,* and *a b, c d,* the half ribs against the walls. The work was now not treated as continued vaulting, but as divided into bays (*travées*), and it was formed by keeping up the ogive or intersecting ribs and their bosses, a sort of construction having some affinity to the dome was formed, which added much to the strength of the groining. Of course, the top of the soffit or ridge of the vault was not horizontal, but rose from the level of the top of the formeret-rib to the boss and fell again, but this could not be perceived from below. This system is illustrated in fig. 4, and more particularly in fig. 5, where the ridge line is

shown by a dotted line. Both these are in isometrical perspective, and the plans are shown below. Indeed, it is a curious fact, that this important circumstance had escaped the notice of any one till a short time ago. As this system of construction got more into use, and as the vaults were required to be of greater span and of higher pitch, the spandrels became larger and wanted more support. To give this another set of ribs was introduced, passing from the springers of the ogive ribs, and going to about half-way between these and the ogive, and meeting on the ridge of the vault; these intermediate ribs are called by the French *tiercerons,* and began to come into use in the transition from early English to Decorated. Thus, in Plate XX. fig. 2, the ribs rising from the vaulting shafts over the centre of the piers are the *arcs doubleaux,* those rising from the same point and meeting over the top of the window are the *ogives,* while the intermediate ribs are the *tiercerons.* About this period a system of vaulting came into use called *hexpartite,* from the fact that every bay is divided into six compartments instead of four, and was invented to cover the naves of churches of unusual width. In every bay there are a set of two arches against each wall, see fig. 5; see also Plate XXXVIII. figs. 6 and 7, which are the elevation and plan of hexpartite groining from the chapel of St Blaise in Westminster Abbey. The filling of the spandrels in this style is very peculiar; and, where the different compartments meet at the ridge, some pieces of harder stone have been used, which have rather a pleasing effect. This will be best understood by an inspection of fig. 7. The arches against the wall being of smaller span than the main arches, cause the centre springers to be perpendicular and parallel for some height, and the spandrels themselves are very hollow. As styles progressed, and the desire for greater richness increased, another series of ribs, called *liernes,* was introduced; these passed crossways from the *ogives* to the *tiercerons,* and thence to the *doubleaux,* dividing the spandrels nearly horizontally (see fig. 11, and Plate XXXIX. fig. 1). These various systems increased in the Perpendicular period, so that the vaults were quite a net-work of ribs, and led at last to the Tudor, or, as it is called by many, *fan tracery* vaulting. In this system the ribs are no part of the real construction, but merely carved upon the voussoirs which form the actual vaulting. In the section of these, the four-centered arch was generally used; the method of finding the lines for which see STONE-MASONRY, MEDIÆVAL, V. 134, &c. FAN TRACERY is so called because the ribs radiate from the springers, and spread out like the sticks of a fan. These later methods are not strictly groins, for the pendentives are not square on plan, but circular, and there is therefore no arris intersection or GROIN POINT (which see). If we imagine a cone, the sides of which are not straight from apex to base, but are of the section of the four-centered curve, and if we suppose this cone cut in half downwards and inverted, and the smaller end placed on the springer, a good representation of the pendentive will be arrived at. A perspective delineation of the under surface of a fan vault, showing the arch stones with their jointing as viewed from below, is given Plate XXXVIII. fig. 9, and another, fig. 10, as viewed from above, shows the outer or upper surface of the same. This method of showing the construction of the voussoirs was first suggested by M. Viollet-le-Duc. Plaet XXXIX. fig. 1, which is from Canterbury, shows the first step to fan vaulting, the spandrel of which is not quite circular in plan, but polygonal; while fig. 2, which is from Winchester, shows the same system in its latest and richest form. Of course they will not cover the whole surface, but will leave an opening of a form somewhat resembling a lozenge with four curved sides (see at *a*, Plate XXXVIII. fig. 9); this is filled in with voussoirs carved much like the other tracery, and full of ribs; if these ribs are disposed, as they often are, in form of stars, this is called STELLAR VAULTING (see Plate XII., which is from Canterbury).

GROINS, WELSH, or UNDERPITCH. When the main longitudinal vault of any groining is higher than the cross or transverse vaults which run from the windows, the system of vaulting is called underpitch groining, or, as termed by the workmen, Welsh groining. A very fine example is at St George's Chapel, Windsor.

GROUND TABLE. (See BASE OF A WALL, BASE TABLE.)

GUILOCHE or GUILOCHOS (Gr. γυιον, a member, and λοχος, a snare). An interlaced ornament like network, used most frequently to enrich the torus. (Plate VIII.)

GUTTÆ (Lat. drops). The small cylindrical drops used to enrich the mutules and regulæ of the Doric entablature are so called.

GUTTER, the channels for carrying off rain-water. Their use and formation is given under the various heads of *Building construction*. The mediæval gutters differed little from others, except they are often hollows sunk in the top of stone cornices, in which case they are generally called *channels* in English, and *cheneau* in French.

HAGIOSCOPE, a name lately coined out of two classic Greek words, very improperly used to describe certain oblique openings in the inside of mediæval buildings for the purpose of seeing the altar. (See SQUINT.)

HALL (Fr. *salle, salon*, Ital. *sala, salone*, Ger. *borsaal*), the principal apartment in the large dwellings of the middle ages, used for the purposes of receptions, feasts, &c. In the Norman castle the hall was generally in the keep above the ground floor (where the retainers lived), the basement being devoted to stores and dungeons for confining prisoners. Later halls, indeed, some Norman halls (not in castles), are generally, on the ground floor, as at Westminster, approached by a porch either at the end, as in this last example, or at the side, as at Guildhall, London, having at one end a raised DAIS or ESTRADE, (which see). The roofs are generally open, and more or less ornamented. (See ROOFS, MEDIÆVAL, figs. 35–47.) In the middle of these was an opening to let out the smoke (see LOUVRE, FUMERELL), though in later times the halls have large chimney places with funnels or chimney shafts for this purpose. At this period there were usually two deeply recessed bay windows at each end of the dais, and doors leading into the withdrawing-rooms or the ladies' apartments; they are also generally wainscoted with oak, in small panels, to the height of five or six feet, the panels often being enriched. Westminster Hall was originally divided into three parts, like a nave and side aisles, as are some on the Continent. Besides the halls above mentioned, the best known to our English readers are those at the Inns of Court and Crosby Hall in London, and at Eltham and Hampton Court, near the same; those of the colleges at Oxford and Cambridge, Ightham in Kent, Mayfield in Sussex, &c.

HAMMER BEAM. (See ROOF, MEDIÆVAL, figs. 42, 43, and 44.)

HAUNCH, HANSE, of a door, a word by some supposed to signify the projecting brackets which support the *sperver*, a species of flat canopy or cover to keep off the weather; by others, with more reason, the spandrel or space between the arch and the square label over the doors in Tudor work. (See Plate XXX. fig. 9. See SPERVER.)

HELIX (Gr. ἑλιξ, a wreath or ringlet), used synonymously with Cauliculus, *q. v.* It forms in the plural Helices.

HEMIGLYPH (Gr. ἡμισυς, half, and γλυφη, an incision or channel). The half-channels, or rather chamfered edges, of a triglyph tablet, may be so called. The two hemiglyphs are included to make the third channel, and complete the triglyph. (See TRIGLYPH.)

HERRING-BONE, bricks or other materials arranged diagonally in building. (See art. BUILDING—*Bricklayer*.)

HEXASTYLE (Gr. ἑξ, six, and στυλος, a column). A portico of six columns in front[1] is of this description. Most of the churches in London which have porticoes have hexaprostyles. (See PROSTYLE.)

HEXPARTITE, a species of groining where each bay is divided into six parts instead of four, as is the more usual plan. (See Plate XXXVIII. figs. 5, 6, and 7—GROINED VAULTING.)

HIGH ALTAR, the principal altar in a cathedral or church. Where there is more than one, it is generally at the end of the choir or chancel, not in the Lady Chapel. At St Albans it stood at the end of the nave, close to the choir screen.

HINGE. (See IRON WORK.)

HIP-KNOB, the finial on the hip of a roof, or between the barge boards of a gable. (See Plate XXXIX. fig. 7 from Ightham. See CROSS, FINIAL.)

HOOD-MOULD, a word used to signify the drip-stone or label over a window or door opening, whether inside or out; but it seems more properly to be applied to the mouldings at the arris of the arch at the inner side of such opening. Sometimes these assume the form of a label, and have jamb-shafts. Frequently the soffit is slightly hollowed and finishes with an arris. (See DRIPSTONE, LABEL.)

HOTEL DE VILLE (Ital. *broletto, palazzo communale*), the town or Guild-hall in France, Germany, and Northern Italy. The building in general serves for all purposes bearing any relation to the administration of justice, the receipt of town dues, the regulation of markets, the residence of magistrates, barracks for police, prisons, and for all other fiscal purposes. As may be imagined, they differ very much in different towns, but it is almost an invariable rule that they have attached to them, or closely adjacent, a large clock-tower (Beffroi), containing one or more bells, intended to call the people together on special occasions. That from the Signoria at Sienna is given, Plate XXIV. fig. 5.

HOTEL DIEU (Fr. *maison dieu*, Ital. *ospedale, lo spedale*), the name for an hospital in mediæval times. In our country there are but few remains of these buildings, one of which is at Dover; abroad there are many. The most celebrated is the one at Angers, lately described by Mr J. H. Parker. They do not seem to differ much in arrangement of plan from those in modern days—the accommodation for the chaplain, medicine, nurses, stores, &c., being much the same in all ages, except that in some of the earlier, instead of the sick being placed in long wards like galleries, as we do now, they are in large buildings, with naves and side aisles like churches. We must refer to the works of Mr Parker, of M. Viollet-le-Duc, and to that of MM. Verdier et Cattois, for further details.

HOUSES, TIMBERED. (See POST AND PAN WORK.)

HYPÆTHRAL. (See HYPÆTHROS.)

HYPÆTHROS (Gr. ὑπο, under, and αιθρα, the air), a temple open to the air, or uncovered. The Greeks frequently made the temples of the supreme divinities hypæthral.

---

[1] The words "in front" are used to prevent the mistake which might arise from the supposition that all the columns in a portico should be counted to designate it. The porticoes of the churches of St Martin-in-the-Fields, and St Mary-le-bone, in London, for instance, have each eight columns, but are hexastyle, nevertheless, there being but six in their front rows respectively.

For instance, those of Jupiter Olympius at Agrigentum in Sicily, of Neptune at Pæstum, and of Minerva Parthenon at Athens, are all of this description. The term may be the more easily understood by supposing the roof removed from over the nave of a church in which columns or piers go up from the floor to the ceiling, leaving the aisles still covered. In that case it would be hypæthral, after the manner of the Greek hypæthros. The Pantheon in Rome having an opening in the centre of the dome, is thereby rendered hypæthral. (See Plates IV. and XI. figs. 4 and 5.)

HYPOGEA (Gr. ὑπο, under, and γη, the earth). Constructions under the surface of the earth, or into the sides of a hill or mountain, are hypogea.

HYPOTRACHELIUM (Gr. ὑπο, under, and τραχηλος, the neck), the moulding or the groove at the junction of the shaft with the capital of a column. In some styles the hypotrachelium is a projecting fillet or moulding, and in others, as the Doric, it is composed of a channel or groove, and sometimes of more than one. (Plate XIII. fig. 1.)

ICHNOGRAPHY (Gr. ιχνος, a footstep or track, and γραφη, a description or representation). A plan, or the representation of the site of an object on a horizontal plane, is its ichnography. The term plan (q. v.) is, however, much more frequently used.

IMPOST, a term in classic architecture for the horizontal mouldings of piers or pilasters, from the top of which the archivolts or mouldings which go round the arch spring. The word is scarcely applicable to mediæval architecture, as the mouldings in general spring from the capital of a shaft, or from a corbel; or they continue without breaking down to the base, or till they are stopped by a chamfer or a regular base moulding, or they die into a plain shaft, or at any rate one of different section. It has been proposed to describe the former as continuous impost and the other as discontinuous. It seems, however, a contradiction in terms to describe an impost, when confessedly there is none. An example of the discontinuous method is given, Plate XVIII. fig. 7. A comparison with fig. 8 will show the difference.

IMPOST (Lat. impositus, laid upon). The horizontal congeries of mouldings forming the capital of a pier, or edge pilaster, which has to support one leg of an arch, is called the impost; sometimes, and more conveniently, this term is used for the pilaster itself, when its capital is called the impost cap or impost mouldings.

INTERCOLUMNIATION (Lat. inter, between, and column, q.v.) The distance from column to column, the clear space between columns is called the intercolumniation.

INTERLACED ARCHES. Arches where one passes over two openings, and they consequently cut or intersect each other. (See Plate XXIII. fig. 2, from St John's, Devizes, and fig. 3, from Canterbury. See ARCH, NORMAN ARCHITECTURE.)

IRON WORK, in mediæval architecture, as an ornament, is chiefly confined to the hinges, &c., of doors and of church chests, &c. That this species of ornamentation was in use among the Saxons, we have the authority of various manuscripts. This people, however, as is well known, were famed as workers in metal. Specimens of Norman iron work are very rare. Of early English there are many, and they are very elaborate. In some instances, not only do the hinges become a mass of scroll work, but the surface of the doors are covered by similar ornaments. In both these periods, the design evidently partakes of the feeling exhibited in the stone or wood carving. In the Decorated period the scroll work is more graceful, and, like the foliage of the time, more natural. As styles progressed, there was a greater desire that the framing of the doors should be richer, and the ledges were

chamfered or raised, then panelled, and at last the doors became a mass of scroll panelling. This, of course, interfered with the design of the hinges, the ornamentation of which gradually grew out of vogue. In almost all styles, the smaller and less important doors had merely plain strap hinges, terminating in a few bent scrolls, and latterly in fleur-de-lys (see DOOR). Escutcheon and ring handles, and the other furniture, partook more or less of the character of the time. On the Continent, the knockers are very elaborate. At all periods doors have been ornamented with nails having projecting heads, sometimes square, sometimes polygonal, and sometimes ornamented with roses, &c. The iron work of windows is generally plain, and the ornament confined to simple fleur-de-lys heads to the stancheons. The iron work, as screens inclosing tombs and chapels, is treated of (see GRILLE) where the method of putting it together is described.

JAMB, the side-post or lining of a doorway or other aperture. The jambs of a window outside the frame are called reveals.

JAMB-SHAFT. Small shafts to doors and windows with caps and bases; when in the inside arris of the jamb of a window they are often called ESCONSON (which see), but apparently without authority.

JAMBETTE, a French term for the upright ashlar piece between the inside of the plate and the rafters. (See ROOFS, MEDIÆVAL.)

JUBÉ, one of the names of the ambo or reading desk in the early Christian church. In later times, a term especially applied to the rood-loft or gallery over the screen, whence the words " Jube Domine benedicere," &c. were read.

KEEL-MOULDING, a round on which there is a small fillet, somewhat like the keel of a ship (see Plate XXXII. figs. 11, 16, and 18). It is common in the early English and Decorated styles.

KEEP (Fr. donjon, Ital. maschio), the inmost and strongest part of a mediæval castle, answering to the citadel of modern times. The arrangement is said to have originated with Gundulf, the celebrated Bishop of Rochester. The Norman keep is generally a very massive square tower; the basements or stories partly below ground being used for stores and prisons. The main story is generally a great deal above ground level, and with a projecting entrance, approached by a flight of steps, and with a drawbridge. This floor is generally supposed to have been the guard-room or place for the soldiery; above this was the hall, which generally extended over the whole area of the building, and is often separated by columns; above this are other apartments for the residents. There are winding staircases in the angles of the buildings, and passages all round in the thickness of the walls. The keep was intended for the last refuge in case the out-works were scaled, and the other buildings stormed. There is generally a well in a mediæval keep, ingeniously concealed in the thickness of a wall, or in a pillar. The most celebrated of Norman times are the White Tower in London, the castles at Rochester, Newcastle, Castle Hedingham, &c. The keep was often circular, as at Conisboro' and Windsor.

KEY-STONE, in classic architecture, the centre voussoir of an arch, often ornamented with carving. In Pointed architecture there is no key-stone. For those to groined arches, see BOSS.

KNOB, KNOT, the bunch of flowers carved on a CORBEL (see Plate XXVIII. figs. 7, 8, and 9), or on a BOSS (see Plate XXV. figs. 2, 3, and 5).

LABEL, the outer projecting moulding over doors, windows, arches, &c., &c., sometimes called Dripstone or Weather Moulding, Hood-Mould. The former terms seem scarcely applicable, as this moulding is often found inside a building where no weather could come, and consequently no drip. The latter term is described, see Hood-Mould. In Norman times the label frequently did not project at all, and when it did it was very little, and formed part of the series of arch mouldings (see at A, Plate XXXII. fig. 4; see also Plate XVIII. figs. 5 to 8; and Plate XXXIV. *Labels*, figs. 1 and 2). In the early English styles they were not very large, sometimes slightly undercut, as Plate XXXIV. fig. 3, sometimes deeply, as fig. 6, sometimes a quarter round and chamfer, fig. 4, and very frequently a " roll " or " scroll moulding " (*ib.* fig. 5), so called because it resembles the part of a scroll when the edge laps over the body of the roll. In the following periods the usual sections are given, *ib.* figs. 8 to 11, and in the Perpendicular, 12 to 17; see also Plate XXXII., and the other plates of doors, windows, arches, &c. Labels generally resemble the string-courses of the periods, and in fact often return horizontally and form strings. They are said to be less common in Continental architecture than in English. (See Dripstone, Hood-Mould, String.)

Label Terminations, carvings on which the labels finished near the springing of the windows. In Norman times these were frequently grotesque heads of fish, birds, &c. Sometimes stiff foliage, as Plate XXIX. fig. 8, from Shoreham. In the early English and Decorated periods, they are often elegant knots of flowers, or heads of kings, queens, bishops, and other persons supposed to be the founders of churches. In the Perpendicular period they often finished with a short, square mitred return or knee, and the foliages are generally leaves of square or octagonal form. See Plate XXIX. figs. 10, 11, and 12; see also the various plates of arches, windows, &c., *passim.*

Lacunar (Lat.), a panelled or coffered ceiling or soffit. The panels or cassoons of a ceiling are more classically called *lacunaria.*

Lancet. A term familiarly applied to the simplest form of the Pointed arch, which is that of the outer end of the surgical instrument, the lancet.

Lancet Arch, a pointed arch of sharper pitch than the equilateral. (See Stone-Masonry, Mediæval, fig. 63.)

Lantern, a name often applied to the *louvre* or *fumerell* on a roof to carry off the smoke. Sometimes to the open construction at the top of towers, as at Ely Cathedral, Boston in Lincolnshire, &c.; probably so called because lights were placed in them at night to serve as beacons.

Lantern (Lat. *lanterna*), a turret raised above a roof or tower, and very much pierced, the better to transmit light. In modern practice this term is generally applied to any raised part in a roof or ceiling, containing vertical windows, but covered in horizontally.

Lantern of the Dead, curious small slender towers, found chiefly in the centre and west of France, having apertures at the top where a light was exhibited at night to mark the place of a cemetery. It is not improbable the round towers in Ireland may have served for this purpose.

Lantern Tower, towers with an open octagonal story at the top, intended to hold a beacon light. There is a very fine Decorated example at Ely, and one of the Perpendicular period at Boston. (See Tower.)

Lavabo (Fr. *lavoir*, Ital. *lavatoio*), the lavatory for washing hands mostly erected in the cloisters of monasteries. Those at Gloucester, Norwich, Lincoln, are best known. A very curious one at Fontenay, surrounding a pillar, is given by Viollet-le-Duc, *q. v.* In general it is a sort of trough, and in some places has an aumbry for towels, &c.

Lich Gate, a covered gate at the entrance of a cemetery, under the shelter of which the mourners rested with the corpse, while the procession of the clergy came to meet them. There is a very fine one at Ashwell, Herts.

Lierne Rib, a rib crossing nearly horizontally from the ogive ribs to the tiercerons or the *arcs doubleaux*, or forming patterns in fan and stellar vaulting. (See Groined Vaulting.)

Light, the portions of a window usually glazed between the mullions, as at A, Plate XXXVI. fig. 13. The upper lights, B, are called *batement lights*; and at C, *angell*, or more probably *angle lights*.

Linen Pattern, a name given to a pattern on the wooden panels of the Tudor period. (See Panel.)

Lip Mould, a moulding of the Perpendicular period like a hanging lip. See the buttress caps, Plate XXV. fig. 12, and the lower mouldings (bases), Plate XXXIV. figs. 9, 10, 11.

Locker, a small aumbry. (See Almery.)

Loft, the highest room in a house, particularly if in the roof; also a gallery raised up in a church to contain the rood, the organ, or singers.

Loop Hole (Fr. *archiére*, *meurtriére*, Ital. *feritore*), openings in the walls of buildings, very narrow on the outside, and splayed within, from which arrows or darts might be discharged on an enemy. They are often in the form of a cross, and generally have round holes at the ends. (See Oylets.)

Louvre, lanterns upon the roofs of halls for the passage of the smoke, when the fire was made on the pavement in the middle. (See Fumerell, Lantern.)

Lucarne, a French term for a garret window; also used to signify the lights or small windows in spires. (See Plate XXXV. figs. 3, 4, and 5.)

Luffers, probably Louvres, pieces of board, slate, or stone, placed slanting so as to exclude the rain, but to allow the passage of smoke, the sound of bells, &c. (See Plate XXXV. fig. 4.)

Lunette, the French term for the circular opening in the groining of the lower stories of towers through which the bells are drawn up.

Machicolation (Fr. *machicollis*) openings between a wall and a parapet, formed by corbelling over the latter, so that the defenders of the building might throw down darts, stones, and sometimes hot sand, melted lead, &c., upon their assailants below.

Manor House, the resident of the suzerain or lord of the manor; in France the central tower or keep of a castle is often called the *manoir*. (See Donjon, Keep.)

Masonry. (See separate article Stone-Masonry.)

Merlon, the solid part of a parapet between the embrasures of a battlement (which see); they are sometimes pierced by loop-holes.

Metope (Gr. μετοπη, a middle space), the square recess between the triglyphs in a Doric frieze. It is sometimes occupied by sculptures. (Plates IV. and V. fig. 4.)

Mezzanine (Ital. *mezzanino*, dim. of *mezzo*, the middle), a low story between two lofty ones. It is called by the French entresol, or inter-story.

Minster, said to be from the German *münster*, but more probably a corruption of monasterium—the large church attached to any ecclesiastical establishment. If the latter be presided over by a bishop, it is generally called a Cathedral, if by an abbot, an Abbey, if by a prior, a Priory (all which see).

Miserere (Fr. *Misericorde*, Ital. *Predella*), seats in the stalls of large churches made to turn up and afford support to a person in a position between sitting and standing. The under side is generally carved with some ornament, and very often with strange grotesque figures and caricatures of different persons. (See Stall.)

Mitre. A moulding returned upon itself at right angles

is said to mitre. In joinery, the ends of any two pieces of wood of corresponding form cut off at 45° necessarily abut upon one another so as to form a right angle, and are said to mitre.

MODILLION (Lat. *modulus*, a measure of proportion), so called because of its arrangement in regulated distances; the enriched block or horizontal bracket generally found under the cornice of the Corinthian entablature. (Plate XIII. fig. 1.) Less ornamented, it is sometimes used in the Ionic. (See also MUTULE.)

MODULE (Lat. *modulus*, from *modus*, a measure or rule). This is a term which has been generally used by architects in determining the relative proportions of the various parts of a columnar ordinance. The semi-diameter of the column is the module, which being divided into thirty parts called minutes, any part of the composition is said to be of so many modules and minutes, or minutes alone, in height, breadth, or projection. The whole diameter is now generally preferred as a module, it being a better rule of proportion than its half.

MONASTERY, a set of buildings adapted for the reception of any of the various orders of monks, the different parts of which are described under the article ABBEY; the word is often used in opposition to that of FRIARY.

MONOPTEROL. (See MONOPTEROS.)

MONOPTEROS (Gr. μονος, one, or single, and πτερον, a wing). This term is incorrectly used by Vitruvius to describe a temple composed of a circular range of columns supporting a tholus, cupola, or dome, but without walls. (See PERIPTERAL.) Such an edifice would be more correctly designated as Cyclostylar, *q. v.*

MONOTRIGLYPH (Gr. μονος, one, or single, and *triglyph, q.v.*) The intercolumniations of the Doric order are determined by the number of triglyphs which intervene, instead of the number of diameters of the column, as in other cases; and this term designates the ordinary intercolumniation of one triglyph. (Plate V. fig. 1.)

MONUMENT, a name given to a tomb, particularly to those fine structures recessed in the walls of mediæval churches.

MOSAIC (Lat. *Opus musivum*, Ital. *Musaico*, Fr. *Mozaique*), pictorial representation, or ornaments formed of small pieces of stone, marble, or enamel of various colours. In Roman houses the floors are often entirely of mosaic, the pieces being cubical. There are several fine specimens in Westminster Abbey, particularly the pavement of the choir.

MOULDING (Lat. *Modulus*, Ital. *Modinatura*, Fr. *Moulure*, Ger. *Seniswerk*). When the face or edge of any work is wrought into long regular channels or projections, the sections of which form various curves or rounds, hollows, ogees, &c., it is said to be *moulded*, and each separate member is called a *moulding*. In mediæval architecture the principal mouldings are those of the arches, doors, windows, piers, &c. The remains of Saxon work are so few that we can tell but little about these mouldings. The arches have sometimes a simple rib on them, as Plate XXXII fig. 1, sometimes chamfered, as fig 2, and sometimes are quite plain, as Plate XXIII. fig. 1. Early Norman work is much the same. By degrees, however, the arrises were finished by a round or bowtell, as Plate XXXII. fig. 4 and fig. 9, which gradually were more numerous, as fig. 3. Later, hollows and rounds together became common, and the arches were set back one behind another, each being frequently supported by a jamb-shaft or column, as Plate XVIII. figs. 5, 6, and 8, though very often the arch mouldings continued down the jambs without any break, as fig. 7. The section is a series of squares set one behind another, as Plate XXXII. fig. 9, where the circles and faint lines show the jamb-shafts, and the

squares they stand in, and the shaded lines show the arch mouldings. Larger arches were more elaborate. See fig. 6 from Shoreham, of which an elevation is given in Plate XXIII. fig. 5 and fig. 9 from Great Grimsby. Fillets and splays are common in early styles, but ogees, whether direct or reversed, are rare for *ornamented mouldings*. (See the next article.) In the early English style, the mouldings, for some time, like those of the preceding period, formed groups set back in squares, as Plate XXXII. figs. 8, 13, and 14; they are smaller, lighter, more graceful, and frequently very deeply undercut. See above, and see also Plate XXXIV. figs. 4 and 5 (capitals), and figs. 6 and 8 (strings). The *scroll* moulding, fig. 5 (strings), is also common. Small fillets now became very frequent in the outer parts of the rounds. (See Plate XXXII. figs. 11, 16, and 18.) This has often been called the *keel* moulding, from its resemblance in section to the bottom of a ship; sometimes also it has a peculiar hollow in each side like two wings, fig. 11. In smaller doorways and windows the mouldings are frequently simply a succession of hollows, as Plate XXXII. fig. 5. As styles progressed, the arch stones seem first to have been chamfered, on which, or rather in which face the mouldings seem to have been sunk, instead of being arranged in squares as before. Later in the Decorated style the mouldings are more varied in design (see fig. 18), though hollows and rounds still prevail. The undercutting is not so deep (compare fig. 3 with figs. 6 and 8, Plate XXXIV. *strings*, also figs. 4 and 5 with figs. 8 and 9 *capitals*), fillets abound, ogees are more frequent, and the *wave* mould, double ogee, or double ressaunt, is often seen (Plate XXXII. fig. 19). In many places the strings and labels are a round, the lower half of which is cut off by a plain chamfer, fig. 17, Plate XXXIV. fig. 4 (strings). The mouldings in the latest styles in some degree resemble those of the Decorated, flattened and extended; they run more into one another, having fewer fillets, and being as it were less grouped. One of the principal features of the change is the substitution of one, or perhaps two (seldom more) very large hollows in the set of mouldings. (Plate XXXII. figs. 24, 25, 26, and 27). These hollows are neither circular nor elliptical, but obovate, like an egg cut across, so that one-half is larger than the other. The wave mould also has a small bead, where the two ogees meet, and is called a *brace* mould. Another sort of moulding which has been called a *lip mould* is common in parapets, bases, and weatherings. (See the sets-off of the buttresses at St Laurence, Evesham, Plate XXV. fig. 12; see also the lower part of the plinth mould, Plate XXXIV. fig. 11).

MOULDINGS, eccentric curves of various kinds, intended to enrich and ornament, by producing light and shade, and obviating the monotony attendant on many flat and angular surfaces. They may be variously carved to increase their efficiency. The most usual forms of mouldings are called the cyma-recta and reversa, cavetto, scotia, torus, astragal or bead, and the echinus or ovalo, *q. v.* Plate VIII. In Pointed architecture, mouldings are not limited either to those names or to the forms they are intended to designate, nor indeed is any other style, except by absurd custom and authority.

MOULDINGS ORNAMENTED. The Saxon and early Norman mouldings do not seem to have been much enriched, but the complete and later styles of Norman are remarkable for a profusion of ornamentation, the most usual of which is what is called the zig-zag. These seem to be to Norman architecture what the meander or fret was to the Grecian, but, however, was probably derived from the Saxons, as it is very frequently found in their pottery. Plate XVIII. figs. 3 and 4, and Plate XXXIII. figs. 2,

3, and 5, show examples on various scales. Bezants, Plate XXXIII. fig. 1; quatrefoils, fig. 2; lozenges, fig. 3; crescents, fig. 4; billets, fig. 6; heads of nails, Plate XXIII. fig. 3, are very common ornaments; besides these, battlements, cables, large ropes, round which smaller ropes are turned, or as our sailors say, "wormed," scallops, pellets, chains, a sort of conical barrels, quaint stiff foliages, beaks of birds, heads of fish, almost every conceivable ornament is sculptured in Norman mouldings; besides this, they are used in such profusion as has been attempted in no other style. (See the doorways at Iffley Church, Plate XVIII. figs. 7 and 8.) The decorations on early English mouldings are chiefly the dog-tooth (Plate XXXIII. figs. 9, 12, and 14), which seems to be one of the great characteristics of this style, though it is to be found in the Transition Norman; and the peculiar stiff foliages of the period either disposed in single leaves or flowers at intervals or in scrolls running along the mouldings. In this period and in the next, the tympanum over doorways, particularly if they are double doors, is highly ornamented. See Plate XXX. fig. 2 from Westminster, fig. 3 from Higham Ferrers, fig. 5 from York, fig. 6, a very curious specimen, from Tournay, &c. Those of the Decorated period re-resemble the former, except that the foliage is more natural (see third article), and the dog-tooth gives way to the *ball-flower* (Plate XXXIII. figs. 10, 13). Some of the hollows also are ornamented with rosettes set at intervals, which are sometimes connected by a running tendril, as the ball-flowers are frequently. Some very pleasing leaf-like ornaments in the labels of windows are often found in Continental architecture, as Plate XXIII. fig. 8, Plate XXXVII. fig. 2 from Lucca, fig. 3 from Le Mans, and fig. 7 from Venice. In the Perpendicular period the mouldings are ornamented very frequently by square four-leaved flowers set at intervals, but the two characteristic ornaments of the time are running patterns of vine leaves, tendrils, and grapes in the hollows, which by old writers are called " vignettes in casements " (Plate XXXIII. fig. 22 from Westminster), and upright stiff leaves, generally called the Tudor leaf (*ib.* fig. 20 from St. Stephen's Hall). On the Continent mouldings partook much of the same character; Plate XXXIII. fig. 7 is from Soissons, fig. 13 and 16 from Rouen, figs. 15 and 19 from Amiens, figs. 17 from Seez, fig. 18 from the Sainte Chapelle at Paris, and fig. 21 from Bourges.

MULLION, MUNION, often corrupted into *munting, monyal,* (Fr. *meneau;* Ital. *regolo;* Ger. *fenster-pfoste*). The perpendicular pieces of stone, sometimes like columns, sometimes like slender piers, which divide the bays or lights of windows or screen work from each other. They originated, no doubt, from the stout shafts or baluster column dividing the lights of windows in the earliest times (Plate XXXV. fig. 2), which is supposed to be Saxon. Later, these became regular columns, with caps and bases, as the lower windows (fig. 3 from St Albans), and sometimes the shaft is attached to a pier, as Plate XVIII. fig. 9 from the School of Pythagoras at Cambridge. In the early English style the divisions became strong piers, sometimes merely plainly chamfered, sometimes moulded, and sometimes ornamented with clustered jamb-shafts, as at Beverley (Plate XIX. fig. 1). These piers became more slender, as Plate XXXVI. fig 4 from Warmington, and began, more or less, to branch, as in fig. 3 from Wenham, and fig. 5 from Chester, where they are regular mullions, the upper parts of which develop themselves in TRACERY (figs. 6 and 7 from the same cathedral). In all styles, in less important work, the mullions are often simply plain chamfered, and more commonly have a very flat hollow on each

side. In larger buildings there is often a bead or bowtell on the edge, and often a single very small column with a capital; these are more frequent in foreign work than in English (Plate XXXVII. fig. 2 from Lucca, fig. 7 from Venice), but these columns are rare in Perpendicular work. Instead of the bowtell they often finish with a sort of double ogee (Plate XXXIV. figs. A and B). As tracery grew richer, the windows were divided by a larger order of mullion, between which came a lesser or subordinate set of mullions which ran into each other, as shown at Plate XXXIV. fig. A, as will be explained hereafter. (See ELEMENT, TRACERY.)

MUTULE (Lat. *mutulus,* a stay or bracket), the rectangular impending blocks under the corona of the Doric cornice, from which gutta or drops depend. Mutule is equivalent to modillion, but the latter term is applied more particularly to enriched blocks or brackets, such as those of Ionic and Corinthian entablatures.

MYNCHERRY, a name for a nunnery still sometimes in use. It is derived from the Anglo-Saxon *minicene, mynecene,* a nun.

NAOS (Gr. ναος, a temple). This term is sometimes used instead of the Latin cella, as applied to the interior; strictly, however, it means the body of the edifice itself, and not merely its interior or cell.

NARTHEX, from the Greek νάρθηξ, a ferula or rule, the long arcaded porch forming the entrance into the Christian basilica. Sometimes there was an inner narthex or lobby before entering the church. When this was the case, the former was salled exo-narthex, and the latter eso-narthex.

NAVE (Lat. *navis,* Ital. *navata,* Fr. *nef,* Ger. *schiff*), the central part between the arches of a church, which formerly was separated from the chancel or choir by a screen. It is so called from its fancied resemblance to a ship. Here in general were the pulpit and font. Abroad, here also is often a high altar, but this is of rare occurrence in England. Instances, however, are to be found at Durham and St Albans. It is also said there was formerly an altar at the west end of the nave in the first cathedral at Canterbury and at St Gall, as there is now at the cathedrals of Worms, Nevers, and some other instances.

NECKING, the annulet or round, or series of horizontal mouldings, which separates the capital of a column from the plain part or shaft. (See the various Plates.) In Norman work they are often corded (Plate XXVI. figs. 9 and 11).

NERVURE. The French name for the ribs or groining. (See GROINING, LIERNE, RIB, &c.)

NEWEL (Fr. *noyau,* Ital. *albero d'una scala,* Ger. *spindel*). In mediæval architecture, the circular ends of a winding staircase which stand over each other, and form a sort of cylindrical column, are called newel or nowell.

NICHE (Fr. *niche,* Ital. *nicchia,* Ger. *nische*), a recess sunk in a wall, generally for the reception of a statue. They sometimes are terminated by a simple label, but more commonly by a *canopy,* and with a *bracket* or *corbel* for the figure, in which case they are often called *tabernacles.* Sometimes they stand singly (as Plate XXII. fig. 1, the upper part), and sometimes form a range like an arcade (see the same fig. in the lower part). (See also figs. 8, 10, and 11; for details of the canopy, pedestal, &c., see also Plate XXIII. fig. 5.) They are very frequently recessed into buttresses (Plate XXI. fig. 1, see also Plate XXV. fig. 11). A curious instance of a sort of projecting niche, forming a bell-cot, is given, Plate XXIII. fig. 16. (See BRACKET, CORBEL, TABERNACLE.)

NORMAN ARCHITECTURE, the style prevailing during the

rule of the Normans in this country, which may be considered as *early*, *complete*, and *transition*, or Anglo-Norman, as it is sometimes called. Early Norman is generally plain and massive, and does not seem to have differed much from that of the Saxons, the arches being quite plain, with square arrises, as Plate XXIII. fig. 1, or one arch behind another, as Plate XVIII. fig. 1. The capitals were of very peculiar shape. (See Plate XXIII. fig. 1 ; see also CAPITAL.) In complete Norman the work became by degrees much richer, till in many instances it was one mass of ornamentation. (See Plate XVIII. figs. 7 and 8.) The doorways then were deeply recessed, and frequently had the peculiar features of a level lintel, and on which was a space or *tympanum*, often richly carved. The piers at first were sometimes square, as Plate XVIII. fig. 1, or cylindrical, figs. 2 and 3, sometimes ornamented with zig-zags, as fig. 4 from Waltham Abbey. (The same thing is also found at Norwich and Durham.) Sometimes they have small shafts set round them. Very frequently the cylindrical piers alternate with octagonal (see Plate XXIII. fig. 9 from Rochester). The groining is generally plain. In the later or transition Norman the arches (which up to this time were invariably circular) are occasionally pointed. Sometimes circular and pointed arches are mixed (see Plate XVIII. fig. 10 from Barfreston). They are also frequently *interlaced* (Plate XXIII. figs. 2 and 3). For fuller information on all these points see ARCH, BUTTRESS, CAPITAL, DOORWAY, GABLE, GROINING, KEEP, MOULDING, ORNAMENTED DITTO, PARAPET, PIER, PORCH, ROOF, TOWER, &c.

OCTASTYLE (Gr. οκτω, eight, and στυλος, a column). A portico of eight columns in front. (See note to HEXASTYLE.) There is no portico in London of this description at present, though the upper one of the west front of St Paul's (Plate XV.) is of eight columns ; but they are coupled, making the arrangement tetrastyle. It may indeed be called a pseudo-octa-prostyle. (See PSEUDO-PROSTYLE.)

OEILLETS, or OYLETS, a name sometimes applied to the arrow slits in towers, &c. ; but it seems more probable its strict meaning is the round hole or circle with which these terminate. (See LOOP HOLE ; see also OYLMENT.)

OGEE (Lat. *cyma*, *cymatium*, Ital. *gola dritta*, *gola a rovescio*, Fr. *cimaise*, *doucine*, *gorge*, *geule*, *geule reversée*, *talon*, Ger. *rehleister*), the name applied to a moulding, partly a hollow, and partly a round, and derived no doubt from its resemblance to an *O* placed at the top of a *G*. It is rarely found in Norman work, and is not very common in early English. In the Decorated it is of frequent use, where it becomes sometimes double, and is called a *wave* moulding, and later still, two waves are connected with a small bead, which is then called a *brace* moulding. In ancient MSS. it is called a *ressaunt* (which see ; also Plates XXXII., XXXIII., XXXIV., &c.).

OGIVE, a term applied by the French to the pointed arch.

OGIVE RIB, the main ribs which cross each other on the intersection of the vaulting. (See GROINED VAULTING.)

OPISTHODOMUS (Gr. οπισθεν, behind, and δομος, a house or other edifice), the part behind a Greek temple corresponding with the pronaos before it. (See PRONAOS.)

ORATORY (Fr. *oratoire*), a place or small chapel for prayer for the use of private individuals, generally attached to a mansion, and sometimes to a church. The name, however, is given abroad to small chapels built to commemorate some miracle or special deliverance. (See CHAPEL.)

ORDER, the name given to the subordinate mullions and tracery which are of smaller size than others in the same window, &c., and whose mouldings fall in with the others as shown. (See MULLION, TRACERY.) Also to

the groups of mouldings arranged on square faces setting back behind one another in Norman and early English work, and not cut in on the splayed faces of the jambs and arch moulds, as in subsequent periods. (See various examples, Plate XXXII., particularly 6 and 9.)

ORDER. A column with its entablature and stylobate is so called. (Plate XIII. fig. 1.) The term is the result of the dogmatic laws deduced from the writings of Vitruvius, and has been exclusively applied to those arrangements which they were thought to warrant.

ORDINANCE, a composition of some particular order or style. It need not, however, be restricted to a columnar composition, for it will apply to any species which is subjected to conventional rules for its arrangement.

ORIEL or ORYEL. (See BAY WINDOW.)

ORTHOGRAPHY (Gr. ορθος, straight or true, and γραφη, a description or representation.) A geometrical elevation of a building or other object, in which it is represented as it actually exists, or may exist, and not perspectively, or as it would appear, is called its orthography.

ORTHOSTYLE (Gr. ορθος, straight or true, and στυλος, a column), any straight range of columns. This is a term suggested to designate what is generally but improperly called a peristyle, *q. v.* ; that is, columns in a straight row or range, but not forming a portico.

OSSUARY. (See CHARNEL HOUSE.)

OVALO (Ital.), egg-formed (see ECHINUS). This is the name most commonly applied to the moulding which appears to have originated in the moulded head of the Doric column, and, with an abacus, forming its capital.

OYLMENTS, a word used in the Beauchamp Roll, signifying the small quatrefoil lights in the head of a Perpendicular window.

PACE, the landing on a broad step in a stair, also any stage raised above the floor. (See DAIS, ESTRADE.)

PAN. (See POST and PAN.)

PANE, probably a diminutive of *panneau*, a term applied to a bay of a window, compartment of a partition, side of a tower, turret, &c. (See BAY, LIGHT.)

PANEL (Fr. *panneau*, Ital. *quadrello*, *formello*, Gr. *feld*), properly the piece of wood framed within the styles and rails of a door, filling up the aperture, but often applied both to the whole square frame and the sinking itself ; also to the ranges of sunken compartments in cornices, corbel tables, groined vaults, ceilings, &c. In Norman work these recesses are generally shallow, and more of the nature of *arcades*. In early English work the square panels are ornamented with quatrefoils, cusped circles, &c. (see under the lower windows, Plate XX. fig. 1), and the larger panels are often deeply recessed, and form *niches* with trefoil heads and sometimes canopies. In the Decorated style the cusping and other enrichments of panels become more elaborate, and they are often filled with shields, foliages, and sometimes figures. Towards the end of this period the walls of important buildings were often entirely covered with long or square panels, the former frequently forming niches with statues. (See the west front of York Minster, Plate XXI. fig. 1.) The use of panels in this way became very common in Perpendicular work, the wall frequently being entirely covered with long, short, and square panels, which latter are frequently richly cusped (see Plate XXII. fig. 7), and filled with every species of ornament, as shields, bosses of foliage, portcullis, lilies, Tudor roses, &c. Wooden panelings partook very much of the character of stone, except in the Tudor period, when the panels were enriched by a varied design, imitating the plaits of a piece of linen, or a napkin folded in a great number of parallel lines. This is generally called the *linen pattern*. Wooden ceilings, which are very common, are composed of thin

ARCHITECTURE.

oak boards nailed on to the rafters, collars, &c., and divided into panels by oak mouldings fixed on them, with carved bosses at the intersections. (See WASCON ROOF.)

PARADISE, PARVISE, PARVYCE, a word of uncertain origin, but supposed to be a corruption of *paradisus*, an inclosed garden. They seem to have been open places surrounded with an enceinte or stone parapet in front of cathedrals or other great buildings, and probably were used to keep the people from pressing on and confusing the marshalling of the public processions. That at Notre Dame, at Paris, is of irregular shape; that at Amiens was round. Nothing of the kind is left in England, though, from a passage in Chaucer, it is supposed there was one in the front of Westminster Hall. The *Promptorium Parvulorum* calls a parvyce *parlatorium*, a place for conversation. In the late fashion for coining names for the various parts of Gothic edifices, and applying old and imperfectly understood terms at random, the small chambers over porches have been called *parvises*, but without any shadow of reason. The irregularly-shaped cloister at Chichester is still called a paradise.

PARAPET (from the Italian *parapetto*, something which comes against the breast, *i. e.*, to lean against, Fr. *parapet*, Ger. *brustwehr*), a dwarf wall along the edge of a roof, or round a lead flat, terrace walk, &c., to prevent persons from falling over, and as a protection to the defenders in case of a siege. Parapets are either plain, embattled, perforated, or panelled. The two last are found in all styles except the Norman. Plain parapets are simply portions of the wall generally overhanging a little, with a *coping* at the top and *corbel* table below. Embattled parapets are fully described. (See BATTLEMENT.) Perforated parapets are those which are pierced in various devices—as circles, trefoils, quatrefoils, and other designs—so that the light is seen through; see the lower part of fig. 11, Plate XXIV., and Plate XXXIX. fig. 3. Panelled parapets are those ornamented by a series of panels, either oblong or square, and more or less enriched, but are not perforated. These are common in the Decorated and Perpendicular periods. (See BATTLEMENT, CORNICE, CORBEL TABLE, &c.)

PARASCENIUM, in a Greek theatre, the wall at the back of the stage.

PARASTAS (Gr. παραστας, standing before), antæ or end pilaster. This is the Greek term for which the Latin antæ is generally used. It has the same meaning, and they may be used alternatively. (See ANTÆ.)

PARCLOSE, a word used for any inclosure to a chantry, tomb, &c. (See SCREEN.)

PAREMENT, a French term for the outside ashlar or casing of a rubble wall, which is tied together by through or bond stones. (See PERPENT.)

PARGETTING, a species of plastering decorated by impressing patterns on it when wet. These seem generally to have been made by sticking a number of pins in a board in certain lines or curves, and then pressing on the wet plaster in various directions, so as to form geometrical figures. Sometimes these devices are in relief, and in the time of Elizabeth represent figures, birds, foliages, &c.; fine examples are to be seen at Ipswich, Maidstone, Newark, &c. The word is Latinised *gypsacio* in the *Promptorium*, and may be derived from the old French *giter*, to cast, to throw, as outside plastering is often thrown against the laths to make it adhere better. (See ROUGH CAST.)

PATIN, PATAND, from the French *patin*, a wooden sole, clog, or patten. The sills in timber-framing are thus named in some old works, though the modern French authors call them *sabliéres*.

PAVEMENTS, MEDIÆVAL (Fr. *carrelage, dallage*). The common old English paving or foot tiles are square, from 4 inches to 12 inches each, of common red pottery clay. These are frequent in our churches of all periods, and make an excellent paving, not so cold as stone, and they never wear so as to be slippery, nor to lose their colour. Where ornamented tiles are too expensive, this is the best material for Gothic churches. In countries where stone abounds, the usual floor is covered with what is called flag-pavement, squares of stone, and in many places large pieces of slate.

PAVEMENTS, ORNAMENTED, of various characters, were much used in the middle ages, and seem to have succeeded the Mosaic or Tessellated pavements of the Romans. Their history seems to have been more easily traced on the Continent than in England. In the admirable *Dictionary* of M. Viollet-le-Duc, he has divided his subject into two heads, that of " Dallage" and of " Carrelage." In the former, he treats chiefly of ornamented pavements of stone, incised by hand, and filled in with coloured resinous mastics or other similar composition. The earliest of them, he says, is in the church of St Menoux, near Moulins. The stone is white, the filling in black, and the design, a succession of circles with a border, is very artistic. It dates from the 12th century. One of the richest examples of this sort of pavement is at St Osmane. As we have very little, if any, of these in England, we must refer our readers to the *Dictionnaire*, vol v. page 9. Our author also treats of ornamented paving of tiles proper, which he calls " carrelage." These are of two kinds, tiles of various shapes, each of one colour only, mixed with those of other colours so as to form a pattern. The earliest known in France is at the Abbey of St Denys, and is of about the 12th century. Here we see clearly the attempt to imitate the *tesseræ* of the Romans, and it is eminently successful. The examples given of squares and lozenges, filled up with minute squares or circles touching each other, and intersected with bold diagonal bands of dark colour, and, more curious than all, greenish-black tiles so made that other tiles of yellow earth in the form of fleur-de-lys, quatrefoils, and other designs, can be inserted in the middle of the dark squares in several pieces, give a high idea of the potter's art in that period. In fact, it is a difficult matter, in the present day, to manufacture such tiles, so as to come together with any degree of truth. The third kind of ornamented tile, which concerns ourselves as being common in this country, is the glazed ornamented tile improperly called *encaustic*. The origin of these is very ancient. M. Viollet-le-Duc cites as the earliest example some now existing in the church of St Colombe, at Sens. This monastery was founded in 630, and our author believes these tiles to be about the same date. They are almost always of a red earth, on which a design is impressed or stamped, the colour being filled in, and no part of the figures above the level of the body of the tile. In fact, the tile seems to have been made, and when partly dry, the subject appears to have been impressed thereon by a stamp, like making an impression by a seal. Once having this form, it seems to have suggested itself to the tilemakers that it would be a great improvement, and a thing very easy to effect, to fill these hollows up with a clay which would burn to a different colour, and thus make a tile of a flat uniform surface which would present a pictorial representation. It was soon found that the effect was improved by covering the whole surface with a glaze, and thus, as it were, varnishing the whole picture. The earliest found in England are said to be those at Castle Acre, but as these are decorated with coats of arms on regular shields, they cannot be earlier than the 13th century. The fact is, that the presence of armorial bearings is a very fallacious test as to dates. As is natural, those who displayed these objects not only gave their own arms, but those of their ancestors, real or supposed; and these, particularly in the 15th and 16th centuries, were often little better than fabulous persons. The chief part of the glazed tiles in England seem to have been made of the ordinary potter's clay, which we know burns to a dull red. On these, when made, the design was impressed, as has before been described, and the hollow filled with some white clay like pipe-clay. The glaze was probably that used now by the red-ware potters of the South of England, which brightens the red ground, and gave a yellow colour to the white insertions. Many of the tiles have a complete pattern,—a quatre-foil, fleur-de-lys, vesica piscis, monogram, interlaced triangles, a star, crest, badge, or single coat of arms, being very general subjects; but in many instances the pattern is formed not by one, but by a *set* of tiles. Thus two, three, or four of a side make assemblages of four, nine, or sixteen tiles, forming one large regular pattern. It is not often we get more in England; but at St Pierre-sur-Dive, as given by M. Didron in the *Annales Archeologiques*, a wonderful *rosace* is delineated, which must have been formed by the union of between 300 and 400 separate tiles. About the end of the 16th and in the 17th century, greenish tiles, with designs in relief, were used, commonly called *galley* tiles. After this time they seem to have gone out of use as a pavement. As a wall decoration, the white glazed tiles, with pictorial subjects in blue, were very common, particularly for lining chimneys. Like the galley tiles, these came from Flanders or Holland. Latterly the manufacture has been revived with great success.

PEDESTAL (Gr. πους, a foot, and στυλος, a column). An insulated stylobate is for the most part so called. The term is, moreover, generally applied to any parallelogramic or cylindrical mass, used as the stand or support of any single object, as a statue or vase.

PEDIMENT, that part of a portico which rises above its entablature to inclose the end of the roof, whose triangular form it takes. The cornice of the entablature, or its corona and part of the bed-mould only, with the addition of a cymatium, bounds its inclined sides, and gives it an obtuse angle at the apex. In Pointed architecture, however, the angle of a pediment is for the most part acute.

PENDENT, a name given to elongated bosses, either moulded or foliated, which hang down from the intersection of groins, especially in fan tracery, or at the end of hammer beams. Sometimes long corbels, like those in Plate XX. fig. 2, under the wall pieces, have been so called. A curious bracket and pendant, from Pietraperzia, in Sicily, is given. (Plate XXV. fig. 7. See also Plate XXII. figs. 8, 9, 11, 12, and 13.)

† PENDENT (Lat. *pendens*, hanging). In some of the later works of the Pointed style, large masses depend from enriched ceilings, and appear to be formed by the other legs of intersecting arches: these are called pendents. They also occur in canopies. (See Plate XXII. fig. 1, 8, 9, 11, and 12.)

PENDENT POST, a name given to those timbers which hang down the side of a wall from the plate, and which receive the hammer braces. (See ROOFS, MEDIÆVAL, fig. 42, letter C. They are now generally called the wall pieces.)

PENDENTIVE, a name given to those arches which cut off, as it were, the corners of square buildings internally, so that the superstructure may become an octagon or a dome. In Mediæval architecture, these arches, when under a spire in the interior of a tower, are called *squinches* (which see).

PERIBOLUS (Gr. περι, around or about, and βαλλω, to throw), an inclosure. Any inclosed space is a peribolus; but the term is applied more particularly to the sacred inclosure about a temple. The wall forming the inclosure is also called the peribolus.

PERIPTERAL. (See PERIPTEROS.)

PERIPTEROS (Gr. περι, around or about, and πτερον, a wing). A temple or other structure with the columns of its end prostyles, or porticoes, returned on its sides as wings, at one intercolumniation distant from the walls. Almost all the Doric temples of the Greeks were peripteral. The term is, however, incorrectly applied by Vitruvius to peristylar structures, though it is clear that a perfectly round building, such as he describes to be peripteral, cannot be said to be winged or to have wings.

PERISTYLAR, having a peristyle. (See PERISTYLE.)

PERISTYLE (Gr. περι, around or about, and στυλος, a column), a range of columns encircling an edifice, such as that which surrounds the cylindrical drum under the cupola of St Paul's. The columns of a Greek peripteral temple form a peristyle also, the former being a circular, and the latter a quadrilateral peristyle. The same term is generally but incorrectly applied to a range of columns in almost any situation when they do not form a portico. (See ORTHOSTYLE.)

PERPENDICULAR STYLE, that which prevailed in England from about 1400 to 1550, and which has no exact parallel on the Continent. It has been so called from the fact that the tracery, which, in the preceding period, ran into a system of curves, representing branches, stems, and even leaves (see the large window, Plate XX. fig. 1) of the greatest delicacy and fancy, now gradually formed a series of upright oblong forms, in tier over tier (see Plate XXII. fig. 1). The cause of this change it is difficult to discover; it may, however, be supposed the intention was to provide better forms for the glass painter to fill up—single whole-length figures being now in vogue, instead of flowing patterns, with subjects in medallions. The later portion of this style, from 1490 to its termination, is generally called Tudor,—by which name we propose to designate its peculiarities. The introduction of perpendicular lines into window tracery, as has been said above, was of slow growth (Plate XIX. figs. 7 and 8); the tower windows (Plate XXI. fig. 1, Plate XXXVI. figs. 11 to 13), and the larger window (Plate XXII. fig. 1), will show various examples of the changes of style. Besides this, the transom was constantly used, which was hardly ever seen in the decorated styles. Plate XXI. fig. 1, the tower windows, and Plate XXII. fig. 1, will show the usual method of treating them in England. A very curious exceptional case from Dorchester, Oxon, is given (Plate XXXVII. fig. 11). In some cases the transoms are multiplied; at Winchester the interior of the west end shows six in succession. (See TRANSOM.) The arches at first differ but little from the Decorated; by degrees the rise becomes less and less, till, in the Tudor period, they became four-centred (see ARCH, DATES OF STYLES; and, for the method of finding the centres, and setting out, see STONE-MASONRY, MEDIÆVAL, 131 to 136). The mouldings in the Perpendicular style are flatter than in that preceding, and their peculiarities (the large shallow hollow, the brace mould, wave mould, lip mould, &c.) are fully described under MOULDING. The doorways are arched as above described; the labels, however, run up and form a square, having a spandril over the arch, the whole of which is described under DOOR. The ROOFS, GROINING (with its peculiarly English feature, the fan-tracery), the TOWERS, SPIRES, BUTTRESSES, FLYING BUTTRESSES, PIERS, PARAPETS, FOLIAGES, &c., will all be found under their respec-

P

tive heads. Shortly after the middle of the 16th century the Perpendicular style became mixed with bad imitations of classic architecture, and declined in every way,—so much so, indeed, that critics have designated it as the *debased style*. A serious mistake has arisen out of this appellation, for many think the phrase applies to the whole Perpendicular period. It is true that at no time can this be considered equal to the palmy periods of the Decorated style, and particularly as regards the change from the fairy-like tracery of that time to what a modern French writer has called the " grille de fer," the inferiority is strikingly apparent. Still, when we consider the lightness of the piers, and the span of the arches, the astonishing boldness of the groining, and of the flying buttresses which support it, the noble central towers, the rich open timber roofs, it must be acknowledged that, though not the most perfect of the mediæval styles, it possesses beauties of its own which have never been rivalled since.

PERPENT STONE (Fr. *purpaing*), bond or " through stones," the διατόνοι of the Greeks and Romans. Long stones going right through walls, and tying them together from face to face. Viollet-le-Duc says these were rarely employed in the middle ages, but that the walls were composed chiefly of *carreaux*, square headers, and *boutisses*. However, there seems some confusion in the terms, as he also describes *boutisse* as stones " which go through the whole thickness of a wall, and tie the outside and inside casing (*parements*) together." The English meaning seems clear, from the old dictionary of Cotgrave, who describes perpent or perpender, " stones just as thick as a wall, and showing their smoothed ends on either side thereof."

PERPENT, or PERPEYN WALL, one banded together with stones as described in the foregoing.

PERRON, the grand flight of external steps entering the mansions of the mediæval nobility or high officials, and considered in itself as a mark of jurisdiction, as it is said from thence sentence was pronounced against criminals, who were afterwards executed at the foot of the steps— as at the Giant Stairs at Venice. When ascended by a double or triple flight of steps, they are called " perrons de deux, ou de trois rampes," &c. One of the finest later examples is the flight in the Horse-shoe Court at Fontainebleau.

PEW, a word of uncertain origin, signifying fixed seats in churches, composed of wood framing, mostly with ornamented ends. They seem to have come into general use early in the reign of Henry VI., and to have been rented and " well payed for " (see Bale's *Image of both Churches*) before the Reformation. Some bench ends are certainly of Decorated character, and some have been considered to be of the early English period. They are sometimes of plain oak board, 2½ to 3 inches thick, chamfered, and with a necking and finial generally called a *poppy head;* others are plainly panelled with bold cappings; in others the panels are ornamented with tracery or with the *linen pattern*, and sometimes with running foliages. The divisions are filled in with thin chamfered boarding, sometimes reaching to the floor, and sometimes only from the capping to the seat.

PIER. The solid parts of a wall between windows and between voids generally, are called piers. The term is also applied to masses of brickwork or masonry which are insulated to from supports to gates or to carry arches.

PIGNON, a French term for the gable of a roof. (See GABLE.)

PILASTER (Lat. *pila*, a pillar, and the Ital. augmentative *astro*, which indicates an inferior quality), an inferior sort of column or pillar; a projection from or against a pier, having the form and decorations of antæ, when used cor-

rectly; but too frequently they have capitals, like those of columns, assigned them.

PILLAR, or PYLLER (Fr. *pilier*, Ital. *pilastro, colonna*, Ger. *pfeiller*), a word generally used to express the round or polygonal piers, or those surrounded with clustered columns, which carry the main arches of a building. Saxon and early Norman pillars are generally stout cylindrical shafts built up of small stones. (See the various examples, Plates XXIII. XXVI.) Sometimes, however, they are quite square, sometimes with other squares breaking out of them (Plate XVIII. fig. 1, this is more common on the Continent), and sometimes with angular shafts, and sometimes they are plain octagons (Plate XXIII. fig. 9). At Norwich, at Durham, and at Waltham Abbey (Plate XVIII. fig. 4), some of the cylindrical pillars are ornamented with large zigzags. In later Norman work the pillar is sometimes square, with two or more semicircular or half columns attached. In the early English period the pillars become loftier and lighter, and in most important buildings are a series of clustered columns, frequently of marble, placed side by side, sometimes set at intervals round a circular centre (see Plate XX. fig. 1 *a*), and sometimes almost touching each other. (See *ibid.* fig. 1 *b;* see also Plate XVIII. fig. 14, Plate XX. fig. 2.) These shafts are often wholly detached from the central pillar, though grouped round it, in which case they are almost always of Purbeck or Bethersden marbles. In Decorated work, the shafts on plan are very often placed round a square set anglewise, or a lozenge, the long way down the nave; the centre or core itself is often worked into hollows or other mouldings, to show between the shafts, and to form part of the composition, much as in Plate XXXII. fig. 11. In this and the latter part of the previous style there is generally a fillet on the outer part of the shaft, forming what has been called a *keel* moulding. (See FILLET, KEEL MOULDING.) They are also often as it were tied together by bands formed of rings of stone and sometimes of metal. (See BAND OF A COLUMN.) About this period, too, these intermediate mouldings run up into and form part of the arch moulds, the impost not being continuous, or rather there is no impost, but the shafts have each their own separate cap. (See IMPOST.) This arrangement became much more frequent in the Perpendicular period, in fact it was almost universal, the commonest section being a lozenge set with the long side from the nave to the aisle, and not towards the other arches, as in the Decorated period, with four shafts at the angles, between which were shallow mouldings, one of which in general was a wide hollow, sometimes with wave moulds, and as the pillar altogether by the arrangement was wider than the wall above the shafts facing the nave, ran up to the roof, and served in place of the vaulting shafts of the previous periods. (See VAULTING SHAFT, see also Plate XX. figs. 1 and 2.) In all periods, in small churches, instances are found of plain round or plain octagon pillars, and sometimes of both placed alternately (see Plate XVIII. fig. 13). The small pillars at the jambs of doors and windows, and in arcades, and also those slender columns attached to pillars, or standing detached, are generally called SHAFTS (which see).

PILLOWED. A swollen or rounded frieze is said to be pillowed or pulvinated.

PINNACLE (Fr. *pinacle, finoison*, Ital. *pinnacolo* (literally a little feather), Ger. *pinnafyl*), an ornament originally forming the cap or crown of a buttress or small turret, but afterwards used on parapets at the corners of towers and in many other situations. Some writers have stated there were no Norman pinnacles; but conical caps to circular buttresses, with a sort of finial, are not uncommon in France at very early periods. Viollet-le-Duc gives examples from St Germer and St Remi, and there is one of similar form at the west front of Rochester Cathedral. In the later Norman period, two examples have been cited, one from Bredon in Worcestershire, and the other from Cleeve in Gloucestershire. In these the buttresses run up, forming a sort of square turret, and crowned with a pyramidal cap, very much like those of the next period, the early English. In this and the following styles, the pinnacle seems generally to have had its absolute uses. It was a weight to counteract the thrust of the groining of roofs, particularly where there were flying buttresses ; it stopped the tendency to slip of the stone copings of the gables, and counterpoised the thrust of spires ; it formed the piers to steady the elegant perforated parapets of later periods ; and in France especially served to counterbalance the weight of overhanging corbel tables, huge gargoyles, &c. In the early English period, the smaller buttresses frequently finished with GABLETS (which see), and the more important with pinnacles supported with clustered shafts. (See Plate XIX. fig. 1, for both instances.) At this period the pinnacles were often supported on these shafts alone, and were open below ; and in larger work in this and the subsequent periods, they frequently form niches and contain statues. About the Transition, and during the Decorated periods, the different faces above the angle shafts often finish with gablets. Those of the last-named period are much richer, and are generally decorated with crockets and finials, and sometimes ball-flowers. (See Plate XIX. figs. 2 and 3.) The former of these is a sort of compound or group of pinnacles ; they are both from the decorated portion of Beverley Minster. Very fine groups are also found at the rise of the spire of St Mary's, Oxford, and are evidently placed there to counteract the thrust. Perpendicular pinnacles differ but little from Decorated, except that the crockets and finials are of later character. They are also often set angle-ways, particularly on parapets, and the shafts are panelled. (See Plate XXV. figs. 11 and 13 ; see also the rich examples on the towers of York Minster, Plate XXI. fig. 1.) In France, pinnacles, like spires, seem to have been in use earlier than in England. There are small pinnacles at the angles of the tower in the Abbey of Saintes. At Roulet there are pinnacles in a similar position, each composed of four small shafts with caps and bases surmounted with small pyramidal spires ; the like at Loches, but these are octagonal in plan. In all these examples the towers have semicircular headed windows.

PISCINÆ, one or more hollows or *cuvettes* near the altars, with drains to take away the water used in the ablutions at the mass. They seem at first to have been mere cups or small basins, supported on perforated stems, placed close to the wall, and afterwards to have been recessed therein and covered with niche heads, which often contain shelves to serve as aumbries. They are rare in England till the 13th century, after which there is scarcely an altar without one. They frequently take the form of a double niche with a shaft between the arched heads, which are often filled with elaborate tracing. Our space will not allow us to give examples, especially as they are now out of use.

PITCH OF A ROOF, its height above the level of the plate. (See GABLE, see also ROOFS, MEDIÆVAL.)

PLAN, a horizontal geometrical section of the walls of a building ; or indications, on a horizontal plane, of the relative positions of the walls and partitions, with the various openings, such as windows and doors,—recesses and projections, as chimneys and chimney-breasts,—

columns, pilasters, &c. This term is often incorrectly used in the sense of DESIGN, *q. v.*

PLANCEER is sometimes used in the same sense as soffit, but incorrectly, as it is from the French *plancher*, to board or floor. It is more particularly applied to the soffit of the corona in a cornice.

PLASTERING (Fr. *plâtre*, Ital. *intonaco*, Ger. *putzarbeit*), a mixture of lime, hair, and sand, to cover lath-work between timbers or rough walling, used from the earliest times, and very common in Roman work. Among other mistakes of the present day is, in the words of M. Viollet-le-Duc, "the prejudice to believe the constructors of the middle ages did not use plastering." He calls it "une excellente matiére," and says, as every one acquainted with ancient buildings must know, that it was used not only in private but in public constructions. On the inside face of old rubble walls it was not only used for purposes of cleanliness, rough work holding dirt and dust, but as a ground for distemper painting (*tempera*, or, as it is often improperly called, *fresco*), a species of ornament much more often used in the middle ages than we are disposed to think. At St Alban's Abbey the Norman work is plastered and covered with lines imitating the joints of stone. The same thing is found in the Perpendicular work at Ash in Kent. On the outside of the like walls, and often of wood-framing, it was used as *rough cast ;* when ornamented in patterns outside, it is called *pargetting*.

PLATE, POLE PLATE. (See also ROOFS as above.)

PLINTH (Gr. πλινθος, a square tile). In the Roman orders the lowest member of the base of a column is square and vertically faced ; this is called a plinth.

PODIUM, strictly something upon or against which the foot may be placed ; and in this sense, probably, it was applied to the wall which bounds the arena of an amphitheatre, and is thereby at the feet of the most advanced of the spectators.

POLYTRIGLYPH (Gr. πογυς, many, and triglyph, *q. v.*) An intercolumniation in the Doric order of more than two triglyphs. (See MONOTRIGLYPH and DITRIGLYPH.)

POMMEL, a name given to any round knob, as a *boss*, a *finial* (which see).

POPPY HEAD, probably from the French *poupée*, the finials or other ornaments which terminate the tops of bench ends, either to pews or stalls. They are sometimes small human heads (from which the name is probably derived), sometimes richly carved images, knots of foliages, or finials, and sometimes fleurs-de-lys simply cut out of the thickness of the bench end and chamfered.

PORCH (Gr. ναρθηξ, Lat. *porticus*, Fr. *porche*, Ital. *portico* Gr. *vorhalle*), a covered erection forming a shelter to the entrance door to a large building. The earliest known are the long arcaded porches in front of the early Christian basilicas, called *narthex* (which see). In later times they assume two forms—one the projecting erection covering the entrance at the west front of cathedrals, and divided into three or more doorways, &c., and the other a sort of covered chambers open at the ends, and having small windows at the sides as a protection from rain. These generally stand on the north or south sides of churches, though in Kent there are a few instances (as Snodland and Boxley) where they are at the west ends. Porches are of very early use ; those of the Norman period generally have but little projection, and are sometimes so flat as to be but little more than outer dressings and hood moulds to the inner door. They are, however, often very richly ornamented, and, as at Southwell, in England, and Kelso, in Scotland, have rooms over, which have been erroneously called Parvises. (See PARADISE, &c.) Early English porches are much longer, or project much further from the faces of the churches to which they are attached, and in larger and more important buildings have very frequently rooms above; the gables are generally bold and high pitched. In larger buildings also, as at Wells, St Albans, &c., the interiors are very rich in design, quite as much so, in fact, as the exteriors. Decorated and Perpendicular porches partake of much the same characteristics, the pitch of roof, mouldings, copings, battlements, &c., being of course influenced by the taste of the time. As a general rule, however, the later porches had rooms over them more frequently than in earlier times; these are often approached from the lower story by small winding stairs, and sometimes have fireplaces, and are supposed to have served as vestries; and sometimes there are the remains of a piscina, and relics of altars, as if they were used as chantry chapels. It is probable there were wooden porches at all periods, particularly in those places where stone was scarce; but, as may be expected from their exposed situation, the earliest have decayed. At Cobham, Surrey, there was one which had ranges of semicircular arches in oak at the sides, of strong Norman character, but which is now unfortunately destroyed. It is said there are several in which portions of early English work still are traceable, one is at Chevington, in Suffolk. In the Decorated and later periods, however, wooden porches are very common, some plain and others with richly carved tracery, and barge boards; these frequently stand on a sort of half story of stone work or *bahut*. A very curious and beautiful example of a wooden renaissance porch from Ry, near Rouen, is given in Viollet-le-Duc, *Diction.* vol. vii. p. 274. The entrance porches at the west end of our cathedrals are generally called *portals*, and where they assume the character of separate buildings, are designated as *galilees*. Both these last are more common on the Continent than in England. A very fine example of these bays, with nave, side aisles, and clerestory, is at Tournus, and another at Vézelay. At Cluny, the porch is in itself a church, about 115 feet long by 58 feet wide. At Dijon and at Vézelay, the upper part of the building looks into the nave, and forms a sort of gallery thereto. In France, the lower part of the tower often forms a sort of open porch, as at St Benoit-sur-Loire, Moissac, and several others cited by M. Viollet-le-Duc. Many of the French cathedrals have the doors so deeply recessed as to be almost like open porches. These are called *portals* or *portes abrités.* Many, however, have detached porches in front of the portals themselves, one of the best known of which is St Germain-l'Auxerrois, at Paris; another very fine example is at St Nicaise de Rheims, and a third at Troyes. The noblest example of an open projecting western porch in England, and probably in the world, is at Peterborough, of the early English period, attached to the early Norman nave.

PORTAL (Fr. *portail*, Ital. *portone*), a name given to the deeply recessed and richly decorated entrance doors to the cathedrals on the Continent.

PORTCULLIS (M. Lat. *cataracta*, Fr. *herse, coulisse*, Ital. *sarasinesca*, Ger. *fallgatter*), a strong-framed grating of oak, the lower points shod with iron, and sometimes entirely made of metal, hung so as to slide up and down in grooves with counter-balances, and intended to protect the gateways of castles, &c.; the defenders having opened the gates and lowered the portcullis, could send arrows and darts through the gratings, and yet the assailants could not enter. One of these constructions was in existence a short time back in a gateway at York; they are said not to be older than the 12th century, and were probably (as their Italian name imports) invented to repulse the sudden attacks of the Saracens on the coasts of that country.

PORTICO (an Italicism of the Lat. *porticus*), an open space

before the door or other entrance to any building, fronted with columns. A portico is distinguished as prostyle, or *in antis*, as it may project from or recede within the building, and is designated with either of these terms by the number of columns its front may consist of. (See DISTYLE, TETRASTYLE, HEXASTYLE, OCTASTYLE, &c.)

PORTICUS (Lat. See PORTICO). In an amphiprostylar or peripteral temple, this term is used to distinguish the portico at the entrance from that behind, which is called the posticum.

POST AND PAN WORK, a name given to the carpentry framing of old wooden houses; *panne*, in old French, signifying any horizontal piece of timber, as a head, sill, or purlin, though its use now is confined to the latter, sills and plates at present being called *sabliéres*. Where timber was abundant, and stone scarce and dear to work, it was not to be wondered at that timber-framed houses should abound. The posts or uprights seem in early times to have been constructed of small oak trees, 6 or 7 inches square, roughly trimmed by the axe; the girders, &c., are larger, but seldom seem to have been sawn. The framing of the lower story generally stands on a sort of plinth or *bahut* of stone or brick, sometimes as high as the window sills, and the other fronts are each framed separately, and as the joists of each story project over those of that below, each story also projects, till, in narrow streets, it is said the houses almost touched each other at the top. To strengthen the framing, it was customary to tie the angles together with circular braces cut out of the crooked boughs of trees, and to fill in under and sometimes over the window openings with cross struts, sometimes like the St Andrew's cross, and sometimes in circles and various designs. The main posts also were strengthened inside and out with a sort of projecting corbels, called in French *liens* or *décharges*, and by us *spervers*, and which helped to carry the projecting plates above. In the better sort of work these timbers are chamfered and sometimes carved, and the gables have rich barge boards; the roofs invariably have great projections to throw off the wet, and the jutting of the stories, one over the other, no doubt was intended for the same purpose. Old post and pan work is put together with mortices and tenons pinned with pins or trunnels of hard wood; very often there is not a nail in the whole construction. The intermediate upright posts or quarters were called *prich posts*. All these houses are plastered, rough cast, or pargetted between the timbers, sometimes in handsome designs, and as the old oak gets black with age, or as the timbers are often rubbed over with oil, and the plaster whited, they are called in England *black and white houses*. (See PARGETTING and PLASTERING.) Several churches in Essex have post and pan work.

POSTERN, a small gateway in the enceinte of a castle, abbey, &c., from which to issue and enter unobserved. They are often called *Sally Ports*.

POSTICUM (Lat.) A portico behind a temple. (See PORTICUS and PORTICO.)

PRECEPTORY, a small establishment of the Knights' Templars, managed by a preceptor, a subordinate officer to a master, in the same way as a priory was by a prior, and not an abbot.

PRESBYTERY (Lat. *presbyterium*, Ital. *presbiterio*, Fr. *presbytére*), a word applied to various parts of large churches in a very ambiguous way. Some consider it to be the choir itself; others, what is now named the *sacrarium.* Traditionally, however, it seems to be applied to the vacant space between the back of the high altar and the entrance to the lady chapel, as at Lincoln and Chichester, in other words, the *Back* or *Retro Choir* (which see).

PRICK POSTS, an old name given sometimes to the queen

posts of a roof, and sometimes to the filling in quarters in framing. (See Post and Pan.)

PRINCIPALS. (See article Roof.)

PRIORY, a monastic establishment, generally in connection with an abbey, and presided over by a prior, who was a subordinate to the abbot, and held much the same relation to that dignitary as a bishop does to a dean. (See Abbey.)

PROCESSION PATH (Lat. *ambitus templi*), the route taken by processions on solemn days in large churches,—up the north aisle, round behind the high altar, down the south aisle, and then up the centre of the nave.

PRONAOS (Gr. προ, before, and ναος, a temple), the inner portico of a temple, or the space between the porticus, or outer portico, and the door opening into the cella. This is a conventional use of the term; for, strictly, the pronaos is the portico itself.

PROPYLÆUM (Gr. προ, before, and πυλη, a portal), any structure or structures forming the entrance to the peribolus of a temple; also the space lying between the entrance and the temple. In common usage this term, in the plural (propylæa), is almost restricted to the entrance to the Acropolis of Athens, which is known by it as a name.

PROPYLON, an alternation in the Greek form of Propylæum (*q. v.*) in the Latin of Vitruvius.

PROSCENIUM, in a Greek theatre, the stage.

PROSTYLE (Gr. προ, before, and στυλος, a column). A portico in which the columns project from the building to which it is attached is called a prostyle. It is tautologous to say a prostyle portico: a prostyle is a portico. Custom, however, seems to warrant the impropriety, for the word portico is always superadded. In determining the number of columns of which a portico consists, the Greek numerals are prefixed to the term Style, *q. v.*, and prostyle is repeated. It would be more concise, and, at the least, equally correct, to put the numeral before prostyle, and say tetra-prostyle, hexa-prostyle, &c., instead of tetrastyle-prostyle, &c., as the custom is; that mode is adopted in this article throughout.

PSEUDO-DIPTERAL (Gr. ψευδης, false, and dipteral, *q. v.*), false, double-winged. When the inner row of columns of a dipteral arrangement is omitted, and the space from the wall of the building to the columns is preserved of the consequent double projection, it is pseudo-dipteral. The portico of University College, London, is pseudo-dipterally arranged, the returning columns on the ends or sides not being carried through behind those in front.

PSEUDO-PERIPTERAL (Gr. ψευδης, false, and peripteral, *q. v.*), false-winged. A temple having the columns on its flanks attached to the walls, instead of being arranged as in a peripteros, is said to be a pseudo-peripteral.

PSEUDO-PROSTYLE (Gr. ψευδης, false, and prostyle, *q. v.*) This is a term not in general use, but is here suggested to designate a portico projecting less than the space from one column to another, as the western porticoes to St Paul's Cathedral, and the portico to the East India House, in London; but that they are recessed also, and therefore may be described as pseudo-prostyle and recessed. The front of Trinity Church in the New Road, near the Regent's Park, London, presents a mere pseudo-prostyle.

PULPIT (Fr. *chaire de l'église*, Ital. *pulpito*, Ger. *kanzel*), a raised platform with inclosed front, from whence sermons, homilies, &c., were delivered. They were probably derived in their modern form from the *ambones* in the early Christian church, though mentioned in Nehemiah viii. 4. There are many old pulpits of stone, though the majority are of wood. Those in churches are generally hexagonal or octagonal; and some stand on stone bases,

and others on slender wooden stems, like columns. The designs vary according to the periods in which they were erected, having panelling, tracery, cuspings, crockets, and other ornaments then in use. Some are extremely rich, and ornamented with colour and gilding. A few also have fine canopies or sounding-boards. Their usual place is in the nave, mostly on the north side against the second pier from the chancel arch. Pulpits for addressing the people in the open air were common in the mediæval period, and stood near a road or cross. Thus there was one at Spital Fields, one at St Paul's, London. An external pulpit still remains at Magdalene College, Oxford, and at Shrewsbury. Pulpits, or rather places for reading during the meals of the monks, are found in the refectories at Chester, Beaulieu, Shrewsbury, &c., in England; and at St Martin des Champs, St Germain des Prés, &c., in France; also in the cloisters at St Dié and St Lu. Shortly after the Reformation the canons ordered pulpits to be erected in all churches where there were none before. It is supposed to this circumstance we owe so many of the time of Elizabeth and James. Many of them are very beautifully and elaborately carved, and are evidently of Flemish workmanship.

PULVINATED (Lat. *pulvinus*, a cushion or bolster), a term used to express the swelling or bolstering of the frieze, which is found in some of the inferior works of the Roman school, and is common in Italian practice. It is used indifferently with pillowed.

PURLINS. (See article Roof.)

PYCNOSTYLE (Gr. πυκνος, dense, and στυλος, a column), columns thickly set. The space or intercolumniation implied by this term is one diameter and a half. (See Eustyle.)

QUARREL, QUARRY (from the French *carré*, square), any square-shaped opening; applied in the Beauchamp Roll to the quatrefoils in perpendicular windows (Plate XIX. fig. 8); sometimes to squares of paving, but most commonly to the lozenge-shaped pieces of glass in lead casements.

QUARTERS, the main upright posts in framing, sometimes called studs; the filling in quarters were formerly named prick posts.

QUATREFOIL, any small panel or perforation in the form of a four-leaved flower. They are sometimes used alone, as in the triforium (Plate XX. fig. 1); sometimes in circles (*ib.* fig. 2, Plate XIX. fig. 5), and over the aisle windows (fig. 1), but more frequently are in square panels. They are generally cusped, and the cusps are often feathered. (See Cinquefoil.)

QUEEN POST (Fr. *poinçons ajoutés*). (See article Roof.)

QUIRK. (See Joinery, Sec. I. Mouldings.)

QUOIN, large squared stones at the angles of buildings, buttresses, &c., generally used to stop the rubble or rough stone work, and that the angles might be true and stronger. (Plate XVII.) Saxon quoin stones are said to have been composed of one long and one short stone alternately (see Plate XXXV. figs. 1 and 2, see also Saxon). Early quoins are generally roughly axed; in later times they had a draft tooled by the chisel round the outside edges, and later still, were worked fine from the saw.

QUOIN (Lat. *ancon*, an elbow or corner, whence the Fr. *coin*), an outer corner. The stones which are made to project from the regular surface of the walls at the outer angles of a building are technically called quoins. The front of the Farnese Palace exemplifies them. (See Plate XVII.)

RAFTERS (Fr. *chevrons*, Ital. *puntoni*). (See Roof.)

RAG-STONE, a name given by some writers to work done with stones which are quarried in thin pieces, such as the Horsham sandstone, Yorkshire stone, the slate stones,

&c., but this is more properly flag or slab work. By rag-stone, near London, is meant an excellent material from the neighbourhood of Maidstone. It is a very hard limestone of bluish-grey colour, and peculiarly suited for mediæval work. It is often laid as uncoursed work, or random work (art. BUILDING, fig. 15), sometimes as random coursed work (fig. 16), and sometimes as regular ashlar. The first method, however, is the more picturesque.

RAISING PLATE, from A. S. Ræsn, the wall plate on which the principals rest, so called to distinguish it from the POLL PLATE (which see, also ROOF).

RANDOM WORK, a term used by the rag-stone masons for stones fitted together at random without any attempt at laying them in courses. (See art. BUILDING, fig. 15.)

RANDOM COURSED WORK, a like term applied to work coursed in horizontal beds, but the stones are of any height, and fitted to one another, as art. BUILDING, fig. 16.

RAYONNANT, a name given to the second Pointed period, or *ogivale secondaire*, on the Continent. It was derived from the circumstance that the tracery resembled in some degree the form of radiation. The word has now gone out of use, except as regards the tracery. It came in sooner than our Decorated, viz., about 1250, and prevailed till about A.D. 1400.

REAR VAULT, a name sometimes applied to the inner hood-mould of a window or doorway, but no ancient authority for the use of such a term has as yet been cited. (See HOOD-MOULD, see also ESCONSON.)

REBATE. (See JOINERY, Sec. I. Mouldings.)

REFECTORY, the hall of a monastery, convent, &c., where the religious took their chief meals together. It much resembled the great halls of mansions, castles, &c., except that there frequently was a sort of ambo, approached by steps, from which to read the legenda sanctorum, &c., during meals. (See PULPIT.)

REGULA (Lat.), a rule or square. The short fillet or rectangular block, under the tænia, on the architrave of the Doric entablature, is so called.

RELIEVING ARCH (see STONE-MASONRY, Sec. V. art. 136, fig. 67), often called *discharging arch;* (also BUILDING, figs. 9 and 10.)

REREDOS, DORSAL, or DOSSEL (Fr. *rétable*), the screen or other ornamental work at the back of an altar. In some large cathedrals, as Winchester, Durham, St Albans, &c., this is a mass of splendid tabernacle work, reaching nearly to the groining. In smaller churches there are sometimes ranges of arcades or panelling behind the altars; but, in general, the walls at the back and sides of them were of plain masonry, and adorned with hangings or paraments. In large churches abroad, the high altar usually stands under a sort of canopy or ciborium, and the sacrarium is hung round at the back and sides with curtains on movable rods. (See DORSAL, CIBORIUM.) In private houses the iron plates behind the fire, where there are andirons, are sometimes called *reredosses.*

RESPOND, the half pier or pillar at the end of a range of piers and arches, or other arcades; they are generally exactly half the other piers, with a short piece of wall finishing at right angles to the end or cross wall.

RESSAUNT, a sort of flat ogee; Plate XXXII. fig. 20, shows two with a hollow, and two fillets between. A *double ressaunt* is the lowest moulding in fig. 19. A *ressaunt lorymer* (or *larmier*) is supposed to be an ogee with a drip, as the upper moulding in Plate XXXIV. fig. 9.

RETRO CHOIR. (See BACK CHOIR, PRESBYTERY.)

RIB (Fr. *nerf d'arrête, nervure*, Ital. *costola*, Ger. *rippe*). (See GROIN RIB, see also GROINED VAULTING.) The earliest groining had no ribs. In early Norman times, plain

flat arches crossed each other. (See Plate XXV. fig. 1, and Plate XXXVIII.) (See also OGIVE RIB.) These ribs by degrees became narrower, had greater projection, and were chamfered. In later Norman work the ribs were often formed of a large roll placed upon the flat band, and then of two rolls side by side, with a smaller roll or a fillet between them, much like the lower member (Plate XXXII. fig. 6, which is from Shoreham). Sometimes they are enriched with zigzags and other Norman decorations, and about this time bosses became of very general use. (See Boss.) As styles progressed, the mouldings were more undercut, richer and more elaborate, and had the dog-tooth or ball-flower or other characteristic ornament in the hollows. (See Plate XXXIV. ribs, let. C, which is from Robertsbridge.) In all instances the mouldings are of similar contours to those of arches, &c., of the respective periods. (See MOULDINGS.) In perpendicular work the ribs are broader and shallower, and almost always have two great hollows of elliptic shape, one on each side. In those churches of the early English and Decorated periods, where there is a groining of wooden ribs filled in between the spandrils with their narrow oak boards, these ribs resemble those of stone, but are slighter, and the mouldings not so bold. (See CEILINGS.) Later, wooden roofs are often formed into cants or polygonal barrel vaults, and in these the ribs are generally a cluster of rounds, and form square or stellar panels with carved bosses or shields at the intersections.

RIDGE (Fr. *faite, faitage*, Ital. *colma, colmello*, Ger. *rücken*), a flat piece of board running from the apex of principal to principal, to which the heads of the common rafters are nailed (see art. ROOFS); also the lead or tile covering to the same. For ornamental ridges, see CRESTING.

ROLL MOULDING or SCROLL MOULDING, a moulding so called because it resembles the section of half a scroll or flexible book rolled up so that the edge projects over the other part. (See Plate XXXIV. Strings, fig. 5.) Figs. 6 to 11 show varieties of roll mouldings which are called roll and fillet, roll and hollow, roll and bead, as the case may be. (See LABEL, STRING, &c.)

ROMANESQUE, a name given to the debased classic architecture which prevailed, according to some, till the Pointed styles came into use, but more correctly till the Norman took its own peculiar form. In general, it is an imitation of Roman architecture more or less classically carried out. The arches are still semicircular, but out of proportion to the columns which sustain them. The pilasters shallow and strip-like. The entablatures have disappeared, or are very poor and meagre. The ornaments and carving are less artistic. The Romanesque may be considered as the transition from classic to mediæval, rather than one of the styles of the latter.

ROOD, a name applied to a crucifix, particularly to those which were placed in the rood-loft or chancel screens. These generally had not only the image of the crucified Saviour, but also those of St John and the Blessed Virgin Mary, standing one on each side. Sometimes other saints and angels are by them, and the top of the screen is set with candlesticks or other decorations.

ROOD-LOFT, ROOD-SCREEN, ROOD-BEAM, JUBE GALLERY, &c. The arrangement to carry the crucifix or rood, and to screen off the chancel from the rest of the church during the breviary services, and as a place from whence to read certain parts of the same. (See JUBE.) Sometimes the crucifix is carried simply on a strong transverse beam, with or without a low screen, with folding doors below, but forming no part of such support. Sometimes the screen runs up to the beam which forms the upper part of the framing. Sometimes there is a gallery across the top of the screen sufficiently wide for a person to pass along, and stand and read the " Jube Domine benedicere," &c. This is gene-rally approached by a small winding staircase which appears as a turret at the junction of the nave and chancel. The general construction of wooden screens is close panelling beneath, about 3 feet to 3 feet 6 inches high, on which stands screen work composed of slender turned ballusters or regular wooden mullions, supporting tracery more or less rich with cornices, crestings, &c., and often painted in brilliant colours, and gilded. These not only inclose the chancels, but also chapels, chantries, and sometimes even tombs. In mansions, and some private houses, the great halls were screened off by a low passage at the opposite end to the dais, over which was a gallery for the use of minstrels or spectators. These screens were sometimes close and sometimes *glazed*. There are many of these in England, but generally more or less mutilated; one of the most perfect galleries is that at Charlton-on-Otmoor, in Oxfordshire. From the Louth accounts, it appears that the organ was sometimes placed in the rood-loft, as it now often is over the choir screens in cathedrals. The jubé galleries on the Continent are often of extraordinary splendour, but they come rather under the head SCREEN.

ROOD-STAIR, a small winding stair or vice leading to the gallery. (See the above article.) In England they generally run up in a small turret in the wall at the west end of the chancel. This also often leads out on the roof. On the Continent these stairs often are ascended out of the interior of churches, and are inclosed with exquisitely perforated tracery, as at Rouen, Strasbourg, &c.

ROOD-TOWER, a name given by some writers to the central tower, or that over the intersection of the nave and chancel with the transepts.

ROOF. (See separate articles, especially ROOF, MEDLÆVAL.)

ROSE WINDOW (Fr. *rosace*), a name given to a circular window with radiating tracery, called also *wheel window*. (Plate XIX. fig. 1.) (See WINDOWS, CIRCULAR.)

ROUGH CAST, a sort of external plastering in which small sharp stones are mixed, and which, when wet, is forcibly thrown or *cast* from a trowel against the wall, to which it forms a coating of pleasing appearance. Some of the rough cast at St Albans is supposed to be coeval with the building itself. This material was also much used in timber houses, and when well executed, is sound and durable.

ROVING, anything following the line of a curve; thus the bowtell or torus going up the side of a bench end and round a finial is called a roving bowtell. (See BOWTELL.)

RUBBLE WORK, a name applied to several species of masonry. One where the stones are loosely thrown together in a wall between waling boards, and grouted with mortar almost like concrete. This is called in Italian *muraglia di getto*, and in French *blocage*, and is thus defined by Viollet-le-Duc : " Blocs de pierre gros ou menus jétés pêle-mêle dans un bain de mortier." Work executed with large stones put together without any attempt at courses, or random work (see art. BUILDING, fig.15), is also called rubble. (See RAG-WORK, RANDOM WORK.)

SACRISTY (Lat. *sacrarium*, Fr. *sacristie*, Ital. *sagrestia*), a small chamber attached to churches, where the chalices, vestments, books, &c., were kept by the officer called the sacristan. In the early Christian basilicæ there were two semicircular recesses or apsides, one on each side of the altar. One of these served as a sacristy, and the other as the bibliotheca or library. Some have supposed the sacristy to have been the place where the vestments were kept, and the vestry where the priests put them on; but we find from Durandus the *sacrarium* was used for both these purposes. Latterly the place where the altar stands inclosed by the rails has been called *sacrarium*, but without authority. (See VESTRY.)

passing from mullion to mullion, and often through the whole window from side to side, to steady the stone work, and to form stays to which the lead work is secured. When the bays of the windows are wide, the lead lights are further strengthened by upright bars passing through eyes forged on the saddle bars, and called *stancheons* (which see, also ARMATURE). When saddle bars pass right through the mullions in one piece, and are secured to the jambs, they are said to have been called *stay bars*, but no authorities are quoted.

SANCTUS BELL-COT or TURRET, a turret or inclosure to hold the small bell sounded at various parts of the service, particularly where the words " Sanctus, &c.," are read. They differ but little from the common bell-cot, except they are generally on the top of the arch dividing the nave from the chancel. At Cleeve, however, the bell seems to have been placed in a cot outside the wall. (See Plate XXIII. fig. 16.) Sanctus bells have also been placed over the gables of porches. On the Continent they run up into a sort of small slender spire, called *flèche* in France, and *guglio* in Italy. (See Plate XXIV. figs. 8 and 9. See BELL-COT.)

SAXON, the architecture of England immediately before the coming of the Normans, and which resembled the latter so much that it is frequently very difficult accurately to discriminate one from the other. There has been much controversy on the subject. Before the late investigations, almost all the early Norman was classed as Saxon, even in work such as at Norwich, Chichester, and Lincoln, where it has been proved from documents the original cathedrals stood in other places, in Saxon times, and the present buildings are entirely of Norman foundation. These proofs have led some antiquaries into a belief that no Saxon work, or if any, the smallest and most inconsiderable quantity possible, remains to us, and others boldly say that all the Saxon churches were of wood, and have perished long ago. The subject is too long to be discussed in this place; it is clear, however, that churches were built of stone as early as the time of Bede, and that cathedrals at Ely, Peterborough, London, Westminster, Worcester, Winchester, and Gloucester, were erected in the century preceding the conquest. The Saxons ruled in England for years, and were a wealthy people, and, if we judge from the ornaments dug up in the barrows and various places, had considerable pretensions to art. Besides this, there are many parts of the Continent where the Normans never penetrated, and yet Romanesque churches erected about our Saxon period are extant, and show signs of an advanced state of architecture. There seems no abstract reason, therefore, why we may not have many remains of Saxon work in this country; indeed, now the controversy has cooled down, this fact is believed by many of our soundest antiquaries. The characteristics attributed to the buildings of the Saxons are generally supposed to be rudeness of design and coarseness of workmanship; the carrying up a number of narrow pilasters of stone, imitating the quarters and braces of timber houses (see Plate XXXV. figs. 1 and 2), or using the same in lieu of buttresses; the formation of the quoins or angle stones, in what is called long and short work—that is, of narrow pieces, one laid flat and the other set on end, and probably derived from the Roman rustics; the like work in the narrow pilasters, and in the jambs of the doors and windows; stones placed over openings, leaning together at the top, forming a sort of triangular pediment; rude imposts moulded horizontally, as if the idea were taken from those of classic times; very peculiar shafts dividing the bays of windows, and strongly resembling turned balusters (see as above), some with capitals evidently imitating Roman Corinthian. The belief that these are Saxon characteristics is further strengthened by the fact, that all these peculiarities have been found in the delineations of buildings in the MSS. of the 11th century. One very curious example is given in ROOFS, MEDIÆVAL, fig. 34, and represents Lot entertaining the two angels. One undoubted example of Saxon architecture is part of an arcade in the wall of the old refectory at Westminster Abbey, built by Edward the Confessor (see Plate XXIII. fig. 1), which does not differ from early Norman work. It is clear, however, that the strip pilasters, long and short work, triangular pediments, and the peculiar imposts, are never found in such work, and as these are evidently not Roman, the natural inference is they must be Saxon. For further particulars of the controversy see Mr Parker's edition of *Rickman's Gothic Architecture*, Mr Bloxam's *Manual*, Mr Ashpitel's paper on *Repton Church*, read before the Archæological Association (see vol. vii. page 7 of their *Transactions*), and many other separate papers.

SCAPPLING, reducing a stone to a rough square by the axe or hammer; in Kent, the rag-stone masons call this knobbling.

SCOTIA (Gr. σκοτια, shadow or darkness), a concave moulding most commonly used in bases, which projects a deep shadow on itself, and is thereby a most effective moulding under the eye as in a base. It is like a reversed ovalo, or rather what the mould of an ovalo would present. (Plate VIII.)

SCREEN, any construction subdividing one part of a building from another—as a choir, chantry, chapel, &c. The earliest screens are the low marble *podia* shutting off the *chorus cantantium* in the Roman basilicas, and the perforated *cancelli* inclosing the bema, altar, and seats of the bishops and presbyters. The chief screens in a church are those which inclose the choir or place at the times when the breviary services are recited. This is done on the Continent not only by doors and screen work, but also, when these are of open work, by curtains, the laity having no part in these services. In England, screens were of two kinds, one of open wood work, generally called *rood-screens* or *jubés* (which see), and which the French call *grilles, clotures du chœur;* the other, massive inclosures of stone work enriched with niches, tabernacles, canopies, pinnacles, statues, crestings, &c., as at Canterbury, York, Gloucester, and many other places both in England and abroad.

SCROLL, synonymous with volute. The term scroll is commonly applied to the more ordinary purposes, whilst volute is generally restricted to the scrolls of the Ionic capital.

SCUTCHEON. (See ESCUTCHEON.)

SECTION, a drawing showing the internal heights of the various parts of a building. It supposes it to be cut through entirely, so as to exhibit the walls, the heights of the internal doors, and other apertures; the heights of the stories, thicknesses of the floors, &c. It is one of the species of drawings necessary to the exhibition of a Design, *q. v.*

SEDILIA, seats used by the celebrants during the pauses in the mass. They are generally three in number, for the priest, deacon, and sub-deacon, and are in England almost always a species of niches cut into the south walls of churches, separated by shafts or species of mullions, and crowned with canopies, pinnacles, and other enrichments more or less elaborate. The piscina and aumbry sometimes are attached to them. Abroad, the sedilia are often movable seats; a single stone seat has rarely been found as at Lenham, but some have considered this to be a confessional chair, and others a frith-stole, or place to which criminals fled for sanctuary.

SEPULCHRE, EASTER, a recess in the wall of a church, generally in the north side, often ornamented with a canopy, finials, &c., for the crucifix to stand in during certain rites from Good Friday to Easter Day.

SET-OFF, the horizontal line shown where a wall is reduced in thickness, and consequently the part of the thicker portion appears projecting before the thinner. In plinths this is generally simply chamfered (See Plate XXV. fig. 9). In other parts of work, the set-off is generally concealed by a projecting string. Where, as in parapets, the upper part projects before the lower, the break is generally hid by a corbel table. The portions of buttress caps which recede one behind another are also called sets-off, *ib.* fig. 10. (See STRING, CORBEL TABLE, BUTTRESS.)

SEVERY (probably from the English *sever*, to divide), any main compartment or division of a building. (See BAY.) It has been supposed this is a corruption of Ciborium, as Gervase of Canterbury uses the word in this sense, but he probably alludes to the *vaulted* form of the upper part of the groining of each severy. (See CIBORIUM.)

SHAFT (Fr. *colonette*, Ital. *colonnetto*, Ger. *schaft*), in classical architecture, that part of a column between the necking and the congé at top of the base. (See Plate XIII. fig. 1.) In later times the term is applied to slender columns, either standing alone or in connection with pillars, buttresses, jambs, vaulting, &c. The earliest known are probably the *baluster shafts* in Saxon work (which see). In Norman work the shafts to doors and windows are generally slender cylinders, disposed in receding squares. (See Plate XXXII. fig. 4, also Plate XVIII. figs. 5, 6, and 8.) They are often enriched with zigzags, &c., as in the last instance; sometimes octagonal, as Plate XXIII. fig. 3, and instances are known where they are fluted (see Plate XXVI. fig 5). In the early English period they are longer in proportion to their height, sometimes attached to, or clustered round pillars, and sometimes standing entirely detached, frequently banded and often of marble; they also are frequently filleted. For particulars as to arrangements, &c., in this and the following styles, see BAND, FILLET, PILLAR, &c. For window shafts, see MULLION. See also BUTTRESS SHAFTS, JAMB SHAFTS, VAULTING SHAFTS.

SHED ROOF or *Lean-to*, a roof with only one set of rafters falling from a higher to a lower wall, like an aisle roof.

SHINGLE (Med. Lat. *scandula, scindula*, Fr. *bardeau, essente*, Ital. *scandola*, Gr. *schindel*), a sort of wooden tile, generally of oak, used in places where timber is plentiful, for covering roofs, spires, &c. In England they are generally plain, but on the Continent the ends are sometimes rounded, pointed, or cut into ornamental form.

SHRINE (Med. Lat. *feretorium, scrinium*, Fr. *châsse, écrin*, Ital. *scrigno*), a sort of ark or chest to hold relics; sometimes they are merely small boxes, generally with raised tops like roofs; sometimes actual models of churches; sometimes large constructions like that of Edward the Confessor, at Westminster, St Geneviève, at Paris, &c. Many are covered with jewels in the richest way; that of San Carlo Borromeo, at Milan, is of beaten silver. (See TABERNACLE.)

SILL or SOLE (Lat. *solum*, a threshold, whence the Fr. *seuil*). The horizontal base of a door or window-frame is called its sill, though in practice a technical distinction is made between the inner or wooden base of the window-frame, and the stone base on which it rests,—the latter being called the sill of the window, and the former that of its frame. This term is not restricted to the bases of apertures; the lower horizontal part of a framed partition is called its sill. It is sometimes incorrectly written cill.

ARCHITECTURE.

SKEW TABLE, any string, *table*, or coping running obliquely up or under the gable of a roof; in contradistinction from those which are horizontal. (See STRING, TABLE, &c.)

SLEEPER (Fr. *dormant*), a piece of timber laid on low cross walls as a plate to receive ground joists.

SLYPE, a name for the covered passage usually found in monasteries between the end of the transept and the chapter-house.

SOFFIT (Ital. *soffito*, a ceiling), the inverted horizontal face of anything. The horizontal face of an entablature resting on, and lying open between, the columns, is its soffit. The underface of an arch, where its thickness is seen, is its suffit.

SOLAR, SOLLER (Med. Lat. *solarium*, Fr. *galetas*, Ital. *solaio*), a room in some high situation, a loft or garret, also an elevated chamber in a church from whence to watch the lamps burning before the altars.

SOLE. (See SILL.)

SOMMER, from the French *sommier*, a girder or main beam of a floor; if supported on two story posts and open below, it is called a *bress-summer*, or more properly *brace-sommer*.

SOUND-BOARD (Fr. *abat-voix*), the covering of a pulpit to deflect the sound into a church. (See TESTER.)

SPAN, the width or opening of an arch between the walls, &c., from which it springs, also the like of a roof between the plates. The word is generally used in contradistinction to *pitch*, which signifies the height above such level. (See PITCH.)

SPAN ROOF, a roof having two sides inclining from a centre or ridge, in contradistinction to a *shed roof* or lean-to (which see).

SPANDRIL or SPANDREL, the space between any arch or curved brace and the level label, beams, &c., over the same. (See ROOFS, figs. 43, 47. See also under the triforium, Plate XX. fig. 2.) The spandrels over doorways in Perpendicular work are generally richly decorated (Plate XXII. fig. 4). At Magdalen College, Oxford, is one which is perforated, and has a most beautiful effect (Plate XXX. fig. 9). The spandrel of doors is sometimes ornamented in the Decorated period (see Plate XXI. fig. 1), but seldom forms part of the composition of the doorway itself, being generally over the label.

SPAR, a word still used in the navy for any piece of timber.

SPUR, SPERVER. The word *spur* is often applied to the carved wooden brackets or *hanses* which support the penthouse of a door, the level part being called a *sperver*.

SPIRE (Fr. *aiguille*, *flèche*, Ital. *guglio*, Ger. *spike*), a sharply-pointed pyramid or large pinnacle, generally octagonal in England, and forming a finish to the tops of towers. In this country, in Norman times, the only attempt at anything like a spire consisted in the termination of some turrets, as those at Rochester, St Peter's, Oxford, &c.; but these are rather PINNACLES (which see) than spires. Later Norman spires are supposed to have been merely low pyramidal roofs. In the early English period they appear at first to have been low, as the remains of the one at Christ Church, Oxford, but afterwards they become much more lofty and sharply pointed (see Plate XXIV. fig. 10; Plate XXXV. fig. 4). The probability is that the sight of the high domes and aspiring minarets of the Holy Land had suggested the erection of these lofty monuments to the Crusaders. At this period the spires generally covered the whole tower top, and had haunchings where the square broke into the octagon. (See BROACH SPIRE; see also the figs. last cited.) In the Decorated period the spires became still slenderer and sharper; the broach spire gradually gave place to those rising at once in octagon form from the flat of the towers surrounded with parapets, often richly perforated,

and with pinnacles at the angles. When there are broach spires, the haunches are less; compare Plate XXXV. figs. 4 and 5. The spires themselves often are decorated with ball-flowers and crockets, and sometimes have broad horizontal bands of tracery at intervals. In both these styles spire lights or *lucarnes* are common (which see). Examples are also given at Plate XXXV. fig. 5. Perpendicular spires partake also of most of these characteristics, except that scarcely an example of a broach spire is to be found in those times. It is remarkable with how little materials some of the loftiest spires have been erected, that at Salisbury being barely 8 inches thick at the bottom. On the Continent the spire seems to have been used earlier than with us. That at Brantôme is a mere low pyramid. At Saintes it is a low carved cone, something of domical character. At Roulet it is a sharp circular cone, with four open pinnacles at the base. At Isomes it is octagonal, and as sharp as many of our early English spires. In all these examples the windows below are semicircular. Abundant instances of spires of this character are given by M. Viollet-le-Duc, under the articles *clocher* and *flèche*. In the later periods on the Continent the spires became very elaborate, and are, in fact, rather masses of tabernacle work in stage above stage than spires. There are very fine examples at Strasbourg and Brussels. In the north of Italy there are constructions very similar; these are called *guglio*, a fine example of which is at Milan; but the prevailing belltower of the churches is the *campanile* (which see). Timber spires are very common in England. Some are covered with lead in flat sheets, others with the same metal in narrow stripes laid diagonally with the hips. Very many are covered with *shingle* (which see). Abroad there are some elegant examples of spires of open timber work covered with lead. (See Plate XXIV. figs. 8 and 11.) The latter (at Paris) is of considerable height, and, as may be seen, extremely rich in its decoration. Generally the flèches are at the intersection of the transepts with the nave and choir. (See SANCTUS, BELL-COT.)

SPIRE-LIGHTS. See LUCARNE.

SPRINGER, the stone from which an arch springs; in some cases this is a *capital*, or *impost* (which see); in other cases the mouldings continue down the pier. The lowest stone of the coping of a gable is sometimes called so. (See SKEW TABLE, GABLET.)

SQUINCH, small arches or corbelled sets-off running diagonally, and, as it were, cutting off the corners of the interior of towers, to bring them from the square to the octagon, &c., to carry a spire. (See PENDENTIVE.)

SQUINT, oblique openings, often mere narrow, square-headed slits piercing the walls of the chancel arch, and evidently intended to afford a view of the high altar. They are often without any ornament, but sometimes arched and occasionally enriched with open tracery. Sometimes they look from the rooms over porches, sometimes from side chapels, but in every instance are so situate that the altar may be seen. The most probable interpretation of their use is to enable the acolyte appointed to ring the sanctus bell to see the performance of mass, and to sound it at the proper time. These openings have lately been called *hagioscopes* (which see), but without a shadow of authority. Indeed it seems most absurd to coin words from ancient Greek to describe parts of mediæval churches.

STAGE, an elevated floor, particularly the various stories of a bell-tower, &c. It is also applied to the plain parts of buttresses between cap and cap where they set back, as Plate XXV. figs. 10 and 12, or where they are divided by horizontal strings and panelling, as in Plate XXI.

fig. 1.   The phrase is also used by William of Worcester to describe the compartments of windows between transom and transom, in contradistinction to the word *bay*, which signifies a division between mullion and mullion. (See STORY.)

STAIR.   See the various articles in JOINERY, STONE-MASONRY, &c., for the nomenclature.

STALL, fixed seats in the choirs for the use of the clergy. In early Christian times the *thronus, cathedra*, or seat of the bishop, was in the centre of the apsis or bema behind the altar, and against the wall; those of the presbyters also were against the wall, branching off from side to side round the hemicircle.   In later times the stalls occupied both sides of the choir, return seats being placed at the ends for the prior, dean, precentor, chancellor, or other officers.   The seats are very peculiar. (See MISERERE).   In general, in cathedrals, each stall is surmounted by tabernacle work, and rich canopies, generally of oak, of which those at Winchester, Henry the Seventh's Chapel, and Manchester, may be quoted as fine instances. (See TABERNACLE, CANOPY).   The word is sometimes used to express any chief seat as in a dining hall.

STANCHEON, a word derived from the French *etançon*, a wooden post, and applied to the upright iron bars which pass through the eyes of the saddle bars or horizontal irons to steady the lead lights.   The French call the latter *traverses*, the stancheons *montans*, and the whole arrangement *armature*.   Stancheons frequently finish with ornamental heads forged out of the iron. (See Plate XXIV. figs. 14 and 15.)

STAY BARS, saddle bars passing through the mullions in one length across the whole window, and secured to the jambs on each side. (See SADDLE BAR.)

STEEPLE (Fr. *clocher*, Ital. *campanile*, Ger. *glockenthurm*), a general name for the whole arrangement of *tower, belfry, spire*, &c. (See those articles.)

STELE (Gr. στηλη, a cippus or small monument).   The ornaments on the ridge of a Greek temple, answering to the antefixæ on the summit of the flank entablatures, are designated stelai.

STEREOBATE (Gr. στερεος, firm or solid, and βασις, a base or fulciment), a basement.   It is here sought to make a distinction between this term and Stylobate, *q. v.*, by restricting the latter to its real import, and applying stereobate to a basement in the absence of columns.

STILTED, anything raised above its usual level.   For stilted arch, see Plate XXXVIII. fig. 6.

STOA (Gr. στοα, a portico).   This is the Greek equivalent for the Latin porticus and the Italo-English portico, *q. v.*

STORY (Lat. *tabulatum*, Fr. *étage*, Ital. *piano*, Ger. *geschoss*).   When a house has rooms one over the other, each set of chambers divided horizontal by the floors is called a story.   They are thus named in the different languages :—

	English.	French.	Italian.	German.
Lowest story.	Basement.	Souterain—Cave.	Sotteraneo.	Keller geschoss.
Ground do.	Ground floor.	Rez de chaussée.	Pianterreno.	Boden geschoss.
Half story or intermediate.	Mezzanine.	Entresol.	Mezzanino.	...
First story.	First floor.	Premier étage, also Bel étage.	Primo piano, also piano nobile.	Haupt geschoss.
Second story.	Second floor.	As their	numbers.	...
Upper story.	Garret.	Mansard.	Solaio.	Dach geschoss.

(See STAGE.)

STOUP (Fr. *bénitier*), a vessel placed close to the entrance of a church to contain the holy water.   They are generally small bowls fixed against a column, or on a stem.   In the north of Italy they are larger, and often carried on the back of a lion, and sometimes they are elegant *tazze* of exquisite workmanship.

STRING COURSES, horizontal mouldings running under windows, separating the walls from the plain part of the parapets, dividing towers into stories or stages, &c. Their section is much the same as the labels of the respective periods; in fact, these last, after passing round the windows, frequently run on horizontally and form strings. In Plate XXXIV., 1 and 2 are Norman; 3 to 6, early English; 7 and 8, Transitional; 9 to 11, Decorated; and 12 to 17, Perpendicular.   Like labels, they are often decorated with foliages, ball-flowers, &c. (See CURS-TABLE, LABEL, TABLE.

STRING or STRING-COURSE, a narrow, vertically-faced, and slightly projecting course in an elevation.   If window-sills are made continuous they form a string-course; but if this course is made thicker or deeper than ordinary window-sills, or covers a set-off in the wall, it becomes a blocking-course.

STRIP PILASTER, very narrow pilasters. (See SAXON ARCHITECTURE.)

STRUTS. (See articles CARPENTRY, ROOFS, &c.)

STUDS, an old name for upright quarters or posts; thus door-studs are door-posts or jambs.

STYLE (Gr. στυλος, a column).   The term style, in architecture, has obtained a conventional meaning beyond its simpler one, which applies only to columns and columnar arrangements.   Eustyle, *q. v.*, is a graceful distribution of columns as to space, and so it may be taken to apply to any good and pleasing distribution of solids and voids in the composition of a structure.   But Eustyle being a condition of style not always found, or not recognised, in all architectural compositions, the term is taken without the qualifying prefix, and is applied with such qualifications and descriptions as may be necessary to make it available to all classes of architectural composition.   Style as a term in architecture must be understood to be in no sense the same word as style in literature.

STYLOBATE (Gr. στυλος, a column, and βασις, a base or fulciment), a basement to columns. (See STEREOBATE.) Stylobate is synonymous with pedestal, but is applied to a continued and unbroken substructure or basement to columns, while the latter term is confined to insulated supports.

SURBASE (Lat. *super*, whence the Fr. *sur*, above or upon, and *base, q. v.*), an upper base.   This term is applied to what, in the fittings of a room, is familiarly called the chair-rail.   It is also used to distinguish the cornice of a pedestal or stereobate, and is separated from the base by the dado or die.

SYSTYLE (Gr. συν, together with, and στυλος, a column), columns rather thickly set.   An intercolumniation to which two diameters are assigned. (See EUSTYLE.)

TABERNACLE, a species of niche or recess in which an image may be placed.   In Norman work there are but few remains, and these generally over doorways.   They are shallow and comparatively plain, and the figures are often only in low relief, and not detached statues.   In early English work they are deeper, and instead of simple arches there is often a canopy over the figure, which was placed on a small low pedestal.   Later in the style the heads of the tabernacles became cusped, either as trefoils or cinquefoils, and they are often placed in pairs, side by side, or in ranges, as at Wells Cathedral, where they are not unlike the arcade at Wells (Plate XXIII. fig. 6). Decorated tabernacles are still deeper and more ornamented, the heads are sometimes richly cusped and surmounted with crocketed gables, as at York (Plate XXI. fig. 1), or with projecting canopies, very much like the arcade at Lichfield (Plate XXIII. fig. 5). In this case the under side of the canopy is carved to imitate groin ribs, and the figures stand either on high pedestals, or on corbels. Perpendicular tabernacles possess much the same features, but the work is generally more elaborate; the figures generally stand on rich pedestals (see Plate XXII. fig. 10), but sometimes on corbels, and the canopies generally project, sometimes in a triangular form (*ib.* fig. 8), and sometimes having a sort of domical top (*ib.* fig. 11); see also the general view, fig. 1. (See CORBEL, CANOPY, NICHE, &c.) The word tabernacle is also often used for the receptacle for relics, which was often made in the form of a small house or church. (See SHRINE.)

TABERNACLE-WORK. The rich ornamental tracery forming the canopy, &c., to a tabernacle, is called tabernacle-work; it is common in the stalls and screens of cathedrals, and in them is generally open or pierced through.

TABLE, TABLET, a name for various mouldings, as string courses, cornices, &c. BENCH TABLES (which see) are low seats against the walls inside churches. CORBEL TABLES (which see), courses of corbels supporting projecting parapets. CURSTABLES or COURSE TABLES (see STRINGS, EARTH TABLES, FOOT TABLES, GRASS TABLES, GROUND TABLES). LEDGEMENT TABLE (see BASE OF A WALL). PIGNUN (*i.e.* pignon) TABLES (see GABLES, SKEW TABLES). WATER TABLES (which see), weatherings, or sets-off, to throw off water.

TERMINAL. Figures of which the upper parts only, or perhaps the head and shoulders alone, are carved, the rest running into a parallelopiped, and sometimes into a diminishing pedestal, with feet indicated below, or even without them, are called terminal figures.

TESSELATED PAVEMENTS, those formed of *tesseræ*, or, as some write it, *tessellæ*, or small cubes from half an inch to an inch square, like dice, of pottery, stone, marble, enamel, &c. (See MOSAIC.)

TESTER, anything placed horizontally over the head, as the sound-board of a pulpit, the flat boards over an old-fashioned bed, the like over the pent-house of a door or SPERVER (which see; also SOUND-BOARD.)

TETRAPROSTYLE. (See TETRASTYLE and PROSTYLE.)

TETRASTOÖN (Gr. τετρα, four, and στοα, a portico). An atrium or rectangular court-yard, having a colonnade or projected orthostyle on every side, is called a tetrastoön.

TETRASTYLE (Gr. τετρα, four, and στυλος, a column), a portico of four columns in front. (See note to HEXA-STYLE.)

THOLOBATE (Gr. θολος, a dome or cupola, and βασις, a base or substructure), that on which a dome or cupola rests. This is a term not in general use, but it is not the less of useful application. What is generally termed the attic above the peristyle and under the cupola of St Paul's, would be correctly designated the tholobate. A tholobate of a different description, and one to which no other name can well be applied, is the circular substructure to the cupola of the Univ. Coll. London.

THOLUS or THOLOS (Gr.), a dome or cupola, or any round edifice. This is the only term used by Greek writers that can apply to the conoidal chambers which approach, in internal form, to that of the modern cupola or dome, and is therefore made the Greek equivalent for those terms.

THROUGH CARVING, a term supposed to signify such as is much undercut, as the tendrils, stalks, &c., in Decorated, and the vignettes, in Perpendicular work. In the Durham roll it clearly means perforated work, as it is "to giue ayre."

THROUGH STONE, stones going through the wall. (See BOND STONES, PERPENT STONES.)

TIE BEAM (Lat. *transtrum*, Fr. *tirant*, Ital. *catena*. See the various articles on ROOF).

TILES, ROOF (Lat. *tegula, imbrex*, Fr. *tuile*, Ital. *tegola, imbrice*), flat pieces of clay burned in flare kilns, to cover roofs in place of slates or lead. In England, in mediæval times, the flat or plane tile seems only to have been used, judging from what we find now left. From MS. and remains abroad, a sort of plane tile, with ornamented ends, forming a sort of scale covering, seems to have been in vogue.

TOOTH ORNAMENT. (See DOG-TOOTH.)

TOPH STONE (Fr. *tuf*, Ital. *tufo*), a very light stone, like pumice, often used in the filling in of vaulting, particularly in Norman times. The tufo of Italy is volcanic, and that deposited in the beds of rivers is called travertine. Both sorts, in the mediæval periods, seem to have been called toph stone.

TORUS (Lat.), a protuberance or swelling, a moulding whose form is convex, and generally nearly approaches a semicircle. It is most frequently used in bases, and is generally the lowest moulding in a base. (Plate VIII.)

TOUCH (Fr. *pierre de touche*, Ital. *pietra di paragone*), a close black stone, much like marble, and capable of bearing a fine polish, much used for monuments in the later styles. It is so called because goldsmiths prove the quality of gold thereon, by the use of aquafortis and *touch needles*.

TOWER (Gr. Πυργος, Lat. *turris*, Fr. *tour, clocher*, Ital. *torre*, Ger. *thor, thurm*), an elevated building originally designed for purposes of defence, and of the remotest antiquity; indeed, those buildings are mentioned in the earliest Scriptures. In mediæval times they are generally attached to churches, to cemeteries, to castles, or are used as bell-towers in public places of large cities. In *churches*, the towers of the Saxon period are generally square, the only round example being supposed to be that at Tasburgh. They are not very lofty, and are of strong, rude workmanship. Two only, Brigstock and Brixworth, have staircases supposed to be original; both these are on the west front of the tower. The masonry partakes of the usual character attributed to Saxon work, as strip pilasters, long and short work, &c. The upper windows are generally circular-headed in two lights, separated by a shaft much resembling a turned baluster sometimes with heavy projecting caps. (See BALUSTER SHAFT. See Plate XXXV. fig. 1, at Sompting, and fig. 2, at Earl's Barton. The roof of the former finishes with four very curious gables. See also SAXON ARCHITECTURE.) Norman towers are also generally square. Many are entirely without buttresses; others have broad, flat, shallow projections, which serve for this purpose. The lower windows are very narrow, with extremely wide splays inside, probably intended to be defended by archers. The upper windows, like those of the preceding style, are generally separated into two lights, but by a shaft or short column, and not by a baluster. Sometimes these towers have arcades round them, and are ornamented, as at St Albans (Plate XXXV. fig. 3), and very richly so, as at Norwich, Winchester, Tewkesbury, Southwell, Sandwich, &c. They frequently have stone staircases at one of the angles. In many of our churches the Norman tower is placed between the chancel and nave, and is of the full width of the latter. For the covering of these towers abroad, see SPIRE, PINNACLE, PARAPET. A few round towers of this period (and also of the next) are found on the coasts of Norfolk and Suffolk; as these mostly have no external doors, and are accessible only from the church, and as some have chimneys, they are supposed to have been built as places of refuge in case of invasion. *Early English towers* are generally taller, and of more elegant proportions. They almost always have large projecting buttresses, and frequently stone staircases. The lower windows, as in the former style, are frequently mere arrow-slits; the upper are in couplets or triplets, and sometimes the tower top has an arcade all round, as at Middleton Stoney. The spires are generally BROACH SPIRES (which see); but sometimes the tower tops finish with corbel courses and plain parapets, and (rarely) with pinnacles. Some of the towers and spires, particularly in the midland counties, are richly ornamented; a very good specimen is at Raunds (Plate XXXV. fig. 4). A few early English towers break into the octagon from the square towards the top, and still fewer finish with two gables, as at Brookthorpe and Ickford. Both these methods of termination, however, are common on the Continent. At Vendôme, Chartres, and Senlis, the towers have octagonal upper stages surrounded with pinnacles, from which elegant spires arise. Decorated towers differ but little from these last, except that they are often lighter in effect; the buttresses, too, are set angularly; the parapets are also frequently embattled, or perforated in elegant designs, and these generally have pinnacles. The spires, also, now generally arise at once from the octagon, and are not broach spires; those that are of this latter character have the haunchings much smaller. (See Plate XXXV. figs. 4 and 5.) The latter is a fine example of a Decorated tower and spire at Ellington, in Huntingdonshire. Most *Perpendicular towers* are very fine, particularly the great central towers, as at Canterbury and Gloucester. They are generally richly panelled throughout; the buttresses project boldly, and are sometimes set anglewise (see Plate XXXV. fig. 6), and sometimes square, not close to each other, but showing a small portion of the angle of the tower where they otherwise would have intersected. The pinnacles are often richly canopied and the battlements panelled, and often perforated; sometimes a pinnacle, and sometimes a canopied niche, is placed in the middle of the parapet. (See Plate XXI. fig. 1.) At Boston, and in several other places, there are fine lanterns at the tops of the towers. Taunton, Evesham, Louth, Magdalen College, Oxford, and very many other places, have very fine Perpendicular towers. Of course, in all towers, the doors, window-strings, mouldings, &c., are in accordance with the features of the respective styles, and all these articles should be referred to. In France the towers and spires resemble our own, with the differences as above stated; but in the north of Italy, and in Rome, they are generally tall, square shafts in four to six stages, with couplets or triplets of semicircular windows in each stage, generally crenellated at top, and covered with a low pyramidal roof. (See CAMPANILE.) The well-known hanging tower at Pisa is cylindrical, in five stories of arcaded colonnades. In Ireland there are in some of the *church-yards* some very curious *round towers*. They are covered with conical caps, have openings at the top, sometimes as many as eight, and the door or lower opening is at a height from the ground, easily reached by a ladder. It is not improbable they served for the same purpose as the Fanal or Lantern of the Dead, which is so often found in the cemeteries of the Celtic portion of France. (See LANTERN OF THE DEAD.) The *towers of castles* may be shortly noticed, though scarcely within the scope of civil architecture. In the Norman period, the fortification consisted of one or more walls, forming an inclosure or enceinte called a Base Court, in the middle of which was the KEEP, fully described under that term.

The towers in the outer walls were either square, or more generally circular, and projecting so as to flank the plain wall or curtain. In the Edwardian castles, and those of later date, the domestic apartments, as the hall, chapel, &c., were generally distinct buildings, situate in convenient parts of the internal courts. The walls were flanked with small towers as before, but the main towers were generally in the various lines of defence, and resembled GATE HOUSES rather than the keep of the Normans (see that art.), those on the external line of works being defended by drawbridges, barbicans, &c.; but as this is a matter which has little to do with civil architecture, we must refer to the various works on the subject, particularly the articles on military architecture in the *Dictionary* of M. Viollet-le-Duc. A very good example of a large square tower standing by itself without outworks may be mentioned as Tattersall Castle. Bell-towers (Fr. *beffroi*, Ital. *campanile*), for the use of the citizens, are common on the Continent in all large towns. They were intended to call the people together at public meetings, or to sound an alarm in cases of danger. They are generally tall, plain shafts, battlemented, and sometimes machicolated. (See Plate XXIV. figs. 4 at Brescia, and 5 at Sienna.) The nobles often erected these towers so as to command the streets by arrows and other missiles. Two very curious examples may be seen at Bologna, generally called the Leaning Towers of the Garisenda and Assinelli.

TRACERY, the ornamental filling in of the heads of windows, panels, circular windows, &c., which has given such characteristic beauty to the architecture of the 14th century. Like almost everything connected with mediæval architecture, this elegant and sometimes fairy-like decoration seems to have sprung from the smallest beginnings. The circular-headed window of the Normans gradually gave way to the narrow-pointed lancets of the early English period, and as less light was afforded by the latter system than by the former, it was necessary to have a greater number of windows; and it was found convenient to group them together in couplets, triplets, &c. (See Plate XVIII. fig. 12; Plate XIX. fig. 1; Plate XX. fig. 1, clerestory, Plate XXXVI. figs. 3, 4, 8, &c.) When these couplets were assembled under one label, as Plate XIX. fig. 4, Plate XXXVI. fig. 3, a sort of vacant space or spandrel was formed over the lancets and under the label. To relieve this, the first attempts were simply to perforate this flat spandrel, first by a simple opening generally of lozenge form, as Plate XIX. fig. 4, but sometimes circular, and afterwards by a quatrefoil, as *ib.* fig. 5, Plate XX. fig. 1. This expedient being a mere perforation of the flat stone, has lately been called plate, or more correctly from its French origin, *plat*, tracery; and although, as yet, the windows had no tracery at all strictly speaking, clearly it was thus this beautiful decoration became developed. As windows were made larger, and had a greater number of lights, the subdivision in the heads became more numerous, and it was necessary, for the purpose of keeping the lead lights into their places, not merely to pierce the flat surfaces, but to bend the mullions themselves into various forms. Thus the simple central shaft, carrying two pointed arches, as Plate XIX. fig. 4, was converted into a mullion branching two ways, as Plate XXXVI. fig. 3; and so by degrees, as these would not sufficiently fill up the heads, the vacancies were supplied by making these mullions branch in various other ways, and so inclose a quatrefoil, as Plate XIX. fig. 5, Plate XX. fig. 2 (the triforium), or plain circles, as Plate XXXVI. fig. 5. Instead, therefore, of the tracery, such as it is, being formed by perforations in the plate or flat stone, the window bar or mullion was twisted into shapes and forms which have been lately called (but not very happily) *bar-tracery*. In a short time it was found that it would add very much to the lightness and beauty of the windows if these additions were filled with trefoiled cusps, so that the plain circles of fig. 5 took the forms as in fig. 6. These afterwards became quatrefoils and cinquefoils, and had even a greater number of cusps, as in the windows and arcaded tracery, Plate XX. fig. 2, and in the lower windows, Plate XXI. fig. 1. We thus have the origin of what is generally called *geometrical tracery* forming the early part of the DECORATED period (which see), and also GEOMETRICAL style. We must now invite the student's attention to the series of examples, Plate XXXVI. figs. 5 to 13, the whole of which are from the choir and eastern part of Chester Cathedral, which was built slowly and gradually as the money came in, from about 1250 to 1400, and which thus form a very curious and instructive consecutive series of windows. We have shown how the plain mullion or circle gradually became enriched with cuspings, and we have now to show how the geometrical forms gradually gave way to what is appropriately called *flowing tracery ;* the mullions twisting and flowing, as has been expressed by our great Scottish poet, as if some fairy had twined wreaths together and then transformed them into stone. An inspection of the Plate XXXVI. figs. 7 and 8, will show the progress of their flowing lines, still, however, retaining the geometrical circles. These, however, are abandoned in fig. 9, and though the regularity of the branching mullion in this latter shows that a feeling for lines obviously struck from regular centres still prevailed, we see in fig. 10 a window which looks as if entirely struck by hand, as if dictated by fancy alone, and as if the art of geometry had nothing whatever to do with its invention. Nevertheless, however fanciful the design may be, the whole element is really geometrical,—that is, formed of portions of circles, the centres of which fall on the intersections of certain geometrical figures. An example is given as worked out, STONE-MASONRY, fig. 67, where a window, apparently drawn by hand, is in fact portions of circles drawn from sixty or seventy centres. The great east window at Carlisle is composed of 86 distinct pieces of stone, and is struck from 263 centres ; and the glorious west window at York (Plate XXI. fig. 1) is probably produced from a still greater number. For some reason or other, of which we have no account, the flowing tracery gradually admitted upright straight lines into its element. These were at first substitutions for curved lines, or rather the vesica piscis-like openings being elongated, a portion of the mullion became straight from mere necessity. (See Plate XXXVI. fig. 10.) It seems difficult to understand why men of such taste permitted such a change, but by degrees the perpendicular lines entirely superseded the curved lines, and such windows as *ib.* figs. 8, 9, and 10, became like those, figs. 12 and 13, and as Plate XIX. figs. 6 and 7 became like fig. 8, or with exception of the two great branching ribs, one mass of horizontal and perpendicular lines, as Plate XXII. fig. 1. It is not improbable this change may have been made to afford, as it were, rectilineal frames to suit the glass painter, the foliages and medallions of the preceding styles having given way to single figures standing on pedestals under rich canopies. Be this as it may, it has given a name to the style of the 15th and 16th centuries. (See PERPENDICULAR STYLE ; also GLASS.) In the later part of the reign of Henry VIII., and in the following century, first of all the cuspings disappeared from the tracery. The mullions then, as at King's College Chapel, St John's, Oxford, and several other

examples, had more flow and fewer perpendicular lines, till at last plain, upright, and transverse bars took their place, and held casement lights, which were at last superseded by our modern sash windows. On the Continent, the windows of the first period, or *ogivale primitive,* were very much like our own early English. So in like manner those of the early part of the *ogivale secondaire* were very much like our own *geometrical decorated* (which see). In Plate XXXVII. fig. 3 is from Le Mans, fig. 2 from Lucca, and fig. 7 from S. Madonna dell' Orto at Venice; while fig. 5, from Amiens, much resembles Plate XXXVI. fig. 9 in character, and the fanciful tracery of Plate XXXVII. fig. 1, that at the house of Jacques Cœur. Later, however, in France and Germany, two styles prevailed, the one having tracery assuming the character of stars or rays, and after this another coeval with our Perpendicular, resembling flames of fire; these are described under the arts. RAYONNANT and FLAMBOYANT.

TRACHELIUM (Gr. τραχηλος, the neck). In Doric and Ionic columns there is generally a short space intervening between the hypotrachelium and the mass of the capital, which may be called the trachelium or neck.

TRANSEPT. In a church of which the plan is in the form of a cross, the arms lying across at the intersection of the nave and chancel form the transept. Commonly each arm is spoken of as a transept, but strictly the transept is one.

TRANSEPT (Med. Lat. *crux,* Fr. *transept,* Ital. *crociata,* Ger. *querschiffe*), that portion of a church which passes transversely between the nave and choir at right angles, and so forms a cross on plan. Its origin seems to have been derived from a wish to have a larger space in front of the bema, or place where the altar stood in the early basilicas. Some of the columns were therefore omitted on the right and left of the upper end, and the building there had a transeptal roof as to the clerestory, but the outer walls were straight, so that, though there was a transept internally, there was no cross on plan. The whole formed, in fact, what is called a false transept. The introduction of domes still further confirmed architects in the use of this internal transept, a most notable example of which is at St Sophia at Constantinople ; and, as in the Greek churches, the limbs of the cross are of equal length, the plan might be said to be two naves or aisles crossing each other. In France the former term, *nef transversal,* is often used, and in England, *cross aisle.* By degrees the transept was made to project, and so form a real transept or cross upon plan. Our Saxon ancestors seem to have used this, as at Worth, and in Norman times transepts were very common—a low square lantern tower filling up the intersection. The form, however, generally was that of the Latin cross, the upper limb or choir (as in the ancient basilica) being very often merely an apse, and the nave larger than the other limbs. Transepts, in almost all instances, exactly match each other, or are in pairs, and consequently are often designated as the north or south transepts. At Exeter there are two noble Norman towers, which are supposed to have been the western towers which flanked the original Norman front. At Chester one transept is of similar Norman work, while the other is elongated and of the Decorated period, being now, and supposed always to have been, built as a separate church. At Soissons one transept is square and the other semicircular. These seem the only examples which are known of those irregular in construction. In England second transepts are found eastward of those under the central tower, as at Canterbury, York, Salisbury, Lincoln, and several other examples. Those at York, however, do not project, but are *false transepts.* A similar arrangement is said to have existed at Cluny, but nothing of the kind is

ARCHITECTURE.

to be found among the French examples given by Viollet-le-Duc, though they are more than thirty in number. Transepts are often without aisles, but in large churches almost always have not only arcades, but triforia and clerestories exactly like those of the nave. In some instances, as Peterbro', Salisbury, Lichfield, Durham, and Lincoln, they have but one aisle and arcade. No instance of a transept with one aisle only is cited by M. Viollet-le-Duc, though he gives several without any. On the other hand, on the Continent the transepts frequently have apses, which are very rare indeed in England. (See AISLE, APSE, CROSS AISLE, &c.)

TRANSITION. A name given to the gradual conversion or blending off of one style into another till the change became complete, and an entirely new style was established. (See DATES OF STYLES.)

TRANSOM (Fr. *traverse*, Ital. *traversa*, Ger. *querbalken*), the horizontal construction which divides a window into heights or stages. They are sometimes simple pieces of mullions placed transversely as cross-bars, and in later times are richly decorated with cuspings, &c. The earliest known are in early English buildings, as in the well-known example in the hall of the castle at Winchester. They then are found as simple bars dividing the long lucarne spire lights. Transoms are very rare in windows of the Decorated period. The earliest, probably, is at Albrighton. At Hull there is a Decorated window in which are two transoms, the upper cusped and the lower plain. In the Perpendicular style they are extremely common, forming stage over stage. At Winchester, the interior of the west end has seven transoms, one over the other. The common form is that in the tower windows. (Plate XXI. fig. 1, and Plate XXII. fig. 1.) A very peculiar arrangement is given from Dorchester (Plate XXXVII. fig. 11), where the mullions are not over each other, but rise from the point of the arch which carries the cusping in the lower stage. Another elegant example is given from Evesham (fig. 12). As the Perpendicular style never prevailed on the Continent, it is not to be wondered at so few transoms are found there. One very curious example, however, is given from Amiens, and a very elegant one exists at the churches of St Madonna dell' Orto and at the Frari at Venice.

TRAYLE. (See VINETTE.)

TREFOIL (Lat. *trifolium*), a cusping, the outline of which is derived from a three-leaved flower or leaf, as the *quatrefoil* and *cinquefoil* are from four and five.

TRIFORIUM, the arcaded story between the lower range of piers and arches and the clerestory. The name is used by Gervase, of Canterbury, in his description of the rebuilding of that Cathedral, but is not found in any other author. It is supposed to be derived from *tres* and *fores*—three doors or openings—that being a frequent number of arches in each bay. The roof over the aisle usually being a shed-roof, or lean-to from the sills of the clerestory windows to the wall of the aisle, the triforium generally has no external windows, and is, therefore, frequently called the *blind-story*. In early works the aisles are often groined at the height of the lower arches. The floor of the triforium is then frequently laid on the top of these, and paved. At Durham and Westminster it is called the nunnery, and the tradition is, the nuns sat there unobserved during the services hidden by curtains. This is not improbable, as a similar arrangement prevails to this day at the Church of St Agnese fuori le mura, at Rome. At other cathedrals, as at Rochester, the forium is a passage-way, or *galerie de service*, looking down into both nave and aisles. Sometimes, however, there is a close wall behind the gallery, and the triforium looks into the nave only. Norman triforia frequently

consist of one large arch of the same span as that below. Such is the arrangement at the chapel in the Tower of London (Plate XVIII. fig. 2); also at Waltham Abbey, Southwell Minster, &c. But more frequently this large arch is filled in with other arches carried on light shafts, the spandrels over which are ornamented with a species of diaper. At the Cathedral, Oxford, and at Romsey, there is a very curious arrangement. The piers and main arches rise to the top of the triforium, which is as it were fixed in between them. In early English work the triforium is often composed of two sets of arches, each subdivided into two others, as Plate XX. fig. 1. At Salisbury all four are under one great arch spanning the whole bay with pierced spandrels. Very frequently, however, the triforium is simply composed of a range or arcade of small arches, as in Plate XVIII. fig. 11. Sometimes the shafts of the rear arches are set alternately with those of the front range, so that they show between them, which has a beautiful effect, as at Wells (Plate XXIII. fig. 3), at Beverly, at Lincoln, and several other places. A very curious arrangement exists as to the triforium in Westminster Abbey. The flying buttresses run through the roof and the upper part of the story, which is lighted with windows in form of a multifoiled circle inclosed in a spherical triangle (Plate XXXVII. fig. 9), circumstances which would lead us to suppose the outer walls and roofs were raised, as an after-thought, to give more accommodation to the *nunnery* within. The Decorated triforium has much the same features as the preceding (see Plate XX. fig. 2, from Lincoln, and Plate XXXIX. fig 8, which is from Lichfield); but as the roofs of the aisles are generally flatter, it is of less height, and in many cases is merely a low arcade, as at Exeter. In the Perpendicular period, when the aisle roofs were generally lead flats, the triforium became of still less importance, the clerestory windows came down to the strings over the lower arches, and a sort of false triforium was formed by blocking up the lights of the lower stage of the windows, till at length in later work the triforium disappeared altogether. On the Continent much the same arrangement prevailed as in England. At Soissons, however, in the rond point is a second arcade over the principal range, forming a sort of double triforium. In several large churches, also, the triforium has been lighted, as at Westminster. Instances are given by M. Viollet-le-Duc at Amiens, St Denys, and Notre Dame at Paris. These are called *triforium ajouré*, or *à claire-voie*.

TRIGLYPH (Gr. τρεις, three, and γλυφη, an incision or channel). The vertically channelled tablets of the Doric frieze are called triglyphs, because of the three angular channels in them, two perfect and one divided; the two chamfered angles or hemiglyphs being reckoned as one. The square sunk spaces between the triglyphs on a frieze are called metopes.

TRUSS, a term in CARPENTRY; but it has been made to apply to consols, or ornamented corbels. (See CONSOL.)

TUDOR STYLE, a word generally applied to the later portion of the *Perpendicular* period (which see).

TUDOR FLOWER, or CRESTING, an ornament much used in the Tudor period on the tops of the cornices of screen work, &c., instead of battlements. It is a sort of stiff, flat, upright leaf standing on stems. (See Plate XXXIII. fig. 20.)

TURRET (Fr. *tourelle*), a small tower, especially those at the angles of larger buildings, sometimes overhanging and built on corbels, and sometimes rising from the ground. These frequently inclose winding stairs, particularly those to towers of churches, and at the junction of naves and chancels. (See JUBÉ, ROOD-STAIR.) Sometimes they hold a single bell. (See SANCTUS BELL.)

ARCHITECTURE.

The name is also often given to large pinnacles, particularly at the main fronts of important buildings. In the Norman period they are often round; in the early English, octagonal, and are surrounded with light shafts, and crowned with a sort of spire. (See those, Plate XIX. fig 1.) In the later periods they finish with perforated battlements or other enrichments, and the spires are crocketed. They sometimes serve as louvres to let light and air into a roof (see FUMERELL, LOUVRE, LANTERN), and sometimes surmount the apex of gables. (See Plate XXII. fig. 1.)

TYMPANUM (Gr. τυμπανον). The triangular recessed space inclosed by the cornice which bounds a pediment. The Greeks sometimes placed sculptures representing subjects in connection with the purposes of the edifice, in the tympana of temples.

UNDER-CROFT, a vaulted chamber under ground. (CRYPT.)

VALURE, VAMURE. (See ALURE.)

VANE (Fr. girouette, Ital. banderuola, Ger. wetterfalme), the weather-cock on a steeple. It is mentioned by Durandus as in form of a cock, gallus, a form it often preserves to this day, and whence its familiar name. They seem in early times to have been of various forms, as dragons, &c.; but in the Tudor period, the favourite design was a beast or bird sitting on a slender pedestal, and carrying an upright rod, on which a thin plate of metal is hung like a flag, ornamented in various ways.

VAULT (Ital. Voltato, turned over). An arched ceiling or roof. A vault is, indeed, a laterally conjoined series of arches. The arch of a bridge is, strictly speaking, a vault. Intersecting vaults are said to be groined. (See GROINING.)

VAULTING. For barrel vaults, spherical vaulting, &c., see STONE-MASONRY; for mediæval vaulting, see ibid., also GROINED VAULTING.

VAULTING SHAFT, a small column or series of clustered shafts, rising from above the capitals of the pillars of an arcade, and generally supported on a corbel, and thence rising and finishing with a capital, from which the various groin ribs spring. (See those at Plate XX. figs. 1 and 2, see also GROINED VAULTING.)

VERGE, the edge of the tiling projecting over the gable of a roof; that on the horizontal portion being called eaves.

VERGE BOARD, erroneously called Barge Board, the board under the verge of gables, sometimes moulded, and often very richly carved, perforated, and cusped, and frequently having pendents and sometimes finials at the apex. They are probably of very early origin, being in fact merely decorations of the outer common rafter; but, as they are of perishable material, there are no remains earlier than the 14th century. Plate XXXIX. fig. 6, gives an example from Caen; and fig. 7 is a very fine specimen from Ightham.

VESICA PISCIS (Fr. amande mystique), panels, windows, and other ornaments of the form of a species of oval with pointed ends, but in reality struck from two centres, and forms part of two circles cutting each other. A window of this kind is given, Plate XIX. fig. 1, from Beverley.

VESTRY. (See SACRISTY.)

VICE, VYS (from the French vis, a screw), a name for a winding staircase. (See STAIRS.)

VIGNETTE, a running ornament, as its name imports, representing a little vine, with branches, leaves, and grapes. It is common in the Tudor period, and runs or roves in a large hollow or casement. The one in Plate XXXIII. fig. 22, is from Westminster. It is also called TRAYLE.

VOLUTE (Lat. volutum, from volvo, rolling up or over, convolving). The convolved or spiral ornament which forms the characteristic of the Ionic capital is so called. The common English term is scroll, q. v. Volute, scroll, helix, and caliculus, are used indifferently for the angular horns of the Corinthian capital.

VOUSSOIR, a name in common use for the various wedge-shaped stones of an arch. (See STONE-MASONRY, see 106.)

WAGGON-CEILING, a boarded roof of the Tudor time, either of semicircular or polygonal section, boarded with thin oak, and ornamented with mouldings forming panels, and with loops at the intersections. (See PANEL.)

WAINSCOTTING. (See PANELLING, &c., ut supra.)

WARD, a name for the inner courts of a fortified place. At Windsor Castle they are called the upper and lower wards. (See BAILEY, BASE COURT, ENCEINTE, &c.)

WATER TABLE, a chamfered weathering to throw the water off from sets-off (which see, also TABLES, WATER.)

WEATHERING, a slight fall on the top of cornices, window sills, &c., to throw off the rain.

WICKET (Fr. guichet, Ital. sportello), a small door opening in a larger. They are common in mediæval doors, and were intended to admit single persons, and guard against sudden surprises.

WIND-BRACES, diagonal braces to tie the rafters of a roof together and prevent racking. In the better sort of mediæval roofs arched, and run from the principal rafters to catch the purlins. (See ROOFS, MEDIÆVAL, fig. 40.)

WINDOW (Lat. fenestra, Fr. fenétre, Ital. finestra, Ger. fenster), openings in walls for admission of light. For those of the Saxon, Norman, and early English periods, see those arts. For those of the Decorated and Perpendicular, see TRACERY. See also MULLION, TRANSOM, CUSP, ELEMENT, LABEL, HOOD-MOULD, ESCONSON, QUARRY, SADDLE BARS, STANCHEON; see also TOWER, SPIRE-LIGHTS, LUCARNE, GLASS—STAINED; see also the art. STONE-MASONRY, MEDIÆVAL.

WINDOW, CIRCULAR, &c. In addition to the usual form of windows, we often find in clerestories, gables, &c., openings of other forms. The most common of these is circular. In early Norman work these are plain; later they were filled with radiations like the spokes of a wheel, from which circumstance they are often called Catherine wheel windows. See the large window in the gable, Plate XIX. fig. 1, which is from Beverley; see also the smaller ones over the aisles, one of which is given to a larger scale, Plate XXXVII. fig. 8. They then became filled with tracery, and were called rose windows (Fr. rosace), of which a very fine example is given from Chichester, fig. 4. Fig. 6 is from S. Antonio, at Padua. In the early English period a form called vesica piscis is common. See the upper window in the gable at Beverley, Plate XIX. fig. 1. Windows in the form of an equilateral spherical triangle are also frequent. One of these, containing a cusped circle from Westminster, is given, Plate XXXVII. fig. 9. See also ROSE WINDOW, VESICA PISCIS.

ZIG-ZAG (Fr. batons rompus), a very common decoration in the Norman style, of which examples are given, St John Devizes, Plate XXIII. fig. 2; Plate XXXIII. figs. 3 and 5. (See also Plate XVIII. figs. 3, 4, 7, and 8.) It is believed that no examples of this ornament are found in the remains of Saxon architecture, unless those at Waltham Abbey (Plate XVIII. fig. 4) be of that period. Here the great nave columns are covered with zig-zag, as they also are at Norwich. The ornament, however, is common in Saxon pottery.

ZOÖPHORUS (Gr. ζωον, an animal, and φερω, to bear). This term is used in the same sense as frieze, and is so called because that part of the entablature is made the receptacle of sculptures which are frequently composed of various animals.

CONSTRUCTION.

THE sense in which the term "Construction" is used in this place, is that in which it is employed in the article ARCHITECTURE; purporting, however, rather fabrication than conformation.

The object of construction is to adapt and combine fit materials in such a manner that they shall retain in use the forms and dispositions assigned to them by the constructor. If an upright wall be properly constructed upon a sufficient foundation, the combined mass will retain its position, and bear pressure acting in the direction of gravity, to any extent that the ground on which it stands and the component materials of the wall can sustain. But pressure acting laterally has a necessary tendency to overturn a wall, and therefore it will be the aim of the constructor to compel, as far as possible, all forces that can act upon an upright wall to act in the direction of gravity; or else to give it permanent means of resistance in the direction opposite to that in which a disturbing force may act. Thus when an arch is built to bear against an upright wall, the constructor applies a buttress or other counterfort in a direction opposed to the pressure of the arch. In like manner the inclined roof of a building, spanning from wall to wall, tends to thrust out the walls; and hence the constructor applies a tie to hold the opposite sides of the roof together at its base, where alone a tie can be fully efficient, and thus compels the roof to act upon the walls wholly in the direction of gravity; or where an efficient tie is inapplicable, he adds buttresses or counterforts to the walls, to enable them to resist the pressure outwards. A beam laid horizontally from wall to wall, as a girder to carry a floor and its load, may sag or bend downwards, and tend thereby to force out the walls; or the beam itself may break. Both these contingencies are obviated by trussing, which renders the beam stiff enough to place its load on the walls in the direction of gravity, and strong enough to carry it safely. Or if the beam be rigid in its nature, or uncertain in its structure, or both (as cast iron is), and will break without bending, the constructor, by the smith's art, will supply a check and ensure it against the possible contingency.

Stability is then the aim of the constructor; but perfect and enduring stability is not to be attained with materials which are subject to influences beyond the control of man, and all matter is subject to certain influences of that nature. The influences with which the constructor has mostly to contend are heat and humidity; the former of which produces movement of some kind, or to some extent, in all bodies; the latter, movement in many kinds of matter; whilst the two acting together tend to disintegration or to decay in all materials available for the purposes of construction. These pervading influences the constructor seeks to counteract, by the selection and disposition of his materials accordingly. From the tenacity of wrought iron, and its almost plastic character in the hands of the smith, the constructor will employ it to tie together other more bulky but less costly and more rigid materials; but on account of its exceeding susceptibility of heat, and its consequent expansion and contraction, he will use wrought iron in short lengths only, unless where protected from great alternations of heat and cold. The rapid decay, too, of wrought iron when exposed to humidity, and especially to alternations of wet and dry, will teach the constructor not to expect enduring stability in his works if he makes them dependent upon wrought iron. Cast iron is brittle, and may not be exposed with impunity to transverse strain, especially if such strain be attended by action tending to induce vibration: it expands and contracts under the influence of heat, but it resists compression also in every direction, and if used in small bodies, is valuable as a means of connecting other materials. Timber, being practically unchangeable in the direction of its length from the mere absorption of either heat or humidity, and at the same time practically both inextensible and incompressible in that direction, and being also readily wrought and easily combined alike with itself and with iron, is a valuable material in the hands of the constructor; but it shrinks and swells in the direction of its thickness, and, in consequence, is subject to rapid decay when exposed to alternations of moisture and dryness; and although in many varieties timber is perdurable and unchangeable in form if it be kept either altogether free from moisture or always wholly wet, its quality of inextensibility is greatly diminished in value to the constructor on account of the comparatively slight resistance it offers to compressing power, and the comparative ease with which its fibrous structure is torn asunder. From this cause it cannot be grasped or otherwise held so that its power of resisting extension may be made available in any degree proportioned to its strength; whilst its quality of incompressibility in the opposite direction is of less value to the constructor for many purposes which require that quality in the material, because it absorbs moisture by the ends of the fibre more readily, and with a far more mischievous effect, than it does in the direction in which it is compressible. Hence timber rots more rapidly by the ends than by the sides.

Stone and brick, the other main available materials in general construction, keep their places in combination by means of gravity. They may be merely packed together, but in general they are compacted by means of mortar; so that although the main constituent materials are wholly incompressible, masses of either, or of both combined in structures, are compressible until the mortar has indurated to the same condition of hardness.

That kind of stone is best fitted for the purposes of general construction which is least absorbent of moisture, and at the same time free to work. Absorbent stone exposed to the weather rapidly disintegrates; and for the most part non-absorbent stone is so hard that it cannot always be used with a due regard to economy. The constructor therefore, when he can command fitting stone of both qualities, exposes a face of harder stone to the weather, or to the action which the softer stone cannot resist, and forms the main body of the structure of the latter so protected. The hard and the soft should be made to bear alike, and should therefore be coursed and bonded together by the mason's art, whether the work be of stone wrought into blocks, or gauged to thickness, or of rough dressed, or otherwise unshaped rubble compacted with mortar.

Brick, if good, is less absorbent of moisture than any stone of the same degree of hardness, and it is a better non-conductor of heat than stone. As the basis of a stable structure, brickwork is more to be relied upon than stone in the form of rubble, when the constituents bear the relation to one another last above referred to, the setting material being the same in both; because the brick by its shaped form seats itself truly, and produces by bonding a more perfectly combined mass; whilst the imperfectly-shaped and variously-sized stone as dressed rubble can neither bed nor bond truly; the inequalities of the form being to be compensated for with mortar, and the irregularity of size of the main constituent accounted for by the introduction of larger and smaller stones.

The most perfect stability is to be obtained nevertheless from truly wrought and accurately seated and bonded blocks of stone, mortar being used to no greater extent than may be necessary to exclude wind or water, to prevent the disintegrating action of both upon even the most durable stone. When water alone is to be dealt with, and especially when it is liable to act with force, mortar is necessary for securing to every block in the structure its own full weight and the aid of every other collateral and superimposed stone in order to resist the loosening effect which water in powerful action is sure to produce.

In the application of construction to any particular object, the nature of the object will greatly affect the character of the constructions and the materials of which they are to be formed. The object of a breakwater is to check the run of the sea when it is acted upon with power. It may be that in some cases piles of timber driven into the bed of the sea, and made rigid by repetition or by combination, might have strength enough to withstand and to break the run of the sea; but there is a range between high and low tide throughout which the piles would be exposed alternately to opposite influences, either of which alone would do the timber little injury, but which acting in rapid succession will in a short time destroy what might otherwise have resisted the extremest force of the sea for ages. Timber failing in enduring usefulness, which it will the more rapidly if it be left in any degree dependent upon wrought iron; or if the depth and the run of the sea be too great for timber, stone is adopted. Large blocks of the densest stone may

be tipped into the sea, and serve as a base to other blocks in succession until a mound, mole, or dike, be formed reaching above high-water level. This dike may have had such slopes given to it as experience had shown that heaped up blocks of rough-hewn stone would take on dry ground; but if, even while the work is yet in progress, a gale of wind occasion a great run of the sea, the half-raised breakwater will be converted into something little better than a shoal of rubble;—it having been overlooked that the blocks of stone, when immersed in water, had lost so much of their weight that far less force than was necessary to move them in air tumbled them about in the water as small pebbles are rolled on the sea-shore. The stones were uncombined, and every surface stone was exposed in a half floating condition to the force of the sea, while its fellows below became exposed as those above them were rolled away. The construction wanted rigidity. The water must be excluded from among the stones, so that the whole shall form one mass, from the level at which the run of the sea in a storm can first be felt, up to and above the highest high-water level at the place. By such a process, labour and skill will perform effectually, with a comparatively small body of stone, what a huge mass of crude material unskilfully dealt with had failed to accomplish.

In a line of road, the object sought in constructing an embankment is a firm foundation to the roadway or the railway, by which the traffic may be carried on with perfect safety. The work that failed as a breakwater may form an enduring embankment on dry land. The loose blocks of stone which, when immersed in water, were too light to resist the force of the sea, will remain undisturbed by the trains rolling over them when heaped up in air, though unarranged by the art of the mason, if the stones be allowed to make such slopes as they will naturally form when dropped down from above as if dropped into the sea; or as a measure of dry sand tipped out upon an earthen floor will form its own slopes, so a rubble tipped embankment will take slopes and maintain them against any pressure from above. Such a construction is rigid in air, but subject to movement in water: the circumstances are different. But construction would nevertheless dictate the propriety of filling in the interstices between the rubble, to prevent water, as rain and snow, from passing through the embankment, and so to soften the ground beneath that the stones would sink into it, and thus produce movement in the otherwise rigid structure.

With these considerations would come also another. A small amount of skill in packing stone rubble will greatly reduce the quantity of material necessary to form such an embankment, by substituting less inclined slopes for the natural inclination of the dropped rubble. From merely packing dry rubble to laying it in courses,—and from the construction of a heavy mound of rubble to the construction of a series of piers and arches,—the steps are gradual; but safe structure and a stable foundation may be obtained by any of the means indicated. Circumstances must dictate the kind of construction proper in each particular case. The quiet traffic of a canal might seem to permit the use of constructions less carefully carried out than those which may be necessary to the carriage road, and more particularly to a railway; but while derangements may occur in the substructure of a railway without stopping its traffic, the canal must be so constructed as to prevent the possibility of defect, for the water is constantly and insidiously working to make defects, the existence of which may involve the safety of the surrounding country, as well as the interruption of its traffic. Hence canal works must be sound and secure constructions;—there is no tampering with water;—and the same rigorous attention to security should be given to railway constructions, which so often verge upon the dangerous.

R

The railway or road embankment is more commonly made of clays and other plastic earths than of stone; and, as usually made, it is a formation rather than a construction. But an earthen embankment ought to be a construction in the sense in which a wall is a construction. It ought to be executed in layers, and not tipped in heaps.

The lateral spread of an earthwork embankment from its crest to its base will depend mainly on the character of the soil. This is to be ascertained by experiments upon the inclination at which it will stand; but as there are soft and yielding places over which earthwork embankments have to be formed, it may be that an embankment will require a wider base because of the yieldingness of the ground under it, than the soil of which the embankment is to be made would require for itself, even as a merely tipped deposit. On the other hand, it will be found in practice that the application of the principle of construction, as above stated, to the formation of earthwork embankments, gives the means of compelling the earth in any case to stand at much steeper slopes than it will without such disposition and working.

The converse of an embankment in a line of road of any kind, is a cutting, the matter cut out being the material of which the constructed embankment is formed. The sides of a cutting must be securely retained, and constructions of some kind may become necessary. They may be necessary from the peculiar character of the soil, or desirable for economy's sake, whether it be on account of the costliness of the site, or of the heavy earth-works involved to give the sides of a cutting such long slopes as the soil in the case may require. Constructions, that is to say, combinations of foreign materials disposed artificially, do become necessary both with or to embankments and in cuttings. Bridges over rivers or other water-courses, in the bottom of a valley so deep that the material from the nearest cuttings will not fill it, or fill it only at a cost greater than the cost of a bridge; or cuttings so deep, or in such a soil, that economy or stability dictate the employment of retaining constructions which shall be wholly independent of the soil in the work or in the slopes;—these involve considerations of the same nature and character as those which arise in the apparently complex design of Cologne Cathedral, or of the Abbey Church of Westminster.

Every piece of construction should be complete in itself, and independent as such of everything beyond it. A door or a gate serves its purpose by an application wholly foreign to itself; but it is a good and effective, or a bad and ineffective piece of construction, independently of the posts to which it may be hung.

Whilst the wheel of a wheel-barrow, comprising fellies, spokes, and axle-tree, is a piece of construction complete in itself, and independent as such of everything beyond it, an arch of masonry, however large it may be, is not necessarily a piece of construction complete in itself,—it would fall to pieces without abutments. Thus, a bridge consisting of a series of arches, however extensive, may be but one piece of construction, no arch being complete in itself without the collateral arches in the series to serve as its abutments, and the whole series being dependent thereby upon the ultimate abutments of the bridge, without which the structure would not stand.[1]

A bridge, of which the bridging way is formed upon arches of masonry, may be thus but one piece of construction; and

in like manner, that paragon of constructive skill, the complete church, whether cathedral or otherwise, as built in the pointed style when that style was practised with perfect knowledge of and in full accordance with true constructive principles, is but one piece of construction. Like the long series of arches in a bridge, viaduct, or other such work, in which the piers are vertical supports to the bridging structure, and may be of no greater substance than is necessary to bear the weight coming directly by vertical pressure from the superincumbent structure and its possible load, but throwing all the pressure arising from weight acting laterally, or as thrust, upon terminal abutments;—nothing may be omitted, as nothing can be removed from the structure of the pointed-arch cathedral, or other church built in that style,—the whole system of which is bridge-like in construction,—without leaving something unsupported or unresisted that requires vertical support or lateral resistance. The western towers of a pointed cathedral form effective abutments to the long series of arches of the inner ranges over the piers which stand between the nave and the aisles on either side, whilst turrets or massive buttresses and deep porches upon the northern and southern transept fronts perform the same duty in respect of the arches of the transepts. The counter-acting east end of the chancel forms a true constructive abutment to the arches of the chancel, whilst the tower, with, it may be, a spire upon it, at the intersection of the four grand compartments of the cross, gives, by its weight, abuttal to them all. The want of this last-named grand and essential body in the system is but too strongly marked in many of the English cathedrals by the iron bars which have been applied to tie in the arches of the nave, transepts, and chancel, and to relieve the piers upon which the transept arches bear at a higher level, from the thrust to which—being without the weight of a tower upon them— they have continually yielded.

Transversely the weight and the thrust of the vaulted ceilings of the nave are brought up to, and thrown against, the piers of the clerestory, which stand upon the main piers or columns of the interior below, and are abutted by flying buttresses, which carry the thrust down to the pinnacle-weighted buttresses of the outer aisle walls which have already received the weight and thrust of the vaulted ceilings of the aisles themselves. Corbels in the walls, and spreading capitals upon shafts take the weight directly, and leave the walls and piers but little encumbered in the middle, so that the vertical structure is continued upwards without bearing upon the springing stones of the arches.

But it is not necessary that the arch employed should be the pointed arch to produce combinations as effective in construction as the most perfectly designed and extensively elaborated work of the kind referred to as models of constructive skill; the skill consists in a full and clear perception of the bearing and leaning of every part, and of the means necessary to support and counteract the bearings and the leanings within the reasonable limits of the work with reference to its object and purpose—to the end that the work may become complete in itself, and independent as a piece of construction of everything beyond it.[2]

An application of the principles of construction exhibited in the most perfect works of constructive skill ever executed, as above indicated, may be made in the rougher operations of mere practical utility. It has been intimated that the sides of cuttings through certain earths in the for-

---

[1] This illustration is not intended to apply to the widely distended masses of the older bridges, by which each pier becomes sufficient to abut the arches springing from it; but which, in providing for a way over a river, chokes up the way by the river itself, or compels the river to throw it down, or otherwise destroy its own banks.

[2] In making reference to the noble works of construction above referred to, and in which the art of the mason is mainly employed, as works exhibiting construction most fully and most truly, the HALL must not be passed over without remark.

Of all the great halls of the class to which Westminster Hall belongs, this hall is itself the most effective as a work of construction; and its effect is wholly produced by the magnificent roof which covers it. This roof is a piece of carpentry admirably designed to resolve it into a compact body to act upon the walls in the direction of gravity alone. But the object was not wholly attained, and of this

Construc-
tion.

mation of lines of inland communication, whether carriage-roads, railways, or canals, are sometimes required to be widened out to an inordinate extent because of the looseness or slipperiness of the soil, or otherwise to be retained or held upright by special constructions.

The expense of the first formation of a cutting under given circumstances is easily calculable, and so is the time within which the work may be effected. Experience has proved that there is for every soil a limit in depth beyond which it becomes more expedient to drift the required way, and construct a vaulted tunnel of sufficient dimensions, than to make an open cutting with the requisite slopes. Even when the first cost would not decide the question, the preference is nevertheless often given to the tunnel because of the greater security of constructed work.

A tunnel is expensive, not from the nature and extent of its constructions, but from the circumstances in which those constructions must be executed. The mere constructions are less than would be consumed by common retaining walls to the sides of a cutting not deeper than the height of an ordinary railway tunnel: the several parts of a tunnel derive support from each other, which is not the case with ordinary retaining walls, whose efficiency depends wholly upon the resistance which their own mass or weight and extent of base enable them to offer to the pressure of the body to be retained. If to two opposite retaining walls be given sufficient means of assisting one another, they may be at once reduced to one-third of the bulk they would otherwise require, and would then be as safe as the sides of a constructed tunnel, the strength of which, supposing the work to be properly executed, is only limited by the power of the setting material employed in the work to resist compression.

Before proceeding to the consideration of the means of enabling opposite retaining walls to assist each other, it may be worth while to consider, whether retaining walls are generally constructed so as best to adapt their components to the duty to be performed.

No one would place a buttress intended to resist the thrust of an arch, within the springing walls, or under the arch whose thrust is to be resisted; yet in the construction of retaining walls, according to the common practice, the counterfort is placed on that side which receives the pressure, where its utility is very questionable, except to keep the retaining wall from falling back against its load, which, from the transverse section generally given to such walls, they would be apt to do, if not so propped up by their counterforts. Wharf and quay walls, and the revetment walls of military works, may require a face unbroken by projections; but this is not the case with retaining walls for roads and railways, where a long line of projecting buttresses would be unobjectionable, the counterforts becoming but tresses and merely changing places with the wall.

On account of the common practice of battering the faces of retaining walls in curved lines and of radiating the beds of the brickwork composing them from the centre of curvature in every part, the back of the wall must contain more setting material than the face, with the same quantity of solid brick; that is, if the work be bonded through. Counterforts must be built in the same courses, and consequently must have still thicker beds of compressible mortar than

the wall; or the bond between the wall and its counterfort must be dropped, and the counterfort thus become utterly inefficient. Construc-
tion.

The retaining walls in the cutting upon the line of the extension of the London and North-Western railway, from Camden Town to Euston Square, are, according to the common practice, built wholly of brickwork in radiating courses and with counterforts following their own contour. In this case the centre of gravity of the wall falls wholly behind its base; and the counterforts not commencing until the wall has reached one-third its height, render it still more dependent for support upon the ground it is intended to retain. It is well known that these extensive walls, though furnished with all the collateral works necessary to protect them from exposure to undue influences, and although set nearly one-fourth of their height in the ground, failed to a considerable extent. A system of strutting with cast-iron beams, across from the opposite walls, to make each aid the other, was applied to meet the emergency; but this is limited to the upper parts of the walls.

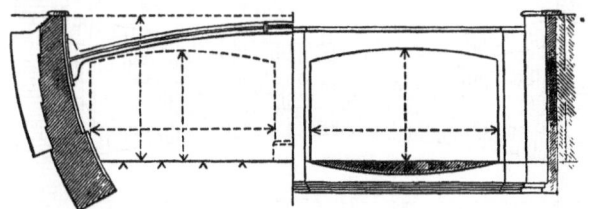

Transverse section of the Euston Incline retaining walls, one-half as executed with cast-iron struts to counterforted and reclining walls, and the other half with the brick-built abutting beam to counter-arched retaining walls strutted at the toes of the springing walls by inverted arches.

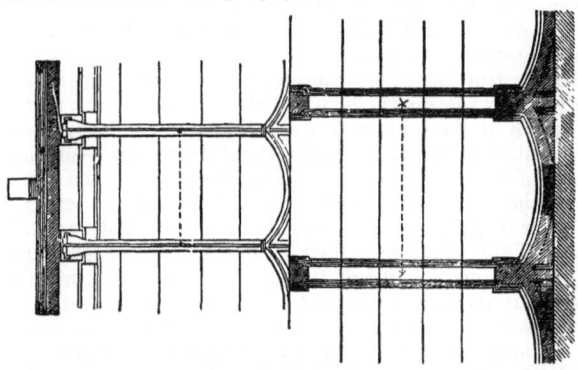

Plan of the above showing the part as executed above the iron struts, with the rails passing underneath, and the other part at the level of the rails, with the inverts in plan under them.

Abutting struts from opposite walls, occurring at intervals only, leave the intermediate portions of the walls exposed to pressure from behind without support, unless these intermediate portions are so disposed as to communicate the pressure upon them to the struts. Hence a common retaining wall, abutted at intervals, would require these intervals to be more or less distant, in proportion to the strength of the wall between them. Instead, therefore, of a continuous wall on each side of the cutting, buttress walls should be placed at intervals, opposite to one another, and strutted apart at their toes by an inverted arch, and above, at a height sufficient for whatever traffic the cutting is to accommodate, by a built beam of brickwork, in vertical

---

the constructor was fully conscious; for whilst erecting massive walls on which to place his elaborate combination of timber, he threw up against the lateral walls a series of flying buttresses to check the tendency of the roof to spread under its own weight in the absence of a thorough transverse tie. These buttresses are supposed still to remain (*and it is to be hoped they will still remain*), though they are mostly incorporated in or encased by the recent erections on the flanks of the hall.

The open or untied roof, of which that of Westminster Hall is so egregious an example, had its origin probably in the want of timber long enough to serve as tie-beams at the higher level of the collars in these roofs, or to reach across from wall to wall where it was sought to dispense with inner ranges of supports as columns and piers. But the great old roofs of this kind are placed upon stout walls well and safely abutted, whilst the puny modern imitations of such roofs are made temporarily safe—*and only temporarily so*—by the aid of straps and screws.

courses, supported on an arch, and prevented from rising under the pressure by an invert upon it. This built beam will then be, as it were, a piece of walling turned down on its vertical transverse section, and will resist any pressure brought upon it through the buttress walls, to the full extent of the power of such a wall built vertically, to bear weight laid upon its summit;—the pressure would be applied in the line of the greatest power of resistance, and there would be no tendency to yield, except to a crushing force. Let such transverse buttress walls, so strutted apart, with the road between them, be the springing walls of longitudinal counter-arched retaining walls, which, being built vertically and in horizontal courses, but arched in plan, against the ground to be retained, will carry all the force exerted against them to their springing walls, and the springing walls or buttresses will communicate, through the struts, the power of resistance of each side to the other, and thus insure the security of both.

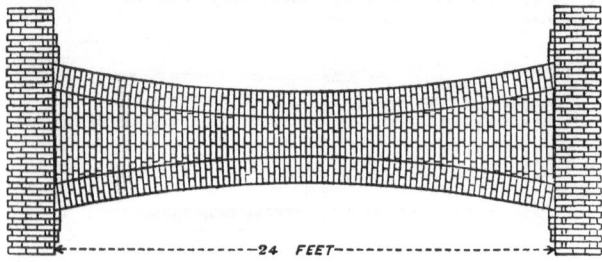

Built abutting Beams.

This arrangement may be carried to any extent in height, by repeating the abutting beam or strut at such intervals as the thrust to be resisted and the strength of the buttress springing-walls may require.

To constructions thus arranged, any requisite power may be given, by altering the quantity of materials in each part ; the length of the buttresses transversely of the cutting,—the number of struts to each pair of buttresses,—or the length of the compartments. The thickness of the buttresses should be in proportion to their height and length, and their length should be in proportion to the flatness and weight of the struts with their arches, and to the space in height between any two of them, as well as to the magnitude of the thrust brought to them by the counter-arched retaining walls. The inverted arch below and the built beam above must, of course, have sufficient substance to enable them to resist, without yielding in any direction, the pressure brought to them through the buttresses ; and the retaining walls themselves must have substance given to them according to their height,—to the pressure they are liable to receive from behind,—to the length of the compartments—and the extent of their flexure ;—subject, of course, as to all these, to the nature of the materials, workmanship, and mode of structure.

The positive strength which such constructions should possess depends much, of course, upon the nature of the soil, and its susceptibility of being affected by external influences ; but it depends, even in a greater degree, upon the manner in which the constructions can be applied to the ground they are intended to retain. A very slight power will retain at rest a body which the exertion of great force could not stop if once in motion ; and a half-brick counter-arch, set in close contact with undisturbed ground, would hold safely up what three times the substance would not stop if there were space and opportunity for motion between the ground and the brickwork. It is impossible, therefore, to state precisely what is the least strength which the retaining constructions must have ; but there can be no question that too much strength is better than too little, and it is generally cheaper to pay in materials than in labour to save materials.

These diagrams represent a cutting 65 feet deep to the level of the rails. It is assumed, that the ground at the top

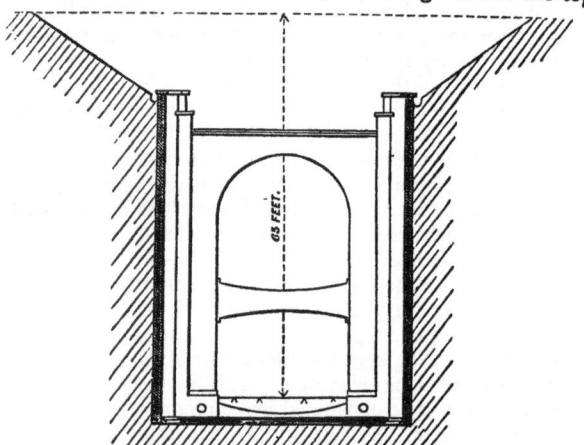

Transverse Section through the centre of a Bay.

may stand for the first 15 feet at less than 2 to 1, and that it may, therefore, be cheaper to run out to that depth with slopes, leaving 50 feet from the rails, or about 52 feet in all, to be retained. As the bricklayer may follow up the excavator with bay after bay, his work lying mostly on the side and out of the way of the excavator, the latter would run out the spoil without interruption, his work being benched onwards and shored as he proceeded. As every compartment, with its buttresses, invert, abutting beams and counter-arches is complete in itself, the ground being backed against the counter-arches as the work rises, the shoring would come out, and be sent on for use on the forward benches.

The invert may be turned upon footings in half-brick rings, to get the largest quantity of solid resisting matter in the curved line. At a height from the surface of the rails sufficient for headway—assumed at 14 feet 6 inches—a 14-inch bonded arch is turned from buttress to buttress, springing from skewbacks on corbelled courses. Upon the back of this arch the abutting beam is built of brick on end and edge, bonded as a wall, with beds vertical and widening over the haunches of the discharging arch and under the similar inverted arch turned upon it; so that although the beam be in the centre but 21 inches deep, it presents an abutment at each end of three times that depth. The ob

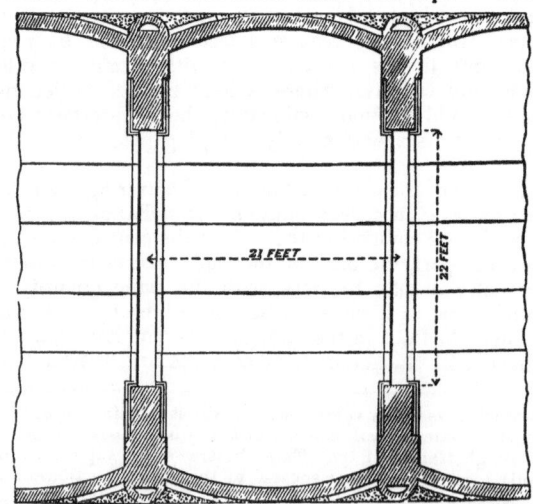

Plan at twice the Scale of the Section.

ject of the invert over the abutting beam is to stiffen it and to bring down and distribute the weight and pressure from the buttresses more effectually.

The built beam, and its sustaining and stiffening arches, should be composed of particularly well-formed bricks of really good quality, set in Roman cement or other quick-setting mortar, that there may be no yielding to the pressure which must be immediately thrown upon this part of the construction.

Another built beam, of greater depth, because of the absence of any inverted arch to stiffen it, is thrown across over the back of a semi-circular arch, with its abutting ends extended in like manner.

To relieve the work from water, a drain being run along over the middle of the inverts, or side-drains being passed by ring culverts through the buttresses, drain-shafts are carried up at the backs of the buttresses against the springings of the counter-arches, to within a few feet of the surface. These shafts being steened with open joints at intervals to admit drainage water and communicating with the drains below, prevent the possibility of water lodging about the backs of the counter-arches, or even in the ground itself. The drain shafts should be semi-domed with bricks set dry and covered in, and the walls also backed up with good clean gravel, through which the surface water might percolate and pass freely down to the shafts.

The constructions are assumed to be of brickwork, for the obvious reason that the cases supposed being clay cuttings, brick is the material which would be most economical. But if masonry be cheaper, it may of course be used with the same effect. Where a cutting intersects loose beds of laminated stone, and particularly strata inclined to the horizon, so as to be unsafe with the ordinary slopes, such constructions are available; and in cases where the sides of the cutting will stand vertically or nearly so, as in chalk, it may be useful to apply similar constructions, though of slight character, to check the separation and fall of masses from the precipitous sides.

It is obvious, too, that these constructions present the means of security, when the stratum forming the base of any cutting is too weak to bear the weight of slopes, or of retained sides, without rising between them. Sheet-piling may be driven to any depth along the backs of the counter-arched walls so as to be retained at the head by the walls; and thus in effect the walls would be carried down to a safe depth, even through the weak stratum; whereas such piling at the toes of slopes is commonly found to be almost if not wholly useless, for the want of a stay to the head.

Embankments formed in the manner already described—that is, by a process of construction—may be, as previously stated, raised higher and with a relatively narrower base than if formed in the usual inartificial way; but there is a limit which may not be passed in heaping up compressible or otherwise yielding materials, how skilfully soever they may be disposed, and this will indicate the limit at which constructions proper may be introduced with economy, as in the converse case of retaining constructions to avoid insecure or expensively long slopes in cuttings.

The height to which the particular soil may be raised upon itself with safety and economy being determined, the greater height required is to be obtained, not by an endeavour to encase the bank by constructions extending laterally as retaining walls, but by bridge-like culverts built under the bank, or of so much of the bank as may be safely built of earth.

Such bridge-like culverts may be composed with great advantage, both in respect of economy and of strength, in the manner suggested by the present writer for the upper works of Westminster Bridge, in the practical treatise on Bridge Building, forming part of the Theory, Practice, and Architecture of Bridges, published by Mr Weale in 1842. This involves a system of groining whereby that important element in construction—rigidity—is obtained, whilst the composition is but a variety of that which is embodied in the structure of the nave and aisles of the model of construction before referred to. Such substructed works may be rough but must be sound as work, and being covered with fifteen or twenty feet in depth of earth, they would not feel the vibrations which act so mischievously upon the lofty and costly viaducts in which railway-makers have exhibited their skill and taste to the cows and the crows, and which react both upon the upper works of the railway and upon the rapidly passing train in the shock felt throughout when the moving load passes the line between earthwork and masonry.

CONSTRUCTION AS APPLIED TO CIVIC STRUCTURES GENERALLY.

The ignitibility of timber, and the rapidity with which it burns when placed in circumstances so favourable to that effect as by its disposition in an erected building, have led to its prohibition for the purposes of the main inclosures of houses, and buildings generally, in London, and in some of our larger provincial towns. It is possible, however, so to protect timber employed in the inclosures and for the internal partitions and floors of buildings as to render mere dwelling-houses practically fire-proof. Whilst, however, the liability of timber to take fire and to burn may in a great measure be counteracted, and notwithstanding that this material combines the advantage of economy with security, stone and brick are undoubtedly better adapted for the main structure of a building. Brick or stone, or brick and stone together, with mortar as a setting material, ought to be employed, but in such manner only as to be free from dependence upon other and less trustworthy materials. The most perfect erections as buildings are those in the composition of which this principle has been understood and fairly practised. If adventitious aid be given to brick or stone walls by foreign materials, the materials ought to be at the least harmless. Iron in bulk is not a proper substance to incorporate with walls, because of its great expansibility by heat; but iron used in thin laminæ, as hoop-iron laid in walls in the bed-joints of the brick or stone, cannot be productive of any bad consequences whilst it is most beneficial in that form as a tie to the structure.

Bricks come ready shaped to the hands of the workman in a form the best adapted for the arrangement in the construction of a wall which, under the designation of bond, gives it such a degree of consistency that a weight placed upon the top is carried by the wall in every part throughout its whole thickness, and throughout a greater or less proportion of the length according to the height of the wall.[1]

---

[1] Bond in brickwork is most conveniently and most effectively formed and maintained by disposing the bricks in their courses either endwise and lengthwise (technically, header and stretcher), alternately brick and brick, or course and course; that is to say, that the bricks in every course should be laid alternate header and stretcher, or that the courses should consist of all headers and all stretchers alternately. The former arrangement—alternate header and stretcher in the same and in every course—is known in this country as Flemish bond; and the latter—alternate courses of header and stretcher—is distinguished by the term English bond. Neat work in face can be produced more easily with Flemish bond, but English bond has the reputation of being the best bond structurally. But why these two arrangements should be distinguished by the names they bear is unknown; at least it is unknown to the present writer, who supposed, in common with most other people with whom he had conversed upon the subject, that alternate header and stretcher in the same course was the practice in Flanders, and generally in the neighbouring countries on the Continent, whilst the term English bond seemed to imply that the arrangement which bears that designation is peculiar to England. A visit made a few years ago to the countries where Flemish bond ought most to abound, if the name be properly applied, enabled the writer to observe what had

Stone, on the other hand, comes to the workman without regular form; and with skill on his part to dispose and arrange the materials, good erections may be produced of rubble; for although the thickness of which walls may be built of rubble with safety will depend in a great degree upon the quality of the mortar, much depends also on the skill of the workman in bedding and bonding the stones. Under any circumstances, however, a wall so composed cannot, safely, be charged with heavy weights, nor be exposed to the vibrating action of floors, until the mortar shall have indurated to some extent; whereas a wall of brickwork is secure by the horizontal bedding of the bricks, and by the effect of the transverse bond which the alternation of header and stretcher almost necessarily produces. Stone, again, may be dressed to any shape, and so as to mould it to every variety of construction with the smallest possible quantity of mortar or cement. From blocks with rough hammer-dressed parallel beds, up to the most complete and perfectly wrought parallelopipeds adapted to any arrangement of bond that may be best adapted to the structure, and with combinations of rudely formed and perfectly formed pieces of stone, walls may be built of stone of greater strength than the best bricks can be made to yield, whilst stone walls are liable to be inferior in every respect to brick-built walls of ordinary quality.

Some combinations of the two kinds of materials have the effect, however, of making a better wall than could be produced by the main constituent in the form employed alone. A stone-rubble or pebble-built wall is greatly improved by one or two bonding courses of brickwork at short intervals; and a brick wall is improved and adapted for a higher purpose by thorough courses, at intervals, of good stone, wrought to bed and joint truly; whilst on the other hand, a wall substantially of stone-rubble or pebble, and faced with brickwork, is essentially an unsound wall; as in like manner a brick wall faced with wrought stone is liable to be weaker than the brickwork would have been without the stone.

With regard to the thicknesses of the walls of buildings, it is generally considered that these should be governed by the height of the structure; but they ought not to be determined by that condition alone. Chimney-breasts, or other buttress-like projections, built up with a wall, and extending to more than the thickness of the wall, make it in fact stronger in its transverse section, and justify less general thickness in the body of the wall, whilst window and other openings in a wall leave piers which ought to be of greater thickness than the mere height would require. But all returns, indeed, whether as chimney-breasts or as cross walls, built and bonded with a wall, tend to render unnecessary the full thickness which the height might require; whilst, as just intimated, the omission of portions of a wall for door and window openings should be compensated for by additional substance to the parts which remain.

Walls subjected to undue action, such as that arising from slight joists tailed into them, or that occasioned by inclined timbers, as under galleries in churches, chapels, and theatres, require to be of greater thickness than they otherwise would; whilst it is quite wonderful to what great heights brick walls may be built with safety, if they are well built, and exposed to no other action than direct vertical weight. When, indeed, such walls stand upon a sufficient foundation, direct vertical weight without motion is a means of security to the walls so long as the weight is reasonably within the power of resistance of the materials to crushing pressure. The object to be looked at, therefore—the walls being honestly built—is, as hereinbefore remarked, to make the weight to be imposed upon any wall act upon its solids vertically and steadily.

Floors upon girders, or framed to strong trimmers—the girders or the trimmer-joists running into and bearing upon the piers or solids of the walls—are far preferable to what are termed single floors, of which each joist runs into the wall. Girders, as the basis of floors, render plates in the walls wholly unnecessary, by depositing the weight in the right places, without requiring plates to carry it on from the weaker to the stronger places; and being of necessity stout and rigid, they form a fair tie and strut to the walls into which their bearing ends are tailed.

Whether girders or trimmer joists be employed for placing the weight of floors upon the walls of a building in the safest manner, the bearing timbers ought to be placed upon pieces of stone as templates built into the walls, and be made to take a cog-hold of the templates, so as to enable them to tie and stay the walls by means of the cogs.[2]

It is by means of the girder bearing upon the solids of the walls, though with bad carpentry, that the French are able to carry up their soft stone rubble walls to heights that would certainly be unsafe if the walls were seamed with wooden plates, and shaken by floors of single joists;[3] and it is by means of the solidity given to the floors by the girders, and the solid bearings which the girders obtain, that the floors are able to carry the dead weight of matter which renders them practically fireproof, as hereinafter described, in addition to the moving weights to which the floors of buildings are necessarily exposed in use.

We can and do frame floors most effectively by carpentry alone; whereas the French do the work so badly, that no important bearing is, or indeed may be, trusted by them to the framed joint—dog-nailed stirrup straps of iron being always brought in aid. But the common practice with us is to use single or unframed floors, which carry the weight and the vibration to which floors are exposed into the walls, over voids as well as over solids; while the French frame their floors to or upon girders, by means of which the floors are brought to bear upon the solids of the walls. The walls are thus not only less exposed to vibratory action, but are both tied together and strutted apart with better effect by the stout girders stiffened by joists, than by joists which themselves require some foreign aid to stiffen them. Moreover, single floors of joists, unless trimmed at frequent

---

never, to his knowledge, been remarked by any person who had published his remarks, and what was quite unknown to every one to whom he has stated, since his return, what he had observed. At Rotterdam and at the Hague, at Antwerp, at Brussels, and at Liège, at Cologne, at Mayence, and at Frankfort, and again throughout the north-eastern parts of France, brick walls are built according to the arrangement distinguished in England as English bond; and Flemish bond is unknown, at least no single example of it fell under the writer's observation in any of the towns and countries indicated, although his attention was called to the subject by the quay walls at Rotterdam before he set foot on shore.

[2] A cog-hold is best obtained through the agency of a chair of cast iron, which should, however, be itself cogged or joggled to a stone template laid in the wall under it, and be capped or covered by another broad flat stone, as an inverted template, with a joggle from the chair running up into it.

[3] The author, being at Paris in 1846, measured the thickness in the ground-floor story of a newly-built coursed-rubble party-wall, in the Rue de la Banque (the Gresham Street of Paris), and found it to be exactly 18 English inches in that part, whilst the total height of the wall was not less than 85 feet. The wall ran up of that same thickness through six stories, a height of not less than 65 feet, and was terminated by a gable of from 12 to 15 feet high, of the same kind of structure; and there was besides a vaulted basement story, throughout which the wall might have been 20 inches thick, as other similar walls then in progress to neighbouring buildings proved to be.

intervals, when, indeed, they may be termed half-framed, are supposed to require plates of timber laid along the inside faces of outer walls and upon internal walls. This defect is avoided by our neighbours, who exclude all timber, except the bearing ends of girders, from their walls, and use framed floors.

When the walls of a building have reached their full height, the wall-plate comes into use legitimately—to cope the walls, in fact, and to form a curb as a base upon which to place the roof, which should deposit its weight, nevertheless, by means of its tie-beams upon the plates over the solids of the wall below, and which should, moreover, oversail, so as to cover and effectually shelter from the weather, the inclosing walls also.

In setting forth the structural advantages derivable from the use of girders as the bases of floors, it may be necessary to repeat the warning already intimated against the use of girders of a material of uncertain strength, and of treacherous character when exposed to transverse strain.

Cast iron is of uncertain strength, mainly because of the imperfections which the most skilful founders, with the best materials and every appliance at command, cannot always avoid, and which are most liable to occur in the production of complex forms in long lengths; whilst careless founding and rapid cooling are contingencies connected with the production of cast-iron girders—which are necessarily long and complex castings. Cast iron is treacherous, inasmuch as it is brittle and liable to be startled into fracture by impact trifling when compared with what it may have borne safely as a dead-weight.

Proving long metal castings by straining them upon their transverse section does but aggravate imperfections, and leave the casting weaker; whilst no dead-weight proof is proof as against blows or other action inducing vibration. It is only under circumstances which do not admit of concussive action upon the beam, or which prevent it from vibrating under any shock that may reach it, that cast iron can be safely used in beams of long lengths to carry heavy weights, without some appliance to mitigate, at least, the imperfections which this substance exhibits.

The application of wrought-iron tension bars as soles to beams and girders of cast iron, would prevent the most serious consequences from attending the failure of the casting, if the beam were also prevented by binders, or by other sufficient means from turning round when the blow produces an oblique fracture.

The foundation of a building of ordinary weight is, for the most part, sufficiently provided for by applying what are technically termed footings to the walls. The reason for a footing is, that the wall obtains thereby a bearing upon a breadth of ground so much greater than its own width or thickness above the footing, as to compensate for the difference between the power of resisting pressure of the wall and of the ground or ultimate foundation upon which the wall is to rest. It will be clear from this, that if a building is to be erected upon rock as hard as the main constituent of the walls, no expanded footings will be necessary; if upon chalk—upon strong or upon weak gravel—upon sand —or upon clay—the footing must be expanded with reference to the power of resistance of the stratum to be used as a foundation; whilst in or upon made-ground, or other loose and badly combined or imperfectly resisting soil, a solid platform bearing evenly over the ground, and wide enough not to sink into it, becomes necessary under the constructed footing. For this purpose the easiest, the most familiar, and, for most purposes, the most effectual and durable, is a layer of concrete, which may be formed so as to cover a surface large enough to obtain from the most yielding soil the amount of resistance to pressure required to support the weight of the intended building. It will be evident that upon a concrete foundation a footing or expanded base may or may not be required to a wall, according to the hardness of the concrete and the kind of wall to be built; but it is perhaps better to give the footing to the wall than to wait for the sufficient induration of the concrete to enable the wall to do without a footing; and better still, to lay the concrete of such height only with reference to the spread or extent of base beyond the toes of the footing, that the gravel of which the concrete is made would stand at in an uncombined condition.[1]

Inasmuch, however, as some soils are liable to change in form, expanding and contracting under meteoric influences, as clays which swell when wetted and shrink when dried, concrete foundations are commonly interposed upon such soils to protect the building from derangement from this cause; or rather, to that effect walls are brought up from a level sufficiently below the ordinary surface of the ground of the cheaper material, concrete, instead of the more expensive brick or stone structure.

When concrete is used to obviate the yieldingness of the soil to pressure, expanse or extent of base is required to answer the end; and to this end the concrete, being widely spread, should be deep or thick as a layer, only with reference to its own power of transmitting to the ground the weight of the wall to be built upon it, without breaking across or being crushed. But when concrete is used as a substitute for a wall, in carrying a wall down to a low level, it is in fact a wall, wide only in proportion to its comparative weakness in the absence of manipulated bond in its construction, and encased by the soil within which it is placed.

Concrete, indeed, is at all times more safely to be regarded as a substance to be placed as a layer, than as a substance to be set up as a wall; for although excellent erections as walls may be made of concrete—as erections in the same form may be made of tempered clay—neither concrete nor tempered clay is to be regarded as a proper substance with which to form the lofty walls of buildings in towns.

SECURITY OF BUILDINGS AGAINST FIRE.

It is seldom that houses take fire from common accidents such as occur to the lighter moveable furniture and to drapery; but, for the most part, from the exposure of timber in or about the structure to the continued action of fire, or of heat capable sooner or later of inducing the combustion of timber; and as the source is most commonly in some stove, furnace, flue, pipe, or tube, for generating or for conveying heat, or for removing the products of combustion, much of the real danger to buildings from fire would be prevented by avoiding that degree of proximity between timber and all such things as can lead to its combustion.

With a view to lessen the danger to which buildings with timber in their structure are exposed from fire, it will be well to consider how far the timber, and wooden fittings commonly used, may be necessary either to the stability of the buildings, or to comfort and convenience.

So long as danger of fire is brought to buildings through pipes and tubes, the necessity must be admitted of guarding the combustible materials used in buildings from any chance of becoming ignited. When heat is produced and passed

---

[1] This is indeed the only safe practice in cases in which the full induration of concrete cannot be obtained before the wall or other structure to be raised upon it is begun to be built. Gravel poured dry upon the ground will resolve itself into a cone, having an angle of inclination to the horizon more or less acute according to the sharpness, or otherwise, of the sand and stones of which it is composed.

through pipes in any manufactory, whether it be to act as power, or for drying or for warming, the fires used may be guarded, and the machinery which regulates the intensity of the heat to be transmitted may be under constant care ; but even in such cases there can be no certain assurance that the heat shall not at some time arrive at the point of danger as it regards the ignition of combustible substances. But when heat is diffused throughout dwelling-houses by means of apparatus which is committed to persons unskilled in its use, and unconscious or careless of the danger which may arise from neglect, it seems impossible to lay down inflexible rules for distances from timber which shall render it safe from heated pipes. Twelve or fifteen inches may not be a greater distance than safety requires under some circumstances, whilst there are many cases in which the actual contact of such pipes with timber is hardly inconsistent with safety. When the air about the heated bodies is not confined, as it is between the joists and the floor and ceilings of an ordinary floor, a distance between timber and the heated surface equal to the longest diameter of any tube or pipe, will be found a safe distance if the temperature of the pipe does not exceed that of boiling water.

It is to be understood, at the same time, that a piece of wood will bear a powerful dead heat upon its sides for an indefinite period without igniting, unless a transverse section of the fibre, as at or around a live knot, or where a branch had been lopped, present itself to the action. It is by the end that a piece of wood exposed to powerful heat most readily ignites. The gases evolved in the substance of the timber by the action of heat applied to its surface expanding as they are evolved, are thrown out by the pores among the fibres at their ends, if the ends are near enough to the action to allow of this effect, with less power than may be enough to obtain vent for the inflammable gases laterally

The legislature in this country, when it has legislated upon such matters, has generally confined itself to making provision that the inclosing walls of buildings should be formed of incombustible materials. In providing of what least thicknesses such walls might be, these were generally determined with reference to the height of the building, and to the area to be inclosed, as an indication of the probable lengths of the walls; and this both for the purpose of promoting safety of structure, and of checking the spread of fire from building to building. As, however, in most cases greater thickness is required in the side wall of an ordinary dwelling-house in a town to render its structure secure, than is necessary to enable it to check the spreading of fire, such walls are frequently made of greater thickness than would be necessary to fulfil the objects which the legislature has had in view, if the walls were not supposed to extend the whole length of the two longer sides of a parallelogram without intermediate cross or return walls. A solid well-built brick wall, one brick or nine inches thick, between two ordinary dwelling-houses of five or six squares in area each, will prevent the communication of fire through it from one to the other.

But, in towns, ordinary dwelling-houses which occupy each an area of five or six squares are generally disposed in plan as parallelograms, having their opposite sides 18 or 20 feet, and 28 or 30 feet respectively in length, and are seldom carried up to less height than 35 or 40 feet ; and walls of such lengths and heights could hardly be deemed safe if not more than one brick thick. Consequently, a greater thickness has been prescribed, as the least thickness of the walls of buildings of the sizes indicated. In the older Metropolitan Building Acts much greater thicknesses were prescribed for the walls likely to be the longer walls ; whilst the only necessity for more than one brick arises from structural requisites, and not from any insufficiency of

a wall of solid brickwork, one brick thick, as a means of preventing the spread of fire. But the requisites of the structure would be as well fulfilled by one-brick walls upon the long sides as by 1½-brick walls, if the ordinary internal cross partition for dividing a house into front and back rooms were built of brickwork, abutting upon, and at right angles to, the longer walls, and carried up coursed and bonded with them. That is to say, party-walls of one brick or nine inches in thickness, connected at their ends by 1½-brick or 13-inch front and back walls, and at or about the middle of their length by other 9-inch cross walls, would be at the least as strong as 1½-brick party-walls, though connected in the same manner at the two ends, but without the abutting and connecting cross-wall of brickwork. Instead, however, of such internal cross walls, hollow partitions of timber are commonly used in all stories above the basement story ; and it is by these partitions, and by the light and highly inflammable wooden stairs, that fire extends itself rapidly throughout ordinary dwelling-houses; whilst the substitution of a brick wall for the cross timber partition would in most cases justify the abatement of a half brick of the thickness otherwise necessary to party-walls, and give an indestructible internal support to the floors, whereby also one of the means by which fire travels rapidly through a house would be removed. It is true that there must be openings as doorways, and fittings in them for doors, in such internal partition wall ; but the wall could not carry fire up from floor to floor through its own heart, as the hollow wood-lathed quartering partition carries it. Doors and shutters, and door and window linings, in and against brick or stone walls, may take fire and burn in any story of an ordinarily built dwelling-house, without carrying it beyond the story in which the fire occurs; for a plastered ceiling of the most common description will resist the action of flame upon its surface for a long time, and plastering of really good quality, though upon wood laths, will keep fire off from the joists by which it is held up, almost without danger, so long as the fire acts upon the face only of the plastering. If, however, fire reach the joists through the agency of hollow quartering partitions, the enemy has turned the flank of the plastering, and the floors and skirtings above and behind it taking fire, the building almost inevitably falls a prey to the flames. Any step, indeed, from the hollow quartering partition towards a solid wall, is a step towards security. A brick wall is, perhaps, the best internal partition for all the purposes of strength and security from fire ; and in small houses, which will not afford the expense of 9-inch walls, half-brick walls with 9-inch jambs at the doors, and short-9 inch piers on alternate sides of the partition, at intervals of three or four feet in length, will give sufficient strength; but even quartering partitions, if based upon brick walls, may be rendered nearly proof against fire by brick-nogging them, especially if care be taken to fill in with brickwork between the joists over the head of one partition and under the sill of another, as well as between the timbers of the partitions. Filling in between the joists, and up as high as the skirtings go, will do something, indeed, towards diminishing the dangerous tendency of even lathed and plastered timber partitions ; whilst the adoption of the plan now commonly practised in Paris, in forming not only internal partitions, but the rearward external inclosures of buildings, would secure to the structure the structural efficiency of timber on end in carrying weight, and give the solid and incombustible character of a brick or stone wall to a partition or inclosure which is structurally of timber.

The plan referred to is, to frame and brace with timber quarterings much in the manner practised in England, except that the timber used in Paris is commonly oak, and is generally seasoned previously. The framed structure being complete, strong oak batten-laths, from two to three inches wide, are nailed up to the quarterings horizontally, at four, six, or even eight inches apart, according to the character of the work, throughout the whole height of the inclosure or partition; and the spaces between the quarterings, and behind the laths, are loosely built up with rough stone rubble, which the laths prevent from falling out until the next process has been effected. This is, to apply a strong mortar, which in Paris is mainly composed of plaster of Paris, which is there of excellent quality, laid on from both sides at the same time, and pressed through from the opposite sides so that the mortar meets and incorporates, embedding the stone rubble by filling up every interstice, and with so much body on the surfaces as to cover up and embed also the timber and the laths—in such manner, indeed, as to render the concretion of stone and plaster, when thoroughly set, an independent body, and giving strength to, rather than receiving support from, the timber.

Our brick-nogged partition is, in point of structure, nothing but through the aid of the timber; the plastering is merely spread out upon the surfaces of brick and wood, and is fragile in the extreme, and always liable to crack and drop off; whilst, on the other hand, according to the French practice, the mortar, meeting through the interstices of the rubble, becomes one consistent mass throughout the whole thickness of the partition. Our lathed and plastered partition is composed of the hollow framework of the timber quarters, with two slight thicknesses of mortar, as plastering, hung upon slighter laths, over and between which the flaccid mortar forms a key for itself; but all necessarily depends upon the timber, and fails with it wherever decay or fire may destroy it.

Only second in importance to the internal partition as a source of danger or as a means of safety, are the stairs; and the stairs are second in importance only when the partitions are made to carry the floors of the several stories. In England, and in London particularly, even when the steps and intermediate landings are of stone, it is but too common to find the passage from the street door to the foot of the stairs, and the floors which connect flight with flight at the several landings, either wholly of wood or of slight stone paving laid upon wooden joists or bearers. Any stone paving upon wooden joists will certainly retard the action of fire upon the joists, especially if assisted by a well-plastered ceiling; but in this, again, if the floors be not formed of wholly incombustible materials, the French practice as to floors would be better than ours.

In Paris stone stairs are far less common in modern houses than they are in London in houses of corresponding character and date; but wooden staircases in Paris are rendered almost as safe as common stone staircases are made with us, by a process similar in character to that applied to partitions and inclosures. The result is an almost incombustible structure. Wooden staircases formed between brick or stone walls, or between partitions of the kind above described, as commonly made in modern buildings in Paris, filled with a solid mass of concreted rubble, may perhaps be set on fire, but they can hardly burn.

It has been remarked that a mere plastered ceiling will resist the action of fire for a long time, although the plastering be upon wooden laths, and the laths nailed to joists of timber; and as fire does not readily act downwards, flooring boards may take fire from above without any immediately serious consequence to the joists under them, so long as there is no access of air from below. But our indoor plastering upon laths is commonly of the most fragile kind, and the slightest weight falling upon the back of a ceiling will make a breach through it, whilst our floors are commonly of deal laid upon fir joists, and are exposed to the action of fire from below directly the lathed and plastered

ceiling has failed; if, indeed, the fire have not found its way to the joists under the flooring boards by the hollow lathed and plastered quartering partitions. In the timber inclosures and partitions, which economy induces the Paris builder to introduce as substitutes for walls, the timber is so embedded in and made part of a solid concrete, as to be protected from almost every casualty of which it is susceptible.

But the French render their floors also so nearly fireproof as to leave but little to desire in that respect, and in a manner attainable with single joists, as well, at the least, as with joists framed into girders. According to their practice, the ceiling *must* be formed before the upper surface or floor is laid, as the ceiling is formed from above instead of from below. The carpenters' work being complete, strong batten-laths are nailed up to the under sides of the joists, as laths are with us; but they are much thicker and wider than our laths, and are placed so far apart that not more, perhaps, than one-half of the space is occupied by the laths. The laths being affixed—and they must be soundly nailed, as they have a heavy weight to carry—a platform, made of rough boards, is strutted up from below parallel to the plane formed by the laths, and at about an inch below them. Mortar is then laid in from above over the platform, and between and over the laths, to a thickness of from two inches and a-half to three inches, and is forced in under the laths, and under the joists and girders. The mortar being gauged, as our plasterers term it, or rather, in great part composed of plaster of Paris, it soon sets sufficiently to allow the platform to be removed onwards to another compartment, until the whole ceiling is formed. The plaster ceiling thus produced is, in fact, a strong slab or table, in the body of which the batten-laths which hold it up are incorporated, and in the back of which the joists, from which the mass is suspended, are embedded. The finishing coat of plastering is then laid on. Such a ceiling will resist any fire that can act upon it from below, under ordinary circumstances; and it would be difficult for fire to take such a hold from above as to destroy the joists to which a ceiling so composed is attached, the laths and the under side of the joists being alike out of its reach; and consequently such a ceiling alone would diminish the danger from fire, although the floor above the joists were laid with deal boards. But a boarded floor in Paris is a luxury not to be found in the dwellings of the labouring classes, nor, indeed, are boarded floors to be found in any dwelling-houses but those of the more costly description. Whether the eventual surface is to be a boarded floor or not, however, the flooring joists are covered by a table of plaster above, as completely as they are covered by a plaster ceiling below. Rough battens, generally split and in short lengths, stout enough to bear the weight of a man without bending, are laid with ends abutting upon every joist, and as close together as they will lie without having been shot or planed on their edges. Upon this rough loose floor mortar of nearly similar consistence to that used for ceilings is spread to a thickness of about three inches; and as it is made to fill in the voids at the ends and sides of the floor-laths upon the joists, the laths become bedded upon the joists, whilst they are to some extent also incorporated with the plaster. The result is a firm floor, upon which, in ordinary buildings, paving-tiles are laid, bedded in a tenacious cement.

It must be clear that the timbers of a floor so encased could hardly be made to burn even if fire were let in between the floor and ceiling. But it has been already stated that the practice of making these almost fire-proof floors is connected with the use of walls which have no timber laid in them bedwise, and that the timber inclosures employed instead of walls, and the internal partitions, are rendered practically fire-proof, whilst the wooden staircase

s

which economy dictates to the Parisian builders—the free-stone which is used in building walls being altogether too soft for the purpose—is also rendered, in the manner already shown, almost unassailable by fire.[1]

It may be added in explanation of the statement that in Paris the practice of forming a table of plaster over the joists when tiles are to be used as the flooring surface, is employed also when a boarded floor is to supervene—that as the surfaces of the true joists lie under the mortar, a base is formed for the boards of what English carpenters would call stout fillets of wood, about $2\frac{1}{2}$ inches square, ranged as joists, and strutted apart to keep them in their places, over the mortar table, to which they are sometimes scribed down, and that to these fillets, or false joists, the flooring boards are secured by nails; so that in truth the boarded floor is not at all connected with the structure of the floor, but is formed upon its upper coat of plaster. The wooden floor thus becomes a mere fitting in an apartment, and not extending beyond the room the floor might burn without communicating fire to the stairs, even if the stairs were readily ignitible.[2]

The necessity which arises with us of dividing the upper stories of houses into more rooms, as bed-rooms, than are commonly required in the lower stories, will be made an objection to any process that would render the partitions heavier; but it is not in the upper stories that the lathed and plastered partition is most dangerous in respect of fire. Generally the stairs may be inclosed by solid partitions throughout almost the whole height of an ordinary dwelling-houses without occasioning any inconvenience as regards the greater weight of such a partition; and generally, too, the partition which divides the front from the back rooms of such houses may be carried up throughout the whole height of a house without removing the bearing, if the house be judiciously disposed. But even if a partition rest upon a beam or girder, a very slight addition to the scantling of the timber will make up for the additional weight which the filling in of the partition would involve, if the materials of the core be well chosen; and it is well known that a piece of timber placed over a void as a brest-summer, and carrying a wall, resists the action of fire for a long time, and the longer if it be of oak or other hard wood. It is not necessary, however, that the timber employed in partitions and inclosures should be of oak; though it is desirable that main bearing timbers, in situations which render them most liable to be exposed to the action of fire, in the event of casualty, should be of such-like timber rather than of fir: but the quarterings, or partition timbers, which the plaster concrete wholly encases, may be of fir as safely almost as of oak.

The core used in Paris consists for the most part of chips and spalls arising in the process of dressing the soft free-stone which is the main constituent of the walls of most buildings in that city. Almost any hard material, however, will furnish rubble fit for the purpose, which must be angular and irregular in form, so as to allow the mortar to pass freely through the rubble, and embed it all. Rubble of brick material, as broken burrs, or even of old bricks freshly broken, will answer very well; but if brickbats or shreds of plain tiles be used, care must be taken in packing not to bring flat beds together, or the mortar will not pass through and

make a perfect concrete. Rubble of almost any kind may be used; but the kinds of stone which are themselves concretions, and present rough surfaces upon the fracture, afford the best, while schistose, or scaling slaty stones, are the worst for the purpose. But there is no better substance for coring partitions upon the plan described than clay burnt into a kind of brick rubble,—an excellent ingredient, indeed, in concrete for any purpose.

The same process applied to external inclosures will justify the use of timber in their structure in situations and under circumstances in which it may be properly prohibited when the timber is merely lathed and plastered, or even brick-nogged, for brick-nogging adds nothing, as already remarked, to the strength of a partition or an inclosure, but rather takes from it, being itself a source of infirmity. But chimneys and their flues ought not under any circumstances to be formed in an inclosure in which timber is employed as a part of the structure. Chimneys—with their congeners, stoves and furnaces—should be confined to walls of brick or stone; and as these almost always occur most conveniently in party-walls when buildings stand together, or in walls which, though not technically party-walls, are so near to other buildings, as to require to be similarly dealt with, inclosures of the kind indicated need not be desired, because it would not be prudent to form flues in them.

Under some circumstances, again,—that is to say, when any street of a town is so wide and the buildings to be built fronting to it are to be of such small elevation, as to make the communication of fire from one side to the opposite side so nearly impossible as, for all the purposes of security, to be so, if the buildings adjoining laterally are effectually separated from one another by sufficient walls, party or otherwise, and these project before the outside faces of the front and back inclosures so as effectually to prevent fire from passing round them,—the temperature of dwelling-houses may be much more easily maintained and regulated if the outside surface be boarded. Weather-boarding is a safe and economical, as well as a neat, wholesome, and equable outside casing for the fronts of a dwelling-house, if the boarding be backed up solidly, and the timber quarterings necessary to secure it be properly filled in between and behind with brick or stone work, or with rubble and concrete in the manner already described. Brickwork builds up badly with the raking braces of timber-framed inclosures, and the concrete described would not be so perfect with weather-boarding on one side as if the mortar were thrown in from both sides; but raking braces are less essential to inclosures which are filled in and backed with a heavy body of brickwork or concrete, than when mere lathing or even brick-nogging is to be employed on the inside. A nine-inch brick wall may, indeed, be very well built up with framed quarterings without raking braces, if the work be built between and around the quarterings, carrying, that is to say, the inner half-brick before the inside faces of the quartering, and so as to show on the inside a plain brick wall.

The foregoing remarks have been written with reference to the articles ARCHITECTURE, BUILDING, CARPENTRY, MASONRY, &c., to which, accordingly, the attention of the reader is directed.   (W. H—G.)

---

[1] It may be remarked here, with reference to the employment of any substance such as cinder, being of the nature of pozzolano, or volcanic scoria, in mortar, to form a floor in the manner above described (about three inches thick), that as all such mortars expand in setting, the walls of buildings may be forced out by the expansion of the plaster floors, if the whole surface of the floor in any story be at once covered with the mortar. A margin of four or five inches on every side should be left void until the expansion has taken place, when the floor may be completed with an assurance of close joints, and without injury to reasonably stable walls.

[2] The most recent practice in Paris, in respect of floors, is to form the structure of slight wrought-iron bars rolled to the form known with us as T and L iron, and to fill in with the same strong plaster between, below, and above the iron, and so to form a slab of plaster from 6 to 8 inches thick, according to the bearing and the depth of the iron bars—the bars being enveloped in the plaster as the bottom laths are when the structure is of timber.

BUILDING.

THE art of building comprises the practice of civil architecture, or the mechanical operations necessary to carry the designs of the architect into effect. It is not unfrequently called practical architecture; but the adoption of this term would have tended only to confuse, by rendering it difficult to make the distinction generally understood between architecture as a fine or liberal art, and architecture as a mechanical art. The execution of works of architecture necessarily includes building; but building is frequently employed when the result is not architectural: a man may be a competent builder without being an architect; but no one can be an accomplished architect unless he be competent to specify and direct all the operations of building. A scientific knowledge of the principles of masonry, carpentry, joinery, &c., and of the qualities, strength, and resistance of materials, though of the utmost importance to an architect, must be attended by a minute acquaintance with a great variety of less ambitious details. Such are those which relate to the arrangement of a plan for the greatest possible degree of convenience on the smallest space, and at the least expense; its transference to the ground; the preparation and formation of foundations; the arrangement and construction of drains, sewers, and ventshafts; the varieties of walling with stone, and of laying bricks in brick-work; the merit of the various modes of bonding and tying walls, both lengthwise and across; the arrangement of gutters on roofs, to get sufficient fall, and to conduct the water to the least inconvenient places for fixing trunks to lead it down; the arrangement and formation of flues; the protection of walls from damp, of timber from moisture and stagnant air, and of metals generally from exciting causes; the cost of materials and labour, and the quantity of each required to produce certain results. Together with these, an architect ought to be practically acquainted with all the modes of operation in all the trades or arts employed in building. Everything must be clearly understood, or it will be impossible properly to specify beforehand, in detail, everything and every operation to be done and performed; and minutely to estimate, beforehand also, the absolute cost involved in the execution of a proposed structure. The power to do the latter necessarily involves that of measuring work, and ascertaining quantities after it is done. These things may certainly be referred to the surveyor or measurer, but they are not the less incumbent on the architect, who cannot be said to be thoroughly master of building, or the practice of his profession, unless he be skilled in these operations.

The architect having furnished the specification and working drawings of his design, the first step in the process is to prepare the foundation. (See article STONE-MASONRY, sect. 60.) Much in this particular, it is evident, must depend on localities. It is not of so much importance that the ground be hard, or even rocky, as that it be compact, and of similar consistence throughout; that it be so constituted as to resist entirely and throughout, or yield equally to the superincumbent weight.

But in the ordinary processes of building, the artificial preparation of foundations hardly need be considered. Common prudence would refer it to professional management, when such is found necessary; and a work of this kind cannot contain sufficient information and instruction to qualify a man to act professionally on any subject, and more particularly on those subjects which demand initiatory practice and experience. We therefore proceed to the ordinary routine of practice.

The artificers whose trades come within the immediate range of the builder's business are the following: Digger or excavator, bricklayer, mason, slater, sawyer, carpenter, joiner, plasterer, modeller, carver and gilder, plumber, smith, glazier, painter and decorator.

Digger or Excavator.—The digger works with a pick-axe and a spade or shovel. With the pick-axe he breaks down the soil if it be hard or very stiff, and throws it out with the shovel; but compacted sand and alluvial soil is spitted and thrown out with the spade alone, without previous breaking down. When rock occurs in a foundation, the assistance of the quarryman is requisite to cut through or blast it, as the occasion may require. The digger should be required to produce a perfect level in every direction, and especially in trenches for walls; nor may this be done by placing again loose matter, but the level must be produced on the solid or undisturbed bed.

Digger's work is valued by the cubic yard, and is generally made to include, besides excavating, the removal of the soil and rubbish. The price per yard is therefore necessarily contingent on the stiffness of the soil, the depth

Building. to which the excavations may reach below the surface, and the distance the stuff is to be removed; so that it is impossible to determine what the cost may be, without reference to each and all of these particulars, most of which must be different in every different place; and all are again affected by the local cost of labour or wages. A good excavator will dig and throw out, of common soil, into a basket or wheelbarrow, eight or ten yards per diem; but of stiff clay or firm gravel, not more than six yards.

Bricklayers.—The manufacture of brick being made the subject of a separate article, we need only refer to that for information on the subject; and in the same manner the components and merits of mortars and cements will be found in sections 20 *et seq.* of the article under the head STONE-MASONRY. A few observations on the composition of mortar for bricklaying will nevertheless be necessary here.

Particular attention must be paid to cleansing the sand to be used for mortar of every particle of clay or mud that may adhere to or be mixed up with it. Sea sand is objectionable for two reasons; it cannot be perfectly freed from a saline taint, and the particles are moreover generally rounded by attrition, caused by the action of the sea, which makes it less efficient for mortar than if they retained their natural angular forms. Lime should not be slaked until the moment it is to be mixed up with the sand in mortar, but the sooner that is done after it is burnt the better. The proportion of lime to sand is generally taken at as much as one-fourth of the whole mass; but if both the materials be of good quality, that is, if the lime slake freely, and become a fine pungent impalpable powder, perfectly clear from argillaceous or any other foreign matter, and the sand clean and sharp, and of variously sized particles, one-fifth of lime to sand is quite enough: more is injurious. The ingredients should be well mixed together, and with water, and as little water used as will suffice to make the compound consistent and paste-like. Rain, or any other soft water, should be used for the purpose of making mortar, and not spring or hard water, though any other may be preferred to what is brackish even in the slightest degree. A quick-setting cement, such as that which is most commonly used in building in this country, and known as Parker's or Roman cement, can only be mixed or gauged as it is required for use. A bricklayer will keep a labourer fully employed in gauging cement for him alone. It is mixed with sand in the same manner that lime is in common mortar, in the proportion of about two or three of sand to one of cement, according to the quality of the latter; and the labourer, as he gauges on one board, supplies the mixture to the bricklayer fit for use on another board, a spadeful at a time; it must then be applied within half a minute, or it sets and is spoiled.

The average size of bricks in this country is a fraction under nine inches long, four and a half wide, and two and a half inches thick; and in consequence of this uniformity of size, a wall of this material is described as of so many bricks in thickness, or of the number of inches which result from multiplying nine inches by any number of bricks; a nine-inch or one-brick wall; a fourteen-inch wall, or one brick and a half (13½ inches would be more correct, in fact, for although a joint of mortar must occur in this thickness, yet the fraction under the given size of the brick is enough to form it); eighteen-inch, or two bricks, and so on.

The great art in bricklaying is to preserve and maintain a bond, to have every course perfectly horizontal, both longitudinally and transversely, and perfectly plumb; which last, however, may not mean upright, though that is the general acceptation of the term, for the plumb-rule may be made to suit any required inclination, as inward against a bank, for instance, or in a tapering tower; and also to make the vertical joints recur perpendicularly over each other: this is vulgarly and technically called keeping the *perpends*.

By bond in brick-work is intended that arrangement which shall make the bricks of every course cover the joints of those in the course below it, and so tend to make the whole mass or combination of bricks act as much together, or dependently one upon another, as possible. The object of this will be understood by reference to the diagram, fig. 1. Here it is evident, from the arrangement of the bricks, any weight placed on *a* would (supposing, as we are obliged to suppose, that every brick feels equally, throughout its whole length, a stress laid on any part of it) be carried down and borne alike in every course from *b* to *c*; in the same manner the brick *d* is upborne by every brick in the line *e f*, and so throughout the structure. But this forms a longitudinal bond only, which cannot extend its influence beyond the width of the brick; and a wall of one brick and a half, or two bricks thick, built in this manner, would, in effect, consist of three or four half-brick-thick walls, acting independently of each other, as shown in the plan at *i*, in the diagram, under fig. 1. If the bricks were turned so as to show their short sides or ends in front, instead of their long ones, certainly a compact wall of a whole brick in thickness would be produced; but the longitudinal bond would be shortened one-half, as at *g c h*, and a wall of any greater thickness, in the same manner, must be composed of so many independent one-brick walls, as at *k* in the plan before referred to. To obviate this, to produce a transverse, and yet preserve a true longitudinal bond, the bricks are laid in alternate courses of headers and stretchers, or of ends and sides, as shown in fig. 2, thus combining the advantages of the two modes of arrangement, *a b c* and *g c h* fig. 1, in *a b c* fig. 2. Each brick in fig. 2 showing its long side in front, or being a stretcher, will have another lying parallel to it, and on the same level, on the other side, to receive the other ends of the bricks showing as headers in front, which in their turn bind, by covering the joint between them, as shown in the end of such a wall at *d*. Thus a well-bonded nine-inch or one-brick wall is produced. The end elevations of the same wall at *e* and *f* show how the process of bonding is pursued in walls of one and a half and two bricks thick, the stretcher being abutted in the same course by a header; thus, in a fourteen-inch wall, inverting the appearance on the opposite sides, as seen at *e*, and producing the same appearance in an eighteen-inch wall, as at *f*. In the diagram under fig. 2, at *g*, is the plan of a fourteen-inch wall, showing the headers on one side, and the stretchers on the other, and at *h* is the plan of the course immediately above it, in which the headers and stretchers are inverted; at *k* and *i* are shown in the same manner, the plans of two courses of an eighteen-inch wall. This is called English bond. Thicker walls are constructed in the same manner by the extension of the same principal.

But a brick being exactly half its length in breadth, it is impossible, commencing from a vertical end or quoin, to make a bond with whole bricks, as the joints must of necessity fall one over the other. This difficulty is obviated by cutting a brick longitudinally into two equal parts, which are called half headers. One of these is placed next to a whole header, inward from the angle, and forms with it a three quarter-length between the stretchers above and below, thus making a regular overlap, which may then be preserved throughout: half headers so applied are technically termed closers. (See the joints in the heading courses next the upright angle of the wall fig. 2, and the first joints inwards from the square ends by the headers in the plans at *g* and *h*.) A three-quarter stretcher is obviously as available for this purpose as a half header, but the latter is preferred, because, by the use of it, uniformity of appearance is preserved, and whole bricks are retained on the returns. In walls of almost all thicknesses above nine inches, to preserve the transverse, and yet not destroy the longitudinal bond, it is frequently necessary to use half bricks; but it becomes a question whether more is not lost in the general firmness and consistence of the wall by that necessity, than is gained in the uniformity of the bond. It may certainly be taken as a general rule, that a brick should never be cut if it can be worked in whole, for a new joint is thereby created in a construction, the difficulty of which consists in obviating the debility arising from the constant recurrence of joints. Great attention should be paid to this, especially in the quoins of buildings, in which half bricks most readily occur; and there it is not only of consequence to have the greatest degree of consistence, but the quarter bricks used as closers are already admitted, and the weakness consequent on their admission would only be increased by the use of other bats, or fragments of bricks.

Another mode of bonding brick-work, which may be supposed to have arisen from the appearance of the ends of a wall according to the former mode of arrangement (see *e* and *f*, fig. 2), instead of placing the bricks in alternate courses of headers and stretchers, places headers and stretchers alternately in the same course, fig. 3. The plans below this at *c* and *d* are of two courses of a fourteen-inch wall, with their bond, showing in what manner the joints are broken in the wall horizontally as well as vertically on its face. This is called Flemish bond. Closers are necessary to both varieties of bond, in the same manner, and for the same purpose; half bricks also will occur in both, but what has been said with reference to the use of them in the former applies even with more force to the latter, for they are more frequent in Flemish than in English, and its transverse tie is thereby rendered less strong. Their occurrence is a disadvantage which every care should be taken to obviate. The arrangement of the joints, however, in Flemish bond, presenting a neater appearance than that of English bond, it is generally preferred for external walls when their outer faces are not to be covered with stucco, or plaster composition of any kind; but English bond should have the preference when the greatest degree of strength and compactness is considered of the highest importance, because it affords, as we have already noticed, a better transverse tie than the other. It is a curious fact, that what is in England called Flemish bond in brick-work is unknown in Flanders, and, so far as the observation of the present writer has extended, is practised in the British Isles alone. In Flanders, Holland, and Rhenish Germany, which are all brick-laying countries, no kind of bond is found but what is known in England as English bond.

It has been attempted to improve the bond in thick walls by laying raking courses in the core between external stretching courses, and reversing the rake when the course recurs. This obviates whatever necessity may exist of using half bricks in the heading courses, but it leaves triangular interstices to be filled up with bats, as the diagram fig. 4 shows. This represents the plan of a thirty-six inch or three-brick wall with raking courses at *a*, between external ranges of stretchers, and lying on a complete course of headers, and at *b* a wall of the same thickness herring-boned; courses of headers would bed and cover this also, and, in the second course above, the raking or herring-boning would be repeated, but the direction of the bricks inverted. It will be seen that the latter demands, in addition to the triangular filling in bats at the outer ends of the diagonally placed bricks, half bricks to fill up the central line of interstices, rendering herring-boning more objectionable in that particular, though it has some advantages over simply raking, or thorough diagonal courses, in some other points. Neither mode should, however, be recurred to for walls of a less thickness than three bricks, and that indeed is almost too thin to admit of any great advantage from it.

Skilful and ingenious workmen are well aware of the necessity of attending to the bond, and are ready both to suggest and to receive and practise an improvement; but generally the workmen themselves are both ignorant of its importance and careless in preserving it, even according to the common modes. Their work should therefore be strictly supervised as they proceed with it; for many of the failures which are constantly occurring may be referred to their ignorance or carelessness in this particular.

Not second in importance to bonding in brick-work is, that it be perfectly plumb, or vertical, and that every course be perfectly horizontal, or level, both longitudinally and transversely. The lowest course in the footings of a brick wall should be laid with the strictest attention to this latter particular; for the bricks being of equal thickness throughout, the slightest irregularity or incorrectness in that will be carried into the superimposing courses, and can only be rectified by using a greater or less quantity of mortar in one part or another, so that the wall will of course yield unequally to the superincumbent weight, as the work goes on, and perpetuate the infirmity. To save the trouble of keeping the plumb-rule and level constantly in his hands, and yet to insure correct work, the bricklayer, on clearing the footings of a wall, builds up six or eight courses at the external angles (see fig. 5), which he carefully plumbs and levels across, and from one to the other. These form a gauge for the intervening parts of the courses, a line being tightly strained from one end to the other, resting on the upper and outer angles of the gauge bricks of the next course to be laid, as at *a* and *b*, fig. 5, and with this he makes his work range. If, however, the length be great, the line will of course sag; and it must therefore be care

fully set and propped at sufficient intervals. Having carried up three or four courses to a level, with the guidance of the line, the work should be proved with the level and plumb-rule, and particularly with the latter at the quoins and reveals, as well as on the face: a smart tap with the end of the handle of the trowel will generally suffice to make a brick yield what little it may be out, while the work is so green, and not injure it. Good workmen, however, take a pride in showing how correctly their work will plumb without tapping. To work which is circular on the plan, both the level and the plumb-rule must be used, together with a gauge-mould or a ranging trammel, to every course, as it must be evident that the line cannot be applied to such in the manner just described. To every wall of more than one brick thick, two men should be employed at the same time, one outside and the other in: one man cannot do justice from one side, even to a fourteen-inch wall. Inferior workmen and apprentices are generally employed as inside men, though the work there is of quite as great importance as exteriorly, except for neatness, and for that only if the brickwork is to show on the outside.

In the operation of bricklaying, the workman holds the trowel in his right hand, and with the left he takes up the bricks from the scaffold, and lays them in their places. Spooning or shovelling up mortar from the board with the trowel, he throws it on the course last laid, and with the point strews it over the surface to form a bed for that which he is about to set; whatever bulges or projects over the outer edge of the work below is struck off, and being caught on the flat face of the trowel, is put against the side or end of the last brick laid in the new course. Then taking up a brick, he presses it down in its place until its upper and outer angle comes exactly to the line; and if this be not readily effected by the hand, a slight drawing blow with the obtuse point of the edge of the trowel does it, or a tap with the end of the handle both draws it and settles it down farther than the hand can press it. The small quantity of mortar that is pressed out in front, by this operation, being struck off, the joints are neatly drawn by compressing the mortar with the point of the trowel, and thus producing a fine smooth surface,—that is, if the work is to be seen; for if it is to be plastered, the rough face is left that the plastering may the more readily attach itself, and the joint is not drawn at all, but the workman proceeds in the same manner with the next brick in advance along the course, or to fill in behind the one he has laid in front to meet the work of his mate on the other side of the same wall. This is the common mode of *laying* bricks. They should not however be merely *laid*; every brick should be rubbed and pressed down in such a manner as to force the slimy matter of the mortar into the pores of the bricks, and so produce absolute adhesion. Moreover, to make brick-work as good and perfect as it may be, every brick should be made damp, or even wet, before it is laid, otherwise it immediately absorbs the moisture of the mortar, and, its surface being covered with dry dust, and its pores full of air, no adhesion can take place; but if the brick be damp, and the mortar moist, the dust is enveloped in the cementitious matter of the mortar, which also enters the pores of the brick, so that when the water evaporates, their attachment is complete, the retention and access of air being thus altogether precluded. To wet the bricks before they were carried on to the scaffold would, by making them heavier, add materially to the labour of carrying: in dry weather they would, moreover, become dry again before they could be used; and for the bricklayer to wet every brick himself would be an unnecessary waste of his time: boys might therefore be advantageously employed to dip the bricks on the scaffold, and supply them in a damp state to the bricklayer's hand. A watering pot with a fine rose to it should also be used to moisten the upper surface of the last laid course of bricks, preparatory to strewing the mortar over it. In bricklaying with quick-setting cements these things are of even more importance; indeed, unless the bricks are quite wet to be set with cement, it will not attach itself to them at all.

As mortar is a more yielding material, used in brick-work merely for the purpose of making the detached portions of the staple adhere, by filling up their interstices and producing exhaustion, and the object being to produce as unyielding and consistent a mass as possible, as much of it should be used as is sufficient to produce the desired result, and no more. No two bricks should be allowed to touch, because of their inaptitude to adhere to each other; and no space between them should be left unoccupied by mortar which may produce adhesion. When the bricks are a fraction under two-and-a-half inches thick, no four courses of bricks and mortar, or brick-work, should exceed eleven inches in height; and if they are fully that thickness, four courses should not reach eleven and a half inches. The result of thick beds of mortar between the bricks is, that the mortar is pressed out after the joint is drawn, on the outside, in front; and being made convex instead of slightly concave, the joints catch every drop of rain that may trickle down the face of the wall, and are thus saturated; the moisture freezes, and in thawing bursts the mortar, which crumbles away, and creates the necessity which is constantly recurring, of pointing the joints to preserve the wall.

The diagram shows the section of a nine-inch wall, with the joints on the side *a* as drawn, and on the side *b* as bulged, in consequence of the quantity of mortar in them yielding to the weight above. This, too, is in addition to the inconvenient settling, which is the consequence of using too much mortar in the beds.

In practice, bricklayers lay the mortar on the course last finished, and spread it over the surface with the trowel, without considering, or caring for it, that they have put no mortar between the bricks of that course, except in the external edges of the outside joints; that the mortar is not, or ought not to be, so thin as to fall into the joints by its own weight; and that unless they press it down, half the height of the space between the bricks remains in every case unoccupied, and the wall is consequently hollow, incompact, and necessarily imperfect. To obviate this, it is common to have thick walls grouted in every course; that is, mortar made liquid, and called grout, is poured on and spread over the surface of the work, that it may run in and fill up the joints completely. This, at the best, is but doing with grout what should be done with mortar; and the difference between the two consisting merely in the difference in the quantity of water they contain, mortar must be considered the best; for the tendency of grout is, by hydrostatic pressure, to burst the wall in which it is employed; and, moreover, it must, by taking a much longer time to dry and shrink than the mortar of the beds and external joints, make and keep the whole mass unstable, and tend to injure rather than benefit it. Filling or flushing up every course with mortar is therefore far preferable, and may be done with very little additional exertion on the part of the workman.

It is a very common thing for two sorts of mortar to be used in the same wall, a finer and whiter for the outside, and a coarser for the inside work; the former made of cleaner and finer sand, and a greater quantity of lime, than the latter, with the intention of exposing a better-looking and more durable material to the view and the weather. The sand, we have already shown, ought to be as clean as it can be made for mortar to be used in bricklaying; therefore there should be no possibility of making a difference in that particular; and the addition of a greater quantity of lime than is necessary to make good mortar makes it less durable, and occasions a sacrifice in an important quality for the sake of an unimportant advantage. Moreover the mortar which contains the greater quantity of lime will yield or settle more than that which has the greater proportion of sand.

All the walls of a building that are to sustain the same floors and the same roof should be carried on simultaneously; under no circumstances should more be done in one part than can be reached from the same scaffold, until all the walls are brought up to the same height, and the ends of the part first built should be racked back, as at *a b*, fig. 2, and not carried up vertically with merely the toothing necessary for the bond, as at *a b*, fig. 3.

Brick-work should never be carried on in frosty weather, nor even when it is likely that frost will occur before the walls can be covered in or become so dry as not to be affected by frost. Covering an unfinished wall with a thick layer of straw, when frost may supervene, is a very useful precaution; on the straw, weather boarding should be laid, to prevent access of moisture from rain or snow. Merely wet weather may be guarded against by following the directions given above as to flushing every course of the work well up with mortar, so that no interstices be left into which water may insinuate itself, and by covering the walls with boards to act as a coping when the men are not actually at work on them; the joints in the face of a wall that is not to be plastered in any way should be protected in this manner with great care.

In ordinary practice the bricklayer's scaffolds are carried up with the walls, and are made to rest on them. Having built up the walls as high as he can reach from the ground, he plants a row of poles, which vary in height from thirty to forty and even fifty feet, parallel to and at a distance of about four feet six inches from the walls, and from ten to twelve feet apart. To these, which are called standards, are attached by means of cords other poles called ledgers, horizontally and on the inside, with their upper surface on a level with the highest course of the wall yet laid; and on the ledgers and wall short transverse poles called putlogs or putlocks are laid as joists to carry the floor of scaffold boards. These putlocks are placed from four to six feet apart, according to the length and strength of the scaffold boards; and the ends which rest on the walls are carefully laid on the middle of a stretcher, so as to occupy the place of a header brick, which is inserted when the scaffolds are struck after the work is finished. On the floor of the scaffold thus formed the bricklayer stands, and the materials are brought to him by labourers, in hods, from the ground below, or they are hoisted up in baskets and buckets by means of a pulley wheel and fall. The

mortar is placed on ledged boards of about three feet square, placed at convenient distances along the scaffold; and the bricks are strewn on the scaffold between the mortar boards, leaving a clear way against the wall for the bricklayers to move along unobstructedly. The workman then recommences the operation of bricklaying, beginning at the extreme left of his course, and advancing to the right until he reaches the angle or quoin in that direction, or the place where his fellow-workman on the same side may have begun. Thus he goes on with course after course until the wall is as high as he can conveniently reach from that scaffold, when another ledger is tied to the poles, another row of putlocks laid, and the boards are removed up to the new level. The ledger and most of the putlocks, however, remain to give steadiness to the temporary structure, and so on to the full height of the wall, piecing out the poles by additional lengths as may be required. If a scaffold be very much exposed, and run to a great height, it must be braced. This is done by tying poles diagonally across on the outside to the standards and ledgers, and it may be further secured by tying the ends of some of the putlocks to the ledgers; but an outside scaffold should never be attached in any way to the building about which it stands. A scaffold should never be loaded heavily, as well on account of the work as of the scaffold itself; for the putlocks resting, as they do, on single bricks, in a green wall, they exert an injurious influence on it, which every additional pound weight on the scaffold must necessarily increase. A constant and steady supply of bricks and mortar on the part of the labourers, without overloading the scaffold at any one time, should be strictly required.

Arches in brick-work are plain, rough, cut, or gauged. Plain arches are built of uncut bricks, and the bricks being parallelopipedons, an arch built of them must be made out with mortar; that is, the difference between the outer and inner periphery of the arch

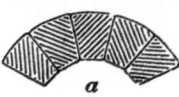 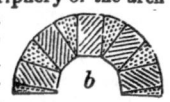

a   *b*

requiring the parts of which an arch is made up to be wedge-formed, as at *a*, which the brick is not, the difference must be made in mortar, as at *b*, so that the inner or lower angles of bricks used for this purpose should all but touch, and the mortar should be more consistent than that used in ordinary walling; nor should the centre on which an arch of this kind is set or built be struck or removed until the work is thoroughly hard, or rather all such arches should be set in cement which will harden immediately. In consequence of this inherent defect in uncut-brick arches, in extensive continuous works, such as sewers, tunnels, vaults, &c., it is advisable to make them in thin independent rings of half-brick or one brick thick, as the case may be; that is, a nine-inch arch should be in two half-brick arches, as at *a*, fig. 6, and an eighteen-inch arch in two one-bricks, as at *b*, each arch in the latter case being bonded in itself as in a common nine-inch wall with headers and stretchers. It is evident that, by this mode of structure, a greater quantity of the solid material comes into the back or outer ring or arch than into the lower one; and if they had been bonded together into one arch, as at *c*, all that difference must have been made up with mortar. Moreover, whatever pressure comes on the outer ring is carried by it directly to the inner or lower, from whose joints, however, the mortar cannot escape or be pressed out, the inner angles of the bricks, by meeting, preventing it below, and the bricks themselves of the upper arch, which conveys the pressure, are themselves opposed to the back of the same joints, so that its power of resistance is made equal to that of the bricks themselves, except at the ends; which, in such works as we have supposed, are remote, and may be protected by the use of cement in their joints, whilst mortar is used in the rest.

Rough arches are those in which the bricks are roughly cut with an axe to a wedge form, and are used over openings, such as doors and windows, when the work is to be plastered on the outside, or in plain back fronts, outhouses, garden-walls, &c., when, however, they are neatly pointed with what is called a tuck or tucked joint. Semicircular and elliptical arches are generally made plain, or without cutting the bricks; but arches composed of a smaller segment of a circle (vulgarly and technically called *scheme* arches), if not gauged, are cut or axed. Very flat arches are technically distinguished from the quicker segment, or scheme, by the term camber, from the French word *cambrer*, to round like an arch. It is arches of this kind which are generally employed over windows and doors in external work, and they too are either cut or gauged.

Gauged arches are composed of bricks which are cut and rubbed to gauges and moulds, so as to form perfectly fitting parts, as in masonry. Gauging is equally applicable to arches and to walling, as it means no more than the bringing every brick exactly to a certain form, by cutting and rubbing, or grinding it to a certain gauge or measure, so that it will exactly fit into its place, as in the finer works of masonry. Gauged brick-work is set in a putty instead of common mortar, but it is seldom used except for arches in the fronts of houses, &c., which are to be neatly finished. These are for the most part straight, and are generally from eleven to twelve inches in depth, or the height of four courses of brick-work. Their value as arches will be best understood by reference to the diagram, fig. 7, by which it appears that all the material between the soffit of the straight arch or head of the opening *b c*, and the dotted line *b f c*, is useless, the intrados or soffit of the really efficient part of the arch being at that dotted line itself. This is the arc of an angle of 60°; its chord, the width of the opening, being the base of an equilateral triangle constructed on it, and the joints are the radii of a circle whose centre is at *a*. *b d* and *c e*, the continuations of the sides of the triangle or radii *a b* and *a c*, are technically termed the skew-back of the arch. Sometimes the arc is made under a more acute angle, in which case the skew-back is less, that is, the external angles *c b d*, and *b c e*, are less obtuse; a smaller unavailable portion of the arch is thus left between the arc and its chord, but that portion is less securely retained under the flatter segment, because the joints or radii diverge less, or are more nearly parallel. These gauged arches being, as they for the most part are, but a half brick in thickness, and not being tied by a bond to anything behind them—for indeed almost the whole, if not the whole, of their height, is occupied behind by the reveal and the wooden lintel—require to be executed with great care and nicety. It is a common fault with workmen to rub the bricks thinner behind than before, to insure a very fine joint in front. This tends to make the work bow outwards; it should rather be inverted, if it be done at all, though the best work is that in which the bricks are gauged to exactly the same thickness throughout. Fig. 8 is a transverse section of fig. 7, and the gauged arch, lintel, &c., in it showing the total disconnection of the gauged arch with any surrounding brick-work to which it might be bonded. The absurdity of constructing arches circular on the plan, especially in a thin unbonded shell of bricks, is so clear as hardly to require notice.

Gauged facing to a wall is exceedingly objectionable, unless the bricks used for the gauged work be originally a little larger than those which are to be worked in behind, whose size should be their gauge, otherwise no bond can be kept between the bulk of the wall and its face; and the same mortar or putty should be used throughout, of equal consistence, and with joints of equal thickness, or the work cannot be sound and compact.

Everything relating to the construction of niches, groins, domes, &c., may be referred to the articles ARCH, BRIDGE, and STONE-MASONRY; the difference between stone and brick, as far as the principle is concerned, being only in the comparative magnitude of the parts; for to make perfect arches, &c., it is clear that the bricks must be cut to the same forms that are required in stone.

It is generally held that nothing but its own components should be admitted into a brick wall, except what is absolutely necessary for its connection with the other parts of a building, such as wall-plates and wood-bricks (and that these should be avoided as much as possible), templates, lintels, &c. Wall-plates are applied to receive the ends of the joists, and distribute the weight of the floor to which they belong equally along the walls. If the joists tailed singly on the naked bricks, their thin edges would crush those immediately under them, and the rest of the brick-work would escape immediate pressure altogether. Wall-plates may be avoided by the use of framed floors, which are carried by a few large beams, under whose ends stout pieces of timber two or three feet in length are placed. These are intended, like a wall-plate, to distribute the weight over a considerable part of the wall, and prevent the necessity of placing the beam on the naked friable bricks, and are called templates. Lintels are used over square-headed windows and doors, instead of arches in brick-work. They are useful to preserve the square form and receive the joiner's fittings, but they should always have discharging arches over them, and should not tail into the wall at either end more than a few inches, that the discharging arch be not wider than is absolutely necessary. Fig. 9 indicates the elevation of the inside of part of an external wall with a window in it, and shows the lintel over the opening with a discharging arch over it, and wood bricks under its ends, on the jambs of the opening. Discharging arches should be turned over the ends of beams, and templates also, as in fig. 10. They may generally be quadrants of a circle, or even flatter, and should be turned in two or more half bricks over doors and windows, and other wide openings, but over the ends of beams they need not be in more than one half brick.

Wood bricks are used to prevent the necessity of driving wedges into the joints of brick-work to nail the joiner's work to. They are pieces of timber generally cut to the size and shape of a brick, and worked in as bricks in the inner face of a wall, where it is known the joiners have occasion for something of the kind. This is principally in the jambs of the windows and doors for their fittings, and along the walls, at proper heights, for the skirtings or wainscotting, as the case may be.

The use of bond timber in brick walls is objectionable, because of its liability to shrink and swell, to decay, and to be consumed by fire, in any of which cases the structure to which it belongs is either injured, endangered, or absolutely destroyed; and in England the use of timber in walls has, since the extension of the manufacture of iron in these countries, been in a great degree superseded by that metal in the form known as hoop iron. Thin and narrow strips of this metal are laid in the bed joints of mortar, at intervals more or less frequent according to the nature and character of the work, with the best effect in respect of compactness and consequent strength.

It will be generally found that a brick wall built with mortar and faced with ashlar has settled inward to a greater or less extent, as the work has been more or less carefully performed. Indeed in the nature of things it cannot be otherwise, unless the brick backing be worked in some cement which sets and hardens at once; for the outer face is composed of a layer of unyielding material, with few and very thin joints, which perhaps do not occupy a fiftieth part of its height, while the back is built up of an infinity of small parts, with fully one eighth its height of joints, which are composed of material that must both yield to pressure and shrink in drying. Some part of the ill effect attendant on this is obviated by the bond-stones, which tail in or run through the wall, and tend to keep the discordant materials together; but still much of it remains: and besides this, the internal or cross walls, which have no stone in them, will either settle down and shrink away from the external walls, or drag them inward, as they happen to be well or ill bonded or tied. For these reasons, brick-work built in this manner with masonry should be executed with exceedingly well-tempered mortar, made with no more lime than is absolutely necessary to cement the particles of sand together, and the sand again to the bricks, worked as stiff as it can be, and laid in as thin courses as may be to answer the purpose required of it. Above all, work of this kind must not be hurried, but allowed time to dry and shrink as it goes on.

Discharging arches over vacuities having been disposed of incidentally, we have now only to speak of them under openings, in which situation their use is to distribute the superincumbent weight equally over the substructure, or along the foundation as the case may be. For this purpose the arch is inverted, as shown in the diagram, fig. 14, and by means of it the weight brought down by the piers is carried along the footings, which are thus equally borne upon throughout their whole length. Arches of two half bricks are indicated here, that being sufficient for ordinary purposes, and to develop the principle; in large and heavy works, arches of three half bricks, and even greater, may be judged necessary. Any arc between a quadrant and a semicircle may be used with advantage; but an arc of less than 45° cannot be recommended for the inverted discharging arch under piers. If it should so happen that an old well or cess-pool, that cannot without great inconvenience and expense be filled up with sound walling, or in some other efficient manner, the ground being sound on either side of it, a second discharging arch may be formed under the pier and over the unsound part, resting its legs on, or springing from, the inverted arch under the opening, and on the sound ground, as indicated by the dotted arch in the last-quoted diagram, fig. 14. For the most part, however, the bonding of the work may be trusted to carry the weight down to the ground under all but very wide openings very low down in the work. Arches require abutments whether they are erect or inverted, and this is often forgotten when inverted arches are used.

Not the least important part of the bricklayer's art is the formation of chimney and other flues. Great tact is required in gathering over properly above the fire-place, so as to conduct the smoke into the smaller flue, which itself requires to be built with great care and precision, that it be not of various capacity in different parts, in one place contracted to a narrow straight, and in another more widely expanded, and so on. With the present imperfect means of cleaning chimney flues, it is absolutely necessary that they be of a certain magnitude, which should be carefully maintained throughout; but it would be better that they were made oval, or with the angles taken off at least, than parallelograms in plan, as the practice is. Chimney flues are plastered or pargetted with a mortar in which a certain proportion of cow-dung is mixed, which prevents it from cracking and peeling off with the heat to which it is exposed. Experiment has proved that a tapering and nearly cylindrical flue of much smaller bore than is now required is the best for carrying away smoke. But the bore should be regulated by the size of the fire-place, or rather by the quantity of smoke to which it is required to give vent.

Sewers and drains which are not cylindrical should be built with concave bottoms, although the sides be parallel and the covering horizontal. The concave channel keeps the stream more together, and enables it the better to carry its impurities along with it: whereas a flat-bottomed drain offers a large surface for the particles of soil to attach themselves to, and the stream of water, being more scattered, is less efficient in force. All drains in houses and in other places where it may be necessary to open them at any time, should be of the form of which *a*, fig. 11, is a section, with a flat covering of stone paving, or large, strong, paving tiles, set and jointed with cement. Gun-barrel drains, as at *b*, are the best in exposed situations, because they are the strongest; but as there is no mode of cleaning but by breaking them up, if they are too long to be raked they should not be employed except with a considerable fall, and a frequent or constant stream of water through them, as from a pump trough, rain-water trunks, &c. They are constructed on a barrelled centre, which the bricklayer drags on as he advances with his work, finishing as he goes. Large sewers, which are accessible from the ends by men to clear or remove any accidental obstructions, are best circular or elliptical; the latter of the two is generally preferred, because, in proportion to its capacity, its height is greater. No drain should have an inclination or fall of less than one-quarter of an inch to a foot; and where the stream is infrequent and dull, as much more would be a great advantage. In building drains it is of great importance that proper traps should be constructed to prevent the return of foul air and the passage of vermin. At every sink there should be a bell-trap, and a well-trap within that, or near the hither end of the drain. Suppose a drain of the form of that shown at *a*, fig. 11, nine inches wide and nine inches deep, leading from a kitchen or scullery to the common drain of the house, in which it meets that which comes from the water-closet and other places. The bell-trap in the sink itself will prevent the return of smell when it is constantly in use, but it is liable to be broken and otherwise injured by the ignorance and impatience of servants and others, or it may become dry by evaporation in some situations; it is therefore necessary to have a trap not so liable to contingencies. Let a well be made eighteen inches or two feet in diameter, square or round, and two feet six inches or three feet deep, across and below the level of the drain, as shown in the plan, fig. 12, and longitudinal section of the same, fig. 13; it must be built around with brick, in cement, and be plastered on the inside with the same material, which will make it capable of retaining fluids. Uprightly across this well, and in the transverse direction of the drain, must be placed a sound piece of paving stone, so long that its ends may be inserted in the sides of the well, as shown in fig. 12, and so wide that its upper edge shall touch the covering of the drain, and that its lower may reach six or nine inches down into the well below the bottom of the drain. Mortar or cement must prevent the passage of air between the upper edge of this trap-stone and the cover of the well and drain, and the trap is complete. The water coming from the sink flows along the drain from *a* to *b* (fig. 13), where it falls into the well, and filling it up to that level, it flows on again from *c* in the direction of *d*, to the cess-pool or common sewer, from which, however, no smell can return; for the trap-stone *e*, the lower half of which is thus immersed in water, completely bars the passage. It is evident, however, that if the well should leak, the water in it may fall below the lower edge of the stone, and the efficiency of the trap be destroyed; but if it be made perfect in the first instance, there can be no danger of any inconvenience that a bucket of water thrown in at the sink will not cure. It is from the drying up of the water in these well-traps (vulgarly called *stink traps*) that uninhabited houses are so frequently offensive. It must be clear, moreover, that these traps form an effectual bar to vermin, and they may therefore be advantageously placed at the entrance of water-closet drains, to prevent rats from getting at the soil-pipes, which they will gnaw and destroy if they can get access to them. Internal drains, or those which go through a house, should always pass under the doorways if possible, in external walls at least. If, however, circumstances should render it absolutely necessary to take a drain through a wall, an arched ring or bull's eye should be made for it to pass by.

Cess-pools should be made cylindrical, and be bricked round, but whether they are made to retain fluids or not, can seldom be a matter of consequence, as they are generally put in secluded places, where, if the object be not to get rid of the waste, there is seldom, at least, any desire to retain it. In towns and cities where the common sewering is as complete as it ought to be, and water-closets are used instead of privies, cess-pools are unnecessary, as the soil becomes so much diluted by the water that goes down with it, that it flows readily enough through the private drains to the common sewer, and so on with the rest, to the common receptacle. Sometimes, indeed, it may be found necessary to clean out the well-traps, but this cannot often occur.

Pipes being hollow cylinders of well-made and well-burnt pottery form the most efficient house-drains. Such pipes may be put together end to end, with great accuracy and sufficient strength, with the aid of collars of the slightest sheet-iron looped together as hoops. The common and bad practice is to form such pipes with sockets, so as to fit spigot and faucet fashion; but the addition to the substance of the pipe to form the socket almost insures a defect in the pipe in or about the socket, whilst the kind of connection which the socket establishes renders it impossible to take out any one length of pipe, and thereby to open a drain in the event of an

obstruction occurring without disturbing many lengths, and makes perfect re-instatement impossible, without taking up and relaying the drain from one or other of its ends. All this is precluded by the use of a collar, but the collar used must not be of the brittle pottery itself. In using pipes for drains, it should be borne in mind that a little larger than large enough, is better than the reverse of this. No pipes should be laid down for a house-drain of less bore than six inches, nor should pottery be used for drains requiring a greater bore than twelve inches; the material is too weak to allow of more. Nor should pottery drain-pipes be laid under any carriage road, for the same reason, that the material is not strong enough to stand more than a dead pressure.

Brick and tile paving is performed by the bricklayer. Brick-paving is either flat or on edge, in sand or in mortar or cement. Brick flat-paving in sand, that is, with the bricks laid on their broadest surfaces, and bedded in and on dry sand, is very slight and fragile, and brick flat-paving set and bedded in mortar is very little better; for if the soil on which the paving is laid be light and sandy, the bricks are easily displaced by being pressed unequally; and if it be clayey it will probably be moist, and the thin porous brick absorbing the moisture, will generally become saturated, and present a damp, unwholesome floor. Paving with bricks on their edges, however, forms a much better floor, and is preferable to a stone paving, if the latter be laid on the ground without the intervention of footings. Brick-on-edge paving in sand is generally used in beer cellars, pantries, dairies, stables, &c., as its numerous open joints allow wasted or discharged fluids readily to escape; and it is both cool and dry under ordinary circumstances. In mortar or cement, bricks on their edges form a sound, dry floor; the smallness of the surface exposed by each brick in this manner leaves them of course less susceptible of partial pressures, and the depth from the soil to the surface is such that damp rarely shows through. The paving brick differs from the common brick only in thickness, its dimension in that direction being rather less than two inches, instead of two inches and a half, and in being rather harder and more compact. Dutch clinkers are paving bricks, smaller and much harder than the English; they are six inches long, three inches wide, and one inch and a half thick, and are always set on edge and herring-boned; that is, instead of being placed in parallel lines, they are set at right angles to each other thus,—with nevertheless a perfectly even face. Paving tiles are made nine inches and a half and eleven inches and a half square, though they are called ten-inch and twelve-inch or foot tiles respectively, the former being one inch, and the latter one inch and a half thick; they are set in courses, as stone paving would be, the alternating courses breaking joint.

Tiling being much less in vogue than formerly, in consequence of the better appreciation of the superior qualities of slate for covering roofs, and the moderate cost at which slates are now furnished to the builder, it no longer maintains its separate artificer, but is performed, when it is required, by the bricklayer. It consists, for the most part, of two sorts—plain tiling and pan tiling. Plain tiles are simple parallelograms, generally about ten inches and a half in length, six inches wide, and five-eighths of an inch thick; and each tile has a hole pierced through it near one end, to receive the wooden pin by which it is hooked on to the lath. The tiles are laid in mortar on the laths, which in this country are of oak or fir, with an overlap of six, seven, or eight inches. The greatest overlap or smallest gauge makes the securest work, though it does not present so good an appearance externally as a longer gauge does; and it requires, moreover, a greater number of tiles and laths, thereby adding materially both to the weight and the cost. The great overlap and the mortar are both necessary, nevertheless, to prevent the rain and snow from driving in between and under the tiles. Plain tiling requires the pitch of the roof to be at an angle of at least 50°, and is one of the heaviest coverings that can be used, though it is at the same time one of the warmest. The tiles, however, readily and rapidly absorb moisture, which they communicate to the laths and rafters under them, to the serious injury of both the latter; and the mortar in which they are set requires to be frequently pointed, the constant atmospheric changes to which it is exposed occasioning it to crumble and fall away in no long time.

Pan tiles are parallelograms of irregular surface, straight in the direction of their length, which is thirteen inches and a half, but twisted to this form ⌒⌒ in the transverse section. Measuring the whole surface across, a tile is nine inches wide, but in a right line from point to point not more than seven, and its thickness is half an inch; a small tongue or lip is bent down at one end from its flatter convexity, on the under side, to hook it on to the lath by, instead of a wooden pin through it, as in a plain tile. Pan tiles are set dry or in mortar, on laths. They are not laid side by side, but overlapping laterally thus; consequently all the overlap they have longitudinally is three or four inches only, or enough to prevent rain and snow from driving up under the upper, over the end of the lower tile; and thence pan tiling is but little more than half the weight of plain tiling. It is nevertheless a much less warm covering for houses, and is more liable to be injured by violent gales or gusts of wind than the latter is; but again, it presents a far more pleasing appearance to the eye. Pan tiling will not bear a much flatter pitch than the other, but it is greatly improved by being pointed on the inside with lime and hair. 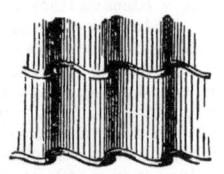 Sometimes indeed the whole of the work is, as we have said, set in mortar; but this mode has disadvantages to which pointing internally is not liable, and its superiority in other respects is questionable. To both pan and plain tiling there is a large concave tile used to cover the hips and ridges of a roof. These are not generally made to overlap each other in any situation, but are set in mortar, and fastened with nails and hooks fitted for the purpose, and driven into the wood-work of the roof.

When the top of a brick wall is not protected by a roof, it must be covered or coped in some manner, or it will soon be destroyed by the weather. Sometimes this is done by means of a course of bricks set across it on their edges in cement, and called a barge course, but it is a very imperfect covering, for water will trickle down the face of the wall on both sides, as the coping brick can be no longer than the thinnest wall is in thickness. Two double courses of plain tiles may be put side by side under the barge course, making a projection over either face of about one inch and a half: thus,—

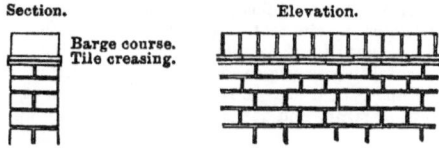

Section.   Elevation.

Barge course.
Tile creasing.

This is much better than the barge course alone: but still the covering receives no inclination outwards to throw the water off; the upper surfaces are all horizontal. The same objection exists to foot-paving tiles, which are also used as a coping; but none of these methods is available for any wall above nine inches in thickness. Stone coping, therefore, which may be made of sufficient width, and be both weathered and throated, is much to be preferred.

One of the greatest faults in the modern practice of building, both architecturally as a matter of taste, and practically as a matter of prudence, is, that these copings, and cornices which serve as such, do not project sufficiently to protect the face of the wall on which they may be placed, from the weather. A bold, massive, and well-projected cornice on a wall serves as a roof or pent-house to it, and, besides imparting great beauty to the plainest structure, protects the wall from the premature decay of its upper part especially, and of the joints generally, if it be unplastered brick-work, which thereby calls for the frequent repetition of pointing. Effective and pleasing cornices and blocking courses may be formed with uncut bricks alone; and these, set in cement, would, with judicious management, add materially both to the appearance and durability of brick-work, without the foreign aid of either the plasterer or the mason.

From the injury which accrues to the joints of brick-work through bad management in its execution, and imperfect protection when executed, arises the necessity so frequent at the present day of pointing.

Sometimes frost will have supervened before the surfaces of the joints in a wall are dry; consequently the mortar bursts and peels away, and the whole then requires to be pointed. Preparatory to this operation the scaffold, if it has been struck, must be re-erected. the mortar raked out of the joints to a depth of about three-eighths of an inch, or deeper if the injury have reached further;—this can be done by a labourer;—a bricklayer then goes over the whole with a hard hair brush and water to cleanse and moisten the joints; and then, with mortar prepared for the purpose, he carefully fills them all up, and neatly draws them with his trowel. This mortar must be of the best quality; it is generally compounded with a certain proportion of forge ashes, which gives it a blue tinge, and greatly aids its power of resisting the action of the weather. Cement is sometimes used instead of this blue mortar. If the wall to be pointed be a front or other important one, in which peculiar neatness is required, every joint is marked with a narrow parallel ridge of a fine white putty, in the composition of which bone lime forms a principal ingredient. The former is called flat-joint, and the latter tuck-pointing. If it be an old wall that requires pointing, a scaffold must be erected before it; and where the putlocks

cannot be rested on window-sills and the like, half bricks are generally drawn from the wall to make rests for them, and restored again when the work is done. The former process is then gone through with a common wall; but if it require tuck-pointing, the whole surface is well washed, and then coloured, to look like new, before the pointing is done. The gauged arches over the windows and doors are always coloured, and the joints drawn with peculiar neatness. If in the original building of the wall the perpends have not been preserved, that is, if the vertical joints have not been made to fall perpendicularly in the alternately recurring courses, the workman in pointing stops up the old joints, which are irregular, with putty of a brick colour, and forms false new ones in the proper places.

The tools and implements mostly employed by the bricklayer are the trowel, the plumb-rule, the level, the square, the bevel, line-pins and lines, the raker, and the hammer, together with a hod and spade for his labourer. Besides these there are sundry others used in cutting and gauging bricks and some which are peculiar to tiling and paving; but the most material operations can be performed with those enumerated here. A pug-mill and screens for mixing and tempering mortar are also auxiliaries of great importance.

Brick-work is valued by the rod. A rod of brick-work is a quantity whose superficies is $272\frac{1}{4}$ feet (taken in practice at the round number 272 without the fraction), and thickness one brick and a half. Reckoning the one brick and a half at thirteen inches and a half,—its average extent,—the cubic foot is to the reduced superficial foot as eight to nine, so that a *cubic* rod of brick-work consists of 306 feet, the result of 272 multiplied by nine and divided by eight. The reduced superficial rod, however, is that commonly used in practice; and the process of measuring, to ascertain the quantities and bring them to a standard, is as follows:—

The exact superficies of so much of a wall as may be of the same thickness is taken, and the number of bricks it is in thickness placed marginally; all the different portions or parts being of the same thickness are taken in like manner, and then deductions, as of window openings and doorways, are taken as such, in superficies, with their respective thicknesses placed marginally also. The dimensions, on being squared, are abstracted in half bricks, the deductions made of like thicknesses from like thicknesses, and the whole reduced by multiplying each quantity by the number of half bricks in the thickness of the parts of the wall which the margin expresses, and dividing the product by three (the number of half bricks in one brick and a half, the standard), the reduced quantity which results, divided again by 272, the number of feet in a rod, gives the quantity of rods and feet in the wall; as, for example,— The front wall of a house is thirty-five feet in length on the ground floor. (Fig. 14.) It has a basement story twelve feet high from the top of the footings to the level of the ground floor, and two and a half bricks thick, which is a half brick more than the wall above. The footings are three spreading courses high, each course a half brick thicker than the one above it. In the basement wall there are a door and two windows, the former seven feet by three feet six inches between the reveals, and the latter five feet by three feet nine inches between the reveals also. The measurement of thus much will show how all the rest must be done.

The footings consisting of three equally spreading courses, the extent of the middle one both in length and breadth will be an average of them all, so that they may be taken in one height. To the length of the ground floor, thirty-five feet, must be added twice three sets-off of one-fourth of a brick at each end of the basement, and of the two first courses of footings for the length of the second of them; this is equal to three half bricks, or thirteen and a half inches, which, added to thirty-five feet, makes thirty-six feet one and a half inch the dimension of length for the footings, by nine inches, their height; their average thickness, to be placed in the margin, is three and a half bricks, the highest course being three bricks, the second three and a half, and the third or lowest four bricks. That is the first quantity. The next is of the wall above. The length (one-half brick, for the two sets-off, added to thirty-five feet, gives) thirty-five feet four and a half inches, by the height twelve feet, two and a half bricks thick. The deductions are seven feet by three feet six inches in one brick for the door, between the reveals, and seven feet four and a half inches by four feet three inches in one and a half brick behind the reveals, the rest of the thickness of the wall, an addition of one-half brick being made to the height, and of two half bricks to the width, because of the reveals. The windows are taken in exactly the same manner, with the same additions; but as the two are of the same size, their number is marked against the one dimension. The dimensions are now to be squared, and the squaring is done by duodecimals, or cross-multiplication. 36 feet $1\frac{1}{2}$ inches $\times$ ·9 inches $=$ 27 feet 1 inch; 35 feet $4\frac{1}{2}$ inches $\times$ 12 feet is $=$ 424 feet 6 inches, and so on with the rest. An abstract is then made of these quantities in two columns, the first is marked " one-half brick," and the second " deductions in that thickness." In the first column is placed the first quantity, multiplied by seven, the number of half bricks in three and a half, which stands marginally to it; 24 feet 6 inches × 7 = 189 feet 7 inches. The second dimension follows in the same column, multiplied by five, the number of half bricks in its thickness; the next quantity is a deduction, that is placed in the second column, multiplied by two, the thickness of the part deducted being one brick, and the rest in the same manner. The abstract being completed, the columns are added, and the amount of the second deducted from that of the first, and the difference divided by three, which brings it to the reduced standard. Dividing now by 272, the number of rods and feet in the given wall appears to be 2 rods, 185 feet, 8 inches. The quantities are more generally abstracted in one-brick and one and a half brick columns, with deductions in other parallel columns, to which thicknesses they are all readily brought. The single column in one-half brick is, however, assumed here as the more simple and the more easily explained.

It must be remembered, that in taking the return or end walls,

3½	36	1½	27	1	Footings.
		9			
2½	35	4½	424	6	Basement wall from the top of footings to the level of ground floor.
	12	0			
1	7	0	24	6	Deduct for door between the reveals.
	3	6			
1½	7	4½	31	4	Do. behind the reveals.
	4	3			
2)					
1	5	0	37	6	Do. for the windows between the reveals.
	3	9			
2)					
1½	5	4½	48	4½	Do. behind the reveals.
	4	6			

Abstract of the above Quantities.

½ brick		Deductions in ½ brick.	
189	7		
2122	6	49	0
		94	0
2312	1	75	0
363	1	145	1½
3)1940	0	363	1½
272)649	8	(2 rods 105 feet 8 inches.	
544			
105	8		

the thickness of that which has been already taken in front is to be deducted from their length, or the angle-pier or quoin will be taken twice. Work which is circular on the plan may be taken separately, and charged at a higher price altogether, or it may be measured as plain, and an extra taken at so much the superficial foot. Chimney breasts are taken as additional quantities, with the thicknesses they project, and the opening for the fire-place is deducted; but the flues are measured as solid, the extra labour and mortar in forming and pargetting them being fully equal in value to the bricks saved.

A rod of brick-work will consume about 4500 bricks, though the number will be a few more or less than this, as the bricks happen to be below or above the average size, and as the joints are made thicker or thinner. The quantity of mortar, it is evident, will be affected by the latter consideration also; but in London it is generally reckoned at from ninety to a hundred striked bushels, or from four to four and a half cart loads, each containing about one cubic yard, to the rod. The labour on a rod of brick-work may be taken on an average at the wages of a bricklayer, and his assistant or labourer, for four days; this, however, does not include making and turning the mortar, nor scaffolding, which latter must be separately considered. Many things will, however, affect the time in which the work may be performed, both of the bricklayer and his labourer; the former can do one-fourth as much more, at the least, in walls which are to be plastered, as in those in which he has to keep the perpends and draw the joints, &c., and more in thick walls than in thin ones; and the capability of the latter will depend, inversely, on the rate at which the former can proceed, on the distance he may have to carry the bricks and mortar to the foot of the ladder, and mainly on the height he has to carry the materials up the ladder. In great heights, however, the materials should always be hoisted.

Gauged arches are taken at so much per foot superficial, in addition to being measured in as brick-work. Both the vertical and horizontal surfaces are measured to obtain the superficies of the arch, or rather of the work upon it. Rough arches are also taken as an extra superficial quantity; but plain arches in vaults, &c. and discharging arches, are not considered extras, though an allowance is made for cutting to moulds, for inverted discharging arches, at per foot run.

If a wall be faced with bricks of a more costly sort than that of which the bulk is composed, or worked in a peculiar manner, it is calculated by the foot superficial, also in addition to its measurement as brick-work. It should be a matter of previous agreement whether or not there shall be an extra charge for plumbing quoins and reveals. Under ordinary circumstances no allowance is made for it; but oblique vertical angles, both internal and external, which require to have bricks neatly cut to form them, are taken at so much per foot running measure. External oblique angles are technically termed *squint-quoins*, and internal, *birds-mouth*. Oblique

angles within a building are taken as run of cut splay. Cuttings to rakes or inclined straight lines are taken by the running foot also, but with reference to the thickness of the wall. Cuttings to ramps or concave lines are measured and valued in the same manner. Sailing or projecting courses, preparations for plaster cornices and brick cornices themselves, are all taken at so much per foot run, according to the labour and materials involved in working them, over and above the regular charge for the brick-work by the rod.

Everything, indeed, which adds to the labour of executing brick-work, and consumes more than the ordinary quantity of materials, is taken in addition, either by the foot superficial, or by the foot running, or in numbers, as the setting of chimney-pots, bedding and pointing door and sash-frames, &c. Bond-timbers, lintels, and wall-plates, are generally measured in with the brick-work, on account of the trouble of bedding them, and the delay generally occasioned to the bricklayer in setting them. If they are not included with the brick-work, bedding them is an extra charge, at so much per foot run; and then filling in between the ends of the joists and beams generally requires to be taken also.

Brick-nogging is measured by the superficial yard, including the quarterings and interstices, and making no deductions but for openings. Drains and sewers are measured by the foot run, according to their form and capacity. The quantity of materials consumed, and labour required in constructing them, may be readily obtained by calculating the one, and observing the quantity a man with a labourer can execute under the circumstances, whatever they may be, within a given time. Digging out the ground, filling in over, and the removing of the spoil, necessarily cause an addition to the charge for the drain, which must vary according to circumstances.

Paving is measured by the superficial yard of nine feet; tiling by the square of one hundred feet;—eaves courses, ridges, and hips, being extra charges, by the foot run. Pointing, whether to old or new work, is measured by the superficial foot; and the scaffolding for it, when scaffolding is required, is either included in the price per foot for pointing, or a charge is made for the use of it, together with the cost of carting, and the men's time in setting up and removing it.

Mason.—We must refer to the separate articles under STONE-MASONRY for information on that subject. It will, however, be necessary to give a few particulars here on masons' work, as it has to do with other artificers' works in the process of building, and especially with reference to various species of walling, or modes of constructing walls of stone.

From the regular and determined form of bricks, modes or systems for setting or arranging them may be formed, and any workman, by habit and an exertion of memory merely, may become competent to build a brick wall as well as it can be built; but it is not so with stone used in common masonry walling. The workman in this material has for the most part to deal with masses of all forms and of all sizes, and a continual exercise of the judgment is required from him beyond the tact or skill which may be acquired by practice. For this reason workmen are generally less to be trusted to themselves, or to their own discretion, in stone than even in bricklaying or walling. The best or highest sort of stone walling is the easiest to set; it is that in which the stones are all tooled and gauged in regular parallelogramic figures, to range in courses and suit the thickness of the wall to which they are to belong; and the most difficult to execute properly is that in which amorphous stones are used,—the mason being allowed merely to dress them roughly with his hammer or axe, and fit them in as he best can to form the most compact mass: this is called rubble walling.

From the brittle nature of stone, great tact is required in setting, to prop or bear up the longer pieces in every part, or they will break across, and thus occasion more injury than could accrue if their whole mass had been made up of small pieces. Very long lengths, therefore, should be avoided, even in regular tooled courses, with which the bearing is or should be perfectly even, and a settling down of the work itself is hardly to be feared. There is a certain medium which may be preserved; and although the object is obviously, in stone as in brick walls, to form a compact mass, as unbroken into parts as possible, a mason will act judiciously in breaking a very long stone into two or more shorter ones, and working them in that state, though he thus makes two or more additional joints, well knowing that he has the power of counteracting to a certain extent the ill effect of joints made by himself; but that those made by accident are irremediable.

The observations made in the section of this article on bricklaying, on the use of mortar, will apply here also. Of whatever quality the stone may be of which a wall is to be built, it should consist as much of stone and as little of mortar as possible. If it be inferior in durability and power of resisting the action of the atmosphere, &c. to the mortar, besides the certain fact that the mortar will yield until it has set hard, and so far act injuriously, no ulterior good is gained; and if the stone be the more durable material, the more of it that enters into the wall the better. Indeed, in rough walling, if the stones be pressed together until the more prominent angles on their faces come into actual contact, the interstices being occupied by mortar, it will be better than if a thick yielding mass were allowed to remain between them. Absolute contact, however, should not be permitted, any more than in brick-work, lest the shrinking of the mortar in drying leave the stones to such unequal bearing as the prominent parts alone would afford. Stone being generally of a less absorbent nature than brick, it is not a matter of so much importance that it be wetted before setting; nevertheless, adhesion on the part of the mortar is more certain and more complete if the stones be worked in at least a damp state. What bond is, and the necessity for it, have also been shown in the preceding section; and bond is of not less importance in stone walling than in bricklaying. We have also hinted above at the greater difficulty of understanding, forming, and preserving it in the former, and can now only add a few observations in addition that can be of any use, and these with reference to rubble walling particularly. Instead of carefully making the joints recur one over the other in alternate courses, as with bricks and gauged stones, the joints should as carefully be made to lock, so as to give the strength of two or three courses or layers between a joint in one course, and one that may occur vertically over it in another. In bonding through a wall, or transversely, it is much better that many stones should reach two-thirds across, alternately from the opposite sides, than that there should be a few thorough stones, or stones extending the whole thickness of the wall. Indeed, one of the many faults of stone-masons is that of making a wall consist of two scales or thin sides, with thorough stones now and then laid across to bind them together, the core being made up of mortar and small rubble merely. This is a mode of structure that should be carefully guarded against. There is no better test of a workman's tact and judgment in rubble walling than the building of a dry wall, or wall without mortar, affords; —walls are frequently built with mortar that without it would have fallen down under their own weight in a height of six feet, in consequence of their defective construction;—thus rendering it evident that they are only held together by the tenacity of the mortar, which is very seldom an equivalent for a proper bond of stone. Masons are very apt to set thin broad stones on their narrow edges to show a good face, by which the wall is injured in two ways; it tends to the formation of a mere case on the surface of a wall, and it for the most part exposes the bed of the stone to the atmosphere, as a stone is more likely to be broad in the direction of its bed than across it.

Rubble walling is either coursed or uncoursed. In the latter sort, fig. 15, the work is carried on with stones of any sizes, as they may occur, and without reference to their heights, somewhat in the manner of the Cyclopæan walling of antiquity; the interstices of the larger being filled up with smaller stones. For this work the mason uses no tool but the trowel to lay on the mortar, the scabling hammer to break off the most repulsive irregularities from the stones,

and the plumb-rule to keep his work perpendicular. The line and level are equally unnecessary, as the work is independent of considerations which are affected by them. An attentive and intelligent workman will, however, make a sound wall with this species of construction, by fitting the stones well together and packing them with as little mortar as possible, yet filling every crevice with it, and carefully bonding through to secure compactness, transversely at the least.

In coursed rubble walling, fig. 16, the line and level are used, the work is laid in courses, each course being carefully brought up to the same level in itself, though no attention is paid to uniformity in the heights of the different courses. For this species of walling the stones are generally roughly dressed by the workman in the gross before he begins building. He is careful to get parallel beds to them, and he brings the best face of each stone to a tolerably even surface at right angles to the beds; the ends, too, receive some little attention, and for this purpose he uses an axe in addition to his scabling hammer. The quoins in coursed rubble walling are generally built with peculiar neatness and precision, and they are set to serve as gauge courses for the rest. This, when well executed, makes a sound and excellent wall. It presents, however, rather a rough and homely appearance, and in finer works must be covered with stucco or cement, or faced with ashlar.

Ashlar is an external rind of gauged stones in equal courses, having tooled or closely-fitting joints to give a wall a neat and uniform appearance; it is axed, tooled, or rubbed, as may be thought most in character with the structure, or that part of it to which it is to belong. Ashlar stones, or ashlars as they are commonly called, are made of various sizes on the surface, as the character of the edifice may require or convenience demand, and vary in thickness from five to eight or nine inches. Some of the ashlar stones must, it is clear, be used transversely as bond stones, or the facing, having nothing to connect it with the wall behind, would soon totter and fall. Bond stones are generally put in alternate courses, with the backing to the jambs of openings, such as windows, and oftener, if these do not recur within a length of five or six feet; the bond stones themselves, too, should not fall in the same vertical chain, except when they are in the jambs of openings, but break in their alternate courses. Ashlar is commonly set in a fine mortar or in putty. It is generally recommended that ashlars should not be made regular parallelopipedons, but run back irregularly to tooth in with the backing, the vertical joints being left open from about an inch within the face of the wall, and the upper surface or bed of the stones made narrower than, though perfectly parallel to, the lower. These things may exert a slightly beneficial influence under some circumstances; but the mode of construction involved is so radically bad, that unless the backing is set in a quick-setting cement, or be so well packed as to be proof against its general tendency to settle away from the ashlar facing, no means of the kind can materially improve it. A well-compacted wall of coursed rubble, the courses being frequently made up of whole stones and faced with ashlar, may be made tolerably sound and trustworthy. Brick backing, with ashlar facing, cannot be considered as good, though it has the advantage of not requiring battening and lathing for inside plastering, as the stone-backed wall does. Uncoursed rubble with ashlar has all the disadvantages of both the preceding, with nothing to recommend it before either of them.

There are, besides, many sorts of walling or modes of structure arising from the nature of the materials furnished in various localities. That of most frequent occurrence, perhaps, is a manner in which either broken or rounded flints are used. These depend almost entirely on the mortar with which they are compacted, and on a coursed chain, which is commonly introduced at short intervals of larger stones or of bricks, to act as a bond; the quoins, too, in this species of structure are generally constructed of dressed stones or of brick.

Whatever objections lie against bond timber in brick-work apply with equal force at least to the use of it in stone walls. Hoop-iron bond is not only available in all kinds of stone walling, including the highly-wrought close-jointed kind, but it is invaluable, as it may be used both longitudinally and transversely as it may in brick-work; whilst it compels the building mason to bring his work up to a true and fair bed as often as the bond is to be laid in it.

Discharging arches, it must be evident, are as necessary in and to stone walls as to walls of brick, and they may be treated much in the same manner.

Rubble walls are scaffolded with single, and ashlar-fronted or other gauged stone walls with double-fronted scaffolding, the former tailing one end of the putlocks in on the wall, and the other having an inner row of standard poles and ledgers parallel to the outer, making the scaffold entirely independent of the wall. In some places, however, it is the custom to dispense altogether with an external scaffold in building stone walls, particularly with gauged stones. With light and plain work this may be done without much inconvenience or retardation; but if the work be heavy or delicate, considerable delay and incorrectness result. Sometimes the finer work, such as that to mouldings, flutes, and foliate or other enrichments, is merely boasted or roughed out before the stones are set, and finished afterwards. This can be done well only from a secure floor or scaffold, on which the workman may move freely.

When walls are not entirely of masonry, in the ordinary course of economic building, stone is frequently used for copings, cornices, string and blocking courses, sills, landings, pavings, curbs, steps, stairs, hearth-stones and slabs, and chimney-pieces; to these may be added quoins and architectural decorations, or dressings for windows, doors, &c., though both the former and latter are not unfrequently executed in plaster composition, or cements. Copings (see Glossary to the article ARCHITECTURE) to cover walls, parapets, &c., are worked with a plain horizontal bed, two vertical faces, and an inclined or weathered back or upper surface; either forming an acute angle with the outer and wider, and an obtuse angle with the inner and narrower face, to throw the water off, as shown at *a*, fig. 19; or to both sides from the middle, as at *b*; the latter is technically termed saddle-back coping. In both cases they are made to project over the wall or parapet on both sides; and in the projected part of the bed under the edge or edges towards which the inclination is given, a channel or groove, called a throat, is cut, to intercept the water in its inclination to run inwards to the wall. On gables or other inclined planes the coping is neither weathered nor throated, as the water is necessarily impelled along its course to the lower end, and not over the sides. To protect the separate stones of a coping course from the danger of being displaced by high winds or other accidental cause, and to form a chain through its whole length, the stones are linked together by cramps of copper or iron let into their backs and run with lead. These metals, however, especially the iron, for the most part act very injuriously, from their exceeding susceptibility of atmospheric changes, and their greater or less tendency to oxidation; indeed, the stone invariably suffers more than the work benefits from the metal cramps. Tenons, dowels, joggles or dovetails of stone, or of hard wood or cast iron, applied so as to be protected from the weather, would be far better, and would answer every desirable purpose sufficiently. Cornices (*vide ut sup.*) are but ramified copings, and are or may be subjected to the same general laws. Care must be taken, however, in arranging them, that their centre of gravity be not brought too far forward, in the anxiety to project them sufficiently, lest they act injuriously on the

wall by pressing unequally, and their own seat be also endangered. String courses (*vide ut sup.*) economically, in contradistinction to architecturally, are meant to protect a set-off in a wall, by projecting over its lower face in the manner of a coping (see fig. 17, at *c*); the beds are worked parallel, and the outer face vertical or at right angles to them, but so much of the upper surface is weathered or sloped off as protrudes from the upper part of the wall to carry the water off; and, for the reason above stated with regard to copings, the lower bed just within the outer face is throated. A stone string course, cramped or dovetailed in the bed, forms an excellent chain round a brick-wall; but the part of it in the wall should be of the exact thickness of one, two, or more courses of brick-work. A blocking course (*vide ut sup.*) is either a very thick string projecting over or flush with the face of the lower part of the wall to cover a set-off, or it is a range of stone over a crowning cornice to bring the centre of gravity more in on the wall than it otherwise would be; in the former case it is treated exactly as a string, excepting that, if it be flush below, there is no occasion for a throat; and in the latter it has a horizontal bed, parallel vertical sides, and a weathered back or upper surface. Sills (*vide ut sup.*) are weathered and throated like the parts of a string course (see fig. 17, at *a* and *b*); they are laid across window openings as a base to the sash-frame; distinct sills in the same line may, indeed, be considered as an intercepted string course. In the ordinary practice of building, window sills are seldom set in brick walls until they are absolutely required to set the sash-frames on; or they are set but not bedded, except at the ends. The object of this is to prevent any settlement that may occur in the piers from breaking the sills across on the unyielding part of the wall under the windows. A necessity for this, however, can only arise from bad construction; for with a good bond in the brick-work, all would settle together, and the sills might be completely bedded across at once. Landings are platforms of stone, either over an area before a door, at the head of a flight of stairs, or as the floor of a balcony. They are made four, five, six, or eight inches in thickness, according to their extent and bearing; if not of one piece of stone, they are of nicely jointed pieces joggled and plugged together, and are worked on the face and edges just as their situation may demand. Stone pavings are of various kinds, and are prepared, shaped, and laid in various ways. Stone paving that is not exposed to the sun and air, if next the ground, should be laid on footings of brick or stone, or it will be constantly damp if the soil be close and clayey; but in yards, open areas, &c., it may be laid on the ground, bedded in sand, and jointed with mortar or cement. Stone-paved floors are either on brick arches, or on a timber floor prepared for the purpose; the latter is a very bad mode of supporting paving, as the impression derived from the presence of the stone is, that the floor is incombustible; but if it be bedded on combustible material, the danger to human life in the event of fire is greater than if the stone paving did not exist at all. It is worked, cut, and set more or less expensively, according to circumstances. A curb is a range or course of thicker and stronger stone to bound a pavement, and is either flush with the paving, showing as a step on its outer edge, or raised above it to receive a balustrade, and shows on the outer side as a blocking course; in the latter situation it is generally joggled and plugged in the joints. The term step or steps alone is generally understood to mean external steps, whether arranged in long or short flights, or the single step in a doorway into which the door-frame is tenoned. A step should have a plain horizontal bed, and a very slightly weathered tread or upper surface; the front or riser worked plain and vertical, or with a moulded nosing, and the back sunk with a joggle or bird's-mouth joint to receive the step or landing above or behind it. Stairs are but a flight or combination of steps used internally; the principles upon which they are constructed will be found under the heads STONE-MASONRY and JOINERY. Hearths are the stone-flooring of fire-places; and a slab is that part of the floor of a room which lies immediately before the fire-place and along the extent of its front. This slab is supported by a flat brick arch called a brick trimmer, which is turned from the chimney-breast under the hearth on one side, to the trimmer joist on the other. (See a section of all these at fig. 18.) Chimney-pieces consist simply of mantle and jambs; that is, the vertical sides, and the architrave or transverse covering with its shelf or cornice. The parts of a chimney-piece are generally put together with an adhesive plaster or cement, and affixed to the wall or chimney-breast behind with cramps, holdfasts, and plugs. The material of which chimney-pieces are composed varies from the coarsest stone to the finest marble; and the labour on them varies to a still greater extent. Quoin-stones are gauged and wrought blocks with parallel beds and vertical faces, placed on the angles of buildings with the intention of adding to their beauty and strength; they are used either with brick or stone walls, and are generally made to project before the face of that to which they are attached, mostly with a weathered angular joint, or with a rectangularly grooved or moulded one. The quoins are coursed with the rest of the wall if it be of stone, and are made to occupy the exact space of a limited number of courses of brick in a brick wall. (See fig. 17.)

Masonry to receive architectural decorations is generally worked into the walls as they are carried up; but as they are seldom homogeneous either in matter or construction, the result is mostly the converse of what it purports to be, for the work is more frequently weakened than strengthened by the decorative masonry. Stones of which columns are to be composed, whether each column is to be of one stone or more, are generally roughly boasted out before they are set, and are finished afterwards to traversing moulds and templets with a plumb-rule, whose sides are cut to the diminution, whatever it may be. Flutes are cut at the same time and in the same manner. The beds of the joints in columns should be worked with the greatest precision, that they may fit firmly and closely together; they must not, however, be worked hollow to make a close joint externally, or the arrises will chip off. It is considered a good plan to put a piece of thin milled lead between the beds, cut circular, and extending to within a short distance of the surface, and that the rest be filled with a fine adhesive putty, made as nearly of the colour of the stone as possible. This makes a solid bed, and protects the arrises effectually; but it will not do so well for slight columns, because it narrows the bed so materially. A joggle or dowel of hard wood or cast iron let into the core might be a sufficient counteraction, and it would certainly add to the stability of a polylithic shaft. The other parts of a columnar composition may be sufficiently cramped and joggled together with wood and metals, according to the situation, though it may be again remarked, that neither wood nor metal should be used, unless it can be protected from access even of the atmosphere.

Stone walling is generally measured by the perch of twenty-one feet superficial, at a standard of eighteen inches in thickness, or a cubic quantity of thirty-one feet six inches. Sometimes it is taken by the rod of 272 feet, like brick-work, but at the eighteen-inch standard instead of the fourteen-inch, or a brick and a half, as in the latter species of walling. The perch, however, as first stated, is the standard of this country. The quantities may be ascertained in the same manner that they are in measuring brick-work, the number of inches the wall is in thickness being substituted in the margin for the number of brick lengths. In abstracting, the superficial quantities may be taken out in columns under the different thicknesses; the amount of each column being multiplied by the thickness in inches, and divided by eighteen, gives the reduced quantity; but if the work be taken in cubic quantities, it is evident that three dimensions of every part multiplied together brings the whole at once to cubic feet, and no further process is necessary, unless it be required to bring the total quantity into reduced perches, which may be done by dividing it by thirty-one and a half.

The custom being different in different places with regard to the double measurement of quoins or angle piers, and as to whether openings, such as windows and doors, shall or shall not be deducted, because of the greater care and trouble required in setting and plumbing quoins and reveals, these particulars should be made matter of previous agreement. Perhaps the best way is to take the quantities exactly, and allow a running measurement extra on the parts requiring more than the usual quantity of labour, or, the nature of the work being of course obvious beforehand, the price per perch, per rod, or per foot cube, on the exact quantity, may be made to include the proposed extras. In the same manner, chisel-dressing (that is, facing the stones neatly and truly with the chisel), whether plain or sunk, may or may not be charged extra, according to agreement, or, in the absence of a previous agreement, to the custom of the place. To ascertain the value of stone walling, the cost of everything that enters into some fixed quantity on the spot must be calculated, for almost everything connected with it varies in almost every place. The original price of the stone at the quarry; the expense of carrying it from thence to the place where it is to be worked up; its texture or comparative hardness, which will materially affect the quantity of walling a mason may execute in a certain time; the cost on the spot, of lime and sand, and the height to which stones must be carried or hoisted from the ground; must all be ascertained and considered, as well as the wages of masons and labourers, and the sort of walling proposed to be executed.

Stone used in string and blocking courses, sills, copings, cornices, steps, quoins, columns, entablatures, &c. is measured by the foot cube, and the work on it is taken as plain, sunk, or moulded, by the foot superficial. The dimensions for the cubic quantities are taken on the unreduced block, or rather on the greatest breadth and thickness which the finished works exhibits; for instance, the string course, which appears in section at c, fig. 17, would be taken as of the thickness throughout which it holds in the wall; and in the same manner, the thickness of the sill at b would be taken under the wooden sill of the sash frame, which must have been the original thickness of the whole scantling. Stone sawed into thin slabs for paving, chimney-pieces, &c., is taken by the superficial foot, at a certain thickness, the value being ascertained from the cubic quantity and the cost of sawing on the surface, whilst some articles, being of a fixed breadth and thickness fitting them to peculiar purposes, are taken by the running foot; but both these latter modes suppose labour included.

Plain work is the even surface produced on stone by the chisel, without the necessity of taking away more than the mere inequalities, and is equivalent to what the joiner calls trying-up, that is, making the surfaces perfectly straight both longitudinally and transversely, and so that it shall be quite out of winding, which indeed is a term to express the result of trying up. Sunk work arises from the necessity of chiseling or hacking away below the level surface of the plain work, such sa the weathering of copings, string courses, cornices, &c.; and mouldings cut in stone produce what is called moulded work. Sunk and moulded work are either straight or circular; circular plain work is certainly spoken of, but incorrectly, for every flexure in stone must be produced by sinking. The joints and beds, that is, the upper and lower horizontal sides, and the vertical ends of stones, are taken as plain work, as well as their faces and edges, if they have been wrought with the chisel to produce the surface; or their superficies are taken as sawing or half plain work, if the surfaces are as the saw left them. An extra charge is made on plain work for rubbing to produce a smooth unchannelled surface; and again, a higher charge is made for plain work if it be equally channelled or furrowed in vertical lines over the surface; this latter operation is technically termed tooling. Whenever any two surfaces meet in an oblique angle, one of them may be taken as sunk work, and it will generally be that which is not parallel to its opposite side. It is valued at about two-sevenths more than plain work; and circular sunk work, that is, circular in the direction of its length, at about one-sixth more than straight sunk. Moulded work is measured by girding the moulding or mouldings with a cord or tape, carrying it into all the quirks, and round all the arrises; the dimension thus given is multiplied by the length for the superficial quantity. This is valued at about one-fifth more than sunk work, and circular-moulded at about one-half more than straight. Narrow jointings, groovings, throatings, jogglings, &c. are taken by the foot run. Mortises, holes, notches, cramps, dovetails, &c. are numbered and charged at so much a piece, according to the labour and cost involved in making them. The common pavings, landings, copings, sills, and steps generally used in London for ordinary purposes, are of a laminated stone from Yorkshire, and they are for the most part worked to size and shape in the quarry, so that there can be very little labour on them beyond the mere fitting and setting, making mortises, fitting coal-plates, traps, &c., when such are required, unless they be rubbed, which occasions, of course, an extra charge. York pavings and landings are taken by the superficial foot, at such a thickness; and copings, sills, steps, &c. by the foot run, according to their size.

Plasterer, &c.—No art in the economy of building contributes more to produce internal neatness and elegance, and no one is less absolutely important, so far as the use and stability of a structure are concerned, than that of the plasterer. Its very general application, too, is of comparatively late date; for wainscotted walls, and boarded or boarded and canvassed ceilings, or naked joists alone, are frequently found in houses of even less than a century old, both in these countries and on the Continent.

The plasterer, as the term imports, works in plastic, adhesive compositions, which are laid on walls, both internally and externally, to stop crevices, reduce inequalities, and produce an even, delicate surface, capable of receiving any decoration that may be applied to it, either in colour or otherwise. These compositions are as various as the modes of applying them, the rudest being a compost of loam, a marly clay, and lime; this is used only for the commonest purposes, and being laid on in one coat, is washed over with a thin mixture of lime and water, which process is termed white-washing; the highest work of the plasterer is the making an imitation of marbles and other costly stones, of the purest calcined gypsum, mixed with a solution of gum and isinglass, and colouring matter to produce the required imitation. For the more common operations of plastering, however, comparatively few tools and few materials are required. The plasterer is attended by a labourer, who supplies his boards with mortar, and by a boy on the scaffold with him to feed his hawk; he is necessarily furnished with a lathing hammer, a laying-on trowel, a hawk, floats, brushes, jointing trowels and rules, moulds and straight edges, together with a screen, spade, rake, and hod, for his labourer, and a feeding-spade or server for his hawk-boy. The lathing hammer is chequered on the face with indented lines, to make it less liable to slip over the head of the nail; the upper or back part of the hammer is made like a hatchet,

but very narrow, and on its inner side or edge there is sometimes a square nick or grove, by means of which the workman is enabled to draw a nail that has gone awry. The laying-on trowel is a thin plate of hardened iron or steel, ten inches long and two and a half inches wide, rounded at one end and square at the other end or heel; it is very slightly convex on the face; and to the back, about the middle of it, the spindle or handle is rivetted in at right angles, which, returning in the direction of the heel parallel to the tool, fits into a rounded wooden handle, by which the workman grasps it. The plasterer is obliged to keep this implement particularly clean and dry when he is not actually using it, lest it rust in the slightest degree, as it is clear that the brown oxide of iron would sadly discolour his finer work on touching it again with the trowel. The hawk is a piece of wood about ten inches square, to receive a small portion of mortar on, for the convenience of carrying it readily up to the wall or ceiling, to be there delivered and spread by the trowel. The hawk is traversed across the back by a dove-tailed piece, into which the wooden handle is fixed at right angles, and by this the workman holds it in his left hand. A hand-float is a piece of board shaped something like a plastering trowel, with a ledge handle to it, and is used to rub over the finished work, to produce a hard, smooth, and even face. A quirk-float is of wood also, and is angularly shaped to work in angles; and a derby is a long two-handed float, which is that principally used in forming the floated coat of lime and hair. The plasterer's brush is broad and thin, with a stout or slight row of coarse or fine hair, as it may be required for rough or fine work. Jointing trowels are thin plates of polished steel, of triangular shape, the point being a very acute angle; the handle is adapted to the heel or base of the tool. They are of three or four different sizes, and are principally used in making good cornices, and joining them at their internal and external angles, which is called mitring. Jointing rules are auxiliary to the jointing trowel. Moulds are pieces of hard wood cut to the contour of cornices or separate mouldings, to assist the workmen in forming them readily. For work of any importance the moulds are cut in copper plates, which are inserted in the wooden stock, and narrow pieces of wood are fixed to the moulds transversely, to guide and steady them along the screeds. A straight edge is a board of considerable length, shot perfectly straight on one edge, to bring the plastering on a wall or ceiling to a perfectly even surface, by traversing it in every direction. A screen is a large parallelogramic wooden frame, on which metal wires are fixed at regulated distances from each other, to act as a sieve. This is propped up in nearly a vertical direction by a counter-frame hinged to it like a common step ladder, and the coarser materials which enter into the composition of plastering mortar are thrown against its outer face, to separate the particles which are too large for the purpose from the finer. The sand and lime, too, are mixed much more efficiently and completely by screening them together than in any other manner. The spade and hod are like those of the bricklayer's labourer. The rake is used to separate the hair used in the mortar, and distribute it throughout the mass. The hawk boy's server is about the size and shape of a common garden hoe, but the handle is in the direction of the instrument. With this the boy rebeats the mortar on the board, to destroy any set it may have taken, and delivers it in small pats or portions on to the plasterer's hawk.

The plasterer's materials are laths and lath nails, lime, sand, hair and plaster, of which are formed coarse stuff or lime and hair, fine stuff, gauge stuff, &c.; and besides these, a variety of stuccoes and cements, together with various ingredients to form colouring washes, &c. are more or less in request.

Laths are narrow strips of some straight-grained wood (in this country they are generally of fir, though oak laths are sometimes used), in lengths of three and four feet, or to suit the distances at which the joists of a floor or the quarterings of a partition are set, and in thickness from one-eighth to three-eighths of an inch; those of the greater thickness are called lath and a half. Lath nails are either wrought or cut; cut nails are in common use in this country with fir laths. Coarse stuff is composed of ox or horse hair from the hide, in addition to the lime and sand mortar of the bricklayer and mason; this is intended to act as a sort of mesh to net or tie it together, and form a coarse but plastic felt. The hair should be as long as it can be procured, and free from grease and filth of every kind. Road drift is unfit to be used for mortar, unless it be completely cleansed from all animal and vegetable matter, and of all mud and clay. Nothing but clean sharp sand should be used with the lime and hair in the composition of this, any more than of brick mortar. Fine stuff is a mortar made of fine white lime, exceedingly well slaked with water, or rather macerated in water to make the slaking complete; for some purposes a small quantity of hair is mixed up with this material. Fine stuff very carefully prepared of the finest powdered lime macerated so completely as to be held in solution by the water, thus forming a mere paste, which is then allowed to evaporate until it is of a sufficient consistence for working, is called putty. Gauge stuff is composed of about three-fourths of putty and one-fourth of calcined gypsum or plaster of Paris; this may be mixed only in small quantities at a time, as the plaster or gauge renders it liable to set very rapidly. Bastard stucco is made of two-thirds fine stuff, without hair, and one-third of very fine and perfectly clean sand (the cleanliness or purity of sand may be determined by the facility with which it may, when in a moist state, be struck off from the hand without leaving a soil); and common stucco is composed of about three-fourths of clean sharp sand and one-fourth of the best lime, well incorporated. This must be protected from the air from the time it is made up until it is required to be laid on the walls. The cement best known and most commonly used in this country is called Parker's, or Parker's Roman cement. This material, when of good quality, with fine clean sharp sand, in the proportion of about three of the former to one of cement, and well executed, forms a very good external coating for walls.

A composition known as Portland cement, because the mortar formed by it when mixed with sand, presents, or is supposed to present, the appearance of stone from the Portland quarries, has grown of late years into repute in London more particularly, and it is in one particular at least preferable for an outside stucco, that the colour to which it dries is sufficiently agreeable to the eye without any colouring wash, whereas Parker's cement is too often of a dark dirty tint, requiring painting or colouring to render it tolerable. Portland cement is also much esteemed as being proof against water when used as a mortar in setting brickwork, and in the composition of concrete for foundations.

The various coatings of plastering are thus designated: On laths, plastering in one coat simply is said to be laid, and in two coats, laid and set. In three-coat plastering on laths, however, the first is called the pricking up, the second is said to be floated, and the third set. On brick or stone walls, without the intervention of laths, plastering in one plain coat is termed rendering; with two coats, a wall is said to be rendered and set; and in three, rendered, floated, and set. Before the plasterer begins to lath a ceiling, he proves the under face of the joists, to which he has to work, by the application of a long straight edge, and makes out any slight inequalities in them, when the work is not to be of a very superior description, by nailing on laths or slips to bring them as nearly even as he can. When the inequalities are great, or if the work is to be of fine quality, he recurs to the carpenter, who takes off inordinate projections with his adze, and nails on properly dressed slips where the joists do not come down low enough, and thus brings the whole to a perfect level. This operation is called firring, that is, putting on pieces of fir, though it is vulgarly termed and frequently spelt *furring*. If it be a framed floor or ceiling joists the plasterer has to work to, it is tolerably sure to be straight; but the carpenter must have firred down on the beams or binders to the level of the ceiling joists, unless the ceiling joists have been nailed to the beams or binders, when nothing of this kind need be necessary. If a ceiling is to be divided into compartments or panels, the projecting or depending portions must be bracketed or cradled down to receive the laths. It is an important point to be attended to in plastering on laths, and in ceilings particularly, that the laths should be attached to as small a surface of timber as possible, because the plastering is not supported or upborne by its adhesion or attachment to the wood, but by the keying of the mortar itself, which passes through between the laths, and bends round over them. If then the laths are in constantly recurring contact with thick joists and beams, the keying is as constantly intercepted, and the plastering in all such places depends entirely on the portions between them which are properly keyed. Under a single floor, therefore, in which the joists are necessarily thick, a narrow fillet should be nailed along the middle under the whole length of them all, to receive the laths and keep them at a sufficient distance from the timber to allow the plastering to key under it; and thus too the surface might be made more perfectly even, by blocking out the fillets, and contrariwise, as it is in single floors that inequalities mostly occur. This being all arranged, the plasterer commences lathing. The laths should be previously sorted, reserving the crooked and knotty, if there be such, for inferior works, and selecting the best for the work of most importance, so that the workman shall find none to his hand that is not fit to be brought in. Taking a lath that will reach across three or four openings, he strikes a nail into it on one of the intermediate joists, at about three-eighths of an inch from the one before it, and then secures the ends of that and the one that it meets of the last row with one nail, leaving the other end of the lath he has just set to be secured in the same manner with that which shall meet it of the next bay in continuation. It is of importance also that he pay attention to the bonding of his work, either by using longer and shorter laths in bays or squares, and in breaking the headings, or with laths of the same length, the first and last courses or bays only having the bond formed by half laths. In lathing on quartering partitions and battened walls, the bonding is not a matter of much importance; nor is the thickness of the timbers behind the latter of so much consequence as in a ceiling, because the toothing which the thickness of the lath itself affords to the plastering is enough to support it vertically; but, nevertheless, the more complete the keying, even in works of this kind, the better, as the toothing above will not protect it from any exciting cause to fall forwards, or away from the laths. The thinner or weaker sort of lath too is generally considered sufficiently strong for partitions, whilst the stronger is used for ceilings. Thin weak laths, if used in a ceiling, are sure to produce inequalities, by sagging with or yielding to the weight attached to them. A chance one or two weak ones in a ceiling of otherwise strong laths may be the ruin of the best piece of work. Care should be taken therefore not to allow a thin lath, or one of unequal thickness, to go on to a scaffold with thicker and more equable ones, lest the workman should, through carelessness or otherwise, put it up with the rest. When the lathing is completed, the work is either laid or pricked up, according as it is to be finished with one, two, or three coats. Laying is a tolerably thick coat of coarse stuff or lime and hair brought to a tolerably even surface with the trowel only; for this the mortar must be well tempered, and of moderate consistence,—thin or moist enough to pass readily through between the laths, and bend with its own weight over them, and at the same time stiff enough to leave no danger that it will fall apart, a contingency, however, that in practice frequently occurs in consequence of badly composed or badly tempered mortar, unduly close lathing, or sufficient force not having been used with properly consistent mortar to force it through and form keys. If the work is to be of two coats, that is, laid and set, when the laying is sufficiently dry, it is roughly swept with a birch broom to roughen its surface, and then the set, a thin coat of fine stuff, is put on. This is done with the common trowel alone, or only assisted by a wetted hog's bristle brush, which the workman uses with his left hand to strike over the surface of the set, while he presses and smooths it with the trowel in his right. If the laid work should have become very dry, it must be slightly moistened before the set is put on, or the latter, in shrinking, will crack and fall away. This is generally done by sprinkling or throwing the water over the surface from the brush. For floated or three-coat work, the first, or pricking up, is roughly laid on the laths, the principal object being to make the keying complete, and form a layer of mortar on the laths to which the next coat may attach itself. It must, of course, be kept of tolerably equal thickness throughout, and should stand about one-quarter or three-eighths of an inch on the surface of the laths. When it is finished, and while the mortar is still quite moist, the plasterer scratches or scores it all over with the end of a lath in parallel lines from three to four inches apart. The scorings should be made as deep as possible without laying bare the laths; and the rougher their edges are the better, as the object is to produce a surface which the next coat will readily attach itself to. When the pricked up coat is so dry as not to yield to pressure in the slightest degree, preparations may be made for the floating. Ledges or margins of lime and hair, about six or eight inches in width, and extending across the whole breadth of a ceiling or height of a wall or partition, must be made in the angles or at the borders, and at distances of about four feet apart throughout the whole extent; these must be made perfectly straight with one another, and be proved in every way by the application of straight edges: technically these ledges are termed *screeds*. The screeds are gauges for the rest of the work; for when they are ready, and the mortar in them is a little set, the interspaces are filled up flush with them; and a derby float or long straight edge being made to traverse the screeds, all the stuff that projects beyond the line is struck off, and thus the whole is brought to a straight and perfectly even surface. To perfect the work, the screeds on ceilings should be levelled, and on walls and partitions plumbed. When the floating is sufficiently set and nearly dry, it is brushed with a birch broom as before described, and the third coat or set is put on. This for a fine ceiling that is to be whitened or coloured must be of putty; but if it is to be papered, ordinary fine stuff, with a little hair in it, will be better. Walls and partitions that are to be papered are also of this latter, or of rough stucco; but for paint the set must be of bastard stucco trowelled. This coat must be worked of exactly the same thickness throughout, to preserve to the external surface the advantage that has been obtained by floating. For all but this last mentioned, the set on floated work, the trowel and brush are considered sufficient to produce fine and even work; but trowelled stucco must moreover be hand-floated. In this operation the stucco is set with the trowel in the usual manner, and brought to an even surface with that tool to the extent of two or three yards. The workman then takes the hand-float in his right hand, and rubs it smartly over the surface, pressing gently to condense the material as much as possible. As he works the float he sprinkles the surface with water from the brush in his left hand, and eventually pro-

duces a texture as fine and smooth almost as that of polished marble. But lathing and plastering on laths as practised in England, is at best a very flimsy affair, and greatly requires improvement. Stronger laths than the laths commonly employed, put on further apart, and with headed wrought nails, and the plastering laid on upon both sides in upright work, or both above and below the ceilings at the same time, two men working against one another will produce work in some degree worthy of the name. But the practice of the French in this respect—a recent practice truly—is well worthy the consideration, and to a great extent the imitation, of English plasterers. The process of plastering on the naked brick or stone wall differs but little, except in name, from that we have described as the mode on lath. The single coat, or equivalent for laying, on lath, is rendering, and it need differ only in the quantity of hair, which may be less than is necessary for laying, and in the consistence of the mortar, which may be made more plastic, to work easier, and because in a moister state it will attach itself more firmly to the wall: the wall, however, must itself be wetted before the rendering is applied. The set is the same, and is put on in the same manner as to two-coat work on lath. For three coat, or floated work, the first or rough rendering should be made to fill up completely whatever crevices there may be in the work behind it, and be incorporated with it as much as possible. As its name imports, its surface may, indeed should, be rough; but it is not scratched or lined as the similar coat on lath is: for this, too, the wall must be previously wetted, that the mortar may the better attach itself to it. For the floating, screeds must be formed as before described, and the consecutive process is exactly the same as on lath, both for the floated and for the set coat. In almost every case in which plastering is to be floated, the workman finds a guide for the feet of his wall screeds in the narrow grounds which the joiner has previously fixed for his skirtings; from these he plumbs upwards, and makes his work perfectly flush with them.

Mouldings and cornices, as large combinations of mouldings and flat surfaces in the angles of rooms, immediately under their ceilings, are called, are formed with running moulds, and are generally executed before the setting coat is put on the walls and ceiling. If the cornice do not project more than about an inch and a half, or two inches, from the ordinary work, a backing of lime and hair will be sufficient; and if any one part only happen to be more than ordinarily protuberant, a row of nails from six to twelve inches apart stuck into the wall or ceiling in the line of that part will give it sufficient support. But if the general mass of the cornice be more than that amounts to, and extend more than six or eight inches along the ceiling, it must be bracketed out, and the bracketing lathed and pricked up, as for ordinary work. This pricking up, or other preparation, must of course be perfectly set before the cornice is run; and there should be one-fourth of an inch at least of clear space between the preparation and the mould in the nearest part. A wooden screed or parallel straight edge is tacked with brads on to the wall, and another on the ceiling, if the cornice be large and heavy, as guides or gauges for the mould, whose rests are chased to fit them; and then one man laying on gauge stuff in an almost fluid state with an angular trowel, another works the mould backwards and forwards over it, which strikes off what is superfluous, and gives the converse of its form to the rest. The mould is never taken down from the work at right angles to the line of it, but is drawn off at the end, so that none of the parts of the moulding or cornice is injured or torn by it, which must otherwise frequently be the case, from the peculiar forms at times given to the details. If a cornice be too large and heavy to be executed at once, it may be done in the same manner at two or more times, in so many parts; and if any part or parts of a moulding or cornice is to be

Building. enriched, the space for it is left vacant by the mould, and the enrichment is afterwards supplied. As a cornice cannot be completed up to the angles by the mould, it is worked by hand in those situations to a joint. The joinings are termed mitres, and in forming them the plasterer uses the jointing tools we have already described. Models for enrichments are made by the modeller, according to the design or drawing submitted to him, and from them the plasterer makes wax moulds, or, as in ordinary practice, the modeller supplies the moulds in which the ornament is cast in plaster of Paris. If the ornament be in recurring lengths or parts, as is usually the case, only one length or part is modelled, and casts of as many as are required are taken from the mould; some single ornaments, again, which are very large, require to be moulded and cast in parts, which are put together by means of cement. When the cast ornaments are sufficiently dry the pieces are scraped and trimmed, the joints made clean and even, and they are set in the cornice with plaster of Paris, with white lead, or with a composition called iron cement, as the case may require. If the castings have something in the cornice to rest upon, the first will do; but if there is nothing to retain or attach them but the cement, one of the two latter must be used. Flowers and other ornaments in ceilings which are too large and heavy to be trusted to adhesive matter alone, must be screwed on to wooden cradling behind and above them.

In plastering a wall with common stucco (and its use is mostly for outside work), the first thing to be done is to remove the dust from it by brushing, and then wetting it very completely with water; if the wall to be stuccoed be an old one, or one of which the joints have been drawn, the mortar of the joints must be chipped or even raked out, and the bricks picked, to expose a new and porous surface to the plastering before brushing and wetting. The wall is then covered with stucco in a fluid state, applied with a broad and strong hog's-bristle brush, like common whitewashing. When this is nearly dry the stucco must be laid on as in common rendering, unless the work is to be floated, when the process is nearly similar to that in floated plastering. Screeds must be formed at the highest and lowest extremities of the wall, or of that part of the wall which is in the same vertical line, and is not intercepted by string courses, and be returned at the angles, putting the whole surface into a sort of frame. These must be made perfectly straight and plumb, so as to be quite out of winding, by the careful application of the plumb-rule and straight edge. Inner vertical screens must then follow at three or four feet apart across the whole surface, and be made to range exactly with the outer ones, and then the interstices must be filled in as before. As the work is made good it must be well rubbed with the hand float, as in the execution of trowelled stucco internally, to compress the material, and produce a hard, even, and glossy surface. Preparations for cornices and other projections from the straight surface of the work must have been previously made in or on the brick or stonework, by the protrusion of bricks, tiles, or whatever may be best suited to form a core, and the mouldings and cornices are run with moulds, in the manner described for the same things internally, only that in work of this kind no plastic material but the stucco itself is used; that is, there is no preparation of any softer material than the stucco itself put under it. In running cornices in this material, workmen are very apt to mix a little plaster of Paris with the stucco to make it set under the mould, and thus give sharpness and fulness to the mouldings; but this should not be permitted; for the plaster is not qualified to stand the weather as the stucco is, and, if mixed with it, will produce premature decay. When the stucco is perfectly dry, it may be painted in oil colours, or be coloured in distemper; and in either case it is generally ruled over the surface with a lead point, to give it the appearance of gauged stone-work.

Rendering in Roman cement is executed almost exactly in the same manner as stucco rendering is, only that it is laid on the saturated wall directly, without the preliminary operation of roughing in, or washing the surface with a solution of the material. The same process, too, is followed in floating this cement, and with the same exceptions; and as, in addition to its superior hardness and capacity for duration, it is a quick-setting cement, it is far preferable to any of the common stuccoes for running cornices, mouldings, &c. Roman cement, or as it is vulgarly called by most persons concerned in the operations of building, compo, a contraction of composition, may, like stucco, be painted in oil or coloured; but instead of a size colour, which is used for almost every other purpose in plastering, the colour for this composition is mixed with diluted sulphuric acid. This too may be lined and tinted to imitate stone and stonework of any description.

It may not be amiss here to refer to some of the causes of the premature decay which takes place in stuccoes and cements when used externally as a coating to walls. The primary cause is the presence of muddy earth and decayed animal and vegetable matter in the sand used with the lime and cement. To this may be added frequent impurities in the limes and cements themselves, particularly of argillaceous matter in the former, and sometimes to the too great proportions of lime or cement to sand. These things might, however, remain quiescent for a long time, if the work were well protected from access of moisture, which is the grand exciting cause. The paint, or distemper wash, on the surface, is generally sufficient to prevent the rain which may beat against a vertical face from penetrating, especially if the work have been well hand-floated and trowelled, to make it close and compact; but the evil arises from exposure above, and from the numberless horizontal unfloated surfaces which are constantly presented. These receive and collect the water, and convey in streams over the vertical surfaces what is not immediately absorbed; and the work thus becoming saturated, frost seizes and bursts it, or warmth calls the vegetative powers of the impurities in it into action, and the whole is covered with a green sward. Let the sand of which a plaster composition is to be formed, whether with lime or cement, be washed until it no longer discolours clean water, and be well compounded with cementitious matter free from the impurities with which it is so frequently charged; let the work be well hand-floated and trowelled, particularly on the backs or upper horizontal surfaces of projections, and protected above by projecting eaves or otherwise; and the work, with common care and attention to paint or distemper at intervals, will last as long as anything of the kind can be expected, or is found, to last anywhere.

A cheap and useful covering for external walls which are protected by projecting eaves, in plain buildings, is rough cast. This is executed in the following manner. The surface is first roughed in, or rendered with lime and hair; and when that is set dry, another coat of the same material is superadded, laid as evenly as it can be without floating, and as soon as a piece of two or three yards in extent is executed, the workman lays on it an almost fluid mixture of fine clean gravel and strong lime, which have been well mixed together. This is immediately washed with any ochreous colour that may be desired, and the whole dries into one compact mass.

In renovating and repairing plastering, the whole surface is first well washed to remove the dirt which may have attached itself, and as much of the earthy matter of the previous coat of whitening or colouring as will come away; any injuries the work may have received, such as cracks and fractures, are then repaired; and when the new stuff is quite dry, the joinings are scraped to produce an even surface, and the whole is again whitened or coloured once or twice, or oftener as may be required, to make it bear

Building. out well. Stuccoed walls which have been painted must be well rubbed with pumice stone, to take off the old paint as much as possible before they are newly painted.

Plastering is measured in feet and inches, and valued by the yard superficial of nine square feet. It is taken under separate heads according to the nature and description of the work, such as rendered; rendered and set; rendered, floated, and set; and with lath, for the lathing and plastering are valued together; lathed and laid; lathed, laid, and set; and lathed, plastered, floated, and set. Whitening and colouring are taken under separate heads, and the quantities of them are reduced to yards also. Work done in narrow slips, such as to the jambs and soffits of doorways and other openings, is measured by the foot superficial, and so are the backs of niches, niche-heads, &c. Arrises, or external angles and quirks, are taken extra by the running foot, and beads and other very small mouldings are measured in the same manner. Larger mouldings, however, and cornices, whether plain or enriched, are taken by the foot superficial, and the quantity is ascertained by multiplying the length, minus once the projection, by the girth, of the moulding or cornice, which is best determined by measuring its mould with a tape or cord. Enrichments are either numbered or taken at so much the running foot, making the modeller's an extra charge, if the design was original and required special modelling and moulding; and mitres are taken at so much a piece beyond a limited number. This number, in an ordinary room, is generally the four which necessarily occur in its four angles, making those which are usually occasioned by the projection of the chimney-breast extra; but it is not an uncommon practice to bring them within the limit, and count only all that may occur above eight, for no difference is made between internal and external angles. Circular work, whether it be convex or concave, of every kind, may be charged about one-fourth higher than straight. Stuccoes and other compositions are also valued by the yard, and according to the description of the work, with almost similar exceptions to those mentioned with regard to common plastering. Used externally, however, all the arrises or external angles, throatings, grooves, chamfers, &c., are taken as extra by the running foot at such a width.

In the practice of measuring plasterer's work, it is customary to take the whole surface at first, and then whatever deductions there may be. Thus the side of a room is measured over all, from the upper edge of the skirting grounds up to the cornice. The windows and doors are deducted by taking to the outside of their framed grounds for the width, and from the skirting grounds up to the top of those of the door or window for the height. If there be more than one of each, or either of them, to deduct, of course the same dimension will serve for all, multiplied by as many times as each deduction occurs. A ceiling also is generally taken over the whole surface, from cornice to cornice, a chimney-breast or other projection being made a deduction. It is a moot point whether the plasterer should not be allowed that part of the ceiling and wall which is covered by the cornice, as he has actually finished the whole except setting. When the cornice is bracketed, however, he may fairly claim up to the brackets.

Scaffolding is not generally made an extra charge with new work; but with old work it is, if scaffolding be necessary; for, under ordinary circumstances, the plasterer is enabled to wash, stop, and whiten the ceilings and walls of rooms from trestles, with boards laid across them. In lofty saloons and halls, churches, &c. scaffolding is indispensable, and must then be charged. A scaffold is necessary, too, to a front that is to be plastered in any way; but it may be afterwards washed, repaired, and coloured, from a ladder, without the intervention of a scaffold.

Slater.—The principle on which slates are laid is that which is employed in plain tiling. To a roof with project-

ing eaves, a wide board is placed over the rafters' feet; but Building. when the eaves tail into gutters, the gutter-board is made wide enough to receive the eaves-course. For light slating it is necessary to board a roof all over. This is done by the carpenter, and is called sound-boarding; but for strong heavy slates, fillets or battens are better; and these are laid by the slater himself, to suit the length of his slates. Three inches wide and one inch thick is a sufficient size for them, if the rafters be not more than twelve inches apart. Against gable or party-walls, a feather-edged board called a tilting fillet is laid to turn the water from the wall.

Before he begins to work on a roof, the slater shapes and trims the slates on the ground. With a large knife or chopper called a saixe, sax, or zax, he strikes off the unevenness on one side of a slate, making it as nearly straight as he can; he then runs a gauge along it, marking the greatest width the slate will bear, and, cutting to that line, makes it perfectly parallel. He next, with a square, brings the thickest and best end to right angles with the sides, generally by chopping, but sometimes by sawing; and then marking upward from the squared foot or tail, makes two nail holes, where, by calculating the gauge the slate in hand will bear, he knows the batten must come. All the slates being thus gauged to width, dressed, and sorted in lengths, they are then taken up to the roof in rotation, beginning with the longest and largest for the lowest courses. The first course the slater lays is little more than half the length of that which is intended to cover it, and is necessary to break the joints at the eaves. This is called the doubling eaves-course; and the covering eaves-course is brought to the same foot line, completely to cover it. Then to ascertain the gauge: From the length of the slate deduct the bond, which should never be less than two inches, and need not be more than three and a half inches, and the half of what remains will be the gauge. Thus, if the bond be fixed at three inches, and the slate is two feet three inches in length, the gauge will be one foot. This gauge or margin is set up from the foot of the eaves-course at each end, and a line strained to mark it along the whole length, and so on, to the ridge or top, where another half-course is required to complete the work, and that is in its turn secured by a covering of sheet lead. To a hipped roof care is taken to complete every course up to the angle, by cutting slates to fit its inclination; and these are also covered by an overlap of sheet lead. In nailing a slate, it must not be strained or bent in the slightest degree, or it will certainly fly in some sudden atmospheric change, to which it is of course constantly liable, even if it escape fracture, from being trodden on by the workmen themselves or by others. Copper, being less liable to oxidize from exposure to common causes than any other metal that will answer the purpose, is generally used for slate nails. Zinc is also used for the purpose; and iron tinned and painted nails are sometimes substituted by dishonesty on the part of the workman or builder, or bad economy on that of the proprietor.

A very light and neat covering is produced, by laying wide slates side by side, and covering their joints with narrow slips bedded in putty, the overlap at the ends being no more than the bond is with the usual mode. It is known as patent slating, and was introduced by the late Mr Wyatt, though he never obtained a patent for it. Indeed it is in principle the mode which was adopted in ancient Greece in covering the roofs of temples. Neither boards nor fillets are used, the slates bearing from rafter to rafter, and to the rafters the slates are screwed. The covering slips are also screwed, as well as bedded in putty. Slating of this kind may be laid at no greater elevation than ten degrees; whereas, for slating in the ordinary way, the angle should never be much less than twenty-five degrees, though large slates with a three and a half inch bond, carefully laid and pointed, may perhaps be trusted at a rise of twenty degrees.

Building. This mode of applying slate is not without the disadvantage attending the fixing of any substance that freely takes up and readily parts with heat. In expanding and contracting, the joints are too often destroyed and leaks are the common consequence.

The mode above described of ascertaining the gauge or margin by the bond, is equally applicable to every sort of roof-covering that is made up of small inflexible parallelogramic slabs or tablets; and it should be borne in mind that the greater the angle is at which the rafters rise, or, in technical language, the higher the pitch of the roof, the less the bond may be, and *vice versa*. With slabs or tablets that vary in length, too, as slates generally do in this country as they are brought to market, it is the bond which it is of importance to observe; but if they are of an invariable length, as tiles are, it is sufficient that the gauge or margin be attended to.

The best slate this country produces is from the quarries of Bangor in Caernarvonshire, and of Kendal in Westmoreland. Good slate is also procured in the neighbourhood of Tavistock in Devonshire, and in some parts of Scotland. The scantlings of slate are cut in the quarries to set sizes, and these are split into tablets, thicker or thinner according to the size of the slab and the capacity of the slate; for the inferior qualities are neither so compact in material, nor so clearly laminated or schistose, as the superior, and will not therefore rend so freely. The sizes of slates best known in the British market are distinguished by the names of ladies, countesses, duchesses, and queens. Ladies measure fifteen inches by eight, countesses twenty inches by ten, duchesses twenty-four inches by twelve, and queens thirty-six inches by twenty-four; and they are esteemed in proportion to their magnitude. Besides these, there is a slate which equals the queen in extent of surface, but is of very much greater thickness; this is called Welsh rag. A smaller slate, again, which is less indeed than the lady, and is cut from the refuse of large scantlings, is called a double. In size it does not often exceed twelve inches by six. Westmoreland slates are thick and heavy like the Welsh rag, but do not generally run so large.

The best slate is of a bluish-gray colour, and breaks before the zax like well-burnt pottery, and will ring in the same manner on being struck. Whitish or light gray-coloured slate is for the most part stony: dark blue or blackish slate, on the other hand, cuts very freely; but it absorbs moisture, and decays rapidly.

Slater's work is measured by the square of a hundred superficial feet. In a parallelogramic piece of slating, as in a gabled roof with projecting eaves, the length along the eaves by the breadth or height from that to the ridge, with the addition to the latter dimension of the gauge or margin for doubling the eaves, will give the quantity of one side. Projections for chimney-shafts or breasts, skylights, &c., must be deducted; but an addition must be made of the run round them by six inches, for cutting and waste. In a hipped roof the length from point to point of the eaves on one of the long sides of a quadrilateral roof, by the breadth or height, with the addition as before, will give that side and half of each of the ends. The other side will, of course, in the same manner, include the other halves of the ends. The length of the hips taken as a superficial dimension in feet, or by twelve inches, is added for cutting and waste, and valleys are taken and added in the same manner when they occur.

Carpenter.—For the scientific principles of carpentry we must refer the reader to the article under that head, and to the articles ROOF, STRENGTH OF MATERIALS, and TIMBER. Here we have merely to speak of the practical details of carpenter's work in the operations of building,—of *carpentering*, indeed, or the practice of carpentry, considering it as a mechanical art.

The carpenter works in wood, which he receives from the sawyer in beams, scantlings and planks, or boards, which he cuts and combines into bond-timbers, wall-plates, floors, and

roofs. He is distinguished from the joiner by his operations being directed to the mere carcass of a building,—to things which have reference to structure only. Almost everything the carpenter does in and to an edifice is absolutely necessary to its stability and efficiency, whereas the joiner does not begin his operations until the carcass is complete; and every article of joiners' work might at any time be removed from a building without undermining it or affecting its most important qualities. Certainly, in the practice of building, a few things do occur which it is difficult to determine to whose immediate province they belong; but the distinction is nevertheless sufficiently broad for general purposes. The carpenter, with the bricklayer or mason, and some of the minor artificers, constructs the frame or hull; and the joiner, with the plasterer and others, decorates and rigs the vessel: on the former the actual existence of the ship depends, and on the latter depends her fitness for use.

The carpenter frames or combines separate pieces of timber by scarfing, notching, cogging, tenoning, pinning, and wedging; and the tools he uses are the rule, the axe, the adze, the saw, the mallet, hammers, chisels, gouges, augers, hook-pins, a square, a bevel, a pair of compasses, and a gauge, together with the level and plumb-rule; besides these, planes, gimlets, pincers, a sledge hammer, a maul or beetle, wedges, and a crow-bar, may be considered useful auxiliaries, though they are not absolutely necessary to the performance of works of carpentry.

To scarf is to cut away equally from the ends, but on the opposite sides, of two pieces of timber, for the purpose of tying or connecting them lengthwise. This is done to wall-plates and bond-timber, and especially to beams when they are required of greater length than can be procured without joining. (See CARPENTRY.) The usual mode of scarfing bond and wall-plates is by cutting about three-fifths through each piece on the upper face of the one and the under face of the other, about six or eight inches from the end, transversely, making what is technically termed a *calf* or *herf*, and longitudinally from the end, from two-fifths down on the same side, so that the pieces lap together with a sort of half dovetail. The heavy supervening weight of the wall and joists renders it impossible that they should be drawn apart without tearing the fibres asunder or lifting the weight. (See fig. 20.) Nevertheless these joints are generally spiked, Pl. XLV. and it is always required that they be made to fall in or under a pier. Notching is either square or dovetailed; it is used in connecting the ends of wall-plates and bond-timber at the angles, in letting joists down on beams or binders, purlines on principal rafters, &c. Nos. 1, 2, 3, 4, and 5, fig. 21, show varieties of notches applied as we have described. No. 1 is a simple square notch or halving of the ends of bond-timbers or wall-plates at a right angle; No. 2, a dovetailed notch. No. 3, the notch most commonly used; it is similar to No. 1, but that the ends are allowed to run on so that the one piece grasps the other, and each forms a cog to the other. No. 4 is an oblique-angled, dovetailed notch; and No. 5 shows how joists are notched or let down on beams and binders, and purlines on principal rafters. A notch is cut into the under edge of the joist or purline an inch or an inch and a half in depth, and considerably shorter than the beam, binder, or rafter is in thickness. Notches are also cut down on the upper angles of the bearing pieces as long as the rider is thick, as deep as the notch before described of the latter is, and so far in as to leave a thickness on its own edge equal to the length of the notch in the riding joist or purline. In the diagram one joist is indicated in its place let down in the notch, and another indicates the notch in its own edge, and leaves exposed the notches in the binder. Cogging, or corking, as it is vulgarly termed, is the last-mentioned species of notch extended on one side, and leaving a narrow tooth or cog alone in the bearing-piece flush with its upper face, No. 1, fig. 22. It is used principally in tailing joists and beams on wall and tem-plates, and the cog is here made narrower, because the end of the joist or rider coming immediately beyond the plate, that part which forms the shoulder of the notch would be liable, on being strained, to be chipped off or torn away, if it were not kept as long as possible; and it is not of so much importance to guard against weakening a wall-plate which is supported along its whole length, as a beam, binder, or principal rafter, which rests on distant points alone. No. 2 of the same diagram shows another mode of tailing on joists and beams by a dovetail notch, which, to distinguish it from the flat notches, Nos. 2 and 4, fig. 21, is called corking, or cogging also, though the operation certainly is not cogging. This is a good mode if the timber be so well seasoned as not to be likely to shrink more; but it would be improved by allowing the rider to take a bearing in a notch like that to No. 1 before the dovetail commenced, as at No. 3, for in the ordinary mode it is weakened in a point of great importance.

Tenoning implies mortising also, as a matter of course. They are the names of the two operations necessary to one result,—that of producing a connection between two pieces by inserting part of the end of one into a hole of similar size cut in the side or face of the other. A tenon is formed by cutting in on each side or edge of a piece of timber, near its end, transversely, to a certain depth, or rather, leaving a certain part of the breadth or depth uncut, and then cutting in longitudinally from the ends as far from each edge as the transverse cuts have been made in depth, thus removing two square prisms and leaving a third undivided. This is the tenon. An excavation in the side of a piece of timber, of a certain depth, in the direction of its thickness, parallel to its edges, and bounded lengthwise by lines at right angles to them, is a mortise. Tenons and mortises are made of exactly corresponding size, and are most frequently at equal distances from one or the other side or edge of the two pieces to be conjoined; and for the most part, too, every angle formed in the process of tenoning, both internal and external, is a right angle. Tenons are called joggles in some situations, when they are not intended to be borne upon; and their use is merely to keep the piece of timber to which they belong steadily in its place, without being liable to slight accidents from lateral pressure or violence. In combining timbers by means of mortises and tenons, to produce as great a degree of strength as possible, it must be obvious that the object to be kept in view is to maintain the end or tenon of the one as large and efficient as it may be, and weaken the other as little as possible in forming the mortise. For the efficiency of the mortised piece in a horizontal bearing, it is clear that as much of its thickness should be below the mortise as possible, as at *a*, fig. 23; for if it be put low, as at *b*, the superincumbent weight on the tenon would more readily split or rend it in the direction of the grain, as indicated; but the case is inverted with the tenoned pieces. With the mortise at *a* the tenon could only have the efficacy of so much of the piece to which it belongs as there is of it above its under surface, which is a very small part of its depth, whereas with the tenon at *b* it would command the power of the greatest part of the depth. To guard as much as possible against the danger of too great a mortise and too small a tenon on one side and the other, and to obviate the difficulty arising from the efficiency of one or the other of the two pieces being affected by putting the tenon too high or too low, a compound, called a tusk tenon, is used for almost all horizontal bearings of any importance, especially to joists and binders, to trimmers, beams, girders, brestsummers, &c. The body of the tenon in this is a little above the middle of the end, and it runs out two, three, or four inches, or more, as the case may require. Below it the tusk protrudes, and above it the shoulder is cut down at an obtuse angle with the horizontal line, giving the strength of the whole depth of the timber above the under tusk to the

tenon, and giving it a bearing in a shallow mortise, whilst a greater depth of the mortised piece than the tusk rests on receives the body of the tenon, and so protects its comparatively narrow margin from undue pressure. The diagram No. 1, fig. 24, shows the tusk tenon, with the section of a beam into which it is mortised; and No. 2 indicates perspectively the appearance of the mortise in front. See also CARPENTRY.

Pinning is the insertion of nearly cylindrical pieces of wood or iron through a tenon, to detain it in the mortise, or prevent it from being drawn out by any ordinary force. For this purpose the pin is inserted either in the body, or beyond the thickness, of the mortised piece, as indicated at *a*, fig. 24, or at *a*, fig. 25. Wedging (see *bb*, No. 2, fig. 25) is the insertion of triangular prisms, whose converging sides are under an extremely acute angle, into or by the end of a tenon, to make it fill the mortise so completely, or bind it so tightly, that it cannot be easily withdrawn. The wedging of tenons also assists in restoring to the mortised piece of timber much of the strength it had lost by the excision of so much of its mass, which indeed the tenon itself does if it fit closely in every direction; but the assistance of the wedge renders the restoration more perfect than the tenon could be made to do of itself, by compressing the fibres of both, longitudinally to those of the one, and transversely to those of the other, thus removing the tendency of the mortised piece to yield in any degree in the weakened part, though it cannot make up the loss in its tenacity occasioned by the scission of its fibres.

In scarfing, cogging, and notching, the shoulders are always cut in with the saw; but the cheek is for the most part struck out with the mallet and chisel, or adze, as may be most convenient. Tenons should be made entirely with the saw: mortises are generally bored at the ends with an auger whose diameter equals their thickness; the intervening part is taken out with a wide chisel, cutting in the direction of the fibre; and the ends are squared down with a chisel whose breadth just equals the thickness of the mortise. Wood pins must be rent to insure the equal tenacity of their whole mass. Wedges are cut with the saw, but straight-grained stuff is always preferred for them.

Bond-timbers and wall-plates should be carefully notched together at every angle and return, and scarfed at every longitudinal joint. The scarf shown at fig. 20 is sufficient for the purpose; and the notch at No. 3, fig. 21, may be preferred where notching is required; neither pinning nor nailing, however, can be of great use to either the notch or the scarf. Bond-timbers are passed along and through all openings, and are not cut out until such openings are to be permanently occupied, that is, windows with their sash-frames, &c., because they assist in preventing irregular settlements, by helping to carry the weight of a heavy part along the substruction generally, instead of allowing it to press unduly upon the part immediately under it.

Whatever notches and cogs for beams and joists are required in wall and tem-plates, should be made before they are set on or in a wall; for, as they are always bedded in mortar, anything that may break the set must be avoided.

It is the duty of the carpenter to supply the bricklayer or mason with wood bricks in sufficient quantity, and to direct him where they should be placed to receive the joiner's fittings, or the battening, which the carpenter himself may have to put up for the plasterer.

The framed quartering partitions which may be required should be set up in every story before the beams and joists of the floors are laid, that their horizontal timbers may be notched on to the wall-plates, and that the joists or binders may be notched on to them if occasion require it; but they should be fixed rather below than above the level of the wall-plates, because they are not liable to settle down so much as the walls, though even that will depend in a great degree on the nature of the walling, and its liability to yield.

The carpenter makes and fixes or sets centres of all kinds, whether for single arches, or niches. The striking out of the centres, in the first instance, is necessarily contingent on the arches to be turned on them, for the forms of which the carpenter must look to the bricklayer or mason, whose instructions for describing arches will be found under the head STONE-MASONRY. Large centres are framed in distinct ribs, and are connected by horizontal ties; whilst small ones are made of mere boards cut to the required sweep, nailed together, and connected by battens notched into or nailed on their edges. Precision and stability are nevertheless equally and absolutely necessary, as it is impossible for an arch to be turned or set correctly on an incorrect or unstable centre.

The timbers or frame-work of floors is called naked flooring, and it is distinguished as single, double, and framed. Of these the first, under ordinary circumstances, is the Pl. XLVI. strongest. Single flooring (See No. 1 and 2, fig. 29) consists of one row or tier of joists alone, bearing from one wall or partition to another, without any intermediate support, receiving the flooring boards on the upper surface or edges of the joists, and the ceiling, if there be one, on the lower. Joists in single floors should never be less than two inches in thickness, because of their liability to be split by the brads or nails of the boards if they are thinner; and they should never be much more, because of the keying of the ceiling, which is injuriously affected by great thickness of the joists. Twelve inches from joist to joist is the distance generally allowed; that dimension, however, from centre to centre of the joists would be better. Strength to almost any extent may be given by adding to the depth of the joists, and diminishing the distance between them; and they may be made firm, and be prevented from buckling or twisting, by putting struts between them. These struts are short pieces of batten, which should not be less than an inch, and need not be more than an inch and a half thick, and three or four inches wide, placed diagonally between the joists, to which they are nailed, in a double series, or crossing, as indicated by the full and dotted lines in the diagram, fig. 26; and they should be made to range in a right line, that none of their effect may be lost; and these ranges or rows should be repeated at intervals not exceeding five or six feet. The struts should be cut at the ends with exactly the same inclination or bevel, to fit closely. Great care should be taken, too, not to split the struts in nailing; but the trouble of boring with a gimlet is saved by making a slight nick or incision with a wide-set saw for each nail, of which there should not be less than two at each end; and the nails used should be clasp-nails. If the struts were notched into the joists, it would add very materially to their efficiency, but perhaps not in proportion to the additional labour it would involve. This strutting should be done to single flooring under any circumstances, as it adds materially to its firmness, and indeed to its strength, by making the joists transmit any stress or pressure from one to another. The efficiency of single flooring is materially affected by the necessity which constantly occurs in practice of trimming round fire-places and flues, and across vacuities. Trimming is a mode of supporting the end of a joist by tenoning it into a piece of timber crossing it, and called a trimmer, instead of running it on or into the wall which supports the ends of the other joists generally. A trimmer requires for the most part to be carried or supported at one or both of its ends by some of the joists, which are called trimming joists, and are necessarily made stouter than if they had to bear no more than their own share of the stress. Commonly it is found enough to make the trimmers and

trimming joists from half an inch to an inch thicker than common joists. In trimming, tusk tenons should be used; and the long tongue or main body of the tenon should run not less than two inches through, and be draw-pinned, and wedged, moreover, if it do not completely fill the mortise in the direction of the length of the latter. The principal objection, however, to single flooring is, that sound readily passes through, the attachment of the boards above and of the ceiling below being to the same joints throughout. Another objection, and one already referred to, is the necessity of making the joists so thin, not to injure the ceilings, that they with difficulty receive the flooring brads in their upper edges without splitting. A partial remedy for both these disadvantages is found in a mode sometimes adopted of making every third or fourth joist an inch or an inch and a half deeper than the intervening joists; and to these, ceiling joists are notched and nailed, or nailed alone, as shown in the diagram, fig. 26. This, by diminishing the number of points of contact between the upper and lower surface, for the ceiling joists must be carefully kept from touching the shallower joists of the floor, is less apt to convey sound from one story to another, and allows conveniently thin joists to be used for the ceiling without affecting those of the floor.

Double flooring (see sections No. 1 and 2, fig. 27, and plan No. 3, fig. 29) consists of three distinct series of joists, which are called binding, bridging, and ceiling joists. The binders in this are the real support of the floor; they run from wall to wall, and carry the bridging joists above and the ceiling joists below them. Binders need not be less and should not be much more than six feet apart, that is, if the bridging or flooring joists are not inordinately weak. The bridging joists form the upper tier, and are notched down on the binders with the notch shown at No. 5, fig. 21. The ceiling joists range under the binders, and are notched and nailed as shown at No. 1, fig. 27; but the notch must be taken entirely out of the ceiling joists, for the lower face or edge of the binder may not be wounded by any means or on any account, and moreover no good would be gained in any other respect by doing so. When it is an object to save height in the depth or thickness of this species of floor, the ceiling joists may be tenoned into the binders, instead of being nailed on to them; in this case the latter must be chase-mortised on one side, for the convenience of receiving the former when they are themselves set and fixed. A chase is a long wedge-formed groove of the breadth or thickness of the mortise, of which it is indeed an elongation, so that the tenon at one end of a ceiling joist being inserted in the regular mortise in the binder prepared for it, that at the other end is driven along the chase up to its place in the mortise in the next binder. When ceiling joists are thus chase-mortised, their lower or under faces are allowed to come a little below the under face of the binders, and the space across is firred down by slips not wider than the ceiling joists are thick. No. 2, fig. 27, shows a transverse compartment, or bay, of a floor in this manner; but it is not so good a one as the preceding; for, besides weakening the binders, by cutting so many mortises and chases in them, it is almost impossible to give the ceiling floor the degree of firmness and consistency it possesses in the other way, besides requiring the firring down on the binders. The same space would be better gained by cutting the bridging joists so much lower down; as they may, with the sort of notch indicated above, be let down fully half their depth without great injury to either bridging joists or binder, for they can always be made to fit tightly or firmly, and very little more labour is involved in notching deeply than slightly.

Flooring is said to be framed when girders are used together with binding, bridging, and ceiling joists. (See sections No. 1 and 2, fig. 28, and plan No. 4, fig. 29.) Girders are large beams, in one or more pieces, according to the length required, and the size and strength of which timber can be procured. They are intended for longer bearings than mere binders may be trusted at, and may be strengthened to almost any extent by trussing; but to be efficient, the height of the truss must always be greater than the depth of the beam itself, and the strength is increased by extending that height as the space or bearing increases. A truss is indeed a wooden arch, whose lateral thrust will of course be greater the smaller the angle subtended by it, and *vice versa*. It has been a commonly received opinion, that a truss within the depth of a girder adds materially to its strength; but experiments have proved that very little advantage is gained by such a one when executed in the best manner, and that, badly executed, the beam or girder is weaker with the truss than without it. Binders are made dependent on the girders by means of double tusk tenons, and on and to them the bridging and ceiling joists are attached in the manner before described. No. 1, fig. 28, shows the transverse section of a compartment or bay of a framed floor; No. 2 the same longitudinally of the girder, and of the bridging and ceiling joists, and transversely of the binders. No. 1, fig. 29, is the plan of a single floor of joists tailing in on wall-plates with two chains of struts, and trimmed to a fire-place. No. 2 is a floor similar to No. 1, with ceiling joists nailed to deeper flooring joists at intervals, as shown in fig. 26. No. 3 is the plan of a double floor; and No. 4 is that of a framed floor of joists, bays of which are shown in section at figs. 27 and 28. It is to be observed, however, with reference to the diagram No. 1, fig. 28, that binders ought not to be framed into the girders opposite to one another, as they are here shown to be as a matter of convenience, since the girder is unduly weakened by being mortised on both sides at the same place. Cast-iron shoes render mortising the one forming a tenon upon the other almost unnecessary; and in like manner cast-iron shoes laid into a wall upon stone tem-plates give a good and safe bearing to the girders; but it is not everywhere that cast-iron shoes are attainable, and mortises and tenons may be made anywhere.

Partitions of timber are called quartering partitions, and they are generally framed. Common quartering partitions which rest on a wall or floor, and have nothing to carry, consist merely of a sill, a head, and common uprights to receive the lath for plastering: these last may be simply joggled or tenoned into the head and sill, in the manner shown at *c*, fig. 23, and stiffened by struts or stretching pieces put between them and nailed. When, however, a quartering partition is over a vacuity, or rests only on certain points, and has, moreover, to sustain a weight, a floor perchance, it is framed and trussed with king or queen posts and trussing pieces as to the tie beam of a roof; and the filling in of common uprights or quarters for the laths is generally performed by joggling them at one end into either head or sill, and nailing them securely to the trussing pieces. In the diagram No. 1, fig. 30, it is supposed that an opening or doorway is to be made in the partition, so that the timbers of the truss are placed around it with queen-posts, and a small internal truss is put over the door-head to prevent it from sagging, and to carry the long part of the partition, which we supposed required to bear a floor, so that the partition acts also, in fact, the part of a trussed girder in the most available form. No. 2 presents another method of framing a similar partition.

Shoring or propping up walls or floors, or it may be a whole building, is done by the carpenter. In appearance it is a simple operation, and under ordinary circumstances it really is so; but nevertheless it often demands the exercise of considerable skill and tact to determine and to counteract the tendency the part or thing to be supported has in one direction or another.

Pugging floors, firring down joists, and bracketing and cradling for plastering, and some other things, are operations performed indifferently by the carpenter or the joiner, as less or greater precision is required in the performance.

The labour of carpenter's work is valued by the square of one hundred superficial feet whenever it will admit of being so measured, and the timber is as generally valued by the cubic foot. It is customary for the carpenter's work to be measured at the same time with the walls and roof covering, or when the carcass of a building is completed, and before the joiner and plasterer commence their operations; for then the work is still exposed, and may be fully and correctly ascertained, whereas much must be taken on trust if the measurement be deferred until the works are completely finished.

Bond timber, wood bricks, and wall and tem-plates, are taken under the same head, and are reduced to cubic feet of timber at so much per foot, including the labour of every kind on it. The naked flooring is taken on the surface from wall to wall, with a description of the nature of it, whether it be single, double, or framed—if trimmed to chimneys, party walls, stairs, or anything else—if notched or cogged to wall-plates and partition heads—the number and size of the large timbers, ceiling joists as notched and nailed to wall-plates, and as framed or notched and nailed to binders or common joists; and everything indeed that affects the quantity of labour required in forming it. The superficial feet are reduced to squares for the labour and nails involved and used in forming and fixing or setting the floors. The timbers of which the flooring is composed are then taken in detail and in cubic quantities, and are said to be without labour, or with no labour. Roofing is measured in the same manner, by the superficial square, for labour and nails, taken on the common rafters from ridge to heel; the length of a rafter by the length of the roof for one side of a common span, and repeated or doubled for the other, noting also a description of the roof, whether it be lean-to or shed roofing, if on purlines and with struts; common span-roofing; curb roofing; span roofing with purlines and collar beams, strutted or otherwise, from walls or partitions; span roofing with framed principals, tie-beams, king-posts or queen-posts, straining beam, straining sill, struts, purlines, pole-plates, and so on or as the case may be, and this too for labour and nails. All the timbers are then taken, measuring every one to the extent of any tenon or tenons at its ends, in cubic quantities also, and as without labour. Bolts, bars, straps, stirrups, &c. are taken separately, and their dimensions noted from which to ascertain their weight. Gutter-boards and bearers are measured and valued by the foot superficial, according to thickness of the former. Rough boarding for lead on flats, and sound boarding for slates or lead, are taken superficially, and reduced into squares. Centring to vaults is measured on the periphery of the arch, or round back of the centre, for the breadth, by the length, and is valued by the square; to apertures in the thickness of walls, by the foot, and to camber-arches, by number, so much a piece. Quartering partitions are measured by the square for labour and nails, and the material is taken by the cubic foot. Battening to walls is also measured by the square, but the stuff is generally included with the labour, as in boarding. Cradling and bracketing is valued by the foot superficial, and with reference to the quantity of stuff required or worked up. Any planing that may have been necessary, and it will happen at times on beams, joists, &c. when it is not intended to have a ceiling under the floor, is charged by the foot on the surface, and any beading or other moulding by the foot running.

It sometimes happens that a superficial quantity for labour and nails on framed timber cannot be obtained; in that case tne timoer is measured by the cubic foot as framed, or with the labour of framing included with its own cost, &c. In this case, however, it is necessary to make a distinction between one quantity and another, as the labour employed upon an equal quantity of stuff in framing some parts of a roof is much greater than is required in most floors. Many things, such as strong door and window frames, that are to be worked into the walls, story-posts, brestsummers, &c. are always taken as framed timber, with any addition that may occur of wrought, rebated, beaded, &c., as the case may be.

The price or value to be attached to the varieties of carpenter's work depends almost as much on the degree of hardness of the timber employed, as on its cost. What the timber itself should be charged at may be thus determined. To its price in the gross at the timber merchant's must be added the cost of carriage to the spot where it is to be employed, which will be so much the load of fifty cubic feet, or so much per foot; then to the cost of each cubic foot of timber add the price of four superficial feet of sawing, which will form a fair average for the variously sized scantlings, and one-eighth of the increased amount to it as an allowance for waste in cutting up and working. This gives the actual cost, to the builder, of the timber as it is worked up; and if it is to be charged as with no labour, his profit and remuneration for his own labour of superintending, &c., alone remain to be included. If, however, labour of any kind is to be charged with the stuff, it should be added first, and the builder's profit, &c., taken on both, or on the increased amount for the price per foot. The cost of labour depends so much upon such a variety of circumstances, that it is impossible to aid the inquirer materially in apportioning prices for the various operations. In this, as in other things, it is well, when the parties are not otherwise qualified to determine a scale of charges, to observe the time a man or a certain number of men are employed in executing so much work of a certain description, and compare the quantity by measurement with the time employed in executing it, or rather with the wages of the workmen for the time. In fixing a price for labour in carpenter's work, the size of the timbers, and the heights they have to be hoisted, together with such scaffolding and machinery for hoisting as may be found necessary, if the timbers be heavy, and the height and expense great, must be considered. As the timber used in shoring is not consumed, a charge is made for use and waste to the amount of one-third of its value if it be much cut up, and one-fourth if but little, in addition to the labour of setting up and taking down, whatever that may be.

Joiner.—The principles of joinery also will be found in an article under that head in another part of this work; here we have merely to do with the modes of operation, and the tools employed by the workman, together with the manner of estimating or ascertaining the value of his work.

The distinction between the operations of the carpenter and the joiner is shown at the beginning of the preceding section on the trade of the former. A man may be a good carpenter without being a joiner at all; but he cannot be a joiner without being competent, at least, to all the operations required in carpentry. It is, indeed, very truly remarked in the article JOINERY, "that the rough labour of the carpenter renders him in some degree unfit to produce that accurate and neat workmanship which is expected from a modern joiner;" but it is no less true that the habit of neatness and the great precision of the joiner, make him a much slower and less profitable workman than the practised carpenter, in works of carpentry.

The joiner operates on battens, boards, and planks, with saws, planes, chisels, gouges, hatchet, adze, gimblets, and other boring instruments, which are aided and directed by chalked lines, gauges, squares, hammers, mallets, and a great many other less important tools: and his operations are principally sawing and planing in all their extensive varieties, setting out, mortising, dovetailing, &c. A great range of other operations, none of which can be called unimportant, such as paring, gluing up, wedging, pinning, fixing, fitting, and hanging, and many things besides which depend on nailing, &c., such as laying floors, boarding ceilings, wainscotting walls, bracketing, cradling, fiering, and the like. In addition to the wood on which the joiner works, he requires also glue, nails, brads, screws, and hinges, and accessorily he applies bolts, locks, bars, and other fastenings, together with pulleys, lines, weights, white-lead, hold-fasts, wall-hooks, &c. &c.

Battens are narrow boards running from half an inch to an inch and a half or two inches thick, and from three to six or seven inches wide. A piece of stuff of too small a scantling to be a batten is called a fillet. The term board is applied to sawed stuff when its width exceeds that of a batten, and its thickness does not exceed two inches or two inches and a half. The term plank is applied to large pieces of stuff whose width is great in proportion to their thickness, and whose thickness nevertheless does not exceed three or four inches. In London these terms are used in much more restricted senses than they are here described to mean, because of the fixed and regular sizes and forms in which stuff for the joiner's use is for the most part brought to market there. A batten, to a London joiner, is a fine flooring board from an inch to an inch and a half in thickness, and just seven inches wide. A board is a piece cut from the thickness of a deal whose width is exactly nine inches; and everything, almost above that width, and not large enough to be called a scantling of timber, is a plank.

The joiners' work for a house is for the most part prepared at the shop, where every convenience may be supposed to exist for doing everything in the best and readiest manner; so that little remains to be done when the carcass is ready, but fit, fix, and hang, that is, after the floors are laid. The sashes and frames, the shutters, back flaps, backs, backs and elbows, soffits, grounds, doors, &c., are all framed and put together, that is, wedged up and cleaned off, at the shop; the flooring boards are prepared, that is, faced, shot, and gauged with a fillister rebate; and all the architraves, pilasters, jamb linings, skirtings, mouldings, &c., are all got out, that is, tried up, rebated, and moulded, at the shop.

When the carcass of a building is ready for the joiner, the first thing to be done is to cut the bond timber out of the openings, set the sash frames, and fill them with old sashes or with oiled paper on frames, to exclude the weather, but admit light. The flooring joists are then proved with straight-edges, and any inequalities in them are removed with the adze; the flooring boards are next cut down to their places, and they are turned with their faces downwards until the ceilings are done; but first the pugging floors, if any are intended, are formed, and the pugged clay is put in on them. Floors are in ordinary cases either straight joint or folding, and are edge or face nailed. Folding floors are those in which three, four, or five boards are laid at a time, with their heading joints all on the same joist, and of course in the same straight line. In laying them, one board being firmly nailed to the joists at the extremity of the floor, another is laid parallel to it at the distance of the width of three or four others, or rather within their width, and these are then forced down and nailed, the forcing having brought all the joints up close. This is a bad mode, however, and should never been used. Straight joint flooring is when every board is laid separately, or one at a time, the heading joint or joints being broken or covered regularly in every case. Straight joint flooring may be with square joints, when it is entirely face nailed, or it may be dowelled or tongued, when it is side or edge nailed only. Dowelling is the driving pins of wood or iron half their length, into the edge of the last laid board, the outer edge of which has been skew-nailed, their other ends running into holes prepared for them in the inner edge of the next board, in the way the head of a cask is held together, and then its outer edge is skew-nailed in the same manner, and so on. Tongueing is effected by grooving both edges of every board, and fitting thin slips and tongues into them, as described in the article JOINERY. The boards are forced together by pressure applied to the outer edge. The nail used in face-nailing floors is called a flooring brad; it has no head, but a mere tongue projecting on one side of the top of the nail, which is put in the direction of the grain, that it may admit of being punched in below the surface level, otherwise the superficial inequalities could not be reduced when the floor was completed, because of the projecting heads of the nails. For side or edge nailing, however, clasp-nails, nails whose heads extend across on two of the opposite sides, are used.

Another early operation the joiner has to attend to, is the fixing of the framed door and window and the narrow skirting grounds (see fig. 35) to which the plasterers may float their work. The skirting grounds are generally dovetailed at the angles, and are well blocked out, so that they may not vibrate on being struck, or yield to pressure when the plasterer's straight-edge passes roughly over their surface; they must also be set with the utmost truth and precision. When the floors are cut down and the grounds fixed, the joiner's operations in a building should be suspended until the plasterers have finished, or nearly so, and then the floors may be laid. By deferring this operation until that period, the workmen of the two different trades are prevented from interrupting each other, and indeed injuring each other's work; and joiners always find employment in the shop preparing, as before intimated.

The preparation flooring boards receive, is planing on the face, shooting on the edges, and gauging to a thickness; the common fillister, or stop rebate plane, being used to work down to the gauge mark, from the back of every board, and about half an inch in on each edge. When a board is to be laid, it is turned on its face in the place it is to occupy, and the workman with his adze cuts away from the back over every joist down to the gauge rebate, so that on being turned over it falls exactly into its place, and takes the same level with all its fellows, which have been brought to the same gauge; then follows the process of laying as before described, and the result must, if the work be done well, be a perfectly even and level surface. The slight inequalities of surface which may occur are reduced with a smoothing-plane, the brads being previously punched below the surface if the floor be face-nailed. See the article JOINERY, sections 35 and 36.

In getting out skirtings, if the work be of a superior description, the boards should be tried up as if for framing in every particular except bringing to a width, which need not be done. The face edges, however, must be worked with great precision, and moulded or rebated as the case may require. Rebating or tongueing will be necessary when the skirting consists of more than one piece, that the different pieces may be made to fit neatly and firmly together; and all but the lowest piece must of necessity be brought to a width, as well as tried up in other particulars. A skirting in a single width is called by that term; but when it is made up of more than one part it is designated a base: the lowest board is then called the skirting board, and the upper the base moulding or mouldings. (Figs. 31 and 35.) The reason why the skirting board is not brought to a width is, that the labour would be lost according to the ordinary mode of fixing it. The board is applied to its place with its lower edge touching the floor; but as the most perfectly wrought floors are found to have some slight unevenness of surface so close to the wall, a straight edge would not fit closely down to it in every part. The board is therefore propped up at one end or the other until the upper or faced edge is perfectly parallel with the average line of the floor, or rather to be perfectly level. A pair of strong compasses, such as those used by the carpenter, is taken, and opened to the greatest distance the lower edge of the skirting board is from the floor throughout its length; the outer edge, near the point of one leg of the compasses, is then drawn along the floor, whilst the point of the other, being kept vertically above it, is pressed against the face of the board, on which it marks a line exactly parallel to the surface of the floor, indicating, of course, every, even the slightest irregularity there may be in it. If the floor be not a very uneven one, the excluded part may be ripped off with the hand or the panel saw, which may generally be made to follow the traced or inscribed line exactly; if, however, the line be a very irregular one, having quick turns in it, the hatchet must be used. This operation is called *scribing*, and the result of it is evidently to make the skirting fit down on the floor with the utmost precision. Care must be taken, in performing the operation, that the upper edges of the skirtings be not only level, but that all which are in immediate connection be scribed to the same height, that their upper edges may exactly correspond. Sometimes skirtings are let into a groove in the floor, as indicated in the diagram, fig. 35, and thus a slight degree

of shrinking is made of less importance, and scribing rendered unnecessary. Before skirtings are fixed, vertical blocks are put at short intervals, extending from the floor to the narrow grounds, and made exactly flush with and true to the latter, and are firmly nailed. These form a sound backing, to which the skirtings may be bradded or nailed; and so prevent them from warping or bending in any manner. If, however, the skirting be not very wide, and be sufficiently stout to stand without a backing, a fillet only is nailed along the floor as a stop for its lower edge; but this is rendered unnecessary if the skirting be tongued into the floor, as the tongue will answer every purpose of a stop. The ends of skirtings should be tongued into each other when it is necessary to piece them in length; and on returns or angles the end of one should be tongued into the returned face of the other in the square parts, and mitred in the oblique-angled or moulded parts.

When a chair-rail or surbase is required, grounds similar to those for the base are fixed to range like them with the face of the plastering; the surbase itself must be wide enough to cover the grounds and the joints formed by them and the plastering, completely; it is in effect a cornice to the stereobate and the space intervening it and the base is generally understood to be wainscotted, though it is more frequently plastered.

In framing or framed work, the outer vertical bars which are mortised are called styles; and the transverse, those on whose ends the tenons are formed, are called rails. (Fig. 32.) In doors, particularly, the open spaces or squares formed internally by the rails and styles are divided in the width by bars parallel to the styles. These are tenoned into the rails, and are called mountings, or, vulgarly, *muntins*. The frame being formed by trying up, setting up, mortising, and tenoning, the inner or face edges of the styles, and of the highest and lowest rails, and both edges of the muntins and of the inner rails, are grooved with the plough to receive the edges and ends of the filling-in parts, or panels of the frame-work. Panels are either flat, raised, or flush. (Fig. 33.) Flat panels are no thicker than the grooves into which they are fitted, and consequently their faces are as much below the surface of the framing as the groove is in from each side of the styles and rails. Raised panels are thicker than the groove in the framing, but are not so thick as to reach the surface; nor is the panel thickened through its whole extent. It fits exactly into the groove, and thickens gradually for an inch or two, and then sets off at a right angle with the surface, increasing suddenly three or four sixteenths of an inch. A panel may be raised on one side only, or on both sides. Flush panels are rebated down from one face to the distance the plough groove is in from the surface of the framing; and the back of a panel thus rebated on one side is worked down to be even with the other edge of the groove, leaving a tongue to fit it exactly; for if it be required to make panels flush on both sides, it is generally effected by filling in on the back or flattened side with an extraneous piece. Framing is not, however, often finished in the manner above described, especially with raised and flush panels; mouldings are generally introduced, and are either struck or worked in the solid substance of the framing, or in separate pieces or slips, and laid in with brads. If a moulding be struck or laid in on one side only, and the other is left plain, the framing is described as moulded and square, a flat panel being in that case understood; if the panel be raised the framing will be described as moulded with a raised panel on one side, and square or flush on the other. It may be moulded with a flat panel, or moulded with a raised panel, on both sides; and the moulding may, as before intimated, be either struck in the solid, or laid in any of the preceding cases. Mouldings which are laid in round the panels of framing are neatly mitred at the angles, and bradded, to appear as much as possible as if they were struck in the solid. In nailing or bradding the mouldings, the brads should be driven into the frame-work, and not into the panels. With a flush panel, however, the moulding is always either a bead, or a series of beads called *reeds*; and is, in the case of a single bead, which is most common, always struck on the solid frame, and the work is called bead-flush; but reeds are generally struck on the panel in the direction of the grain, and laid in on the panel across it, or along the ends; this is termed reed-flush. Flush panels in inferior works have a single bead struck on their

Plate XLVII.

sides in the direction of the grain alone, the ends abutting plainly, as in the first diagram of a flush panel, and this is termed bead-butt, the fact that the panels are flush being inferred. The plainest quality of framing, in which it is square on both sides, is used in the fittings of inferior bed-rooms, inner closets, and the plainer domestic offices, but always internally; framing moulded on one or both sides, in rooms and places of a greater degree of importance, and in places where the work may be more generally seen; in some cases a flat panel may be enriched by a small moulding laid on its surface, leaving a margin between it and the larger moulding at its extremities; this may be done in drawing-rooms and apartments of that class, especially if they be in an upper story; and raised panels should be confined to the framed fittings of dining-rooms and other apartments on a ground or principal story. Framing with flush panels is almost restricted to external doors, &c., one side of a door being bead-flush, and the other flat and moulded, perhaps, or the face may be moulded with a raised panel, and the back-bead flush; and this for principal entrances. Bead-butt framing is found in external doors to offices, &c. Doors are made four-panelled for the most part when the panels are flat and the framing square, six-panelled when the latter is moulded, and six, eight, or even ten panelled when the framing is of the superior descriptions. Doors which are hung in two equal widths to occupy the doorway, and are hung to the opposite side posts or jambs of the frame, are said to be double-margined; that is, the styles or margins are repeated necessarily in the middle where they meet. Doorways are fitted with jamb linings, and architraves or pilasters. Jamb linings may be framed to correspond with the door on the outer faces; and when they exceed nine or ten inches in width they should always be so, or they may be solid. Narrow and plain jamb linings to inferior rooms are rebated on one side only, and the rebate forms the frame into which the door is fitted. To superior work they are rebated on both sides, as if it were intended to put a door on each side. The jambs are fixed to the inner edges of the grounds; and if they are wide, and not framed, backings are put across to stiffen them; and these backings are dovetailed into the edges of the grounds. Architraves and pilasters are variously sunk and moulded, according to the fancy of the designer. They are fixed to the grounds with their internal edges exactly fitting to the rebates in the jambs, and they form the enriched margin or moulding of the frame in which the door is set. Architraves are mitred at the upper angle, but pilasters have generally a console or an enriched block or cap resting on them, to which they fit with a square joint; both the one and the other either run down and are scribed to the floor, or rest on squared blocks or bases, which may be the height of the skirting board, or of the whole base.

The parts of the outside frame of a sash are distinguished by the terms applied to the similar parts of common framing. The upright sides are styles, and the transverse or horizontal ones, which are tenoned into the ends of the styles, are rails; but the inner frame-work or divisions for the panes are called merely upright and cross bars; the upright being the mortised, and the cross bars the tenoned, nevertheless, as with the outer frame-work. (Fig. 31.) Sashes are got out like common framing; the parts are tried up, set out, mortised and tenoned, exactly in the same manner, allowance being made in the length of the rails and all tenoned pieces, in the setting out, as in common framing also, for the portions of the mortised styles and upright bars, which are worked away in forming the moulding and rebate. The meeting rails of sashes which are in pairs, to be hung with lines, are made thicker than the other parts by the thickness of the parting bead, and they are bevelled or splayed off, the one from above and the other from below, that they may meet and fit closely. When the framework is completed, although it cannot be put together because of what has just been referred to, the rebate is formed by the sash fillister on the further part of the face edge, and the moulding struck on its hither angle. These things being done, the moulded edges are either mitred or scribed at the shoulders and haunches, and the sash may be put together. If sash bars are mitred at the joints, they require dowels in the cross bars to act as tenons; but if they can be scribed, dowelling is not necessary. Sashes are either hung upon hinges or hung with lines, pulleys, and weights. Fixed sashes are put into frames, of which every part may be solid but the stop, which must be put in behind the sash to detain it. Sashes hung with hinges require solid rebated frames; but there can be no stops to them except their own moveable fastenings, and the outer stop, which of course the rebate furnishes. Sashes hung with lines require cased frames to receive the pulleys and weights. The sill of the frame is made, as in the former cases, solid, is sunk and weathered, and is generally made of a more durable material than the rest of the frame; the sides in the direction of the thickness of the frame are of one and a quarter or one and a half inch board, very truly tried up, and grooved to receive a parting bead; for it must be obvious that sashes hung with lines to run vertically up and down within the height of the frame must be themselves in two heights, and must pass each other in two separate and distinct channels. The ends of these boards are fixed into the upper face of the solid sill below, and into a similar board parallel to the sill which forms a head above; and they are called pulley pieces, or styles, because they receive the pulleys, which are let into them near their upper ends. Linings from four to six inches in width, and from three-fourths of an inch to an inch in thickness, are nailed on to the edges of the pulley pieces, and to the sill and head above and below, inside and outside in the direction of the breadth of the sash frame, and are returned along the head in the direction of its length. The outside linings are made to extend within the pulley pieces about half an inch, to form a stop for the upper and outer sash; and the inside linings are made exactly flush with their inner faces. The casing is completed by fixing thin linings on to the outer edges of the outside and inside linings, parallel to the pulley pieces, to prevent any thing from impeding the weights. Thin slips called parting beads are fitted tightly into the grooves previously noticed in the pulley pieces, but they are not fixed, as the upper sash can be put in or taken out only by the temporary removal of the parting bead. An inner or stop bead is mitred round on the inside to complete the groove or channel for the lower sash; the stop bead covers the edge of the inside linings on the sides and head, and is fixed by means of screws, which may be removed without violence when it is required to put in or take out the sashes. A hole covered with a moveable piece large enough to allow the lead or iron weight to pass in and out, is made in each of the pulley pieces, so that the sashes may be hung after the frames are set, and to repair any accident that may occur to the hangings in after-use. (Fig. 34.) It may be here remarked, that sash-frames require greater truth and precision from the workman than anything else in the joiners' work of a building; and unless the stuff employed be quite sound and perfectly seasoned, all the workman's care in operating will be thrown away. The fittings of a window which has boxed shutters consist of back linings, grounds, back, elbows and soffit, together with shutters and back flaps, and architraves or pilasters round on the inside to form a moulded frame. (Figs. 31 and 34.) Back linings are generally framed with flush panels; they fit in between the inside lining of the sash frame and the framed ground, to both of which they are attached, and form the back of the boxing into which the shutters fall back. They are tongued into the inside lining by their inner edge, and on their outer edge the ground is nailed, and they are set at right angles to the sash-frame, or obtusely outwards, as the shutters may be splayed or not. The back is the continuation of the window fittings from the sash-sill to the floor on the inside; the elbows are its returns on either side under the shutters, and the soffit is the piece of framing which extends from one side of the window to the other, across the head, or from back lining to back lining. These are all framed to correspond with the shutters on the face; but, as they are fixed, their backs are left unwrought. Window shutters are framed in correspondence with the door and other framed work of the room to which they belong, in front, and generally with a flush panel behind: the back flaps are in one or two separate breadths to each shutter, according to the width of the window and the depth of the recess; they are made lighter than the shutters themselves; and they should, when shut to, present faces exactly corresponding with those of the shutters, both internally and externally. The shutters are hung to the sash-frame with butt hinges, and the back flaps are hung to their outer styles with a hinge called a back-flap, from its use. The shutters and their back flaps are hung in one, two, or more heights, as may be found convenient. The moulded margin round the boxings of a window on the inner face are made to harmonize generally with the similar parts of the doors of the room or place to which it belongs. The fixing and hanging of window fittings or dressings are hardly less important, for the accuracy required, than the making and fixing of the sash-frame itself; the slightest infirmity or inaccuracy in any part will be likely to derange some essential operation. Sashes, it may be remarked, are never fitted until the frames are immoveably fixed, so that if there be any inaccuracy in the latter, the sashes are cut away or pieced out to make them fit; but, as they are intended to traverse, the fitting in that case can only apply to one particular position, and in every one but that there must be something wrong. Any incorrectness in the sash-frame, again, must throw the shutters and their back flaps out; indeed the sash-frame, though apparently a secondary part of the arrangement, is that which affects all the rest beyond anything else. When sashes have been fitted, a plough groove, wide and deep enough to receive the sash-line, is made in the outer edges of the styles, for about two-thirds of their length, at their upper ends. They are then primed and glazed, and when the putty is sufficiently set the joiner hangs them. He is furnished with sash-line, tacks, and iron or lead weights, which are generally made cylindrical, with a ring at one end, to which the line may be attached. A sash is weighed, and two weights are selected which together amount to within a few ounces of a counterpoise. The line is then passed through the pulley, which was previously fixed in the pulley style; the end is knotted to a weight which is passed in at the hole left for the purpose, and at a sufficient distance, which a common degree of intelligence will readily determine; the line is cut off and the end tacked into the groove in the style of the sash.

Glue is used principally in putting framed work together, but not at all in fixing; and even for the former purpose it is much less used by good workmen than by bungling hands. When the stuff is well seasoned, and the trying up, setting out, mortising, and tenoning, are well and accurately executed, there is no necessity for glue on the tenons and shoulders; the wedges alone need be glued, to attach them to the sides of the tenons, that their effect may not depend on mere compression. Joiners are generally furnished with a cramp, with which to force the joints of framing into close contact; it is either of wood acting by means of wedges, or of iron with a screw. This, too, is unnecessary with good work, every joint of which may be brought perfectly close without great violence of any kind. The cramp will sometimes give bad work the semblance of good, but it cannot make it really so. If any cracking and starting be heard in the joiner's work of a new building, it generally indicates one of two things; either the cramp has been required in putting the framing together, or, having been put together, it has been forced out of winding in fixing, and the constrained fibres are seeking to regain their natural position. A good workman does not require a cramp, nor will his work, if he has been supplied with seasoned stuff, ever require to be strained; and consequently the cracking and starting of joiner's work indicates unfit stuff or bad work, or perhaps both. It is true that glued joints will sometimes fly; but when they do, there need be no hesitation in determining the presence of both bad work, and stuff in an improper state.

Floors are measured and valued by the square of a hundred superficial feet; but anything beyond the mere flooring, such as the mitred borders generally put as a margin to the stone slab of a fire-place, is taken extra by the foot superficial, or running, as the additional work may be above or below three inches in width. The first important thing to note in measuring a floor is the thickness of the boards, by which to determine the cost of the principal material. A floor of boards unplaned on the face, and shot on the edges, laid folding, is the roughest that can be supposed; with the boards wrought or planed on the face, and laid in the same manner, will be the next in advance; and straight-joint flooring, in all its varieties, is the most troublesome, and consequently the most expensive in common and general use. Whether the boards be wide or narrow is a consideration to be noted, an equal surface being of course more rapidly covered with wide than with narrow boards; whether they be gauged, and if brought to a thickness throughout, or only rebate gauged, and cut down on the joints with the adze; in what manner the heading joints are formed and secured; how the longitudinal joints are executed, whether square, ploughed and tongued, or dowelled;

and whether the boards are face or edge nailed. Solid frames, as for outside doors, &c., are measured and valued by the cubic foot, labour being calculated upon the stuff according to the nature and extent of what may have been applied to it.

With trifling and unimportant exceptions, everything else in joiners' work that exceeds three inches in width is taken by the superficial foot; and the dimensions are taken on the finished and fixed work, so that allowances must be made for whatever waste may have been of necessity made. The stuff worked up by the joiner is always supposed to have been in planks and boards a certain number of quarters of an inch in thickness, so that whatever the finished work may stand, it is taken as of the thickness which in quarters of an inch it is next below; thus, if the styles of a door stand at even less than an inch and seven-eighths, it is taken as a two-inch door; for a piece of framing is always considered to be of the thickness of its outer frame-work, the description determining the substance of the panels. Framed grounds are measured round on the outside for the length; their width is not that of the frame, but of the styles and head as they actually are; and their thickness that of the stuff before it was planed at all. Narrow grounds are taken by the foot running, their width being noted in the description of them. Jamb linings are measured to the full length they may be of by their width, the thickness being noted, together with a description of the work on them,—if they are single or double rebated, if framed, and in what manner, &c.

The dimensions of a door are generally taken within the rebates in which it is to hang, with its thickness and description noted,—as of four, six, or eight panels, moulded on one or both sides, with flat or raised panels, &c.; if it be double-margined, that is stated, and the amount of the lap or rebate in their meeting styles is added to the width, to increase the superficies by so much. The hinges with which a door is hung, and the lock or other fastenings which may be on it, are taken, with a description of their sizes and qualities, immediately after the door itself. If sashes are in a solid frame they are taken alone, but sashes in cased frames are measured in and with the frames. To the clear height between the sill and the head, three inches are added for the thickness of the sill, and four inches for the depth of the case at the head, for the height; and to the width between the pulley-styles is added eight, nine, or ten inches, as the case may be, for the breadth of the casing on each side, for the width; these give the superficies of the sashes and frame. The sashes and frame are described, with the thickness of the former, which determines that of the latter; the sill is described as sunk or merely weathered; the pulley-styles as of such a thickness; the pulleys, line, and stuff employed in the different parts of the frame, as of such and such qualities and sorts; and whether the sashes be single or double hung, with what fastenings, &c. The boxings for the shutters are taken in a superficial quantity, as square or splayed, if circular on plan, whether with a flat or quick sweep, or if circular-headed, and straight on plan. The back linings, the backs, elbows, and soffits, the shutters and the back flaps, are all measured by the superficial foot, according to their thicknesses and descriptions, the hinges and fastenings of the shutters and back flaps being numbered and noted independently of them. The capping to backs is taken by the running foot; and elbow cappings are numbered. Moulded architraves are taken superficially, the length by their girt, or by the run at such a girt. Skirtings are measured superficially at such a thickness, as scribed or tongued, as square or moulded, or rebated for base moulding, as the case may be. Base and surbase, and indeed all other moulding which girds at four inches and above, should be taken superficially; and mouldings which are of less girt may be taken by the run if they be taken independently of the other work, or that to which they belong, at all. A moulding projecting from the face of the work to which it belongs may be assumed as independent of it; whereas a receding one, if it be small, will merely add the character of moulded to the work, and if large will qualify all in immediate connection with it to be taken as a superficial quantity of moulding. All circular work, or work which diverges from a straight line, is noted and charged proportionally to the additional labour and waste of stuff involved; the shorter the radius of the arc, or quicker the sweep, the higher must be the proportioned charge. Things which have been bent to their flected form are less costly in proportion than those which must have been worked in the solid or glued up in thicknesses.

Stairs are measured by the superficial foot, the length of one step being taken by the breadth of a step and riser, increased by once the thickness of the former for a quantity, and this multiplied by the number of steps there may be of the same kind; that is, when the steps are flyers; for in winding steps the treads and risers are taken in separate dimensions, for greater accuracy. The thicknesses of the steps and risers are noted, as well as the mode in which they are worked; they have either rounded or moulded nosings, are housed into the string, or have returned nosings, the riser being mitred to the string or to cut brackets on the ends of the steps. Curtail ends to steps are numbered. The frame-work or bearers on which the stairs rest is included with the stairs themselves. String-boards are taken according to their thickness and the quantity of work on them; the grooves or housings in them are numbered. The capping on a close string is taken by the run; but when the nosings of the steps are returned, the strings are said to be cut; and if there are any cut and mitred blocks, they are numbered. Stair skirting is taken as raking and scribed, and as straight, circular, ramped, or wreathed, by the foot superficial; wooden balusters are taken by the run, and the mortises or dovetails in which they are set are numbered; newels are taken by the run for the stuff and the fixing, and the turnings on them are numbered. Hand-rails are said to be merely rounded, or moulded; they are measured by the running foot; and a distinction is kept up between the straight, the circular, the ramps, the wreaths, and the scroll; nuts and screws in their joints are numbered.

All sorts of framing, whether it be fixed or hung—all linings above three inches in width—all sorts of ledged work, such as plain doors and shutters, partitions in lofts and stables, bracketing, cradling, &c.—must be measured superficially. All narrow linings, very narrow skirtings, staff beads, fillets, water trunks and spouts, legs, rails, and runners to dressers, groovings, flutings, reedings, cappings, &c., and any work on superficial quantities that does not pervade the whole, but is in itself peculiar, should be taken lineally, or by the running foot. Insulated parts, such as short, interrupted grooves, blocks, pateras, brackets, trusses, cantilevers, holes, mortises for articles taken lineally, mitres to cornices, heads and feet to flutes and reeds, &c., are numbered and charged at so much a piece. Ironmongery goods employed by the joiner are numbered under their different heads, and charged as fixed; that is, to the price of a lock is added a charge for the labour employed in fitting and fixing it, and whatever accessories it may have required which are not included in its own cost, such as screws, &c., to a rim or dead lock. To the price of hinges, however, only the cost of screws should be added, as the fixing of them is usually included in hanging the work to which they are attached.

The cost at which joiners' work can be executed can only be determined by calculation and observation. The cost of the materials employed may be readily determined by dissecting a piece of work and reckoning its contents; but the labour depends on so many contingencies, that very accurate observation indeed is necessary to determine the quantity that may have been required to produce a certain result. In carpenters' work, the material forms the principal part of the charge; but in joiners' work the materials are for the most part of far less importance than the labour which has been expended on them. The stuff employed in a sash must be costly indeed to amount to as much as the labour of making the sash; whereas, in most doors, under ordinary circumstances, the materials may cost as much as the labour.

Sawyer.—The labour of the sawyer is applied to the division of large pieces of timber or logs into forms and sizes to suit the purposes of the carpenter and joiner. His working place is called a saw-pit, and his almost only important tool a pit-saw. A cross-cut saw, axes, dogs, files, compasses, lines, lamp-black, black-lead, chalk, and a rule, are all accessories which may be considered necessary to him.

Unlike most other artificers, the sawyer can do absolutely nothing alone: sawyers are therefore always in pairs; one of the two stands on the work, and the other in the pit under it. The log or baulk of timber being carefully and firmly fixed on the pit, and lined for the cuts which are to be made in it, the top-man standing on it, and the pit-man below or off from its end, a cut is commenced, the former holding the saw with his

two hands by the handle above, and the other in the same manner by the box handle below. The attention of the top-man is directed to keeping the saw in the direction of and out of winding with the line to be cut upon, and that of the pit-man to cut down in a truly vertical line. The saw being correctly entered, very little more is required than steadiness of hand and eye in keeping it correctly on throughout the whole length. It is the custom to project so much of the log over the first transverse bearer as can be done without rendering it liable to vibrate or be insecure; and when all the cuts proposed are advanced up to that bearer, the end is slightly raised to allow the bearer to be passed out beyond the termination of the advanced cuts. The advantage of, or rather the necessity for, the moveable handle at the lower end of the saw is now evident, the top-man removing the saw readily from cut to cut from above, his mate having merely to strike the wedge in the box one way or the other, to fix or loosen it.

It is absolutely necessary that the top-man should stand in such a manner on the log or piece operated on, that a line down the centre of his body should fall exactly upon the line of the cut he is to work on, and be as exactly perpendicular to it and to the plane of the horizon. He must, therefore, when the cut is near the outer edge, be provided with a board or plank, one end of which may rest on something firm at a short distance from the log, and the other on or against it, to put the outer foot on, and so keep himself in such a position that he may always, and without constraint, see his saw out of winding, and so that a spectator standing on the fore-end of the pit may see the saw an imaginary line passing down the centre of the workman's body, and the line of the cut in exactly the same vertical plane. The labour of the top-sawyer should consist solely in lifting the saw up by the handle as high as his arms can carry it, and that of the pit-man in drawing it down with a slight pressure or tendency onward, sufficient to make it bite into the timber as much as his strength will enable him to make it cut away. The only assistance the pit-man should give in lifting the saw, is in holding it back that the teeth may not drag against the cut in the ascent; and all the top-man should do in cutting downward is to keep the teeth steadily and firmly in contact with the part to be eroded. Good workmen may work with a narrower or closer set to their saw than bad ones can, though the wider or more open set saw is more liable to make bad work. It works more slowly and consumes more stuff than the close set; but it is not so likely to hang in the cut with unnecessary pushing up of the pit-man and jerking down of the other, as if it were set more closely. A good top-man, nevertheless, is of much more importance, though he be badly mated, than the converse. Indeed the best possible pit-man could not work satisfactorily with a bad top-man, and therefore the latter is always considered the superior workman, and on him devolves the care of sharpening and setting the saw, &c. In the operations of the carpenter and joiner much depends on the manner in which the sawyers have performed their part. The best work on the part of the carpenter cannot retrieve the radical defects in his materials from bad sawing; and although the joiner need not allow his work to suffer, bad sawing causes him great loss of stuff and immense additional and otherwise unnecessary labour. Planks or boards, and scantlings, on coming from the saw-pit, should be as straight and true in every particular, except mere smoothness of surface, as if they had been tried upon the joiner's bench; and good workmen actually produce them so. Saw-mills, too, by the truth and beauty with which they operate, show the sawyer what may be effected; for though he can hardly hope to equal their effect, he may seek to approach it.

Sawyers' work is valued at so much the hundred superficial feet; the sawing on a board or squared scantling being once its length, by a side and an edge, or half the amount of its four sides. In squared timber, however, it is generally valued at so much per load of fifty cubic feet, four cuts to the load, any cuts exceeding that number being paid for at so much per hundred feet; in this case the length of the cut by its depth gives the superficial quantity of sawing in it. Pieces again of determined and equal length and breadth, such as the deals and planks commonly used for joiners' work in this country, admitting of a regulated scale, the sawing that may be required in them is valued at so much the dozen cuts.

Modeller.—The modeller copies, in a solid material, the drawings of designs which may have been prepared for enrichments, in whatever material they are to be cast, whether in plaster, in metals, or in compostition of any kind, for the plasterer, smith, or decorator. The model is made in a finely tempered and plastic clay, or in wax; and the modeller works with his fingers, assisted by a few ivory or bone tools for finishing off neatly and sharply, and for working in parts which he cannot reach with his fingers. He is generally the best workman who can do most towards producing the required forms with his fingers unassisted by artificial tools, as a greater degree of ease and freedom almost always results from the use of the hands alone. The model being completed, it is moulded, that is, moulds are made fitting it exactly in every part, and fitting exactly to each other at the edges, and in these, casts are made to any extent that may be required.

The modeller having some pretensions to be considered an artist rather than a mere artificer, is for the most part paid according to his merits as such, rather than for so much time, according to the ordinary mode of determining the value of artificers' works.

Carver and Gilder.—The carver is strictly an independent artist, whose business it is to cut ornaments and enrichments in solid and durable material, such as wood and stone, so that, like the modeller, he must be paid according to the taste and power he may exhibit in his works, rather than as a common artificer. Carving has, however, been in a great measure superseded by modelling and casting, so that the carver is hardly known in economic building except in connection with the gilder. Gilding may indeed be applied to castings as well as to carvings; but the former being, almost as a matter of course, less sharp and spirited in their flexures and details, as well as less firm in substance than the latter, castings can less bear to be further subdued by the application of foreign matters to their surfaces than carvings may.

Gilding is the application of gold leaf to surfaces, which require, however, to be previously prepared for its reception. The work is first primed with a solution of boiled linseed oil and carbonate of lead, and then covered with a fine glutinous composition called gold size, on which, when it is nearly dry, the gold leaf is laid in narrow slips with a fine brush, and pressed down with a piece of cotton wool held in the fingers. As the slips must be made to overlap each other slightly, to insure the complete covering of the whole surface, the loose edges will remain unattached; these are readily struck off with a large sable or camel-hair brush, fitted for the purpose; and the joints, if the work be dexterously executed, will be invisible. This is called oil gilding, and it is by far the best fitted for the enrichment of surfaces in architecture, because it is durable, and is easily cleaned, and does not destroy or derange the forms under it so much as burnished gilding does. This latter requires the work to be covered with various laminæ of gluten, plaster, and bole, which last is mixed with gold size, to procure the adhesion of the leaf. The most durable mode of gilding metals in common use is by amalgamation.

The surfaces generally operated on by the gilder are so diverse, that the real value of his work can be determined satisfactorily only by taking his time and the materials employed and consumed in executing a piece of work.

Plumber.—Lead, as the name imports, is the material in and with which the plumber operates. The previous preparation, casting and milling of lead into sheets, pipes, &c., and the composition and uses of solder, will be found described under the head PLUMBERY.

The principal operations of the plumber are directed to the covering of roofs and flats, laying gutters, covering hips, ridges, and valleys, fixing water trunks, making cisterns and reservoirs, and laying on the requisite pipes and cocks to them, fixing water-closet apparatus, setting up pumps, and applying indeed all the hydraulic machinery required in economic building. His tools are knives, chisels, and gouges for cutting and trimming, rasps or files and planes for fitting and jointing, a dressing and flatting tool for the purposes its name expresses, iron hammers and wooden mallets for driving and fixing, ladles in which to melt solder, grozing irons to assist in soldering, a hand-grate or stove which may be conveniently moved from place to place, for melting solder and heating the grozing irons, a stock and bits for boring holes, and a rule, compasses, lines and chalk for setting out and marking, together with weighing apparatus, as the quantities of most of the materials used by the plumber must be either proved or determined by weight. A plumber is always attended by a labourer, who does the more

laborious work of carrying the materials from place to place, helps to move them when they are under operation, melts the solder and heats the grozing irons, attends to hold the one or the other, as neither may be set down or put out of hand when in use, and assists in some of the minor and coarser operations. In boarding roofs, flats, and gutters for lead, clasp-nails or flooring brads should be used; and the first care of the plumber should be to punch them all in from an eighth to a quarter of an inch below the surface, and stop the holes carefully and completely with putty, or a chemical process will ensue on the slightest access of moisture if the iron heads of the nails come in contact with the lead, and the latter will, in the course of no long period, be completely perforated over every one of them. Neither should lead in surfaces of any extent be soldered, or in any manner fastened at the edges, without being turned up so as to make sufficient allowance for the expansion and contraction which it is constantly undergoing during the various changes in the temperature of the atmosphere. It may be taken, indeed, as a general rule, that solder should be dispensed with as much as possible. Like glue to the joiner, it is indispensable in many cases; but like glue also, it is in common practice made to cover many defects, and much bad work, that ought not to exist.

Sheet lead, whether cast or milled, is supplied of various weight or thickness; and it is always described as of such a weight in pounds to the superficial foot. This varies from four to ten or twelve, so that the weight to the foot being ascertained, the whole weight of any quantity of the same thickness may be determined by admeasurement. There are very few purposes, indeed, in building, in which lead of less than six pounds to the foot should be used, and very few in which the weight need to exceed ten. For roofs, flats, and gutters, under ordinary circumstances, eight pounds lead is a very fair and sufficient average; for hips and ridges, lead of six pounds to the foot is thick enough; and for flashings five-pound lead need not be objected to. Cast lead has been preferred for the former purposes, because its surface is harder, but milled lead is of more even thickness throughout, it bends without cracking, which is not always the case with cast lead, and it makes neater work. Sheets of cast lead run from sixteen to eighteen feet long and six feet wide; milled sheets are made of about the same width, and six or eight feet longer than cast sheets. Neither the one nor the other may be safely used on flats, or in gutters exposed to the wide range of temperature of our climate, in pieces of more than half the length and half the breadth of a sheet; that is to say, from eight to twelve feet long, and three feet wide, are the limits within which sheet lead will expand and contract without puckering and cracking, and to allow it to move freely it is laid with rolls and drips in such a manner that any extent of surface may be covered with the effect of continuity, though the pieces of lead forming the covering be of such small sizes as above stated. But all *fixing*, whether by soldering, or otherwise, is to be carefully avoided. A roll is a piece of wood made about two inches thick and two and a half inches wide, rounded on one edge, and fixed with that edge uppermost, so as to come four inches within half the width of a sheet, that the edges may be turned up and folded round and over it, being lapped by, or lapping the similar edge of the adjoining sheet (fig. 37). Lead sufficiently stout, dressed neatly and closely down to the boards under it, and over the rolls at its edges, will require no fastening of any kind, unless it be so light as to be moveable by the wind. Rolls occur for the most part in roofs and flats, and drips principally in gutters. The drip is formed in the first instance by the carpenter in laying the gutter boards according to an arrangement with the plumber. It is a difference made in the height of the gutter of two or three inches, where one sheet terminates in length, and meets another in continuation. The end of the lower is turned up against the drip, and that of the upper is dressed down over it, so as effectually to prevent the water from driving up under it. Gutters should have a current of at least an eighth of an inch to the foot, and in flats it should be rather more; ends and sides which are against a wall should turn up against it from five to seven inches, according to circumstances; and the turning up under the slates, tiles, or other roof covering, to a gutter, should be to the level of that against the wall at the least. The turning up against the wall should be covered by a flashing. This is a piece of lead let into one of the joints of the wall above the edge of the gutter lead, and dressed neatly down over, to prevent water from getting in behind it. (Fig. 36.) Lead on

Building. ridges and hips not being in sufficient masses to be secured by its own weight, must be held down by nails.

In making cisterns and reservoirs, unless they be cast, the sheets of lead must of necessity be joined by soldering; but the water they are intended to contain protects the lead from the frequent and sudden changes to which in other and more exposed situations it is exposed.

Water trunks and pipes are made of a certain number of pounds weight to the yard in length, to every variety of bore or calibre that can be required. Water trunks or pipes are fitted with large case heads above. to receive the water from the gutter spouts, and with shoes to deliver the water below; they are fixed or attached to the walls of buildings with flanges of lead, which are secured by means of spike nails. Service and waste pipes to cisterns, &c., are generally supported and attached by means of iron holdfasts.

Plumbers' work is for the most part estimated by the hundredweight of a hundred and twelve pounds, though there are of course many things which must be taken in detail, by the pound weight, by number, and even by size. It has been already shown in what manner the quantity of lead consumed may be determined, whether it be in sheets or in pipes; the weight per superficial foot of the one, and per lineal foot or yard of the other, being known, and it is always ascertainable, the dimensions of the various parts or portions of the work readily give the total amount in hundredweights or tons. The waste of lead in working is very trifling, as cuttings all go to the melting pot again with little or no loss but that of refounding or casting; and even old lead is taken by the lead merchant in exchange for new, at a very trifling allowance for tare and the cost of reworking. Water-closet apparatus, pumps, cocks, bosses, ferules, washers, valves, balls, grates, traps, funnels, &c., can all readily be counted and noted according to their sizes and peculiarities; and so may the various requisite joints in pipes, and attachments of cocks, &c., to the pipes, which must also be taken in addition to the articles themselves. The prices of all these goods, from the sheets of lead and the pipes, to the smallest articles used by the plumber, may be ascertained from the wholesale merchants and manufacturers; an addition of thirty per cent. to these prime costs will, under ordinary circumstances, afford the builder or tradesman an ample profit, and payment with sufficient profit on them also, for labour, solder, and nails, excepting cost of carriage, and any other contingent expense, which must be added to the gross. The materials may, however, be taken with a recognised profit added to the prime costs and the actual labour expended; and solder and nails worked up may be reckoned from observation, or account kept of the workman's time, &c.

These things are mentioned more particularly, because a nefarious custom has obtained in this country, and is still allowed to a very great extent, by which the plumber is permitted to take not only an extortionate profit on his goods, but actually to charge twice for labour and the accessories. There is nothing more common than to find in a plumber's account a charge for lead (meaning sheet-lead) *and labour*, at so much per hundredweight,—charges for pipe of a certain bore or diameter at so much per foot,—for so many joints in pipe of such a size,—that is, for the labour and solder consumed and expended in making them,—and so on through all sorts of things, the account winding up at length, or being interspersed from time to time, with so many pounds or hundredweights of solder, and so many days' work of plumber and labourer! The now prevalent custom of artificers' work being done by general builders by tender and contract, has considerably lessened the injury to the public from this abuse, and proved it to be really so by the moderate profits the same men will content themselves with if they make a tender, who would persist in charging at the old rate if they were instructed to do the work without being bound by a contract. Such too is the effect of custom on the courts of justice in England, that the abuse referred to has been protected by them, and probably would be so still, because it was the custom and had been allowed!

Smith and Founder.—The goods supplied by the smith are charged by the pound according to the quantity of labour on them, and the founder has generally an average charge for iron castings at so much per hundredweight or per ton. The working up or fitting and fixing of smiths'-work devolves for the most part on the carpenter in whose favour it is taken, generally, however, in combination with some of his own peculiar works; but founders' work commonly requires to be fitted and fixed by the smith.

Glazier.—The business of the glazier may be confined to the mere fitting and setting of glass; even the cutting of the plates up into squares being generally an independent art, requiring a degree of tact and judgment not necessarily possessed by the building artificer. (See the articles GLASS, Manufacture of, and GLASS-CUTTING.) The glazier is supplied with a diamond cutting tool, laths, or straight-edges of various lengths, a square, a glazing-knife, a hacking-knife, hammer, duster, sash-tool, and rule; and his materials are simply glass, putty, and priming or paint.

The glass is supplied by the glass-cutter in squares or plates, of the sizes and qualities required for the particular work to be executed. The putty is made by the glazier himself or by a labourer, of fine clean powdered chalk or whitening, and linseed oil, well mixed and combined, and kneaded to the consistence of dough. No more putty should be made at once than is likely to be worked up in the course of a day, as, the oil drying out, it becomes hard and partially set, and is therefore less available for its purposes. Priming is a thin solution of white, with a little red, lead in linseed oil. When the sashes come to the glazier from the joiner, they have been fitted into their places, and only require to be glazed before they may be permanently set or hung. Supposing that no preliminary process is required, such as stopping (the result of bad joiners' work) and knotting (and knotty stuff should not be admitted in sashes), the sashes require to be primed. The priming is laid on every part of the sash except the outer edges of the styles and of the bottom and top rails, with the sash tool or painting brush, that is, if the sashes are intended to be painted; for if not, the rebates only must be primed. The object of this is to prepare the material of which the sash is composed for the reception of the putty, which would not otherwise attach itself so firmly as it does after this preparation. The priming being sufficiently dry, the workman cuts the panes of glass down into their places, making every one fall readily into the rebates without binding in any part; indeed the glass should fit so nicely as not to touch the wood with its edges any where, and yet hardly allow a fine point to pass between it and the sash-bar or rebate, the object being to encase it completely in putty, and yet that the putty should not be in greater quantity than is absolutely necessary. The glass being fitted or cut down, the workman takes the glazing-knife in his right hand, and a lump of putty in the palm of his left, the sash being laid on its face, that is, with the rebates upward, before him; with the knife he lays a complete bedding of putty on the returning narrow stops of the rebates, all round to every pane. This being done, the panes of glass are put in on it as they have been fitted, and every one is carefully rubbed down with the fingers, forcing the putty out below and around the edges of the glass, until they are nearly brought into contact with the wood or other material of the sash. The rebates are then filled in with putty behind, the mass forming exactly a right-angled triangle, its base being the extent of the stop of the rebate, and its perpendicular the depth from the glass to the outer edge of the rebate; the third side or hypothenuse is neatly smoothed off, and the sash being then turned on its edge and held uprightly by the left hand, the protruded putty of the bedding is struck off with the knife, and the section of it neatly drawn. The sashes are now deposited on their faces, to allow the putty to set, and then they may be hung and painted. To very large squares, and to plate-glass, small nails called sprigs are used to retain the glass securely while the putty is still soft and yielding.

Lead-work, as it is termed, is the glazing of frames rather than of sashes with small squares or quarries of glass, which are held together by reticulations of lead; and these are secured to stout metal bars, which are fixed to the window frames. The leaden reticulating bars are grooved on their edges to receive the quarries, and are tied by means of leaden ribands or wires to the saddle bars, which, in their turn, are affixed to the stouter bars before mentioned, if the bay or frame be so large as to require both.

Glazing is valued by the superficial foot, the squares or panes being measured between the rebates in which they are set. The value of plate-glass is very much effected by the sizes of the panes, every additional inch in extent of surface adding materially to the cost of production of the whole piece or plate ; it must therefore be carefully noted according to its magnitude. Common window glass is divided into best, seconds, and thirds, and is charged higher as the panes increase in size, because for large panes the table cuts to waste more than in cutting small ones. In ordinary practice, panes containing two superficial feet and under are classed together ; then from two feet to two feet six inches, and so on ; and according to the quality of the article. Flatting, bending, grinding, staining, &c., are all subjects of separate and independent charge.

Lead lights are taken by the superficies generally of a hundred feet, lead and glass being included in the same charge, which, however, depends on the size of the quarries. Stay and saddle bars are taken separately, according to their number and magnitude.

Painter.—For economical painting, see article Painting in Encyclopædia Britannica. The real object of painting is to protect wood, metals, and stuccoes from being readily acted upon by the atmosphere, by covering their surfaces with a material which is capable of resisting it. A continued succession of moisture and dryness, and of heat and frost, soon effects the decomposition of woods, causes oxidation in most of the metals used for economical purposes, and destroys the generality of stuccoes if their surfaces be exposed nakedly to it. A solution of ceruse or white lead in linseed oil spread over them prevents these injuries in a great measure, and for a considerable period of time ; and as the application of such an unction can be repeated without much trouble or expense as often as occasion may require, it may be said to furnish a protection against the cited contingencies. In addition to the utility of painting, it is also available as an ornament, by bringing disagreeably or diversely coloured surfaces to a pleasing and uniform tint, or by diversifying a disagreeable monotony of tint, to suit the taste and fancy ; and this is done in a great measure by the addition of various pigments to the solution before mentioned.

The painter works with hog's-bristle brushes of various sizes, which, with the exception of pots to hold his colours, a grinding-stone and grinder or muller for grinding or triturating them, a pallet and a pallet knife, are almost his only implements. His materials are comparatively few also ; but for some purposes these require a great variety of ingredients, the preparation and combination of which, however, devolves principally on the manufacturer or colourman, and not on the painter himself.

The first thing the workman has to attend to in painting wood-work, is to prepare its surface for the reception of paint, by counteracting the effect of anything that may tend to prevent it from becoming identified with the material. Thus, in painting pine-woods of any kind, the resin contained in the knots which appear on the surface must be neutralized, or a blemish will appear in the finished work over every resinous part. Inequalities or unevennesses of surface, too, must be reduced with sand-paper or pumice-stone, or made up with putty. The necessary process for killing knots, just referred to, will generally leave a film, which must be rubbed down ; and the heads of nails and brads having been punched in, will present indentations, which should be stopped. In painting or laying on the colour, the brush must be constantly at right angles to the face of the work, only the ends of the hairs, in fact, touching it, for in this manner the paint is at the same time forced into the pores of the wood and distributed equally over the surface ; for if the brush be held obliquely to the work, it will leave the paint in thick masses wherever it is first applied after being dipped for a fresh supply into the pot, and the surface will be daubed but not painted. Painting, when properly executed, will not present a shining, smooth, and glossy appearance, as if it formed a film or skin, but will show a fine and regular grain, as if the surface were natural, or had received a mere stain without destroying the original texture. Imitative grainings, however, and the varnishes which are intended to protect them, and make them bear out, necessarily produce a new and artificial texture ; and for this reason they are all to a greater or less extent disagreeable, how well soever the imitations may be effected.

As it must be presumed that all the wood submitted to the operations of the painter, which has passed through the hands of the joiner, was already well seasoned and properly dry, it is only necessary to say generally, that work should be free from moisture of any and every kind before paint is applied to it, or it will at the least prove useless, and probably injurious rather than beneficial. This remark applies alike to wood and to plastered work, both internal and external ; that is, whether they be subjected to the more violent changes of the weather or not. Dampness or moisture in woods, and stopped in or covered up with paint, will, under ordinary circumstances, tend to their destruction ; and in stuccoes it will spoil the paint, and most probably injure the plastering itself too.

Painters' work, on extended surfaces, is valued by the yard superficial, according to the number of coats, or the number of times the paint has been applied to the surface, and to the manner in which, and matter with which, it is finished. On skirtings, surbases, narrow cornices, reveals, single mouldings, sills, string courses, &c., it is measured by the foot run ; sash-frames and the squares or panes of sashes, are numbered, the latter by the dozen ; and so are other things which do not readily admit of being measured. Rich cornices, expensive imitations, &c., are taken by the foot superficial ; and preparations before the work can be commenced are most fairly charged for by the time they occupy and the materials they consume. The work is taken, as one, two, three, four, or more times in oil, common colour ; or so many times finished of a certain colour that is more expensive than what is called common ; or as so many times, and flatted of such a colour, the flatting being an extra coat ; or as painted so many times, and grained and varnished. Common colours are those which are produced by the addition of lamp-black, red-lead, or any of the common ochres to white-lead and oil ; blues, greens, rich reds, pinks, and yellows, &c., being more costly, are taken as such. Unflatted white is a common colour ; flatted, it classes with the rich colours. If the same surface be painted of two different tints, it is said to be in party colours, and an allowance is made in the price for the additional trouble of finishing in that manner. Carved mouldings and other enrichments having to be picked in with a pencil or small brush, that the quirks, &c., be not choked up, must be taken extra, by the run or by number ; and if the picking be in party colours, the labour is necessarily greater than if the work be plain.

What is termed decorating, is divided between the painter and the paper-hanger. Decorations must necessarily depend upon the taste and skill required or employed in producing them ; and the remuneration must also of course be contingent. Decorative papers are paid for by the piece or yard, a piece being made in this country twelve yards long and twenty inches wide, and the hanging is charged at so much the piece. Borders are charged by the yard for the material, and by the dozen for hanging. Sizing and otherwise preparing the walls are considered beyond the charges for hanging. (w. h—g.)

STONE-MASONRY.

History. 1. STONE-MASONRY is the art of building in stone. The word *mason* is derived directly from the French *maçon*, which signifies indifferently a bricklayer or mason. Du *Definition and derivation.* Cange attributes the origin of the word to the low Latin *maceria*, a wall; but by far the most probable derivation is that of the French *maison*, a house; thus *maisonner* is to build houses, and the *maçon*, the man who builds them. Among ourselves, at present, we reckon three sorts of artificers—rubble or rag-stone masons, freestone masons, and marble masons. This last branch, however, is rather that of the carver or statuary. The art of working or reducing stone to the proper shape for the mason to *set*, or place them in the walls, &c., has generally been called *stone-cutting*, and depends very much on the nature of the stone for its details.

Origin and history. 2. The art of building with stone is of very great antiquity; it was originally, no doubt, suggested by the holes in the rocks, or natural caves, in which our forefathers sheltered themselves from the inclemencies of the weather. The perishable nature of their wooden huts afterwards suggested an imitation of them in stone as a durable material; and the trunks of trees, and the beams laid across them, were probably the prototypes of columns and architraves.

Cromlechs, &c. 3. Among rude and barbarous people there seems to have been always a great desire to erect huge masses of stones, either as memorials of some event, or for the purposes of religion, and the early history of almost every country treats of some of these structures.

Cyclopæan masonry. 4. The necessity for defence against predatory tribes seems to have given the next impulse to building with stone, and to this we probably owe those extraordinary walls, commonly called Cyclopean or Pelasgic. These are huge polygonal blocks of stone, carefully cut so as to fit exactly to each other without mortar, and forming walls which must have been impregnable at that time. An idea of their size may be gathered from the fact that in the Etruscan walls at Rusellæ, Mr Dennis (*Cities and Cemeteries of Etruria*, vol. ii.) measured a stone 12 feet 8 inches long by 2 feet 10 inches high. Most of the blocks forming these walls would weigh from 6 to 8 tons. It seems very difficult to understand in that state of civilization how they were hoisted and set. Pausanias (ii. 25), describing those of Argolis, says,—"The walls, the only remains of the city left, are the work of the Cyclops, and are made of rough blocks of such size that a yoke of mules would be unable to move the smallest."

Nineveh, Persepolis, &c. 5. At Nineveh the walls seem mostly of unburned brick, but they are lined with huge slabs of marble, or rather a species of alabaster, the working and carving of which show a very great advance in art. The architecture of the Hindus, Persians, Phœnicians, and Jews, will be found under their respective heads. But there is nothing about their *masonry* differing from that of other similar structures on which it would be profitable to dwell.

Egyptian masonry. 6. The Egyptians, however, seem not only to have used gigantic masonry, but also to have had the power of working, carving, and polishing granite in a way which we certainly cannot at present attain to. The most marvellous fact connected with their masonry seems to be that the whole work was executed with copper or rather bronze *History.* tools, which seem to have answered their purpose better than even our best and hardest steel. Such seems to have been the facility with which they worked this, to us untractable material, that they were not content to cut and polish huge slabs and masses of granite, but they covered them all over with the most delicate and sharp-cut hieroglyphical inscriptions.

7. The masonry of the Greeks yet remaining is chiefly *Greek* of beautiful marble. The workmanship is of the most *masonry.* exquisite character, the joints, &c., of the greatest truth. The artistic beauties of the carving, &c., have been unrivalled at any period. It seems difficult to believe that so enlightened a people were ignorant of the use of the arch, especially as it was clear it was known not only to the Egyptians, but was used at Nineveh. However, no example of a Greek arch exists at this time, as an architectural feature, although, for necessary purposes (as covering drains), and concealed in the walls (as discharging arches), examples are to be found in Greek works. It is probable that, as they had plenty of marble in blocks of almost any size, they preferred to use it in horizontal bearings, to working it into arch forms.

8. It was, however, the contrary with Roman masons; *Roman* although it is true many of their temples were Greek in *masonry.* character, and most of them rivalled those of that nation in size and in the vastness of their material.[1] But in general there was less of that ponderous strength that characterized the Egyptian and the Grecian Doric; and much more science in the construction, particularly as regarded economy; though in point of artistic beauty they were far below the Greeks. To what nation or race the invention of the arch may be attributed, it is clear the Romans were the first to bring it into general use; and though we read of a species of dome among the Greeks called θολος, and though the Hindus delight in domical construction, it is clear the Romans were the first in Europe to use the true dome in covering their temples. Besides this, they had not only good lime but plenty of pozzolano, and therefore their mortar and cement were of first-rate quality. To these advantages we may attribute the vast works which to the present day amaze the spectator, who cannot view their cloacæ, aqueducts, amphitheatres, basilicæ, walls, towers, tombs, domes, harbours, without wonder at the enterprise of the people and the skill of their masons.

9. After the ruin of the Roman empire, and the irruption *Mediæval* of the savage hordes over the whole of civilized Europe, the *masonry.* art of masonry, like all others, declined to the lowest ebb. In fact, except for the erection of rude forts and towers, it became almost extinct. In England we owe its first revival to the works of the Norman invaders; and next, no doubt, to the return of the crusaders, who had witnessed with admiration the marvellous lightness of the buildings in the East, and who brought back with them the arts and learning of the Arabians, especially their mathematical science. From these sources, no doubt, pointed architecture took its rise, and massive cylindrical pillars, composed of many small pieces of stone; small circular-headed windows; walls of vast thickness, with very shallow buttresses; and plain groining without ribs, gradually became changed to

[1] There are pieces of the architrave of the Temple of the Sun now lying on the Quirinal at Rome, measuring 16 ft. 6 in. long by 9 ft. 6 in. high, and about 6 ft. thick, or nearly 50 tons in weight.

History.

light shafted piers, and delicately moulded arches ; windows rich with varied tracery ; pannelled walls, with bold buttresses, surmounted by niches, and crowned with pinnacles ; and groined roofs, fretted with a net-work of ribs, and studded with richly floriated bosses.

Masonry of the sixteenth and seventeenth centuries.

10. The revival of classic architecture threw the arts into another channel, and the masons had hardly forgotten their old traditions when Jones and Wren introduced them to new details. The latter, in particular, formed an excellent school of masonry. The works at St Paul's, and his other public buildings, are executed in a very superior manner. He seems to have been very choice in the selection of his stone. Among all his buildings is scarcely a failure, or a defective block. Besides this, by the assistance of Gibbons, he formed an excellent school of architectural carvers.

Eighteenth and nineteenth centuries.

11. The art of masonry was well upheld by Hawksmoor, Vanburgh, and Chambers ; but shortly began to languish, when the inordinate use of cements, first introduced by the Adamses, came into vogue. Besides this, the heavy duties imposed on the transport of stone by sea, and the high prices which all materials bore during the war, threatened to reduce masonry to its lowest. The revival of Gothic architecture has renewed the use of freestone, and has taught our masons the art of working tracery, groined roofs, flying buttresses, and such use of stone as was supposed, scarce a century ago, to be one of the lost arts. Besides this, the abolition of duties, and the introduction of many facilities of transport by steam, both by land and water, has so reduced the price of stone, that in many places the use of cement is a false economy. Again, our intercourse with the Continent has brought us into more familiar acquaintance with the great works of classic antiquity, and of the Italian Renaissance. In addition to these causes, the vast engineering works, our docks, harbours, lighthouses, bridges, and, above all, our railroads, which have lately been constructed all over the country, have given a vast impetus to the study and practice of constructive masonry on the largest scale ; and the consequence is, we now have in Great Britain a body of masons of higher and more varied skill then perhaps ever was known in this country at any time, both as regards constructive ability and elegant taste.

State on the continent. Italy.

12. In Italy the old customs and traditions are still closely followed, and masonry is extremely well but slowly executed. The fine viaduct lately built at Albano is, however, a favourable example of Italian engineering in stone ; and as railways gradually spread over the country we may expect still greater progress. In France masonry has always flourished, stone being so abundant, especially in the neighbourhood of Paris. The late works at the Great Exposition, particularly the beautiful bridge over the Seine, and the noble buildings which unite the Tuileries and Louvre, form probably the finest palace in the world, and speak highly of the state of the art in France. In Germany we may note the fine works at Munich, particularly the Valhalla, the Pinacothek, and the Glyptothek ; at Berlin, the beautiful buildings of Shinkel ; and in Russia the vast improvements at St Petersburg, particularly the cathedral of St Isaac, the dome of which is surrounded by twenty-four columns, each of one single piece, and each weighing more than 60 tons, and standing at the height of 150 feet above the ground. (See art. ST PETERSBURG, vol. xvii., p. 490.)

France.

Germany.

Russia.

Divisions of the subject.

13. It is now proposed to treat this subject under the following heads :—

I. Materials used in masonry.—1. The various kinds of stone. 2. Their durability and causes of decay. 3. Mortars and cements.

II.—Of the principles of stability and strength in masonry. Under this head it is proposed to give the whole

of the scientific article written by the late Professor Robison expressly for this work.

III.—Of foundations.—1. On land. 2. In the water.

IV.—On stone cutting and setting.—1. General. 2. Mediæval.

V.—On artificial stones, and on the induration of soft stones.

Materials used in Masonry.

I.—OF MATERIALS USED IN MASONRY.

Materials.

14. In our article on MINERALOGICAL SCIENCE, vol. xv., p. 155, we have given an epitome of the various rocks which compose the fabric of the globe. We purpose to go through this list, showing those used in building as they occur *seriatim*, with some practical remarks upon each, which will give a complete epitome of the materials used by masons in all ages.

Igneous rocks of volcanic origin.

15. Of *Igneous Rocks* of volcanic origin, the varieties which have been used are those light stones called tufa and pumice, and that stone called peperino. The two former of these were extensively employed by the Romans in the filling in of vaulting, on account of their great lightness. The latter stone, which is obtained in large quantities near Rome ; and which, though of volcanic origin, resembles a sort of coarse oolitic conglomerate, was used by that people extensively, particularly for substructures, for which it is fitted, being obtained in large blocks.

Trappean.

Of the second division of igneous rocks, the *trappean* ; porphyry, and serpentine have been used, but chiefly as ornamental coloured stones, and have been generally classed as marbles.

Super-silicated stones. Granite.

Of this third division the granite alone is in use, and is now very extensively employed, not only in bridges and engineering works, but in public buildings and dwellings. It is got from the quarries by splitting the blocks with wedges, and is so hard it cannot be cut by any ordinary saws. It can only be worked first with large hammers, and then reduced by pointed chisels, and consequently is very expensive in building. Some very good specimens come from Cornwall and Devonshire, but by far the best are from Dundee and Aberdeen. A variety of the latter, called Peterhead, is only to be equalled by the finest oriental granites.

Aqueous rocks.

16. *Of Aqueous Rocks—Mechanically formed,* and of the *Arenaceous* varieties.—Gravel is used by masons for concrete, and sand in making mortar. Sandstones and gritstones are very extensively used. These are either laminated, as the York stone, and used generally for paving, as it can readily be split into large surfaces of small relative thickness ; or compact as old red sandstones, which stand very well internally, but perish sadly with the weather, as may be seen at Chester cathedral. The new sandstones, the best of which is the Calverley stone, got near Tunbridge Wells. These stones are easily quarried, but, if sawn, the wet saw and sand must be used. The finer grained compact sandstones, which are comparatively free from iron, and form very good building-stones, are very numerous. We name a few. These are the Bramley Fall, used by engineers for bridge copings, &c. ; the Park Spring, Elland Edge, and Whitby, all from Yorkshire. Scotland can boast of some of the finest quarries of sandstone, the best of which, perhaps, is the Craigleith, much used at Edinburgh. The college, courts of law, registry, custom-house, royal exchange, national monument, and many churches and private residences there, are built of this excellent material, which has also been extensively exported to Hamburg, Altona, Gottenburg, and the continent. Humbie stone has also been extensively used, both at Edinburgh and at Glasgow, where it forms the Royal Exchange and bank. It is easier to work than Craigleith. Glammis is also a fine sandstone. The castle there, as well as those at Inverquharity and Cortachy, and Lindertis House, are built of this. In Fifeshire, at Culello, are

Gravel and sand. York paving. Old red sandstone. Tunbridge sandstone. Compact sandstone. Humbie. Glammis. Culello.

Materials used in Masonry. quarries from whence the stones for the monument to Lord Melville at Edinburgh, and that to Lord Nelson at Yarmouth, were obtained. In addition to beauty and durability, these stones have the merit of being capable of receiving the finest and smoothest forms from the chisel of the workman. Another class of sandstones are commonly called *firestones*, as they endure the action of fire better than most others. Of these the best known is the Reigate stone, which is the principal material used at Windsor Castle, Hampton Court, and in many old buildings round London.

Reigate firestone.

Argillaceous clunch. Of *aqueous stones* classed as *argillaceous*, the Clunch only is used in building. It may be seen in Ely and Peterborough cathedrals, and many other mediæval buildings, and is a beautiful material for carving, but will not stand the weather.

Chemically formed. Of those *aqueous stones*, classed as *chemically formed*, and of the subdivision, the *calcareous*, we have none of note but the Travertine, or, properly speaking, Tiburtine. This is a coarse grained stone, of warm colour, found in large blocks, and extensively used at Rome, both in ancient and modern buildings; the great cathedral of St Peter's may be cited as an instance; but it is unknown in England.

Travertine.

Calcareous. Of *aqueous rocks, organically derived*, and of the first subdivision, the *calcareous* claims our principal attention. The chief of these are the *limestones*, which are generally considered by architects as compact, magnesian, or oolitic limestones. Of the former the best, in the south of England, is that called Chilmark, of which Salisbury Cathedral and Wilton Abbey, and many other fine buildings, have been erected. In the midland counties the Tottenhoe stone, of which Dunstable Priory, Woburn Abbey, Luton church, &c., are built, is an excellent stone. There is also a stone of high quality got near Wirksworth in Derbyshire, used at Chatsworth, Belvoir, Drayton Manor &c., &c. Of magnesian limestones we may name the Anstone, or Bolsover Moor stone, used formerly at Southwell Minster, and lately at the houses of Parliament; the Tadcaster stones, used at York, Beverley, and Rippon Minsters, and very many other buildings; and the Roche Abbey, used at the building of that name, and very many other churches in Yorkshire and Lincolnshire. These stones contain a great deal of carbonate of magnesia, from which they take their name, and are of beautiful texture, and stand well in the country as building-stones; but fail in London, from a cause which will shortly be stated.

Limestones.

Chilmark.

Tottenhoe.

Magnesian.
Anstone.
Tadcaster.

Roche Abbey.

Oolitic stones. We now come to the most important subdivision of the limestones used in masonry—the *oolitic*—so called because they resemble, when broken, a conglomerate of globular eggs, also frequently called roe-stones, because they resemble what is called the hard roe of a fish. (See art. GEOLOGY, vol. xv., p. 146.) Very good examples of these are the Barnack stone from Northamptonshire, of which Peterborough Cathedral, Croyland Abbey, Burleigh House, &c., &c., are erected. Ketton stone, used at most of the colleges in Cambridge, and at Bury St Edmunds, Bedford, Stamford, &c., &c. But the principal English oolites used in masonry are the Bath and Portland. The former, as its name imports, is found in the neighbourhood of Bath. The chief quarries are the Box, Combe Down, Farleigh Down, and Corsham Down; all these quarries vary in quality at different depths. The Corsham Down is said to produce the finest in quality, and the Box Ground stone to be the hardest; but everything in the use of this stone depends on the judgment in selection. Large quantities of a similar stone are imported from Caen, in Normandy. They are more compact in texture than Bath, and therefore fitter for carving; but do not appear to stand the weather of our climate so well. The best variety of this stone is said to be D'Aubigny stone. Almost all these oolites can be sawn with a common dry saw, which saves a great deal in the labour of conversion. But, without doubt, the best of

Barnack.

Ketton.

Bath.

Caen.

all this class of stones is that from the Island of Portland; for beauty of texture and for durability, it perhaps exceeds any stone in the world. It seems the only stone unaffected by the smoke of London; and therefore the greater number of its buildings, St Paul's among the rest, are of this stone. It must, however, be sawn by the use of sand and water; and, being of hard texture, is much more expensive to work than the softer oolites. There are between fifty and sixty quarries on the island. The best are said to be those on the north-eastern side; but, like all stone, there is good and bad in every quarry, and everything depends on the selection. It is said, when Sir Christopher Wren built St Paul's, he had this stone quarried, and exposed to the weather on the sea-beach for three years, before he suffered it to be used. A very excellent limestone for walling, especially for Gothic work, is that called *Kentish Rag*. It is found in large quantities in the neighbourhood of Maidstone, and is very hard, and worked, like granite, with large hammers instead of the saw. Jambs, strings, and mouldings are sometimes worked out of it, but the hardness makes the work expensive. Of siliceous stones, flint is sometimes used for rough walling; but in England this work is done by the bricklayer, and not the mason. (See BUILDING, p. 146.)

Material used in Masonry.
Portland.

Kentish Rag.

Flint.

17. The only remaining class is that of the *metamorphic rocks*, of which the crystalline and saccharine, and the serpentinous limestone are used; but these are all species of marbles, and used more as ornamental than as constructive building-stones, and need not be dilated upon in this article.

Metamorphic rocks.

Marble.

The Durability and Causes of Decay in Stones.

18. The causes of durability of stone, and the correspondent causes of failure and decay, are either chemical or mechanical, and may be described either as decomposition or disintegration. Durability also depends much on the power of resistance to wear.

Durability of stones.

19. Decomposition is caused by some of the elements of the stone entering into such new combinations with water, gases, or acids as render them soluble either by the air or water. Granite, though the hardest of building-stones, is liable to serious decomposition when the feldpars are alkaline (see art. GEOLOGY, vol. xv. p. 136), and will unite with water or acids. Some qualities of this stone are rapidly decomposed by the sea. The same is the case with many of the limestones, as is described in the article above quoted, page 151. Stones containing iron are also liable to decay. In its native state it is usually in a low state of oxidation, and is liable to be acted upon by additional quantities of oxygen or carbonic acid. This sort of decomposition is much increased by being alternately wet and dry, or by frequent changes of temperature. Stones, however, containing iron in a high state of oxidation, as rosso-antico, porphyry, &c., do not readily become decomposed. The most curious discovery of modern times is with regard to the magnesian limestones and dolomites. These were chosen for the new Houses of Parliament on account of their durability. The work at Southwell Minster, 800 years old, bears every mark of the tool to the present day, and every circumstance seemed to justify its selection. It appears, however, that magnesia has a great affinity for sulphur; and the consequence is, the sulphurous acid which is present in such quantities in the smoke of London, has already caused serious decomposition in that building, as well as in the Lincoln's Inn Hall. This acid has also so much effect on the softer limestones, that the fronts of several important buildings, Buckingham Palace among the rest, have been obliged to be painted, to save them from decay.

Decomposition.

Alkalies.

Oxides of iron.

Magnesia.

Sulphurous acids.

20. Disintegration, as has before been said, is the separation of parts of stone by mechanical action. The chief cause is the freezing of minute portions of water which get into the

Disintegration.

Freezing.

Y

Materials used in Masonry.

pores, or fissures, or between the laminæ of stones, and swell slowly as crystals of ice are gradually formed, and consequently burst open the pores, or split the grain of the stone. The south sides of buildings, in northern climates, suffer more than others, as their surface becomes thawed and filled with wet in the day, and frozen again at night, more frequently than the others. A very common error in the present day, is the not taking care to set stones with their laminæ, or grain, or, as the workmen call it, "bed," in a horizontal direction. If work be "face-bedded," the action of the weather will cause the laminæ to scale off, one after the other, just as the leaves of a book fall over, if the volume be placed on its back in an upright position. For fuller illustration of the subject of decay and decomposition of rocks, we must refer our readers to the article Mineralogical Science.

Face-bedded work.

Resistance to wear.

21. Resistance to wear is another obvious cause of durability; but this depends rather on the toughness than the mere hardness of material, a quality often attended with brittleness, as also on its situation. The crushing weight of Portland is about 10,000, while that of York is about 12,000, or one-fifth more; but in many situations Portland steps will last much longer than York. Again, the crushing weight of Peterhead granite is about 18,000, or not quite double that of Portland; whereas, if used as street-paving, it would outlast six sets of the latter.

Of Mortars and Cements.

History of mortars and cements.

22. The use of some material, not only to cause stones to adhere together in building, but to fill up crevices between them, and irregularities in bedding them, is of the remotest antiquity. The earliest mention of mortar is in Genesis xii. 3, where the builders of the Tower of Babel are said to have had "slime for mortar," which the LXX. called ἄσφαλτος (bitumen) and πηλός. Herodotus (Clio, 179) tells us the walls of Babylon were built of bricks, cemented together by hot asphalt. The Egyptians used mortar of lime and sand, almost exactly in the proportions we do, as was proved from an analysis of some taken from the pyramid of Cheops. The Grecian mortars and cements are very fine and strong. Vitruvius gives careful directions how to make mortar (lib. 2, ch. 5), and his instructions are probably the best, and his observations the most sound, of any author, at least till the time of the researches of Vicat and the French chemists.

Definition.

23. Mortar is generally considered under two heads: as common mortar, or that mixture of lime and sand ordinarily used in building; and hydraulic mortar, or that which will set under water. Cement is a name given to the produce of certain argillaceous stones, after calcination, which will set rapidly in the air, and become a hard, adhesive substance in a short time; and which will also set under water, both without admixture of any other substance. The name is also given to certain artificial imitations of these substances, possessing the same properties; and besides, to various bituminous, or oleaginous compositions, used in building for similar purposes.

Lime.

24. Pure lime is an oxide of a metal called calcium, but does not exist in a natural state. It is, however, found abundantly in the conditions of carbonates and sub-carbonates, in chalk, and the various other descriptions of limestones. Its chemical qualities and analysis will be found under the proper heads. The first thing is to drive off the water, which all limestones contain in a greater or less degree, and the carbonic acid gas, which is done by calcining or burning in a kiln at red heat, which must be kept up for several hours, taking care, however, to avoid any approach to vitrification. By this process it is slightly

Materials used in Masonry.

Classification.

diminished in bulk, and it loses nearly half its weight and becomes caustic lime.

25. Limes are generally classed, since the publication of the work of Vicat, as—1. rich limes; 2. poor limes; 3. limes slightly hydraulic; 4. hydraulic limes; and 5. eminently hydraulic limes. In treating of mortar, we have to deal with the first two of this division.

Production of hydrates

26. The lime must next be converted into a hydrate. This is done by a process called "slacking," or throwing water over it from time to time till it hisses and cracks with considerable force, and some noise, gives off a large quantity of hot vapour, and falls into a powder. The rich limes, which are the purest oxides of calcium, increase to double their bulk in the process. The poor limes swell to a much less degree. The hydrates thus formed absorb water, and easily take the form of a paste. They contain rather less than one-third water to two-thirds lime. In this state, if treated with pure water, frequently renewed, every particle of rich limes, and very nearly the whole of the poor limes, will be taken up in solution. In the process of slacking, too much water should not be used, as it "drowns" the lime, according to the expression of the workmen. When in the form of paste it begins to absorb the carbonic acid, which, though no component part of air, is always present in it, in large quantities; and gradually to crystallize again, and so to harden. If the air be excluded from the hydrate of pure lime, it may be kept for almost any length of time. Alberti (lib. 2, cap. 2) says that he once discovered some in an old ditch, which, from certain indications, must have been there 500 years, and was as soft as honey or marrow, and as fit for use as it could be.

Rich limestones.

27. The rich limestones give a white lime, which easily slacks, and increases in bulk; but it is curious that, though the stones differ so much in outward appearance and in texture, the lime, if they be well burned, is the same. The softest chalk, and the hardest rag-stone, or marble, yield an equally good lime, the calcium which they contain being the same mineral. But as chalk generally contains water, irregularly distributed in some places and not in others, and as it is does not exhibit the change that marble or stone does, it is frequently unequally burned, and therefore slacks imperfectly. It is said, however,[1] that lime, made from chalk absorbs the carbonic acid more rapidly than that made from stone; but our own experience does not warrant this conclusion.

Poor limestones.

28. Poor limestones are those which contain silica, magnesia, manganese, or metallic oxides. In consequence of this they are more liable to vitrify in burning, and do not slack so freely. The lime is generally of a browner colour than that from rich limestones, which is said to be a proof of the presence of the above-named metallic oxides. If, however, they be ground so as to facilitate the slacking of every particle, and if used immediately it is made up, poor limes produce a mortar which becomes harder than that from the rich limes, and which resists water better. In fact, works where the latter have been used, have been found to fail entirely by the action of running water, which, as before has been said, will continue to remove the whole of a rich lime, particle by particle.

Use of sand.

29. It is found that the mixture of some kind of hard matter in particles or granules facilitates the setting of mortar, renders it harder and more adhesive than when used alone, besides the saving of limestone and expense of burning. The harder this material, and the sharper the particles the better, as the brick or stone has always some irregularities on the surface, into which these angles or sharp points may enter, and form what is called a key. The substance most generally used is sand, which is generally classed as river-sand or pit-sand. The former is

---

[1] Higgins *On Mortars and Cements*, p. 29.

Materials used in Masonry.

generally preferred, as it is more free from any earthy matters, particularly soft loams or clay. If pit-sand be used it should be well washed. Scarcely any material is better than crushed quartz, or flint, from the sharpness of the angles of the particles; in fact, it is said, that very sharp sand, with an inferior lime, will make a more adhesive mortar than soft sand with the best lime. For observations on the practical mixing of mortar, see BUILDING, p. 140.

Other materials. Burned clay.

30. Where sand is scarce, other materials may be used, the principal and cheapest of which is burned clay. The Romans used this extensively in the form of pounded brick. At present the custom is to throw up clay mixed with any fuel in loose heaps, and burn it slowly. The French writers at one time asserted that burned clay, if not equal to pozzolano, was very nearly so; and large quantities were used as hydraulic mortars at various public works. Where the water was fresh, as at Strasbourg, the work stood very well; but where these mortars were exposed to the action of sea-water, they failed and went to powder in three or four years. Vicat gave great attention to the subject; and though he attributed much of the fault to the imperfect carbonization of the materials, it appears with but little doubt there is some inherent difference between the pozzolanos and other volcanic products, which will be treated of shortly, and those produced artificially.[1]

Slag and cinders.

31. The vitrified refuse of furnaces, called slag, and the scoriæ from the iron-works, have also been crushed and used instead of sand; and with lime slightly hydraulic, produce good mortar. The former is preferred to the latter, as having sharper and harder particles, and containing much less iron. Coal cinders have been used, and seem to have some hydraulic properties; they should, however, be employed with caution, for it is considered they make the lime "short." Wood cinders are too alkaline to be used with safety. A very excellent mortar, much used by engineers in tunnels, is composed of one part of moderately hydraulic lime, one part of coal ashes, one part of burned clay, and two parts of sharp sand.

Volcanic products.

Pozzolano and trass.

32. The vitrified and calcined products of volcanoes make most excellent materials for mortars, particularly where required to be eminently hydraulic. The principal of them is the Pozzolano, which abounds in Italy. It is called so from being found in great abundance at Pozzuoli, near Naples, and is, in fact, the basis of all the best Roman mortars, ancient as well as modern. It varies in colour from reddish brown to violet red, and is sometimes grayish. It is usually sent to England from Civita Vecchia. It has a roughly granulated appearance, and sometimes resembles a cinder in texture, and has frequently a spongy appearance. Acids have little effect on it, and it is not soluble in water. A similar earth is found in the centre of France. But one more familiarly known in this country comes from the village of Brohl, near Andernach, on the Rhine; this is called terrass or trass. These materials have a wonderful effect in rendering even the rich limes eminently hydraulic, and in less proportions improving the hydraulic limes. Vicat says, these mortars begin to set under water the first day, grow hard in the third, and in twelve months are as hard as the bricks themselves. The mixture of common lime with these materials, according to the French writers, should be 1 of pounded lime to 2½ of pozzolano, or to 2 of terrass; or 1 of lime to 1 of sand and 1 of pozzolano. The analysis given by them is nearly as follows:—

	Terrass.	Pozzolano.
Silica, per cent.	57	44
Alumina	12	15
Lime	3	8
Magnesia	1	4
Oxide of iron	5	12

Materials used in Masonry.

Cements

Natural cements.

33. In addition to those which we have called hydraulic limes, there is a peculiar class of stones, which, when burned and pulverized, may be used as a species of mortar, without admixture of sand or any similar substance; and which will not only set rapidly under water, but will acquire an unusual degree of hardness and tenacity.

34. These are called natural cements. The inventor is supposed to have been a Mr Parker, at any rate that gentleman took out a patent about sixty years ago for what he called Roman cement. His material were those argillo-calcareous nodules, or septaria, which are found in the Isle of Sheppey, and commonly called bald-pates. They contain about 70 per cent. of carbonate of lime, about 4 per cent. oxide of iron, 18 per cent. of silica, and 6 or 7 per cent. of alumina. The process is simply to break the stones into small pieces, and burn them in running kilns with coal or coke; they are then ground to a powder, and headed up into casks for use. The success of Parker's cement led to experiments in other places, and the same process was carried on with other argillo-calcareous nodules, as the septaria at Hawick; those in Yorkshire, which produce the cement called Atkinson's; and those in the Isle of Wight, which produce the Medina cement. Similar substances were also discovered, and the same processes carried on in France and in Russia. All these cement-stones effervesce with acids, and lose about one-third of their weight in burning. Parker considered the more the stones were burned short of absolute vitrification the better; but this is not the practice in the present day, though no doubt sound in theory. When taken from the kiln these stones will not slack without being pulverized; and if kept dry, and not exposed to the air, the cement will be good almost any length of time; but it rapidly absorbs both water and carbonic acid if not carefully closed, and falls back into a state of subcarbonate, from which it is said it may be recovered by fresh burning, but it is doubted whether it is ever so good as on the first calcination.

Artificial cements.

35. The great utility of these cements, and the expense of obtaining the stone, induced the manufacturers to endeavour to discover some method of making an article by artificial means which should resemble the natural cements. Mr Frost seems to have been the first who attempted it on a large scale; but though assisted by the talent and science of General Pasley, the results did not come up to the expected standard. Of course, the object was to produce an argillo-calcareous substance, containing the same chemical qualities as the natural nodules, and which might be burned in kilns as they are. The attempt to combine argil in the form of burned clay, to be mixed with lime instead of pozzolano, had partially failed, as has been stated before. Our space will not allow us to relate the various experiments by General Pasley here, nor by Vicat at Meudon in France. They were all based on the principle of mixing together, in a mill, with a quantity of water, masses of chalk and clay, just as the brickmakers do for the production of malm bricks, but in the proportion of about four of the former to one of the latter. The fluid mixture is run out into shallow receivers, and when dry is cut into small blocks or lumps, and burned exactly as the natural nodules are. The difficulty seems to have been to give the materials the full degree of calcination short of vitrification. This seems to have been at last attained by the inventors of the Portland cement, so called from its near resemblance to Portland stone in its colour. It not only possesses the property of setting more quickly, and has greater powers of cohesion than the natural cements, but it may be used with a superabundance of water in the

Portland cement.

---

[1] After long investigation, Vicat was of opinion that this failure was due to the quantity of hydro-chloride of magnesia always present in sea-water; but in what way this affected the burned clay and not the volcanic products, he was unable to explain.

Principles of Stability and Strength.

form of grout, which they cannot; and, above all, it seems to resist the action of sea-water beyond all others. It also forms a very superior cement, as is described hereafter. From a series of trials made on the Portland cement, manufactured by Messrs J. B. White of Millbank, it appeared it required more than two tons to separate two blocks of stone 6 inches cube, which had been put together with a joint of Portland stone, one-eighth to one-fourth inch thick. This gives a resistance of 146 lb. per inch of sectional area, that of Bramley Fall being 76 lb., and Whitby stone 57 lb. Its power to resist compression was tested on blocks 9 inches square and 18 inches long, the pressure being exerted on their ends.

		Tons.	
Roman cement, and two parts sand, bore....	5·33 per sq. foot.	
Portland do. and three do. do....	44	"
Roman do. pure	50	"
Portland do. and two parts sand...........	80	"
Do. do. pure.............................	146	"

Marble cements.

36. A class of cements capable of taking a brilliant polish resembling marble, and consequently very suitable for internal decoration, had lately been invented. The chief of these is Keene's marble cement, and the Parian cement. They become excessively hard in a short time, and are capable of being painted in a few days. The principal component part is said to be obtained by the precipitation of alum by an alkali, which gives a white powder of great brilliancy. It is, however, more matter for the plasterer than the other building trades.

Oleaginous cements.

37. Cements made by the mixture of oil with various substances were formerly much used both here and abroad. The best known in England was called Hamelin's mastic, that in France the mastic de Dhil. These cements being very expensive, and requiring to be constantly painted, have now gone nearly out of use. For outside plastering they form a very fine clean surface, as may be seen in the quadrant in Regent Street.

Bituminous cements.

38. The asphaltum, or mineral pitch (see ASPHALTUM), have lately come into extensive use for pavings, and for covering the backs of arches, or rendering the walls of basements where wet is likely to soak through. The best is said to be that from Seyssel, in France. For their chemical properties see BITUMEN. It is used thus—A bed of concrete, made of the best hydraulic lime, is first prepared, and made fair at top by a rendering of similar mortar. The asphalte will not dissolve with heat by itself, but will calcine in the caldron. A small quantity of pure mineral pitch is therefore first put in; when hot the asphalte is added, and soon dissolves; into this is stirred a quantity of powdered stone-dust, and a small portion of quick-lime. The mixture is then laid hot on the bed of concrete (which must be quite dry), and spread close and fair, some sand being sprinkled over the top and well trowelled in. The best proportions are said to be about 2 pints of mineral pitch to 10 lb. of asphalte, and one-fourth bushel of stone-dust. A very inferior imitation is made by mixing a quantity of sharp sand with gas-tar, heated in a caldron, and then adding some quick-lime. This may do for rendering walls, &c., to keep out wet, but is of very little use in paving.

II.—OF THE PRINCIPLES OF STABILITY AND STRENGTH IN MASONRY.

39. The strength and the stability of stone-work depends partly on its mass or weight, and partly on the resistance of the materials. And, since we cannot imagine incompressible fulcra, nor that the materials of masonry are infinitely hard and inflexible, as writers on elementary mechanics consider them to be, therefore, it is essential that the resistance of materials should be considered, and the effect of their weight allowed for in estimating the power of the straining force.

The resistance of stones being dependent on their state of aggregation, and not on the hardness or density of their elementary parts, their comparative strength cannot be judged of by these qualities; indeed there are few kinds of materials of which the resistance is so uncertain as that of stone, and hence, it is not at all adapted for any support where its resistance depends on its cohesion only, unless it be very carefully examined, and abundant strength be allowed. The resistance of stone to compression is less affected by its irregular nature, particularly as it is usually employed in blocks of inconsiderable height; and, in general, there is scarcely any reason to be sparing of a material which it is often more expensive to reduce than to employ in large blocks. When, however, works of great magnitude are to be constructed, the weight of the materials themselves forms the chief part of the straining force; and, consequently, in such cases it becomes desirable to form a tolerably accurate estimate of their power.

40. This power is limited by a property of bodies that has not received that degree of attention which its importance would lead us to expect. We shall in this place make it the basis of an investigation of the power of materials to resist a force applied in any given direction, and show its application to some of the cases where a mason is most likely to need the assistance of calculation.

When any material is strained beyond a certain extent, every time the strain is increased to the same degree, there is a permanent derangement of the structure of the material produced; and a frequent repetition will increase the derangement till the parts actually separate. (See the article CARPENTRY.) When a small base rests upon a considerable mass of matter, as a pier on the ground, the quantity of derangement will increase only till the mass be compressed to that degree which renders the increase insensible; but in many cases a number of years will elapse before the settlement becomes insensible.

41. The strain which produces permanent derangement in the structure of a material varies from one-fourth to two-fifths of that which would destroy its direct cohesion. In stone the lower value should be taken, on account of its being subject to so many defects; and, for the present, let this strain be denoted by f lbs. upon a superficial foot.

42. Imagine ABCD to be a block, fig. 1, strained either in the direction EF or FE, by a force W; and let BDF be a line drawn in the same

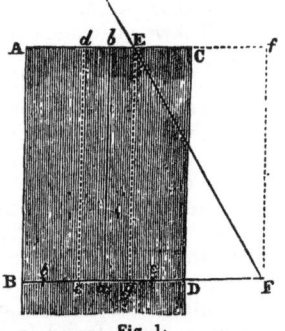

Fig. 1.

plane as the direction of the straining force, and perpendicular to the axis ab of the block. Now, if we consider the resistance of the block to be collected at the centres of resistance t and c, then tF will represent a lever acted upon by three forces; that is, the resistance at t and c, and the straining force at F.

If the angle FEg be denoted by a, then the effect of the force W, reduced to a direction perpendicular to the lever, will be expressed by cos aW . . . . (1.)

Also, if T be the resistance at t, and C the resistance at c, we shall, in the case of equilibrium, have C − T = cos aW . . . (2.)

And, by the property of the lever,

$$\frac{\overline{tc} \times T}{\overline{cF}} = W \cos a \quad . \quad . \quad . \quad . \quad . \quad . \quad . \quad (3.)$$

Hence, $\dfrac{\overline{tc}}{\overline{cF}} = \dfrac{C}{T} - 1$ . . . . (4.)

43. Without stopping to notice some maxims furnished by this equation (see the article BRIDGE), we will proceed to explain the notation used in the investigation which follows:—

l = the length AB.

d = the depth BD, measured in the same plane as the direction of the strain.

b = the breadth.

z = the distance of the neutral point e from the axis ab.

y = the distance of the point g from the axis ab.

$p\left(\tfrac{1}{2}d-z\right)$, and $p\left(\tfrac{1}{2}d+z\right)$ the respective distances of the centres of resistance t and c from the neutral point; and, consequently,

pd = the distance, ct, between them. And,

$g\left(\tfrac{1}{2}d-z\right)$, and $g\left(\tfrac{1}{2}d+z\right)$ the respective distances of the centres of gravity of the sections into which the neutral axis divides the block, counted from the neutral point.

The leverage cF, expressed in this notation, will be

$$\frac{l\sin a}{\cos a}+y-\tfrac{1}{2}pd+(-p)z\,;$$ consequently, equation (4) becomes

$$\frac{pd}{\dfrac{l\sin a}{\cos a}+y-\tfrac{1}{2}pd+(1-p)z}=\frac{C}{T}-1\qquad\ldots\ldots\quad(5.)$$

44. Now, it has been shown, by writers on the resistance of solids, that the resistance of any section, collected at its centre of pressure, is equal to its cohesive force multiplied by the distance of its centre of gravity from the neutral axis, and divided by the distance of the point of greatest strain from the neutral axis. (Emerson's *Mechanics*, prop. 77, 4to ed.)

Accordingly we have $\quad C=\dfrac{fbg\left(\tfrac{1}{2}d+z\right)^2}{\tfrac{1}{2}d+z}\qquad\ldots\ldots\quad(6.)$

And $\qquad\qquad\qquad T=\dfrac{fbg\left(\tfrac{1}{2}d-z\right)^2}{\tfrac{1}{2}d+z}\qquad\ldots\ldots\quad(7.)$

Therefore $\dfrac{C}{T}=\dfrac{\left(\tfrac{1}{2}d+z\right)^2}{\left(\tfrac{1}{2}d-z\right)^2}$; which being substituted in equation (5), we obtain

$$(3p-2)z^2-2z\left(\frac{\sin a.\,l}{\cos a}+y\right)+\tfrac{1}{2}pd^2=o\qquad\ldots\ldots\quad(8.)$$

If the block be rectangular $p=\dfrac{2}{3}$, and therefore the distance of the neutral point from the axis is

$$z=\frac{d^2}{12\left(\dfrac{l\sin a}{\cos a}+y\right)}.\qquad\ldots\ldots\ldots\quad(9.)$$

45. The value of z for a rectangular section being determined, the magnitude of the straining force is easily found, so that it may not exceed the power of the material; for, by the properties of the lever—

$$\text{W}\cos a=\frac{pd\text{C}}{\dfrac{l\sin a}{\cos a}+y+\tfrac{1}{2}pd+z(1-p)}\,;\text{ and since }\text{C}=fbg\left(\tfrac{1}{2}d+z\right)$$

by equation (6), and

$$z=\frac{d^2}{12\left(\dfrac{l\sin a}{\cos a}+y\right)}\text{ by equation (9), and }g=\tfrac{1}{2}\text{ by the form of}$$

the section, we have

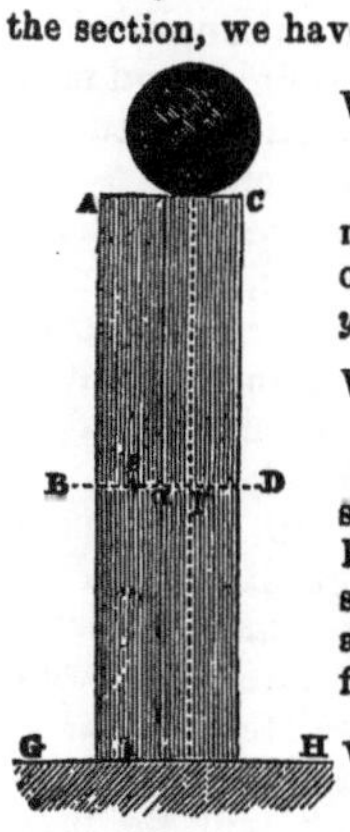

$$\text{W}=\frac{fbd^2}{d\cos a+6l\sin a+6y\,\cos a}\quad\cdots\quad(10.)$$

46. In particular cases this formula becomes more simple; as, for example, when the distance of the point g from the axis ab is o, that is, when $y=o$,

$$\text{W}=\frac{fbd^2}{d\cos a+6l\sin a}\quad\ldots\ldots\quad(11.)$$

In a column or pillar the section of greatest strain will be at the middle of the length, as at BD in fig. 2; and the direction EF of the straining force is usually parallel to the axis; and then $\sin a=o$, and $\cos a=1$, and therefore,

$$\text{W}=\frac{fbd^2}{d+6y}\quad\ldots\ldots\ldots\quad(12.)$$

Fig. 2.

When the direction of the straining force coincides with the axis, or when $y=o$, the strain on a column or pillar is expressed by the equation $\text{W}=fbd$ (13.) These equations apply also to tensile forces.

When the strain becomes transverse, or when EF is perpendicular to the axis, as in fig. 3, then $\cos a=o$, and $\sin a=1$, hence $\text{W}=\dfrac{fbd^2}{6l}\quad\ldots\quad(14.)$

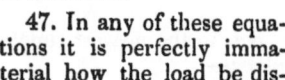
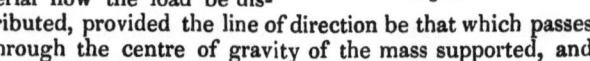

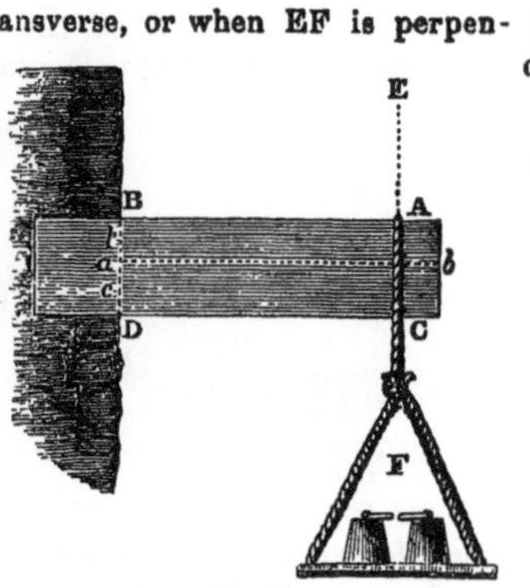

If the block be supported at the ends, and the load be applied in the middle of the length, as in fig. 4, the fracture will take place at BD; and W in equation (14) will be the pressure on either support, which is obviously half the load in the middle.

Fig. 3.

47. In any of these equations it is perfectly immaterial how the load be distributed, provided the line of direction be that which passes through the centre of gravity of the mass supported, and the weight be the whole weight of that mass; or, if the strain be the combined effect of several pressures, then the direction must be that of the resultant of these pressures, as determined by the principles of mechanics. (See the article CARPENTRY.)

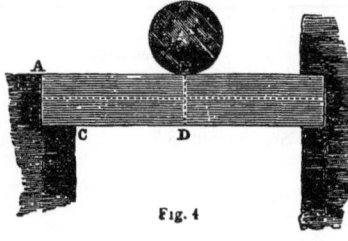

Fig. 4.

48. If a slab of equable thickness and width be supported along two of its sides, as at AC, AB in fig. 5, and it be strained by a force acting at D, in a direction perpendicular to its surface, and DE be made equal to DB, then the fracture will take place in the direction EB; for it may be shown, by the principles of the maxima and minima of quantities, that the resistance, according to that line, is a minimum. And since, in that case, EB = 2 FD, we shall have, by equation (14), $\text{W}=\dfrac{fd^2}{3}\quad\ldots\ldots\ldots\quad(15.)$

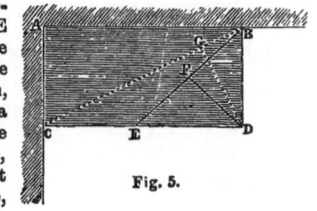

Fig. 5.

A force uniformly diffused over the surface of the slab would fracture it in the direction CB, shown by a dotted line in the figure, and if w be the load in pounds upon a square foot of the surface, then the proper values being substituted for the leverage and breadth in equation (14), $w=\dfrac{fd^2\left(\overline{\text{CD}^2}+\overline{\text{DB}^2}\right)}{\overline{\text{CD}^2}+\text{DB}^2}\quad\ldots\ldots\quad(16.)$

The strength of a series of steps bearing upon one another, as in the perspective sketch (fig. 6), may be determined with sufficient

Fig. 6.

accuracy by the last equation, supposing the depth to be the mean vertical depth of any one step; as, for example, taken at GH in fig. 7, the figure showing the ends of the steps.

49. The case to which equation (14) applies, affords the most convenient, as well as the most accurate, means of determining the value of f for any material; and, suppos-

Principles ing it to be one-fourth of the absolute cohesion (§ 34), the
of Stability last column of the following table of experiments gives its

Strength

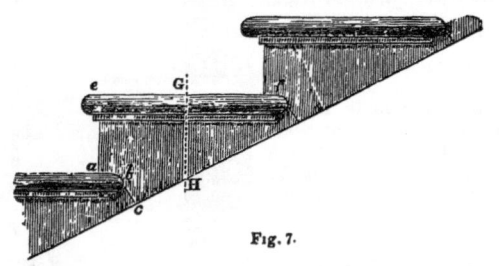

Fig. 7.

value for various stones, mortars, &c., in the nearest simple
numbers under the calculated value :—

TABLE I.—*Experiments on the Transverse Strength of Stones, &c., to the Case Equation* (14).

No. of Expts.	Substance tried.	Weight of a Cubic Foot.	Length, l.	Depth, d.	Breadth, b.	Breaking Weight.	Values of $f=\frac{1}{4}$ of the absolute Strength of a Sq. Foot.
		lb.	in.	in.		lb.	lb.
1	Statuary marble	169·12	15	1·075	1·075	25	65,000
2	" "	...	7·5	1·08	1·05	55	73,000
3	" "	...	7	1·075	1·075	65	78,000
4	Dundee stone	163·80	7	1·5	1·45	207	95,000
5	Portland stone	132	21	1·5	2	28	28,000
6	" "	...	12	1·45	2	50	30,000
7	" "	...	6	1·55	2 07	135	35,000
8	" "	...	15	1·25	1·2	30	26,000
9	Craigleith white sandstone	147·6	7	1·55	1·55	68·5	26,000
10	White sandstone from Hailes } Quarry	134·8	7	1·5	1·55	61·5	26,000
11	White sandstone from Longannet	138·25	9	1·525	1 45	46	26,000
12	" "	...	4·5	1·45	1·525	80	23,000
13	" "	...	3·5	1·55	1·55	116·5	24,000
14	Bath stone	...	2·75	1	1	29	17,000
15	Red porphyry	...	2·5	0·4	1	60	101,000
16	Welsh roof slate	...	6	0·25	1½	30	414,000
17	Scotch roof slate	...	6	0·25	1½	25	345,000
18	Common brick, new	...	4	2·5	4	201·5	6,900
19	" old...	...	4	2·5	4	171·5	5,900
20	Best stock brick	...	4	2·5	4	222	7,600
21	Mortar from an old castle in } Sussex	...	3	1	1	18·5	12,000
22	Mortar, common	...	0 75	0·35	1⅘	11·5	9,300
23	Mortar in the joints of two } inch cubes of stone, one } month after being joined....	...	pou. 8	pou. 2	pou. 2	liv. 6·75	1,400

Numbers 18, 19, and 20 are from Barlow's *Essay on the Strength of Timber* (p. 250), each being a mean of three trials. Number 23 is from Rondelet's *Traité de l'Art de Bâtir* (tome iii. p. 377), lowest result; the rest of the experiments were made by the writer of this article.

TABLE II.—*Experiments on the Direct Resistance of Stones, &c., to the Case Equation* (13).

No. of Expts.	Substance tried.	Weight of a Cubic Foot.	Area of Specimen.	Weight that pulled it Asunder.	Value of $f=\frac{1}{4}$ of the absolute Strength of a Square Foot.
1	Hard stone of Givry	147 lb.	96 lines	164 livres	8,400 lb.
2	Tender stone of Givry	130 "	324 "	183 "	1,400 "
3	Mortar of sand and lime sixteen years old	...	1 pouce	53 "	1,800 "
4	Plaster of Paris...	...	1 "	76 "	2,500 "
5	Adhesion of mortar to lias stone, joined six months	...	4 "	64 "	547 "
6	Adhesion of mortar to brick, joined six months.....	...	4 "	138 "	1,180 "
7	Adhesion of mortar to tile, joined six months	...	4 "	141 "	1,200 "

We have not here availed ourselves of the experiments of Gauthey (Rozier's *Journal de Physique*, tome iv.) on the transverse strength of stones; because those he fixed at

one end appear to have been injured in fixing, and only a
calculated result is given for the other specimens supported
at both ends. As to this, see the article on the STRENGTH
OF MATERIALS.

50. Several experiments have also been made, with the intention of measuring the direct resistance to extension or compression; but theory indicates so nice an adjustment of the direction of the straining force as necessary in these experiments, that the reader may expect the results to differ as widely amongst themselves as they are found to differ from theoretical calculation.

The experiments, Nos. 3, 4, 5, 6, and 7, are extracted from Rondelet's *L'Art de Bâtir* (tome i. p. 312). Nos. 1 and 2 are by Gauthey (Rozier's *Journal de Physique*, tome iv. p. 414.)

51. In the resistance to actual fracture, from a compressive force, the joint effect of cohesion and friction is concerned, and, therefore, a much greater force is required to crush than to tear asunder the same quantity of material. The resistance to fracture might be investigated on principles analogous to those we intend to employ in determining the pressure of earth against retaining walls, &c. (See the article CARPENTRY.) But we conceive that it is neither prudent, nor useful, nor necessary, to load the parts of a structure beyond that limit we have made the basis of our investigation. (See § 34.) Rondelet has observed, that the load under which a stone began to split was nearly always two-thirds of that which crushed it; but that stone of some kinds began to split with half the load that crushed it (*L'Art de Bâtir*, iii. 86 et 101). The value of f should, therefore, not exceed one-fourth of the force which splits stone; and, supposing the splitting force to be always half the crushing one, we shall have $f=$one-eighth of the crush-

ing force.

52. In this, as in the preceding tables, the reader will observe, that the results of all experiments are given in the original weights and measures; but that the value of f and the weight of a cubic foot are in English pounds avoirdupois, and for an English foot. The foreign weights and measures are distinguished by their foreign names.

The experiments, Nos. 1, 21, 22, and 36, were made by Gauthey (Rozier, *Journal de Physique*, tome iv. p. 406). Those numbered 3, 4, 5, 8, 9, 11 to 20, 30, 31, 32, 34, 37, 38, and 39, were made by Mr George Rennie (*Philosophical Transactions* for the year 1818). The others were made by Rondelet (*Traité de l'Art de Bâtir*, tome i. and tome iii.) We have selected those which will be most useful, with others of a more interesting and curious nature; such are Rondelet's experiments on the effect of

Principles of Stability and Strength. beating mortar, the strength and density of ancient mortar, and the resistance of stones used in ancient and in modern structures.

53. It was observed by M. Rondelet, in the course of his very numerous experiments, that it was not the heaviest stones which offered the greatest degree of resistance to compression, but those of a fine even grain and close texture, with a deep colour; that of granites, the most compact and perfectly crystallized was the strongest (*L'Art de Bâtir*, tom. i. 213, 215); and that, when all other qualities were the same, the strength was in proportion to some function of the specific gravity.

The writers who have contributed to our experimental knowledge of the strength of stones are not numerous. The chief are Emerson, in his *Mechanics*, 4to ed., p. 115; Gauthey, in his *Mémoire sur la Charge que peuvent porter les Pierres* in Rozier's *Journal de Physique*, tome iv., 1774, and in his *Construction des Ponts*, tome i., p. 267; Coulomb, in his *Mémoires presentés à l'Académie*, 1773; Rondelet, in his *Traité de l'Art de Bâtir*, tome i. et iii. (the latter volume contains the experiments made by Perronet and Soufflot); Rennie, in the *Philosophical Transactions* for 1818, or *Philosophical Magazine*, vol. liii.; and Tredgold, in the *Philosophical Magazine*, vol. lvi., p. 290.

Actual load put on stone in practice. 54. The last column in each of the three tables of experiments shows the greatest load that we suppose should be borne by a superficial foot of the different kinds of stone contained in those tables. We now propose to give the results of some calculations respecting the extent to which stone has in practice been loaded. The foreign ones are reduced to our own weights and measures, and the whole stated in round numbers.

The pillars of the Gothic church of All-Saints at Angers, of the stone, No. 24, Table III., support on each superficial foot a pressure of 86,000 lb.[1] The pillars of the dome of the Pantheon at Paris, the lower part of which are of Bagneux stone (No. 2151, Table III.), support on each superficial foot 60,000 lb.[2] The pillar in the centre of the chapter-house at Elgin, which is of red sandstone, supports on each superficial foot 40,000 lb.[3] The piers which support the dome of St Paul's in London sustain a pressure on

TABLE III.—*Experiments on the Resistance to Crushing.*

No. of Expts.	Substance Tried.	Weight of a Cubic Foot.	Area of Specimen.	Weight that Crushed it.	Value of $f=\frac{1}{3}$ of the Crushing Force for a Square Foot.
		lb.			lb.
1	Porphyry	179·44	20 lines	5,208 livres	640,000
2	„	174·9	4 pouces	119,808 „	500,000
3	Granite, Aberdeen blue	164·06	2·25 inch.	24,556 b.	196,000
4	„ Peterhead, hard and close grained	...	2·25 „	18,636 „	149,000
5	„ Cornish	166·37	2·25 „	14,302 „	114,000
6	„ gray	171·06	4 pouces	39,168 livres	165,000
7	„ rose oriental	166·32	4 „	52,704 „	220,000
8	Marble, white statuary	172·5	2·25 inch.	13,632 „	109,000
9	„ „	...	1 „	3,216 „	57,000
10	„ „	168·37	4 pouces	19,584 „	83,000
11	„ veined white, Italian	170·37	2·25 inch.	21,783 lb.	174,000
12	„ variegated red, Devonshire	...	2·25 „	16,172 „	129,000
13	Dundee stone	158·12	2·25 „	14,918 „	119,000
14	Craigleith stone, with strata	153·25	2·25 „	15,550 „	124,000
15	„ „	...	2·25 „	12,346 „	98,000
16	Bramley Fall sandstone near Leeds, with the strata	156·62	2·25 „	13,632 „	109,000
17	Portland stone	151·43	2·25 „	10,284 „	82,000
18	„	...	4 „	14,918 „	67,000
19	Culello white sandstone	151·43	2·25 „	10,264 „	82,000
20	Yorkshire paving stone	156·68	2·25 „	12,856 „	102,000
21	Hard stone of Givry	147·31	324 lines	11,208 livres	85,000
22	Tender stone of Givry	129·43	576 „	5,880 „	25,000
23	Saillancourt stone arches of bridge of Neuilly	141·31	4 pouces	7,280 „	30,000
24	Fourneaux stone pillars of All Saints at Angers	160·68	4 „	62,600 „	110,000
25	Bagneux stone pillars of the Pantheon at Paris	137·12	25 centim.	6,125 kilog.	62,000
26	Stone bridge of St Maxence	156·25	4 pouces	23,380 livres	97,000
27	Caserta stone, in Italy	169·87	4 „	36,142 „	150,000
28	Stone of temples at Pæstum	140·87	4 „	13,720 „	58,000
29	Travertino, ancient buildings at Rome	147·37	4 „	18,112 „	77,000
30	Derbyshire grit, a friable red sandstone	144·75	2·25 inch.	7,070 lb.	56,000
31	„ from another quarry	151·75	2·25 „	9,776 „	78,000
32	Roe stone, Gloucestershire	...	2·25 „	1,449 „	11,500
33	Tufa, from Rome	76·00	4 pouces	3,520 „	15,000
34	Chalk	...	2·25 inch.	1,127 „	9,000
35	Pumice-stone	37·81	4 pouces	2,100 „	8,900
36	Brick, hard and well burned	97·31	378 lines	5,280 „	34,000
37	„ pale red	130·31	2·25 inch	1,265 „	10,100
38	„ red, mean of two trials	135·5	2·25 „	1,817 „	14,500
39	„ Stourbridge fire	...	2·25 „	3,864 „	30,900
40	Mortar of lime and sand 18 months old	118·31	4 pouces	2,552 livres	10,900
41	„ „ 16 years old	...	4 „	2,864 „	12,000
42	„ „ not beaten 18 months old	101·56	4 „	1,866 „	7,900
43	Mortar of lime and pit-sand. 18 months old	99·25	4 „	2,475 „	10,600
44	„ beaten together, 18 months old	118·93	4 „	3,420 „	14,600
45	„ of lime and pounded tiles, 18 months old	91·06	4 „	2,896 „	12,300
46	„ beaten together, 18 months old	103·93	4 „	3,970 „	16,900
47	„ „ 16 years old	...	4 „	4,948 „	21,000
48	„ from an ancient wall at Rome	89·37	4 „	4,248 „	18,000
49	„ from the Pont du Gard	93·75	4 „	3,090 „	13,000
50	Lastrico, brought from Naples	62·5	4 „	4,664 „	19,400

each superficial foot of 39,000 lb.[2] The piers which support the dome of St Peter's at Rome sustain a pressure on each superficial foot of 33,000 lb.[2] The pressure on the key-stone (No. 23, Table III.) of the Bridge of Neuilly has been estimated for each superficial foot at 18,000 lb.[4]

In regard to these examples we have to remark that the calculators of them have considered the pressures as uniformly distributed over the pressed surface; but this can only be true when the direction of the resultant of the straining force coincides with the axis of the pier or pillar; besides, stones cannot be wrought absolutely level, nor bedded in perfect contact. From these circumstances, the strength of piers, columns, pillars, and arch-stones,

[1] Gauthey, Rozier's *Journal de Physique*, tom. iv., p. 409; and *Construction des Ponts*, tom. i., p. 273.
[2] Rondelet, *L'Art de bâtir*, tom. iii., p. 74.
[3] Telford, *Edinburgh Encyclopædia*. Art. "Bridge," p. 505.
[4] Gauthey, *Construction des Ponts*, tom. i., p. 260.

Principles of Stability and Strength.

should be estimated by equation (12), and when the line of direction falls within the pier, always making $y=$ half the least dimension of the section, an allowance which will include the effect of the greatest possible inequality of action. We shall, in that case, have

$$\frac{fbd^2}{d+\frac{6d}{2}}=\frac{fbd}{4}=W \quad \ldots \ldots \quad (17.)$$

If the pressure on the Bagneux stone in the piers of the dome of the Pantheon at Paris be estimated by this formula, it will be found that it is sufficient to split the stones, and this has actually happened.[1]

Principles of arches, domes, &c.

55. The chief elements of the theory of arches have already been given in the article BRIDGE (sect. ii.), to which we refer the reader, at the same time expressing a hope that the excellent article referred to will be useful in correcting some absurd notions respecting catenarian and other curves, which are too commonly entertained. The conical support of the lantern of St Paul's is a fine example of an appropriate form, whilst the catenarian dome of the French Pantheon exemplifies a scientific blunder of the first magnitude.[2]

The principles of domes, of groins, and of vaulting of every kind, are the same as those of arches, excepting that each kind has its peculiar manner of distributing the load on the different parts. See prop. M and N, art. BRIDGE.

Of the pressure of Earth, Fluids, &c., against walls.

Pressure of earth against walls.

56. When a high bank of earth, or a fluid, is to be sustained by a wall, as it is often necessary to do in forming bridges, locks, quays, reservoirs, docks, and military works, the construction is very expensive, however economical the means employed may be; hence it is desirable to devote some space to an object of which the importance is manifest.

Let EC, fig. 8, be the line according to which the earth would

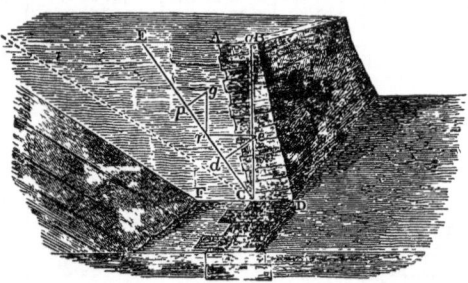

Fig. 8.

separate, if the wall were to yield in a small degree; then AEC will represent the section of the prism of earth, the pressure of which causes the wall to yield.

Put W = the weight of the prism AEC, when its length is unity.
R = the resistance of the wall, when its length is unity.
a = the angle ECa, which the plane of fracture makes with a vertical line.
c = the angle ACa, which the back of the wall makes with a vertical line.
F = the friction of the earth when the pressure is unity.
h = the vertical height of the wall a C in feet.
and S = the weight of a cubic foot of earth, water, or other matter to be supported.

If g be the centre of gravity of the prism of earth, the triangles rpg, CaE, being similar, the effort of the prism to slide in the direction EC, reduced for the friction, will be

$$=\frac{W(1-F \tan a)}{\sec a} \quad \ldots \ldots \quad (18.)$$

This effort is to be opposed by the resistance of the wall, which et us suppose to be collected at c, the centre of pressure, and, educing to the direction CE, the effect of friction being allowed or, it becomes

$$\frac{R(F+\tan a)}{\sec a} \quad \ldots \ldots \quad (19)$$

Hence, in the case of equilibrium,

$$\frac{W(1-F \tan a)}{\sec a}=\frac{R(F+\tan a)}{\sec a};$$

$$\text{Or, } R=W\left(\frac{1-F \tan a}{F+\tan a}\right) \quad \ldots \ldots \quad (20.)$$

57. But, in the case now considered the radius being unity,

$$W=\frac{h^2 S}{2}\left(\tan a-\tan c\right) \quad \text{Therefore, R} =$$

$$\frac{h^2 S}{2}\left[\frac{(\tan a-\tan c)\times(1-F \tan a)}{F+\tan a}\right] \quad \ldots \ldots \quad (21.)$$

And, from the state of the variable quantities in this equation, it is obvious that it has a maximum value, which determines the angle of fracture. By the principles of maxima and minima, the maximum pressure takes place when

$$\tan a=-F+\left(1+F \tan c+\frac{\tan c}{F}+F^2\right)^{\frac{1}{2}} \quad \ldots \ldots \quad (22.)$$

If the angle which the plane of repose (BRIDGE) makes with a vertical plane be denoted by i, then

$$F=\frac{1}{\tan i}; \text{ and } \tan a =$$

$$\frac{-1+(\tan c \tan {}^3 i+\tan {}^2 i+\tan c \tan i+1)^{\frac{1}{2}}}{\tan i} \quad \ldots \ldots \quad (23.)$$

If the back of the wall be vertical, tan $c = o$, and then this equation reduces to the simple form, which Prony obtained, of $\tan a = \tan \frac{1}{2}i$. . . . . (24.)

58. When we substitute in equation (21) the value of the tan a, which has been found in equation (23), it becomes R =

$$\frac{h^2 S}{2 \tan i}\left\{\tan i+\tan c+\frac{2}{\tan i}\right.$$

$$\left.-2\left(\tan c \tan i+\frac{\tan c+1}{\tan i}+1\right)^{\frac{1}{2}}\right\} \quad \ldots \quad (25.)$$

And when the back of the wall is vertical, it becomes

$$R=\frac{h^2 S}{2}(\tan {}^2 \tfrac{1}{2}i). \quad \ldots \ldots \quad (26.)$$

The tan i being the co-tangent of the angle of repose, if the matter to be supported be of so fluid a nature that it naturally assumes a sensibly level surface when at rest, the tan $\frac{1}{2}i$ becomes equal to unity, and consequently,

$$R=\frac{h^2 S}{2} \quad \ldots \ldots \quad (27.)$$

The same result may be obtained from the common principles of hydrostatics in the case of fluids.

Since the only variable quantity which enters into the calculation of the distance of the centre of pressure is the height h, whatever the nature of the supported material may be; therefore that distance counted from the base will always be $\frac{1}{3}h$, as in the pressure of fluids.

59. TABLE IV.—*Table of Constant Quantities necessary for calculating the Pressure of some Materials.*

Substance.	Angle of Repose.	Weight of a Cubic Foot = S	Value of R in Equation (26).	Value of R in Equa. (25) when $c=10°$.
Water..............	0°	62·5 lb.	R=31¼h^2	R=31¼h^2
Fine dry sand ...	33°	92 ,,	R=13·8h^2	R=4·84h^2
Do. moist	...	119 ,,	R=17·85h^2	R=6·2h^2
Quartz sand (dry)	35°	102 ,,	R=13·77h^2	R=4·64h^2

In sand, clay, and earthy bodies, the natural slopes should be taken when the material is dry, and the clay and earth pulverised. When any of these bodies are in a moist state, the parts cohere, and the angle of repose is greater, though the friction be actually less. The preceding table shows that the pressure of water is greater than that of any of the other kinds of matter, and from the nature of fluids it is evident, that if water be suffered to collect behind a retaining wall, calculated to sustain common earth only, it will

[1] Gauthey, *Construction des Ponts*, tom. i., p. 273.
[2] La charge considérable que cette voûte devait porter à son sommet, a déterminé à choisir pour la courbe de son ceintre *la chaînette. Traité l'Art de bâtir* ii. 308.)

most likely be overturned. Such accidents may be prevented by making proper drains.

60. The preceding analysis will apply, without sensible error to the curved walls which have lately become fashionable. Fig. 9 is a section of one of these walls, as executed from a design by Rennie. The vertical height, AB, 21 feet; the wall of uniform thickness, with counterforts 15 feet apart; and the front of the walls described by a 69 feet radius, with the centre in

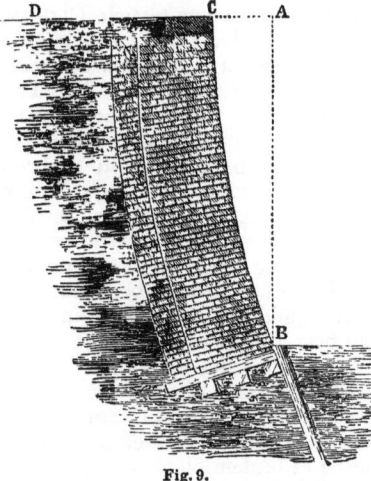

Fig. 9.

the horizontal line DA produced. The wall is built of brick, and the uniform part is 4·5 feet thick. The radius is usually thrice the vertical height of the wall; when this proportion is adhered to, the angle c will be ten degrees, for which the value of R is calculated in the table.

Resistance of Walls.

61. In the first place, we propose to investigate the resistance a wall offers to being overturned; and, in so doing, it appears desirable that the resistance of the mortar in the joints should be considered one of the elements of the strength of the wall. Good mortar adds much to the firmness of walls, and still more to their durability, and, all things considered, its first cost is less than that of bad; besides, the resistance of mortar to compression must be considered, for, in practice, we have no perfectly hard arrises to fulfil the conditions of common mechanical hypotheses.

Put A = the area of the wall.

  w = the weight of a cubic foot of masonry.

  y = the horizontal distance, $g\,a$, between the vertical passing through the centre of gravity of the wall, and the point where the axis cuts the plane of fracture, the same notation being applied to the other quantities as in the foregoing equations.

Let G, fig. 10, be the centre of gravity of the wall; and on the vertical Gg set off gI, the height of the centre of pressure; also, let IK represent A × w = the weight of the wall, and HI the force R of the earth.

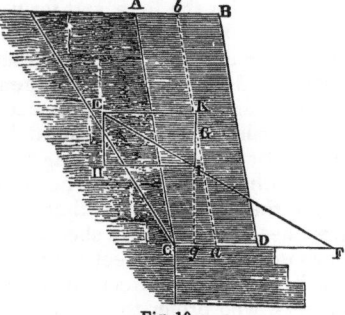

Fig. 10.

Then, completing the parallelogram, EI will represent the direction and intensity of the straining force; consequently, $\dfrac{R}{Aw} = \tan a$ . . . . (28.)

Which determines its direction, and its intensity is

$$W = \frac{Aw}{\cos a} \qquad \text{. . . . . (29.)}$$

But, we have found,

$$W = \frac{fbd^2}{d\cos a + 6l\sin a + 6y\cos a};$$ and as $W = \dfrac{Aw}{\cos a}$ (equa. (29.)); $\tan a = \dfrac{R}{Aw}$; equa. (28.); $l = \tfrac{1}{3}h$, art. 51; and b = unity; the equation reduces to $fd^2 - Adw - 6\,Awy = 2Rh.$ . . . . (30.)

If the section of the wall be a parallelogram, then $A = hd$, and $-\tfrac{1}{3}h\tan c = y$; these values of A and y being substituted in equa. (30), it becomes $-whd^2 + fd^2 + 3h^2wd\tan c = 2Rh.$ . . . (31.)

Or, $d = \dfrac{h^2w}{f-hw}\left(\dfrac{-3\tan c}{2} + \dfrac{\sqrt{2R(f-hw)}}{h^3w^2}\right.$

$$\left. + \frac{9\tan^2 c}{4}\right) \text{. . . . . (32.)}$$

When the section of the wall is a rectangle $y = o$, therefore equa. (31) reduces to

$$d = \sqrt{\frac{2Rh}{f-hw}} \qquad \text{. . . . . (33.)}$$

This last equation is also correct for a wall of which the back is vertical, and the front sloping. We suppress the investigation, to afford the young student an opportunity of proving that the diminution of weight is exactly counterbalanced by the alteration of the distance of the centre of gravity from the axis.

The tendency of a wall to slide forward may be easily prevented, by giving an inclination to the joints.

62. To illustrate these rules we shall give two examples, and in these show the construction of a table, which the reader may enlarge at his pleasure.

Example I. Let it be required to determine the thickness of a rectangular wall for supporting the front of a wharf 10 feet in height, the earth being a loose sand, and the wall to be built of brick.

The weight of a cubic foot of brick-work may be estimated at 100 lbs., and the resistance of mortar being valued at 5000 lbs. *per* superficial foot, the experimental value being 7900 lbs., Table III., Experiment 42, and the difference an allowance for any irregularity in building, consequently, $f = 5000$; $w = 100$; and by Table IV., $R = 13\cdot8h^2$; hence equa.

(33), $d = \sqrt{\dfrac{2Rh}{f-hw}} = \sqrt{\dfrac{2\times13\cdot8\times h^3}{5000-100h}} =$

$\sqrt{\dfrac{h^3}{181-3\cdot62h}}$. When $h = 10$ feet, then the thickness of the wall

$d = 2\cdot644$ feet. If h be made successively 10, 20, 30, 40, &c., feet, the numbers under the head of dry sand in the following table will be obtained, observing that they are only calculated to the nearest tenth of a foot.

The proper thickness being found for supporting one kind of material, that for any other may be easily determined; as the thickness varies as the square root of R, equa. (33). Let the thickness for dry sand be d, then

$\sqrt{13\cdot8} : \sqrt{31\cdot25} :: d : 1\cdot5d$ the thickness for supporting water.

$\sqrt{13\cdot8} : \sqrt{17\cdot85} :: d\ 1\cdot14d$ the thickness for supporting moist sand.

In this manner, by means of Table IV. the thicknesses for other kinds are easily calculated.

Example II. If a retaining wall be intended to support a sandy and loose kind of earth, to be constructed of brick, and to be inclined 10 degrees from the vertical, the thickness being uniform; it is required to determine that thickness for any given height.

By equa. (32), $d = \dfrac{h^2w}{f-hw}\left(-\dfrac{3\tan c}{2}+\right.$

$\left.\sqrt{\dfrac{2R(f-hw)}{h^3w^2}} + \left(\dfrac{3\tan c}{2}\right)^2\right)$ and as $c = 10°$, $\tan c = \cdot18$, hence

$\dfrac{3\tan c}{2} = \cdot27$, and its square $= \cdot0729$. Also $f = 5000$, and $w =$

100, consequently, $d = \dfrac{h^2}{50-h}$

$$\left(-\cdot27 + \sqrt{\frac{2R(\cdot5-\cdot01h)}{h^3}+\cdot0729}\right)$$

For sandy earth $R = 4\cdot8h^2$, therefore $d = \dfrac{h^2}{50-h}$

$$\left(-\cdot27 + \sqrt{\frac{4\cdot8}{h}-\cdot0113}\right);$$ and making h successively 10, 20,

&c., feet, the numbers obtained will be the same within one-tenth of a foot, as those in the following Table, column fifth, in which the thickness of leaning and curved walls for supporting dry sand is shown, at an inclination of 10 degrees.

z

Founda-
tions.

63. TABLE V.—*A Table of the Thicknesses for Retaining Walls, Revetments, Dock-walls, &c.*

Height of Wall.	Thickness of Rectangular Walls to support.			Thickness of Leaning and Curved Walls for supporting Dry Sand, the angle of inclination being 10°.
	Water.	Dry Sand.	Moist Sand.	
10 feet	4·0 feet	2·7 feet	3·1 feet	1 1 feet
20	12·9	8 6	9·8	2·8
30	29·2	19·4	22·2	5·2
40	62·5	41·7	47·5	9·2

Our investigation informs us that the mortar of high walls must be of a superior strength; indeed, we know that when its consolidation takes place, under considerable pressure, it is of much greater strength. According to what function of the pressure the strength increases, we have not experiments to determine, and we therefore point out the circumstance to the notice of experimental inquirers.

Construction of walls.

For further particulars as to construction of walls, particularly of railway embankments, see article CONSTRUCTION.

64. The proper quantity of mortar to be employed in stone-work is another point to which it will be useful to direct the mason's attention. A stone cannot be very firmly bedded upon a very thin layer of mortar; and if the stone be of an absorbent nature, the mortar will dry too rapidly to acquire any tolerable degree of hardness (Vitruvius, lib. ii., cap. viii.), however well it may have been prepared. On the other hand, if the bed of mortar be thicker than is neces--ary to bed the stone firmly, the work will be a long time in settling, and will never be perfectly stable.

When the internal part of a wall is built with fragments of stone, they should be closely packed together, so as to require as little mortar as possible. Walls are often bulged by the hydrostatic pressure of mortar, when it is too plentifully thrown into the interior, to save the labour of filling the spaces with stones.

The walls of houses are frequently built with hewn stone on the outside, and rubble stone on the inside. The settlement of these two kinds of stone-work during the setting of the mortar are so different, that the walls often separate; or where this separation is prevented by bond stones, the wall bulges outwards, and bears unequally on its base. These evils are best prevented by using as little mortar as possible in the joints of the interior part of the wall, and not raising the wall to a great height at one time.

III.—OF FOUNDATIONS.

On Land.

Foundations.

65. Having considered the nature of the materials to be used, and the scientific principles on which we should proceed in their disposition, we must turn to practical results, and first consider the foundations, or the base, upon which these superstructures are to be placed, so as to stand safely. When a good hard soil is easily accessible, as solid gravel, chalk, or rock, we have nothing to do but to excavate the surface mould to the sound bottom, and to build at once,

Footings.

first putting in the footings, which are one or more courses, forming a sort of steps, each a little wider than the other, and the wall that stands on them (see BUILDING, Pl. 43, fig. 5), according to the judgment of the architect. On hard ground one course of masonry, about half as wide again as the superincumbent wall, is ample. On softer ground it was usual to employ footings at least double the width of the wall, and frequently more; but since the invention, or rather revival, of the use of concrete, this is seldom or

Depth of footings.

never done. In this case, or when the ground is a deep clay, be the material used what it may, it should at least go so deep as not to be affected by change of temperature,

or the rising and falling of springs, as the alternate shrinking and swelling of the ground must affect the building. As has been shown (article CLIMATE, p. 768), frost seldom penetrates a foot into the ground in this country; but in clay soils, fissures, the consequences of drought, are found three feet and more in depth. The basis should, therefore, be below this point in such a stratum. If the ground be springy, it should be drained, if possible; if not, a foundation should be made with concrete as low as the lowest level of the water; or if very deep or boggy, piles must be used. The plan of building on sleepers and planking, so common a few years ago, is very bad, as they soon rot, and the building settles in all directions, as the greater weights crush the decayed timbers sooner than the lighter portions of the building. Where ground is alternately wet and dry, the best timber soon decays; even piles should always be wholly below the water.

Foundations.

Definition of concrete.

66. The use of concrete, except in very peculiar occasions, has entirely superseded every other artificial foundation. It may be defined as a sort of rough masonry, composed of broken pieces of stone or gravel, not laid by hand, but thrown at random into the trenches, cemented together with lime prepared in various ways, and thoroughly mixed with it before it is so thrown in. In this country, the lime is generally ground and mixed, when hot, with the stones; in France, the lime is first made into a paste, and the mixture is then called *béton*, not concrete.

History of concrete.

67. The use of this material is of very remote antiquity. It is no doubt the *signinum opus* of Vitruvius, and is described by Alberti. It is very common in mediæval buildings, walls and even arches frequently being made of it. In Rochester castle the staircases are composed of it; the under sides, or soffites, show to this day the marks of the boards which sustained it till it was set. Smeaton states that he was induced to employ it from the observation of the ruins of Corfe castle. Dance employed a sort of concrete in rebuilding Newgate, 1770–1778. The foundation of part of the new structure was a deep bog, and it was rendered available by shooting a quantity of broken bricks into the holes, mixed with occasional loads of mortar, in the proportions of four to one, and suffering them to find their bed.

Materials of concrete.

68. Any hard substance, broken into small pieces, will make good concrete. That most used is gravel, or ballast. This should not be sifted too fine, as the sand which is left will mix with the lime, and form a sort of mortar, and assist to cement the stones together (see I., sec. 29). If broken stones, or masons' chips, are used, it is well to mix some sharp sand with them. The general rule is, that no piece should exceed a hen's egg in size. In this country the lime is generally ground, and used hot. It is mixed with the ballast by scattering it among the stones, and turning them over with a shovel, water being at the same time thrown upon the mass.

Laying concrete.

It is then immediately filled into the trenches, sometimes by shooting from stages erected for the purpose, six or eight feet above the work. But this process has been very justly censured as uncertain by eminent engineers, who prefer to put it in layers of not more than 1 foot in thickness, and to level each course, and ram it down thoroughly. About one-sixth part of lime is generally used.

Swelling of concrete.

When too hastily put into the trenches, the lime, which has not had time thoroughly to be slacked, will continue to do so, and the mass will puff or swell, and sometimes cause considerable mischief. The author has seen the wing walls of bridges thrust out by this means. From some experiments made by the Architectural Publication Society, where the materials were carefully mixed, no change took place in the bulk. The lime, if it can be procured, should be hydraulic; and concrete is much improved by the addition of the volcanic sands. The French authors recommend, as good proportions, one-fifth hydraulic lime, one-fourth pozzolano, one-eighth sharp sand, and the rest

Founda-
tions.

broken stone or gravel; or 20 per cent. hydraulic lime, the same of terrass, the same of sharp sand, 15 per cent. of gravel, and 25 per cent. of broken stone. Perhaps, after all, the very best concrete is made of a simple mixture of gravel with Portland cement.

Foundations in Water—Ancient Systems.

Loose stones shot in.

69. These are often made by shooting quantities of loose rough stones, &c., into water, at hazard, till the mass finds its own bottom, and becomes solid by degrees. When a sufficient quantity has been shot, so as to appear above the surface, the material is levelled, and the superstructure erected upon it. Most of the *break-waters* (see Art.) have been thus constructed. This method is called by the French, foundations *à pierre perdue* (see BREAKWATER, HARBOUR, &c.) Concrete has been used in the same way with great success. Where it is practicable, it is very desirable first to bring the bottom to a level by *dredging*. (See NAVIGATION, INLAND, p. 68.)

Piling.

70. Another method is to drive a number of piles, side by side, through the mud or other soft soil till they reach a sound bottom, the heads are then cut off to a level, and a platform of timber, or better still, of flat stones, is laid on them, and the superstructure erected. The heads of the piles should always be under water. For small operations they may be driven without dams; but for larger a more elaborate system must be pursued, the most effectual of which are coffer-dams.

Coffer-dams.

71. These are of as great antiquity as the time of Vitruvius, and most probably much older. That author, however (lib. 5, cap. *ult.*), describes the method of forming them, and calls them *arcæ*. Like those of later times, they were composed of two parallel rows of piles driven into the water, and kept together by strong horizontal timbers, and continued around the place where the proposed work is to be erected, so as to form a sort of box or coffer entirely round it. The two rows are kept in their places by other timbers, and the vacant space between such double row stuffed full of clay and weeds, till the whole is quite tight, the water is then pumped out by proper engines till the ground appears, which is then levelled and excavated to a solid stratum, if such is readily accessible. The foundations of the pier, &c., are then laid, and the superstructure carried up to above the water-level, when the dam is removed. If solid ground is not easily accessible, structural piling is resorted to. In large works these are of whole timber, pointed at the end, and shod with iron, to facilitate their penetrating the earth.

Pile-engine.

They are driven by a weight called a monkey, which is raised by a machine called a pile-engine, worked by horse or other power, now frequently by steam. When the monkey is raised to a sufficient height, it is suddenly liberated by a contrivance much like a double pair of scissors, and falls with great impetus on the head of the pile, and of course forces it downward into the bed of the stream. When driven to a proper depth, the heads are cut off to a level line, cross timbers are bolted on these, and the superstructure erected, as shown in BRIDGE (Plate CXLV. fig. 3.) The piles of the dam should not be drawn, as that would allow the water to form holes, and to work under the foundations, but they should be cut off close to the bottom of the stream. Fig. 11 shows a plan of the coffer-dam and pier of a bridge erected by Rennie on the Thames. The outer lines are the parallel piles which keep out the water, and form the coffer; these, as will be seen, are strongly bolted together, both across and lengthwise, and also braced diagonally. The general plan is elliptical, the better to resist the pressure of the water,—a course afterwards adopted in the coffer-dams of London Bridge. The plan of the pier is within this, and is shown in four quarters:—A shows the

plan of the great piles, driven down to the solid; B shows the heads of the same piles, when cut off and tied together by strong cross timbers; C shows the planking laid thereon; and D the first courses of the masonry.

Founda-
tions

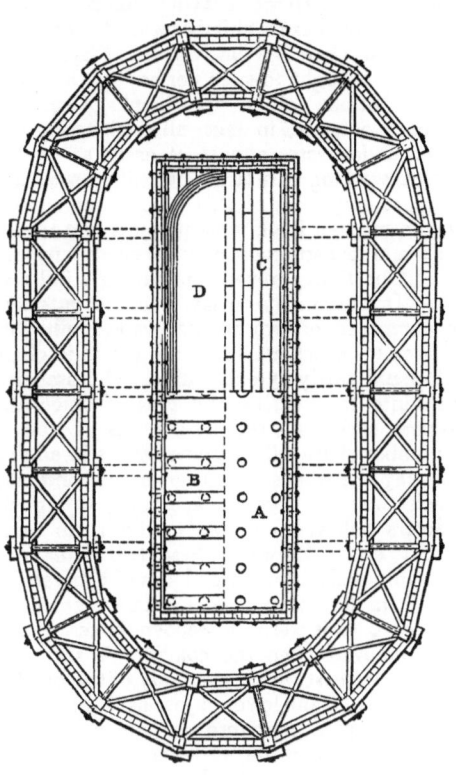

Fig. 11.

Caissons.

72. As the system above described is extremely expensive, especially before the introduction of steam-power for pumping, &c., a very ingenious method was introduced into this country by a Swiss named Labelye, and first used at Westminster Bridge. The bed of the stream was first carefully levelled by dredging. (See NAVIGATION, INLAND, page 68.) Strong frames of timber were then constructed, having upright sides like those of a box, and being about the same area as a coffer-dam. These were floated over the place where the piers were to be built, and the masonry of each pier commenced inside these large cases (the word caisson meaning a large box or *caisse*). It was, in fact, like building in the bottom of a large flat-bottomed barge. Of course, as the weight increased this barge or caisson would gradually sink. The sides were somewhat deeper than the river, and well caulked and pitched to keep the water out and enable the men to work. When the first course of stone was laid and cramped together, the water was let into the caisson by sluices, and the whole sunk to the bottom. It being found this was not sufficiently level, the sluices were closed, the water pumped out, when the whole floated again, and the bottom was again dredged and levelled. This operation was performed three times before the work settled to a level bed. The pier was then built up to a height above water-level, when the sides of the caisson were removed, and used again for the next pier. Blackfriars Bridge was afterwards built on the same plan, but in consequence of the removal of old London Bridge, the scour of the river increased so much as to work under the piers; these directly began to settle in all ways, and the bridges must both be rebuilt. The system of caissons might, perhaps, nevertheless, be used in still waters, but it is manifestly improper for a sharp current, and still worse for a tidal river.

Foundations.

New systems.

Diving-dress.

Dover harbour.

Foundations in Water—Modern Systems.

73. The great advance in all objects of engineering, and the desire to avoid expense, has led to some very ingenious and novel systems of laying foundations in water.

74. The first of these is chiefly owing to the invention of the diving dress. (See DIVING, vol. viii., Plate CCV.) This has now been brought to such perfection, that excavating, levelling the bottom, setting large blocks of stones, cutting off the heads of piles, in fact, all engineering operations are effected under water almost as easily as on land; the men, in fact, working in a large bag filled with air. The most extraordinary work of this kind is now being carried on at Dover, where a huge mole projects into the ocean, of an extent and construction that exceeds any thing yet achieved. Our limits prevent our giving a full account of this work. It must suffice to say, that the outsides of the pier are composed of two parallel walls, built with large blocks of granite, which are sunk into the solid chalk partly by dredging and partly by excavation. A number of piles[1] are driven into the sea, on the tops of which are strong cross-sleepers, each traversed by a series of iron rails, on which a number of travelling cranes move in all directions. These convey the stones exactly over their intended beds, on to which they are lowered, according to signals given by the divers below, who then cramp them together as well as they can. Between these two outer walls is a filling-in, composed of immense blocks of concrete, made of the ballast from the beach and Portland cement. These blocks are cast, as it were, in wooden boxes, the sides of which are removed when the concrete is set. Pieces of rope or chain are cast within the body of the block, and by them they are raised and lowered just as if they were masses of stone. The pier, therefore, is composed of concrete faced with granite; and the work stands extremely well.

Iron piles.

75. The next important change in building in the water is the substitution of iron piles for those of timber. Their success emboldened engineers, and from small piles,

Cylinders.

driven in the usual way, large cast-iron cylinders came into use. These vary, according to the nature of the work, from 3 or 4, to 6 or 7 feet in diameter. They are first lowered into the water in a vertical position, and driven down as far as they will go without much difficulty. A quantity of clay is then thrown in round the outside of each, to keep the water from coming in under the bottom as little as possible. That inside is then pumped out, and workmen descend and excavate the bottom, sending up the stuff in buckets, just as a well is excavated. The cylinder then sinks partly by its own weight, and partly from weights above, as the earth is excavated beneath to the depths required; each cylinder has a series of flanches, on which another is screwed from time to time as the former sinks into the bottom. When the cylinders are sunk to a proper depth, they are brought to a level at the top, and a platform of girders and planking is fixed for permanent use. In many instances the cylinders have been filled in solid with concrete after they have been thus submerged.

Rochester new bridge

76. The most extraordinary operation of this kind has, however, just been executed in the piers of the new bridge at Rochester. The ground here was very difficult to work, being composed partly of rag rock and partly of very hard chalk, and it was found almost impracticable to sink the piles in the manner last described, which is successful enough in clay. To overcome this difficulty the following plan was devised:—a proper stage of piles, sleepers, &c., was first erected, and a number of cast-iron cylinders, each 9 feet long and 7 feet in diameter were bolted together in proper lengths. As it was necessary to employ diving-bells to dredge the bottom, it was considered that each

cylinder might easily be converted into a sort of bell by putting on it an air-tight cap. This was done, and an ingenious method contrived by which the men could pass in and out of the cylinder without admitting the air. It would exceed our limits to go into all the details of this invention, which was called an air-lock, probably from its permitting or hindering the passage of air, much as the lock-gates on a canal do with that of water. But there was this great difference between the method described in [sec. 75] and this new method. In the former the water was pumped *out* of the cylinder, but in this air was forced *into* the cylinder below the air-tight cap and locks, and the water *driven out* by its pressure. The men then entered through the air-locks, and excavated the ground under the cylinder, which descended by its being heavily weighted above. As each cylinder sunk, the cap was removed and another cylinder screwed on, so that some of the piles (as in fact they may be called), consisted of seven pieces, and measured over 60 feet in length, one half of which was buried in the bed of the river. The girders, tying the heads together on the top, as well as the skew-backs, &c., from whence the arches spring, were then fixed in the usual way. (See IRON BRIDGES.)

Foundations.

Screw-piles.

77 A still more curious method of employing iron piles is the invention of Mr Mitchell, and succeeds admirably in soft ground and even in sands. These are hollow, and of wrought-iron, varying in diameter from 5 or 6 inches to about a foot. They terminate at the end with screws of various shapes (see figs. 12, 13, and 14), and are screwed into the bed of the river or the bottom of the sea, as the case may be, to such a depth as to hold the pile firm, their heads are then connected together with sleepers, &c., and the intended superstructure erected. The lighthouse on the Chapman Sand, in the mouth of the River Thames, is built on piles 7 inches in diameter and about 40 feet long, the screw part is of cast-iron about 4 feet in diameter. They are screwed down till only 2 or 3 feet remain above the sand; on their heads are cast-iron cylinders, braces, &c., which support the lighthouse, which is entirely of wrought-iron. The piles are only seven in number; one driven in the centre, and the others at equal distances around it.

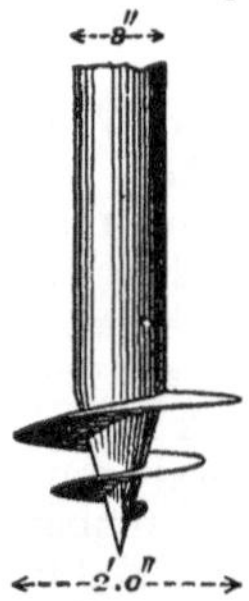

Fig. 12.

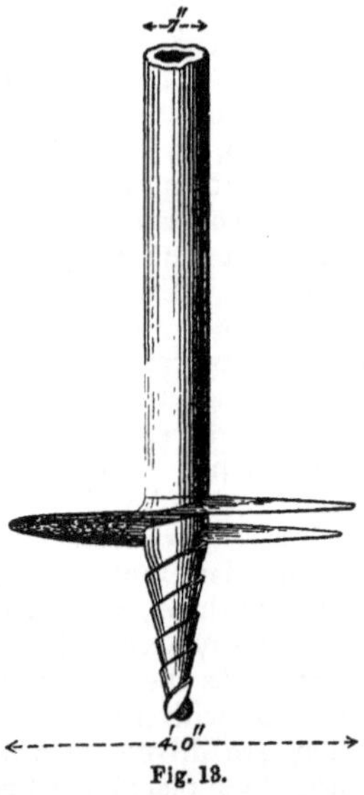

Fig. 13.

78. Several very ingenious adaptations of iron for coffer-dams have been tried with success in various places; they are all, however, more or less expensive, not only in construction and removal, but because they entail constant expenses in pumping.

Iron coffer-dams.

79. The method lately invented by Mr Page for getting in the piers of the new bridge now in course of erection at Westminster is, however, so novel and so important, we feel our work would not be complete without a short description of it. Figs. 15, 16, and 17 show the plan, the long and cross-sections of half each pier. Rows of

Westminster new bridge

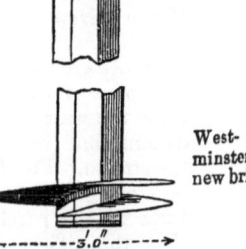

Fig. 14.

---

[1] These piles form a scaffold, and are removed when the work is done.

strong elm piles, AA,[1] about 30 feet long, are driven into the bed of the river, as shown in the plan, passing first through the gravel, which is about 4 or 5 feet thick, and then going about 20 feet into the London clay. There are about 140 or 150 piles to each pier, ranged alternately in threes and fives, around these a range of cast-iron piles

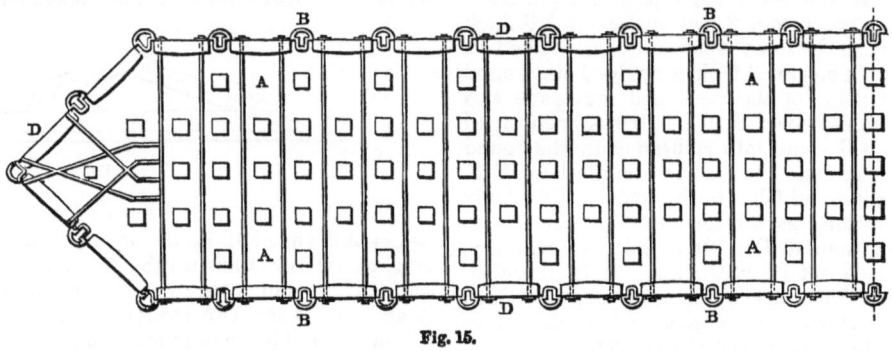

Fig. 15.

BB, are driven about 4 feet apart, as shown in the figure. These are round, 15 inches in diameter, and have strong grooves cast on each side of them. They, however, go into the clay only 10 or 12 feet. Into these grooves large plates of iron, which the engineer calls "plate-piles," are fitted and driven down between the piles, BB. They are marked CC on the cross-section (fig. 16), and, as will be seen, go about 10 feet into the blue clay, and extend to about a

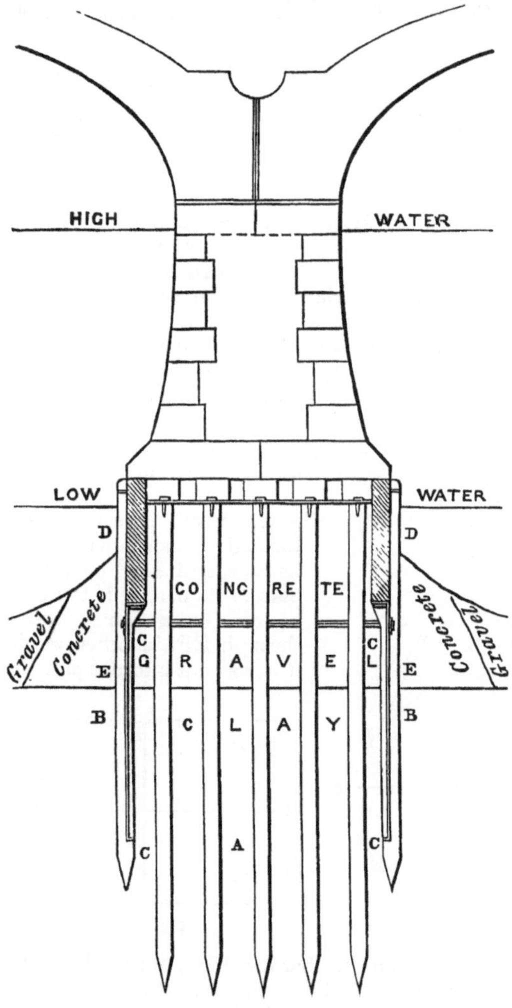

Fig. 16.

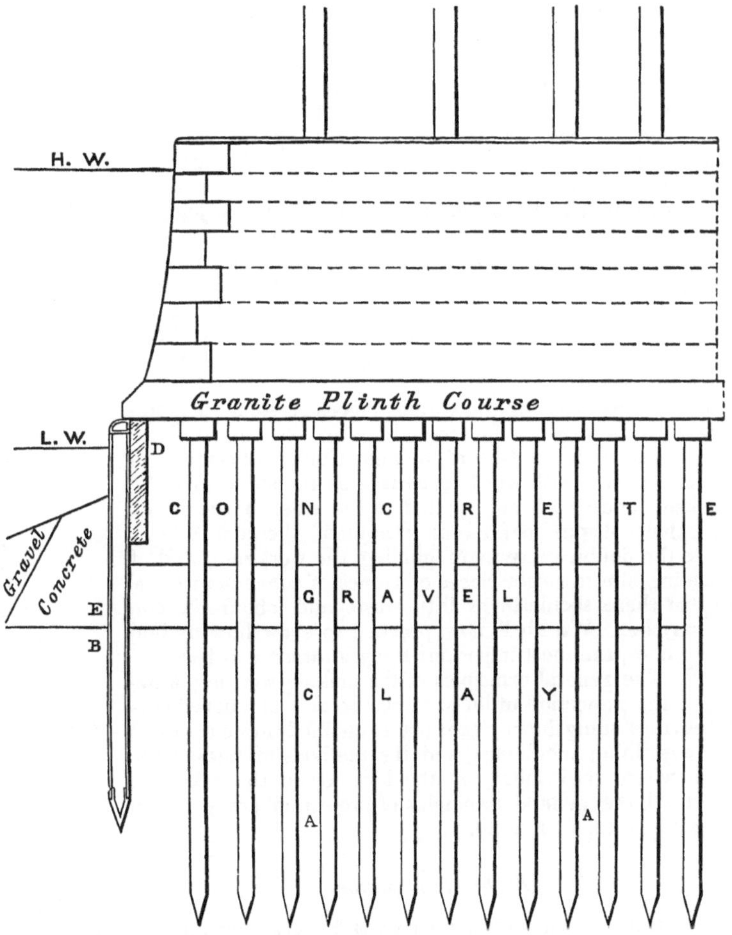

Fig. 17.

foot or two above the natural bed of gravel. Upon these are a series of slabs of granite, DD, placed edgewise, retained in their places thus,—The bottom rests on the "plate-pile," CC, the edges are secured to the round iron-pile A, and the tops to the other masonry. The plate-piles are secured together by two sets of ranges of iron-rods passing through the pier and tying them together. These are all fixed by the divers. It will be seen, therefore, a sort of case or box is made which surrounds the wooden piles AA on all sides. The loose sand and mud

---

[1] The same letters apply to all three figures.

is then dredged out, and the case filled up solid with hydraulic concrete, in which, of course, the piles are embedded, and the whole forms one solid mass to about a foot above low water-mark. At this level the tops of the piles are cut off, and on each top a stone 2 feet square and 1·6 thick is bedded, the spaces between which are again filled in solid with concrete. The gravel is then dredged out around the pier on the outside of the case, and the space also filled with concrete, as shown at EE. It has been urged that the steamers will come into collison with the round piles, BB, and break them, so that the granite slabs, DD, will escape, as it were, and fall into the river. This, however, cannot be as long as the concrete E remains in its place, as the top of the slab D is secured by the masonry, and the bottom would not be accessible. It is, however, intended to protect the piles by floating booms, which would prevent the chance of collison, and would act as safeguards for the steamers as well as the bridge.

For the action of waves, running water, &c., on walls, piers, &c., see HARBOUR.

IV.—STONE-CUTTING.

Previous descriptions. 80. The different methods of reducing stones to shape by the axe or scabbling-hammer, the saw, or the tool; of dressing by the chisel or point; the nature and value of plain work, sunk work, moulded work, and beds and joints; the various sorts of bond, and of rubble, coursed, or ashlar, with their proper backings and quoin-stones; the mortar, and such methods of working and setting the beds of stones, that the faces may not chip or flush; the mode of securing and strengthening work, by discharging and relieving arches; by hoop-iron bond, cramps, dowels, joggles, plugs, &c.; the descriptions of copings, cornices, string-courses, blocking-courses, sills, landings, balconies, paving, curbs, steps, hearths, chimney-pieces, &c.; and of columns, with their beds, joggles, flutes, &c.; in short, all that relates to the mechanical part of stone-cutting, has been already given, article BUILDING, page 146, to which we refer our readers.

Moulds, and to find the lines for. 81. Before working the various pieces of stone, it is necessary to prepare certain moulds, which are generally of thin metal, wood, mill-board, or some similar substances, of the exact form with which each face of the stone is to be worked; and which are applied to the sides of the stone, and their shapes marked or "scribed" thereon. They serve the double purpose of guiding the workmen, and of enabling him to select pieces of stones of convenient sizes, so that there should be as little waste and labour thereon as may be. We shall now proceed to show how to find these lines, the most important thing a mason can learn.

Making working drawings. 82. The general principles of the making working drawings; the projection of lines, of planes, and of curved surfaces; the finding the angles of planes inclined to one another; the describing mouldings, and the methods of finding the lines where they mitre, on the level or on the rake; in short, all the general principles of projection are given in JOINERY, sect. I., 1 to 20.

Arches, to Describe.

Lines for arches. 83. But as arches form no part of joiners' work, and as our articles ARCH, ARCHITECTURE, BRIDGE, &c., though containing full scientific developments on the subject, necessarily involve the highest branches of pure mathematics, we shall refer the readers of abstract science to these articles, and shall proceed to give a few problems in descriptive geometry for the use of masons, as we have done before for joiners.

84. First, of circular arches. These, if of moderate size,

Stone-Cutting. may be set out by a long lath, one brad awl as a centre, and another to trace, or by beam compasses; but if they are flat, the centre is frequently at such a distance as to render this inconvenient, if not impossible. The best way *Circular arches, to draw.* *Flat arches.*

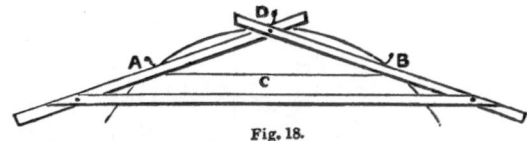

Fig. 18.

to proceed, far better than most of the cyclographs, is thus —Let AB (fig. 18), be the width of the arch, and CD its height; set out their width and height on a floor, or on some boards joined together for the purpose; drive pins at A and B. Take two straight rods AD, DB, place them so that their sides may touch A and B, and their intersection coincide with D. Tack them together, and also a third lath across, to keep them at the same angle; place a pencil at D, which will trace the curve if the rods are moved to the right and left, and are kept pressed against the pins A and B.

85. *Another way* (fig. 19), let the letters represent the *Another way.*

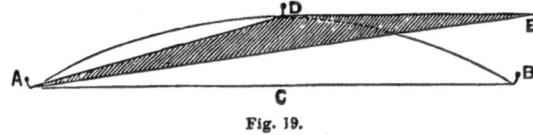

Fig. 19.

same): join AD, draw DE parallel to AB, and make DE equal to AD, then cut out a triangle in wood, or form one as above with three laths; put a pin into the board at D, and a pencil at the same point of the triangle, and it will trace half the curve AD; reverse it, and it will trace the other half DB.

86. Next to circular forms, the most common are those from the sections of a cone, of which the most usual is the *Arches from conic sections.* ellipse; the parabola is sometimes used, and so is the hyperbola, but very seldom. As these forms are necessary in setting out Grecian mouldings, as well as in almost all problems in masonry, it is proposed to treat of all three as shortly as possible.

87. Let ABC (fig. 20) be the section of a cone, or one cut *The ellipsis.* through its axis downwards to its base, and ABD be half the plan of its base. If the cone be cut by a plane passing through EF, in other words, if it was cut into two parts by

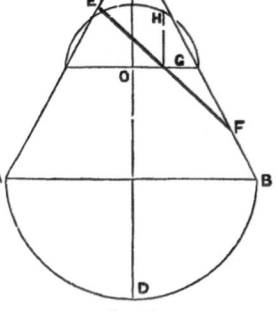

Fig. 20.

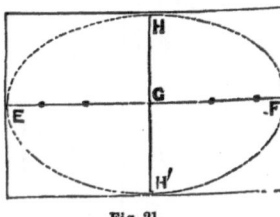

Fig. 21.

a large flat knife passing in that direction, but wholly above the base, its section would be an ellipse. Then EF would be its length, or axis-major. To find its height or axis-minor, bisect EF in G, through G draw a line parallel to AB, touching the sides of the cone; bisect this in O, and on it describe the semicircle as shown; then draw GH parallel to the axis of the cone CD, and this line GH will be the half the height, or the semi axis-minor, and the ellipse will be described within the parallelogram EF, HH' (fig. 21).

Stone-Cutting.

Various ways of describing the ellipsis.

88. This may be done, first, by finding the foci,[1] and describing the ellipsis by a cord or string, thus:—Take the distance FG (fig. 22) in your compasses, and from H as a centre strike the portion of a circle 1, 2, 3, then 1 and 3, where it cuts EF, are the foci. Stick two pins, or brad awls, in their points, and strain a string round H, 1 and 3; place a pencil at H, and move it round, keeping the string tight. The pencil will draw the ellipse, Or it may be done by a trammel. (See vol. viii., Plate CCXXXI. fig. 1; and see article ELLIPTOGRAPH.)

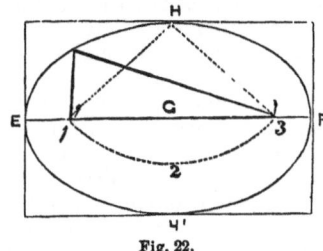

Fig. 22.

To describe an ellipsis by ordinates.

89. Let the parallelogram be, as before (fig. 23). Divide EG, E 4, each into any number of equal parts; here they are divided into 4; from HH' draw lines through 1, 2, 3, as shown, and where they intersect are points in an ellipsis; mark these points, bend a thin lath round them, and strike the curve. The same repeated for the other quarters will complete the ellipse. These methods are mathematically true; but as it is difficult to get laths to bend round a large curve, and very difficult to tie a string exactly to the proper length; and also to prevent its stretching when tied, other means have been taken as approximations.

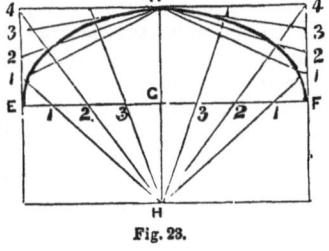

Fig. 23.

To describe an ellipsis by compasses.

90. It is of the utmost importance to the mason, whether in Gothic or other work, to remember, that in all cases where two circles touch each other, the two centres from which they each are struck, and the point in which they touch, must lie in the same straight line; in other words, a line drawn from one centre to the other must pass through the point of contact, a point where they each touch without cutting each other. If this rule be not strictly attended to, any curve coming out of another will not flow freely, but must be crippled. If we attempt to draw an ellipse with the compasses, we must strictly attend to that rule. Now, if we take a diagram similar to fig. 23, but instead of four parts, we suppose the diameter EG and the side E divided into two parts, as fig. 24. Now, joining the lines H 1, H' 1, we get at their intersection, a point K, which, as has been shown before, is a point within the true curve of the ellipse. To prevent the confusion of so many lines, we will now suppose a similar point found in the right hand quarter, and will call it L. Now, we have to draw two portions of circles, one through L and F, the other through LH and K, and they are, of course, to touch each other in L; then this point of contact L and the centres of the circles must be all in the same straight line. To do this we have to join HL by a straight line, and bisect it by a perpendicular line. This is done by taking

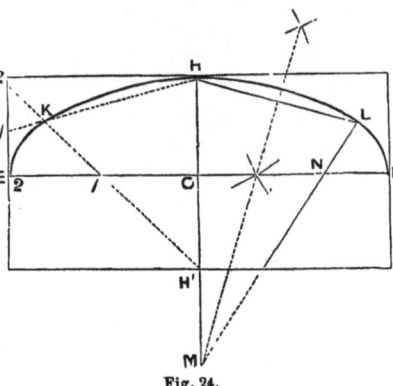

Fig. 24.

any convenient opening of the compasses, and from the points H and L cross two small segments, as shown, and draw a line through them on till they meet HG produced to M. Join ML, cutting EF in N. From N as a centre strike the segment of a circle LF; and as LN and M are in the same straight line from the centre M, strike the segment KHL, which will pass through H, and touch LF at L. We have, therefore, a curved line passing through the points KHLF, all of which points are in the curve of a true ellipse, though the curve itself is not so, but parts of circular arcs. Proceed in the same way for the other side of the parallelogram, and the ellipsis is complete. If greater accuracy is required, divide EG, E 2, each into three or four or even more points, instead of two, and proceed on the same principle given above, viz., join the first pair of points so formed (beginning at the top) by a straight line, bisect this by a perpendicular projected till it cuts the line HG, and then join the next pair of points, and bisect again as before.

The parabola.

91. Let ABC (fig. 25) be the section of a cone as before, and ABD half the plan of its base; if the cone be cut through by a flat straight cut or plane in the direction EF, but it always must be parallel to one of the sides (as it is here to the side CB), the section of the cone thus cut will be a parabola, and its height or axis will be EF. From F draw FG at right angles to the base AB, and FG is half the base of the same.

Fig. 25.

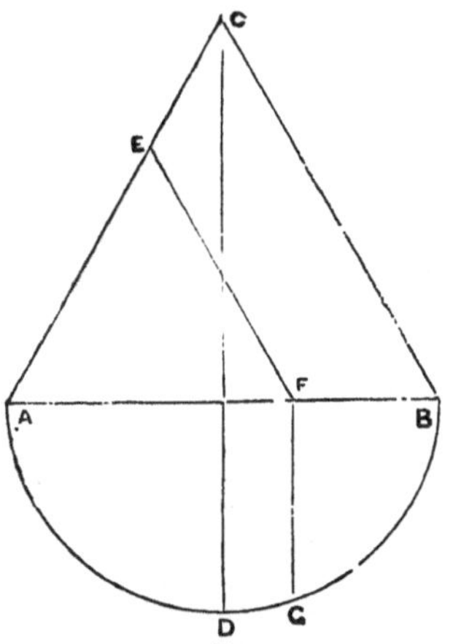

To describe the parabola.

92. Draw a parallelogram (fig. 26) of which the height shall be equal to FE, and the base equal to twice FG. Divide the height and each half the base into any number of equal parts (in this case they are divided into 4), draw co-ordinates crossing each other as shown, and a b c will be points in the curve; bend a lath round, and strike the curve, which will be a parabola. In a similar way a parabola may be drawn, any height and width being given.

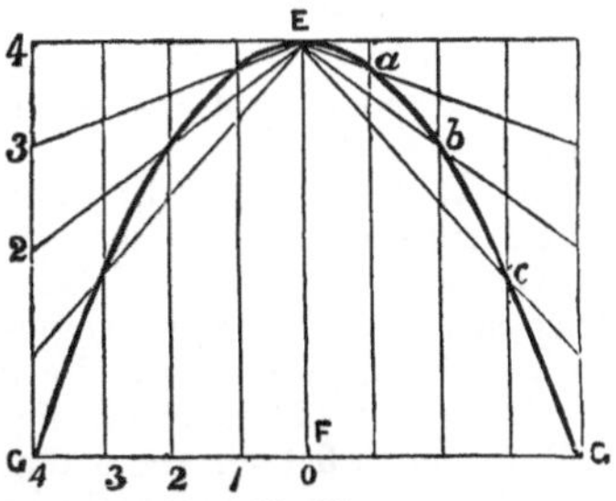

Fig. 26.

The hyperbola.

93. Let ABC, &c. (fig. 27), be the cone and its plan as before, and let it be cut by a plane at EF, falling within the base, but not parallel to the side. The curve of the section is a hyperbola, and FE is its height. From F draw FG at right angles to AB; then twice FG is the base of a parallelogram, within which the hyperbola lies.

To draw the hyperbola.

94. Construct a parallelogram (fig. 28) of the height EF, and width twice FG; then in fig. 27 produce the side AC and the line EF till they meet in H, and make EH (fig. 28) equal EH (fig. 27), divide the sides as shown into any number of equal parts, cross the co-ordinates as before, and through their intersections draw the curve.

The regular solids.

95. It is now necessary to say a few words on the regular solid figures with which the mason has most to do. These are the cone, the cylinder, the globe, and the spheroid. The cone may be considered to be formed

by taking a right angle triangle (ABC, fig. 29), holding it upright, and turning it round as if the perpendicular side AB was an axis. The surface traced by the hypothenuse AC as it turns, would be the surface of a cone, and that

Fig. 27.   Fig. 29.

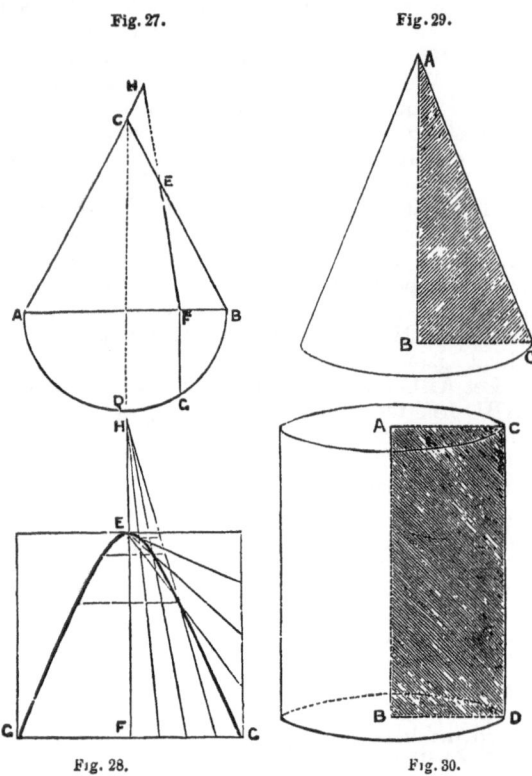

Fig. 28.   Fig. 30.

by the third side BC the base. In the like manner the parallelogram ABCD (fig. 30), turned round on one of its sides, as AB, would describe a cylinder; a semicircle (ABC, fig. 31) turned on its chord AB, a globe; and a semi-ellipsis, on its axis, a spheroid.

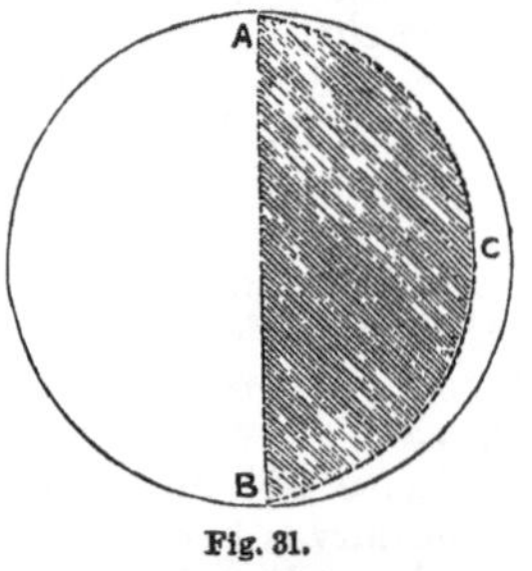

96. If a cone be cut by any plane surface passing through its vertex, the section will be a triangle; if cut at right angles to its axis, the section will be a circle; if cut obliquely (as fig. 20), but at so flat an angle that the plane does not cut into the base, the section is an ellipsis; if by a plane parallel to one of its sides (figs. 25 and 26), which, of course, must cut the base, it is a parabola; if of less angle, or perpendicular (figs. 27 and 28) to the axis, it is a hyperbola.

Fig. 31.

97. If a piece of paper could be wrapped round any figure so as to cover its side exactly, and this then be spread out flat, it is called its development. Let ABC (fig. 32) be a right cone, the plan of which is drawn beneath it. Suppose this plan divided into any number of equal parts, as 12,

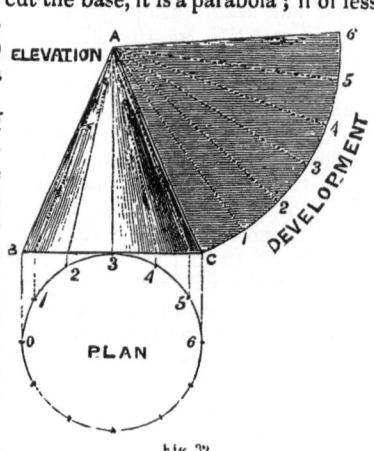

Fig. 32.

and the cone divided in a similar way by lines drawn to the apex A; suppose these lines drawn in wet ink, or paint, and the cone then rolled on its side along a flat surface, A 1, A 2, A 3, &c., would mark the different portions of the cone; and C 1 2 3–6 will be the stretch out of half its base, and AC 6 the development of half its surface. The other half will be merely a continuation of the same. To draw this development from the centre A with the radius AC, describe the segment of a circle, and set off upon it the openings 0, 1–6, equal to the circumference of the half-circle 0–6 on the plan, and join A 1, &c. Each division will be the development of a portion of the cone—thus, AC 3 will be a quarter of the cone, AC 4 one third, &c.

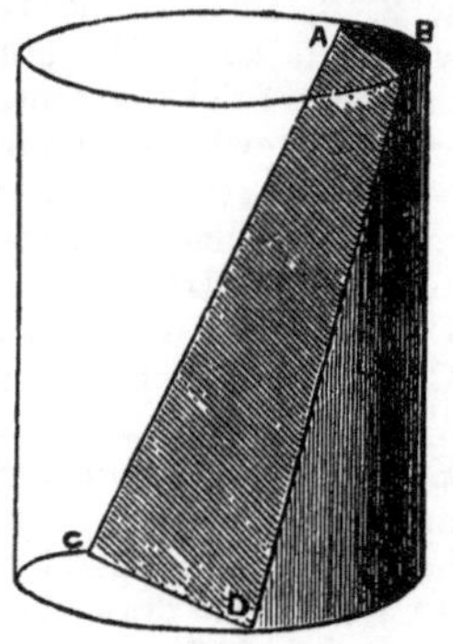

98. If a cylinder be cut parallel to its axis, the section is a parallelogram; if the plane be inclined from the axis so as not to pass outside the ends or bases (fig. 33), a trapezium, ABCD; if at right angles to the axis, a circle; and if cut obliquely to its axis, the plane not cutting into the ends, the section is an ellipsis (fig. 34), of which AB is the larger and CD the lesser axis.

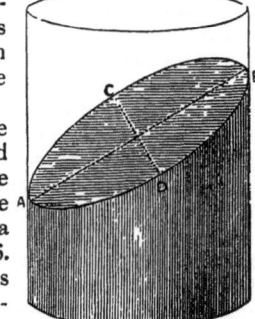

Fig. 33.

99. Let ABCD (fig. 35) be the elevation of a cylinder, and its plan be below it; and suppose it divided as directed for the cone, and then rolled along a plain surface from E o to F 6. 0–6 is the stretch out of half its base, and E o–F 6 is the development of half its surface. To draw this proceed as directed for the cone.

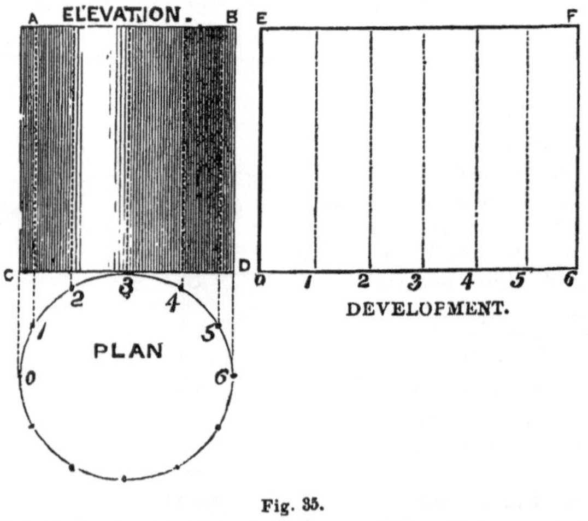

Fig. 34.

ELEVATION.   DEVELOPMENT.

PLAN

Fig. 35.

100. Let ABCD (fig. 36) be a cylinder cut square at the base BC (of which the plan is below), but obliquely at the top, as AD.

The upper section (sec. 98), will be an ellipse, of which AD will be the length, and the radius of the circle (on the plan) equal to half the height EF. This may be described by any of the foregoing methods. Divide the circle at the base as before, and draw lines on to the surface of the cylinder; set out the stretch out from 0 to 6 as before directed, and draw the perpendiculars—as 0 0′, 1 1′, 2 2′.

&c.   Again, draw perpendiculars from 0 1 2, &c., on the stretch out, and cross them by horizontal lines drawn from

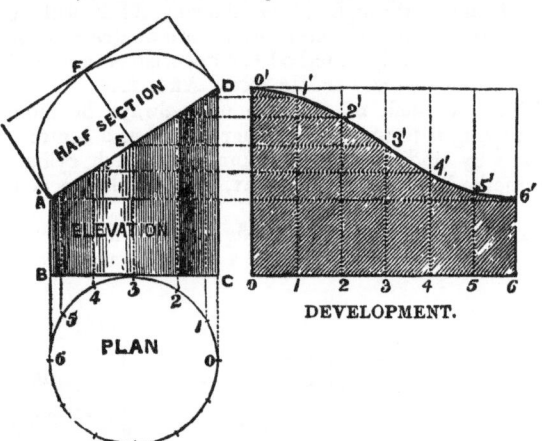

Fig. 36.

similar points in the elevation, cutting the lines on the development at 1′ 2′ 3′, &c., the surface between the curved and the straight line is its development.   In like manner the other half may be drawn.

Sections of a globe.

101. If a globe be cut through by any plane, the section is a circle.   If the plane passes through the centre of the globe it is called a great circle, being the largest that can be cut out of that solid.   For the development of a globe see 117, 118.

Sections of a spheroid.

102. If a spheroid be cut by a plane at right angles to its larger axis, the section is a circle; all other sections are oval, of which all parallel sections are similar.

These foregoing problems are indispensable to a mason, and should be carefully studied and gone over till they are thoroughly understood.   Several other mathematical curves have been used for arches, as cycloidal, catenary, cassinoidal, &c.; but the trouble of setting them out, and the difficulty and confusion the workmen often get into has led engineers and architects to the use of circular and elliptical arches almost exclusively.   Indeed, in many cases where the latter were formerly used, it has been found better and cheaper practically to substitute segments of circles.

To find the Joints of Arches.

Circular arches.

103. In all cases the joints of all arches should be at right angles to the tangents of the curve.   For this reason those of circular arches should simply be drawn to the centre of the circle of which they form a part.   If that be very distant, as is frequently the case with very flat arches (fig. 37), divide the arc into as many equal parts as may be.

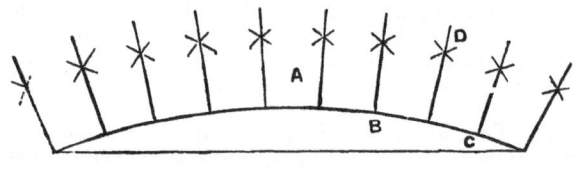

Fig. 37.

convenient, having relation to the size of the arch-stones or voussoirs, and taking care that the middle of the key-stone A shall be exactly in the centre of the curve.   Then, from every alternate centre as B and C, with any convenient opening of the compasses, strike two small arcs, as at D, crossing each other; these will give the angles for the joints.   In fact, this way is very nearly true for all flat arches, whether circular or elliptical.

104. Let EHF be an ellipse, and *a* the point from which the arch-joint should be struck; find the foci 1 3 (sec. 88, fig. 22), and draw lines from each through the point *a*, as shown towards *b* and *c*.   Bisect the angle *b a c*, by the segments as shown; draw *b a*, which is the arch-joint.   In cases of arches drawn by segments of circles, as fig. 39, the arch-joints are simply parts of radii from the respective centres ABC.   This figure

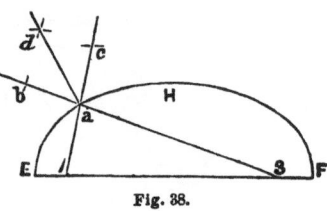

Fig. 38.

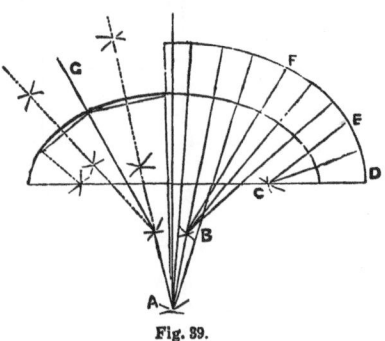

Fig. 39.

shows the case of an arch drawn from three centres in each quadrant instead of two.   The arch-joints from D to E are to be ruled from C, from E to F they are to be drawn from B, and F to G from A.   The method of finding these points is shown on the other half of the diagram.   (See also sec. 90.)

105. The same rule holds good with parabolic arches as with others, that the arch-joint shall be at right angles to a tangent, passing through the point whence it rises.   Then let ABC be a parabola, drawn as shown, sec. 92, figs. 25 and 26.   It is required to find the arch-joint at D, draw DE

Joints to a parabolic arches.

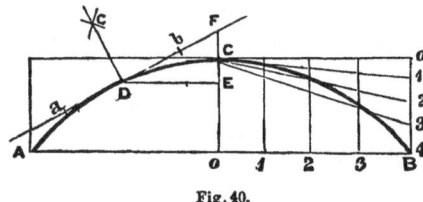

Fig. 40.

parallel to AB, and make CF equal to CE; through D, draw FD *a* as shown; take *a*, *b* equal distances from D, and raise the perpendicular D *e*, which is the arch-joint.

To find the Arch-Stones or Voussoirs.

106. Let fig. 41 be the elevation of the arch, below which is its plan.   Then, as the stones are generally longer than they are high, the workman will take a block of about the size required, and will work any long side, say the bed *a b e f* (see plan).   He will then work the ends square, and of course they will be parallel with the face of the wall; he will then, by means of bevels, set off the angles *a b c*, *b a d*, *b c d*, and work these sides; he will then apply a mould the exact shape of the arch-stone, as shown by the shaded lines, to each end of the stone he is working, and describe its exact shape, taking care to keep the moulds out of winding, and work off the waste between the dotted lines *a d*, *b c*, reducing them to the curve shewn by the mould, and the work is complete.

107. If the arch be splayed inside, as on the right hand

Voussoirs of right arches.

2 A

side of the plan, fig. 41, and *l m* be the half of the width of the square part, and *l n* that of the splayed, then the

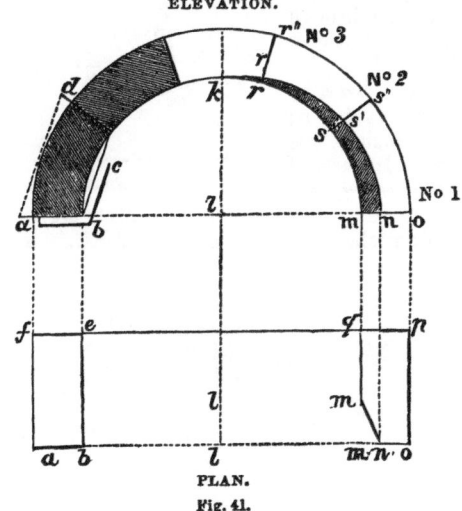

ELEVATION.

PLAN.

Fig. 41.

line marked *r'*, *s'*, *n'*, is an ellipsis. Draw this by any of the methods given above, taking *l n* as the length and *l k* as the height. Then the plan of the bed (No. 1) at the springing will be *m n o p q* (see plan, fig. 41). The next bed *s s' s''*, fig. 42, No. 2 ; the third, *r*, *r₁*, *r₂*, &c., &c.—the

No. 3. No. 2. No. 1.

PLANS OF BEDS OF ARCH-JOINTS.

Fig. 42.

depth of the square part, and the entire thickness of the arch being always the same.

108. Let AB represent the plan of an opening (as a door or window) which cuts obliquely into a wall. If the

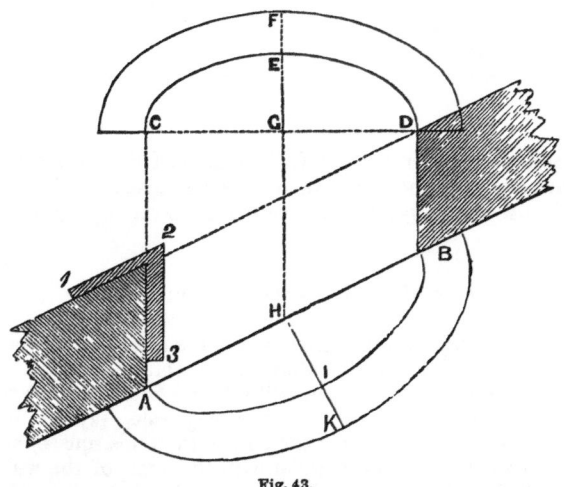

Fig. 43.

face of the arch be a circle, proceed as in 41 ; but instead of working the stones square at the end, they must be worked according to the bevel 1 2 3. The face bevels are obtained as shown before. But if the section of the arch be an ellipse, as CED (fig. 43), and the depth of the voussoir EF, the delineation of the face will be found by drawing HI at right angles to and bisecting AB, and making HI

equal to GE as the height, and AB the length, and describ- ing an ellipse ; also make IK equal to EF, and describe a second ellipse through K as shown ; AKB will be the face of the arch. If these arches are of great length, or very oblique, another method has been lately used, which is given in the separate article SKEW ARCHES.

109. The soffit of an arch may always be found by considering it to touch a cylinder placed as a centre under it, and by finding the development of such cylinder by the methods given in JOINERY, and in the sections 99, 100.

110. Let ABC..F, *a b c*, *f* (fig. 44) be the section of the arch of a tunnel, or of one in a terrace wall, the face of

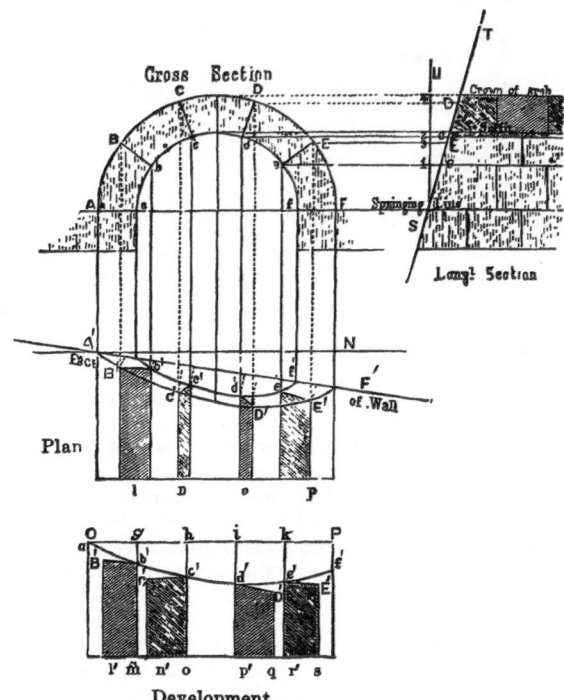

Cross Section

Longl. Section

Plan

Development.

Fig. 44.

which is not square with the section, but inclined on the plan at the angle NAF', and which also batters or falls back out of the perpendicular at the angle UST on the longitudinal section, SU being perpendicular to the springing line. From the intrados *d e f*, &c., draw the faint lines to ST, as 1*e*, 2*d*, &c., and from the extrados draw the dotted lines 3E, 4D, &c. ; also let fall the perpen- dicular *b b'*, *c c'*, *d d*, on A'F'. Then take the several divisions intercepted between ST and the perpendicular US, and transfer them to the line A'F', and set them off at right angles to it as shown ; then *a' b' c'..f*, will be the line of the intrados on plan, and A', B'.. F', that of the extrados ; and the portion between those lines will represent the battering face of the arch. From B'*b'*, C'*c'*, &c., let fall perpendiculars as shown, and the shaded portions will re- present the places of the arch-joints on the plan.

111. Take any line OP equal to the stretch out of the inner circle of the arch *a b .. f*, and divide it into as many parts as there are arch-stones. Draw perpendiculars from these points as shown in the figure, and set out on them *g b'*, *h c'*, &c., equal to the same on the plan. Then *a' b'*, *c'*, &c., is the development of the line of the intrados ; and each portion, as *c' d' o p*, *d' p r e'*, is the mould for the soffit of each respective stone. Next make *l m*, *n o*, &c., equal to A*a*, or D*d*, in the section which will represent the depth of each arch-joint, and make B'*l*, C'*n*, &c., in the development equal to B'*l*, &c., on the plan. B'*b l m*,

d'D$p\,q$, &c., are the moulds for the arch-joints, as shown by the shaded surfaces.

112. Let A$a\,b\,c$, &c. (fig. 45), be a circular-headed arch

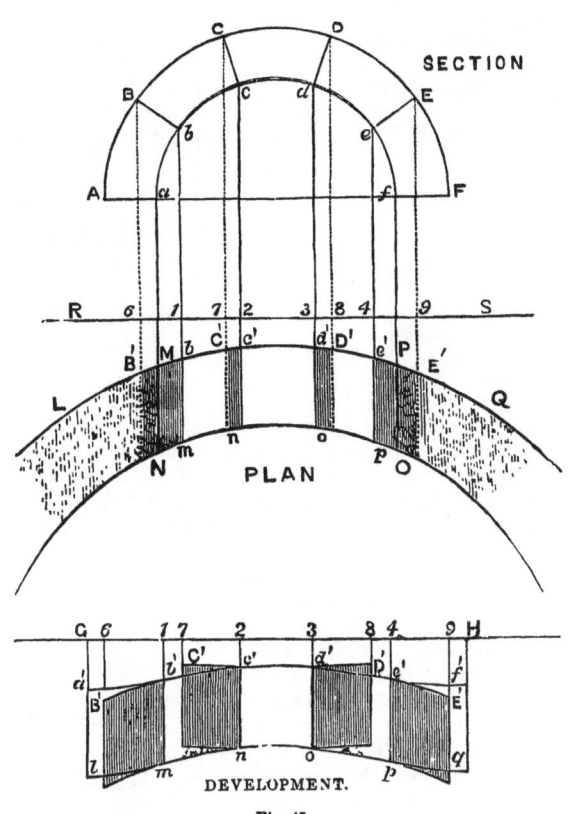

SECTION

PLAN

DEVELOPMENT.

Fig. 45.

over a door or window, in the wall of a circular bow or tower; the plan of which is LMNOPQ. Draw the arch-joints bB, cC, &c., in the section, and let them fall on the plan in b'B', c'C', then the shaded part will represent the lines of the arch-joints on the plan.

113. Draw any line in the plan RS perpendicular to the lines let fall. Also draw GH in the development, and make it equal to the stretch out $a\,b\,.\,.\,f$, and divide it into as many parts as there are arch-stones at 1, 2, 3, &c.; draw the perpendiculars, and set down 1, b', 2, c', on the development equal to 1b', 2c', &c., on the plan. Do the same with 1m, 2n, 3o, &c., and the line $a'\,b'\,c'$, &c., is the front line of the soffit, and $l\,m$, q, the inner line of the same; also $c'd'\,n\,o$, $b'c'\,m\,n$, will represent the soffit of each arch-stone.

114. From the lines $b'm$, $c'n$, &c., set out the depth of the arch-stone equal to Bb, &c., on the cross section, and on the line GH set down 6B', 7C', &c., equal to the same on the plan. The shaded portions will then be the moulds for the arch-joints.

115. The above problem may be used for any sort of arch, or any form of wall, whether cylindrical or conical, care being taken that all section lines be first carefully transferred to the plan and then to the development. Perhaps the method of finding the arch-joints will be better understood if the reader will suppose the stones of an arch to be transparent or made of glass; and the joints or surfaces where they touch to be blackened. If the eye were placed exactly above such an arch, it would present the appearance as shown in our plans.

Spherical Vaults, Domes, Niche Heads, &c.

116. Spheres or globes may be developed in two ways,—

first, in horizontal sections or zones; and, second, in vertical sections or gores.

117. Let Ao, 1, 4 (fig. 46) be the section of a dome, and 1, 2, 3, 4 the places of the arch-joints; AB 4, the springing line; B, o, the line through the vertex; produce this line towards C, and through 2 1, 3 2, 4 3; draw chords to the vertical line, cutting it in a, b, C. Then o is the centre of the eye of the globe, or the key-stone of the dome; a the centre of the development of the first zone, 1, 2; b, the centre of the second 2, 3, C, of the third 3, 4; and so on for as many divisions as you may think proper to divide the globe into.

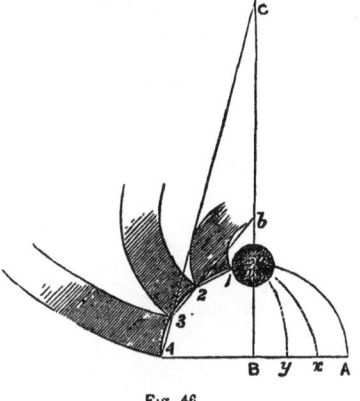

Fig. 46.

118. Let $x\,y$, in fig. 46, be a gore or section of globe, made by two vertical planes passing through its centre. Then those planes are portions of great circles 101, and their sections are quadrants. Let ADB (fig. 47) be one of these quadrants, draw DC parallel to AB, and make it equal to half the gore required. Draw CF parallel to DB; produce AB to F, and CF to E, and join DF. Divide AD into any number of parts, as 0, 1, 2, 3, &c., and draw ordinates parallel to AB, cutting DF and CF. On CE set out 0, 1, 2, &c., so that CE will be the stretch out of AD. Draw ordinates through these points, and make 1, a; 2, b; &c., in CE, equal to the same lines

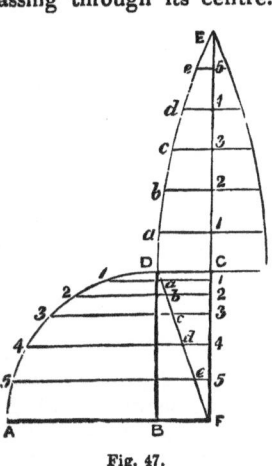

Fig. 47.

intercepted between DF and CF. The curved line which passes through D $a\,b\,c$, &c., will be the development of half the gore. Transfer these points to the other side, and the whole is completed.

By these two methods any portion of the surface of a globe may be drawn, and if cut out of any thin material will form a mould for any of the curved surfaces.

To find the
joints of a
dome.

119. Let fig. 48 be the section, and fig. 49 the plan of a semi-circular dome, or a niche head. For greater clearness this is drawn with only two ranges of arch-stones besides the key, but the theory is the same for any number. Then $a\,b\,c\,d$ upon the plan will be the bed; $a\,b\,e\,f$ on section, the end mould of the first voussoir; the face moulds will be found by the preceding problems. It remains to find the upper bed at $e\,f$. With the radius Of strike part of a circle (fig. 50), and set out on it the line $f\,p\,g$, equal to $f\,g$ stretched out on the plan. If the thickness of the dome be parallel, as on the left side of the diagram, set out such thickness $e\,f$ (fig. 48), and then describe the circle $e\,h$, and $e\,f\,g\,h$ is the mould. The bevels can easily be set from the plan and section. If the arch-stones are intended to bond in with the level masonry, as shown on the right side of the section (fig. 48), make $p\,q$ (fig. 50), equal to $p\,q$ in the section, and proceed as before. In this latter case it will be better to work the arch-stones thus: from the point 1 on section let fall (through p) the perpendicular 1 p 2, and draw 4, 3, parallel thereto; then 2, r, 3, 5, 6, on the plan

(fig. 49), will be the mould of the bed : 1, 4, 2, 3, on the section, fig. 50, of the end : and the plane 1, 2, 8, a square angle

SECTION.

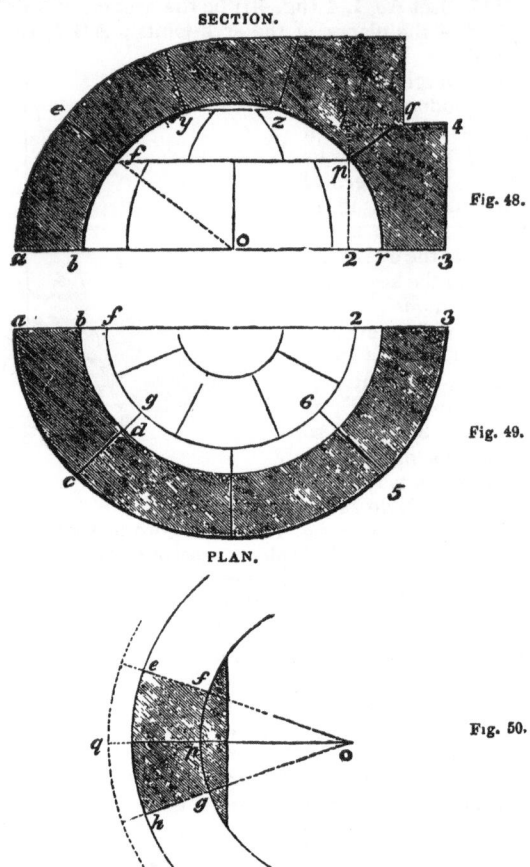

Fig. 48.

Fig. 49.

PLAN.

Fig. 50.

with the plan. The block will be the portion of an upright cylinder shown by 1, 2, 3, 4, 5, 6, 7, fig. 51. Then make q 4, 2 r, 2 p, &c., equal to the same on the section, and work off the waste as shown. As the radii of a globe are all equal, the curve is alike in all parts. A radius rule is therefore very convenient, this is found (fig. 50) by simply making $f g$ a chord to the curve, and the section included by the lines forms the rule. The key is simply part of a cone, the widest diameter of which is $w x$, and narrowest $y z$.

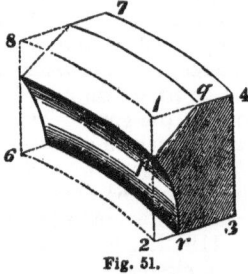

Fig. 51.

Arched Vaults which intersect each other or Groins.

120. Almost all groins (except the mediæval) are either circular or elliptical, and their crowns $a\,b$, $c\,d$ (fig. 52), form straight lines. If the openings and heights be equal, the two sections (Nos. 1 and 2, fig. 52), will be alike, and the centers will intersect on the groin lines. Then (99) the intersection of two cylinders will be an ellipse of the length $a\,b$ on the plan, fig. 53, and of the height of $c\,d$ equal to that of either of the sections. Having thus the width and height, the ellipse may be drawn by any of the methods given above.

121. If arches are of equal height, and level both at crown and springing, but of unequal widths, one arch at least must be the portion of an ellipse, and the intersection an ellipse more or less regular. Let the main arch, No. 1, be semi-circular ABC (fig. 54), and the cross arch, No. 2, of the width DE, but of the same height OC as

No. 1. Join AB on plan No. 3, and draw OC perpendicular to AB, and equal to OC, Nos. 1 and 2. Then the

Fig. 52.

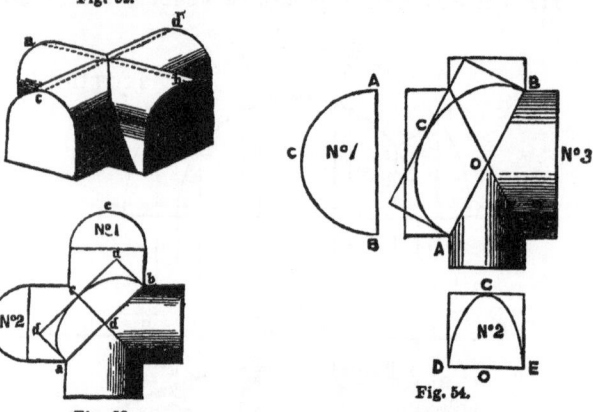

Fig. 54.

Fig. 53.

line of intersection will be an ellipse, the length of which is AB, and the height OC. The section, No. 2, will be an upright ellipse, the greater axis of which is OC and the less DE, all which may be drawn as before. If the cross vault be a semicircle, the main vault will be a flat ellipse, and the groins found as before.

122. These are either from the same springing, in which case the crowns of the intersecting arches are not level ; or the crowns are level, in which case the springing of the lesser arch is higher than that of the greater. In fig. 55, No. 1 is the section of the main arch, No. 2 of the cross arch, No. 3 is the plan of a groin in which both arches are semicircular, but No. 2 is less than No. 1, and of course of different heights. In the smaller arch, No. 2, take any number of points, 0, 1, 2, 3, 4, and draw horizontal lines from them to the perpendicular 0, 1, 2, 3, transfer them to No. 1, return them to the curve, and draw co-ordinates from these points, meeting in the line $a\,b\,c\,d$, which is the curve of intersection. These are called under-pitch groins, and sometimes Welsh arches. In the same way the lines may be formed where one arch intersects another obliquely (fig. 56), where the respective numbers and letters refer to the same things.

Fig. 55.

Fig. 56.

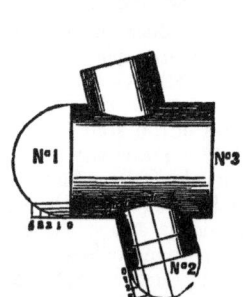

123. In this case the springing of the smaller arch must be higher than the other. This often occurs in Roman work, and almost always in Norman groining. No. 2 (fig. 57) is the smaller semicircle, the springing line of which is AB. Make OC in No. 2 equal to OC in No. 1, which is the height from the crown to the springing of the main arch. Take any points in the curve 0, 1, 2, 2, 3, &c., of No. 2, draw them to the perpendicular A 4, and transfer them to No. 1, and cross the co-ordinates as before, and the line of intersection will be found. But it must be noticed this is a curved line, and not a straight groin point, as will be the case in the next problem.

124. This is often the case at the end of a building with a canted bow, or a church that ends with a hexagonal apsis. Let ABCD (fig. 58, No. 3), be the plan of cylindrical vault, the section of which is DEF, No. 1, and let it be pierced by the portion of a conical vault GHIK. Produce

To find the
lines when
a cylinder
is inter-
sected by a
portion of
a cone.

GH, IK, O, till they meet in O the vertex of the cone, and till *d e* in No. 2 is made equal to DE in No. 1,—*d e*, of course, being parallel to GI. Draw the semicircle *d e f* in No. 2 equal to DEF in No. 1, and divide each into the

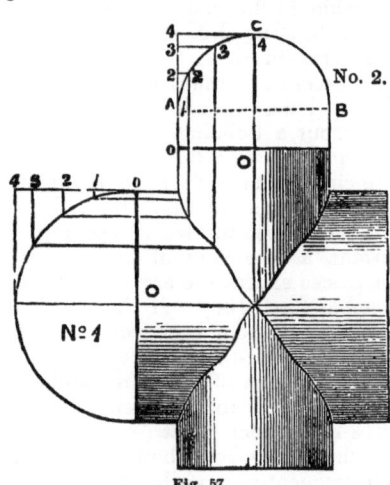

Fig. 57.

same number of equal parts. In No. 2, draw ordinates, first perpendicular to *d e*, and thence radiating to the point

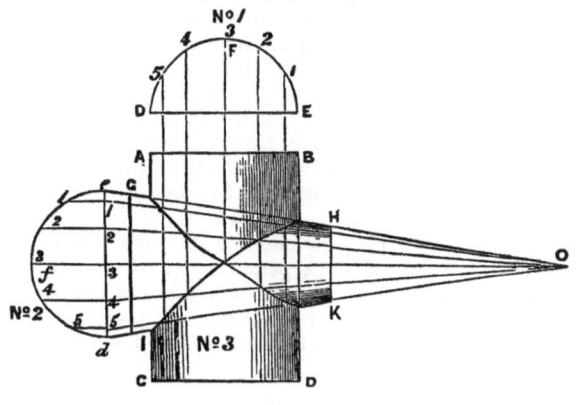

Fig. 58.

O ; cross these by the co-ordinates from No. 1, all as shown in the diagram, and the curved lines GK, IH, will show the intersections.

125. The lines for these may be formed on exactly the same principles, viz., from a plan, two sections, and a double set of ordinates. In descending vaults, however, we have to remark, if it is intended the cross arches should be cylindrical, the groined points will be curved, as in 123 and 124. If it be intended that the groined points shall be straight, and should intersect in the middle of each bay, then the section of the cross arch will be an oblique oval.

If a mason will carefully master these problems, he will find very little difficulty in any methods of stone-cutting.

The same problems give the lines for centering, the practical method of executing which is found in CARPENTRY, &c., &c.

Mouldings.

126. All the sections for Roman mouldings are given in JOINERY, p. 807, but as those used in stone work, particularly in Grecian architecture, are parts of Conic Sections, and not struck by compasses, we give a short problem by which they may all be easily set out.

127. Let (fig. 59) the moulding required, be an ovolo, the height of which (to the point where the moulding curves backward) is AC or BD, and the greatest projection AB

or CD ; and let CE be a tangent line, or line which the curve must touch but not cut. Produce CA to F, and make AF equal to AC, and AG to ED. Divide GB BE each into the same number of equal parts as 5. Draw the co-ordinates from F and C to the respective numbers, their intersections will trace the curve. If BE be more than Stone-
Cutting.

To draw a
Grecian
moulding,
having its
height, its
greatest
projection,
and a tan-
gent to the
curve.

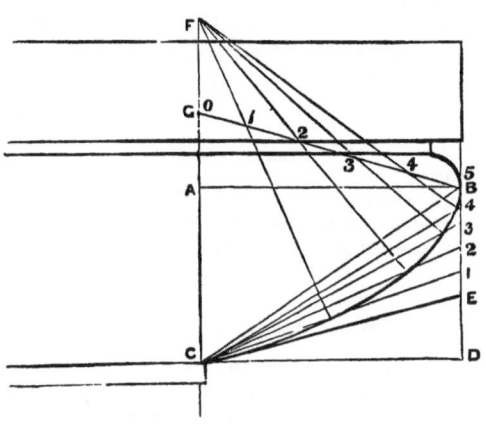

Fig. 59.

half the whole height, the curve is an ellipsis ; if exactly half the height, it is a parabola ; and if BE be less than half BD, the curve will be a hyperbola. All other moulding can be drawn by this method, remembering that cymas, ogees, and all reflex curves, must be divided and drawn in two separate portions.

Staircases.

128. The general principle of designing staircases, as re- gards the rise and tread of steps, setting out curves, curtails, landings, &c., are given in the article JOINERY (35, &c.) The chief difference between these and other staircases consists in the fixing, the one being framed with wooden strings, while the other have no strings, but are supported entirely by the walls. If there be a wall at each end, they are simply built in at the time the work goes up ; but if they are supported at one end only, they are called geometrical stairs, and depend entirely on their being securely wedged into the wall ; on which, and on the support each derives at one edge from the step below, they wholly rely. If they are square in section, they are called solid steps ; but as the under side or soffit, then, is irregular, it is usual to make the steps of somewhat a triangular shape, so as to present a regular soffit. In this case they are called arris, or feather edge steps. Care should be
taken that there are no sudden or irregular changes in the curves. These may be easily avoided by the method shown ·for the easing of the curves and ramps in handrails. (See JOINERY.)

129. Landings should also be very carefully pinned into the walls. Fig. 60 will show the danger, should they not be so, through the full length of their insertion. If the front edge be pinned up, as at A, but a vacancy be left, as at B, the point C will become the fulcrum of a lever, and the landing have a tendency to turn at that point, and to break at the edge C. Every step and landing should have 8 inches hold in a brick wall.

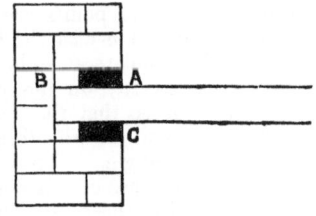

Fig. 60.

(130.) All landings should be well joggled ; the joint

Stone-Masonry.

made as at *a* (fig. 61) is called by workmen a he, and that at *b* a she, joggle. The late accident at the Polytechnic Institution in London arose no doubt from the careless-

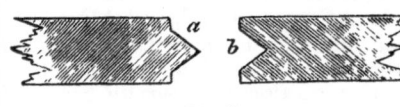

Fig. 61.

ness of the workmen, who put two landings together, on which two she joggles were worked (fig. 62), and filled the open space with plaster. There happened to be a large fossil in the stone close to the wall in the landing *b*, which

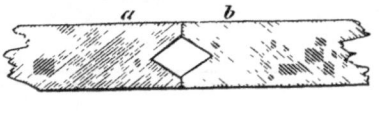

Fig. 62.

having no support from the other landing *a*, gave way, and caused the destruction of the staircase below, upon which it fell.

V.—STONE-MASONRY—MEDIÆVAL.

Materials.

131. It has already been stated (secs. 67, 68) that many of the early buildings of the middle ages were entirely constructed of masses of concrete, often faced with a species of rough cast. The early masonry seems to have been for the most part worked with the axe and not with the chisel. A very excellent example of the contrast between the earlier and later Norman masonry may be seen in the choir of Canterbury cathedral. In those times the groining was frequently filled-in with a light tufa stone, said by some to have been brought from Italy, but more probably from the Rhine. The Normans imported a great quantity of stone from Caen, it being easily worked, and particularly fit for carving. The freestones of England were also much used; and in the first pointed period, Purbeck and Bethersden marbles were employed for column shafts, &c. As time went on the art of masonry advanced with us, till in point of execution it at length rivaled that of any country. The methods of working and setting stone were much the same as at present, except that, as the roads were then in a very bad state, and in many places the only means of conveyance was by pack-horses, the stones were used in much smaller sizes than at present. The methods of setting out work were, however, different from those of other styles, as might be expected from the difference of forms.

Arches.

132. The earliest arches were circular (see ARCHITECTURE,[1] page 64, and figs. 1 to 9, Plate XVIII.), and of course easily set out. But as the pointed styles came in, several methods were used for describing them. Pointed

Lancet arches.

arches may be classed as—1st, lancet; 2d, equilateral; 3d, depressed; and 4th, four-centred or Tudor. In the first the centres (1, 2, fig. 63) are without the arch *a b*. At Westminster Abbey the arches of the choir are so acutely pointed, that the distance from 1 *a*–2 *b* is nearly two-thirds of the entire opening *a b*. In the nave at York the points are without the arch at a distance of about one-fifth the open-

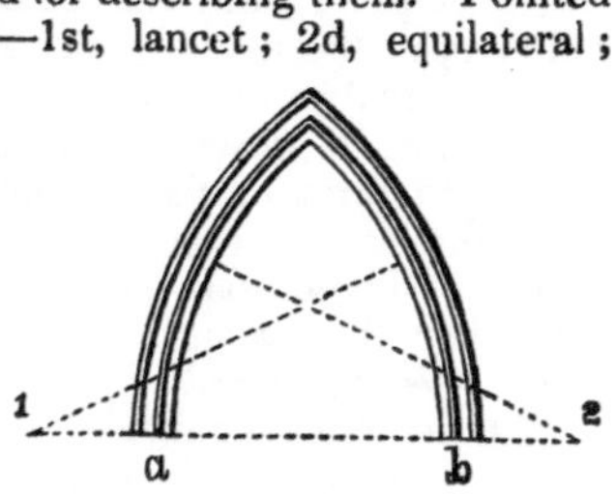

Fig. 63.

Equilateral arches.

ing *a b*. In equilateral arches the centres are exactly on the points *a b* (fig. 64), so that the apex *c*, joined to *a* and *b*, will form an equilateral triangle. The nave arches at Wells are of this description, and also those at Lincoln (Plate XX., figure 1). In later

Depressed arches.

times the arches were of lower

Fig. 64.

pitch, and then of course the centres 1, 2 (fig. 65), were within the arch *a b*. At Salisbury Cathedral (Plate XVIII., fig. 14), the distance *a* 1 is one-sixth of *a b*, while in the choir at Lincoln (fig. 2, Plate XX.) it is as much as two-fifths.

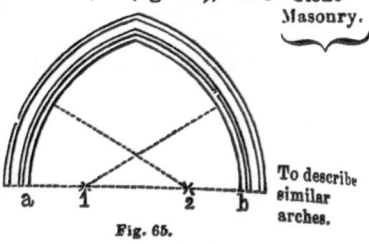

To describe similar arches.

Fig. 65.

133. To describe arches which shall be similar to one another throughout a building, however the openings may differ, this principle must always be borne in mind: that the centres shall always be distant from the points *a b* by some aliquot portion of the whole opening. This is the more important, as the lines of tracery will not fall into their proper places except the arches are set out upon some regular principle (sec. 136). If the arches are not equilateral, some distance from each point, *a b* should be first determined on (say one-third the opening *a b*), and after this, whatever the span of the other arches may be, one-third its own opening is to be taken from the points *a b*, as the centres from which to strike its curves. The only exception is, that in mediæval buildings, the arches to the doorways are frequently somewhat flatter than those of the windows.

To describe four-centred arches.

134. In the Tudor period the arches are very frequently drawn from four centres instead of two. It must be remembered that it has already been stated (sec. 90) the point where two circles touch each other must always be in the same straight line that is drawn through both their centres. As there has been great misapprehension as to four-centred arches, some persons treating them as parts of conic sections, whereas they are really parts of segments of circles, it is thought well to give two methods of describing these arches. First, when the width AB, and the apex height OC, are given, and a tangent to the upper circle CD. In this case draw AD perpendicular to AB, and set out A 1 equal to AD; draw C 3 perpendicular to CD, and make CE equal to AD or A 1; join 1 E and bisect the same as shown by a perpendicular meeting CE produced in 3; join 3 1 and produce towards F, then 1 and 3 will be the

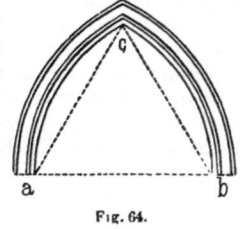

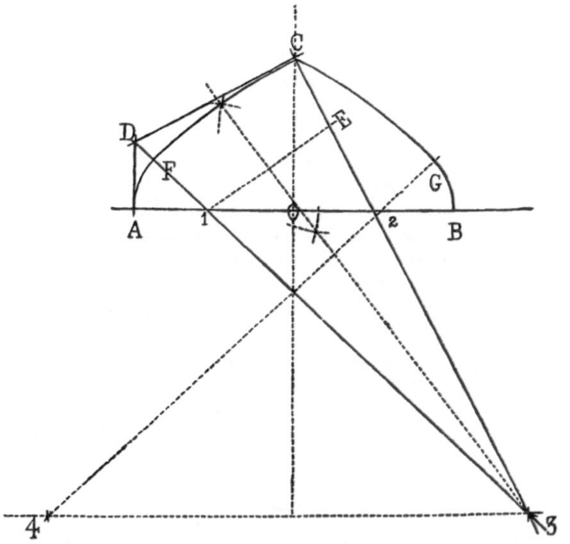

Fig. 66.

centres for half the arch; and, transferring the points across, 2 and 4 will be the centres for the other half. In the second case, when the width AB and the height OC, and the centres of the small circles 1, 2, are given. Make AD equal to A 1, join CD (which will be a tangent to the upper curve), draw C 3 at right angles thereto, make CE

[1] See *Glossary*, "Roman," "Romanesque," and "Norman."

equal to A 1, join 1 E, bisect the same, and proceed as before. The points FG, as has before been explained, are the points where the circles will touch each other. The joints to these arches will all radiate to their respective centres, as has before been explained in secs. 103, 104.

135. The mouldings of mediæval architecture are almost infinite in variety, and even a short description of those used in each style would exceed our limits. They are sometimes set out with the compasses, and many often appear to have been drawn by eye. We must refer our readers to the works of Willis, Paley, and particularly of J. H. Parker, for their description. A very curious treatise was published by the former gentleman called the *Architectural Nomenclature of the Middle Ages*, which goes at great length into the subject. A bead or astragal seems to have been called a bowtelle; a torus, a grete bowtelle; a hollow or scotia, a casement; an ogee, a ressaunte, &c., &c.; but the subject is too long to be discussed in our pages.

136. The various sorts of tracery which adorn the windows of the mediæval periods, and are in fact their greatest glory, are treated of in the art. ARCHITECTURE, and specimens given in the different plates, particularly Plate XIX., figs. 4 to 8. The designs for these are almost infinite, and the various methods of setting them out would fill a volume. But although they display such ingenuity and fancy that one would think the design to be quite arbitrary, it is a curious fact they are all, or very nearly all, set out on the principle of geometrical intersections. An example, therefore, is given (fig. 67) to show the principles on which the mediæval architects proceeded to describe the tracery, and also the method of finding the joints of the various pieces of stone. Let *a b* be the opening of the arch; as there are to be two mullions, divide the same into three equal parts, as *a c*, *c d*, *d b*; then determine the points from which to strike the arch. In this instance, for the sake of simplicity, we make it equilateral (sec. 132 and fig. 64); *a* and *b* then are the centres for striking the main arch *a e g*, *b f g*, and the height *o g* is that of an equilateral triangle. Produce the springing line, and the same opening of the compasses through *c*, and *d* will give the principal inner branches of the tracery *c e*, *d f*. From the centre *o*, with an opening extending to the middle of the lights *a c*, *d b*, strike a semicircle, raise perpendiculars from *c* and *d* to 1 and 2; draw a line through 1 and 2; on this and the springing line will be found the centres of the

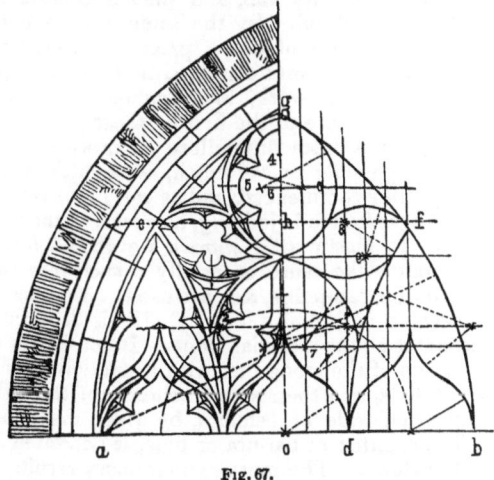

Fig. 67.

lower ogees; bisect the line from the intersection 1, 2, in *h*, which is in fact the same thing as dividing the whole height *o g* into three; divide *h g* into three parts, as 3, 4; through 3 draw a horizontal line, and set off from 3 equal to one-third of the width *o d*, or draw the perpendicular

lines as shown, which is better; then 5, 6, will be the centres of the upper quatre-foil. From the line 1, 2, on the same perpendicular as last, set down similar points as at 7. These will be the centres for the lower sub-division as shown. Next draw *e h f* and sub-divide by similar perpendiculars, and where the lines intersect, as at 8, 9, will be the centres for the upper sub-divisions. The lines thus drawn will form a species of skeleton diagram, as shown on the right side of fig. 67, which is called the *element of the* *tracery*, and is in fact the centre line of the mullion, as shown by *a*, fig. 68. On each side of this, using always the same centres for the same branches, draw lines, showing the face (or what the workmen call the *nose*) of the mullion, and answering to *b c*; and then others answering to the sides of the mullion as *d e*. Any other mouldings upon their sides or faces may be drawn in like manner. Put in the cuspings as shown, and the tracery is complete. The practical stone-mason will take care never to make a joint where there is an angle of any sort, as the point of a cusp. In all cases the joints must tend to the centres of the circles from which they are struck, and where the lines branch off in two directions, the joints must not be in one line, but must tend in two, or as many directions as there are branches, and each to the centres of such respective branch. When the lines are perpendicular, as at *c d*, and at the joint below *h*, the joints are horizontal. A close inspection of fig. 67, where they are carefully drawn, will elucidate the matter more than any number of words can do. Our readers would scarcely believe that the elaborate west window at York is entirely set out on this principle; and so is the still more remarkable instance, the eastern window at Carlisle, which is composed of 86 pieces of stone, and the design for which is drawn from 263 centres. On no account should iron be used as cramps or dowels in Gothic work, as it rusts and breaks pieces out of the stone. The best material is slate run with the Portland cement. Lead is often used; but any metal will expand and contract with heat and cold, and its use is much better avoided altogether. All the upper construction of windows and doors, and of aisle arches, should be protected from superincumbent pressure by strong relieving arches above the labels (see fig. 67), which should be worked in with the ordinary masonry of the walls, and so set that the weight above should not press on the fair work, in which case the joints of the tracery, &c., will sometimes flush or break out.

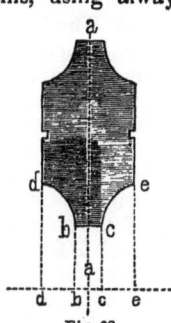

Joints of tracery.

Fig. 68.

137. Mediæval vaults differ much from those before described, principally that the crowns *ab*, *cd*, are not level, as shown in fig. 52, but all have a slight curve or spring, and the filling-in between them also is slightly curved, so as to partake in some degree of the character of the dome as well as of the groined arch. Bearing this carefully in mind, and setting the lines out thus on the sections, the rules we have given for finding the various lines for groins (120–125) will apply as well to Gothic groins as to those of ordinary character; the principle of working from plan, section, and stretch-out being the same, though for the most part the ribs in early vaulting are not true segments of ellipses, but approximations drawn by the compasses. The triumph of mediæval stone-masonry, however, is that species of

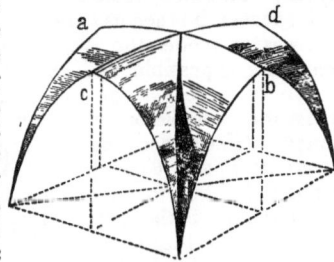

Fig. 69.

Artificial Stones.

groin known as fan vaulting. This is unlike that of any other age or time. The roofs of King's College Chapel, Cambridge, and of Henry VII.'s Chapel at Westminster, are eminent examples. It is impossible in our limited space to give demonstrations of them, and we must refer our readers to the admirable treatises on the subject by Professor Willis, published in the first volume of the *Transactions of the Institute of British Architects*. The filling-in between the ribs of mediæval groins is generally of clunch or some soft stone, over which some concrete is placed in such manner as to bind all together, and to resist the thrust.

Fan vaulting.

Spires

138. The bold and beautiful termination to mediæval towers, which the French call flêches and we call spires, is another proof of the skill of the mediæval masons. These are generally octagonal, and rise partly from the walls of the tower and partly from arches thrown angle-wise from wall to wall, to cut off the corners, as it were, and afford a springing to the spire. The wonder of these constructions is their extreme thinness and lightness. The top of the spire at Salisbury is 411 feet from the ground, of which 207 is taken up by the tower, leaving, of course, 204 feet for the height of the spire itself; this is only 9 inches thick at the bottom, diminishing to 7 inches, or on an average only about the three-hundredth part of its height. It has been attempted to show mathematically that the joints of a spire would be stronger if at right angles to its face; but they would then slope inwards and hold the wet, which in sudden frosts would do most serious injury; practically, therefore, it is found best to lay the courses on a level bed. They should, however, be frequently dowelled and cramped together, but not with metal, as above stated.

VI.—ON ARTIFICIAL STONES, AND ON THE INDURATION OF SOFT STONES.

Cements and terra cottas.

139. The great expense of obtaining and working Portland has driven our masons into the use of the softer free-stones. These (as has before been described, sec. 19, &c.), are liable to rapid decay from the action of frost and wet, and in towns from the action of the sulphurous acid in the coal-smoke. To avoid these inconveniences several contrivances have been resorted to within the last century to obtain a cheap and durable material in which to execute external architectural ornament; or to invent some process by which soft stones may be so protected by an outer coat of impervious matter as to throw off the wet; or may be so indurated as to resist the ordinary wear to which Portland or the harder stones may be subjected. The earliest attempts were to manufacture a hard hydraulic cement, which might be modelled with the tool, or cast into moulds. The earliest use of this is said to have been by the Adam's, it suiting very well with the low relief of their style of ornamentation. (For an account of cements, see this article, sec. 22, *ut supra*.) Another method, which has been frequently tried with success, but seems too expensive to come into general use, was to model the ornaments in good plastic clay, as colourless as possible, and burn them in kilns. The best known of these systems was that called Coade's artificial stone, the manufactory of which, however, is now discontinued. (For an account of this species of manufacture, see POTTERY, TERRA COTTA, &c.) A method, however, which seems to bid fair to excel them all in beauty, cheapness, and durability is that patented by Mr Ransome of Ipswich. It appears, the idea suggested itself to him as long back as 1844, that if he could mix sand, or pounded flint with anything that would make a sort of fluid glass, and stick it together, as it were, in a pasty state, it might be pressed into moulds, and when dry it would form a sort of strong glass. After a variety of experiments, and trying to accomplish his object through a number of changing theories, he at length

Coade's artificial stone.

Ransome's.

succeeded in the following process. He first obtains a strong caustic alkali, which is purified by a most ingenious process. This is made, by the assistance of steam and heat, to act upon some broken flints, which it does, and the alkaline solution is then drawn off and evaporated till it becomes the thickness of treacle. One part of this is mixed with ten parts of sand, one of flint, and one of clay. The whole is kneaded up till like putty, so as to be readily pressed into moulds, and so as to take the sharpest forms. Our space will not permit us to give the full details; but it may suffice to say that the objects, after being dried in close stoves, are then submitted to a strong red-heat in a kiln like that of a potter, which drives off all the alkaline and other chemical agents in the process, and leaves the granules of the sand and flint enveloped in, and, as it were, stuck together by a sort of glass. Time alone, however, can only ultimately show the results. At present the material seems of beautiful colour, texture, and sharpness; and unless some unknown or unforeseen chemical agent should act upon it, Ransome's stone appears to be indestructible.

Artificial Stones.

140. The rendering soft stones hard, and the protecting the surfaces from the weather when worked and set, has been the subject of great investigation lately. The idea of the latter seems to have originated with the late well-known John Sylvester, who tried the method of washing over the faces of stone walls with first a solution of soap and then of alum. Another method was that of washing with what was called water-glass, or silicate of potash, both of which are said to have failed. The next idea was to soak the stone, or in some way to cause the surface to imbibe a quantity of oily or fatty matters to throw off the wet, as well as to harden the stone itself. The first patent was taken out by Mr Hutchison, at Tunbridge Wells, in 1847, and was applied to the new sandstone there. The stones, when worked, were boiled in a solution of resin, turpentine, wax, oil, &c., and sometimes, we believe, pitch, till they were impregnated a sufficient depth from the surface. In 1851 Mr Barrett took out a patent something like the preceding, but far more elaborate; in fact, too long to be described in our pages. The main elements, however, were resins, fats, and tallows, some of which were mixed with gutta percha, unslacked lime, copperas, and a number of other ingredients. In April 1856 Mr Daines took out a patent, not so much to indurate stone, but to preserve stone, or cement walls from damp and efflorescence. His process was to apply, first, a solution of sulphate of zinc, or solution of alum, to the wall, and then a composition of sulphur dissolved in oil. In the same year, and in the next month, Mr Page took out a patent for a similar purpose; his material was wax dissolved in coal-tar, naphtha, or, for more delicate work, in camphine. We are informed the manufacture of the first of these patents is discontinued, but not from any failure of process. Of the others it is impossible to say much, as so little time has elapsed since they commenced, and as early experiments in all manufactures often fail; judging, however, on the grounds expressed as to mastic (37, *supra*), we should fear they would fail from a like cause, especially as such very volatile media as naphtha, camphine, &c., are used. Mr Ransome's, however, seems to promise better. His is deduced from his experiments on the artificial stone. It consists of treating the surface of the stone first with a solution of silicate of potash or soda, and then with a solution of the chloride of barium, or chloride of calcium, by which means an insoluble silicate, either of barium or lime, is deposited in the pores of the stone. The most extraordinary results, however, are promised by Mr Szerelmey's process. The author of this article has been informed by that gentleman that it will not only entirely protect the surface of stone or brick, or cement, but of iron; as a proof, he states that an anchor coated with it was sunk in the sea many months, and raised

Induration of stone.

Sylvester's.

Water-glass.

Oleaginous processes.

Hutchison's.

Barrett's.

Daines's.

Page's.

Ransome's.

Szerelmey's.

again without trace of oxidation. As his process is a profound secret, and his experiments are now in progress, it is impossible to pronounce an opinion on the subject.

The principal works on Stone-Masonry are as follows :—

Foreign.—Jousse de la Fléche, *Secrets d'Architecture*, fol. 1642. Bosse, *La Pratiqué du Trait pour la Coupe des Pierres*, fol. 1643. Francois Derrand, *Des Traits et Coupe des Voutes*, fol. 1643. De la Rue, *Traité de la Coupe des Pierres*, fol. 1728. Frézier, *Traité de Stéreotomie*, 4to, 1737–57; *Elements de Stéreotomie*, 1759. Simonin et Delagardette, *Traité de la Coupe des Pierres*, 4to, 1792. Douliot, *Traité Speciale du Coupe des Pierres*, 4to, 1825. *Vorlegeblatter, per Maurer*, fol. 1835. Adhémar, *Traité de la Coupe des Pierres*, fol. 1836–40. Normand, *Epures d'Escaliers en Pierre*, 4to, 1838. Le Roy, *Traité de Géometrie Descriptive*, 4to, 1850. Claudel et Laroque, *Maçonnerie Pratique*, 8vo, no date. Besides these see the article Maçonnerie in the various Encyclopédiès, and the famous general treatise by Rondelet, *L'Art de Bien Bâtir*, and the older work of De L'Orme, fol. 1643.

In English.—Moxon, *Mechanick Exercises*, 4to, 1677–93, 1700. Batty Langley, *Ancient Masonry*, fol. 1736. Nicholson, *Practical Builder*, 4to, 1823, &c.; *Practical Treatise on Masonry*, 8vo, 1828; *Guide to Railway Masonry*, 8vo, 1839–46. *Practical Masonry, Bricklaying*, &c., 4to, 1830. Dobson, *Rudimentary Treatise on Masonry*, 4to, 1849–56. Besides the valuable articles in Gwilt's well-known *Dictionary of Architecture*, and Cresy's on *Engineering*.   (A. A.)

ARCH.

WE are disposed to give the Greeks the merit of discovering the arch; for we observe arches in the most ancient buildings of Greece, such as the temple of the sun at Athens, and of Apollo at Didymos—not indeed as roofs to any apartment, nor as parts of the ornamental design, but concealed in the walls, covering drains or other necessary openings; and we have not found any *real* arches in any monuments of ancient Persia or Egypt. Sir John Chardin speaks of numerous and extensive subterranean passages at Tchelminar, built of the most exquisite masonry, the joints so exact, and the stones so beautifully dressed, that they look like one continued piece of polished marble; but he nowhere says that they are arched—a circumstance which we think he would not have omitted: no arched door or window is to be seen. Indeed one of the tombs is said to be arch-roofed, but it is all of one solid rock. No trace of an arch is to be seen in the ruins of ancient Egypt; even a wide room is covered with a single block of stone. In the pyramids, indeed, there are two galleries whose roofs consist of many pieces; but their construction puts it beyond doubt that the builder did not know what an arch was; for it is covered in the manner represented in fig. 5, where every projecting piece is more than balanced behind. Yet there are perfect arches, both circular and pointed, in the pyramidal remains at Djebel-el-Berkel, the ancient Napata, in Meroe, the cradle of Egyptian art. The arched dome, however, seems to have arisen in Etruria, and originated in all probability from the employment of the augurs, whose business it was to observe the flight of birds. Their stations for this purpose were *templa*, so called *a templando*, on the summits of hills. To shelter such a person from the weather, and at the same time allow him a full prospect of the country around him, no building was so proper as a dome set on columns; which accordingly is the figure of a temple in the most ancient monuments of that country. We do not recollect a building of this kind in Greece except that called the *Lanthorn of Demosthenes*, but this is covered by a single stone. In the later monuments and coins of Italy or of Rome we commonly find the Etruscan dome and the Grecian temple combined; and the famous Pantheon was of this form, even in its most ancient state.

9. It does not appear that the arch was considered as a part of the *ornamental* architecture of the Greeks during the time of their independency. It is even doubtful whether it was employed in roofing their temples. In none of the *ancient* buildings where the roof is gone can there be seen any rubbish of the vault or mark of the spring of the

arch. It is not unfrequent, however, after the Roman conquest, and may be seen in Athens, Delos, Palmyra, Balbeck, and other places. It is very frequent in the magnificent buildings of Rome, such as the Coloseum, the baths of Diocletian, and the triumphal arches, where its form is evidently made the object of attention. But its chief employment was in bridges and aqueducts; and it is in these works that its immense utility is the most conspicuous: for by this happy contrivance a canal or a road may be carried across any stream, where it would be almost impossible to erect piers sufficiently near to each other for carrying lintels. Arches have been executed 130 feet wide, and their execution demonstrates that they may be made four times as wide.

10. As such stupendous arches are the greatest performances of the masonic art, so they are the most difficult and delicate. When we reflect on the immense quantity of materials thus suspended in the air, and compare this with the small cohesion which the firmest cement can give to a building, we shall be convinced that it is not by the force of the cement that they are kept together; they stand fast only in consequence of the proper balance of all their parts. Therefore, in order to erect them with a well-founded confidence of their durability, this balance should be well understood and judiciously employed. We doubt not but that this was understood in some degree by the engineers of antiquity; but they have left us none of their knowledge. They must have had a great deal of mechanical knowledge before they could erect the magnificent and beautiful buildings whose ruins still enchant the world; but they kept it among themselves. We know that the Dionysiacs of Ionia were a great corporation of architects and engineers, who undertook and even monopolized the building of temples, stadiums, and theatres, precisely as the fraternity of masons in the middle ages monopolized the building of cathedrals and conventual churches. Indeed the Dionysiacs resembled the mystical fraternity now called Free-masons in many important particulars. They allowed no strangers to interfere in their employment; they recognised each other by signs and tokens; they professed certain mysterious doctrines, under the tuition and tutelage of Bacchus, to whom they built a magnificent temple at Teos, where they celebrated his mysteries as solemn festivals; and they called all other men profane because not admitted to these mysteries. But their chief mysteries and most important secrets seem to be their mechanical and mathematical sciences, or all that academical knowledge which forms the regular education of a civil engineer.

2 B

Arch.

We know that the temples of the gods and the theatres required an immense apparatus of machinery for the celebration of some of their mysteries; and that the Dionysiacs contracted for those jobs, even at far distant places, where they had not the privilege of building the edifice which was to contain them. This is the most likely way of explaining the very small quantity of mechanical knowledge that is to be met with in the writings of the ancients. Even Vitruvius does not appear to have been of the fraternity, and speaks of the Greek architects in terms of respect next to veneration. The *Collegium Murariorum*, or incorporation of masons at Rome, does not seem to have shared the secrets of the Dionysiacs.

11. The art of building arches has been most assiduously cultivated by the associated builders of the middle ages of the Christian church, both Saracens and Christians, and they seem to have indulged in it with fondness: they multiplied and combined arches without end, placing them in every possible situation.

12. Having studied this branch of the art of building with so much attention, they were able to erect the most magnificent buildings with materials which a Greek or Roman architect could have made little or no use of. There is infinitely more scientific skill displayed in a Gothic cathedral than in all the buildings of Greece and Rome. Indeed these last exhibit very little knowledge of the mutual balance of arches, and are full of gross blunders in this respect; nor could they have resisted the shock of time so long, had they not been almost solid masses of stone, with no more cavity than was indispensably necessary.

13. Anthemius and Isidorus, whom the Emperor Justinian had selected as the most eminent architects of Greece, for building the celebrated church of St Sophia at Constantinople, seem to have known very little of this matter. Anthemius had boasted to Justinian that he would outdo the magnificence of the Roman Pantheon, for he would hang a greater dome than it aloft in the air. Accordingly he attempted to raise it on the heads of four piers, distant from each other about 115 feet, and about the same height. He had probably seen the magnificent vaultings of the temple of Mars the Avenger, and the temple of Peace at Rome, the thrusts of which are withstood by two masses of solid wall, which join the side walls of the temple at right angles, and extend sidewise to a great distance. It was evident that the walls of the temple could not yield to the pressure of the vaulting without pushing these immense buttresses along their foundations. He therefore placed four buttresses to aid his piers. They are almost solid masses of stone, extending at least ninety feet from the piers to the north and to the south, forming as it were the side walls of the cross. They effectually secured them from the thrusts of the two great arches of the nave which support the dome; but there was no such provision against the push of the great north and south arches. Anthemius trusted for this to the half dome which covered the semicircular east end of the church, and occupied the whole eastern arch of the great dome. But when the dome was finished, and had stood a few months, it pushed the two eastern piers with their buttresses from the perpendicular, making them lean to the eastward, and the dome and half dome fell in. Isidorus, who succeeded to the charge on the death of Anthemius, strengthened the piers on the east side by filling up some hollows, and again raised the dome. But things gave way before it was closed; and while they were building in one part, it was falling in in another. The pillars and walls of the eastern semicircular end were much shattered by this time. Isidorus seeing that they could give no resistance to the push which was so evidently direct-

ed that way, erected some clumsy buttresses on the east wall of the square which surrounded the whole Greek cross, and was roofed in with it, forming a sort of cloister round the whole. These buttresses, spanning over this cloister, leaned against the piers of the dome, and thus opposed the thrusts of the great north and south arches. The dome was now turned for the third time, and many contrivances were adopted for making it extremely light It was made offensively flat, and, except the ribs, was roofed with pumice stone; but, notwithstanding these precautions, the arches settled so as to alarm the architects, and they made all sure by filling up the whole from top to bottom with arcades in three stories. The lowest arcade was very lofty, supported by four noble marble columns, and thus preserved in some measure the church in the form of a Greek cross. The story above formed a gallery for the women, and had six columns in front, so that they did not bear fair on those below. The third story was a dead wall filling up the arch, and pierced with three rows of small, ill-shaped windows. In this unworkmanlike shape it has stood till now, and is the oldest church in the world; but it is an ugly mis-shapen mass, more resembling an overgrown potter's kiln, surrounded with furnaces pierced and patched, than a magnificent temple. We have been thus particular in our account of it, because this history of the building shows that the ancient architects had acquired no distinct notions of the action of arches. Almost any mason of our time would know, that as the south arch would push the pier to the eastward, while the east arch pushed it to the southward, the buttress which was to withstand these thrusts must not be placed on the south side of the pier, but on the southeast side, or that there must be an eastern as well as a southern buttress.

14. No such blunders are to be seen in a Gothic cathedral. Some of them appear, to a careless spectator, to be very massive and clumsy; but when judiciously examined, they will be found very bold and light, being pierced in every direction by arcades; and the walls are divided into cells like a honeycomb, so that they are very stiff, while they are very light.

15. About the middle, or rather towards the end, of last century, when the Newtonian mathematics opened the road to true mechanical science, the construction of arches engrossed the attention of the first mathematicians. The first hint of a principle that we have met with is Dr Hooke's assertion, that the figure into which a chain or rope, perfectly flexible, will arrange itself when suspended from two hooks, is, when inverted, the proper form for an arch composed of stones of uniform weight. This he affirmed on the principle, that the figure which a flexible festoon of heavy bodies assumes, when suspended from two points, is, when inverted, the proper form of an arch of the same bodies, touching each other in the same points; because the force with which they mutually press on each other in this last case are equal and opposite to the forces with which they pull at each other in the case of suspension.

This principle is strictly just, and may be extended to every case which can be proposed. We recollect seeing it proposed in very general terms in 1759, when plans were forming for Blackfriars Bridge in London; and since it is perhaps equal in practical utility to the most elaborate investigations of the mathematicians, our readers will not be displeased with a more particular account of it in this place.

16. Let ABC (fig. 6) be a parcel of magnets of any size and shape, and let us suppose that they adhere with great force by any points of contact. They will compose such a flexible festoon as we have been speaking of if

(margin right) Arch.

(margin right) Dr Hooke's principle of arches

(margin right) explained Plate XLVIII.

suspended from the points A and C. If this figure be inverted, preserving the same points of contact, they will remain in equilibrio. It will indeed be that kind of equilibrium which will admit of no disturbance, and which may be called a *tottering equilibrium*. If the form be altered in the smallest degree, by varying the points of contact (which indeed are points in the *figure of equilibrium*), the magnets will no more recover their former position than a needle, which we had made to stand on its point, will regain its perpendicular position after it has been disturbed.

But if we suppose planes de, fg, hi, &c. drawn through the points of mutual contact a, b, c, each bisecting the angle formed by the lines that unite the adjoining contacts (fg, for example, bisecting the angle formed by ab, bc), and if we suppose that the pieces are changed for others of the same weights, but having flat sides, which meet in the planes de, fg, hi, &c., it is evident that we shall have an arch of equilibration, and that the arch will have some stability, or will bear a little change of form without tumbling down: for it is plain that the equilibrium of the original festoon obtained only in the points a, b, c, of contact, where the pressures were perpendicular to the touching surfaces; therefore, if the curve a, b, c, still passes through the touching surfaces perpendicularly, the conditions that are required for equilibrium still obtain. The case is quite similar to that of the stability of a body resting on a horizontal plane. If the perpendicular through the centre of gravity falls within the base of the body, it will not only stand, but it will require some force to push it over. In the original festoon, if a small weight be added in any part, it will change the form of the curve of equilibration a little, by changing the points of mutual contact. This new curve will gradually separate from the former curve as it recedes from A or C. In like manner, when the festoon is set up as an arch, if a small weight be laid on any part of it, it will bring the whole to the ground, because the shifting of the points of contact will be just the contrary to what it should be to suit the new curve of equilibration; but if the same weight be laid on the same part of the arch now constructed with flat joints, it will be sustained if the new curve of equilibration still passes through the touching surfaces.

17. These conclusions, which are very obviously deducible from the principle of the festoon, show us, without any further discussion, that the longer the joints are, the greater will be the stability of the arch, or that it will require a greater force to break it down. Therefore it is of the greatest importance to have the arch-stones as long as economy will permit; and this was the great use of the ribs and other apparent ornaments in the Gothic architecture. The great projections of those ribs augmented their stiffness, and enabled them to support the unadorned compartments of the roof, composed of very small stones, seldom above six inches thick. Many old bridges are still remaining, which are strengthened in the same way by ribs.

Having thus explained, in a very familiar manner, the stability of an arch, we proceed to give the same popular account of the general application of the principle.

18. Suppose it be required to ascertain the form of an arch which shall have the span AB (fig. 7), and the height F 8, and which shall have a road-way of the dimensions CDE above it. Let the figure ACDEB be inverted, so as to form a figure $AcdeB$. Let a chain of uniform thickness be suspended from the points A and B, and let it be of such a length that its lower point will hang at, or rather a little below, f, corresponding to F. Divide AB into a number of equal parts, in the points 1, 2, 3, &c. and draw vertical lines, cutting the chain in the corre-

Plate XLVIII.

sponding points 1, 2, 3, &c. Now take pieces of another chain, and hang them on at the points 1, 2, 3, &c. of the chain AfB. This will alter the form of the curve. Cut or trim these pieces of chain, till their lower ends all coincide with the inverted road-way cde. The greater lengths that are hung on in the vicinity of A and B will pull down these points of the chain, and cause the middle point f (which is less loaded) to rise a little, and will bring it near to its proper height.

It is plain that this process will produce an arch of perfect equilibration; but some further considerations are necessary for making it exactly suit our purpose. It is an arch of equilibration for a bridge that is so loaded that the weight of the arch-stones is to the weight of the matter with which the haunches and crown are loaded, as the weight of the chain AfB is to the sum of the weights of all the little bits of chain very nearly. But this proportion is not known beforehand; we must therefore proceed in the following manner:—Adapt to the curve produced in this way a thickness of the arch-stones as great as are thought sufficient to insure stability; then compute the weight of the arch-stones, and the weight of the gravel or rubbish with which the haunches are to be filled up to the road-way. If the proportion of these two weights be the same with the proportion of the weights of chain, we may rest satisfied with the curve now found; but if different, we can easily calculate how much must be added equally to or taken from each appended bit of chain, in order to make the two proportions equal. Having altered the appended pieces accordingly, we shall get a new curve, which may perhaps require a very small trimming of the bits of chain to make them fit the road-way. This curve will be very near to the curve wanted.

We have practised this method for an arch of 60 feet span and 21 feet high, the arch-stones of which were only two feet nine inches long. It was to be loaded with gravel and shivers. We made a previous computation, on the supposition that the arch was to be nearly elliptical. The distance between the points 1, 2, 3, &c. were adjusted, so as to determine the proportion of the weights of chain agreeable to the supposition. The curve differed considerably from an ellipse, making a considerable angle with the verticals at the spring of the arch. The real proportion of the weights of chain, when all was trimmed so as to suit the road-way, was considerably different from what was expected. It was adjusted. The adjustment made very little change in the curve. It would not have changed it two inches in any part of the real arch. When the process was completed, we constructed the curve mathematically. It did not differ sensibly from this mechanical construction. This was very agreeable information; for it showed us that the first curve, formed by about two hours' labour, on a supposition considerably different from the truth, would have been sufficiently exact for the purpose, being in no place three inches from the accurate curve, and therefore far within the joints of the intended arch-stones. Therefore this process, which any intelligent mason, though ignorant of mathematical science, may go through with little trouble, will give a very proper form for an arch subject to any conditions.

19. The chief defect of the curve found in this way is a want of elegance, because it does not spring at right angles to the horizontal line; but this is the case with all curves of equilibration, as we shall see by and by. It is not material; for, in the very neighbourhood of the piers, we may give it any form we please, because the masonry is solid in that place; nay, we apprehend that a deviation from the curve of equilibration is proper. The construction of that curve supposes that the pressure on every part of the arch is vertical; but gravel, earth, and rub-

bish, exert somewhat of a hydrostatical pressure laterally in the act of settling, and retain it afterwards. This will require some more curvature at the haunches of an arch to balance it; but what this lateral pressure may be, cannot be deduced with confidence from any experiments that we have seen. We are inclined to think, that if, instead of dividing the horizontal line AB in the points 1, 2, 3, &c. we divide the chain itself into equal parts, the curve will approach nearer to the proper form.

20. After this familiar statement of the general principle, it is now time to consider the theory founded on it more in detail. This theory aims at such an adjustment of the position of the arch-stones to the load on every part of the arch, that all shall remain in equilibrio, although the joints be perfectly polished and without any cement. The whole may be reduced to two problems. The first is to determine the vertical pressure or load on every point of a line of a given form, which will put that line in equilibrio. The second is to determine the form of a curve which shall be in equilibrio when loaded in its different points, according to any given law.

The whole theory is deducible from one principle, which will be found fully developed in the article Roof. It is this: when an assemblage of beams or other pieces of solid matter AB, BC, CD, DE, fig. 8, freely movable about its angles as so many joints, is retained in equilibrium by the joint effect of the pressures produced by the weight of its parts, the thrust at any angle, if estimated in a horizontal direction, is the same throughout, and may be represented by any horizontal line BT; and that if a vertical line QS be drawn through T, the thrust exerted at any angle D by the piece CD, in its own direction, will then be represented by BR, drawn parallel to CD; and in like manner, that the thrust in the direction ED is represented by BS, &c.; and, lastly, that the vertical thrust or loads at each angle B, C, D, by which all these other pressures are excited, are represented by the portions QC, CR, RS, of the vertical intercepted by those lines; that is, all these pressures are to the uniform horizontal thrust as the lines which represent them are to BT. The horizontal thrust, therefore, is a very proper unit, with which we may compare all the others. Its magnitude is easily deduced from the same proposition; for QS is the sum of all the vertical pressures of the angles, and therefore represents the weight of the whole assemblage. Therefore as QS is to BT, so is the weight of the whole to the horizontal thrust.

21. To accommodate this theory to the construction of a curvilinear arch vault, let us first suppose the vault to be polygonal, composed of the chords of the elementary arches. Let AVE (fig. 12) be a curvilinear arch, of which V is the vertex, and VX the vertical axis, which we shall consider as the axis or abscissa of the curve, while any horizontal line, such as HK, is an ordinate to the curve. About any point C of the curve, as a centre, describe a circle BLD, cutting the curve in B and D. Draw the equal chords CB, CD. Draw also the horizontal line CF, cutting the circle in F. Describe a circle BCDQ passing through B, C, D. Its centre O will be in a line COQ, which bisects the angle BCD; and C c, which touches this circle in C, will bisect the angle b C d, formed by the equal chords BC, CD. Draw CLP perpendicular to cb, and DP perpendicular to CD, meeting CL in P. Through L draw the tangent GLM, meeting CD in G, and the vertical line CM in M. Draw the tangent F a, cutting the chords BC, CD, in b and d, and the tangent to the circle BCDQ in c. Lastly, draw d N parallel to bc.

From what will be demonstrated in the article Roof, it appears that if BC, CD be two pieces of an equilibrated heavy polygon, and if CF represent the horizontal

thrust in every angle of the polygon, C d and C b will severally represent the thrusts exerted by the pieces DC, BC, and that bd, or CN, will represent the weight lying on the angle BCD, by which those thrusts are balanced. In the mean time the reader may, without that article, understand the nature of the equilibrium in the following manner. Produce d C to o, so that C o may be equal to C d. Draw bn to the vertical parallel to d C, and join no. It is evident that bnoC is a parallelogram, and that n C (= bd) = CN. Now the thrust or support of the piece BC is exerted in the direction C b, while that of DC is exerted in the direction C o. These two thrusts are equivalent to the thrust in the diagonal C n; and it is with this compound thrust that the load or vertical pressure CN is in immediate equilibrium.

22. Because bCL, NCF are right angles, and FCL is common to both, the angles bCF and MCL are equal; therefore the right-angled triangles bCF and MCL are similar. And since CF is equal to CL, Cb is equal to CM. It is evident that the triangles GCM and dCN are similar. Therefore CG : Cd = CM : CN = C b : C N. Therefore we have $CN = \dfrac{Cb \times Cd}{CG}$. But because CDP and CLG are right angles, and therefore equal, and the angle GCP is common to the two triangles GCL, PCD, and CD is equal to CL, we have CG equal to CP; therefore $CN = \dfrac{Cb \times Cd}{CP}$. Also, since CDP is a right angle, DP meets the diameter in Q, the opposite point of the circumference, and the angle DQC is equal to DCc or cCb (because bCd is bisected by the tangent), that is, to PCQ (because the right angles bCP, cCO are equal, and cDP is common). Therefore PQ is equal to PC; and if PO be drawn perpendicular to CQ, it will bisect it, and O is the centre of the circle BCDQB.

Now let the points B and D continually approach to C (by diminishing the radius of the small circle), and ultimately coincide with it. It is evident that the circle BCDQ is ultimately the equicurve circle, and that PC ultimately coincides with OC, the radius of curvature. Also Cb × Cd becomes ultimately Cc². Therefore CN, the vertical load on any point of a curve of equilibration, is $= \dfrac{Cc^2}{\text{Rad. Curv.}}$.

It is further evident that CF is to Cc as radius to the secant of the elevation of the tangent above the horizon. Therefore we have the load on any point of the curve always proportional to $\dfrac{\text{Sec.}^2 \text{ Elev.}}{\text{Rad. Curv.}}$.

This load on every elementary arch of the wall is commonly a quantity of solid matter incumbent on that element of the curve, and pressing it vertically; and it may be conceived as made up of a number of heavy lines standing vertically on it. Thus, if the element Ee of the curve were lying horizontally, a little parallelogram REer standing perpendicularly on it would represent its load. But as this element Ee has a sloping position, it is plain that, in order to have the same quantity of heavy matter pressing it vertically, the height of the parallelogram must be increased till it meets in ϵ, the line R ϵ drawn parallel to the tangent EG. It is evident that the angle REϵ is equal to the angle AEG. Therefore we have ER : Eϵ = Rad. : Sec. Elev.

If therefore the arch is kept in equilibrio by the vertical pressure of a wall, we must have the height of the wall above any point proportional to $\dfrac{\text{Sec.}^3 \text{ Elev.}}{\text{Rad. of Curv.}}$.

23. Corol. 1. If OS be drawn perpendicular to the

vertical CS, CS will be half the vertical chord of the equicurve circle. The angle OCS is equal to cCF, that is, to the angle of elevation. Therefore $1 : \text{Sec. Elev.} = \text{CS} : \text{CO}$, and the secant of elevation may be expressed by $\dfrac{\text{CO}}{\text{CS}}$,

and its cube by $\dfrac{\text{CO}^3}{\text{CS}^3}$. Therefore the height of wall is

proportional to $\dfrac{\text{CO}^3}{\text{CS}^3 \times \text{CO}}$, or to $\dfrac{\text{CO}^2}{\text{CS}^3}$, or $\dfrac{\text{CO}^2}{\text{CS}^2 \times \text{CS}}$, or to

$\dfrac{\text{Sec.}^2 \text{ of Elev.}}{\text{Vert. Chord of Curve}}$.

Corol. 2. If we make the arch VC $= z$, the abscissa VH $= x$, the ordinate HC $= y$, the radius osculi CO $= r$, and the $\frac{1}{2}$ vertical chord CS $= s$, the height of wall pressing on any point is proportional to $\dfrac{dz^3}{r dy^3}$; or to $\dfrac{dz^2}{s dy^2}$, or

$\dfrac{dx^2 + dy^2}{s dy^2}$. Therefore, when the equation of the curve is given, and the height of wall on any one point of it is also given, we can determine it for any other point; for the equation of the curve will always give us the relation of dz, dx, dy, and the value of r or s. This may be illustrated by an example or two. For this purpose it will generally be most convenient to assume the height above the vertex V for the unit of computation. The thickness of the arch at the crown is commonly determined by other circumstances. At the vertex the tangent to the arch is horizontal, and therefore the cube of the secant is unity or 1. Call the height of wall at the crown H, and let the radius of curvature in that point be R, and its half-chord R (it being then coincident with the radius), and the height on any other point h; we have $\dfrac{1}{R} : \dfrac{dz^3}{r dy^3} = \text{H} : h$, and

$h = \text{H} \times \dfrac{dz^3}{dy^3} \times \dfrac{\text{R}}{r}$. The other formula gives

$h = \text{H} \times \dfrac{dz^2}{dy^2} \times \dfrac{\text{R}}{s}$.

24. *Ex.* 1. Suppose the arch to be a segment of a circle, as in fig. 10, where AE is the diameter, and O the centre. In this arch the curvature is the same throughout, or $\dfrac{\text{R}}{r}$

$= 1$. Therefore $h = \text{H} \times \dfrac{dz^3}{dy^3}$, or $= \text{H} \times$ Cube Sec. Elev.

This gives a very simple calculus. To the logarithm of H add thrice the logarithm of the secant of elevation. The sum is the logarithm of h.

It gives also a very simple construction. Draw the vertical CS, cutting the horizontal diameter in S. Draw ST, cutting the radius OC perpendicularly in T. Draw the horizontal line Tz, cutting the vertical in z. Join zO. Make Cu = Vv, and draw ux parallel to zO : Cc must be made = Cx. The demonstration is evident.

It is very easy to see that if CV is an arch of 60°, and Vv is $\frac{1}{14}$th of VO, the points v and c will be on a level; for the secant of CV is twice CO, and therefore Cc is eight times Vv, which is $\frac{4}{7}$th of VH.

The dotted line $vgcf$ is drawn according to this calculus or construction. It falls considerably below the horizontal line in the neighbourhood of c; and then, passing very obliquely through c, it rises rapidly to an immeasurable height, because the vertical line through A is its asymptote. This must evidently be the case with every curve which springs at right angles with a horizontal line.

It is plain that if vV be greater, all the other ordinates

of the curve $vgcf$, resting on the circumference AVE, will be greater in the same proportion, and the curve will cut the horizontal line drawn through v in some point nearer to v than c is. Hence it appears that a circular arch cannot be put in equilibrio by building on it up to a horizontal line, whatever be its span, or whatever be the thickness at the crown. We have seen that when this thickness is only $\frac{1}{14}$th of the radius, an arch of 120 degrees will be too much loaded at the flanks. This thickness is much too small for a bridge, being only $\frac{1}{23}$th of the span CM, whereas it should have been almost double of this, to bear the inequalities of weight that may occasionally be on it. When the crown is made still thinner, the outline is still more depressed before it rises again. There is therefore a certain span, with a corresponding thickness at the crown, which will deviate least of all from a horizontal line. This is an arch of about 45 degrees, the thickness at the crown being about one fourth of the span, which is extravagantly great. It appears in general, therefore, that the circle is not a curve suited to the purposes of a bridge or an arcade, which requires an outline nearly horizontal.

Ex. 2. Let the curve be a parabola AVE (fig. 14), of which V is the vertex, and DG the directrix. Draw the diameters DCF, GVN, the tangents CK, VP, and the ordinates VF and CN. It is well known that GV is to DC as VP^2 to CK^2, or as CN^2 to CK^2. Also 2 GV is the radius of the osculating circle at V, and 2 DC is one half of the vertical chord of the osculating circle at C.

Therefore CN^2 : CK^2 (or $dy^2 : dz^2$) = R : s; and s $= \dfrac{dz^2}{dy^2}$ R. But Cc, or $h = \text{H} \times \dfrac{dz^2 \text{R}}{dy^2 s}$. Therefore $h = \text{H}$

$\times \dfrac{dz^2 \text{R}}{\frac{dz^2}{dy^2} \text{R}} = \text{H} \times \dfrac{dz^2 \text{R}}{dz^2 \text{R}} = \text{H}$. Therefore C$c$ = vV.

It follows from this investigation, that the back or extrados of a parabolic arch of equilibration must be parallel to the arch or soffit itself; or that the thickness of the arch, estimated in a vertical direction, must be equal throughout; or that the extrados is the same parabola with the soffit or intrados.

We have selected these two examples merely for the simplicity and perspicuity of the solutions, which have been effected by means of elementary geometry only, instead of employing the analytical value of the radius of the osculatory circle, viz. $\dfrac{dz^3}{dy d^2 x - dx d^2 y}$, which would have involved us at last in the doctrine of second fluxions. We have also preferred simplicity to elegance in the investigation, because we wish to instruct the practical engineer who may not be a proficient in the higher mathematics.

25. The converse of the problem, namely, to find the form of the arch when the figure of the back of it is given, is the most usual question of the two, at least in cases which are most important and most difficult. Of these, perhaps, bridges are the chief. Here the necessity of a road-way, of easy and regular ascent, confines us to an outline nearly horizontal, to which the curve of the arch must be adapted. This is the most difficult problem of the two; and we doubt whether it can be solved without employing infinite approximating series instead of accurate values.

Let ave (fig. 9) be the intended outline or extrados of the arch AVE, and let vQ be the common axis of both curves. From c and C, the corresponding points, draw the ordinates ch, CH. Let the thickness vV at the top be a, the abscissa vh be $= u$, and VH $= x$, and let the equal ordinates ch, CH, be y, and the arch VC be z

Then, by the general theorem, $cC = \frac{dz^3}{rdy^3}$, r being the radius of curvature. This, by the common rules, is $= \frac{dz^3}{dyd^2x - dxd^2y}$. This gives us $cC \doteq \frac{dyd^2x - dxd^2y}{dy^3}$, or $= \frac{dyd^2x - dxd^2y}{dy^3} \times C$; where C is a constant quantity, found by taking the real value of cC in V, the vertex of the curve. But it is evident that it is also $= a + x - u$.

Therefore $a + x - u = \frac{dyd^2x - dxd^2y}{dy^3} \times C = \frac{C}{dy} \times$ fluxion of $\frac{dx}{dy}$.

If we now substitute the true value of u (which is given because the extrados is supposed to be of a known form), expressed in terms of y, the resulting equation will contain nothing but x and y, with their first and second fluxions, and known quantities. From this equation the relation of x and y must be found by such methods as seem best adapted to the equation of the extrados.

Fortunately the process is more simple and easy in the most common and useful case than we should expect from this general rule; we mean the case where the extrados is a straight line, especially when this is horizontal. In this case u is equal to o.

Pl. XLIX. *Ex.* To find the form of the balanced arch AVE, having the horizontal line cv for its extrados.
Fig. 6.

Keeping the same notation, we have $u=o$, and therefore

$$a + x = \frac{C}{dy} \times \text{fluxion of } \frac{dx}{dy}.$$

Assume $dy = \frac{dx}{v}$; then $\frac{dx}{dy} = v$, and $\frac{C}{dy} \times$ fluxion of $\frac{dx}{dy} = \frac{Cvdv}{dx}$, that is, $a + x = \frac{Cvdv}{dx}$. Therefore $adx + xdx = Cvdv$; and by taking the fluents, we have $2ax + x^2 = Cv^2$; and $v = \sqrt{\frac{2ax + x^2}{C}}$. Consequently, $dy = \frac{\sqrt{C}dx}{\sqrt{2ax + x^2}} \left(\text{being} = \frac{dx}{v}\right)$. Taking the fluent of this, we have $y = \sqrt{C} \times L (2a + 2x + 2\sqrt{2ax + x^2})$. But at the vertex, where $x = o$, we have $y = \sqrt{C} \times L (2a)$. The corrected fluent is therefore $y = \sqrt{C} \times L\frac{a + x + \sqrt{2ax + x^2}}{a}$.

It only remains to find the constant quantity C. This we readily obtain by selecting some point of the extrados where the values of x and y are given by particular circumstances of the case. Thus, when the span $2s$ and height h of the arch are given, we have

$$s = \sqrt{C} \times L \left(\frac{a + h + \sqrt{2ah + h^2}}{a}\right), \text{ and conse-}$$

quently $\sqrt{C} = \dfrac{s}{L\left(\dfrac{a + h + \sqrt{2ah + h^2}}{a}\right)}$. Therefore

the general value of $y = s \times \dfrac{L\left(\dfrac{a + x + \sqrt{2ax + x^2}}{a}\right)}{L\left(\dfrac{a + h + \sqrt{2ah + h^2}}{a}\right)}$

$$= \frac{s}{L\frac{a + h + \sqrt{(2ah + h^2)}}{a}} \times L\frac{a + x + \sqrt{2ax + a^2}}{a}.$$

26. As an example of the use of this formula, we subjoin a table calculated by Dr Hutton of Woolwich, for an arch, the span of which is 100 feet, and the height 40, which are nearly the dimensions of the middle arch of Blackfriars Bridge in London.

y	x	y	x	y	x
0	6,000	21	10,381	36	21,774
2	6,035	22	10,858	37	22,948
4	6,144	23	11,368	38	24,190
6	6,324	24	11,911	39	25,505
8	6,580	25	12,489	40	26,894
10	6,914	26	13,106	41	28,364
12	7,330	27	13,761	42	29,919
13	7,571	28	14,457	43	31,563
14	7,834	29	15,196	44	33,299
15	8,120	30	15,980	45	35,135
16	8,430	31	16,811	46	37,075
17	8,766	32	17,693	47	39,126
18	9,168	33	18,627	48	41,293
19	9,517	34	19,617	49	43,581
20	9,934	35	20,665	50	46,000

The figure for this proposition is exactly drawn according to these dimensions, that the reader may judge of it as an object of sight. It is by no means deficient in gracefulness, and is abundantly roomy for the passage of craft; so that no objection can be offered against its being adapted on account of its mechanical excellency.

The reader will perhaps be surprised that we have *Defects of* made no mention of the celebrated Catenarean curve, *the Cate-* which is commonly said to be the best form for an arch; *narean* but a little reflection will convince him, that although *curve.* it is the only form for an arch consisting of stones of equal weight, and touching each other only in single points, it cannot suit an arch which must be filled up in the haunches, in order to form a road-way. He will be more surprised to hear, after this, that there is a certain thickness at the crown, which will put the Catenarea in equilibrio, even with a horizontal road-way; but this thickness is so great as to make it unfit for a bridge, being such that the pressure at the vertex is equal to the horizontal thrust. This would have been about 37 feet in the middle arch of Blackfriars Bridge. The only situation, therefore, in which the Catenarean form would be proper, is an arcade carrying a height of dead wall; but in this situation it would be very ungraceful. Without troubling the reader with the investigation, it is sufficient to inform him, that in a Catenarean arch of equilibration the abscissa VH is to the abscissa vh in the constant ratio of the horizontal thrust to its excess above the pressure on the vertex

27. Thus much will serve, we hope, to give the reader a clear notion of this celebrated theory of the equilibrium of arches, one of the most delicate and important applications of mathematical science. Volumes have been written on the subject, and it still occupies the attention of mechanicians. But we beg leave to say, with great deference to the eminent persons who have prosecuted this theory, that their speculations have been of little service, and are little attended to by the practitioner. Nay, we may add, that Sir Christopher Wren, perhaps the most accomplished architect that Europe has seen, seems to have thought it of little value; for, among the fragments which have been preserved of his studies, there are to be seen some imperfect dissertations on this very subject, in which he takes no notice of this theory, and considers the balance of arches in quite another way. These are collected by the author of the account of Sir Christopher Wren's family. This man's great sagacity,

and his great experience in building, and still more his experience in the repairs of old and crazy fabrics, had shown him many things very inconsistent with this theory, which appears so specious and safe. The general facts which occur in the failure of old arches are highly instructive, and deserve the most careful attention of the engineer; for it is in this state that their defects, and the process of nature in their destruction, are most distinctly seen. We venture to affirm, that a very great majority of these facts are irreconcilable to the theory. The way in which circular arches commonly fail, is by the sinking of the crown and the rising of the flanks. It will be found by calculation, that in most of the cases it ought to have been just the contrary. But the clearest proof is, that arches very rarely fail where their load differs most remarkably from that which this theory allows. Semicircular arches have stood the power of ages, as may be seen in the bridges of ancient Rome, and in the numerous arcades which the ancient inhabitants have erected. Now, all arches which spring perpendicularly from the horizontal line require, by this theory, a load of infinite height; and even to a considerable distance from the springing of the arch, the load necessary for the theoretical equilibrium is many times greater than what is ever laid on those parts; yet a failure in the immediate neighbourhood of the spring of an arch is a most rare phenomenon, if it ever was observed. Here is a most remarkable deviation from the theory; for, as is already observed, the load is frequently not the fourth part of what the theory requires.

28. Many other facts might be adduced which show great deviation from the legitimate results of the theory. We hope to be excused, therefore, by the mathematicians for doubting of the justness of this theory. We do not think it erroneous, but defective, leaving out circumstances which we apprehend to be of great importance; and we imagine that the defects of the theory have arisen from the very anxiety of the mechanicians to make it perfect. The arch-stones are supposed to be perfectly smooth or polished, and not to be connected by any cement, and therefore to sustain each other merely by the equilibrium of their vertical pressure. The theory insures this equilibrium, and this only, leaving unnoticed any other causes of mutual action.

The authors who have written on the subject say expressly that an arch which thus sustains itself must be stronger than another which would not; because when, in imagination, we suppose both to acquire connection by cement, the first preserves the influence of this connection unimpaired; whereas in the other, part of the cohesion is wasted in counteracting the tendency of some parts to break off from the rest by their want of equilibrium. This is a very specious argument, and would be just, if the forces which are mutually exerted between the parts of the arch in its settled state were merely vertical pressures, or, where different, were inconsiderable in comparison with those which are really attended to in the construction.

But this is by no means the case. The forms which the uses for which arches are erected oblige us to adopt, and the loads laid on the different points of the arch, frequently deviate considerably from what are necessary for the equilibrium of vertical pressures. The varying load on a bridge, when a great waggon passes along it, sometimes bears a very sensible proportion to the weight of that point of the arch on which it rests. It is even very doubtful whether the pressures which are occasioned by the weight of the stuff employed for filling up the flanks really act in a vertical direction, and in the proportion which is supposed. We are pretty certain that this is not

the case with sand, gravel, fat mould, and many substances in very general use for this purpose. When this is the case, the pressures sustained by the different parts of the arch are often very inconsistent with the theory; a part of the arch is overloaded and tends to fall in, but is prevented by the cement. This part of the arch, therefore, acts on the remoter parts by the intervention of the parts between, employing those intermediate parts as a kind of levers to break the arch in a remote part, just as a lintel would be broken. We apprehend that a mathematician would be puzzled how to explain the stability of an arch cut out of a solid and uniform mass of rock. His theory considers the mutual thrusts of the arch-stones as in the direction of the tangents to the arch. Why so? Because he supposes that all his polished joints are perpendicular to those tangents. But in the present case he has no existing joints; and there seems to be nothing to direct his imagination in the assumption of joints, which, however, are absolutely necessary for employing his theory, because, without a supposition of this kind, there seems no conceiving any mutual abutment of the arch-stones. Ask a common but intelligent mason, what notion he forms of such an arch? We apprehend that he will consider it as no arch, but as a lintel, which may be broken like a wooden lintel, and which resists entirely by its cohesion. He will not readily conceive that, by cutting the under side of a stone lintel into an arched form, and thus taking away more than half of its substance, he has changed its nature of a lintel, or given it any additional strength. Nor would there be any change made in the way in which such a mass of stone would resist being broken down, if nothing were done but forming the under side into an arch. If the lintel be so laid on the piers that it can be broken without its parts pushing the piers aside (which will be the case if it lies on the piers with horizontal joints), it will break like any other lintel; but if the joints are directed downwards, and converging to a point within the arch, the broken stone (suppose it broken at the crown by an overload in that part) cannot be pressed down without forcing the piers outwards. Now, in this mode of acting, the mind cannot trace any thing of the statical equilibrium that we have proceeded on in the foregoing theory. The two parts of the broken lintel seem to push the piers aside in the same manner that two rafters push outwards the walls of a house when their feet are not held together by a tie-beam. If the piers cannot be pushed aside (as when the arch abuts on two solid rocks), nothing can press down the crown which does not crush the stone.

This conclusion will be strictly true if the arch is of such a form that a straight line drawn from the crown to the pier lies wholly within the solid masonry. Thus. if the vault consist of two straight stones, as in Plate XLVIII. fig. 1, or if it consist of several stones, as in Plate XLIX. fig. 7, disposed in two straight lines, no weight laid on the crown can destroy it in any other way than by crushing it to powder.

29. But when straight lines cannot be drawn from the overloaded part to the firm abutments through the solid masonry, and when the cohesion of the parts is not able to withstand the transverse strains, we must call the principles of equilibrium to our aid; and, in order to employ them with safety, we must consider how they are modified by the excitement of the cohering forces.

The cohesion of the stones with each other by cement or otherwise has in almost every situation a bad effect. It enables an overload at the crown to break the arch near the haunches, causing those parts to rise, and then to spread outwards, just as a Mansarde or Kirb roof would do if the truss-beam which connects the heads of the lower rafters were sawn through. This can be prevented

Arch.

only by loading that part more than is requisite for equilibrium. It would be prudent to do this to a certain degree, because it is by this cohesion that the crown always becomes the weakest part of the arch, and suffers more by any occasional load.

We expect that it will be said in answer to all this, that the cohesion given by the strongest cement that we can employ, nay the cohesion of the stone itself, is a mere nothing in comparison with the enormous thrusts that are in a state of continual exertion in the different parts of an arch. This is very true; but there is another force which produces the same effect, and which increases nearly in the proportion that those thrusts increase, because it arises from them. This is the friction of the stones on each other. In dry freestone this friction considerably exceeds one half of the mutual pressure. The reflecting reader will see that this produces the same effect, in the case under consideration, that cohesion would do; for while the arch is in the act of failing, the mutual pressure of the arch-stones is acting with full force, and thus produces a friction more than adequate to all the effects we have been speaking of.

Process of the breaking of an arch.

30. When these circumstances are considered, we imagine that it will appear that an arch, when exposed to a great overload on the crown (or indeed on any part), divides of itself into a number of parts, each of which contains as many arch-stones as can be pierced (so to speak) by one straight line, and that it may then be considered as nearly in the same situation with a polygonal arch of long stones abutting on each other like so many beams in a Norman roof, but without their braces and ties. It tends to break at all those angles; and it is not sufficiently resisted there, because the materials with which the flanks are filled up have so little cohesion, that the angle feels no load except what is immediately above it; whereas it should be immediately loaded with all the weight which is diffused over the adjoining side of the polygon. This will be the case, even though the curvilinear arch be perfectly equilibrated. We recollect some circumstances in the failure of a considerable arch, which may be worth mentioning. It had been built of an exceedingly soft and friable stone, and the arch-stones were too short. About a fortnight before it fell, chips were observed to be dropping off from the joints of the arch-stones, about ten feet on each side of the middle, and also from another place on one side of the arch, about twenty feet from its middle. The masons in the neighbourhood prognosticated its speedy downfall, and said that it would separate in those places where the chips were breaking off. At length it fell; but it first split in the middle, and about fifteen or sixteen feet on each side, and also at the very springing of the arch. Immediately before the fall a shivering or crackling noise was heard, and a great many chips dropped down from the middle, between the two places from whence they had dropped a fortnight before. The joints opened above at those new places above two inches, and in the middle of the arch the joints opened below, and in about five minutes after this the whole came down. Even this movement was plainly distinguishable into two parts. The crown sunk a little, and the haunches rose very sensibly, and in this state it hung for about half a minute. The arch-stones of the crown were hanging by their upper corners: when these splintered off, the whole fell down.

We apprehend that the procedure of nature was somewhat in this manner. Straight lines can be drawn within the arch-stones from A (Plate XLIX. fig. 8) to B and D, and from these points to C and E. Each of the portions ED, DA, AB, BC, resist as if they were of one stone, composing a polygonal vault EDABC. When this is overloaded at A,

A can descend in no other way than by pushing the angles B and D outwards, causing the portions BC, DE, to turn round C and E. This motion must raise the points B and D, and cause the arch-stones to press on each other at their *inner* joints b and d. This produced the copious splintering at those joints immediately preceding the total downfall. The splintering which happened a fortnight before arose from this circumstance, that the lines AB and AD, along which the pressure of the overload was propagated, were tangents to the soffit of the arch in the points F, H, and G, and therefore the strain lay all on those corners of the arch-stones, and splintered a little from off them till the whole took a firmer bed. The subsequent phenomena are evident consequences of this distribution and modification of pressure, and can hardly be explained in any other way, at least not on the theoretical principles already set forth; for in this bridge the loads at B and D were very considerably greater than what the equilibrium required; and we think that the first observed splintering at H, F, and G, was most instructive, showing that there was an extraordinary pressure at the inner joints in those places, which cannot be explained by the usual theory.

Not satisfied with this single observation after this way of explaining it occurred to us, and not being able to find any similar fact on record, the writer of this article got some small models of arches executed in chalk, and subjected them to many trials, in hopes of collecting some general laws of the internal workings of arches which finally produce their downfall. He had the pleasure of observing the above-mentioned circumstances take place very regularly and uniformly when he overloaded the models at A. The arch always broke at some place B considerably beyond another point F, where the first chipping had been observed. This is a method of trial that deserves the attention both of the speculatist and the practitioner.

If these reflections are any thing like a just account of the procedure of nature in the failure of an arch, it is evident that the ingenious mathematical theory of equilibrated arches is of little value to the engineer. We ventured to say as much already, and we rested a good deal on the authority of Sir Christopher Wren. He was a good mathematician, and delighted in the application of this science to the arts. He was a celebrated architect, and his reports on the various works committed to his charge show that he was in the continued habit of making this application. Several specimens remain of his own methods of applying them. The roof of the theatre of Oxford, the roof of the cupola of St Paul's, and in particular the mould on which he turned the inner dome or that cathedral, are proofs of his having studied this theory most attentively. He flourished at the very time that it occupied the attention of the greatest mechanicians of Europe; but there is nothing to be found among his papers which shows that he had paid much regard to it. On the contrary, when he has occasion to deliver his opinion for the instruction of others, and to explain to the dean and chapter of Westminster his operations in repairing that collegiate church, this great architect considers an arch just as a sensible and sagacious mason would do, and very much in the way that we have just now been treating it. (See *Account of the Family of Wren*, p. 356.) Supported therefore by such authority, we would recommend this way of considering an arch to the study of the mathematician; and we would desire the experienced mason to think of the most efficacious methods for resisting this tendency of arches to rise in the flanks. Unfortunately there seems to be no precise principle to point out the place where this tendency is most remarkable.

31 We are therefore highly pleased with the ingenious

Arch.

Construction of Blackfriars Bridge Pl. XLIX.

contrivance of Mr Mylne, the architect of Blackfriars Bridge in London, by which he determines this point with precision, by making it impossible for the overloaded arch to spring in any other place. Having thus confined the failure to a particular spot, he with equal art opposes a resistance which he believes to be sufficient; and the present condition of that noble bridge, which does not in any place show the smallest change of shape, proves that he was not mistaken. Looking on this work as the first, or at least the second, specimen of masonic ingenuity that is to be seen in the world, we imagine that our readers will be pleased with a particular account of its most remarkable circumstances.

The span $k\,a$ (fig. 1) of the middle arch is 100 feet, and its height OV is 40, and the thickness KV of the crown is six feet seven inches. Its form is nearly elliptical; the part AVZ being an arch of a circle whose centre is C, and radius 56 feet, and the two lateral portions A k B and Z a E being arches described with a radius of 35 feet nearly. The thickness of the pier at $a\,b$ is 19 feet. The thickness of the arch increases from the crown V to Y, where it is eight or nine feet. All the arch-stones have their joints directed to the centres of their curvature. The joints are all joggled, having a cubic foot of hard stone let half-way into each. By this contrivance the joints cannot slide, nor can any weight laid on the crown ever break the arch in that part if the piers do not yield; for a straight line from the middle of KV to the middle of the joint YI is contained within the solid masonry, and does not even come near the inner joints of the arch-stones; therefore the whole resists like one stone, and can be only broken by crushing it. The joint at Z is very nearly perpendicular to a line ZF drawn to the outer edge of the foundation of the pier. By this it was intended to take off all tendency of the pressure on the joint d Z to overset the pier; for if we suppose, according to the theory of equilibration, that this pressure is necessarily exerted perpendicularly to the joint, its direction passes through the fulcrum at F, round which it is thought that the pier must turn in the act of oversetting. This precaution was adopted in order to make the arch quite independent of the adjoining arches; so that although any of them should fall, this arch should run no risk.

Still farther to secure the independence of the arch, the following construction was practised to unite it into one mass, which should rise all together. All below the line $a\,b$ is built of large blocks of Portland stone, dovetailed with sound oak. Four places in each course are interrupted by equal blocks of a hard stone called *Kentish rag*, sunk half-way in each course. These act as joggles, breaking the courses, and preventing them from sliding laterally.

The portion a Y of the arch is joggled like the upper part. The interior part is filled up with large blocks of Kentish rag, forming a kind of coursed rubble-work, the courses tending to the centres of the arch. The under corner of each arch-stone projects over the one below it. By this form it takes fast hold of the rubble-work behind it. Above this rubble there is constructed the inverted arch I e G of Portland stone.[1] This arch shares the pressure of the two adjoining arches, along with the arch-stones in Y a and in G b. Thus all tend together to compress and keep down the rubble-work in the heart of this part of the pier. This is a very useful precaution; for it

often happens, that when the centres of the arches are struck before the piers are built up to their intended heights, the thrust of the arches squeezes the rubble-work horizontally, after the mortar has set, but before it has dried and acquired its utmost hardness. Its bond is broken by this motion, and it is squeezed up, and never acquires its former firmness. This is effectually prevented by the pressure exerted by the back of the inverted arch.

Above this counter-arch is another mass of coursed rubble, and all is covered by a horizontal course of large blocks of Portland stone, abutting against the back of the arch-stone ZI and its corresponding one in the adjoining arch. This course connects the feet of the two arches, preserves the rubble-work from too great compression, and protects it from soaking water. This last circumstance is important; for if the water which falls on the road-way is not carried off in pipes, it soaks through the gravel or other rubbish, rests on the mortar, and keeps it continually wet and soft. It cannot escape through the joints of good masonry, and therefore fills up this part like a funnel.

Supposing the adjoining arch fallen, and all tumbled off that is not withheld by its situation, there will still remain in the pier a mass of about 3500 tons. The weight of the portion VY is about 2000 tons. The directions of the thrusts VY and YF are such, that it would require a load of 4500 tons on VY to overturn the pier round F. This exceeds VY by 2500 tons—a weight incomparably greater than any that can ever be laid on it.

Such is the ingenious construction of Mr Mylne. It evidently proceeds on the principles recommended above— principles which had occurred to his experience and sagacious mind during the course of his extensive practice. We have seen attempts by other engineers to withstand the horizontal thrusts of the arch by means of counter-arches inserted in the same manner as here, but extending much farther over the main arch; but they did not appear to be well calculated for producing this effect. A counter-arch springing from any point between Y and V has no tendency to hinder that point from rising by the sinking of the crown; and such a counter-arch will not resist the precisely horizontal thrust so well as the straight course of Mr Mylne.

32. The great incorporation of architects who built the cathedrals of Europe departed entirely from the styles of ancient Greece and Rome, and introduced another, in which arcades made the principal part. Not finding in every place quarries from which blocks could be raised in abundance of sufficient size for forming the far-projecting cornices of the Greek orders, they relinquished those proportions, and adopted a style of ornament which required no such projections; and having substituted arches for the horizontal architrave or lintel, they were now able to erect buildings of vast extent with spacious openings, and all this with very small pieces of stone. The form which had been adopted for a Christian temple occasioned many intersections of vaultings, and multiplied the arches exceedingly. Constant practice gave opportunities of giving every possible variety of these intersections, and taught the art of balancing arch against arch in every variety of situation. An art so multifarious, and so much out of the road of ordinary thought, could not but become an object of fond study to the architects most eminent for ingenuity and invention. Becoming thus the dupes of

Origin of the Gothic arches.

___

[1] We know from good authority that the counter-arch here spoken of, although originally intended, was never executed, because it was not thought necessary. The notion was, however, excellent, and it has, we believe, been actually executed in the Strand Bridge. We rather think the joggling was also abandoned, and, as far as we can judge, was not likely to be of any use.

their own ingenuity, they were fond of displaying it even when not necessary. At last arches became their principal ornament, and a wall or ceiling was not thought dressed out as it should be till filled full of mock arches, crossing and abutting on each other in every direction. In this process in their ceilings they found that the projecting mouldings, which we now call the Gothic tracery, formed the chief supports of the roofs. The plane surfaces included between those ribs were commonly vaulted with very small stones, seldom exceeding six or eight inches in thickness. This tracery, therefore, was not a random ornament. Every rib had a position and direction that was not only proper, but even necessary. Habituated to this scientific arrangement of the mouldings, they did not deviate from it when they ornamented a smooth surface with mock arches; and in none of the highly ornamented *ancient* buildings will we find any false positions.

33. This is by no means the case in many of the modern imitations of Gothic architecture, even by our best architects. Ignorant of the directing principle, or not attending to it, in their stucco-work they please the unskilled eye with pretty radiated figures; but in these we frequently see such abutments of mouldings as would infallibly break the arches, if these mouldings were really performing their ancient office, and supporting a vaulting of considerable extent. Nay, this began even before the Gothic style was finally abandoned. Several instances are to be found in the highly enriched vaultings of New College and Christ Church in Oxford, in St George's Chapel at Windsor, and Henry VII.'s Chapel in Westminster.

We call the middle ages rude and barbarous; but there was surely much knowledge in those who could execute such magnificent and difficult works. The working drafts which were necessary for such varieties of oblique intersections must have required considerable skill, and would at present occupy many very expensive volumes of *Masons' Jewels, Carpenters' Manuals,* and the like. All this knowledge was kept a profound secret by the corporation, and on its breaking up we had all to learn again.

34. There is no appearance, however, that those architects had studied the theory of equilibrated arches. They had adopted an arch which was very strong, and permitted considerable irregularities of pressure—we mean the pointed arch. The very deep mouldings with which it was ornamented made the arch-stones very long in proportion to the span of the arch. But they had studied the mutual thrust of arches on each other with great care; and they contrived to make every invention for this purpose become an ornament, so that the eye *required* it as a necessary part of the building. Thus we frequently see small buildings having buttresses at the sides. These are necessary in a large vaulted building, for withstanding the outward thrust of the vaulting; but they are useless when we have a flat ceiling within. Pinnacles on the heads of the buttresses are now considered as ornaments, but originally they were put there to increase the weight of the buttress: even the great tower in the centre of a cathedral, which now constitutes its greatest ornament, is a load almost indispensably necessary, for enabling the four principal columns to withstand the combined thrust of the aisles, of the nave, and transepts. In short, the more closely we examine the ornaments of this architecture, the more shall we perceive that they are essential parts, or derived from them by imitation; and the more we consider the whole style of it, the more clearly do we see that it is all deduced from the relish for arcades, indulged in the extreme, and pushed to the limit of possibility of execution.

35. There is another species of arch which must not be overlooked, namely, the DOME or CUPOLA, with all its varieties, which include even the pyramidal steeple or spire.

It is evident that the erection of a dome is also a scientific art, proceeding on the principles of equilibration, and that these principles admit and require the same or similar modifications, in consequence of the cohesion and friction of the materials. At first sight, too, a dome appears a more difficult piece of work than a plain arch; but when we observe potters kilns, and glasshouse domes, and cones of vast extent, erected by ordinary bricklayers, and with materials vastly inferior in size to what can be employed in common arches of equal extent, we must conclude that the circumstance of curvature in the horizontal direction, or the abutment of a circular base, gives some assistance to the artist. Of this we have complete demonstration in the case of the cone. We know that a vaulting in the form of a pent roof could not be executed to any considerable extent, and would be extremely hazardous, even in the smallest dimensions; while a cone of the greatest magnitude can be raised with very small stones, provided only that we prevent the bottom from flying out, by a hoop, or any similar contrivance.

36. When we think a little of the matter, we see plainly, that if the horizontal section be perfectly round, and the joints be all directed to the axis, they all equally endeavour to slide inwards, while no reason can be offered why any individual stone should prevail. They are all wedges, and operate only as wedges. When we consider any single course, therefore, we see that it cannot fall in, even though it may be part of a curve which cannot stand as a common arch; nay, we see that a dome may be constructed having the convexity of the curve, by the revolution of which it is formed, turned towards the axis, so that the outline is concave. We shall afterwards find that this is a stronger dome by far than if the convexity were outwards, as in a common arch. We see also that a cone may be loaded on the top with the greatest weight, without the smallest danger of forcing it down, so long as the bottom course is firmly kept from bursting outwards. The stone lanthorn on the top of St Paul's cathedral in London weighs several hundred tons, and is carried by a brick cone of 18 inches thick, with perfect safety, as long as the bottom course is prevented from bursting outwards. The reason is evident: The pressure on the top is propagated along the cone in the direction of the slant side; and, so far from having any tendency to break it in any part, it tends rather to prevent its being broken by any irregular pressure from foreign causes.

37. For the same reasons the octagonal pyramids, which form the spires of Gothic architecture, are abundantly firm, although very thin. The sides of the spire of Salisbury cathedral are not eight inches thick after the octagon is fully formed. It is proper, however, to direct the joints to the axis of the pyramid, and to make the coursing joints perpendicular to the slant side, because the projecting mouldings which run along the angles are the abutments on which the whole panel depends. A considerable art is necessary for supporting those panels or sides of the octagon which spring from the angles of the square tower. This is done by beginning a very narrow pointed arch on the square tower at a great distance below the top; so that the legs of the arch being very long, a straight line may be drawn from the top of the keystone of the arch through the whole arch-stones of the legs. By this disposition the thrusts arising from the weight of these four panels are made to meet on the massive masonry in the middle of the sides of the tower, at a great

distance below the springing of the spire. This part, being loaded with the great mass of perpendicular wall, is fully able to withstand the horizontal thrust from the legs of those arches. In many spires these thrusts are still farther resisted by iron bars which cross the tower, and are hooked into pieces of brass firmly bedded in the masonry of the sides.

38. There is much nice balancing of this kind to be observed in the highly ornamented open spires; such as those of Brussels, Mecklin, Antwerp, &c. We have not many of this sort in Britain. In those of great magnitude, the judicious eye will discover, that parts, which a common spectator would consider as mere ornaments, are necessary for completing the balance of the whole. Tall pinnacles, nay even pillars carrying entablatures and pinnacles, are to be seen standing on the middle of the slender leg of an arch. On examination we find that this is necessary, to prevent the arch from springing upwards in that place by the pressure at the crown. The steeple of the cathedral of Mecklin was the most elaborate piece of architecture in this taste in the world, and was really a wonder; but it was not calculated to withstand a bombardment, which destroyed it in 1578.

Such frequent examples of irregular and whimsical buildings of this kind show that great liberties may be taken with the principle of equilibration without risk, if we take care to secure the base from being thrust outwards. This may always be done by hoops, which can be concealed in the masonry; whereas in common arches these ties would be visible, and would offend the eye.

39. It is now time to attend to the principle of equilibrium as it operates in a simple circular dome, and to determine the thickness of the vaulting when the curve is given, or the curve when the thickness is given. Therefore, let B b A (Plate XLIX fig. 2) be the curve which produces the dome by revolving round the vertical axis AD. We shall suppose this curve to be drawn through the middle of all the arch-stones, and that the coursing or horizontal joints are everywhere perpendicular to the curve. We shall suppose (as is always the case) that the thickness KL, HI, &c. of the arch-stones is very small in comparison with the dimensions of the arch. If we consider any portion HA h of the dome, it is plain that it presses on the course, of which HL is an arch-stone, in a direction bC perpendicular to the joint HI, or in the direction of the next superior element βb of the curve. As we proceed downwards, course after course, we see plainly that this direction must change, because the weight of each course is superadded to that of the portion above it, to complete the pressure on the course below. Through B draw the vertical line BCG, meeting βb, produced in C. We may take bC to express the pressure of all that is above it, propagated in this direction to the joint KL. We may also suppose the weight of the course HL united in b, and acting on the vertical. Let it be represented by bF. If we form the parallelogram bFGC, the diagonal bG will represent the direction and intensity of the whole pressure on the joint KL. Thus it appears that this pressure is continually changing its direction, and that the line, which will always coincide with it, must be a curve concave downward. If this be precisely the curve of the dome, it will be an equilibrated vaulting; but so far from being the strongest form, it is the weakest, and it is the limit to an infinity of others, which are all stronger than it. This will appear evident, if we suppose that bG does not coincide with the curve A b B, but passes without it. As we suppose the arch-stones to be exceedingly thin from inside to outside, it is plain that this dome cannot stand, and that the weight of the upper part will press it down, and spring the vaulting outwards at the joint KL. But

let us suppose, on the other hand, that bG falls within the curvilinear element bB. This evidently tends to push the arch-stone inward towards the axis, and would cause it to slide in, since the joints are supposed perfectly smooth and slipping. But since this takes place equally in every stone of this course, they must all abut on each other in the vertical joints, squeezing them firmly together. Therefore, resolving the thrust bG into two, one of which is perpendicular to the joint KL, and the other parallel to it, we see that this last thrust is withstood by the vertical joints all around, and there remains only the thrust in the direction of the curve. Such a dome must therefore be firmer than an equilibrated dome, and cannot be so easily broken by overloading the upper part. When the curve is concave upwards, as in the lower part of the figure, the line bC always falls below bB, and the point C below B. When the curve is concave downwards, as in the upper part of the figure, b'C' passes above, or without bB. The curvature may be so abrupt, that even b'G' shall pass without b'B', and the point G' is above B'. It is also evident that the force which thus binds the stones of a horizontal course together, by pushing them towards the axis, will be greater in flat domes than in those that are more convex; that it will be still greater in a cone, and greater still in a curve whose convexity is turned inwards, for in this last case the line bG will deviate most remarkably from the curve. Such a dome will stand (having polished joints) if the curve springs from the base with any elevation, however small; nay, since the friction of two pieces of stone is not less than half of their mutual pressure, such a dome will stand although the tangent to the curve at the bottom should be horizontal, provided that the horizontal thrust be double the weight of the dome, which may easily be the case if it do not rise high.

40. Thus we see that the stability of a dome depends on very different principles from that of a common arch, and is in general much greater. It differs also in another very important circumstance, viz. that it may be open in the middle; for the uppermost course, by tending equally in every part to slide in toward the axis, presses all together in the vertical joints, and acts on the next course like the keystone of a common arch. Therefore an arch of equilibration, which is the weakest of all, may be open in the middle, and carry at top another building, such as a lanthorn, if its weight do not exceed that of the circular segment of the dome that is omitted. A greater load than this would indeed break the dome, by causing it to spring up in some of the lower courses; but this load may be increased if the curve is flatter than the curve of equilibration: and any load whatever, which will not crush the stones to powder, may be set on a truncate cone, or on a dome formed by a curve that is convex toward the axis; provided always that the foundation be effectually prevented from flying out, either by a hoop, or by a sufficient mass of solid pier on which it is set. *Stability of a dome depends on principles different from that of a common arch.*

41. We have mentioned the many failures which happened to the dome of St Sophia in Constantinople. We imagine that the thrust of the great dome, bending the eastern arch outwards as soon as the pier began to yield, destroyed the half-dome which was leaning on it, and thus almost in an instant took away the eastern abutment. We think that this might have been prevented without any change in the injudicious plan, if the dome had been hooped with iron, as was practised by Michael Angelo in the vastly more ponderous dome of St Peter's at Rome, and by Sir Christopher Wren in the cone and the inner dome of St Paul's at London. The weight of the latter considerably exceeds 3000 tons, and they occasion a horizontal thrust which is nearly half this quantity, the elevation of the cone being about 60°. This being distributed

Arch.

round the circumference, occasions a strain on the hoop $= \frac{7}{2 \times 22}$ of the thrust, or nearly 238 tons. A square inch of the worst iron, if well forged, will carry 24 tons with perfect safety; therefore a hoop of 7 inches broad and $1\frac{1}{2}$ inch thick will completely secure this circle from bursting outwards. It is, however, much more completely secured; for, besides a hoop at the base of very nearly these dimensions, there are hoops in different courses of the cone, which bind it into one mass, and cause it to press on the piers in a direction exactly vertical. The only thrusts which the piers sustain are those from the arches of the body of the church and the transepts. These are most judiciously directed to the entering angles of the building, and are there resisted with insuperable force by the whole lengths of the walls, and by four solid masses of masonry in the corners. Whoever considers with attention and judgment the plan of this cathedral, will see that the thrusts of these arches, and of the dome, are incomparably better balanced than in St Peter's church at Rome. But to return from this sort of digression.

Theory of the curves proper for domes.

42. We have seen that if $b\,G$, the thrust compounded of the thrust $b\,C$, exerted by all the courses above HILK, and of the force $b\,F$, or the weight of that course, be everywhere coincident with $b\,B$, the element of the curve, we shall have an equilibrated dome: if it falls within it, we have a dome which will bear a greater load; and if it falls without it, the dome will break at the joint. We must endeavour to get analytical expressions of these conditions. Therefore draw the ordinates $b\,\delta\,b''$, BDB'', CdC''. Let the tangents at b and b'' meet the axis in M, and make MO, MP, each equal to bc, and complete the parallelogram MONP, and draw OQ perpendicular to the axis, and produce $b\,$F, cutting the ordinates in E and e.[1] It is plain that MN is to MO as the weight of the arch HAb to the thrust $b\,$C which it exerts on the joint KL (this thrust being propagated through the course HILK); and that MQ, or its equal $b\,e$, or $\delta\,d$, may represent the weight of the half HA.

Let AD be called x, and DB be called y. Then $be = dx$, and $e\,$C $= dy$ (because $b\,$C is in the direction of the element $\beta\,b$). It is also plain that if we make dy constant, BC is the second fluxion of x, or BC $= d^2x$, and be and bE may be considered as equal, and taken indiscriminately for dx. We have also $b\,$C $= \sqrt{dx^2 + dy^2}$. Let h be the depth or thickness HI of the arch-stones. Then $h\sqrt{dx^2 + dy^2}$ will represent the trapezium HL; and since the circumference of each course increases in the proportion of the radius y, $hy\sqrt{dx^2 + dy^2}$ will express the whole course. If $\int$ be taken to represent the sum or aggregate of the quantities annexed to it, the formula will be analogous to the fluent of a fluxion, and $\int hy\sqrt{dx^2 + dy^2}$ will represent the whole mass, and also the weight of the vaulting, down to the joint HI. Therefore we have this proportion, $\int hy\sqrt{dx^2 + dy^2} : hy\sqrt{dx^2 + dy^2} := be : b\,\text{F}, = b\,e : \text{CG},$

$= \delta\,d : \text{CG}, = dx : \text{CG}.$ Therefore $\text{CG} = \dfrac{hydx\sqrt{dx^2 + dy^2}}{\int hy\sqrt{dx^2 + dy^2}}.$

If the curvature of the dome be precisely such as puts it in equilibrium, but without any mutual pressure in the vertical joints, this value of CG must be equal to CB or to d^2x, the point G coinciding with B. This condition will be expressed by the equation $\dfrac{hydx\sqrt{dx^2 + dy^2}}{\int hy\sqrt{dx^2 + dy^2}} = d^2x,$

or, more conveniently, by $\dfrac{hy\sqrt{dx^2 + dy^2}}{\int hy\sqrt{dx^2 + dy^2}} = \dfrac{d^2x}{dx}.$ But this form gives only a tottering equilibrium, independent of

the friction of the joints and the cohesion of the cement. An equilibrium, accompanied by some firm stability, produced by the mutual pressure of the vertical joints, may be expressed by the formula $\dfrac{hy\sqrt{dx^2 + dy^2}}{\int hy\sqrt{dx^2 + dy^2}} > \dfrac{d^2x}{dx},$ or

by $\dfrac{hy\sqrt{dx^2 + dy^2}}{\int hy\sqrt{dx^2 + dy^2}} = \dfrac{d^2x}{dx} + \dfrac{dt}{t}$, where t is some variable positive quantity, which increases when x increases. This last equation will also express the equilibrated dome, if t be a constant quantity, because in this case $\dfrac{dt}{t}$ is $= 0$.

Since a firm stability requires that $\dfrac{hydx\sqrt{dx^2 + dy^2}}{\int hy\sqrt{dx^2 + dy^2}}$ shall be greater than d^2x, and CG must be greater than CB; hence we learn that figures of too great curvature, whose sides descend too rapidly, are improper. Also, since stability requires that we have $\dfrac{hydx\sqrt{dx^2 + dy^2}}{d^2x}$ greater than $\int hy\sqrt{dx^2 + dy^2}$, we learn that the upper part of the dome must not be made very heavy. This, by diminishing the proportion of $b\,$F to $b\,$C, diminishes the angle CbG, and may set the point G above B, which will infallibly spring the dome in that place. We see here also, that the algebraic analysis expresses that peculiarity of dome-vaulting, viz. that the weight of the upper part may even be suppressed.

The fluent of the equation $\dfrac{hy\sqrt{dx^2 + dy^2}}{\int hy\sqrt{dx^2 + dy^2}} = \dfrac{d^2x}{dx} + \dfrac{dt}{t}$ is most easily found: it is L$\int hy\sqrt{dx^2 + dy^2} = \text{L}\,dx +$ Lt, where L is the hyperbolic logarithm of the quantity annexed to it. If we consider dy as constant, and correct the fluent so as to make it nothing at the vertex, it may be expressed thus, L$\int hy\sqrt{dx^2 + dy^2} - \text{L}\,a = \text{L}\,dx$ $- \text{L}\,dy + \text{L}\,t.$ This gives us L$\dfrac{\int hy\sqrt{dx^2 + dy^2}}{a} = \text{L}\dfrac{dx}{dy}t,$

and therefore $\dfrac{\int hy\sqrt{dx^2 + dy^2}}{a} = t\dfrac{dx}{dy}.$

This last equation will easily give us the depth of vaulting, or thickness h of the arch, when the curve is given. For its fluxion is $\dfrac{hy\sqrt{dx^2 + dy^2}}{a} = \dfrac{dtdx + td^2x}{dy},$

and $h = \dfrac{adtdx + atd^2x}{ydy\sqrt{dx^2 + dy^2}}$, which is all expressed in known quantities; for we may put in place of t any power or function of x or of y, and thus convert the expression into another, which will be applicable to all sorts of curves.

Instead of the second member $\dfrac{d^2x}{dx} + \dfrac{dt}{t}$, we might employ $\dfrac{pd^2x}{dx}$, where p is some number greater than unity. This will evidently give a dome having stability, because the original formula $\dfrac{hydx\sqrt{dx^2 + dy^2}}{\int hy\sqrt{dx^2 + dy^2}}$ will then be greater than d^2x. This will give $h = \dfrac{pax^{r-1}d^2x}{ydy\sqrt{dx^2 + dy^2}}.$ Each of these forms has its advantages when applied to particular cases. Each of them also gives $h = \dfrac{ad^2x}{ydy\sqrt{dx^2 + dy^2}}$ when the curvature is such as is in precise equilibrium.

---

[1] The letters e and d are wanting in the plate; e ought to be at the intersection of b E and C c'', and d at that of AD and C c''

Arch.

And, lastly, if h be constant, that is, if the vaulting be of uniform thickness, we obtain the form of the curve, because then the relation of d^2x to dx and to dy is given.

The chief use of this analysis is to discover what curves are improper for domes, or what portions of given curves may be employed with safety. Domes are generally built for ornament, and we see that there is great room for indulging our fancy in the choice. All curves which are concave outwards will give domes of great firmness: they are also beautiful. The Gothic dome, whose outline is an undulated curve, may be made abundantly firm, especially if the upper part be convex and the lower concave outwards.

The chief difficulty in the case of this analysis arises from the necessity of expressing the weight of the incumbent part, or $\int hy \sqrt{dx^2 + dy^2}$. This requires the measurement of the conoidal surface, which, in most cases, can be had only by approximation by means of infinite series. We cannot expect that the generality of practical builders are familiar with this branch of mathematics, and therefore will not engage in it here; but content ourselves with giving such instances as can be understood by such as have that moderate mathematical knowledge which every man should possess who takes the name of engineer.

The surface of any circular portion of a sphere is very easily had, being equal to the circle described with a radius equal to the chord of half the arch. This radius is evidently $= \sqrt{dx^2 + dy^2}$.

In order to discover what portion of a hemisphere may be employed (for it is evident that we cannot employ the whole) when the thickness of the vaulting is uniform, we may recur to the equation or formula $\dfrac{hy\,dx\sqrt{dx^2 + dy^2}}{dx^2}$ $= \int hy \sqrt{dx^2 + dy^2}$. Let a be the radius of the hemisphere. We have $dx = \dfrac{ay\,dy}{\sqrt{a^2 - y^2}}$, and $d^2x = \dfrac{a^2 dy^2}{(a^2 - y^2)^{\frac{3}{2}}}$. Substituting these values in the formula, we obtain the equation $y^2 \sqrt{a^2 - y^2} = \int \dfrac{a^2 y\,dy}{\sqrt{a^2 - y^2}}$. We easily obtain the fluent of the second member $= a^3 - a^2 \sqrt{a^2 - y^2}$, and $y = a\sqrt{-\frac{1}{2} + \sqrt{\frac{5}{4}}}$. Therefore, if the radius of the sphere be 1, the half breadth of the dome must not exceed $\sqrt{-\frac{1}{2} + \sqrt{\frac{5}{4}}}$, or 0·786, and the height will be ·618. The arch from the vertex is about 51° 49'. Much more of the hemisphere cannot stand, even though aided by the cement, and by the friction of the coursing joints. This last circumstance, by giving connection to the upper parts, causes the whole to press more vertically on the course below, and thus diminishes the outward thrust; but it at the same time diminishes the mutual abutment of the vertical joints, which is a great cause of firmness in the vaulting. A Gothic dome, of which the upper part is a portion of a sphere not exceeding 45° from the vertex, and the lower part is concave outwards, will be very strong, and not ungraceful.

Dome of St Peter's at Rome.

43. But the public taste has long rejected this form, and seems rather to select more elevated domes than this portion of a sphere, because a dome, when seen from a small distance, always appears flatter than it really is. The dome of St Peter's is nearly an ellipsoid externally, of which the longer axis is perpendicular to the horizon. It is very ingeniously constructed. It springs from the base perpendicularly, and is very thick in this part. After rising about 50 feet, the vaulting separates into two thin vaultings, which gradually separate from each other.

Arch

These two shells are connected together by thin partitions, which are very artificially dovetailed in both, and thus form a covering which is extremely stiff, while it is very light. Its great stiffness was necessary for enabling the crown of the dome to carry the elegant stone lanthorn with safety. It is a wonderful performance and has not its equal in the world; but it is an enormous load in comparison with the dome of St Paul's, and this even independent of the difference of size. If they were of equal dimensions, it would be at least five times as heavy, and is not so firm by its gravity; but as it is connected in every part by iron bars (lodged in the solid masonry, and well secured from the weather by having lead melted all around them), it bids fair to last for ages if the foundations do not fail.

If a circle be described round a centre, placed anywhere in the transverse axis AC (Plate XLVIII. fig. 11) of an ellipse, so as to touch the ellipse in the extremities B, b of an ordinate, it will touch it internally, and the circular arch B a b will be wholly within the elliptical arch BAb. Therefore, if an elliptical and a spherical vaulting spring from the same base, at the same angle with the horizon, the spherical vaulting will be within the elliptical, will be flatter and lighter, and therefore the weight of the next course below will bear a greater proportion to the thrust in the direction of the curve; therefore the spherical vaulting will have more stability. On the contrary, and for similar reasons, an oblate elliptical vaulting is preferable to a spherical vaulting springing with the same inclination to the horizon. (Fig. 13.)

Dimensions of the best form of a dome.

44. Persuaded that what has been said on the subject convinces the reader that a vaulting perfectly equilibrated throughout is by no means the best form, provided that the base is secured from separating, we think it unnecessary to give the investigation of that form, which has a considerable intricacy, and shall content ourselves with merely giving its dimensions. The thickness is supposed uniform. The numbers in the first column of the table express the portion of the axis counted from the vertex, and those of the second column are the lengths of the ordinates.

AD	DB	AD	DB	AD	DB
0,4	100	610,4	1080	2990	1560
3,4	200	744	1140	3442	1600
11,4	300	904	1200	3972	1640
26,6	400	1100	1260	4432	1670
52,4	500	1336	1320	4952	1700
91,4	600	1522	1360	5336	1720
146,8	700	1738	1400	5756	1740
223,4	800	1984	1440	6214	1760
326,6	900	2270	1480	6714	1780
465,4	1000	2602	1520	7260	1800

The curve delineated in fig. 15 is formed according to these dimensions, and appears destitute of gracefulness, because its curvature changes abruptly at a little distance from the vertex, so that it has some appearance of being made up of different curves pieced together. But if the middle be occupied by a lanthorn of equal or of smaller weight, this defect will cease, and the whole will be elegant, nearly resembling the exterior dome of St Paul's in London.

Advantages of dome-vaulting.

45. It is not a small advantage of dome-vaulting, that it is lighter than any that can cover the same area. If, moreover, it be spherical, it will admit considerable varieties of figure by combining different spheres. Thus, a dome may begin from its base as a portion of a large hemisphere, and may be broken off at any horizontal

Arch.

course, and then a similar or a greater portion of a smaller sphere may spring from this course as a base. It also bears being intersected by cylindrical vaultings in every direction, and the intersections are exact circles, and always have a pleasing effect. It also springs most gracefully from the heads of small piers, or from the corners of rooms of any polygonal shape; and the arches formed by its intersections with the walls are always circular and graceful, forming very handsome spandrels in every position. For these reasons Sir Christopher Wren employed it in all his vaultings, and he has exhibited many beautiful varieties in the transepts and the aisles of St Paul's, which are highly worthy of the observation of architects. Nothing can be more graceful than the vaultings at the ends of the north and south transepts, especially as furnished off in the fine inside view published by Gwynn and Wale.

Effects of cement and friction in dome-vaulting.

46. The connection of the parts arising from cement and from friction has a great effect on dome-vaulting. In the same way as in common arches and cylindrical vaulting, it enables an overload on one place to break the dome in a distant place. But the resistance to this effect is much greater in dome-vaulting, because it operates all round the overloaded part. Hence it happens that domes are much less shattered by partial violence, such as the falling of a bomb or the like. Large holes may be broken in them without much affecting the rest; but, on the other hand, it greatly diminishes the strength which should be derived from the mutual pressure in the vertical joints. Friction prevents the sliding in of the arch-stones, which produces this mutual pressure in the vertical joints, except in the very highest courses, and even there it greatly diminishes it. These causes make a great change in the form which gives the greatest strength; and as their laws of action are as yet but very imperfectly understood, it is perhaps impossible, in the present state of our knowledge, to determine this form with tolerable precision. We see plainly, however, that it allows a greater deviation from the best form than the other kind of vaulting, and domes may be made to rise perpendicular to the horizon at the base, although of no great thickness; a thing which must not be attempted in a plain arch. The immense addition of strength which may be derived from hooping largely compensates for all defects; and there are hardly any bounds to the extent to which a very thin dome-vaulting may be carried when it is hooped or framed in the direction of the horizontal courses. The roof of the Halle du Bled at Paris is but a foot thick, and its diameter is more than 200, yet it appears to have abundant strength. It is, on the whole, a noble specimen of architecture.

The iron bridge at Sunderland described.

47. We must not conclude this article without taking notice of that magnificent and elegant arch erected in cast iron at Wearmouth, near Sunderland, in the county of Durham. The inventor and architect was Rowland Burdon, Esq.

This arch (of which a view is given at the article BRIDGE) is a segment of a circle whose diameter is about 444 feet. The span or chord of the arch is 236 feet, and its versed sine or spring is 34 feet. It springs at the elevation of 60 feet from the surface of the river at low water, so that vessels of 200 or perhaps 300 tons burden may pass under it in the middle of the stream, and even 50 feet on each side of it.

The sweep of the arch consists of a series of frames of cast iron, which abut on each other, in the same manner as the voussoirs of a stone arch. One of these frames or blocks (as we shall call them in future) is represented in Plate XLIX. fig. 3, as seen in front. It is cast in one piece, and consists of three pieces or arms BC, BC, BC, the middle

one of which is two feet long, the upper being somewhat more, and the lower somewhat less, because their extremities are bounded by the radius drawn from the centre of the arch. These arms are four inches square, and are connected by other pieces KL, of such length that the whole length of the block is five feet in the direction of the radius. Each arm has a flat groove on each side, which is expressed by the darker shading, three inches broad and three fourths of an inch deep. A section of this block, through the middle of KL, is represented by the light-shaded part BBB, in which the grooves are more distinctly perceived. These grooves are intended for receiving flat bars of malleable iron, which are employed for connecting the different blocks with each other. Fig. 4 represents two blocks united in this manner. For this purpose each arm has two square bolt-holes. The ends of the arms being nicely trimmed off, so that the three ends abut equally close on the ends of the next block; and the bars of hammered iron being also nicely fitted to their grooves, so as to fill them completely, and have their bolt-holes exactly corresponding to those in the blocks; they are put together in such a manner that the joints or meetings of the malleable bars may fall on the middle between the bolt-holes in the arms. Flat-headed bolts of wrought iron are then put through, and keys or forelocks are driven through the bolt-tails, and thus all is firmly wedged together, binding each arm between two bars of wrought iron. These bars are of such length as to connect several blocks.

In this manner a series of about 125 blocks are joined together, so as to form the precise curve that is intended. This series may be called a rib, and it stands in a vertical plane. The arch consists of six of these ribs, distant from each other five feet. These ribs are connected together so as to form an arch of 32 feet in breadth, in the following manner.

Fig. 5 represents one of the bridles or cross pieces which connect the different ribs, as it appears when viewed from below. It is a hollow pipe of cast iron four inches in diameter, and has at each end two projecting shoulders, pierced with a bolt-hole near their extremities, so that the distance between the bolt-holes in the shoulders of one end is equal to the distance between the holes in the arms of the blocks, or the holes in the wrought iron bars. In the middle of the upper and of the under side of each end may be observed a square prominence, more lightly shaded than the rest. These projections also advance a little beyond the flat of the shoulders, forming between them a shallow notch about an inch deep, which receives the iron of the arms, where they abut on each other, and thus give an additional firmness to the joint. The manner in which the arms are thus grasped by these notches in the bridles is more distinctly seen in fig. 4, at the letter H, in the middle of the upper rail.

The rib having been all trimmed and put together, so as to form the exact curve, the bolts are all taken out, and the horizontal bridles are then set on in their places, and the bolts are again put in and made fast by the forelocks. The bolts now pass through the shoulders of the bridles, through the wrought iron bars, and through the cast iron arm that is between them, and the forelocks bind all fast together. The manner in which this connection is completed is distinctly seen in fig. 4, which shows in perspective a double block in front, and a single block behind it. The abutting joints of the two front blocks are at the letters E, E, E; the holes in the shoulders of the horizontal cross pieces are at H.

48. This construction is beautifully simple, and very judicious. A vast addition of strength and of stiffness is procured by lodging the wrought iron bars in grooves

Arch.

formed in the cast iron rails; and for this purpose it is of great importance to make the wrought iron bars fill the grooves completely, and even to be so tight as to require the force of the forelocks to draw them home to the bottom of the grooves. There can be no doubt but that this arch is able to withstand an enormous pressure, as long as the abutments from which it springs do not yield. Of this there is hardly any risk, because they are masses of rock, faced with about four or five yards (in some places only) of solid block masonry. The mutual thrusts of the frames are all in the direction of the rails, so that no part bears any transverse strain. We can hardly conceive any force that can overcome the strength of those arms by pressure or crushing them. The manner in which the frames are connected into one rib effectually secures the abutting joints from slipping; and the accuracy with which the whole can be executed secures us against any warping or deviation of a rib from the vertical plane.

But when we consider the prodigious span of this arch, and reflect that it is only five feet thick, it should seem that the most perfect equilibration is indispensably necessary. It is but like a film, and must be so supple, that an overload on any part must have a great tendency to bend it, and to cause it to rise in a distant part; and this effect is increased by the very firmness with which the whole sticks together. The overloaded part acts on a distant part, tending to break it with all the energy of a long lever. This can be prevented only by means of the stiffness of the distant part. It is very true, the arch cannot break in the extrados except by tearing asunder the wrought iron bars which connect the blocks along the upper rail, and each of these requires more than a hundred tons to tear it asunder; yet an overload of five tons on any rib at its middle will produce this strain at twenty feet from the sides, supposing the sides held firm in their position. It were desirable, therefore, that something were done to stiffen the arch at the sides, by the manner of filling up the spandrels or space between the arch and the road-way. This is filled up in a manner that is extremely light and pleasing to the eye, namely, by large cast iron circles, which touch the extrados of the arch and touch the road-way. The road-way rests on them as on so many hoops, while they rest on the back of the arch, and also touch each other laterally. We cannot think that this contributes to the strength of the arch; for these hoops will be easily compressed at the points of contact, and, changing their shape, will oppose very little resistance. We think that this part of the arch might have been greatly stiffened and strengthened by connecting it with the road-way by trussed frames, in the same way that a judicious carpenter would have framed a roof. If a strong cast iron pillar had been made to rest on the arch at about 20 feet from the impost, and been placed in the direction of a radius, the top of this pillar might have been connected by a diagonal bar of wrought iron with the impost of the arch, and with the crown of the arch by another string or bar of the same materials. These two ties would cause the radial pillar to press strongly on the back of the arch, and they must be torn asunder before it could bend in that place in the smallest degree. Supposing them of the same dimensions as the bars in the arms, their position would give them near ten times the force for resisting the strain produced by an overload on the crown.

This beautiful arch contains only 260 tons of iron, of which about 55 are wrought iron. The superstructure is of wood, planked over a top. This floor is covered with a coating of chalk and tar, on which are laid the materials for the carriage road, consisting of marl, limestone, and gravel, with foot-ways of flag-stones at the sides. The

weight of the whole did not exceed 1000 tons, whereas the lightest stone arch which could have been erected would have weighed 15,000. It was turned on a very light but stiff scaffolding, most judiciously constructed for the preservation of its form, and for allowing an uninterrupted passage for the numerous ships and small craft which frequent the busy harbour of Sunderland. The mode of framing the arch was so simple and easy that it was put up in ten days, without an accident; and when all was finished, and the scaffolding removed, the arch did not sensibly change its form. The whole work was executed in three years, and cost about L.26,000.

APPENDIX.

49. The excellence of the preceding article, written by the late Dr John Robison, Professor of Natural Philosophy in the University of Edinburgh, for the Supplement to the third edition of this work, may be inferred from the fact, that almost every later writer on mechanics has spoken of it with approbation, and borrowed more or less from it. (See Gregory's *Mechanics*, book i. chap. 6; Hutton's *Tracts*, vol. i., Bridges; Whewell's *Mechanics*, art. 72, &c.) There is however one part of the article, viz. the purely mathematical part, which must have been quite unintelligible as it originally appeared, because of its numerous typographical errors. These are here corrected, we believe for the first time. Even with this advantage, we fear it has rather a forbidding aspect to the student of modern mathematics, because of the very complicated diagram (see Plate XLVIII. fig. 12) from which the differential equation of the equilibrated arch has been deduced. We shall therefore, as an appendix, treat of the equilibrated polygon and equilibrated arch upon the general principles of statics, and employ in this investigation only the most simple geometrical figures. The theory of the equilibrated arch cannot be delivered without employing the principles of the differential and integral calculus; but we shall endeavour to pass from the finite equation of the equilibrated polygon to the differential equation of the arch by the shortest and most direct road.

Equilibrated Polygon.

50. In Plate XLIX. fig. 9, let A B C D E D' C' B' A' be a polygon formed by beams or rods AB, BC, CD, DE, &c. of any length, movable about the points B, C, D, E, &c. as joints, and forming an equilibrium in a vertical plane by the mutual thrusts at the joints and by the weight of the beams, the extreme sides of the polygon being supposed supported or attached to fixed points. Let AB, BC, CD, be any three consecutive sides of the polygon: produce AB, DC, the extreme sides, until they meet in H. The beam BC is kept in its position by the thrusts of the adjoining beams AB, CD, in the directions BH, CH, and its own weight, which is equivalent to pressures or loads on the joints B, C, acting in a vertical direction. Let G be the point in BC, which is the centre of gravity of weights proportional to loads at B and C. By the principles of statics, the directions of these forces must pass through the same point; therefore G must be in a vertical line passing through H.

51. Draw BL, CK, perpendicular to the vertical H G. Let φ, φ', φ'', denote the angles which the lines AB, BC, CD, in their order, make with any horizontal line in the plane of the polygon; then

φ = angle HBL, φ' = GBL = GCK, φ'' = HCK:

also let w denote the load on the joint B, and w' the load on the joint C.

By the nature of the centre of gravity,

$$w : w' = CG : BG = CK : BL,$$

therefore $\quad w : w' = \dfrac{1}{BL} : \dfrac{1}{CK} = \dfrac{HG}{BL} : \dfrac{HG}{CK}:$

Now $\dfrac{HG}{BL} = \dfrac{HL}{BL} - \dfrac{GL}{BL} = \tan. \varphi - \tan \varphi'$

and $\dfrac{HG}{CK} = \dfrac{GK}{CK} - \dfrac{HK}{CK} = \tan. \varphi' - \tan. \varphi''$;

hence it appears that

$$w : w' = \tan. \varphi - \tan. \varphi' : \tan. \varphi' - \tan. \varphi''.$$

If w'' denote the load on the next joint D, and φ''' the angle which the line DE makes with the horizontal plane, it will appear in the same way that

$$w' : w'' = \tan. \varphi' - \tan. \varphi'' : \tan. \varphi'' - \tan. \varphi''',$$

and so on throughout the whole polygon: whence we have this important proposition:

The vertical pressures on any two joints of the polygon, whether adjoining or remote, are to one another as the differences of the tangents of the angles which the sides about the joints make with the plane of the horizon.

52. From this proposition we may infer, that if φ, φ' denote the angles which any two adjoining sides of the polygon make with the horizontal plane, and w the load on the joint at their intersection, then,

$$w = C (\tan. \varphi - \tan. \varphi')\dots\dots\dots\dots(A);$$

and in this formula, C denotes some constant quantity, which is the same for all the angles of the polygon. This is the general equation of the equilibrated polygon; and it shows that the loads at the joints depend entirely on the angles which the beams make with the horizontal plane, and are independent of the lengths of the beams themselves.

Equilibrated Arch.

53. We shall next investigate the differential equation of an equilibrated arch, deducing it from the finite equation of the equilibrated polygon.

Let us suppose an equilibrated polygon of a very great number of sides (fig. 10), and that a curve ABC passes through all the joints: the sides of the polygon will then be chords in the curve; and as the number of these may be conceived to be greater than any assigned number, and each shorter than any given line, they may be regarded as elements of the curve, and as constituting it.

54. Let us suppose the curve ABC formed in this manner to be the intrados of an arch of a bridge, and that the extrados is the line MDN, which may be curved or straight. Let AC be the *span*, or greatest horizontal width of the arch, and BH, which bisects AC at right angles, its *rise* or height; also let BD be the thickness of the arch of the crown: draw a straight line EDF perpendicular to DH, and draw AE, CF perpendicular to EF.

Let Pp be any indefinitely small part of the intrados ABC. Draw PRQ perpendicular to the horizontal line EF, meeting the extrados in R, and rp parallel to RP; also draw PK perpendicular to DH, meeting rp in q, and PT touching the intrados in p; and, referring the two curves to the same axes DE, let us put

$x = DQ$ the common abscissa,
$y = PQ$ the ordinate of the intrados,
$v = RQ$ the ordinate of the extrados,
$\varphi = $ angle TPK made by the tangent and horizontal line PK,
$c = DB$ the given value of y at the crown,
$c' = AE$ the given value of y at either haunch.

Then, because Pp is a differential of the arc BP, we have P$q = dx$, and $pq = dy$.

55. The load on the arc BP is the gravitating matter between it and DR, the arc of the extrados immediately over it. This is expressed by the area BDRP; therefore the load on Pp, the differential of the arc, is $(y - v)dx$, the differential of that area.

We have put φ to denote the angle which a tangent at P makes with any horizontal line; similarly, let φ' denote the angle which a line touching the curve at p makes with a horizontal line. If now the curve be considered as a polygon of an infinite number of infinitely short sides, and the lines which touch the curve at P and p as the prolongations of two adjoining sides, then, by formula (A), sect. 52, the vertical pressure at their intersection will be C $(\tan. \varphi - \tan. \varphi')$. But these infinitely short touching lines may be regarded as forming Pp, the differential of the curve, the load on which we have found to be $(y - v)dx$; and, moreover, $\tan. \varphi - \tan. \varphi' = d(\tan. \varphi)$, the differential of the trigonometrical tangent of φ; therefore, the relation between the intrados and extrados will be expressed by this equation,

$$C \, d(\tan. \varphi) = (y - v)dx.$$

Now in all curves, $\tan. \varphi = \dfrac{dy}{dx}$, and making dx constant,

$d(\tan. \varphi) = \dfrac{d^2y}{dx}$, let us, to make the members of the equation homogeneous, put c^2 instead of C, which is always positive, and the above equation becomes

$$c^2 \frac{d^2y}{dx} = (y - v)dx;$$

and hence

$$\frac{d^2y}{dx^2} - \frac{y}{c^2} = - \frac{v}{c^2}\dots\dots\dots\dots\dots\dots\dots(B).$$

This differential equation, which is of the second order, and *linear*, or of the first degree, comprehends in it the whole theory of the equilibrated arch. We may now deduce from it the resolution of two problems.

PROB. 1. Having given the form of the intrados ABC, to find that of the extrados.

PROB. 2. Having given the nature of the extrados, to determine the intrados.

56. The first problem is easy, and may be resolved by the differential calculus.

Ex. Let the intrados ABC be a segment of a circle whose centre is O. Draw PO to the centre. Let $\varphi = $ angle POB, and $a = $ PO the radius of the circle; then,

$x = DQ = PK = a \sin. \varphi,$
$y = PQ = BD + BK = c' + a (1 - \cos. \varphi),$
$dx = a \cos. \varphi \, d\varphi, \qquad dy = a \sin. \varphi d\varphi,$

$$\frac{dy}{dx} = \frac{\sin. \varphi}{\cos. \varphi} = \tan. \varphi \, ; \quad \frac{d^2y}{dx} = \frac{d\varphi}{\cos.^3\varphi},$$

$$\frac{d^2y}{dx^2} = \frac{1}{a \cos.^3 \varphi} = \frac{\sec.^3 \varphi}{a}.$$

Now, by formula (B),

$$\frac{y - v}{c^2} = \frac{d^2y}{dx^2} = \frac{\sec.^3 \varphi}{a};$$

therefore $y - v = \dfrac{c^2}{a} \sec.^3 \varphi.$

But when $x = 0$, then $v = 0$, $y = c'$, $\varphi = 0$, and $\sec. \varphi = 1$

therefore $c' = \dfrac{c^2}{a}$ and $c^2 = ac'$; hence we have

$$v - y = c' \sec.^3 \varphi.$$

This shows that the vertical line between the intrados and extrados is always proportional to the cube of the secant of the angle which the radius OP makes with the perpendicular; a conclusion which agrees with section 25 of the preceding article.

57. The second problem, viz. having given the nature

of the ext ados, to find that of the intrados, is the more important of the two, and more difficult. Its solution requires the integral calculus; but the difficulty is no greater than that of finding the area of a curve whose equation is given. This can always be accomplished, if not in finite terms, at least by infinite series. We shall now give a general solution of the problem.

Lagrange has shown (*Théorie des Fonctions Analytiques*, chap. viii.) that the integration of the equation

$$\frac{d^2y}{dx^2} - \frac{y}{c^2} = X,$$

(where X denotes an explicit function of the variable x), can always be accomplished when two particular integrals of this other equation, viz.

$$\frac{d^2y}{dx^2} - \frac{y}{c^2} = 0,$$

are known. Now $y = pe^{\frac{x}{c}}$ and $y = qe^{-\frac{x}{c}}$ are such integrals (p and q being arbitrary constants, and $e = 2\cdot7182818$, the number whose Napierian logarithm is unity), as may be proved by differentiation; therefore, following Lagrange, to integrate the equation (B), viz.

$$\frac{d^2y}{dx^2} - \frac{y}{c^2} = -\frac{v}{c^2},$$

we assume

$$y = pe^{\frac{x}{c}} + qe^{-\frac{x}{c}}$$

for the complete integral equation; but now p and q are to be considered as functions of the variable x. To determine these, we differentiate, and get

$$\frac{dy}{dx} = \frac{1}{c}\left\{pe^{\frac{x}{c}} - qe^{-\frac{x}{c}}\right\} + e^{\frac{x}{c}}\frac{dp}{dx} + e^{-\frac{x}{c}}\frac{dq}{dx}:$$

Since p and q are indeterminate functions of x, we may assume that

$$e^{\frac{x}{c}}\frac{dp}{dx} + e^{-\frac{x}{c}}\frac{dq}{dx} = 0,$$

and then we have

$$\frac{dy}{dx} = \frac{1}{c}\left\{pe^{\frac{x}{c}} - qe^{-\frac{x}{c}}\right\}:$$

Again, by differentiating, we find

$$\frac{d^2y}{dx^2} = \frac{1}{c^2}\left\{pe^{\frac{x}{c}} + qe^{-\frac{x}{c}}\right\} + \frac{1}{c}\left\{e^{\frac{x}{c}}\frac{dp}{dx} - e^{-\frac{x}{c}}\frac{dq}{dx}\right\}$$

$$= \frac{y}{c^2} + \frac{1}{c}\left\{e^{\frac{x}{c}}\frac{dp}{dx} - e^{-\frac{x}{c}}\frac{dp}{dx}\right\}.$$

This result, compared with equation (B), gives

$$\frac{1}{c}\left\{e^{\frac{x}{c}}\frac{dp}{dx} - e^{-\frac{x}{c}}\frac{dq}{dx}\right\} = -\frac{v}{c^2}.$$

From this and the equation

$$e^{\frac{x}{c}}\frac{dp}{dx} + e^{-\frac{x}{c}}\frac{dq}{dx} = 0,$$

we obtain $\left(\text{putting } X = -\frac{v}{c^2}\right)$

$$\frac{dp}{dx} = \frac{c}{2}e^{-\frac{x}{c}}X, \qquad \frac{dq}{dx} = -\frac{c}{2}e^{\frac{x}{c}}X,$$

$$p = \frac{c}{2}\int e^{-\frac{x}{c}}Xdx + b; \quad q = -\frac{c}{2}\int e^{\frac{x}{c}}Xdx + b'.$$

Here b and b' are arbitrary constants, and the integrals are supposed to be taken so as to vanish when $x = 0$. The complete integral equation is now

(C)

$$y = be^{\frac{x}{c}} + b'e^{-\frac{x}{c}} + \frac{c}{2}\left\{e^{\frac{x}{c}}\int e^{-\frac{x}{c}}Xdx - e^{-\frac{x}{c}}\int e^{\frac{x}{c}}Xdx\right\}.$$

To determine the arbitrary constants b and b', we must consider that a tangent to the intrados at the vertex is perpendicular to the vertical DH; therefore, when $x = 0$, then $\frac{dy}{dx} = 0$; but from the equation just found we get

$$\frac{dy}{dx} = \frac{1}{c}\left\{be^{\frac{x}{c}} - b'e^{-\frac{x}{c}}\right\} + \frac{1}{2}\left\{e^{\frac{x}{c}}\int e^{-\frac{x}{c}}Xdx + e^{-\frac{x}{c}}\int e^{\frac{x}{c}}Xdx\right\}.$$

Now, when $x = 0$, then $e^{\frac{x}{c}} = 1$, and $e^{-\frac{x}{c}} = 1$; and by hypothesis the integrals $\int e^{\frac{x}{c}}Xdx, \int e^{-\frac{x}{c}}Xdx$ in this particular case vanish; therefore, when $x = 0$, the last equation becomes $0 = b - b'$, so that $b' = b$. But again, in the general equation (C), when $x = 0$, then $y = c'$; therefore we have also $c' = 2b$ and $b = \frac{1}{2}c'$. On the whole, the equation of the extrados is

(D)

$$y = \frac{c'}{2}\left\{e^{\frac{x}{c}} + e^{-\frac{x}{c}}\right\} + \frac{c}{2}\left\{e^{\frac{x}{c}}\int e^{-\frac{x}{c}}Xdx - e^{-\frac{x}{c}}\int e^{\frac{x}{c}}Xdx\right\};$$

the integrals being taken so as to vanish when $x = 0$. We have thus brought the solution to depend on the integration of the two differentials

$$e^{\frac{x}{c}}Xdx, \quad e^{-\frac{x}{c}}Xdx,$$

which in fact will only differ in their sign, because the branches of the extrados on opposite sides of the vertical are exactly alike, and therefore the substitution of $-x$ for $+x$ will not change the sign of u nor of X. Now this integration can always be effected b known methods, therefore the second problem may be regarded as completely resolved.

57. *Example.* Let us suppose that the extrados is a horizontal straight line EF.

The line PT being supposed to touch the curve, let us as before put

$$c' = DB, \quad c'' = EA, \quad s = DE,$$
$$x = DQ, \quad y = PQ, \quad \varphi = \text{angle TPK.}$$

In this case $v = 0$ and $X = 0$, and the equation of the curve is simply

$$y = \frac{c'}{2}\left\{e^{\frac{x}{c}} + e^{-\frac{x}{c}}\right\}.$$

This case of the general problem has been resolved in sect. 25; the equation of the curve is, however, here given under a different form. We shall now deduce from it a formula for logarithmic calculation.

In all curves, $\tan.\varphi = \frac{dy}{dx}$; in the present case

$$\tan.\varphi\left(= \frac{dy}{dx}\right) = \frac{c'}{2c}\left\{e^{\frac{x}{c}} - e^{-\frac{x}{c}}\right\}:$$

Let ψ be such an angle that

$e^{\frac{x}{c}} = \tan. (45° + \frac{1}{2}\psi)$, then $e^{-\frac{x}{c}} = \tan. (45 - \frac{1}{2}\psi)$.

Hence, by the arithmetic of sines (ALGEBRA, sect. 244, (K) and sect. 240, (C) No. 1),

$$e^{\frac{x}{c}} + e^{-\frac{x}{c}} = \frac{1}{\cos.(45° + \frac{1}{2}\psi)\cos.(45 - \frac{1}{2}\psi)} = \frac{2}{\cos.\psi} = 2\sec.\psi$$

$$e^{\frac{x}{c}} - e^{-\frac{x}{c}} = \frac{\sin.\psi}{\cos.(45° + \frac{1}{2}\psi)\cos.(45° - \frac{1}{2}\psi)} = \frac{2\sin.\psi}{\cos.\psi} = 2\tan.$$

2 D

Also, by the theory of logarithms (ALGEBRA, sect. xix.),

$$\frac{x}{c} \log. e = \log. \frac{\tan. (45° + \frac{1}{2}\psi)}{\text{rad.}} = \log. \tan. (45° + \frac{1}{2}\psi)$$
$$- 10.$$

From these expressions we obtain the relation of the three principal elements of the curve, viz. φ, x, y, as follows:

$$\tan. \psi = \frac{c}{c'} \tan. \varphi \dots\dots\dots\dots\dots\dots\dots(1)$$

$$x = \frac{c}{\log. e} \left\{ \log. \tan. (45° + \frac{1}{2}\psi) - 10 \right\} \dots(2)$$

$$y = \frac{c'}{\cos. \psi} = d \sec. \psi \dots\dots\dots\dots\dots(3)$$

But before these formulæ can be applied, the value of c must be known. To find this, let α denote the value of ψ when $x = s$ and $y = c'$; then equations (3) and (2) become

$$c'' = \frac{c'}{\cos. \alpha}$$

$$s = \frac{c}{\log. e} \left\{ \log. \tan. (45° + \frac{1}{2}\alpha) - 10 \right\}$$

From these we obtain

$$\cos. \alpha = \frac{c'}{c''} \dots\dots\dots\dots\dots\dots\dots\dots(4)$$

$$c = \frac{\log. e}{\log. \tan.(45° + \frac{1}{2}\alpha) - 10} s \dots\dots\dots(5)$$

These formulæ determine c, and this known, the values of x and y corresponding to any proposed value of φ may be readily found from formulæ (1) (2) and (3).

58. We may also determine y directly from x without φ by eliminating $\frac{c}{\log. e}$ by formulæ (2) and (5); we have then, to determine y from x, these formulæ,

$$\cos. \alpha = \frac{c'}{c''} \dots\dots\dots\dots\dots\dots\dots\dots(a)$$

$$\log.\tan.(45° + \frac{1}{2}\psi) = 10 + \frac{x}{s} \left\{ \log.\tan.(45° + \frac{1}{2}\alpha) - 10 \right\} (b)$$

$$y = \frac{c'}{\cos. \psi} \dots\dots\dots\dots\dots\dots\dots(c)$$

1. As an example, let us take the case of Blackfriars Bridge, for which a table of corresponding values of x and y has been given, sect. 26. Here the span is 100 feet, the height or rise forty feet, and the thickness at the crown six feet; and first, let it be required to find the ordinate y, when $x = 20$ feet; we have now

$$s = 50, c' = 6, c'' = 46, x = 20, \frac{x}{s} = \frac{20}{50}.$$

Logarithmic Calculation.

$c' = 6 \dots\dots\dots\dots\dots\dots\dots\dots\dots\dots0.7781512$
$c'' = 46 \dots\dots\dots\dots\dots\dots\dots\dots\dots\dots1.6627578$

$\cos. (\alpha = 82° 30' 19'') \dots\dots\dots\dots\dots9.1153934$

$\log. \tan. (45° + \frac{1}{2}\alpha = 86° 15' 9''.5) - 10 = 1.1837773$

To $\dots\dots\dots\dots\dots\dots\dots\dots\dots\dots\dots10.0000000$
add $\frac{20}{50} \times 1.1837773 \dots\dots\dots\dots = 0.4735109$

$\tan. (45° + \frac{1}{2}\psi = 71° 25' 18'') \dots\dots\dots10.4735109$

$\cos. (\psi = 52° 50' 36'') \dots \dots\dots\dots\dots 9.7810344$

$y = \frac{c'}{\cos. \psi} = 9.9338$ feet $\dots\dots\dots\dots 0.9971168$

The greater part of the above calculation serves for all the values of x, and need not be repeated in constructing a table of the ordinates.

2. Let it be required to find y when $x = 32$ feet.
To $\dots\dots\dots\dots\dots\dots\dots\dots\dots10.0000000$
add $\frac{32}{50} \times 1.1837773 = \dots\dots\dots\dots 0.7576175$

$\tan. (45° + \frac{1}{2}\psi = 80° 5' 18'') \dots\dots10.7576175$

$c' = 6 \dots\dots\dots\dots\dots\dots\dots\dots\dots 0.7781512$
$\cos. (\psi = 70° 10' 36'') \dots\dots\dots\dots 9.5303431$

$y = 17.6933$ feet $\dots\dots\dots\dots\dots 1.2478081$

In this way may all the numbers in the table of sect. 26. be computed with greater expedition than by the formula given there.

The value of φ to each value of x may be found from formulæ (5) and (1) of last article.   (J. R.)   (W. W.)

CARPENTRY.

WE must begin by informing our readers, that the bulk of the present article was written by the late Professor Robison, in order to form, with those on ROOF, and STRENGTH OF MATERIALS, also written by him for the Encyclopædia, a uniform system of the most useful departments of practical mechanics, deduced, in the same familiar and elementary manner, from the simple principles of the composition of forces. In here reprinting his contribution, we shall premise some introductory observations, which may be considered as a retrospective summary of the doctrine of Passive Strength, accompanied by some of the most useful propositions respecting the resistance of elastic substances, derived from the principles which have been already laid down in our article BRIDGE; and subjoining a few notes on such passages as may appear to require further illustration or correction. Some of the demonstrations will be partly borrowed from a work which has been published since the death of Professor Robison, but others will be more completely original; and of the remarks the most important will probably be those which relate to the form and direction of the abutment of rafters; a subject which seems to have been very incorrectly treated by former writers on Carpentry.[1]

I.—ABSTRACT OF THE DOCTRINE OF PASSIVE STRENGTH.

The effects of forces of different kinds, on the materials employed in the mechanical arts, require to be minutely examined in the arrangement of every work dependent on them; and of these effects, as exhibited in a solid body at rest, we may distinguish seven different varieties; the extension of a substance acting simply as a tie; the compression of a block supporting a load above it; the detrusion of an axis resting on a support close to its wheel, and resisting by its lateral adhesion only; the flexure of a body bent by a force applied unequally to its different parts; the torsion or twisting, arising from a partial detrusion of the external parts in opposite directions, while the axis retains its place; the alteration or permanent change of a body which settles, so as to remain in a new form, when the force is withdrawn; and lastly, the fracture, which consists in a complete separation of parts before united, and which has been the only effect particu-

larly examined by the generality of authors on the strength of materials.

The analogy of the laws of extension and compression has been demonstrated in the article BRIDGE, and their connection with flexure has been investigated; but it is not easy to compare them directly with the resistance opposed to a partial detrusion, the effects of which are only so far understood as they are exhibited in the phenomena of twisting; and these appear to justify us in considering the resistance of lateral adhesion as a primitive force, deduced from the rigidity or solidity of the substance, and proportional to the deviation from the natural situation of the particles. The resistance exhibited by steel wire, when twisted, bears a greater proportion to that of brass than the resistance to extension or compression, but the forces agree in being independent of the hardness produced by tempering.

Flexure may be occasioned either by a transverse or by a longitudinal force. When the force is transverse, the extent of the flexure is nearly proportional to its magnitude; but when it is longitudinal, there is a certain magnitude which it must exceed in order to produce, or rather to continue, the flexure, if the force be applied exactly at the axis. But it is equally true that the slightest possible force applied at a distance from the axis, however minute, or with an obliquity however small, or to a beam already a little curved, will produce a certain degree of flexure; and this observation will serve to explain some of the difficulties and irregularities which have occurred in making experiments on beams that are exposed to longitudinal pressure.

Stiffness, or the power of resisting flexure, is measured by the force required to produce a given minute change of form. For beams similarly fixed, it is directly proportional to the breadth and the cube of the depth, and inversely to the cube of the length. Thus a beam or bar two yards long will be equally stiff with a beam one yard, provided that it be either twice as deep or eight times as broad. If the ends of a beam can be firmly fixed, by continuing them to a sufficient distance, and keeping them down by a proper pressure, the stiffness will be four times as great as if the ends were simply supported. A hollow substance of given weight and length, has its stiffness

---

[1] These introductory observations to Professor Robison's article, and the notes subjoined to it, were written by the late Dr Thomas Young.

nearly proportional to the square of the diameter; and hence arises the great utility of tubes when stiffness is required, this property being still more increased by the expansion of the substance than the ultimate strength. It is obvious that there are a multiplicity of cases in carpentry where stiffness is of more importance than any other property, since the utility as well as beauty of the fabric might often be destroyed by too great a flexibility of the materials.

If we wish to find how much a beam of fir will sink when it is loaded in the middle, we may multiply the cube of the length in inches by the given weight in pounds, and divide by the cube of the depth, and by ten million times the breadth; but, on account of the unequal texture of the wood, we must expect to find the bending somewhat greater than this in practice, besides that a large weight will often produce an alteration, or permanent settling, which will be added to it : a beam of oak will also sink a little more than a beam of fir with the same weight.

With respect to torsion, the stiffness of a cylindrical body varies directly as the fourth power of the diameter, and inversely in the simple proportion of the length : it does not appear to be changed by the action of any force tending to lengthen or to compress the cylinder; and it may very possibly bear some simple relation to the force of cohesion, which has not yet been fully ascertained; but it appears that, in an experiment of Mr Cavendish, the resistance of a cylinder of copper to a twisting force, acting at its surface, was about $\frac{1}{100}$ of the resistance that the same cylinder would have opposed to direct extension or compression.

Alteration is often an intermediate step between a temporary change and a complete fracture. There are many substances which, after bending to a certain extent, are no longer capable of resuming their original form; and in such cases it generally happens that the alteration may be increased without limit, until complete fracture takes place, by the continued operation of the same force which has begun it, or by a force a little greater. Those substances which are the most capable of this change are called ductile; and the most remarkable are gold, and a spider's web. When a substance has undergone an alteration by means of its ductility, its stiffness, in resisting small changes on either side, remains little or not at all altered. Thus, if the stiffness of a spider's web, in resisting torsion, were sufficient at the commencement of an experiment to cause it to recover itself, after being twisted in an angle of ten degrees, it would return ten degrees, and not more, after having been twisted round a thousand times. The ductility of all substances capable of being annealed is greatly modified by the effects of heat. Hard steel, for example, is incomparably less subject to alteration than soft, although in some cases more liable to fracture; so that the degree of hardness requires to be proportioned to the uses for which each instrument is intended; although it was proved by Coulomb, and has since been confirmed by other observers, that the primitive stiffness of steel, in resisting small flexures, is neither increased nor diminished by any variation in its temper.

The strength of a body is measured by the force required completely to overcome the corpuscular powers concerned in the aggregation of its particles, and it is jointly proportional to the primitive stiffness and to the toughness of the substance, that is, to the degree in which it is capable of a change of form without permanent alteration. It becomes, however, of importance in some cases to consider the measure of another kind of strength, which has sometimes been called resilience, or the power of resisting a body in motion, and which is proportional to the strength and the toughness conjointly, that is, to the

stiffness and the square of the toughness. Thus, if we Carpentry. double the length of a given beam, we reduce its absolute strength to one half, and its stiffness to one eighth; but since the toughness, or the space through which it will continue to resist, is quadrupled, the resilience will be doubled, and it would require a double weight to fall from the same height, or the same weight to fall from a double height, in order to overcome its whole resistance. If we wish to determine the resilience of a body from an experiment on its strength, we must measure the distance through which it recedes or is bent previously to its fracture; and it may be shown that a weight which is capable of breaking it by pressure, would also break it by impulse if it moved with the velocity acquired by falling from a height equal to half the deflection. Thus, if a beam or bar were broken by a weight of 100 pounds, after being bent six inches without alteration, it would also be broken by a weight of 100 pounds falling from a height of three inches, or moving in a horizontal direction with a velocity of four feet in a second, or by a weight of one pound falling from a height of 300 inches. This substitution of velocity for quantity of matter has, however, one limit, beyond which the velocity must prevail over the resistance, without regard to the quantity of matter; and this limit is derived from the time required for the successive propagation of the pressure through the different parts of the substance, in order that they may participate in the resistance. Thus, if a weight fell on the end of a bar or column with a velocity of 100 feet in a second, and the substance could only be compressed $\frac{1}{200}$ of its length, without being crushed, it is obvious that the pressure must be propagated through the substance with a velocity of 20,000 feet in a second, in order that it might resist the stroke; and, in general, a substance will be crushed or penetrated by any velocity exceeding that which is acquired by a body falling from a height, which is to half that of the modulus of elasticity of the substance, as the square of the greatest possible change of length is to the whole length. From the consideration of the effect of rigidity in lessening the resilience of bodies, we may understand how a diamond, which is capable of resisting an enormous pressure, may be crushed with a blow of a small hammer, moving with a moderate velocity. It is remarkable that, for the same substance in different forms, the resilience is in most cases simply proportional to the bulk or weight, while almost every other kind of resistance is capable of infinite variation by change of form only.

The elaborate investigations of M. Lagrange, respecting the strength and the strongest forms of columns, appear to have been conducted upon principles not altogether unexceptionable; but it is much easier to confute the results than to follow the steps of the computations. One great error is the supposition that columns are to be considered as elastic beams, bent by a longitudinal force; while, in reality, a stone column is never slender enough to be bent by a force which it can bear without being crushed; and even for such columns as are capable of being bent by a longitudinal force, M. Lagrange's determinations are in several instances inadmissible. He asserts, for example, that a cylinder is the strongest of all possible forms, and that a cone is stronger than any conoid of the same bulk; but it appears to be demonstrable in a very simple manner, and upon incontestable principles, that a conoidal form may be determined, which shall be stronger than either a cone or a cylinder of the same bulk.

When a column is crushed, its resistance to compression seems to depend in great measure on the force of lateral adhesion, assisted by a kind of internal friction, dependent on the magnitude of the pressure; and it commonly gives way by the separation of a wedge in an

Carpentry. oblique direction. If the adhesion were simply proportional to the section, it may be shown that a square column would be most easily crushed when the angle of the wedge is equal to half of a right angle; but if the adhesion is increased by pressure, this angle will be diminished by half the angle of repose appropriate to the substance. In a wedge separated by a direct force from a prism of cast iron, the angle was found equal to $32\frac{1}{2}°$, consequently the angle of repose was $2 \times 12\frac{1}{4}° = 25°$, and the internal friction to the pressure as 1 to ·466, the tangent of this angle; there was, however, a little bubble in the course of the fracture, which may have changed its direction in a slight degree. The magnitude of the lateral adhesion is measured by twice the height of the wedge, whatever its angle may be. In this instance the height was to the depth as 1·57 to 1, consequently the surface, affording an adhesion equal to the force, was somewhat more than three times as great as the transverse section, and the lateral adhesion of a square inch of cast iron would be equal to about 46,000 pounds; the direct cohesive force of the same iron was found by experiment equal to about 20,000 pounds for a square inch. It is obvious that experiments on the strength of a substance in resisting compression ought to be tried on pieces rather longer than cubes, since a cube would not allow of the free separation of a single wedge so acute as was observed in this experiment; although, indeed, the force required to separate a shorter wedge on each side would be little or no greater than for a single wedge. The same consideration of the oblique direction of the plane of easiest fracture would induce us to make the outline of a column a little convex externally, as the common practice has been; for a circle cut out of a plank possesses the advantage of resisting equally in every section, and consequently of exhibiting the strongest form, when there is no lateral adhesion; and in the case of an additional resistance proportional to the pressure, the strongest form is afforded by an oval consisting of two circular segments, each containing twice the angle formed by the plane of fracture with the horizon. If we wish to obtain a direct measure of the lateral adhesion, we must take care to apply the forces concerned at a distance from each other not greater than one sixth of the depth of the substance, otherwise the fracture will probably be rather the consequence of flexure than of detrusion. Professor Robison found this force in some instances twice as great as the direct cohesion, or nearly in the same proportion as it appears to have been in the experiment on the strength of cast iron; Mr Coulomb thinks it most commonly equal only to the cohesion; and in fibrous substances, especially where the fibres are not perfectly straight, the repulsive strength is generally much less than would be inferred from this equality, and sometimes even less than the cohesive strength.

It is well known that the transverse strength of a beam is directly as the breadth and as the square of the depth, and inversely as the length; and the variation of the results of some experiments from this law can only have depended on accidental circumstances. If we wish to find the number of hundredweights that will break a beam of oak supported at both ends, supposing them to be placed exactly on the middle, we may multiply the square of the depth in inches by 100 times the breadth, and divide by the length; and we may venture in practice to load a beam with at least an eighth as much as this, or, in case of necessity, even a fourth. And if the load be distributed equally throughout the length of the beam, it will support twice as much; but for a beam of fir the strength is somewhat less than for oak. A cylinder will bear the same curvature as the circumscribing prism, and it may be shown that its strength, as well as its stiffness, is to that of the prism as one fourth of its bulk is to one third of the bulk of the prism. The strength of a beam supported at its extremities may be doubled by firmly fixing the ends where it is practicable; and we have already seen that the stiffness is quadrupled: but the resilience remains unaltered, since the resistance is doubled, and the space through which it acts is reduced to a half. It is therefore obviously of importance to consider the nature of the resistance that is required from the fabric which we are constructing. A floor, considered alone, requires to be strong; but in connection with a ceiling, its stiffness requires more particular attention, in order that the ceiling may remain free from cracks. A coach-spring requires resilience for resisting the relative motions of the carriage, and we obtain this kind of strength as effectually by combining a number of separate plates, as if we united them into a single mass, while we avoid the stiffness, which would render the changes of motion inconveniently abrupt.

In all calculations respecting stiffness, it is necessary to be acquainted with the modulus of elasticity, which may be found for a variety of substances in the annexed table.

Height of the Modulus of Elasticity in Thousands of Feet.

Iron and steel	10,000	Fir wood	10,000
Copper	5,700	Elm	8,000
Brass	5,000	Beech	8,000
Silver	3,240	Oak	5,060
Tin	2,250	Box	5,050
Crown glass	9,800	Ice	850

II.—PROPOSITIONS RELATING TO FLEXURE.

A. *The stiffness of a cylinder is to that of its circumscribing rectangular prism, as three times the bulk of the cylinder is to four times that of the prism.*

We may consider the different strata of the substance as acting on levers equal in length to the distance of each from the axis; for although there is no fixed fulcrum at the axis, yet the whole force is the same as if such a fulcrum existed, since the opposite actions of the opposite parts would relieve the fulcrum from all pressure. Then the tension of each stratum being also as the same distance x, and the breadth of the stratum being called $2y$, the fluxion of the force on either side of the axis will be $2x^2ydx$, while that of the force of the prism, the radius being r, is $2rx^2dx$. Now z being the area of half the portion included between the stratum and the axis, of which the fluxion is ydx, the fluxion of $z - \dfrac{y^3x}{rr}$ will be

$$ydx - \frac{y^3dx}{r^2} - \frac{3y^2xdy}{r^2}$$

$$= ydx \left(1 - \frac{y^2}{r^2}\right) - \frac{3yx}{r^2}\, ydy.$$

But $1 - \dfrac{y^2}{r^2} = \dfrac{x^2}{r^2}$, and $-ydy = xdx$, therefore the fluxion is

$$\frac{x^2ydx}{r^2} + \frac{3x^2ydx}{r^2} = \frac{4x^2ydx}{r^2};$$

consequently the fluent of $2x^2ydx$ is $\frac{1}{2}r^2z - \frac{1}{2}y^3x$, which when $y = 0$, becomes $\frac{1}{2}r^2z$, or one fourth of the product of the square of the radius by the area of the section while the fluent of $2rx^2dx$, that is, $\frac{2}{3}rx^3$, the force of the prism becomes $\frac{2}{3}r^4$ or $\frac{1}{3}r^2 \times 2r^2$, one third of the product of the same square into the area of the section of the prism.

Hence the radius of curvature of a cylindrical column, instead of $\dfrac{Maa}{12fy}$ (Art. BRIDGE, Prop. G), will be $\dfrac{Maa}{16fy}$, the weight of the modulus M decreasing in the same propor

Carpentry. tion as the bulk when the prism is reduced to a cylinder. The force is supposed in this proposition to be either transverse or applied at a considerable distance from the axis; but the error will not be material in any other case.

B. *When a longitudinal force f is applied to the extremities of a straight prismatic beam, at the distance b from the axis, the deflection of the middle of the beam will be* $b\left(\text{SECANT}\left[\sqrt{\left(\frac{3f}{M}\right)}\cdot\frac{e}{a}\right]-1\right)$; M *being the weight of the modulus, e the length of the beam, and a its depth.*

The curvature being proportional to the distance from the line of direction of the force, or to the ordinate, when that line is considered as the absciss, the elastic curve must in this case initially coincide with a portion of the harmonic curve, well known for its utility in the resolution of a variety of problems of this kind. Now if the half length of the complete curve be called k, corresponding to a quadrant of the generating circle, and the greatest ordinate y, c being the quadrant of a circle of which the radius is unity, the radius of curvature r corresponding to y will be $\frac{kk}{ccy}$, that is, a third proportional to y and $\frac{k}{c}$ the radius of the generating circle; consequently $\frac{Maa}{12fy}=\frac{kk}{ccy}$, $kk=\frac{Maacc}{12f}$, and $k=\frac{1}{2}\sqrt{\frac{M}{3f}}\cdot ac$; but by the nature of the curve, $y:b=1:\cos.\frac{ec}{2k}=\text{sec}.\frac{ec}{2k}:1$, and $y=b\ \text{sec}.\frac{ec}{2k}=b\ \text{sec}.\sqrt{\frac{3f}{M}}\cdot\frac{e}{a}$, which is the ordinate at the middle; and the deflection from the natural situation is $y-b$.

It follows that, since the secant of the quadrant is infinite, when $\sqrt{\frac{3f}{M}}\cdot\frac{e}{a}$ becomes equal to c, the deflection will be infinite, and the resistance of the column will be overcome, however small the distance b may be taken, provided that it be of finite magnitude; and since in this case $\frac{3fee}{Maa}=cc$, $f=\frac{Maacc}{3ee}=\cdot8225\ M\frac{aa}{ee}$, which is the utmost force that the column will bear: and for a cylinder we find, by the same reasoning, $f=\frac{Maacc}{4ee}=\cdot6169\ M\frac{aa}{ee}$. If b be supposed to vanish, we shall have in theory an equilibrium without flexure; but since it will be tottering, it cannot exist in nature.

By applying this determination to the strength of wood and iron, compared with the modulus of elasticity, it appears that a round column or a square pillar of either of these substances cannot be bent by any longitudinal force applied to the axis, which it can withstand without being crushed, unless its length be greater than twelve or thirteen times its thickness respectively; nor a column or pillar of stone, unless it be forty or forty-five times as long as it is thick. Hence we may infer, as a practical rule, that every piece of timber or iron intended to withstand any considerable compressing force, should be at least as many inches in thickness as it is feet in length, in order to avoid the loss of force which necessarily arises from curvature.

C. *When a beam, fixed at one end, is pressed by a force in a direction deviating from the original position of the axis in a small angle, of which the tangent is t, the deflection becomes* $d=at\frac{M}{12f}\ \text{TANG}.\left(\sqrt{\frac{12f}{M}}\cdot\frac{e}{a}\right).$

The inclination of the curve to the absciss being inconsiderable, it will not differ sensibly from a portion of a har-

monic curve; and supposing the quadrantal length of Carpentry this curve k, we have again, as in the last proposition, $k=\frac{1}{2}\sqrt{\frac{M}{3f}}\cdot ac$, or, for a cylinder, $k=\frac{1}{4}\sqrt{\frac{M}{f}}\cdot ac$. Now, the tangent of the inclination of the harmonic curve varies as the sine of the angular distance from the middle; consequently, as $\text{SIN}.\frac{k-e}{k}\cdot c$, or $\cos.\frac{ec}{k}$, is to the radius, so is the tangent t, expressing the difference of inclination of the end of the beam and the direction of the force, which is also that of the middle of the supposed curve, to the tangent of the extreme inclination of the curve to its absciss, which will therefore be $t\ \text{SEC}.\frac{ec}{k}$; consequently the greatest ordinate will be $\frac{kt}{c}\ \text{SEC}.\frac{ec}{k}$, and since the ordinates are as the sines of the angular distances from the origin of the curve, the ordinate at the fixed end of the beam, corresponding to the angle $\frac{ec}{k}$, that is, the deflection, will be $\frac{kt}{c}\ \text{SEC}.\frac{ec}{k}.\text{SIN}.\frac{ec}{k}=\frac{kt}{c}\ \text{TANG}.\frac{ec}{k}=\frac{1}{2}\sqrt{\frac{M}{3f}}\cdot at$ TANG. $\frac{2e}{a}\sqrt{\frac{3f}{M}}$, or, for a cylinder, $\frac{1}{4}\sqrt{\frac{M}{f}}\cdot at$ TANG. $\frac{4e}{a}\sqrt{\frac{f}{M}}.$

By means of this proposition we may determine the effect of a small lateral force in weakening a beam or pillar which is at the same time compressed longitudinally by a much greater force, considering the parts on each side of the point to which the lateral force is applied, as portions of two beams, bent in the manner here described, by a single force slightly inclined to the axis.

D. *A bar fixed at one end, and bent by a transverse force applied to the other end, assumes initially the form of a cubic parabola, and the deflection at the end is* $d=\frac{4e^3f}{Maa}.$

The ordinate of a cubic parabola varying as x^3, its second fluxion varies as $6x\ (dx)^2$, or since the first fluxion of the absciss is constant, simply as the absciss x, measured from the vertex of the parabola, which must therefore be situated at the end to which the force is applied, and the absciss must coincide with the tangent of the bar. But if we begin from the other end, we must substitute $e-x$ for x, and the second fluxion of the ordinate will be as $6(e-x)(dx)^2$, the first as $6exdx-3x^2dx$, and the fluent as $3ex^2-x^3$, which, when $x=e$, becomes $2e^3$, while it would have been $3e^3$ if the curvature had been uniform, and the second fluxion had been everywhere $6e(dx)^2$. Now the radius of curvature at the fixed end being $r=\frac{Maa}{12ef}$, and the versed sine of a small portion of a circle being equal to $\frac{ee}{2r}$, this versed sine will be expressed by $\frac{6e^3f}{Maa}$; and two thirds of this, or $\frac{4e^3f}{Maa}$, will be the actual deflection.

E. *The depression of a bar, fixed horizontally at one end, and supporting only its own weight, is* $d=\frac{3e^4}{2maa}$; *m being the height of the modulus of elasticity.*

The curvature here varies as the square of the distance from the end, because the strain is proportional to the weight of the portion of the bar beyond any given point, and to the distance of its centre of gravity conjointly, that is, to $(e-x)\frac{1}{2}(e-x)$, so that if the second fluxion

Carpentry. at the fixed end be as $e^2(dx)^2$, it will elsewhere be as $(e-x)^2(dx)^2$; and the corresponding first fluxions being e^2xdx and $e^2xdx - ex^2dx + \frac{1}{3}x^3dx$, the fluents will be $\frac{1}{2}e^2x^2$, and $\frac{1}{2}e^2x^2 - \frac{1}{3}ex^3 + \frac{1}{12}x^4$, or, when $x = e$, $\frac{1}{12}e^4$, and $(\frac{1}{2} - \frac{1}{3} + \frac{1}{12})e^4 = \frac{1}{4}e^4$; consequently the depression must be half the versed sine in the circle of greatest curvature. Now the radius of curvature $\frac{Maa}{12fy}$ becomes here $\frac{Maa}{6ef}$, the force being applied at the distance $\frac{1}{2}e$; and since the weight of the bar is to that of the modulus of elasticity in the proportion of the respective lengths, we have $\frac{f}{M} = \frac{e}{m}$, and $r = \frac{maa}{6ee}$, and the versed sine for the ordinate e will be $\frac{3e^4}{maa}$, half of which is the actual depression.

F. *The depression of the middle of a horizontal bar, fixed at both ends, and supporting its own weight only, is* $d = \frac{5e^4}{32maa}$.

The transverse force at each point of such a bar, resisted by the lateral adhesion, is as the distance x from the middle (*see* BRIDGE, under Prop. L); but this force is proportional to the first fluxion of the strain or curvature, consequently the curvature itself must vary as the corrected fluent of $\pm\, xdx$, taking here the negative sign, because the curvature diminishes as x increases; and the corrected fluent will be $\frac{1}{4}e^2 - x^2$, since it must vanish when $x = \frac{1}{2}e$; the first fluxion of the ordinate will then be $\frac{1}{4}e^2xdx - \frac{1}{3}x^3dx$, and the fluent $\frac{1}{8}e^2x^2 - \frac{1}{12}x^4$, or for the whole length $\frac{1}{2}e$, $\frac{1}{192}e^4$, instead of $\frac{1}{32}$, or $\frac{6}{192}$, which would have been its value if the curvature had been equal throughout. Now the strain at the middle is the difference of the opposite strains produced by the forces acting on either side; and these are the half weight acting at the mean distance $\frac{1}{4}e$, and the resistance of the support, which is equal to the same half weight, but acts at the distance $\frac{1}{2}e$, the difference being equivalent to the half weight, acting at the distance $\frac{1}{4}e$, so that the curvature at the middle is the same as if the bar were fixed there, and loose at the ends; that is, as in the last proposition, substituting $\frac{1}{2}e$ for e, $r = \frac{2maa}{3ee}$; and the versed sine at the distance $\frac{1}{2}e$ being $\frac{e^2}{8r}$, or $\frac{3e^4}{16maa}$, $\frac{5}{6}$ of this will be $\frac{5e^4}{32maa}$. This demonstration may serve as an illustration of two modes of considering the effect of a strain, which have not been generally known, and which are capable of a very extensive application.

It follows that where a bar is equally loaded throughout its length, the curvature at the middle is half as great as if the whole weight were collected there, the strain derived from the resistance of the support remaining in that case uncompensated. The depression produced by the divided weight will be $\frac{5}{8}$ as great as by the single weight, since $\frac{5}{6} \times \frac{1}{2}$ is to $\frac{3}{8}$ as 5 to 8. M. Dupin found the proposition, by many experiments, between $\frac{2}{3}$ and $\frac{5}{7}$; and $\frac{5}{8}$ is a very good mean for representing these results.

III.—ELEMENTS OF CARPENTRY.

Definition. " Carpentry is the art of framing timber for the purposes of architecture, machinery, and, in general, for all considerable structures."

It is not intended in this article to give a full account of carpentry as a mechanical art, or to describe the various Carpentry ways of executing its different works, suited to the variety of materials employed, the processes which must be followed for fashioning and framing them for our purposes, and the tools which must be used, and the manner in which they must be handled. This would be an occupation for volumes, and, though of great importance, must be entirely omitted here. Our only aim at present will be to deduce, from the principles and laws of mechanics, and the knowledge which experience, and judicious inferences from it, have given us concerning the strength of timber, in relation to the strain laid on it, such maxims of construction as will unite economy with strength and efficacy.

This object is to be attained by a knowledge, 1*st*, of the strength of our materials, and of the absolute strain that is to be laid on them; 2*dly*, of the modifications of this strain, by the place and direction in which it is exerted, and the changes that can be made by a proper disposition of the parts of our structure; and, 3*dly*, having disposed every piece in such a manner as to derive the utmost advantage from its relative strength, we must know how to form the joints and other connections in such a manner as to secure the advantages derived from this disposition.

This is evidently a branch of mechanical science which An import- makes carpentry a *liberal* art, constitutes part of the learn- ant branch ing of the ENGINEER, and distinguishes him from the of mecha- workman. Its importance in all times and states of civil nical sci- society is manifest and great. In the present condition of ence. these kingdoms, raised by the active ingenuity and energy of our countrymen to a pitch of prosperity and influence unequalled in the history of the world, a condition which consists chiefly in the superiority of our manufactures, attained by prodigious multiplication of engines of every description, and for every species of labour, the Science (so to term it) of carpentry is of immense consequence. We regret therefore exceedingly that none of our celebrated artists have done honour to themselves and their country, by digesting into a body of consecutive doctrines the results of their experience, so as to form a system from which their pupils might derive the first principles of their education. The many volumes called Complete Instructors, Manuals, &c. take a much humbler flight, and content themselves with instructing the mere workman; or sometimes give the master builder a few approved forms of roofs and other framings, with the rules for drawing them on paper, and from thence forming the working draughts which must guide the saw and the chisel of the workman. Hardly any of them offer any thing that can be called a principle, applicable to many particular cases, with the rules for this adaptation. We are indebted for Principally the greatest part of our knowledge of this subject to the indebted to labours of literary men, chiefly foreigners, who have pub- foreigners lished in the memoirs of the learned academies disserta- for a know. tions on different parts of what may be termed the *Science* ledge of *of Carpentry*. It is singular that the members of the ject. Royal Society of London, and even of that established and supported for the encouragement of the arts, have contributed so little to the public instruction in this respect. We have observed some beginnings of this kind, such as the last part of Nicholson's *Carpenter's and Joiner's Assistant*; and it is with pleasure we can say, that we were told by the editor this work was prompted in a great measure by what has been delivered in our articles ROOF and STRENGTH OF MATERIALS. It abounds more in important and new observations than any book of the kind that we are acquainted with. We again call on such as have given a scientific attention to this subject, and pray that they would render a meritorious service to their country by imparting the result of their researches. The very limited nature of this work does not allow us to treat the subject in de-

Carpentry. tail; and we must confine our observations to the fundamental and leading propositions.

The theory, so to term it, of carpentry is founded on two distinct portions of mechanical science, namely, a knowledge of the strains to which framings of timber are exposed, and a knowledge of their *relative* strength.

We shall therefore attempt to bring into one point of view the propositions of mechanical science that are more immediately applicable to the art of carpentry, and are to be found in various articles of our work, particularly ROOF and STRENGTH OF MATERIALS. From these propositions we hope to deduce such principles as shall enable an attentive reader to comprehend distinctly what is to be aimed at in framing timber, and how to attain this object with certainty; and we shall illustrate and confirm our principles by examples of pieces of carpentry which are acknowledged to be excellent in their kind.

Composition and resolution of forces.

The most important proposition of general mechanics to the carpenter is that which exhibits the composition and resolution of forces; and we beg our practical readers to endeavour to form very distinct conceptions of it, and to make it very familiar to their minds. When accommodated to their chief purposes, it may be thus expressed:

1. If a body, or any part of a body, be at once pressed in the two directions AB, AC (fig. 1, Plate L.), and if the intensity or force of those pressures be in the proportion of these two lines, the body is affected in the same manner as if it were pressed by a single force acting in the direction AD, which is the diagonal of the parallelogram ABDC formed by the two lines, and whose intensity has the same proportion to the intensity of each of the other two that AD has to AB or AC.

Such of our readers as have *studied* the laws of motion, know that this is fully demonstrated. Such as wish for a very accurate view of this proposition will do well to read the demonstration given by D. Bernoulli, in the first volume of the *Comment. Petropol.*, and the improvement of this demonstration by D'Alembert in his *Opuscules* and in the *Comment. Taurinens.* The practitioner in carpentry will get more useful confidence in the doctrine, if he will shut his book, and verify the theoretical demonstrations

Illustrated by experiment.

by actual experiments. They are remarkably easy and convincing. Therefore it is our request that the artist, who is not so habitually acquainted with the subject, do not proceed further till he has made it quite familiar to his thoughts. Nothing is so conducive to this as the actual experiment; and since this only requires the trifling expense of two small pulleys and a few yards of whipcord, we hope that none of our practical readers will omit it: they will thank us for this injunction.

2. Let the threads A*d*, AF*b*, and AE*c* (fig. 2), have the weights *d*, *b*, and *c*, appended to them, and let two of the threads be laid over the pulleys F and E. By this apparatus the knot A will be drawn in the directions AB, AC, and AK. If the sum of the weights *b* and *c* be greater than the single weight *d*, the assemblage will of itself settle in a certain determined form: if you pull the knot A out of its place, it will always return to it again, and will rest in no other position. For example, if the three weights are equal, the threads will always make equal angles, of 120 degrees each, round the knot. If one of the weights be three pounds, another four, and the third five, the angle opposite to the thread stretched by five pounds will always be square, &c. When the knot A is thus in equilibrio, we must infer that the action of the weight *d*, in the direction A*d*, is in direct opposition to the combined action of *b* in the direction AB, and of *c* in the direction AC. Therefore, if we produce *d*A to any point D, and take AD to represent the magnitude of the

force, or pressure exerted by the weight *d*, the pressures Carpentry. exerted on A by the weights *b* and *c*, in the directions AB, AC, are in fact equivalent to a pressure acting in the direction AD, whose intensity we have represented by AD. If we now measure off by a scale on AF and AE the lines AB and AC, having the same proportions to AD that the weights *b* and *c* have to the weight *d*, and if we draw DB and DC, we shall find DC to be equal and parallel to AB, and DB equal and parallel to AC; so that AD is the diagonal of a parallelogram ABDC. We shall find this always to be the case, whatever are the weights made use of; only we must take care that the weight which we cause to act without the intervention of a pulley be less than the sum of the other two; if any one of the weights exceeds the sum of the other two, it will prevail, and drag them along with it.

Now since we know that the weight *d* would just balance an equal weight *g*, pulling directly upwards by the intervention of the pulley G; and since we see that it just balances the weights *b* and *c*, acting in the directions AB, AC; we must infer that the knot A is affected in the same manner by those two weights, or by the single weight *g*; and therefore that *two pressures, acting in the directions and with the intensities AB, AC, are equivalent to a single pressure having the direction and proportion of AD.* In like manner, the pressures AB, AK, are equivalent to AH, which is equal and opposite to AC. Also AK and AC are equivalent to AI, which is equal and opposite to AB.

We shall consider this combination of pressures a little more particularly.

Suppose an upright beam BA (fig. 3), pushed in the direction of its length by a load B, and abutting on the ends of two beams AC, AD, which are firmly resisted at their extreme points C and D, which rest on two blocks, but are nowise joined to them; these two beams can resist no way but in the directions CA, DA, and therefore the pressures which they sustain from the beam BA are in the directions AC, AD. We wish to know how much each sustains: Produce BA to E, taking AE from a scale of equal parts, to represent the number of tons or pounds by which BA is pressed. Draw EF and EG parallel to AD and AC; then AF, measured on the same scale, will give us the number of pounds by which AC is strained or crushed, and AG will give the strain on AD.

It deserves particular remark here, that the length of AC or AD has no influence on the strain arising from the thrust BA, while the directions remain the same. The effects, however, of this strain are modified by the length of the piece on which it is exerted. This strain compresses the beam, and will therefore compress a beam of double length twice as much. This may change the form of the assemblage. If AC, for example, be very much shorter than AD, it will be much less compressed: the line CA will turn about the centre C, while DA will hardly change its position; and the angle CAD will grow more open, the point A sinking down. The artist will find it of great consequence to pay a very minute attention to this circumstance, and to be able to see clearly the change of shape which necessarily results from these mutual strains He will see in this the cause of failure in many very great works. By thus changing shape, strains are often produced in places where there were none before, and frequently of the very worst kind, tending to break the beams across.

The dotted lines of this figure show another position of the beam AD'. This makes a prodigious change, not only in the strain on AD', but also in that on AC. Both of them are much increased; AG is almost doubled, and AF' is four times greater than before. This addition was

Carpentry. made to the figure to show what enormous strains may be produced by a very moderate force, AE, when it is exerted on a very obtuse angle.[1]

The fourth and fifth figures will assist the most uninstructed reader in conceiving how the very same strains, AF, AG, are laid on these beams, by a weight simply hanging from a billet resting on A, pressing hard on AD, and also leaning a little on AC; or by an upright piece, AE, joggled on the two beams AC, AD, and performing the office of an ordinary king-post. The reader will thus learn to call off his attention from the means by which the strains are produced, and learn to consider them abstractedly merely as strains, in whatever situation he finds them, and from whatever cause they arise.

We presume that every reader will perceive, that the proportions of these strains will be precisely the same if every thing be inverted, and each beam be drawn or pulled in the opposite direction. In the same way that we have substituted a rope and weight in fig. 4, or a king-post in fig. 5, for the loaded beam BA of fig. 3, we might have substituted the framing of fig. 6, which is a very usual practice. In this framing, the batten DA is stretched by a force AG, and the piece AC is compressed by a force AF. It is evident that we may employ a rope or an iron rod hooked on at D, in place of the batten DA, and the strains will be the same as before.

This seemingly simple matter is still full of instruction; and we hope that the well-informed reader will pardon us, though we dwell a little longer on it for the sake of the young artist.

By changing the form of this framing, as in fig. 7, we produce the same strains as in the disposition represented by the dotted lines in fig. 3. The strains on both the battens AD, AC, are now greatly increased.

The same consequences result from an improper change of the position of AC. If it is placed as in fig. 8, the strains on both are vastly increased. In short, the rule is general, that the more open we make the angle against which the push is exerted, the greater are the strains which are brought on the struts or ties which form the sides of the angle.

The reader may not readily conceive the piece AC of fig. 8 as sustaining a compression; for the weight B appears to hang from AC as much as from AD. But his doubts will be removed by considering whether he could employ a rope in place of AC. He cannot; but AD may be exchanged for a rope. AC is therefore a strut, and not a tie.

In fig. 9, Plate LI., AD is again a strut, butting on the block D, and AC is a tie; and the batten AC may be replaced by a rope. While AD is compressed by the force AG, AC is stretched by the force AF.

If we give AC the position represented by the dotted lines, the compression of AD is now AG', and the force stretching AC' is now AF'; both much greater than they were before. This disposition is analogous to fig. 8, and to the dotted lines in fig. 3. Nor will the young artist have any doubts of AC' being on the stretch, if he consider whether AD can be replaced by a rope. It cannot, but AC' may; and it is therefore not compressed, but stretched.

In fig. 10 all the three pieces, AC, AD, and AB, are ties, on the stretch. This is the complete inversion of fig. 3; and the dotted position of AC induces the same changes in the forces AF, AG', as in fig. 3.

Thus have we gone over all the varieties which can happen in the bearings of three pieces on one point. All calculations about the strength of carpentry are reduced to this case; for when more ties or braces meet in a point (a thing that rarely happens), we reduce them to three, by substituting for any two the force which results from their combination, and then combining this with another; and so on.

The young artist must be particularly careful not to mistake the kind of strain that is exerted on any piece of the framing, and suppose a piece to be a brace which is really a tie. It is very easy to avoid all mistakes in this matter by the following rule, which has no exception. (See Note AA.)

Take notice of the direction in which the piece acts from which the strain proceeds. Draw a line in that direction *from* the point on which the strain is exerted, and let its length (measured on some scale of equal parts) express the magnitude of this action in pounds, hundreds, or tons. From its *remote* extremity draw lines parallel to the pieces on which the strain is exerted. The line parallel to one piece will necessarily cut the other, or its direction produced. If it cut the piece itself, that piece is compressed by the strain, and it is performing the office of a strut or brace; if it cut its direction produced, the piece is stretched, and it is a tie. In short, the strains on the pieces AC, AD, are to be estimated in the direction of the points F and G *from* the strained point A. Thus, in fig. 3, the upright piece BA, loaded with the weight B, presses the point A in the direction AE; so does the rope AB in the other figures, or the batten AB in fig. 5.

In general, if the straining piece is within the angle formed by the pieces which are strained, the strains which they sustain are of the opposite kind to that which it exerts. If it be pushing, they are drawing; but if it be within the angle formed by their directions produced, the strains which they sustain are of the same kind. All the three are either drawing or pressing. If the straining piece lie within the angle formed by one piece and the produced direction of the other, its own strain, whether compression or extension, is of the same kind with that of the most remote of the other two, and opposite to that of the nearest. Thus, in fig. 9, where AB is drawing, the remote piece AC is also drawing, while AD is pushing or resisting compression.

In all that has been said on this subject, we have not spoken of any joints. In the calculations with which we are occupied at present, the resistance of joints has no share; and we must not suppose that they exert any force which tends to prevent the angles from changing The joints are supposed perfectly flexible, or to be like compass joints, the pin of which only keeps the pieces together when one or more of the pieces draws or pulls. The carpenter must always suppose them all compass joints when he calculates the thrusts and draughts of the different pieces of his frames. The strains on joints, and their power to produce or balance them, are of a different kind, and require a very different examination.

Seeing that the angles which the pieces make with each other are of such importance to the magnitude and the proportion of the excited strains, it is proper to find out some way of readily and compendiously conceiving and expressing this analogy.

In general the strain on any piece is proportional to the straining force. This is evident.

Rule for distinguishing the cases compression and extension.

General expression of the magnitude of the strain.

---

[1] The reader is requested to add accents to the extreme letters D and F of fig. 3, which correspond to the position of the beam AGD indicated by the dotted lines. Accents are also wanted to the upper F and the lower C and G in fig. 9; also to the upper F and lower G in fig. 10; and in this *b* should be C. In fig. 11, the *i* towards the left should be *t*, and an accent is wanting over the upper *f*. In fig. 12, the dotted line CK should be continued upward and marked I. In fig. 16, the letters should stand thus, A C E *e* D *f* F B.

2 E

Secondly, the strain on any piece AC is proportional to the sine of the angle which the straining force makes with the other piece directly, and to the sine of the angle which the pieces make with each other inversely.

For it is plain that the three pressures AE, AF, and AG, which are exerted at the point A, are in the proportion of the lines AE, AF, and FE (because FE is equal to AG). But because the sides of a triangle are proportional to the sines of the opposite angles, the strains are proportional to the sines of the angles AFE, AEF, and FAE. But the sine of AFE is the same with the sine of the angle CAD, which the two pieces AC and AD make with each other; and the sine of AEF is the same with the sine of EAD, which the straining piece BA makes with the piece AC. Therefore we have this analogy, Sin. CAD : Sin.

$$EAD = AE : AF, \text{and } AF = AE \times \frac{\text{Sin. EAD}}{\text{Sin. CAD}}.$$ Now the sines of angles are most conveniently conceived as decimal fractions of the radius, which is considered as unity. Thus. *Sin.* 30° is the same thing with 0·5, or $\frac{1}{2}$; and so of others. Therefore, to have the strain on AC, arising from any load AE acting in the direction AE, multiply AE by the sine of EAD, and divide the product by the sine of CAD.

This rule shows how great the strains must be when the angle CAD becomes very open, approaching to 180 degrees. But when the angle CAD becomes very small, its sine (which is our divisor) is also very small; and we should expect a very great quotient in this case also. But we must observe, that in this case the sine of EAD is also very small; and this is our multiplier. In such a case, the quotient cannot exceed unity.

But it is unnecessary to consider the calculation by the tables of sines more particularly. The angles are seldom known any otherwise but by drawing the figure of the frame of carpentry. In this case, we can always obtain the measures of the strains from the same scale, with equal accuracy, by drawing the parallelogram AFCG.

Strains propagated to the points of support.

Hitherto we have considered the strains excited at A only as they affect the pieces on which they are exerted. But the pieces, in order to sustain, or be subject to, any strain, must be supported at their ends C and D; and we may consider them as mere intermediums, by which these strains are made to act on those points of support : Therefore AF and AG are also measures of the forces which press or pull at C and D. Thus we learn the supports which must be found for these points. These may be infinitely various. We shall attend only to such as somehow depend on the framing itself.

Such a structure as fig. 11 very frequently occurs, where a beam BA is strongly pressed to the end of another beam AD, which is prevented from yielding, both because it lies on another beam HD, and because its end D is hindered from sliding backwards. It is indifferent from what this pressure arises: we have represented it as owing to a weight hung on at B, while B is withheld from yielding by a rod or rope hooked to the wall. The beam AD may be supposed at full liberty to exert all its pressure on D, as if it were supported on rollers lodged in the beam HD; but the loaded beam BA presses both on the beam AD and on HD. We wish only to know what strain is borne by AD.

All bodies act on each other in the direction perpendicular to their touching surfaces; therefore the support given by HD is in a direction perpendicular to it. We may therefore supply its place at A by a beam AC, perpendicular to HD, and firmly supported at C. In this case, therefore, we may take AE, as before, to represent the pressure exerted by the loaded beam, and draw EG perpendicular to AD, and EF parallel to it, meeting the

perpendicular AC in F. Then AG is the strain compressing AD, and AF is the pressure on the beam HD.

Carpentry.

The form of the abutting joint of n great importance.

It may be thought, that since we assume as a principle that the mutual pressures of solid bodies are exerted perpendicular to their touching surfaces, this balance of pressures, in framings of timbers, depends on the directions of their butting joints; but it does not, as will readily appear by considering the present case. Let the joint or abutment of the two pieces BA, AD, be mitred in the usual manner, in the direction *fAf'*. Therefore, if A*c* be drawn perpendicular to A*f*, it will be the direction of the actual pressure exerted by the loaded beam BA on the beam AD. But the re-action of AD, in the opposite direction A*t*, will not balance the pressure of BA; because it is not in the direction precisely opposite. BA will therefore slide along the joint, and press on the beam HD. AE represents the load on the mitre joint A. Draw E*c* perpendicular to A*c*, and E*f* parallel to it. The pressure AE will be balanced by the re-actions *c*A and *f*A; or, the pressure AE produces the pressures A*c* and A*f*, of which A*f* must be resisted by the beam HD, and A*c* by the beam AD. The pressure A*f* not being perpendicular to HD, cannot be fully resisted by it; because (by our assumed principle) it re-acts only in a direction perpendicular to its surface. Therefore draw *fp, fi*, parallel to HD, and perpendicular to it. The pressure A*f* will be resisted by HD with the force *p*A; but there is required another force *i*A, to prevent the beam BA from slipping outwards. This must be furnished by the re-action of the beam DA. (See Note BB.) In like manner, the other force A*c* cannot be fully resisted by the beam AD, or rather by the prop D, acting by the intervention of the beam; for the action of that prop is exerted through the beam in the direction DA. The beam AD, therefore, is pressed to the beam HD by the force A*c*, as well as by A*f*. To find what this pressure on HD is, draw *cg* perpendicular to HD, and *co* parallel to it, cutting EG in *r*. The forces *g*A and *o*A will resist, and balance A*c*.

Thus we see that the two forces A*c* and A*f*, which are equivalent to AE, are equivalent also to A*p*, A*i*, A*o*, and A*g*. But because A*f* and *c*E are equal and parallel, and E*r* and *fi* are also parallel, as also *cr* and *fp*, it is evident, that *if* is equal to *r*E, or to *o*F, and *i*A is equal to *rc*, or to G*g*. Therefore the four forces A*g*, A*o*, A*p*, A*i*, are equal to AG and AF. Therefore AG is the compression of the beam AD, or the force pressing it on D, and AF is the force pressing it on the beam HD. The proportion of these pressures, therefore, is not affected by the form of the joint.

This remark is important; for many carpenters think the form and direction of the butting joint of great importance; and even the theorist, by not prosecuting the general principle through *all* its consequences, may be led into an error. The form of the joint is of no importance, in as far as it affects the strains in the direction of the beams; but it is often of great consequence, in respect to its own firmness, and the effect it may have in bruising the piece on which it acts, or being crippled by it.

Origin of the strain on a tie-beam.

The same compression of AB, and the same thrust on the point D by the intervention of AD, will obtain, in whatever way the original pressure on the end A is produced. Thus, supposing that a cord is made fast at A, and pulled in the direction AE, and with the same force, the beam AD will be equally compressed, and the prop D must re-act with the same force.

But it often happens that the obliquity of the pressure on AD, instead of compressing it, stretches it; and we desire to know what tension it sustains. Of this we have a familiar example in a common roof. Let the two rafters AC, AD (fig. 12), press on the tie-beam DC. We may

Carpentry. suppose the whole weight to press vertically on the ridge A, as if a weight B were hung on there. (See Note CC.) We may represent this weight by the portion A*b* of the vertical or plumb line, intercepted between the ridge and the beam. Then drawing *bf* and *bg* parallel to AD and AC, A*g* and A*f* will represent the pressures on AC and AD. Produce AC till CH be equal to A*f*. The point C is forced out in this direction, and with a force represented by this line. As this force is not perpendicularly across the beam, it evidently stretches it; and this extending force must be withstood by an equal force pulling it in the opposite direction. This must arise from a similar oblique thrust of the opposite rafter on the other end D. We concern ourselves only with this extension at present; but we see that the cohesion of the beam does nothing but supply the balance to the extending forces. It must still be supported externally, that it *may resist*, and by resisting obliquely, be stretched. The points C and D are supported on the walls, which they press in the directions CK and DO, parallel to A*b*. If we draw HK parallel to DC, and HI parallel to CK (that is to A*b*), meeting DC produced in I, it follows from the composition of forces, that the point C would be supported by the two forces KC and IC. In like manner, making DN = A*g*, and completing the parallelogram DMNO, the point D would be supported by the forces OD and MD. If we draw *go* and *fk* parallel to DC, it is plain that they are equal to NO and CI, while A*o* and A*k* are equal to DO and CK, and A*b* is equal to the sum of DO and CK (because it is equal to A*o* + A*k*). The weight of the roof is equal to its vertical pressure on the walls.

Thus we see, that while a pressure on A, in the direction A*b*, produces the strains A*f* and A*g*, on the pieces AC and AD, it also excites a strain CI or DM in the piece DC. And this completes the mechanism of a frame; for all derive their efficacy from the triangles of which they are composed, as will appear more clearly as we proceed.

External action of a frame.

But there is more to be learned from this. The consideration of the strains on the two pieces AD and AC, by the action of a force at A, only showed them as the means of propagating the same strains in their own direction to the points of support. But, by adding the strains exerted in DC, we see that the frame becomes an intermedium, by which exertions may be made on other bodies in certain directions and proportions, so that this frame may become part of a more complicated one, and, as it were, an element of its constitution. It is worth while to ascertain the proportion of the pressures CK and DO, which are thus exerted on the walls. The similarity of triangles gives the following analogies:

DO : DM = A*b* : *b*D
CI, or DM : CK = C*b* : A*b*
Therefore DO : CK = C*b* : *b*D.

Or, *the pressures on the points C and D, in the direction of the straining force A*b*, are reciprocally proportional to the portions of DC intercepted by A*b*.*

Also, since A*b* is = DO + CK, we have
A*b* : CK = C*b* + *b*D (or CD) : *b*D, and
A*b* : DO = CD : *b*C.

In general, any two of the three parallel forces A*b*, DO, CK, are to each other in the reciprocal proportion of the parts of CD, intercepted between their directions and the direction of the third.

And this explains a still more important office of the frame ADC. If one of the points, such as D, be supported, an external power acting at A, in the direction A*b*, and with an intensity which may be measured by A*b*, may be set in equilibrio with another acting at C, in the direction CL, opposite to CK or A*b*, and with an intensity

represented by CK; for since the pressure CH is partly withstood by the force IC, or the firmness of the beam DC supported at D, the force KC will complete the balance. When we do not attend to the support at D, we conceive the force A*b* to be balanced by KC, or KC to be balanced by A*b*. And, in like manner, we may neglect the support or force acting at A, and consider the force DO as balanced by CK.

Thus our frame becomes a lever, and we are able to trace the interior mechanical procedure which gives it its efficacy: it is by the intervention of the forces of cohesion, which connect the points to which the external forces are applied with the supported point or fulcrum and with each other.

These strains or pressures A*b*, DO, and CK, not being in the directions of the beams, may be called *transverse*. We see that by their means a frame of carpentry may be considered as a solid body: but the example which brought this to our view is too limited for explaining the efficacy which may be given to such constructions. We shall therefore give a general proposition, which will more distinctly explain the procedure of nature, and enable us to trace the strains as they are propagated through all the parts of the most complicated framing, finally producing the exertion of its most distant points.

It becomes a lever.

We presume that the reader is now pretty well habituated to the conception of the strains as they are propagated along the lines joining the points of a frame, and we shall therefore employ a very simple figure.

General proposition.

Let the strong lines ACBD (fig. 13) represent a frame of carpentry. Suppose that it is pulled at the point A by a force acting in the direction AE, but that it rests on a fixed point C, and that the other extreme point B is held back by a power which resists in the direction BF: It is required to determine the proportion of the strains excited in its different parts, the proportion of the external pressures at A and B, and the pressure which is produced on the obstacle or fulcrum C.

It is evident that each of the external forces at A and B tend one way, or to one side of the frame, and that each would cause it to turn round C if the other did not prevent it; and that if, notwithstanding their action, it is turned neither way, the forces in actual exertion are in equilibrio by the intervention of the frame. It is no less evident that these forces concur in pressing the frame on the prop C. Therefore, if the piece CD were away, and if the joints C and D be perfectly flexible, the pieces CA, CB, would be turned round the prop C, and the pieces AD, DB, would also turn with them, and the whole frame change its form. This shows, by the way, and we desire it to be carefully kept in mind, that the firmness or stiffness of framing depends entirely on the triangles bounded by beams which are contained in it. An open quadrilateral may always change its shape, the sides revolving round the angles. A quadrilateral may have an infinity of forms, without any change of its sides, by merely pushing two opposite angles towards each other, or drawing them asunder. But when the three sides of a triangle are determined, its shape is also invariably determined; and if two angles be held fast, the third cannot be moved. It is thus that, by inserting the bar CD, the figure becomes unchangeable; and any attempt to change it by applying a force to an angle A, immediately excites forces of attraction or repulsion between the particles of the stuff which form its sides. Thus it happens, in the present instance, that a change of shape is prevented by the bar CD. The power at A presses its end against the prop; and in doing this it puts the bar AD on the stretch, and also the bar DB. Their places might therefore be supplied by cords or metal wires. Hence it is evident that

DC is compressed, as is also AC; and, for the same reason, CB is also in a state of compression; for either A or B may be considered as the point that is impelled or withheld. Therefore DA and DB are stretched, and are resisting with attractive forces. DC and CB are compressed, and are resisting with repulsive forces; and thus the support of the prop, combined with the firmness of DC, puts the frame ADBC into the condition of the two frames in fig. 8 and fig. 9. Therefore the external force at A is really in equilibrio with an attracting force acting in the direction AD, and a repulsive force acting in the direction AK. And since all the connecting forces are mutual and equal, the point D is pulled or drawn in the direction DA. The condition of the point B is similar to that of A, and D is also drawn in the direction DB. Thus the point D, being urged by the forces in the directions DA and DB, presses the beam DC on the prop, and the prop resists in the opposite direction. Therefore the line DC is the diagonal of the parallelogram, whose sides have the proportion of the forces which connect D with A and B. This is the principle on which the rest of our investigation proceeds. We may take DC as the representation and measure of their joint effect. Therefore draw CH, CG, parallel to DA, DB. Draw HL, GO, parallel to CA, CB, cutting AE, BF, in L and O, and cutting DA, DB, in I and M. Complete the parallelograms ILKA, MONB. Then DG and AI are the equal and opposite forces which connect A and D; for GD = CH = AI. In like manner DH and BM are the forces which connect D and B.

The external force at A is in immediate equilibrio with the combined forces, connecting A with D and with C. AI is one of them, therefore AK is the other; and AL is the compound force with which the external force at A is in immediate equilibrium. This external force is therefore equal and opposite to AL. In like manner, the external force at B is equal and opposite to BO; and AL is to BO as the external force at A to the external force at B. The prop C resists with forces equal to those which are propagated to it from the points D, A, and C. Therefore it resists with forces CH, CG, equal and opposite to DG, DH; and it resists the compressions KA, NB, with equal and opposite forces Ck, Cn. Draw kl, no, parallel to AD, BD, and draw ClQ, CoP: It is plain that kCHl is a parallelogram equal to KAIL, and that Cl is equal to AL. In like manner Co is equal to BO. Now the forces Ck, CH, exerted by the prop, compose the force Cl; and Cn, CG, compose the force Co. These two forces Cl, Co, are equal and parallel to AL and BO; and therefore they are equal and opposite to the external forces acting at A and B. But they are, primitively, equal and opposite to the pressures, or at least the compounds of the pressures, exerted on the prop, by the forces propagated to C from A, D, and B. Therefore the pressures exerted on the prop are the same as if the external forces were applied there in the same directions as they are applied to A and B. Now if we make Cv, Cz, equal to Cl and Co, and complete the parallelogram Cvyz, it is plain that the force yC is in equilibrio with lC and oC. Therefore the pressures at A, C, and B are such as would balance if applied to one point.

Lastly, in order to determine their proportions, draw CS and CR perpendicular to DA and DB. Also draw Ad, Bf, perpendicular to CQ and CP; and draw Cg, Ci, perpendicular to AE, BF.

The triangles CPR and BPf are similar, having a common angle P, and a right angle at R and f.

In like manner, the triangles CQS and AQd are similar. Also the triangles CHR, CGS, are similar, by reason of the equal angles at H and G, and the right angles at R and S. Hence we obtain the following analogies:

$$Co : CP = on : PB, = CG : PB$$
$$CP : CR = \qquad\qquad PB : fB$$
$$CR : CS = \qquad\qquad CH : CG$$
$$CS : CQ = \qquad\qquad Ad : AQ$$
$$CQ : Cl = AQ : hl, = AQ : CH.$$

Therefore, by equality,

$$Co : Cl = \qquad\qquad Ad : fB$$
$$\text{or } BO : AL = \qquad\qquad Cg : Ci.$$

That is, the external forces are reciprocally proportional to the perpendiculars drawn from the prop on the lines of their direction.[1]

This proposition, sufficiently general for our purpose, is fertile in consequences, and furnishes many useful instructions to the artist. The strains LA, OB, CY, that are excited, occur in many, we may say in all, framings of carpentry, whether for edifices or engines, and are the sources of their efficacy. It is also evident that the doctrine of the transverse strength of timber is contained in this proposition; for every piece of timber may be considered as an assemblage of parts, connected by forces which act in the direction of the lines which joined the strained points on the matter which lies between those points, and also act on the rest of the matter, exciting those lateral forces which produce the inflexibility of the whole. See STRENGTH OF MATERIALS.

Thus it appears that this proposition contains the principles which direct the artist to frame the most powerful levers; to secure uprights by shores or braces, or by tiers and ropes; to secure scaffoldings for the erection of spires; and many other more delicate problems of his art. He also learns from this proposition how to ascertain the strains that are produced, without his intention, by pieces which he intended for other offices, and which, by their transverse action, puts his work in hazard. In short, this proposition is the key to the science of this art.

We would now counsel the artist, after he has made the tracing of the strains and thrusts through the various parts of a frame familiar to his mind, and even amused himself with some complicated fancy framings, to read over with care the articles STRENGTH OF MATERIALS and ROOF. He will now conceive its doctrine much more clearly than when he was considering them as abstract theories. The mutual action of the woody fibres will now

---

[1] " The learned reader will perceive that this analogy is precisely the same with that of forces which are in equilibrio by the intervention of a lever. In fact, this whole frame of carpentry is nothing else than a *built* or *framed lever* in equilibrio. It is acting in the same manner as a solid, which occupies the whole figure compressed in the frame, or as a body of any size and shape whatever that will admit the three points of application A, C, and B. It is always in equilibrio in the case first stated; because the pressure *produced* at B by a force applied to A is always such as balances it. The reader may also perceive, in this proposition, the analysis or tracing of those internal mechanical forces which are indispensably requisite for the functions of a lever. The mechanicians have been extremely puzzled to find a legitimate demonstration of the equilibrium of a lever ever since the days of Archimedes. Mr Vince has the honour of first demonstrating, most ingeniously, the principle assumed by Archimedes, but without sufficient ground for *his* demonstration; but Mr Vince's demonstration is only a putting the mind into that perplexed state which makes it acknowledge the proposition, but without a clear perception of its truth. The difficulty has proceeded from the abstract notion of a lever, conceiving it as a mathematical line—inflexible, without reflecting how it is inflexible; for the very source of this indispensable quality furnishes the mechanical connection between the remote pressures and the fulcrum; and this supplies the demonstration (without the least difficulty) of the desperate case of a straight lever urged by parallel forces." See the article ROTATION.

Carpentry. be easily comprehended, and his confidence in the results will be greatly increased.

Decision of a disputed and very important question. There is a proposition (see article Roof) which has been called in question by several very intelligent persons; and they say that Belidor has demonstrated, in his *Science des Ingenieurs*, that a beam firmly fixed at both ends is not twice as strong as when simply lying on the props; and that its strength is increased only in the proportion of two to three; and they support this determination by a list of experiments recited by Belidor, which agree *precisely* with it. Belidor also says that Pitot had the same result in his experiments. These are respectable authorities, but Belidor's reasoning is any thing but demonstration, and his experiments are described in such an imperfect manner that we cannot build much on them. It is not said in what manner the battens were secured at the ends, any further than that it was by *chevalets*. If by this word is meant a trestle, we cannot conceive how they were employed; but we see it sometimes used for a wedge or key If the battens were wedged in the holes, their resistance to fracture may be made what we please; they may be made loose, and therefore resist little more than when simply laid on props. They may be (and probably were) wedged very fast, and bruised or crippled.

Our proposition mentioned distinctly the security given to the ends of the beams. They were mortised into remote posts. Our *precise* meaning was, that they were simply kept from rising by these mortises, but at full liberty to bend up at E and I, and between G and K. Our assertion was not made from theory alone (although we think the reasoning incontrovertible), but was agreeable to numerous experiments made in those precise circumstances. Had we mortised the beams firmly into two very stout posts which could not be drawn nearer to each other by bending, the beam would have borne a *much* greater weight, as we have verified by experiments. We hope that the following mode of conceiving this case will remove all doubts.

Let LM be a long beam (fig. 14) divided into six equal parts, in the points D, B, A, C, E. Let it be firmly supported at L, B, C, M. Let it be cut through at A, and have compass joints at B and C. Let FB, GC, be two equal uprights, resting on B and C, but without any connection. Let AH be a similar and equal piece, to be occasionally applied at the seam A. Now let a thread or wire AGE be extended over the piece GC, and made fast at A, G, and E. Let the same thing be done on the other side of A. If a weight be now laid on at A, the wires AFD, AGE, will be strained, and may be broken. In the instant of fracture we may suppose their strains to be represented by A*f* and A*g*. Complete the parallelogram, and A*a* is the magnitude of the weight. It is plain that nothing is concerned here but the cohesion of the wires; for the beam is sawed through at A, and its parts are perfectly movable round B and C.

Instead of this process, apply the piece AH below A, and keep it there by straining the same wire BHC over it. Now lay on a weight. It must press down the ends of BA and CA, and cause the piece AH to strain the wire BHC. In the instant of fracture of *the same* wire, its resistances H*b* and H*c* must be equal to A*f* and A*g*, and the weight *h*H which breaks them must be equal to A*a*.

Lastly, employ all the three pieces FB, AH, GC, with the same wire attached as before. There can be no doubt but that the weight which breaks all the four wires must be = *a*A + *h*H, or twice A*a*.

The reader cannot but see that the wires perform the very same office with the fibres of an entire beam LM held fast in the four holes D, B, C, and E, of some upright posts.

In the experiments for verifying this, by breaking slender bars of fine deal, we get complete demonstration, by measuring the curvatures produced in the parts of the beam thus held down, and comparing them with the curvature of a beam simply laid on the props B and C; and there are many curious inferences to be made from these observations, but we have not room for them in this place.

We may observe by the way, that we learn from this The best manner of framing purlins. case that purlins are able to carry twice the load when notched into the rafters that they carry when mortised into them, which is the most usual manner of framing them. So would the bending joists of floors; but this would double the thickness of the flooring. But this method should be followed in every possible case, such as brest summers, lintels over several pillars, &c. These should never be cut off and mortised into the sides of every upright; numberless cases will occur which show the importance of the maxim.

We must here remark, that the proportion of the spaces BC and CM, or BC and LB, has a very sensible effect on the strength of the beam BC; but we have not yet satisfied our minds as to the *rationale* of this effect. It is undoubtedly connected with the serpentine form of the curve of the beam before fracture. This should be attended to in the construction of the springs of carriages. These are frequently supported at the middle point (and it is an excellent practice); and there is a certain proportion which will give the easiest motion to the body of the carriage. We also think that it is connected with that deviation from the best theory observable in Buffon's experiments on various lengths of the same scantling. The force of the beams diminished much more than in the inverse proportion of their lengths.

We have seen that it depends entirely on the position In what case ties are better than struts. of the pieces in respect of their points of ultimate support, and of the direction of the external force which produces the strains, whether any particular piece is in a state of extension or of compression. The knowledge of this circumstance may greatly influence us in the choice of the construction. In many cases we may substitute slender iron rods for massive beams, when the piece is to act the part of a tie. But we must not invert this disposition; for when a piece of timber acts as a strut, and is in a state of compression, it is next to certain that it is not equally compressible in its opposite sides through the whole length of the piece, and that the compressing force on the abutting joint is not acting in the most equable manner all over the joint. A very trifling inequality in either of these circumstances (especially in the first) will compress the beam more on one side than on the other. This cannot be without the beam's bending, and becoming concave on that side on which it is most compressed. When this happens, the frame is in danger of being crushed, and soon going to ruin. It is, therefore, indispensably necessary to make use of beams in all cases where struts are required of considerable length, rather than of metal rods of slender dimensions, unless in situations where we can effectually prevent their bending, as in trussing a girder internally, where a cast-iron strut may be firmly cased in it, so as not to bend in the smallest degree. In cases where the pressures are enormous, as in the very oblique struts of a centre or arch frame, we must be particularly cautious to do nothing which can facilitate the compression of either side. No mortises should be cut near to one side; no lateral pressures, even the slightest, should be allowed to touch it. We have seen a pillar of fir twelve inches long, and one inch in section, when loaded with three tons, snap in an instant when pressed on one side by sixteen pounds, while another bore four and a half tons without hurt, because it was inclosed (loosely) in a stout pipe of iron. (See Note DD.)

Carpentry. In such cases of enormous compression it is of great importance that the compressing force bear equally on the whole abutting surface. The German carpenters are accustomed to put a plate of lead over the joint. This prevents, in some measure, the penetration of the end fibres. M. Perronet, the celebrated French architect, formed his abutments into arches of circles, the centre of which was the remote end of the strut. By this contrivance the unavoidable change of form of the triangle made no partial bearing of either angle of the abutment. This always has a tendency to splinter off the heel of the beam where it presses strongest. It is a very judicious practice. (See Note EE.)

When circumstances allow it, we must rather employ ties than struts for securing a beam against lateral strains. When an upright pillar, such as a flag-staff, a mast, or the uprights of a very tall scaffolding, are to be shoared up, the dependence is more certain on those braces that are stretched by the strain than on those which are compressed. The scaffolding of the iron bridge near Sunderland had some ties very judiciously disposed, and others with less judgment.

We should proceed to consider the transverse strains as they affect the various parts of a frame of carpentry; but we have very little to say here in addition to what will be found in the articles STRENGTH OF MATERIALS and ROOF. What we shall add in this article will find a place in our occasional remarks on different works. It may, however, be of use to recal to the reader's memory the following propositions.

General theorems concerning the relative strength of beams.

1. When a beam AB (fig. 15) is firmly fixed at the end A, and a straining force acts perpendicularly to its length at any point B, the strain occasioned at any section C between B and A is proportional to CB, and may therefore be represented by the product $w \times$ CB; that is, by the product of the number of tons, pounds, &c. which measure the straining force, and the number of feet, inches, &c. contained in CB. As the loads on a beam are easily conceived, we shall substitute this for any other straining force.

2. If the strain or load is uniformly distributed along any part of the beam lying beyond C (that is, farther from A), the strain at C is the same as if the load were all collected at the middle point of that part; for that point is the centre of gravity of the load.

3. The strain on any section D of a beam AB (fig. 16) resting freely on two props A and B, is $w \times \dfrac{\text{AD} \times \text{DB}}{\text{AB}}$.

(See ROOF, No. 19, and STRENGTH OF MATERIALS, No. 92, &c.) Therefore,

4. The strain on the middle point, by a force applied there, is one fourth of the strain which the same force would produce if applied to one end of a beam of the same length having the other end fixed.

5. The strain on any section C of a beam, resting on two props A and B, occasioned by a force applied perpendicularly to another point D, is proportional to the rectangle of the exterior segments, or is equal to $w \times \dfrac{\text{AC} \times \text{DB}}{\text{AB}}$. Therefore,

The strain at C occasioned by the pressure on D is the same with the strain at D occasioned by the same pressure on C.

6. The strain on any section D, occasioned by a load uniformly diffused over any part EF, is the same as if the two parts ED, DF, of the load were collected at their middle points e and f. Therefore,

The strain on any part D, occasioned by a load uniformly distributed over the whole beam, is one half of the strain that is produced when the same load is laid on at D; and

The strain on the middle point C, occasioned by a load uniformly distributed over the whole beam, is the same which half that load would produce if laid on at C.

7. A beam supported at both ends on two props B and C (fig. 14), will carry twice as much when the ends beyond the props are kept from rising, as it will carry when it rests loosely on the props.

8. Lastly, the transverse strain on any section, occasioned by a force applied obliquely, is diminished in the proportion of the sine of the angle which the direction of the force makes with the beam. Thus, if it be inclined to it in an angle of thirty degrees, the strain is one half of the strain occasioned by the same force acting perpendicularly.

On the other hand, the RELATIVE STRENGTH of a beam, or its power in any particular section to resist any transverse strain, is proportional to the absolute cohesion to the section directly, to the distance of its centre of effort from the axis of fracture directly, and to the distance from the strained point inversely.

Thus, in a rectangular section of the beam, of which b is the breadth, d the depth (that is, the dimension in the direction of the straining force), measured in inches, and f the number of pounds which one square inch will just support without being torn asunder, we must have $f \times b \times d^2$, proportional to $w \times$ CB (fig. 15). Or, $f \times b \times d^2$, multiplied by some number m, depending on the nature of the timber, must be equal to $w \times$ CB. Or, in the case of the section C of fig. 16, that is strained by the force w applied at D, we must have $m \times fbd^2 = w \times \dfrac{\text{AC} \times \text{DB}}{\text{AB}}$. Thus, if the beam is of sound oak, m is very nearly $= \frac{1}{9}$ (see STRENGTH OF MATERIALS, No. 116.) Therefore we have $\dfrac{fbd^2}{9} = w \times \dfrac{\text{AC} \times \text{CB}}{\text{AB}}$. (See Note FF.)

Hence we can tell the precise force w which any section C can just resist when that force is applied in any way whatever; for the above-mentioned formula gives $w = \dfrac{fbd^2}{9\text{CB}}$ for the case represented by fig. 15. But the case represented in fig. 16, having the straining force applied at D, gives the strain at C $(= w) = f \times \dfrac{bd^2 \times \text{AB}}{9\text{AC} \times \text{CB}}$.

Example. Let an oak beam, four inches square, rest freely on the props A and B, seven feet apart, or eighty-four inches. What weight will it just support at its middle point C, on the supposition that a square inch rod will just carry 16,000 pounds, pulling it asunder?

The formula becomes $w = \dfrac{16000 \times 4 \times 16 \times 84}{9 \times 42 \times 42}$,

or $w = \dfrac{86016000}{15876}$, $= 5418$ pounds. This is very near what was employed in Buffon's experiment, which was 5312.

Had the straining force acted on a point D, half way between C and B, the force sufficient to break the beam at C would be $= \dfrac{16000 \times 4 \times 16 \times 84}{9 \times 42 \times 21} = 10836$ lbs.

Had the beam been sound red fir, we must have taken $f = 10,000$ nearly, and m nearly 8; for although fir be less cohesive than oak in the proportion of five to eight nearly, it is less compressible, and its axis of fracture is therefore nearer to the concave side.

Having considered at sufficient length the strains of Of joint different kinds which arise from the form of the parts of a frame of carpentry, and the direction of the external forces which act on it, whether considered as impelling or as supporting its different parts, we must now proceed to con sider the means by which this form is to be secured, and the connections by which those strains are excited and communicated.

The joinings practised in carpentry are almost infinitely various, and each has advantages which make it preferable in some circumstances. Many varieties are employed merely to please the eye. We do not concern ourselves with these: nor shall we consider those which are only employed in connecting small works, and can never appear on a great scale; yet even in some of these, the skill of the carpenter may be discovered by his choice; for in all cases, it is wise to make every, even the smallest, part of his work as strong as the materials will admit. He will be particularly attentive to the changes which will necessarily happen by the shrinking of timber as it dries, and will consider what dimensions of his framings will be affected by this, and what will not; and will then dispose the pieces which are less essential to the strength of the whole, in such a manner that their tendency to shrink shall be in the same direction with the shrinking of the whole framing. If he do otherwise, the seams will widen, and parts will be split asunder. He will dispose his boardings in such a manner as to contribute to the stiffness of the whole, avoiding at the same time the giving them positions which will produce lateral strains on truss beams which bear great pressures; recollecting, that although a single board has little force, yet many united have a great deal, and may frequently perform the office of very powerful struts.

Our limits confine us to the joinings which are most essential for connecting the parts of a single piece of a frame when it cannot be formed of one beam, either for want of the necessary thickness or length; and the joints for connecting the different sides of a trussed frame.

Of building up beams. Much ingenuity and contrivance has been bestowed on the manner of building up a great beam of many thicknesses, and many singular methods are practised as great nostrums by different artists; but when we consider the manner in which the cohesion of the fibres performs its office, we will clearly see that the simplest are equally effected with the most refined, and that they are less apt to lead us into false notions of the strength of the assemblage.

Building up a girder or lever.
Joggling preferable to scarfing.
Thus, were it required to build up a beam for a great lever or a girder, so that it may act nearly as a beam of the same size of one log, it may either be done by plain joggling, as in Plate LII. fig. 17, A, or by scarfing, as in fig. 17, B or C. If it is to act as a lever, having the gudgeon on the lower side at C, we believe that most artists will prefer the form B and C; at least this has been the case with nine tenths of those to whom we have proposed the question. The best informed only hesitated; but the ordinary artists were all confident in its superiority, and we found their views of the matter very coincident. They considered the upper piece as grasping the lower in its hooks; and several imagined, that by driving the one very tight on the other, the beam would be stronger than an entire log; but if we attend carefully to the internal procedure in the loaded lever, we shall find the upper one clearly the strongest. If they are formed of equal logs, the upper one is thicker than the other by the depth of the joggling or scarfing, which we suppose to be the same in both; consequently, if the cohesion of the fibres in the intervals is able to bring the uppermost filaments into full action, the form A is stronger than B, in the proportion of the greater distance of the upper filaments from the axis of the fracture. This may be greater than the difference of the thickness if the wood is very compressible. If the gudgeon be in the middle, the effect, both of the joggles and the scarfings, is considerably diminished; and if it is on the upper side the scarfings act in a very

different way. In this situation, if the loads on the arms Carpentry. are also applied to the upper side, the joggled beam is still more superior to the scarfed one. This will be best understood by resolving it in imagination into a trussed frame. But when a gudgeon is thus put on that side of the lever which grows convex by the strain, it is usual to connect it with the rest by a powerful strap, which embraces the beam, and causes the opposite point to become the resisting point. This greatly changes the internal actions of the filaments, and in some measure brings it into the same state as the first, with the gudgeon below. Were it possible to have the gudgeon on the upper side, and to bring the whole into action without a strap, it would be the strongest of all; because in general the resistance to compression is greater than to extension. In every situation the joggled beam has the advantage, and it is the easiest executed. (See Note GG.)

We may frequently gain a considerable accession of strength by this building up of a beam, especially if the part which is stretched by the strain be of oak, and the other part be fir. Fir being so much superior to oak as a pillar (if Muschenbroeck's experiments may be confided in), and oak so much preferable as a tie, this construction seems to unite both advantages. But we shall see much better methods of making powerful levers, girders, &c. by trussing.

Observe that the efficacy of both methods depends entirely on the difficulty of causing the piece between the cross joints to slide along the timber to which it adheres. Therefore, if this be moderate, it is wrong to make the notches deep; for as soon as they are so deep that their ends have a force sufficient to push the slice along the line of junction, nothing is gained by making them deeper; and this requires a greater expenditure of timber.

Scarfings are frequently made oblique, as in fig. 18; but we imagine that this is a bad practice. It begins to yield at a point where the wood is crippled and splintered off, or at least bruised out a little. As the pressure increases, this part, by squeezing broader, causes the solid parts to rise a little upwards, and gives them some tendency, not only to push their antagonists along the base, but even to tear them up a little. For similar reasons, we disapprove of the favourite practice of many artists to make the angles of their scarfings acute, as in fig. 19. This often causes the two pieces to tear each other up The abutments should always be perpendicular to the directions of the pressures. Lest it should be forgotten in its proper place, we may extend this injunction also to the abutments of different pieces of a frame, and recommend it to the artist even to attend to the shrinking of the timbers by drying. When two timbers abut obliquely, the joint should be most full at the obtuse angle of the end; because, by drying, that angle grows more obtuse, and the beam would then be in danger of splintering off at the acute angle.

It is evident that the nicest work is indispensably ne-We must not wedge too hard.cessary in building up a beam. The parts must abut on each other completely, and the smallest play or void takes away the whole efficacy. It is usual to give the butting joints a small taper to one side of the beam, so that they may require moderate blows of a maul to force them in; and the joints may be perfectly close when the external surfaces are even on each side of the beam. But we must not exceed in the least degree, for a very taper wedge has great force; and if we have driven the pieces together by very heavy blows, we leave the whole in a state of violent strain, and the abutments are perhaps ready to splinter off by a small addition of pressure. This is like too severe a proof for artillery; which, though not sufficient to burst the pieces, has weakened them to such a

Carpentry. degree, that the strain of ordinary service is sufficient to complete the fracture. The *workman* is tempted to exceed in this, because it smooths off and conceals all uneven seams; but he must be watched. It is not unusual to leave some abutments open enough to admit a thin wedge reaching through the beam. Nor is this a bad practice, if the wedge is of material which is not compressed by the driving or the strain of service. Iron would be preferable for this purpose, and for the joggles, were it not that, by its too great hardness, it cripples the fibres of timber to some distance. In consequence of this it often happens, that in beams which are subjected to desultory and sudden strains (as in the levers of reciprocating engines), the joggles or wedges widen the holes, and work themselves loose; therefore skilful engineers never admit them, and indeed as few bolts as possible, for the same reason; but when resisting a steady or dead pull, they are not so improper, and are frequently used.

Beams are built up, not only to increase their dimensions in the direction of the strain (which we have hitherto called their depth), but also to increase their breadth, or the dimensions perpendicular to the strain. We sometimes double the breadth of a girder which is thought too weak for its load, and where we must not increase the thickness of the flooring.

Building of masts.
The mast of a great ship of war must be made bigger athwartship, as well as fore and aft. This is one of the nicest problems of the art; and professional men are by no means agreed in their opinions about it. We do not presume to decide, and shall content ourselves with exhibiting the different methods.

The most obvious and natural method is that shown in fig. 20. It is plain that (independent of the connection of cross bolts, which are used in them all when the beams are square) the piece C cannot bend in the direction of the plane of the figure without bending the piece D along

Method used in the French navy.
with it. This method is much used in the French navy; but it is undoubtedly imperfect. Hardly any two great trees are of equal quality, and swell or shrink alike. If C shrinks more than D, the feather of C becomes loose in the groove wrought in D to receive it; and when the beam bends, the parts can slide on each other like the plates of a coach-spring; and if the bending is in the direction *cf*, there is nothing to hinder this sliding but the bolts, which soon work themselves loose in the bolt-holes.

Another method.
Fig. 21 exhibits another method. The two halves of the beam are tabled into each other in the same manner as in fig. 17. It is plain that this will not be affected by the unequal swelling or shrinking, because this is insensible in the direction of the fibres; but when bent in the direction *ab*, the beam is weaker than fig. 20 bent in the direction *cf*. Each half of fig. 20 has, in every part of its length, a thickness greater than half the thickness of the beam. It is the contrary in the alternate portions of the halves of fig. 21. When one of them is bent in the direction AB, it is plain that it drags the other with it by means of the cross butments of its tables, and there can be no longitudinal sliding. But unless the work is accurately executed, and each hollow completely filled up by the table of the other piece, there will be a lateral slide along the cross joints sufficient to compensate for the curvature; and this will hinder the one from compressing or stretching the other in conformity to this curvature.

Its imperfection.
The imperfection of this method is so obvious that it has seldom been practised; but it has been combined with the other, as is represented in fig. 22, where the beams are divided along the middle, and the tables in each half are alternate, and alternate also with the tables of the other half. Thus 1, 3, 4, are prominent, and 5, 2, 6, are depressed. This construction evidently puts a stop to

both slides, and obliges every part of both pieces to move Carpentry together. *ab* and *cd* show sections of the built-up beam corresponding to AB and CD.

No more is intended in this practice by any intelligent artist, than the causing the two pieces to act together in all their parts, although the strains may be unequally distributed on them. Thus, in a built-up girder, the binding joists are frequently mortised into very different parts of the two sides. But many seem to aim at making the beam stronger than if it were of one piece; and this inconsiderate project has given rise to many whimsical modes of tabling and scarfing, which we need not regard.

British method.
The practice in the British dock-yards is somewhat different from any of these methods. The pieces are tabled as in fig. 22, but the tables are not thin parallelopipeds, but thin prisms. The two outward joints or visible seams are straight lines, and the table No. 1 rises gradually to its greatest thickness in the axis. In like manner, the hollow, 5, for receiving the opposite table, sinks gradually from the edge to its greatest depth in the axis. Fig. 23, No. 1, represents a section of a round piece of timber built up in this way, where the full line EFGH is the section corresponding to AB of fig. 22, and the dotted line EGFH is the section corresponding to CD.

This construction, by making the external seam straight, leaves no lodgment for water, and looks much fairer to the eye; but it appears to us that it does not give so firm a hold when the mast is bent in the direction EH. The exterior parts are most stretched and most compressed by this bending; but there is hardly any abutment in the exterior parts of these tables. In the very axis, where the abutment is the firmest, there is little or no difference of extension and compression.

But this construction has an advantage, which, we imagine, much more than compensates for these imperfections, at least in the particular case of a round mast; it will draw together by hooping incomparably better than any of the others. If the cavity be made somewhat too shallow for the prominence of the tables, and if this be done uniformly along the whole length, it will make a somewhat open seam; and this opening can be regulated with the utmost exactness from end to end by the plane. The heart of those vast trunks is very sensibly softer than the exterior circles; therefore, when the whole is hooped, and the hoops hard driven, and at considerable intervals between each spell, we are confident that all may be compressed till the seam disappears; and then the whole makes one piece, *much* stronger than if it were an original log of that size, because the middle has become, by compression, as solid as the crust, which was naturally firmer, and resisted farther compression. We verified this beyond a doubt by hooping a built stick of a timber which has this inequality of firmness in a remarkable degree, and it was nearly twice as strong as another of the same size.

Our mast-makers are not without their fancies and whims; and the manner in which our masts and yards are generally built up is not near so simple as fig. 23; but it consists of the same essential parts, acting in the very same manner, and derives all its efficacy from the principles which are here employed.

Attended with peculiar advantages.
This construction is particularly suited to the situation and office of a ship's mast. It has no bolts; or, at least, none of any magnitude, or that make very important parts of its construction. The most violent strains perhaps that it is exposed to, is that of twisting, when the lower yards are close braced up by the force of many men acting by a long lever. This form resists a twist with peculiar energy; it is therefore an excellent method for building up a great shaft for a mill. The way in which they are usually built up is by reducing a central log to a poly-

Carpentry. gonal prism, and then filling it up to the intended size by *planting* pieces of timber along its sides, either spiking them down, or cocking them into it by a feather, or joggling them by slips of hard wood sunk into the central log and into the slips. *N.B.* Joggles of elm are sometimes used in the middle of the large tables of masts; and when sunk into the firm wood near the surface, they must contribute much to the strength. But it is very necessary to employ wood not much harder than the pine, otherwise it will soon enlarge its bed, and become loose, for the timber of these large trunks is very soft.

The most general reason for piecing a beam is to increase its length. This is frequently necessary, in order to procure tie-beams for very wide roofs. Two pieces must be scarfed together. Numberless are the modes of doing this, and almost every master carpenter has his favourite nostrum. Some of them are very ingenious; but here, as in other cases, the most simple are commonly Various the strongest. We do not imagine that any, the most in-methods of genious, is equally strong with a tie consisting of two scarfing. pieces of the same scantling laid over each other for a certain length, and firmly bolted together. We acknowledge that this will appear an artless and clumsy tie-beam, but we only say that it will be stronger than any that is more artificially made up of the same thickness of timber. This, we imagine, will appear sufficiently certain.

The simplest and most obvious scarfing, after the one now mentioned, is that represented in fig. 24, No. 1 and 2. If considered merely as two pieces of wood joined, it is plain that, as a tie, it has but half the strength of an entire piece, supposing that the bolts (which are the only connections) are fast in their holes. No. 2 requires a bolt in the middle of the scarf to give it that strength, and in every other part is weaker on one side or the other. (See Note HH.)

But the bolts are very apt to bend by the violent strain, and require to be strengthened by uniting their ends by iron plates; in which case it is no longer a wooden tie. The form of No. 1 is better adapted to the office of a pillar than No. 2, especially if its ends be formed in the manner shown in the elevation No. 3. By the sally given to the ends, the scarf resists an effort to bend it in that direction. Besides, the form of No. 2 is unsuitable for a post; because the pieces, by sliding on each other by the pressure, are apt to splinter off the tongue which confines their extremity.

Fig. 25 and **26** exhibit the most approved form of a scarf, whether for a tie or for a post. The key represented in the middle is not essentially necessary; the two pieces might simply meet square there. This form, without a key, needs no bolts (although they strengthen it greatly); but, if worked very true and close, and with square abutments, will hold together, and will resist bending in any direction. But the key is an ingenious and a very great improvement, and will force the parts together with perfect tightness. The same precaution must be observed that we mentioned on another occasion, not to produce a constant internal strain on the parts by overdriving the key. The form of fig. 25 is by far the best; because the triangle of 26 is much easier splintered off by the strain, or by the key, than the square wood of 25. It is far preferable for a post, for the reason given when speaking of fig. 24, No. 1 and No. 2. Both may be formed with a sally at the ends equal to the breadth of the key. In this shape fig. 25 is vastly well suited for joining the parts of the long corner posts of spires and other wooden towers. Fig. 25, No. 2, differs from No. 1 only by having three keys. The principal and the longitudinal strength are the same. The long scarf of No. 2, tightened by the three keys, enables it to resist a bending much better.

None of these scarfed tie-beams can have more than Carpentry. one third of the strength of an entire piece, unless with the assistance of iron plates; for if the key be made thinner than one third, it has less than one third of the fibres to pull by.

We are confident, therefore, that when the heads of the bolts are connected by plates, the simple form of fig 24, No. 1, is stronger than those more ingenious scarfings. It may be strengthened against lateral bending by a little tongue, or by a sally, but cannot have both.

The strongest of all methods of piecing a tie-beam would be to set the parts end to end, and grasp them between other pieces on each side, as in fig. 27, Plate LIII.— This is what the ship-carpenter calls *fishing* a beam, and Fishing is a frequent practice for occasional repairs. M. Perronet beam. used it for the tie-beams or stretchers, by which he connected the opposite feet of a centre, which was yielding to its load, and had pushed aside one of the piers above four inches. Six of these not only withstood a strain of 1800 tons, but, by wedging behind them, he brought the feet of the truss $2\frac{1}{2}$ inches nearer. The stretchers were 14 inches by 11 of sound oak, and could have withstood three times that strain. M. Perronet, fearing that the great length of the bolts employed to connect the beams of these stretchers would expose them to the risk of bending, scarfed the two side pieces into the middle piece. The scarfing was of the triangular kind (*Trait de Jupiter*), and only an inch deep, each face being two feet long, and the bolt passed through close to the angle.

In piecing the pump-rods and other wooden stretchers of great engines, no dependence is had on scarfing; and the engineer connects every thing by iron straps. We doubt the propriety of this, at least in cases where the bulk of the wooden connection is not inconvenient. These observations must suffice for the methods employed for connecting the parts of a beam; and we now proceed to consider what are more usually called the joints of a piece of carpentry.

Where the beams stand square with each other, and the Square strains are also square with the beams, and in the plane of joints. the frame, the common mortise and tenon is the most perfect junction. A pin is generally put through both, in order to keep the pieces united, in opposition to any force which tends to part them. Every carpenter knows how to bore the hole for this pin, so that it shall draw the tenon tight into the mortise, and cause the shoulder to butt close, and make neat work; and he knows the risk of tearing out the bit of the tenon beyond the pin, if he draw it too much. We may just observe, that square holes and pins are much preferable to round ones for this purpose, bringing more of the wood into action, with less tendency to split it. The ship-carpenters have an ingenious method Fox-tail of making long wooden bolts, which do not pass complete-wedging. ly through, take a very fast hold, though not nicely fitted to their holes, which they must not be, lest they should be crippled in driving. They call it *fox-tail wedging*. They stick into the point of the bolt a very thin wedge of hard wood, so as to project a proper distance; when this reaches the bottom of the hole by driving the bolt, it splits the end of it, and squeezes it hard to the side. This may be practised with advantage in carpentry. If the ends of the mortise are widened inwards, and a thin wedge be put into the end of the tenon, it will have the same effect, and make the joint equal to a dove-tail. But this risks the splitting the piece beyond the shoulder of the tenon, which would be unsightly. This may be avoided as follows: Let the tenon T, fig. 28, have two *very* thin wedges *a* and *c* struck in near its angles, projecting equally; at a very small distance within these, put in two shorter ones *b*, *d*, and more within these if necessary. In driving this tenon,

Carpentry. the wedges *a* and *c* will take first, and split off a thin slice, which will easily bend without breaking. The wedges *b*, *d*, will act next, and have a similar effect, and the others in succession. The thickness of all the wedges taken together must be equal to the enlargement of the mortise towards the bottom.

When the strain is transverse to the plane of the two beams, the principles laid down in No. 85, 86, of the article STRENGTH OF MATERIALS, will direct the artist in placing his mortise. Thus the mortise in a girder for receiving the tenon of a binding joist of a floor should be as near the upper side as possible, because the girder becomes concave on that side by the strain. But as this exposes the tenon of the binding-joist to the risk of being torn off, we are obliged to mortise farther down. The form (fig. 29) generally given to this joint is extremely judicious. The sloping part *a b* gives a very firm support to the additional bearing *e d*, without much weakening of the girder. This form should be copied in every case where the strain has a similar direction.

Oblique mortise and tenon. The joint that most of all demands the careful attention of the artist, is that which connects the ends of beams, one of which pushes the other very obliquely, putting it into a state of extension. The most familiar instance of this is the foot of a rafter pressing on the tie-beam, and thereby *drawing* it away from the other wall. When the direction is very oblique (in which case the extending strain is the greatest), it is difficult to give the foot of the rafter such a hold of the tie-beam as to bring many of its fibres into the proper action. There would be little difficulty if we could allow the end of the tie-beam to project to a small distance beyond the foot of the rafter; but, indeed, the dimensions which are given to tie-beams for other reasons, are always sufficient to give enough of abutment when judiciously employed. Unfortunately this joint is much exposed to failure by the effects of the weather. It is much exposed, and frequently perishes by rot, or becomes so soft and friable that a very small force is sufficient either for pulling the filaments out of the tie-beam, or for crushing them together. We are therefore obliged to secure it with particular attention, and to avail ourselves of every circumstance of construction.

One is naturally disposed to give the rafter a deep hold by a long tenon; but it has been frequently observed in old roofs that such tenons break off. Frequently they are observed to tear up the wood that is above them, and push their way through the end of the tie-beam. This in all probability arises from the first sagging of the roof, by the compression of the rafters and of the head of the king-post. The head of the rafter descends; the angle with the tie-beam is diminished by the rafter revolving round its step in the tie-beam. By this motion the heel or inner angle of the rafter becomes a fulcrum to a very long and powerful lever much loaded. The tenon is the other arm, very short; and being still fresh, it is therefore very powerful. It therefore forces up the wood that is above it, tearing it out from between the cheeks of the mortise, and then pushes it along. Carpenters have therefore given up long tenons, and give to the toe of the tenon a shape which abuts firmly, in the direction of the thrust, on the solid bottom of the mortise, which is well supported on the under side by the wall-plate. This form has the further advantage of having no tendency to tear up the end of the mortise. This form is represented in fig. 30. The tenon has a small portion *ab* cut perpendicular to the surface of the tie-beam, and the rest *bc* is perpendicular to the rafter. (See Note CC.)

But if the tenon is not sufficiently strong (and it is not so strong as the rafter, which is thought not to be stronger than is necessary), it will be crushed, and then the raf-

ter will shade out along the surface of the beam. It is therefore necessary to call in the assistance of the whole rafter. It is in this distribution of the strain among the various abutting parts that the varieties of joints and their merits chiefly consist. It would be endless to describe every nostrum, and we shall only mention a few that are most generally approved of.

The aim in fig. 31 is to make the abutments exactly perpendicular to the thrusts. (See Note CC.) It does this very precisely; and the share which the tenon and the shoulder have of the whole may be what we please, by the portion of the beam that we notch down. If the wall-plate lie duly before the heel of the rafter, there is no risk of straining the tie across or breaking it, because the thrust is made to direct to that point where the beam is supported. The action is the same as against the joggle on the head or foot of a king-post. We have no doubt but that this is a very effectual joint. It is not, however, much practised. It is said that the sloping seam at the shoulder lodges water; but the great reason seems to be a secret notion that it weakens the tie-beam. If we consider the direction in which it acts as a tie, we must acknowledge that this form takes the best method for bringing the whole of it into action.

Fig. 32 exhibits a form that is more general, but certainly worse. Such part of the thrust as is not borne by the tenon acts obliquely on the joint of the shoulder, and gives the whole a tendency to rise up and slide outward.

The shoulder joint is sometimes formed like the dotted line *abcdcfg* of fig. 32. This is much more agreeable to the true principle, and would be a very perfect method, were it not that the intervals *bd* and *df* are so short that the little wooden triangles *bcd*, *dcf*, will be easily pushed off their bases *bd*, *df*.

Fig. 33, No. 1, seems to have the most general approbation. It is the joint recommended by Price, and copied into all books of carpentry as the *true joint* for a rafter foot. The visible shoulder-joint is flush with the upper surface of the tie-beam. The angle of the tenon at the tie nearly bisects the obtuse angle formed by the rafter and the beam, and is therefore somewhat oblique to the thrust. The inner shoulder *ac* is nearly perpendicular to *bd*. The lower angle of the tenon is cut off horizontally, as at *ed*. Fig. 34 is a section of the beam and rafter foot, showing the different shoulders.

We do not perceive the peculiar merit of this joint. The effect of the three oblique abutments, *ab*, *ac*, *ed*, is undoubtedly to make the whole bear on the outer end of the mortise, and there is no other part of the tie-beam that makes immediate resistance. Its only advantage over a tenon extending in the direction of the thrust is, that it will not tear up the wood above it. Had the inner shoulder had the form *eci*, having its face *ic* perpendicular, it would certainly have acted more powerfully in stretching many filaments of the tie-beam, and would have had much less tendency to force out the end of the mortise. The little bit *ci* would have prevented the sliding upwards along *ec*. At any rate, the joint *ab* being flush with the beam, prevents any sensible abutment on the shoulder *ac*.

Fig. 33, No. 2, is a simpler, and in our opinion a preferable, joint. We observe it practised by the most eminent carpenters for all oblique thrusts; but it surely employs less of the cohesion of the tie-beam than might be used without weakening it, at least when it is supported on the other side by the wall-plate.

Fig. 33, No. 3, is also much practised by the first carpenters.

Fig. 35, No. 1, is proposed by Mr Nicholson as preferable to fig. 33, No. 3, because the abutment of the inner

Carpentry. part is better supported. This is certainly the case; but it supposes the whole rafter to go to the bottom of the socket, and the beam to be thicker than the rafter. Some may think that this will weaken the beam too much, when it is no broader than the rafter is thick; in which case they think that it requires a deeper socket than Nicholson has given it. Perhaps the advantages of Nicholson's construction may be had by a joint like fig. 35, No. 2.

Circumstance to be attended to. Whatever is the form of these butting joints, great care should be taken that all parts bear alike; and the artist will attend to the magnitude of the different surfaces. In the general compression, the greater surfaces will be less compressed, and the smaller will therefore change most. When all has settled, every part should be *equally* close. Because great logs are moved with difficulty, it is very troublesome to try the joint frequently to see how the parts fit; therefore we must expect less accuracy in the interior parts. This should make us prefer those joints whose efficacy depends chiefly on the visible joint.

It appears from all that we have said on this subject, that a very small part of the cohesion of the tie-beam is sufficient for withstanding the horizontal thrust of a roof, even though very low pitched. If therefore no other use is made of the tie-beam, one much slenderer may be used, and blocks may be firmly fixed to the ends, on which the rafters might abut, as they do on the joggles on the head and foot of a king-post. Although a tie-beam has commonly floors or ceilings to carry, and sometimes the workshops and store-rooms of a theatre, and therefore requires a great scantling, yet there frequently occur in machines and engines very oblique stretchers, which have no other office, and are generally made of dimensions quite inadequate to their situation, often containing ten times the necessary quantity of timber. It is therefore of importance to ascertain the most perfect manner of executing such a joint. We have directed the attention to the principles that are really concerned in the effect. In all hazardous cases, the carpenter calls in the assistance of iron straps; and they are frequently necessary, even in roofs, notwithstanding this superabundant strength of the tie-beam. But this is generally owing to bad construction of the wooden joint, or to the failure of it by time. Straps will be considered in their place.

There needs but little to be said of the joints at a joggle worked out of solid timber; they are not near so difficult as the last. When the size of a log will allow the joggle to receive the whole breadth of the abutting brace, it ought certainly to be made with a square shoulder; or, which is still better, an arch of a circle, having the other end of the brace for its centre. (See Note EE.) Indeed this in general will not sensibly differ from a straight line perpendicular to the brace. By this circular form, the settling of the roof makes no change in the abutment; but when there is not sufficient stuff for this, we must avoid bevel joints at the shoulders, because these always tend to make the brace slide off. The brace in fig. 36, No. 1, must not be joined as at *b*, but as at *a*, or in some equivalent manner. Observe the joints at the head of the main posts of Drury Lane theatre, fig. 44, Plate LV.

Butting joints. When the very oblique action of one side of a frame of carpentry does not extend, but compress, the piece on which it abuts (as in fig. 11), there is no difficulty in the joint. Indeed a joining is unnecessary, and it is enough that the pieces abut on each other; and we have only to take care that the mutual pressure be equally borne by all the parts, and that it do not produce lateral pressures, which may cause one of the pieces to slide on the butting joint. A very slight mortise and tenon is sufficient at the joggle of a king-post with a rafter or straining beam. It is best, in general, to make the butting plain, bisecting the angle formed by the sides, or else perpendicular to one of the pieces. In fig. 36, No. 2, where the straining beam, *ab*, cannot slip away from the pressure, the joint *a* is preferable to *b*, or indeed to any uneven joint, which never fails to produce very unequal pressures on the different parts, by which some are crippled, others are splintered off, &c.

Directions for placing iron straps. When it is necessary to employ iron straps for strengthening a joint, considerable attention is necessary, that we may place them properly. The first thing to be determined is the direction of the strain. This is learned by the observations in the beginning of this article. We must then resolve this strain into a strain parallel to each piece, and another perpendicular to it. Then the strap which is to be made fast to any of the pieces must be so fixed that it shall resist in the direction parallel to the piece. Frequently this cannot be done; but we must come as near to it as we can. In such cases we must suppose that the assemblage yields a little to the pressures which act on it. We must examine what change of shape a small yielding will produce. We must now see how this will affect the iron strap which we have already supposed attached to the joint in some manner that we thought suitable. This settling will perhaps draw the pieces away from it, leaving it loose and unserviceable (this frequently happens to the plates which are put to secure the obtuse angles of butting timbers, when their bolts are at some distance from the angles, especially when these plates are laid on the inside of the angles); or it may cause it to compress the pieces harder than before, in which case it is answering our intention. But it may be producing cross strains, which may break them, or it may be crippling them. We can hardly give any general rules; but the reader will do well to read what is written in No. 36 and 41 of the article ROOF. In No. 36 he will see the nature of the strap or stirrup, by which the king-post carries the tie-beam. The strap that we observe most generally ill placed is that which connects the foot of the rafter with the beam. It only binds down the rafter, but does not act against its horizontal thrust. It should be placed farther back on the beam, with a bolt through it, which will allow it to turn round. It should embrace the rafter almost horizontally near the foot, and should be notched square with the back of the rafter. Such a construction is represented in fig. 37. By moving round the eye-bolt, it follows the rafter, and cannot pinch and cripple it, which it always does in its ordinary form. We are of opinion that straps which have eye-bolts in the very angles, and allow all motion round them, are of all the most perfect. A branched strap, such as may at once bind the king-post and the two braces which butt on its foot, will be more serviceable if it have a joint. When a roof warps, those branched straps frequently break the tenons, by affording a fulcrum in one of their bolts. An attentive and judicious artist will consider how the beams will act on such occasions, and will avoid giving rise to these great strains by levers. A skilful carpenter never employs many straps, considering them as auxiliaries foreign to his art, and subject to imperfections in workmanship which he cannot discern or amend. We must refer the reader to Nicholson's *Carpenter and Joiner's Assistant* for a more particular account of the various forms of stirrups, screwed rods and other iron work for carrying tie-beams, &c.

As for those that are necessary for the turning joints of great engines constructed of timber, they make no part of the art of carpentry. (See Note II.)

Examples of different pieces of carpentry. After having attempted to give a systematic view of the principles of framing carpentry, we shall conclude by giving some examples which will illustrate and confirm the foregoing principles.

Carpentry.

Roof of Greenwich chapel.

Fig. 38, Plate LIV. is the roof of the chapel of the Royal Hospital at Greenwich, constructed by Mr S. Wyatt.

	Inches Scantling.
AA is the tie-beam, 57 feet long, spanning 51 feet clear..	14 by 12
CC, queen-posts..	9 × 12
D, braces...	9 × 7
E, straining beam..	10 × 7
F, straining piece..	6 × 7
G, principal rafters..	10 × 7
H, a cambered beam for the platform.............	9 × 7
B, an iron string, supporting the tie-beam.......	2 × 2

The trusses are seven feet apart, and the whole is covered with lead, the boarding being supported by horizontal ledgers *h, h*, of six by four inches.

This is a beautiful roof, and contains less timber than most of its dimensions. The parts are all disposed with great judgment. Perhaps the iron rod is unnecessary, but it adds great stiffness to the whole.

The iron straps at the rafter feet would have had more effect if not so oblique. Those at the head of the post are very effective.

We may observe, however, that the joints between the straining beam and its braces are not of the best kind, and tend to bruise both the straining beam and the truss beam above it.

St Paul's, Covent Garden.

Fig. 39, the roof of St Paul's, Covent Garden, designed by Mr Hardwick, and constructed by Mr Wapshot in 1796.

AA, tie-beam spanning fifty feet two inches...	16 × 12
BB, queen-posts..	9 × 8
C, straining beam..	10 × 8
D, king-post (fourteen at the joggle).............	9 × 8
EE, struts...	8 × 7½
FF, auxiliary rafters (at bottom)...................	10 × 8½
HH, principal rafter (at bottom)...................	10 × 8½
gg, studs supporting the rafter......................	8 × 8

The trusses are about ten feet six inches apart, and the dotted lines in the middle compartment show the manner in which the roof is framed under the cupola.

This roof far excels the original one put up by Inigo Jones. One of its trusses contains 198 feet of timber. One of the old roof had 273, but had many inactive timbers, and others ill disposed. The internal truss FCF is admirably contrived for supporting the exterior rafters, without any pressure on the far projecting ends of the tie-beam. The former roof had bent them greatly, so as to appear ungraceful. (See Note KK.)

We think that the camber (six inches) of the tie-beam is rather hurtful, because, by settling, the beam lengthens; and this must be accompanied by a *considerable* sinking of the roof. This will appear by calculation. (See Note LL.)

Birmingham theatre.

Fig. 43, Plate LV., the roof of Birmingham theatre, constructed by Mr George Saunders. The span is eighty feet clear, and the trusses are ten feet apart.

A is an oak corbel....................................	9 × 5
B, inner plate...	9 × 9
C, wall-plate..	8 × 5½
D, pole-plate..	7 × 5
E, tie-beam..	15 × 15
F, straining beam....................................	12 × 9
G, oak king-post (in the shaft)..................	9 × 9
H, oak queen-post (in the shaft)...............	7 × 9
I, principal rafters...................................	9 × 9
K, common ditto.....................................	4 × 2½
L, principal braces........................... 9 and 6	× 9
M, common ditto.....................................	6 × 9
N, purlins..	7 × 5
Q. straining sill......................................	5½ × 9
S, ridge piece...	

This roof is a fine specimen of British carpentry, and is one of the boldest and lightest roofs in Europe. The straining sill, Q, gives a firm abutment to the principal braces, and the space between the posts is 19½ feet wide, affording roomy workships for the carpenters and other workmen connected with a theatre. The contrivance for taking double hold of the wall, which is very thin, is excellent. There is also added a beam (marked R), bolted down to the tie-beams. The intention of this was to prevent the total failure of so bold a trussing, if any of the tie-beams should fail at the end by rot.

Drury-Lane theatre.

Akin to this roof is fig. 44, Plate LV., the roof of Drury Lane theatre, eighty feet three inches in the clear, and the trusses fifteen feet apart, constructed by Edward Grey Saunders.

A, beams...	10 by 7
B, rafters...	7 × 7
C, king-posts..	12 × 7
D, struts..	5 × 7
E, purlins..	9 × 5
G, pole-plates...	5 × 5
H, gutter plates framed into the beams..	12 × 6
I, common rafters....................................	5 × 4
K, tie-beam to the main truss..................	15 × 12
L, posts to ditto......................................	15 × 12
M, principal braces to ditto............... 14 and 12	× 12
N, struts..	8 × 12
P, straining beams....................................	12 × 12

The main beams are trussed in the middle space with oak trusses five inches square. This was necessary for its width of thirty-two feet, occupied by the carpenters, painters, &c. The great space between the trusses afford good store-rooms, dressing-rooms, &c.

It is probable that this roof has not its equal in the world for lightness, stiffness, and strength. The main truss is so judiciously framed, that each of them will safely bear a load of three hundred tons; so it is not likely that they will ever be quarter loaded. The division of the whole into three parts makes the exterior roofings very light. The strains are admirably kept from the walls, and the walls are even firmly bound together by the roof. They also take off the dead weight from the main truss one third.

Remarks

The intelligent reader will perceive that all these roofs are on one principle, depending on a truss of three pieces and a straight tie-beam. This is indeed the great principle of a truss, and is a step beyond the roof with two rafters and a king-post. It admits of much greater variety of forms, and of greater extent. We may see that even the middle part may be carried to any space, and yet be flat at top; for the truss-beam may be supported in the middle by an inverted king-post (of timber, not iron), carried by iron or wooden ties from its extremities; and the same ties may carry the horizontal tie-beam K; for till K be torn asunder, or M, M, and P be crippled, nothing can fail.

The roof of St Martin's church in the Fields is constructed on good principles, and every piece properly disposed. But although its span does not exceed forty feet from column to column, it contains more timber in a truss than there is in one of Drury-Lane theatre. The roof of the chapel at Greenwich, that of St Paul's, Covent-Garden, those of Birmingham and Drury-Lane theatres, form a series gradually more perfect. Such specimens afford excellent lessons to the artist. We therefore account them a useful present to the public.

Project by Mr Nicholson.

There is a very ingenious project offered to the public by Mr P. Nicholson. (*Carpenter's Assistant*, p. 68.) He proposes iron rods for king posts, queen-posts, and all

Carpentry. other situations where beams perform the office of ties. He receives the feet of the braces and struts in a socket very well connected with the foot of his iron king-post; and he secures the feet of his queen-posts from being pushed inwards, by interposing a straining sill. He does not even mortise the foot of his principal rafter into the end of the tie-beam, but sets it in a socket like a shoe, at the end of an iron bar, which is bolted into the tie-beam a good way back.[1] All the parts are formed and disposed with the precision of a person thoroughly acquainted with the subject; and we have not the smallest doubt of the success of the project, and the complete security and durability of his roofs. We abound in iron; but we must send abroad for building timber. This is therefore a valuable project; at the same time, however, let us not overrate its value. Iron is about twelve times stronger than red fir, and is more than twelve times heavier; nor is it cheaper, weight for weight, or strength for strength.

Our illustrations and examples have been chiefly taken from roofs, because they are the most familiar instances of the difficult problems of the art. We could have wished for more room even on this subject. The construction of dome roofs has been, we think, mistaken, and the difficulty is much less than is imagined; we mean in respect of strength; for we grant that the obliquity of the joints, and a general intricacy, increases the trouble of **Wooden** workmanship exceedingly. Wooden bridges form another **bridges.** class equally difficult and important; but our limits are already overpassed, and will not admit them. The principle on which they should all be constructed, without exception, is that of a truss, avoiding all lateral bearings on any of the timbers. In the application of this principle we must further remark, that the angles of our truss should be as acute as possible; therefore we should make it of as few and of as long pieces as we can, taking care to prevent the bending of the truss beams by bridles, which embrace them, but without pressing them to either side. When the truss consists of many pieces, the angles are very obtuse, and the thrusts increase nearly in the duplicate proportion of the number of angles.

Framing of With respect to the frames of carpentry which occur **great le-** in engines and great machines, the varieties are such that **vers.** it would require a volume to treat of them properly. The principles are already laid down; and if the reader be really interested in the study, he will engage in it with seriousness, and cannot fail of being instructed. We recommend to his consideration, as a specimen of what may be done in this way, the working beam of Hornblower's steam-engine. (See STEAM-ENGINE.) When the beam must act by chains hung from the upper end of arch-heads, the framing there given seems very scientifically constructed; at the same time we think that a strap of wrought iron reaching the whole length of the upper bar (see the figure) would be vastly preferable to those partial plates which the engineer has put there, for the bolts will soon work loose.

But when arches are not necessary, the form employed by Mr Watt is vastly preferable, both for simplicity and for strength. It consists of a simple beam, AB (fig. 45, Plate LV.), having the gudgeon C on the upper side. The two piston rods are attached to wrought-iron joints, A and B. Two strong struts, DC, EC, rest on the upper side of the gudgeon, and carry an iron string, ADEB, consisting of three pieces, connected with the struts by proper joints of wrought iron. A more minute description is not needed for a clear conception of the principle. No part of this is exposed to a cross strain; even the beam AB might be sawed through at the middle. The iron string is the only part which is stretched; for AC, DC, EC, BC, are all in a state of compression. We have made the angles equal, that all may be as great as possible, and the pressure on the struts and strings a minimum. Mr Watt makes them much lower, as A*de*B, or A*d*B. But this is for economy, because the strength is almost insuperable. It might be made with wooden strings; but the workmanship of the joints would more than compensate the cheapness of the materials.

We offer this article to the public with deference, and we hope for an indulgent reception of our essay on a subject which is in a manner new, and would require much study. We have bestowed our chief attention on the strength of the construction, because it is here that persons of the profession have the most scanty information. We beg them not to consider our observations as too refined, and that they will study them with care. One principle runs through the whole; and when that is clearly conceived and familiar to the mind, we venture to say that the practitioner will find it of easy application, and that he will improve every performance by a continual reference to it.

IV.—NOTES.

AA, p. 217. This rule may be somewhat more accurately expressed in these words: From the point at which any three forces meet and balance each other, draw a line in the actual direction of any one of them, and from the extremity of this line draw two others, parallel to the directions of the other two forces respectively; then supposing the pieces affording these two forces to be produced indefinitely at their remoter ends, either of them which is cut by one of the two lines will be compressed, and act as a brace, and either of them which is not cut will be stretched, and act as a tie.

BB, p. 218. It is, however, difficult to imagine how the beam DA can furnish a force iA, to prevent the force Af from carrying the beam BA towards H, when DA only affords a repulsive abutment. The true resolution of the force AE is found by considering the intersection of GE with Ae, which are the directions of the separate forces composing it: these lines meeting in a point a little above r, we may call their intersection r^*: then in the triangle AEr^*, the side Ar^* will represent the pressure on the mitred joint, and r^*E the pressure on the beam HD; and the former being again resolved into AG and Gr^*, we have ultimately AG and G$r^* + r^*$E = GE = AF, for the horizontal and vertical forces, however they may be modified by intermediate combinations.

CC, p. 219. The reasoning contained in this and some of the subsequent articles may serve as an approximation to the truth in many cases of common occurrence; but the supposition on which it is founded is by no means generally admissible as affording a result mathematically accurate; for, in reality, the distribution of the weight of a roof over the whole extent of the rafters, or the concentration of the whole weight in the point where they meet, is far from being an indifferent alternative, either with respect to the magnitude of the thrusts, or to the proper directions of the abutments or joints. In the case here discussed, where there is no king-post, it is clear that the centre of gravity of the whole roof must be much nearer to the middle of the figure than the angular point, and that consequently the weights supported by the two walls will be very different from those which would be support-

---

[1] See figures 40, 41, 42, Plate X., and Mr Nicholson's work, p. 68, where these figures are particularly described

Carpentry. ed if the whole load were placed at the summit; although, where there is a heavy king-post, supporting also, as it ought to do, about half the weight of the tie-beam, with its floors or ceiling, the case will approach much nearer to the supposition here assumed.

For a common light roof, without a king-post, the calculation or construction is very simple. When two rafters only meet at the summit, they must support each other by a horizontal thrust (see Art. BRIDGE, Prop. Y); and this thrust, acting on each rafter as a lever, of which the lower end is the fulcrum, must be equivalent to the weight, acting at the horizontal distance of the centre of gravity from the fulcrum, which is a quarter of the whole span; consequently the thrust must be to the weight as a quarter of the span to the height, and the compound oblique thrust on the abutment will be represented by the hypotenuse of the triangle of which those lines are the sides; so that if we had a roof of the same height, and of half the breadth, the direction of its rafters would exactly represent the actual direction of the compound thrust on the end of the tie-beam, and would consequently indicate the proper form for the abutment of the given structure.

But in the case of the unequal rafters represented in the figure, the determination becomes more complicated, and we must first find the direction of the mutual thrust of the rafters, which must evidently be such, that the perpendiculars falling on it from each end of the tie-beam may be in the inverse proportion of the motive powers of the weights of the rafters, that is, of the products of those weights into the horizontal distances of the centres of gravity from the respective fulcrums, or into the segments of the tie-beam made by a vertical line passing through the summit, which are proportional to these distances; and if we produce the base of the triangle, and find in it a point, of which the distance is to the length of the tie-beam as the smaller product to the difference of the products, a line drawn from the summit to this point will show the true direction of the thrust; and its magnitude may then be readily determined by dividing either of the products by the respective perpendicular falling on this line.

Where, however, there is a king-post supporting a heavy tie-beam, it is necessary to determine the centre of gravity of the half roof, together with this addition; and the distance of the centre of gravity from the middle will then be to the half span, as the weight of one of the rafters with its load is to the weight of the whole roof, including the tie-beam and ceiling; and if we erect a perpendicular passing through the centre of gravity thus found, and equal to the height, the oblique thrust on the abutment will be in the direction of the line joining the upper end of this perpendicular and the end of the tie-beam.

DD, p. 221. In order to obtain a distinct idea of the operation of the forces concerned in this experiment, we must have recourse to proposition C of this article, and substitute in the formula for the deflection

$$d = a\,t\,\sqrt{\frac{M}{16f}}\;\text{TANG.}\left(\sqrt{\frac{16f}{M}\cdot\frac{e}{a}}\right),$$

$a = 1$, $t = \frac{8}{6720} = \frac{1}{840}$, M = 1,900,000 pounds, the specific gravity of fir being ·56, $f = 6720$, and $e = 6$, the middle of the pillar being considered as the fixed point: we then find $\sqrt{\frac{16f}{M}\cdot\frac{e}{a}} = 1\cdot427$, which is the length of an arc of 81° 45', and the tangent becomes 6·9, whence we have $d = \frac{1}{840}\times 4\cdot2\times 6\cdot9 = \cdot0345$, or somewhat more than the thirtieth of an inch: consequently the strength

must have been reduced in the proportion of 1·207 to 1. Carpentry (Art. BRIDGE, Prop. E.) But considering how near the arc thus determined approaches to a quadrant, it is obvious that any slight variations of the quantities concerned in the calculation must have greatly affected the magnitude of the tangent; so that the loss of strength may easily have been considerably greater than this, as it appears to have been found in the experiment. It would, however, scarcely have been expected that such a pillar, however supported, could withstand the pressure of ninety hundredweight, since Emerson informs us that the cohesive strength of a pillar of fir an inch in diameter is only about thirty-five: but supposing the facts correct, the coincidence tends to show the near approach to equality of the forces of cohesion and lateral adhesion, as explained in the introduction to this article.

EE, p. 222. A similar remark of the author has already been noticed in the article BRIDGE, at the end of the fifth section. In the form in which it is here expressed, it becomes still more objectionable; for with whatever part of a circular abutment a rafter equal to the radius may be brought into contact, it is very plain that its opposite end can never be either higher or lower than the original centre of curvature: and even if the curvature were made twice as great, so that the rafter might be equal to the diameter of the circle, it would be necessary that the lower end should slide upwards on the abutment as much as the upper end fell, in order to preserve the contact; and there would obviously be no force in the structure capable of producing such a change as this. Any general curvature of the joint must therefore be totally useless; but a judicious workman will make it somewhat looser below than above, when there is any probability that the rafters will sink, taking care, however, to avoid all bearing too near the surface, lest it should splinter, and, for these reasons combined, making the end a little prominent somewhat above the middle of the surface which rests on the abutment.

With this precaution, the direction of the joint between a rafter and a tie-beam ought to be made precisely perpendicular to the true thrust of the rafter, determined as already explained (Note CC); for, in the first place, unless we trust either to the friction, or to straps, the bearing cannot be more nearly horizontal than this, without danger of the rafters sliding outwards; and, in the second place, if we made it more nearly vertical, we should lessen the vertical pressure on the end of the tie-beam, immediately beyond the joint; a pressure which gives firmness to the wood, by pressing its fibres more closely together, and increasing their lateral adhesion, or rather internal friction. If, however, the tie-beam were not deep enough to receive the whole of the rafter so terminated, without too great a reduction of its depth, it would be proper to make the joint a little flatter, or more horizontal, and to restrain the end from sliding upwards by an iron strap fixed in a proper direction. We should preserve the end of the rafter as little diminished in breadth as possible, when the tie-beam is wide enough to receive it; a moderate thickness, left on each side of the mortise in the tie-beam, being sufficient to assist in securing the connection of the ends of the beam with the intermediate parts.

FF, p. 222. The doctrine of the initial equality of the resistances to compression and extension, as stated in the article BRIDGE, enables us to demonstrate that the transverse strength can never exceed one sixth of that which would be derived from the resistance of all the fibres, co-operating at the distance of the whole depth from a fixed fulcrum, and acting with the weaker of the two powers appropriate to the body. It is true that the results of some direct experiments seem to favour the opinion that the cohesive power is the weaker; but where the flexure is already considerable, it is probable that this circumstance materially diminishes the primitive power of resisting compression, so that the principles on which the calculation proceeds are by no means strictly applicable to the case of a bar so broken.

GG, p. 223. There seems to be a little confusion in the idea of the possibility of altering the nature of the action of the fibres of a beam by altering the place of the gudgeon in this manner; but the author has very properly abstained from making any practical application of the supposed modification thus introduced. With respect to the strength required for scarfing or joggling, it may be observed, that the whole of the compressed fibres of the concave side may be considered as abutting against the whole of the extended fibres on the convex side; and this abutment is equally divided throughout the length of the beam; so that if the scarfings or joggles in the whole length of the arm of a lever, taken together, are as strong as one half of the depth of the lever, exerting half its powers, from the inequality of tension, there will be no danger of the failing of these joints; and from this principle it will be easy to determine the depth to which the joints ought to extend in any particular case. Hence also we may understand how a beam may become so short as to be incapable of transverse fracture in its whole extent; for the lateral adhesion between the different fibres of wood is generally far inferior to the longitudinal strength of the fibres: and if, for example, it were only one fourth as great, a beam less than twice as long as it is deep would separate, if urged in the middle by a transverse force, into two strata, from its incapacity of affording sufficient abutment, before its longitudinal fibres would give way.

HH, p. 225. If the bolts were sufficiently numerous and sufficiently firm, so as to produce a great degree of adhesion or of friction between the parts, this joint might be made almost as strong as the entire beam, since there is nothing to prevent the co-operation of each side with the other throughout its extent; but much of the strength would be lost if the bolts became loose, even in an inconsiderable degree.

II, p. 227. The author has reasoned upon the direction of straps, as if it were universally necessary to economize their immediate strength only, without regard to the effect produced on the tightness of the joint; but it may happen that the principal purpose of the strap will be answered by its pressing the rafter firmly upon the beam, and this effect may be produced by a certain deviation from the horizontal position, with but little diminution of the strength of the strap; a deviation which has also the advantage of allowing the strap to embrace the whole of the beam, without weakening it by driving a bolt through it. We must not, however, run the risk of crippling the end of the beam, and the straps represented in fig. 38 may be allowed to be somewhat too erect.

KK, p. 228. It does not appear to be desirable that the ends of the rafters should be supported without any pressure on the ends of the beams, since these ends would bear a small weight without any danger of bending, and would thus lessen the pressure on the king-post.

LL, p. 228. The half length being 25 feet, and the camber 6 inches, the excess of the oblique length will be $\sqrt{625\cdot25} - 25$, or $\frac{1}{200}$ of a foot, that is, $\frac{1}{16}$ of an inch, which is all that the beam would appear to lengthen in sinking; nor would the settling of the roof be more "considerable" than about a quarter of an inch. But there seems to be no advantage in this deviation of the tie-beam from the rectilinear direction; and the idea, which appears to be entertained by some workmen, that a bent beam partakes of the nature of an arch, is one of the many mischievous fallacies which it is the business of the mathematical theory of carpentry to dispel. (T. Y.)

ROOF.

Definition. ROOF, the covering of any building by which its inhabitants and contents are protected from the injuries and inclemencies of the weather. So essential is it, that the word is frequently used for the house itself. To " come under the roof" is a Hebrew phrase; and the word " tectum" had the same meaning among the Romans. It is derived from the Anglo-Saxon *hrof*, who thought so much of its importance, that they called the carpenter *hrof-wyrhta*, or the " roof-worker."

Varieties of covering. Roofs may be considered as to their covering, and the framing which carries such covering. The former is either of metal, as lead, copper, zinc, corrugated or galvanized iron, &c.; or of tile, either Italian, pan or Flemish tile, plain tile, &c.; or of slate, and sometimes of stone. The Greek temples were covered with long thin pieces of marble sunk or worked hollow by the mason, so that the wet could not run back under the next, and consequently these roofs shot off the water easily, and were very flat. Both in ancient and modern times, in all countries, the poorer classes of roofs are covered or thatched with straw, reed, heather, or some similar material. In most hot climates, and also in many parts of Italy, the roofs are flat, and covered with a sort of concrete or cement, which is carried on joists like a floor; the object being to form a sort of terrace to walk on early in the morning or late in the evening, to enjoy the cool air, which can only be felt in elevated situations.

Pitch. The elevation of a roof, which governs the angle its rafters make with the horizon, is called its pitch. On this subject there has been a great deal of controversy. Some have considered, as they find the farther we go south the flatter the roofs are, that the pitch must be governed by climate; and most elaborate calculations have been made of certain angles at which it is proposed that roofs should be constructed in various latitudes. But it should be remembered that in hot climates the rains come all at once; in such floods our roofs could not resist; and it would be poor economy, because for months together there were no rain, if, when it does come, the house should be daily drenched. Others have considered the whole a mere matter of taste, and the pitch is chosen as we wish more or less of a roof to be seen. The Greeks made their roofs very flat, and placed large antefixes along the eaves, so that the roof could not be seen from below except from a great distance. (See the restored view of the Parthenon, ARCHITECTURE, Plate III.) The angle is about 16°, the pitch or height at the apex being about a seventh of the width. Roman roofs average about 22°, or a fifth pitch. That the mediæval builders had no rule, is shown from the extreme variety of the height of roofs in their different edifices. In the Lombardic cathedral of Pisa, erected 1063 (ARCHITECTURE, Plate XXI.), the roof is about 27°, or nearly quarter pitch. The Norman roofs are seldom more than 40°, or less than half pitch; while in the early English period they suddenly sprung up to whole pitch,—*i.e.*, the height equal to the entire width (see Beverley Minster, ARCHITECTURE, Plate XIX., fig. 1), being an angle of about 64°. They then gradually were less in height till the perpendicular period, when many roofs were nearly flat; that of Henry the Seventh chapel, for example, being but about 16°, or as flat as a Greek roof. Now, that this variety was matter of taste,—we had almost said caprice,—is evidenced by this fact: these examples are all covered with lead (which might have been laid quite flat, and yet have been perfectly sound), and all have a stone groined roof below them, which has nothing whatever to do with the upper covering, and which, after all, is the real roof or cover which protects the building from the weather. Much has been said of the propriety of always showing the roof of a building, and the Gothic architects have been eulogised for so doing. The facts stated above, however, prove this was not always the case. We cannot, however, justify the going out of the way to conceal a roof by false attics, stilted balustrades, &c.; and the screen wall at St Paul's at London must always be considered a defect in that fine building. Still, a wide expanse of plain roof is as ugly in itself as a bare wall; and we cannot approve of such roofs as some of the modern imitations of early English work are, where the wall is so low that we could touch the eaves with a walking-stick, and there is three or four times as much roof as wall. The roof of a house has not inaptly been likened to a man's hat. There is no need to try and hide or disguise it if you are obliged to wear it; and if the weather is warm, and you do not require it, it would be folly to wear one without a crown. If on board ship, you would wear as low a hat as possible to avoid striking it against the beams; but, above all, it should bear some reasonable proportion to the height of a man, as the roof should to the wall. It would be absurd to wear a hat as tall as the man himself.

Pitch dependent on materials used. After all, although much latitude must be given to taste, it is probable the pitch of a roof mainly depends on the material with which it is covered. The largest number of buildings are erected with a view to utility and strict economy, and without any regard to æsthetics. Everybody knows that if slates or tiles are laid at too flat a pitch, the wind will drive the rain up under them, and the roof will leak; and everybody also knows that if the same covering be taken off and re-laid to a steeper pitch, the roof will be sound. Practice teaches what is the safe minimum pitch. Let us suppose it to be quarter pitch, and for considerations of taste we make it three-quarter pitch. Now, it is quite clear we waste not only the rafters and covering, but our whole roof must be constructed of stronger timbers, and our walls also must be thicker and stronger, inasmuch as they have more weight to bear. We therefore pay dear in more ways than one for our liking for high-pitched roofs.

Pitch not due to climate. Although it happens that both Greek, Roman, and Italian roofs are flatter than ours, and the climate is warmer, the same material used in our climates would answer perfectly well. An inspection of the elaborate plate 97, in the *Architettura Antica Greca* of the celebrated Canina will show this: and the frequent failure in our climate of Italian

Roof.

tiles (which are exactly like the ancient Roman) arises from the fact, that the tegole and imbrici only have been used, our builders being ignorant of the use of the mattoni, which in Italy are a very essential part of the soundness of those roofs. In one respect climate must be considered, and that is, where there are long winters, and the snow is likely to lie on them; in this case they should be sharper in pitch, and stronger in framing.

Pitch required for different materials.

If covered with lead or other metals, roofs may be made nearly flat, with only so much fall, in fact, as to prevent the water flowing back under the drips. (See BUILDING, PLUMBERS' WORK, &c.) Italian tiles, to be sound, should have a fifth pitch, or 22°. Slates with extra lap may be laid at quarter pitch, about 27°, if it be necessary the roof should be flat; a third pitch (34°) is rather too much: the mean between a third and fourth (31°) is a good rule. Pantiles should be laid rather sharper still, and plain tiles from about 35° to 40°; but of course very much will depend on the gauge they are laid to, or the length of the part of the slate or tile which overlaps the other, as the larger this lap is, the less likely the rain is to drive under. Thatched roofs should be somewhat sharper in pitch than plain tiles.

Qualities of various roof coverings.

Lead or copper, in an economical point of view, are the best materials for roofs. They may be laid nearly flat, and so save all the framing and roof timbers; and the metal, should it be worn into holes, is nearly as valuable as when first laid down; the only objection is, that the first expense is so great. Zinc, though very cheap and light, and though it can be laid flat, is apt to go into holes with the action of acids. Slating is both light and very cheap, and will lie at a flat pitch; and consequently requires much lighter walls and timbers than tiles. It will not decay with the weather. It is apt to break under the feet; and if not very well done will lift with heavy winds. Each slate should be nailed with two copper nails, as iron rusts and breaks them. (See BUILDING, SLATING, &c.) Pan-tiles are dearer than slates, but not much heavier; they also break if trodden on, and the snow will drift under if the pointing comes out. Plain tiles are very durable, but they require a steep pitch, and are very heavy: thus in two ways distressing the walls and the roof timbers.

Weight of different roof coverings.

This also depends on the gauge; but the following may be taken as the ordinary average:—

A square (100 feet superficial, or 11 yards superficial nearly) of zinc will weigh about...	1 cwt.
A square of lead, according to thickness, from...5 to 7	,,
A square of slating from............5½ to 7½	,,
A square of pan-tiling............7½	,,
A square of plain tiling............14 to 16	,,

All roofs, till very lately, except some which have been arched or domed, were framed with timber; no other material being known at that time which possessed such lengths with such qualities of tension. Later years, however, and more extended requirements, have developed the advantages of the use of iron. As everything must depend on the soundness of both design and execution of framing, whether in wood or iron, it is proposed to divide this subject, one of the most important in architecture, into the following sections:—

I. *Theory of Roof* will comprehend the whole of the scientific part of the celebrated essay of Professor Robison, which was originally written for this work, and which is acknowledged to be the best yet given to the public.

II. *Causes of Failure of Roofs*, given in terms that are intelligible to those unacquainted with the higher branches of mathematical analysis.

III. *Mediæval Roofs.*

IV. *Account of Roofs of great span* (*à grande portée.*)
—1. Those trussed with straight timbers (*en bois plat*).
2. Those trussed with curved timbers—*a*, With timbers side by side, breaking joint (*système en planches*

de champ); *b*, Those bent in thickness (*courbes sur leur plat*).

V. *Roofs constructed of Iron*

Roof.

The late Professor Robison's Theory of Roof.

Purpose of this article.

We shall attempt in this article to give an account of the leading principles of this art, in a manner so familiar and palpable, that any person who knows the common properties of the lever, and the composition of motion, shall so far understand them as to be able, on every occasion, so to dispose his materials, with respect to the strains to which they are to be exposed, that he shall always know the effective strain on every piece, and shall, in most cases, be able to make the disposition such as to derive the greatest possible advantage from the materials which he employs.

Principles which regulate the strength of the materials.

It is evident that the whole must depend on the principles which regulate the strength of the materials, relative to the manner in which this strength is exerted, and the manner in which the strain is laid on the piece of matter. With respect to the first, this is not the proper place for considering it, and we must refer the reader to the article STRENGTH OF MATERIALS IN MECHANICS. We shall just borrow from that article two or three propositions suited to our purpose.

The force with which the materials of our edifices, roofs, floors, machines, and framings of every kind, resist being broken or crushed, or pulled asunder, is immediately or ultimately the cohesion of their particles. When a weight hangs by a rope, it tends either immediately to break all the fibres, overcoming the cohesion amongst the particles of each, or it tends to pull one parcel of them from amongst the rest, with which they are joined. This union of the fibres is brought about by some kind of gluten, or by twisting, which causes them to bind each other so hard that any one will break rather than come out, so much is it withheld by friction. The ultimate resistance is therefore the cohesion of the fibre; and the force or strength of all fibrous materials, such as timber, is exerted in much the same manner. The fibres are either broken or pulled out from among the rest. Metals, stone, glass, and the like, resist being pulled asunder by the simple cohesion of their parts.

The force which is necessary for breaking a rope or wire is a proper measure of its strength. In like manner, the force necessary for tearing directly asunder any rod of wood or metal, breaking all its fibres, or tearing them from amongst each other, is a proper measure of the united strength of all these fibres; and it is the simplest strain to which they can be exposed, being just equal to the sum of the forces necessary for breaking or disengaging each fibre. And, if the body is not of a fibrous structure, which is the case with metals, stones, glass, and many other substances, this force is still equal to the simple sum of the cohesive forces of each particle which is separated by the fracture. Let us distinguish this mode of exertion of the cohesion of the body by the name of its *absolute strength*.

When solid bodies are, on the contrary, exposed to great compression, they can resist only a certain degree. A piece of clay or lead will be squeezed out; a piece of freestone will be crushed to powder; a beam of wood will be crippled, swelling out in the middle, and its fibres lose their mutual cohesion, after which it is easily crushed by the load. A notion may be formed of the manner in which these strains are resisted, by conceiving a cylindrical pipe filled with small shot, well shaken together, so that each sphericle is lying in the closest manner possible, that is, in contact with six others in the same vertical plane, this being the position in which the shot will take the least room. Thus each touches the rest in six points. Now suppose them all united, in these six points only, by some cement. This assemblage will stick together and form a cylindrical pillar, which may be taken out of its mould. Now suppose this pillar standing upright, and loaded above. The supports arising from the cement act obliquely, and the load tends either to force them asunder laterally, or to make them slide on each other : either of these things happening, the whole is crushed to pieces. The resistance of fibrous materials to such a strain is a little more intricate, but may be explained in a way very similar.

A piece of matter of any kind may also be destroyed by wrenching or twisting it. We can easily form a notion of its resistance to this kind of strain by considering what would happen to the cylinder of small shot if treated in this way.

And, lastly, a beam, or a bar of metal, or piece of stone or other matter, may be broken transversely. This will happen to a rafter or joist supported at the ends when overloaded, or to a beam having one end stuck fast in a wall and a load laid on its projecting part. This is the strain to which materials are most commonly exposed in roofs ; and, unfortunately, it is the strain which they are the least able to bear ; or rather it is the manner of application which causes an external force to excite the greatest possible immediate strain on the particles. It is against this that the carpenter must chiefly guard, avoiding it when in his power, and in every case diminishing it as much as possible. It is necessary to give the reader a clear notion of the great weakness of materials in relation to this transverse strain. But we shall do nothing more, referring him to the articles STRAIN, and STRESS, and STRENGTH.

Let ABCD (fig. 1) represent the side of a beam projecting horizontally from a wall in which it is firmly fixed, and let it be loaded with a weight W appended to its extremity. This tends to break it ; and the least reflection will convince any person, that if the beam is equally strong throughout, it will break in the line CD, even with the surface of the wall. It will open at D, while C will serve as a sort of joint, round which it will turn. The cross section through the line CD is for this reason called the *section of fracture* ; and the horizontal line drawn through

Fig. 1.

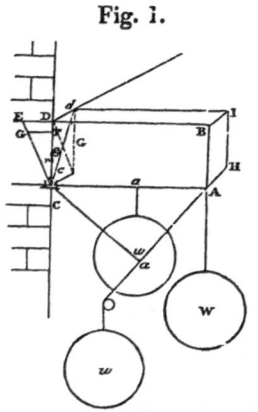

C on its under surface is called the *axis of fracture*. The fracture is made by tearing asunder the fibres, such as DE or FG. Let us suppose a real joint at C, and that the beam is really sawed through along CD, and that in place of its natural fibres, threads are substituted all over the section of fracture. The weight now tends to break these threads, and it is our business to find the force necessary for this purpose.

It is evident that DCA may be considered as a bended lever, of which C is the fulcrum. If f be the force which will just balance the cohesion of a thread when hung on it so that the smallest addition will break it, we may find the weight which will be sufficient for this purpose when hung on at A, by saying, $AC : CD = f : \varphi$, and φ will be the weight which will just break the thread, by hanging φ by the point A. This gives us $\varphi = f \times \dfrac{CD}{CA}$. If the weight be hung on at a, the force just sufficient for breaking the same thread will be $= f \times \dfrac{CD}{Ca}$. In like manner, the force φ, which must be hung on at A in order to break an equally

strong or an equally resisting fibre at F, must be $= f \times \dfrac{CF}{CA}$

And so on of all the rest.

If we suppose all the fibres to exert equal resistances at the instant of fracture, we know, from the simplest elements of mechanics, that the resistance of all the particles in the line CD, each acting equally in its own place, is the same as if all the individual resistances were united in the middle point g. Now this total resistance is the resistance or strength f of each particle, multiplied by the number of particles. This number may be expressed by the line CD, because we have no reason to suppose that they are at unequal distances. Therefore, in comparing different sections together, the number of particles in each are as the sections themselves. Therefore DC may represent the number of particles in the line DC. Let us call this line the depth of the beam, and express it by the symbol d. And since we are at present treating of roofs whose rafters and other parts are commonly of uniform breadth, let us call AH or BI the breadth of the beam, and express it by b, and let CA be called its length l. We may now express the strength of the whole line CD by $f \times d$, and we may suppose it all concentrated in the middle point g. Its mechanical energy, therefore, by which it resists the energy of the weight w, applied at the distance l, is $f \times CD \times Cg$, whilst the momentum of w is $w \times CA$. We must therefore have $f \times CD \times Cg = w \times CA$, or $fd \times \frac{1}{2}d = wl$, and $fd : w = l : \frac{1}{2}d$, or $fd : w = 2l : d$. That is, twice the length of the beam is to its depth as the absolute strength of one of its vertical planes to its relative strength, or its power of resisting this transverse fracture.

It is evident, that what has been now demonstrated of the resistance exerted in the line CD, is equally true of every line parallel to CD in the thickness or breadth of the beam. The absolute strength of the whole section of fracture is represented by fdb, and we still have $2l : d = fdb : w$; or twice the length of the beam is to its depth as the absolute strength to the relative strength. Suppose the beam twelve feet long and one foot deep ; then whatever be its absolute strength, the twenty-fourth part of this will break it if hung at its extremity.

But even this is too favourable a statement. All the fibres are supposed to act alike in the instant of fracture. But this is not true. At the instant that the fibre at D breaks, it is stretched to the utmost, and is exerting its whole force. But at this instant the fibre at g is not so much stretched, and it is not then exerting its utmost force. If we suppose the extension of the fibres to be as their distance from C, and the actual exertion of each to be as their extensions, it may easily be shown (see STRENGTH and STRAIN), that the whole resistance is the same as if the full force of all the fibres were united at a point r distant from C by one third of CD. In this case we must say, that the absolute strength is to the relative strength as three times the length to the depth ; so that the beam is weaker than by the former statement in the proportion of two to three.

Even this is more strength than experiment justifies, and we can see an evident reason for it. When the beam is strained, not only are the upper fibres stretched, but the lower fibres are compressed. This is very distinctly seen, if we attempt to break a piece of cork cut into the shape of a beam. This being the case, C is not the centre of fracture. There is some point c which lies between the fibres which are stretched and those that are compressed. This fibre is neither stretched nor squeezed, and this point is the real centre of fracture ; and the lever by which a fibre D resists, is not DC, but a shorter one Dc, and the energy of the whole resistances must be less than by the second statement. Till we know the proportion between the dilatability and compressibility of the parts, and the relation between the dilatations of the fibres and the resistances which they exert in this state of dilatation, we cannot positively say where the point c is situated, nor what is the sum of the actual resistances, or the point where their action may be supposed concentrated. The firmer woods, such as oak and chestnut, may be supposed to be but slightly compressible; we know that willow and other soft woods are very compressible. These last must therefore be weaker; for it is evident, that the fibres which are in a state of compression do not resist the fracture. It is well known, that a beam of willow may be cut through from C to g without weakening it in the least, if the cut be filled up by a wedge of hard wood stuck in.

We can only say, that very sound oak and red fir have the centre of effort so situated, that the absolute strength is to the relative strength in a proportion of not less than that of three and a half times the length of the beam to its depth. A square inch of sound oak will carry about 8000 pounds. If this bar be firmly fixed in a wall, and project twelve inches, and be loaded at the extremity with 200 pounds, it will be broken. It will just bear 190, its relative strength being $\frac{1}{42}$ of its absolute strength; and this is the case only with the finest pieces, so placed that their annual plates or layers are in a vertical position. A larger log is not so strong transversely, because its plates lie in various directions round the heart.

Practical
inferences. These observations are enough to give us a distinct notion of the vast diminution of the strength of timber when the strain is across it; and we see the justice of the maxim which we inculcated, that the carpenter, in framing roofs, should avoid as much as possible the exposing his timbers to transverse strains. But this cannot be avoided in all cases. Nay, the ultimate strain arising from the very nature of a roof is transverse. The rafters must carry their own weight, and this tends to break them across. An oak beam a foot deep will not carry its own weight if it project more than sixty feet. Besides this, the rafters must carry the lead, tiling, or slates. We must therefore consider this transverse strain a little more particularly, so as to know what strain will be laid on any part by an unavoidable load, imposed either at that part or at any other.

Effect when
beams are
supported
at the ends
and loaded
in the mid-
dle. We have hitherto supposed, that the beam had one of its ends fixed in a wall, and that it was loaded at the other end. This is not an usual arrangement, and was taken merely as affording a simple application of the mechanical principles. It is much more usual to have the beam supported at the ends, and loaded in the middle. Let the beam FEGH (fig. 2) rest on the props E and G, and be loaded at its middle point C with a weight W. It is required to determine the strain at the section CD. It is plain that the beam will receive the same support, and suffer the same strain, if, instead of the blocks E and G, we substitute the ropes Ffe, Hhg, going over the pulleys f and g, and loaded with proper weights e and g. The weight e is equal to the support given by the block E; and g is equal to the support given by G. The sum of e and g is equal to W; and on whatever point W is hung, the weights e and g are to W in the proportion of DG and DE to GE. Now, in this state of things, it appears that the strain on the section CD arises immediately from the upward action of the ropes Ff and Hh, or the upward pressions of the blocks E and G; and that the office of the weight W is to oblige the beam to oppose this strain. Things are in the same state in respect of strain as if a block were substituted at D for the weight W, and the weights e and g were

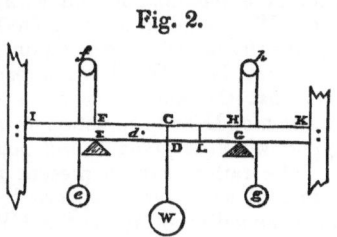

Fig. 2.

hung on at E and G, only the directions will be opposite. The beam tends to break in the section CD, because the ropes pull it upwards at E and G, whilst a weight W holds it down at C. It tends to open at D, and C becomes the centre of fracture. The strain therefore is the same as if the half ED were fixed in the wall, and a weight equal to g, that is, to the half of W, were hung on at G.

Hence we conclude, that a beam supported at both ends, but not fixed there, and loaded in the middle, will carry four times as much weight as it can carry at its extremity, when the other extremity is fast in a wall.

The strain occasioned at any point L by a weight W, hung on at any other point D, is $= W \times \dfrac{DE}{EG} \times LG$. For EG is to ED as W to the pressure occasioned at G. This would be balanced by some weight g acting over the pulley h; and this tends to break the beam at L, by acting on the lever GL. The pressure at G is $W \times \dfrac{DE}{EG}$, and therefore the strain at L is $W \times \dfrac{DE}{EG} \times LG$.

In like manner, the strain occasioned at the point D by the weight W hung on there, is $W \times \dfrac{DE}{EG} \times DG$; which is therefore equal to $\frac{1}{4}$ W when D is the middle point.

Hence we see that the general strain on the beam arising from one weight, is proportional to the rectangle of the parts of the beam (for $\dfrac{W \times DE \times DG}{EG}$ is as $DE \times DG$), and is greatest when the load is laid on the middle of the beam.

We also see, that the strain at L by a load at D, is equal to the strain at D by the same load at L. And the strain at L from a load at D is to the strain by the same load at L as DE to LE. These are all very obvious corollaries, and they sufficiently inform us concerning the strains which are produced on any part of the timber by a load laid on any other part.

If we now suppose the beam to be fixed at the two ends, that is, firmly framed or held down by blocks at I and K, placed beyond E and G, or framed into posts, it will carry twice as much as when its ends were free. For suppose it sawn through at CD, the weight W hung on there will be just sufficient to break it at E and G. Now restore the connection of the section CD, it will require another weight W to break it there at the same time.

Therefore, when a rafter, or any piece of timber, is firmly connected with three fixed points, G, E, I, it will bear a greater load between any two of them than if its connection with the remote point were removed; and if it be fastened in four points, G, E, I, K, it will be twice as strong in the middle part as without the two remote connections.

One is apt to expect from this that the joist of a floor will be much strengthened by being firmly built in the wall. It is a little strengthened; but the hold which can thus be given to it is much too short to be of any sensible service, and it tends greatly to shatter the wall, because, when it is bent down by a load, it forces up the wall with a momentum of a long lever. Judicious builders therefore take care not to bind the joists tight in the wall. But when the joists of adjoining rooms lie in the same direction, it is a great advantage to make them of one piece. They are then twice as strong as when made in two lengths.

It is easy to deduce from these premises the strain on Inferences. any point which arises from the weight of the beam itself, or from any load which is uniformly diffused over the whole or any part. We may always consider the whole of the weight which is thus uniformly diffused over any part as united in the middle point of that part; and if the load is not uniformly diffused, we may still suppose it united at its centre of gravity. Thus, to know the strain at D arising from the weight of the whole beam, we may suppose the whole weight accumulated in its middle point D. Also the strain at L, arising from the weight of the part ED, is the same as if this weight were accumulated in the middle point d of ED; and it is the same as if half the weight of ED were hung on at D. For the real strain at L is the upward pressure at G, acting by the lever GL. Now, calling the weight of the part DE e, this upward pressure will be $\frac{e \times dE}{EG}$, or $\frac{\frac{1}{2} e \times DE}{EG}$.

Therefore the strain on the middle of a beam, arising from its own weight, or from any uniform load, is the weight of the beam or its load $\times \frac{ED}{EG} \times DG$; that is, half the weight of the beam or load multiplied or acting by the lever DG; for $\frac{ED}{EG}$ is $\frac{1}{2}$.

Also the strain at L, arising from the weight of the beam, or the uniform load, is $\frac{1}{2}$ the weight of the beam or load acting by the lever LG. It is therefore proportional to LG, and is greatest of all at D. Therefore a beam of uniform strength throughout, uniformly loaded, will break in the middle.

It is of importance to know the relation between the strains arising from the weights of the beams, or from any uniformly diffused load, and the relative strength. We have already seen, that the relative strength is $\int \frac{db\,d}{ml}$, where m is a number to be discovered by experiment for every different species of materials. Leaving out every circumstance but what depends on the dimensions of the beam, viz. d, b, and l, we see that the relative strength is in the proportion of $\frac{d^2 b}{l}$, that is, as the breadth and the square of the depth directly, and the length inversely.

Now, to consider, first, the strain arising from the weight of the beam itself, it is evident that this weight increases in the same proportion with the depth, the breadth, and the length of the beam. Therefore its power of resisting this strain must be as its depth directly, and the square of its length inversely. To consider this in a more popular manner, it is plain that the increase of breadth makes no change in the power of resisting the actual strain, because the load and the absolute strength increase in the same proportion with the breadth. But, by increasing the depth, we increase the resisting section in the same proportion, and therefore the number of resisting fibres and the absolute strength; but we also increase the weight in the same proportion. This makes a compensation, and the relative strength is yet the same. But, by increasing the depth, we have not only increased the absolute strength, but also its mechanical energy. For the resistance to fracture is the same as if the full strength of each fibre was exerted at the point which we called the centre of effort; and we showed that the distance of this from the under side of the beam was a certain portion (a half, a third, a fourth, &c.) of the whole depth of the beam. This distance is the arm of the lever, by which the cohesion of the wood may be supposed to act. Therefore this arm of the lever, and consequently the energy of the resistance, increases in the proportion of the depth of the beam, and this remains uncompensated by any increase of the strain. On the whole, therefore, the power of the beam to sustain its own weight increases in the proportion of its depth. But, on the other

hand, the power of withstanding a given strain applied at its extremity, or to any aliquot part of its length, is diminished as the length increases, or is inversely as the length; and the strain arising from the weight of the beam also increases as the length. Therefore the power of resisting the strain actually exerted on it by the weight of the beam is inversely as the square of the length. On the whole, therefore, the power of a beam to carry its own weight varies in the proportion of its depth directly and the square of its length inversely.

As this strain is frequently a considerable part of the whole, it is proper to consider it apart, and then to reckon only on what remains for the support of any extraneous load.

In the next place, the power of a beam to carry any load which is uniformly diffused over its length, must be inversely as the square of the length; for the power of withstanding any strain applied to an aliquot part of the length (which is the case here, because the load may be conceived as accumulated at its centre of gravity, the middle point of the beam) is inversely as the length; and the actual strain is as the length, and therefore its momentum is as the square of the length. Therefore the power of a beam to carry a weight uniformly diffused over it, is inversely as the square of the length.

It is here understood, that the uniform load is of some determined quantity for every foot of the length, so that a beam of double length carries a double load.

We have hitherto supposed that the forces which tend to break a beam transversely are acting in a direction perpendicular to the beam. This is always the case in level floors loaded in any manner; but in roofs, the action of the load tending to break the rafters is oblique, because gravity always acts in vertical lines. It may also frequently happen, that a beam is strained by a force acting obliquely. This modification of the strain is easily discussed. Suppose that the external force, which is measured by the weight W in fig. 1, acts in the direction Aw' instead of AW. Draw Ca' perpendicular to Aw. Then the momentum of this external force is not to be measured by $W \times AC$, but by $W \times a'C$. The strain therefore by which the fibres in the section of fracture DC are torn asunder, is diminished in the proportion of CA to Ca', that is, in the proportion of radius to the sine of the angle CAa', which the beam makes with the direction of the external force.

To apply this to our purpose in the most familiar manner, let AB (fig. 3) be an oblique rafter of a building, loaded with a weight W suspended to any point C, and thereby occasioning a strain in some part D. We have already seen, that the immediate cause of the strain on D is the re-action of the support which is given to the point B. The rafter may at present be considered as a lever, supported at A, and pulled down by the line CW. This occasions a pressure on B, and the support acts in the opposite direction to the action of the lever, that is, in the direction Bb, perpendicular to BA. This tends to break the beam in every part. The pressure exerted at B is $\frac{W \times AE}{AB}$, AE being a horizontal line. Therefore the strain at D will be $\frac{W \times AE}{AB} \times BD$. Had the beam been lying horizontally, the strain at D, from the weight W suspended at C, would have been $\frac{W \times AC}{AB} \times BD$.

Fig. 3.

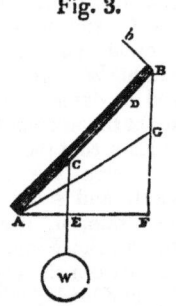

It is therefore diminished in the proportion of AC to AE, that is, in the proportion of radius to the cosine of the elevation, or in the proportion of the secant of elevation to the radius.

It is evident, that this law of diminution of the strain is the same whether the strain arises from a load on any part of the rafter, or from the weight of the rafter itself, or from any load uniformly diffused over its length, provided only that these loads act in vertical lines.

Strength of roofs having different elevations compared.
We can now compare the strength of roofs which have different elevations. Supposing the width of the building to be given, and that the weight of a square yard of covering is also given. Then, because the load on the rafter will increase in the same proportion with its length, the load on the slant-side BA of the roof will be to the load of a similar covering on the half AF of the flat roof, of the same width, as AB to AF. But the transverse action of any load on AB, by which it tends to break it, is to that of the same load on AF as AF to AB. The transverse strain therefore is the same on both, the increase of real load on AB being compensated by the obliquity of its action. But the strengths of beams to resist equal strains, applied to similar points, or uniformly diffused over them, are inversely as their lengths, because the momentum or energy of the strain is proportional to the length. Therefore the power of AB to withstand the strain to which it is really exposed, is to the power of AF to resist its strain as AF to AB. If, therefore, a rafter AG of a certain scantling is just able to carry the roofing laid on it, a rafter AB of the same scantling, but more elevated, will be too weak in the proportion of AG to AB. Therefore steeper roofs require stouter rafters, in order that they may be equally able to carry a roofing of equal weight per square yard. To be equally strong, they must be made broader, or placed nearer to each other, in the proportion of their greater length, or they must be made deeper in the subduplicate proportion of their length. The following easy construction will enable the artist not familiar with computation to proportion the depth of the rafter to the slope of the roof.

Let the horizontal line af (fig. 4) be the proper depth of a beam whose length is half the width of the building; that is, such as would make it fit for carrying the intended tiling laid on a flat roof. Draw the vertical line fb, and the line ab having the elevation of the rafter; make ag equal to af, and describe the semicircle bdg; draw ad perpendicular to ab, then ad is the required depth. The demonstration is evident.

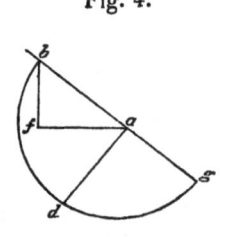

Fig. 4.

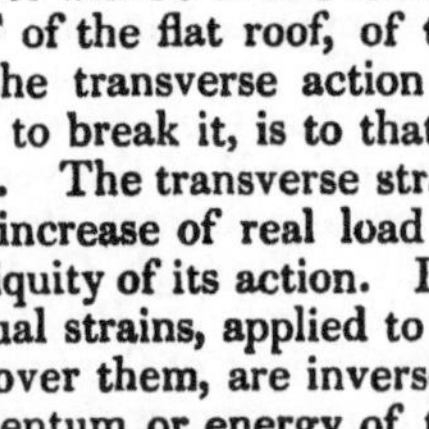

We have now treated in sufficient detail what relates to the chief strain on the component parts of a roof, namely, what tends to break them transversely; and we have enlarged more on the subject than what the present occasion indispensably required, because the propositions which we have demonstrated are equally applicable to all framings of carpentry, and are even of greater moment in many cases, particularly in the construction of machines. These consist of levers in various forms, which are strained transversely; and similar strains frequently occur in many of the supporting and connecting parts. We shall give, in another article, an account of the experiments which have been made by different naturalists, in order to ascertain the absolute strength of some of the materials which are most generally framed together in buildings and engines. The house-carpenter will derive from them absolute numbers, which he can apply to his particular purposes by means of the propositions which we have now established.

We proceed, in the next place, to consider the other strains to which the parts of roofs are exposed, in consequence of the support which they mutually give each other, and the pressures, or *thrusts*, as they are called in the language of the house-carpenter, which they exert on each other, and on the walls or piers of the building.

Let a beam or piece of timber AB (fig. 5) be suspended Effect of other strains, &c by two lines AC, BD; or let it be supported by two props AE, BF, which are perfectly moveable round their remote extremities E, F, or let it rest on the two polished planes KAH, LBM. Moreover, let G be the centre of gravity of the beam, and let GN be a line through the centre of gravity perpendicular to the horizon. The beam will not be in equilibrio unless the vertical line GN either passes through P, the point in which the directions of the two lines AC, BD, or the directions of the two props EA, FB, or the perpendiculars to the two planes KAH, LBM intersect each other, or is parallel to these directions. For the supports given by the lines or props are unquestionably exerted in the direction of their lengths, and it is well known in mechanics that the supports given by planes are exerted in a direction perpendicular to those planes in the points of contact; and we know that the weight of the beam acts in the same manner as if it were all accumulated in its centre of gravity G, and that it acts in the direction GN perpendicular to the horizon. Moreover, when a body is in equilibrio between three forces, they are acting in one plane, and their directions are either parallel or they pass through one point.

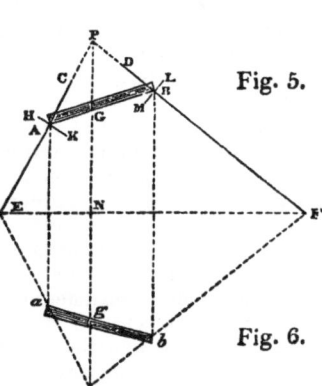

Fig. 5.

Fig. 6.

The support given to the beam is therefore the same as if it were suspended by two lines which are attached to the single point P. We may also infer, that the points of suspension C, D, the points of support E, F, the points of contact A, B, and the centre of gravity G, are all in one vertical plane.

When this position of the beam is disturbed by any external force, there must either be a motion of the points A and B round the centres of suspension C and D, or of the props round these points of support E and F, or a sliding of the ends of the beam along the polished planes KAH and LBM; and in consequence of these motions the centre of gravity G will go out of its place, and the vertical line GN will no longer pass through the point where the directions of the supports intersect each other. If the centre of gravity rises by this motion, the body will have a tendency to recover its former position, and it will require force to keep it away from it. In this case the equilibrium may be said to be *stable*, or the body to have *stability*. But if the centre of gravity descends when the body is moved from the position of equilibrium, it will tend to move still farther; and so far will it be from recovering its former position, that it will now fall. This equilibrium may be called a *tottering equilibrium*. These accidents depend on the situations of the points A, B, C, D, E, F; and they may be determined by considering the subject geometrically. It does not much interest us at present; it is rarely that the equilibrium of suspension is tottering, or that of props is stable. It is evident, that if the beam were suspended by lines from the point P, it would have stability, for it would swing like a pendulum round P, and therefore would always tend towards the position of equilibrium. The intersection of the lines of support would still be at P, and the vertical line drawn through the centre of gravity, when in any other situation, would be on that side of P towards which this centre has been moved. Therefore, by the rules of pendulous bodies, it tends to come back. This would be more remarkably the case if the points of suspension C and D were on the same side of the point P with the points of attachment A and B; for in this case the new point of intersection of the lines of support would shift to the opposite side, and be still further from the vertical line through the new position of the centre of gravity. But if the points of suspension and of attachment are on opposite sides of P, the new point of intersection may shift to the same side with the centre of gravity, and lie beyond the vertical line. In this case the equilibrium is tottering. It is easy to perceive, too, that if the equilibrium of suspension from the points C and D be stable, the equilibrium on the props AE and BF must be tottering. It is not necessary for our present purpose to engage more particularly in this discussion.

It is plain that, with respect to the mere momentary equilibrium, there is no difference in the support by threads, props, or planes, and we may substitute the one for the other. We shall find this substitution extremely useful, because we easily conceive distinct notions of the support of a body by strings.

Observe farther, that if the whole figure be inverted, and strings be substituted for props, and props for strings, the equilibrium will still obtain. For by comparing fig. 5 with fig. 6, we see that the vertical line through the centre of gravity will pass through the intersection of the two strings or props; and this is all that is necessary for the equilibrium; only it must be observed in the substitution of props for threads, and of threads for props, that if it be done without inverting the whole figure, a stable equilibrium becomes a tottering one, and *vice versa*.

Examples. This is a most useful proposition, especially to the unlettered artisan, and enables him to make a practical use of problems which the greatest mechanical geniuses have found it no easy task to solve. An instance will show the extent and utility of it. Suppose it were required to make a mansard or kirb roof whose width is AB (fig. 7), and consisting of the four equal rafters AC, CD, DE, EB. There can be no doubt but that its best form is that which will put all the parts in equilibrio, so that no ties or stays may be necessary for opposing the unbalanced thrust of any part of it. Make a chain *acdeb* (fig. 8) of four equal pieces, loosely connected by pin-joints, round which the parts are perfectly moveable. Suspend this from two pins *a*, *b*, fixed in a horizontal line. This chain or festoon will arrange itself in such a form that its parts are in equilibrio. Then we know that if the figure be inverted, it will compose the frame or truss of a kirb-roof *aγδεb*, which is also in equilibrio, the thrusts of the pieces balancing each other in the same manner that the mutual pulls of the hanging festoon *acdeb* did. If the proportion of the height *df* to the width *ab* is not such as pleases, let the pins *a*, *b* be placed nearer or more distant, till a proportion between the width and height is obtained which pleases, and then make

Fig. 7.

Fig. 8.

the figure ACDEB, fig. 7, similar to it. It is evident that this proposition will apply in the same manner to the determination of the form of an arch of a bridge; but this is not a proper place for a further discussion.

We are now enabled to compute all the thrusts and other pressures which are exerted by the parts of a roof on each other and on the walls. Let AB (fig. 9) be a beam standing anyhow obliquely, and G its centre of gravity. Let us suppose that the ends of it are supported in any directions AC, BD, by strings, props, or planes. Let these directions meet in the point P of the vertical line PG passing through its centre of gravity. Through G draw lines G*a*, G*b* parallel to PB, PA. Then

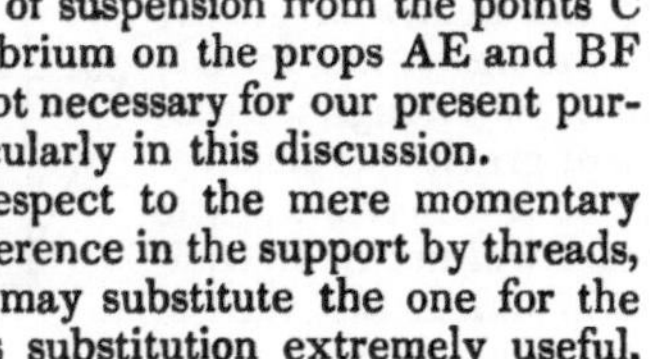

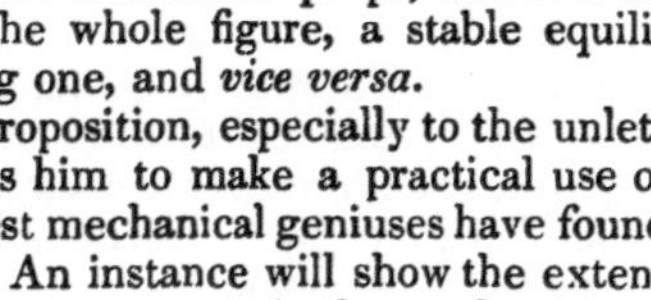

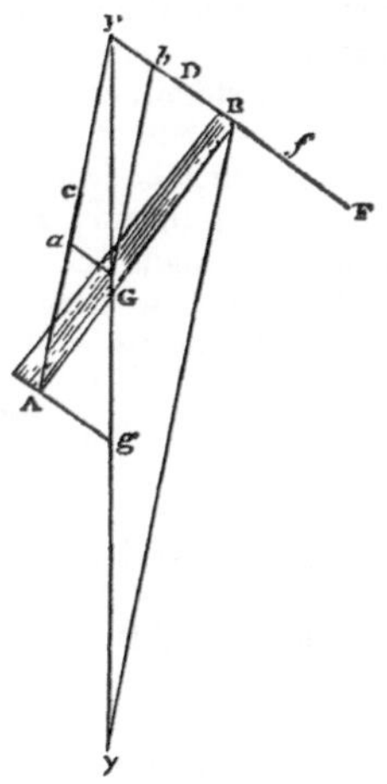

$$\left.\begin{array}{l}\text{The weight of the beam}\\ \text{The pressure or thrust at A}\\ \text{And the pressure at B}\end{array}\right\}\text{is proportional to}\left\{\begin{array}{l}\text{PG}\\ \text{P}a\\ \text{P}b.\end{array}\right.$$

For when a body is in equilibrio between three forces, these forces are proportional to the sides of a triangle which have their directions.

In like manner, if A*g* be drawn parallel to P*b*, we shall have

$$\left.\begin{array}{l}\text{Weight of the beam}\\ \text{Thrust on A}\\ \text{And thrust on B}\end{array}\right\}\text{proportional to}\left\{\begin{array}{l}\text{P}g\\ \text{PA}\\ \text{A}g\end{array}\right.$$

Or, drawing B*γ* parallel to P*a*,

$$\left.\begin{array}{l}\text{Weight of the beam}\\ \text{Thrust at A}\\ \text{And thrust at B}\end{array}\right\}\text{proportional to}\left\{\begin{array}{l}\text{P}\gamma\\ \text{B}\gamma\\ \text{PB}.\end{array}\right.$$

It cannot be disputed that, if strength alone be considered, the proper form of a roof is that which puts the whole in equilibrio, so that it would remain in that shape although all the joints were perfectly loose or flexible. If it has any other shape, additional ties or braces are necessary for preserving it, and the parts are unnecessarily strained. When this equilibrium is obtained, the rafters which compose the roof are all acting on each other in the direction of their lengths; and by this action, combined with their weights, they sustain no strain but that of compression, the strain of all others that they are the most able to resist. We may consider them as so many inflexible lines having their weights accumulated in their centres of gravity. But it will allow an easier investigation of the subject, if we suppose the weights to be at the joints, equal to the real vertical pressures which are exerted on these points. These are very easily computed; for it is plain, that the weight of the beam AB (fig. 9) is to the part of this weight that is supported at B as AB to AG. Therefore, if W represent the weight of the beam, the vertical pressure at B will be $W \times \dfrac{AG}{AB}$, and the vertical pressure at A will be $W \times \dfrac{BG}{AB}$. In like manner, the prop BF being considered as another beam, and *f* as its centre of gravity and *w* as its weight, a part of this weight, equal to $w \times \dfrac{fF}{BF}$, is supported at B, and the whole vertical pressure at B is $W \times \dfrac{AG}{AB} + w \times \dfrac{fF}{BF}$. And thus we greatly simplify the consideration of the mutual thrusts of roof frames. We need hard-

The proper form of a roof is that which puts the whole in equilibrio.

Fig 9.

R O O F.

ly observe, that although these pressures by which the parts of a frame support each other in opposition to the vertical action of gravity, are always exerted in the direction of the pieces, they may be resolved into pressures acting in any other direction which may engage our attention.

All that we propose to deliver on this subject at present may be included in the following proposition.

Let ABCDE (fig. 10) be an assemblage of rafters in a

Fig. 10.

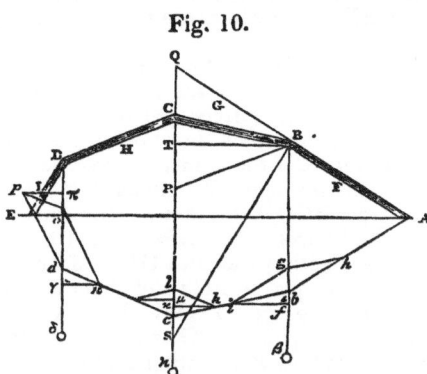

vertical plane, resting on two fixed points A and E in a horizontal line, and perfectly moveable round all the joints A, B, C, D, E; let it be further supposed to be in equilibrio, and let us investigate what adjustment of the different circumstances of weight and inclination of its different parts is necessary for producing this equilibrium.

Let F, G, H, I, be the centres of gravity of the different rafters, and let these letters express the weights of each. Then, by what has been said above, the weight which presses B directly downwards is $F \times \dfrac{AF}{AB} + G \times \dfrac{CG}{BC}$.

The weight on C is in like manner $G \times \dfrac{BG}{BC} + D \times \dfrac{DH}{CD}$,

and that on D is $H \times \dfrac{CH}{CD} + I \times \dfrac{EI}{DE}$.

Let $AbcdE$ be the figure ABCDE inverted, in the manner already described. It may be conceived as a thread fastened at A and E, and loaded at b, c, and d with the weights which are really pressing on B, C, and D. It will arrange itself into such a form that all will be in equilibrio. We may discover this form by means of this single consideration, that any part bc of the thread is equally stretched throughout in the direction of its length. Let us therefore investigate the proportion between the weight β, which we suppose to be pulling the point b in the vertical direction $b\beta$, to the weight δ, which is pulling down the point d in a similar manner. It is evident, that since AE is a horizontal line, and the figures $AbcdE$ and ABCDE equal and similar, the lines Bb, Cc, Dd, are vertical. Take bf to represent the weight hanging at b By stretching the threads bA and bc it is set in opposition to the contractile powers of the threads, acting in the directions bA and bc, and it is in immediate equilibrio with the equivalent of these two contractile forces. Therefore make bg equal to bf, and make it the diagonal of a parallelogram $hbig$. It is evident that bh, bi, are the forces exerted by the threads bA, bc. Then, seeing that the thread bc is equally stretched in both directions, make ck equal to bi; ck is the contractile force which is excited at c by the weight which is hanging there. Draw kl parallel to cd, and lm parallel to bc. The

force lc is the equivalent of the contractile forces ck, cm, and is therefore equal and opposite to the force of gravity acting at C. In like manner, make $dn = cm$, and complete the parallelogram $ndpo$, having the vertical line od for its diagonal. Then dn and dp are the contractile forces excited at d, and the weight hanging there must be equal to od.

Therefore, the load at b is to the load at d as bg to do. But we have seen that the compressing forces at B, C, D may be substituted for the extending forces at b, c, d. Therefore the weights at B, C, D which produce the compressions, are equal to the weights at b, c, d which produce the extensions. Therefore

$$bg : do = F \times \frac{AF}{AB} + G \times \frac{CG}{BC} : H \times \frac{CH}{CD} + I \times \frac{EI}{DE}.$$

Let us inquire what relation there is between this proportion of the loads upon the joints at B and D, and the angles which the rafters make at these joints with each other, and with the horizon or the plumb-lines. Produce AB till it cut the vertical Cc in Q; then draw BR parallel to CD, and BS parallel to DE. The similarity of the figures ABCDE and $AbcdE$, and the similarity of their position with respect to the horizontal and plumb lines, show, without any further demonstration, that the triangles QCB and gbi are similar, and that $QB : BC = gi : ib = hb : ib$. Therefore QB is to BC as the contractile force exerted by the thread Ab to that exerted by bc; and therefore QB is to BC as the compression on BA to the compression on BC.[1] Then, because bi is equal to ck, and the triangles CBR and ckl are similar, $CB : BR = ck : kl = ck : cm$, and CB is to BR as the compression on CB to the compression on CD. And, in like manner, because $cm = dn$, we have BR to BS as the compression on DC to the compression on DE. Also $BR : RS = nd : do$, that is, as the compression on DC to the load on D. Finally, combining all these ratios,

$$QC : CB = gb : bi = gb : kc,$$
$$CB : BR = kc : kl = kc : dn,$$
$$BR : BS = nd : no = dn : no,$$
$$BS : RS = no : do = no : do, \text{ we have finally}$$
$$QC : RS = gb : od = \text{load at B : load at D.}$$

Now

$$QC : BC = \sin. QBC : \sin. BQC = \sin. ABC : \sin. AB b,$$
$$BC : BR = \sin. BRC : \sin. BCR = \sin. CD d : \sin. bBC,$$
$$BR : RS = \sin. BSR : \sin. RBS = \sin. dDE : \sin. CDE.$$

Therefore

$$QC : RS = \sin. ABC \sin. CD d \sin. dDE : \sin. CDE \sin. AB b \sin. bBC.$$

Or

$$QC : RS = \frac{\sin. ABC}{\sin. AB b \sin. CB b} : \frac{\sin. CDE}{\sin. dDC \sin. dDE}.$$

That is, the loads on the different joints are as the sines of the angles at these joints directly, and as the products of the sines of the angles which the rafters make with the plumb-lines inversely.

Or, the loads are as the sines of the angles of the joints directly, and as the products of the cosines of the angles of elevation of the rafters inversely.

Or, the loads at the joints are as the sines of the angles at the joints directly, and as the products of the secants of the angles of elevation of the rafters jointly; for the secants of angles are inversely as the cosines.

Draw the horizontal line BT. It is evident, that if this be considered as the radius of a circle, the lines BQ, BC, BR, BS are the secants of the angles which these lines

---

[1] This proportion might have been shown directly without any use of the inverted figure, or consideration of contractile forces; but the substitution gives distinct notions of the mode of acting, even to persons not much conversant in such disquisitions; and we wish to make it familiar to the mind, because it gives an easy solution of the most complicated problems, and furnishes the practical carpenter, who has little science, with solutions of the most difficult cases by experiment. A festoon, as we called it, may easily be made; and we are certain that the forms into which it will arrange itself are models of perfect frames.

Roof.

make with the horizon; and they are also as the thrusts of those rafters to which they are parallel. Therefore, the thrust which any rafter makes in its own direction is as the secant of its elevation.

The horizontal thrust is the same at all the angles. For $ii = kx = mu = nv = p\pi$. Therefore both walls are equally pressed out by the weight of the roof. We can find its quantity by comparing it with the load on one of the joints. Thus, QC : CB = sin. ABC : sin. ABb

        BC : BT = rad. : sin. BCT = rad. : sin. CBb.

Therefore, QC : BT = rad. × sin. ABC : sin. bBA × sin. bBC.

The length of the beams depends on the weights at the angles.

It deserves remark, that the lengths of the beams do not affect either the proportion of the load at the different joints, or the position of the rafters. This depends merely on the weights at the angles. If a change of length affects the weight, it indeed affects the form also; and this is generally the case. For it seldom happens, indeed it never should happen, that the weight on rafters of longer bearing is not greater. The covering alone increases nearly in the proportion of the length of the rafter.

If the proportion of the weights at B, C, and D is given, as also the position of any two of the lines, the position of all the rest is determined. If the horizontal distances between the angles are all equal, the forces on the different angles are proportional to the verticals drawn on the lines through these angles from the adjoining angle, and the thrusts from the adjoining angles are as the lines which connect them. If the rafters themselves are of equal lengths, the weights at the different angles are as these verticals and as the secants of the angles of elevation of the rafters jointly.

Practical inferences.

This proposition is very fruitful in its practical consequences. It is easy to perceive that it contains the whole theory of the construction of arches; for each stone of an arch may be considered as one of the rafters of this piece of carpentry, since all is kept up by its mere equilibrium. We may have an opportunity in a future article of exhibiting some very elegant and simple solutions of the most difficult cases of this important problem; and we now proceed to make use of the knowledge we have acquired for the construction of roofs.

To determine the best form of a kirb-roof.

We mentioned by the by a problem which is not unfrequent in practice, to determine the best form of a kirb-roof. M. Couplet of the Royal Academy of Paris has given a solution of it in an elaborate memoir in 1726, occupying several lemmas and theorems.

Let AE (fig. 11) be the width, and CF the height; it

Fig. 11.

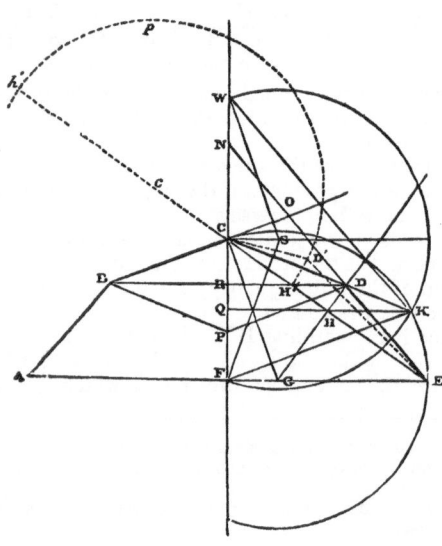

is required to construct a roof ABCDE, whose rafters AB, BC, CD, DE, are all equal, and which shall be in equilibrio.

Roof.

Draw CE, and bisect it perpendicularly in H by the line DHG, cutting the horizontal line AE in G. About the centre G, with the distance GE, describe the circle EKC. It must pass through C, because CH is equal to HE and the angles at H are equal. Draw HK parallel to FE, cutting the circumference in K; draw CK, cutting GH in D; and join CD, ED. These lines are the rafters of half of the roof required.

We prove this by showing that the loads at the angles C and D are equal; for this is the proportion which results from the equality of the rafters, and the extent of surface of the uniform roofing which they are supposed to support. Therefore produce ED till it meet the vertical FC in N; and having made the side CBA similar to CDE, complete the parallelogram BCDP, and draw DB, which will bisect CP in R, as the horizontal line KH bisects CF in Q. Draw KF, which is evidently parallel to DP. Make CS perpendicular to CF, and equal to FG; and about S, with the radius SF, describe the circle FKW. It must pass through K, because SF is equal to CG, and CQ = QF. Draw WK, WS, and produce BC, cutting ND in O.

The angle WKF at the circumference is one half of the angle WSF at the centre, and is therefore equal to WSC or CGF. It is therefore double of the angle CEF or ECS. But ECS is equal to ECD and DCS, and ECD is one half of NDC, and DCS is one half of DCO or CDP. Therefore the angle WKF is equal to NDP, and WK is parallel to ND, and CF is to CW as CP to CN; and CN is equal to CP. But it has been shown above that CN and CP are as the loads upon D and C. These are therefore equal, and the frame ABCDE is in equilibrio.

A comparison of this solution with that of M. Couplet will show its great advantage in respect of simplicity and perspicuity; and the intelligent reader can easily adapt the construction to any proportion between the rafters AB and BC, which other circumstances, such as garret-rooms, &c. may render convenient. The construction must be such that NC may be to CP as CD to $\dfrac{CD + DE}{2}$. Whatever proportion of AB to BC is assumed, the point D′ will be found in the circumference of a semicircle H′D′h', whose centre is in the line CE, and having AB : BC = CH′ : H′E $= ch'$: h'E. The rest of the construction is simple.

In buildings which are roofed with slate, tile, or shingles, the circumstance which is most likely to limit the construction is the slope of the upper rafters CB, CD. This must be sufficient to prevent the penetration of rain, and the stripping by the winds. The only circumstance left in our choice in this case is the proportion of the rafters AB and BC. Nothing is easier than making NC to CP in any desired proportion when the angle BCD is given.

The truss for a roof should always be in equilibrio.

We need not repeat that it is always a desirable thing to form a truss for a roof in such a manner that it shall be in equilibrio. When this is done, the whole force of the struts and braces which are added to it is employed in preserving this form, and no part is expended in unnecessary strains. For we must now observe, that the equilibrium of which we have been treating is always of that kind which we call the tottering, and the roof requires stays, braces, or hanging timbers, to give it stiffness, or keep it in shape. We have also said enough to enable any reader acquainted with the most elementary geometry and mechanics, to compute the transverse strains and the thrusts to which the component parts of all roofs are exposed.

General maxims for all roofs.

It only remains now to show the general maxims by which all roofs must be constructed, and the circumstances which determine their excellence. In doing this we shall be exceedingly brief, and almost content ourselves with exhibiting the principal forms, of which the endless variety of roofs are only slight modifications. We shall not trouble the reader with any account of such roofs as receive part of their support from the interior walls, but confine ourselves to the more difficult problem of throwing a roof over a wide building, without any intermediate support; because when such roofs are constructed in the best manner, that is, deriving the greatest possible strength from the materials employed, the best construction of the others is necessarily included. For all such roofs as rest upon the middle walls are roofs of smaller bearing. The only exception deserving notice is the roofs of churches, which have aisles separated from the nave by columns. The roof must rise on these. But if it is of an arched form internally, the horizontal thrusts must be nicely balanced, that they may not push the columns aside.

Simplest notion of a roof.

The simplest notion of a roof-frame is, that it consists of two rafters AB and BC (fig. 12), meeting in the ridge.

Fig. 12.

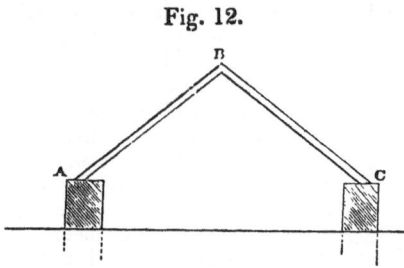

But even this simple form is susceptible of better and worse. We have already seen, that when the weight of a square yard of covering is given, a steeper roof requires stronger rafters, and that when the scantling of the timbers is also given, the relative strength of a rafter is inversely as its length.

Best form of rafters.

But there is now another circumstance to be taken into the account, viz. the support which one rafter leg gives to the other. The best form of a rafter will therefore be that in which the relative strength of the legs, and their mutual support, give the greatest product. Mr Muller, in his Military Engineer, gives a determination of the best pitch of a roof, which has considerable ingenuity, and has been copied into many books of military education both in this island and on the continent. Describe on the width AC (fig. 13) the semicircle AFC, and bisect it by the radius FD. Produce the rafter AB to the circumference in E, join EC, and draw the perpendicular EG. Now $AB : AD$

Fig. 13.

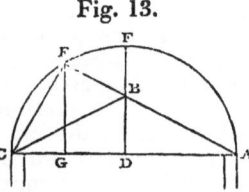

$= AC : AE$, and $AE = \dfrac{AD \times AC}{AB}$, and AE is inversely

as AB, and may therefore represent its strength in relation to the weight actually lying on it. Also the support which CB gives to AB is as CE, because CE is perpendicular to AB. Therefore the form which renders $AE \times EC$ a maximum seems to be that which has the greatest strength. But $AC : AE = EC : EG$, and $EG = \dfrac{AE \times EC}{AC}$, and is therefore proportional to $AE \times EC$. Now EG is a maximum when B is in F, and a square pitch is in this respect the strongest. But it is very doubtful whether this construction is deduced from just principles. There is another strain to which the leg AB is exposed, which is not taken into the account. This arises from the curvature which it unavoidably acquires by the transverse pressure of its load. In this state it is pressed in its own direction by the abutment and load of the other leg. The relation be-

tween this strain and the resistance of the piece is not very distinctly known. Euler has given a dissertation on this subject, which is of great importance, because it affects posts and pillars of all kinds; and it is very well known that a post of ten feet long and six inches square will bear with great safety a weight which would crush a post of the same scantling and twenty feet long in a minute; but his determination has not been acquiesced in by the first mathematicians. Now it is in relation to these two strains that the strength of the rafter should be adjusted. The firmness of the support given by the other leg is of no consequence, if its own strength is inferior to the strain. The force which tends to crush the leg AB, by compressing it in its curved state, is to its weight as AB to BD, as is easily seen by the composition of forces; and its incurvation by this force has a relation to it, which is of intricate determination. It is contained in the properties demonstrated by Bernoulli of the elastic curve. This determination also includes the relation between the curvature and the length of the piece. But the whole of this seemingly simple problem is of much more difficult investigation than Mr Muller was aware of; and his rules for the pitch of a roof, and for the sally of a dock-gate, which depends on the same principles, are of no value. He is, however, the first author who attempted to solve either of these problems on mechanical principles susceptible of precise reasoning. Belidor's solutions, in his *Architecture Hydraulique*, are below notice.

Reasons of economy have made carpenters prefer a low pitch; and although this does diminish the support given by the opposite leg faster than it increases the relative strength of the other, it is not of material consequence, because the strength remaining in the opposite leg is still very great; for the supporting leg is acting against compression, in which case it is vastly stronger than the supported leg acting against a transverse strain.

Thrust on the walls.

But a roof of this simplicity will not do in most cases. There is no notice taken, in its construction, of the thrust which it exerts on the walls. Now this is the strain which is the most hazardous of all. Our ordinary walls, instead of being able to resist any considerable strain pressing them outwards, require, in general, some ties to keep them on foot. When a person thinks of the thinness and height of the walls of even a strong house, he will be surprised that they are not blown down by any strong puff of wind. A wall three feet thick, and sixty feet high, could not withstand a wind blowing at the rate of thirty feet per second (in which case it acts with a force considerably exceeding two pounds on every square foot), if it were not stiffened by cross walls, joists, and roof, which all help to tie the different parts of the building together.

How this is avoided.

A carpenter is therefore exceedingly careful to avoid every horizontal thrust, or to oppose them by other forces. And this introduces another essential part into the construction of a roof, namely, the *tie* or *beam* AC (fig. 14), laid from wall to wall, binding the feet A and C of the rafters together. This is the sole office of the beam; and it should be considered in no other light than as a string to prevent the roof from pushing out the walls. It is indeed used for carrying the ceiling of the apartments under it, and it is even made to support a flooring. But, considered as making part of a roof, it is merely a string; and the strain which it withstands tends to tear its parts asunder. It therefore acts with its whole absolute force, and a very small scantling would suffice if we could contrive to fasten it firmly enough to the foot of the rafter. If it is of oak, we may safely subject it to a strain of three tons for every square inch of its sec-

Fig. 14.

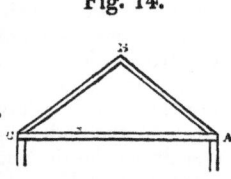

tion. And fir will safely bear a strain of two tons for every square inch. But we are obliged to give the tie-beam much larger dimensions, that we may be able to connect it with the foot of the rafter by a mortise and tenon. Iron straps are also frequently added. By attending to this office of the tie-beam, the judicious carpenter is directed to the proper form of the mortise and tenon, and of the strap. We shall consider both of these in a proper place, after we become acquainted with the various strains at the joints of a roof.

These large dimensions of the tie-beam allow us to load it with the ceilings without any risk, and even to lay floors on it with moderation and caution. But when it has a great bearing or span, it is very apt to bend downwards in the middle, or, as the workmen term it, to sway or swag; and it requires a support. The question is, where to find this support. What fixed points can we find with which to connect the middle of the tie-beam? Some ingenious carpenter thought of suspending it from the ridge by a piece of timber BD (fig. 15), called by our carpenters the *king-post*. It must be acknowledged, that there was very great ingenuity in this thought. It was also perfectly just. For the weight of the rafters BA, BC tends to make them fly out at the foot. This is prevented by the tie-beam, and this excites a pressure, by which they tend to compress each other. Suppose them without weight, and that a great weight is laid on the ridge B. This can be supported only by the abutting of the rafters in their own directions AB and CB, and the weight tends to compress them in the opposite directions, and, through their intervention, to stretch the tie-beam. If neither the rafters can be compressed, nor the tie-beam stretched, it is plain that the triangle ABC must retain its shape, and that B becomes a fixed point very proper to be used as a point of suspension. To this point, therefore, is the tie-beam suspended by means of the king-post. A common spectator unacquainted with carpentry views it very differently, and the tie-beam appears to him to carry the roof. The king-post appears a pillar resting on the beam, whereas it is really a string; and an iron rod of one sixteenth of the size would have done just as well. The king-post is sometimes mortised into the tie-beam, and pins put through the joint, which gives it more the look of a pillar with the roof resting on it. This does well enough in many cases. But the best method is to connect them by an iron strap like a stirrup, which is bolted at its upper ends into the king-post, and passes round the tie-beam. In this way a space is commonly left between the end of the king-post and the upper side of the tie beam. Here the beam plainly appears hanging in the stirrup; and this method allows us to restore the beam to an exact level, when it has sunk by the unavoidable compression or other yielding of the parts. The holes in the sides of the iron strap are made oblong instead of round; and the bolt which is drawn through all is made to taper on the under side; so that driving it farther draws the tie-beam upwards. A notion of this may be formed by looking at fig. 16, which is a section of that post and beam.

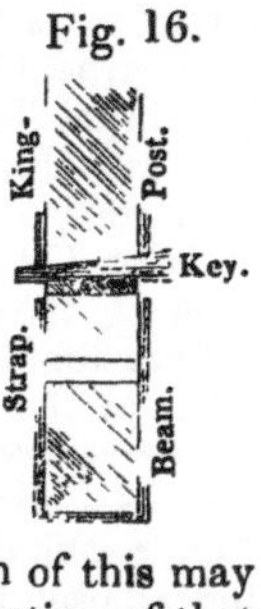

It requires considerable attention, however, to make this suspension of the tie-beam sufficiently firm. The top of the king-post is cut into the form of the arch-stone of a bridge, and the heads of the rafters are firmly mortised into this projecting part. These projections are called joggles, and are formed by working the king-post out of a much larger piece of timber, and cutting off the unnecessary wood from

Fig. 15.

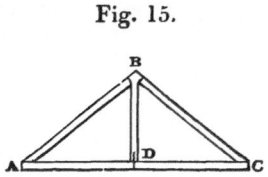

Fig. 16.

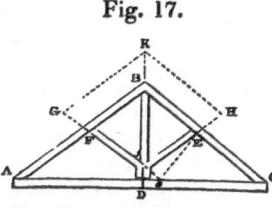

the two sides; and, lest all this should not be sufficient, it is usual in great works to add an iron plate or strap of three branches, which are bolted into the heads of the king-post and rafters.

The rafters, though not so long as the beam, seem to stand as much in need of something to prevent their bending, for they carry the weight of the covering. This cannot be done by suspension, for we have no fixed points above them. But we have now got a very firm point of support at the foot of the king-post. *Braces*, or rather *struts*, ED, FD (fig. 17), are put under the middle of the rafters, where they are slightly mortised, and their lower ends are firmly mortised into joggles formed on the foot of the king-post.

Fig. 17.

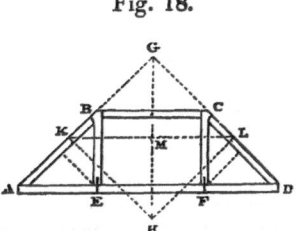

As these braces are very powerful in their resistance to compression, and the king-post equally so to resist extension, the points E and F may be considered as fixed; and the rafters being thus reduced to half their former length, have now four times their former relative strength.

Roofs do not always consist of two sloping sides meeting in a ridge. They have sometimes a flat on the top, with two slopping sides. They are sometimes formed with a double slope, and are called *kirb* or *mansarde roofs*. They sometimes have a valley in the middle, and are then called M roofs. Such roofs require another piece, which may be called the *truss-beam*, because all such frames are called *trusses*, probably from the French word *trousse*, because such roofs are like portions of plain roofs *troussés* or shortened.

A flat-topped roof is thus constructed. Suppose the three rafters AB, BC, CD (fig. 18), of which AB and CD are equal, and BC horizontal. It is plain that they will be in equilibrio, and the roof have no tendency to go on either side. The tie-beam AD withstands the horizontal thrusts of the whole frame, and the two rafters AB and CD are each pressed in their own directions in consequence of their abutting with the middle rafter or truss-beam BC. It lies between them like the key-stone of an arch. They lean towards it, and it rests on them. The pressure which the truss-beam and its load excites on the two rafters is the very same as if the rafters were produced till they meet in G, and a weight were laid on these equal to that of BC and its load. If therefore the truss-beam is of a scantling sufficient for carrying its own load, and withstanding the compression from the two rafters, the roof will be equally strong, whilst it keeps its shape, as the plain roof AGD, furnished with the king-post and braces. We may conceive this another way. Suppose a plain roof AGD, without braces to support the middle B and C of the rafters. Then let a beam BC be put in between the rafters, abutting upon little notches cut in the rafters. It is evident that this must prevent the rafters from bending downwards, because the points B and C cannot descend, moving round the centres A and D, without shortening the distance BC between them. This cannot be without compressing the beam BC. It is plain that BC may be wedged in, or wedges driven in between its ends B and C and the notches in which it is lodged. These wedges may be driven in till they even force out the rafters GA and GD. Whenever this happens, all the mutual pressure of the heads of these rafters at G is taken away, and the parts GB and GC may be cut away, and the

Fig. 18.

Construc-
tion of flat-
topped
roofs.

roof ABCD will be as strong as the roof AGD furnished with the king-post and braces, because the truss-beam gives a support of the same kind at B and C as the brace would have done.

But this roof ABCD would have no firmness of shape. Any addition of weight on one side would destroy the equilibrium at the angle, would depress that angle, and would cause the opposite one to rise. To give it stiffness, it must either have ties or braces, or something partaking of the nature of both. The usual method of framing is to make the heads of the rafters abut on the joggles of two side-posts BE and CF, whilst the truss-beam, or strut as it is generally termed by the carpenters, is mortised square into the inside of the heads. The lower ends E and F of the side-posts are connected with the tie-beam either by mortises or straps.

This construction gives firmness to the frame; for the angle B cannot descend in consequence of any inequality of pressure, without forcing the other angle C to rise. This it cannot do, being held down by the post CF. And the same construction fortifies the tie-beam, which is now suspended at the points E and F from the points B and C, whose firmness we have just now shown.

They are not so strong as the plain roofs.
But although this roof may be made abundantly strong, it is not quite so strong as the plain roof AGD of the same scantling. The compression which BC must sustain in order to give the same support to the rafters at B and C that was given by braces properly placed, is considerably greater than the compression of the braces. And this strain is an addition to the transverse strain which BC gets from its own load. This form also necessarily exposes the tie-beam to cross strains. If BE is mortised into the tie-beam, then the strain which tends to depress the angle ABC presses on the tie-beam at E transversely, whilst a contrary strain acts on F, pulling it upwards. These strains, however, are small; and this construction is frequently used, being susceptible of sufficient strength, without much increase of the dimensions of the timbers; and it has the great advantage of giving free room in the garrets.

Were it not for this, there is a much more perfect form represented in fig. 19. Here the two posts BE, CF are united below. All transverse action on the tie-beam is now entirely removed. We are

Fig. 19.

almost disposed to say that this is the strongest roof of the same width and slope. For if the iron strap which connects the pieces BE, CF with the tie-beam have a large bolt G through it, confining it to one point of the beam, there are five points, A, B, C, D, G, which cannot change their places, and there is no transverse strain in any of the connections.

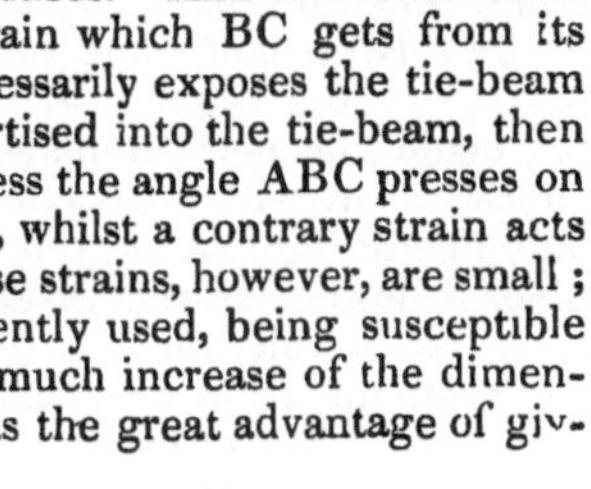

When the dimensions of the building are very great, so that the pieces AB, BC, CD, would be thought too weak for withstanding the cross strains, braces may be added as is expressed in fig. 18 by the dotted lines. The reader will observe, that it is not meant to leave the top flat externally; it must be raised a little in the middle, to shoot off the rain. But this must not be done by incurvating the beam BC. This would soon be crushed, and spring upwards. The slopes must be given by pieces of timber added above the strutting-beam.

Members of which the frame of a roof consists.
And thus we have completed a frame of a roof. It consists of these principal members: the rafters, which are immediately loaded with the covering; the tie-beam, which withstands the horizontal thrust by which the roof tends to fly out below and push out the walls; the king-posts, which hang from fixed points and serve to uphold the tie-beam, and also to afford other fixed points on which we may rest the braces which support the middle of the rafters; and,

lastly, the truss or strutting-beam, which serves to give mutual abutment to the different parts which are at a distance from each other. The rafters, braces, and trusses are exposed to compression, and must therefore have not only cohesion, but stiffness. For if they bend, the prodigious compressions to which they are subjected would quickly crush them in this bended state. The tie-beams and king-posts, if performing no other office but supporting the roof, do not require stiffness; and their places might be supplied by ropes, or by rods of iron of one-tenth part of the section that even the smallest oak stretcher requires. These members require no greater dimensions than what is necessary for giving sufficient joints, and any more is a needless expense. All roofs, however complicated, consist of these essential parts; and if pieces of timber are to be seen which perform none of these offices, they must be pronounced useless, and they are frequently hurtful, by producing cross strains in some other piece. In a roof properly constructed there should be no such strains. All the timbers, excepting those which immediately carry the covering, should be either pushed or drawn in the direction of their length. And this is the rule by which a roof should always be examined.

They are susceptible of numberless combinations and varieties.
These essential parts are susceptible of numberless combinations and varieties. But it is a prudent maxim to make the construction as simple, and consisting of as few parts, as possible. We are the less exposed to the imperfections of workmanship, such as loose joints, &c. Another essential harm arises from many pieces, by the compression and the shrinking of the timber in the cross direction of the fibres. The effect of this is equivalent to the shortening of the piece which abuts on the joint. This alters the proportions of the sides of the triangle on which the shape of the whole depends. Now, in a roof such as fig. 18, there is twice as much of this as in the plain pent-roof, because there are two posts. And when the direction of the abutting pieces is very oblique to the action of the load, a small shrinking permits a great change of shape. Thus, in a roof of what is called pediment pitch, where the rafters make an angle of thirty degrees with the horizon, half an inch compression of the king-post will produce a sagging of an inch, and occasion a great strain on the tie-beam, if the posts are mortised into it.

This method of including a truss within the rafters of a pent-roof is a very considerable addition to the art of carpentry. But to insure its full effect, it should always be executed with abutting rafters under the principal ones, abutting on joggles in the heads of the posts. Without this the strut-beam is hardly of any service. We would therefore recommend fig. 20 as a proper construction of a trussed roof; and the king-post which is placed in it may be employed to support the

Fig. 20.

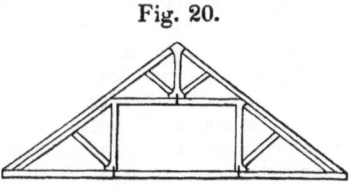

upper part of the rafters, and also for preventing the strut-beam from bending in their direction in consequence of its great compression. It will also give a suspension for the great burdens which are sometimes necessary in a theatre. The machinery has no other firm points to which it can be attached; and the portions of the single rafters which carry this king-post are but short, and therefore may be considerably loaded with safety.

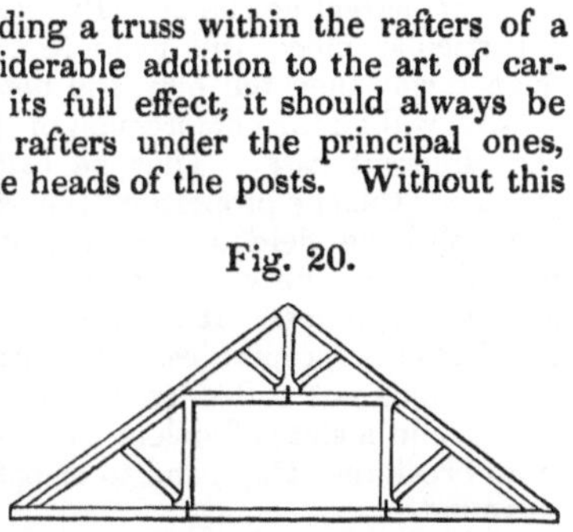

We observe in the drawings which we sometimes have of Chinese buildings, that the trussing of roofs is understood by them. Indeed they must be very experienced carpenters. We see wooden buildings run up to a great height, which can be supported only by such trussing. One of these is sketched in fig. 21. There are some very excellent specimens to be seen in the buildings at Deptford, belonging to the victualling-office, commonly called the *Red-house*, which were erected about the year 1788, and we believe are the performance of Mr James Arrow of the Board of Works, one of the most intelligent artists in this kingdom.

Fig. 21.

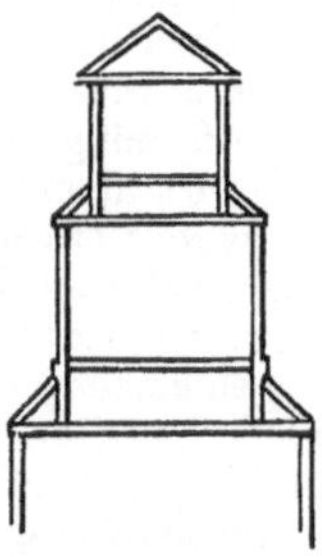

Remarks addressed to practical carpenters.

Thus have we given an elementary, but a rational or scientific, account of this important part of the art of carpentry. It is such, that any practitioner, with the trouble of a little reflection, may always proceed with confidence, and without resting any part of his practice on the vague notions which habit may have given him of the strength and supports of timbers, and of their manner of acting. That these frequently mislead, is proved by the mutual criticisms which are frequently published by the rivals in the profession. They have frequently sagacity enough, for it seldom can be called science, to point out glaring blunders; and any person who will look at some of the performances of Mr Price, Mr Wyatt, Mr Arrow, and others of acknowledged reputation, will readily see them distinguishable from the works of inferior artists by simplicity alone. A man without principles is apt to consider an intricate construction as ingenious and effectual; and such roofs sometimes fail merely by being ingeniously loaded with timber, but still more frequently by the wrong action of some useless pieces, which produces strains that are transverse to other pieces, or which, by rendering some points too firm, cause them to be deserted by the rest in the general subsiding of the whole. Instances of this kind are pointed out by Price in his British Carpenter. Nothing shows the skill of a carpenter more than the distinctness with which he can foresee the changes of shape which must take place in a short time in every roof. A knowledge of this will often correct a construction which the mere mathematician thinks unexceptionable, because he does not reckon on the actual compression which must obtain, and imagines that his triangles, which sustain no cross strains, invariably retain their shape till the pieces break. The sagacity of the experienced carpenter is not, however, enough without science for perfecting the art. But when he knows how much a particular piece will yield to compression in one case, science will then tell him, and nothing but science can do it, what will be the compression of the same piece in another very different case. Thus he learns how far it will now yield, and then he proportions the parts so to each other, that when all have yielded according to their strains, the whole is of the shape he wished to produce, and every joint is in a state of firmness. It is here that we observe the greatest number of improprieties. The iron straps are frequently in positions not suited to the actual strain on them; and they are in a state of violent twist, which both tends strongly to break the straps, and to cripple the pieces which they surround.

In like manner, we frequently see joints or mortises in a state of violent strain on the tenons, or on the heels and shoulders. The joints were perhaps properly shaped for the primitive form of the truss; but by its settling, the bearing of the push is changed. The brace, for example, in a very low-pitched roof, comes to press with the upper part of the shoulder, and, acting as a powerful lever on the tenon, breaks it. In like manner, the lower end of the brace, which at first abutted firmly and squarely on the joggle of the king-post, now presses with one corner in prodigious force, and seldom fails to splinter off on that side. We cannot help recommending a maxim of M. Perronet, the celebrated hydraulic architect of France, as a golden rule, viz. to make all the shoulders of abutting pieces in the form of an arch of a circle, having the opposite end of the piece for its centre. Thus, in fig. 18, if the joggle-point B be of

this form, having A for its centre, the sagging of the roof will make no partial bearing at the joint; for in the sagging of the roof the piece AB turns or bends round the centre A, and the counter-pressure of the joggle is still directed to A, as it ought to be. We have just now said *bends* round A. This is too frequently the case, and it is always very difficult to give the tenon and mortise in this place a true and invariable bearing. The rafter pushes in the direction BA, and the beam resists in the direction AD. The abutment should be perpendicular to neither of these, but in an intermediate direction, and it ought also to be of a curved shape. But the carpenters perhaps think that this would weaken the beam too much to give it this shape in the shoulder; they do not even aim at it in the heel of the tenon. The shoulder is commonly even with the surface of the beam. When the bearing therefore is on this shoulder, it causes the foot of the rafter to slide along the beam till the heel of the tenon bears against the outer end of the mortise (See Price's British Carpenter, plate C, fig. IK). This abutment is perpendicular to the beam in Price's book; but it is more generally pointed a little outwards below, to make it more secure against starting. The consequence of this construction is, that when the roof settles, the shoulder comes to bear at the inner end of the mortise, and it rises at the outer, and the tenon, taking hold of the wood beyond it, either tears it out or is itself broken. This joint therefore is seldom trusted to the strength of the mortise and tenon, and is usually secured by an iron strap, which lies obliquely to the beam, to which it is bolted by a large bolt quite through, and then embraces the outside of the rafter foot. This strap is very frequently not made sufficiently oblique, and we have seen some made almost square with the beam. When this is the case, it not only keeps the foot of the rafter from flying out, but it binds it down. In this case, the rafter acts as a powerful lever, whose fulcrum is in the inner angle of the shoulder, and then the strap never fails to cripple the rafter at the point. All this can be prevented only by making the strap very long and very oblique, and by making its outer end (the stirrup part) square with its length, and making a notch in the rafter foot to receive it. It cannot now cripple the rafter, for it will rise along with it, turning round the bolt at its inner end. We have been thus particular on this joint, because it is here that the ultimate strain of the whole roof is exerted, and its situation will not allow the excavation necessary for making it a good mortise and tenon.

Similar attention must be paid to some other straps, such as those which embrace the middle of the rafter, and connect it with the post or truss below it. We must attend to the change of shape produced by the sagging of the roof, and place the strap in such a manner as to yield to it by turning round its bolt, but so as not to become loose, and far less to make a fulcrum for any thing acting as a lever. The strains arising from such actions, in framings of carpentry which change their shape by sagging, are enormous, and nothing can resist them.

We shall close this part of the subject with a simple method, by which any carpenter, without mathematical science, may calculate with sufficient precision the strains or thrusts which are produced on any point of his work, whatever be the obliquity of the pieces.

Mode of calculating strains or thrusts.

Let it be required to find the horizontal thrust acting on the tie-beam AD of fig. 18. This will be the same as if the weight of the whole roof were laid at G on the two rafters GA and GD. Draw the vertical line GH. Then, having calculated the weight of the whole roof that is supported by this single frame ABCD, including the weight of the pieces AB, BC, CD, BE, CF themselves, take the number of pounds, tons, &c. which expresses it from any scale of equal parts, and set it from G to H. Draw HK,

ROOF.

HL parallel to GD, GA, and draw the line KL, which will be horizontal when the two sides of the roof have the same slope. Then ML measured on the same scale will give the horizontal thrust, by which the strength of the tie-beam is to be regulated. GL will give the thrust which tends to crush the rafters, and LM will also give the force which tends to crush the strut-beam BC.

In like manner, to find the strain of the king-post BD of fig. 17, consider that each brace is pressed by half the weight of the roofing laid on BA or BC, and this pressure, or at least its hurtful effect, is diminished in the proportion of BA to DA, because the action of gravity is vertical, and the effect which we want to counteract by the braces is in a direction Ee perpendicular to BA or BC. But as this is to be resisted by the brace fE acting in the direction fE, we must draw fe perpendicular to Ee, and suppose the strain augmented in the proportion of Ee to Ef.

Having thus obtained in tons, pounds, or other measures, the strains which must be balanced at f by the cohesion of the king-post, take this measure from the scale of equal parts, and set it off in the directions of the braces to G and H, and complete the parallelogram GfH K; and fK measured on the same scale will be the strain on the king-post.

Strength of the truss. The artist may then examine the strength of his truss upon this principle, that every square inch of oak will bear at an average 7000 pounds compressing or stretching it, and may be safely loaded with 3500 for any length of time; and that a square inch of fir will in like manner securely bear 2500. And, because straps are used to resist some of these strains, a square inch of well-wrought tough iron may be safely strained by 50,000 pounds. But the artist will always recollect, that we cannot have the same confidence in iron as in timber. The faults of this last are much more easily perceived; and when the timber is too weak, it gives us warning of its failure by yielding sensibly before it breaks. This is not the case with iron; and much of its service depends on the honesty of the blacksmith.

Sketch of some trussed roofs, &c. In this way may any design of a roof be examined. We shall here give the reader a sketch of two or three trussed roofs, which have been executed in the chief varieties of circumstances which occur in common practice.

Fig. 22 is the roof of St Paul's Church, Covent Garden,

Fig. 22.

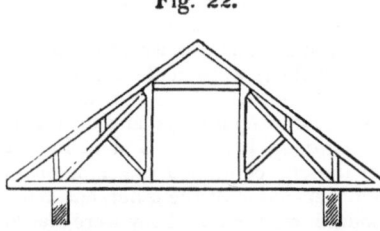

London, the work of Inigo Jones. Its construction is singular. The roof extends to a considerable distance beyond the building, and the ends of the tie-beams support the Tuscan corniche, appearing like the mutules of the Doric order. Such a roof could not rest on the tie-beam. Inigo Jones has therefore supported it by a truss below it; and the height has allowed him to make this extremely strong with very little timber. It is accounted the highest roof of its width in London. But this was not difficult, by reason of the great height which its extreme width allowed him to employ without hurting the beauty of it by too high a pitch. The supports, however, are disposed with judgment.[1]

Fig. 23 is a kirb or mansarde roof by Price, and supposed to be of large dimensions, having braces to carry the middle of the rafters. It will serve exceedingly well for a church having pillars. The middle part of the tie-beam being taken away, the strains are very well balanced, so that there is no risk of its pushing aside the pillar on which it rests.

Fig. 23.

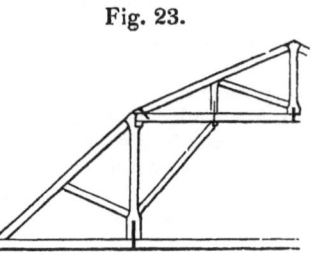

Fig. 24 is the celebrated roof of the Theatre of the Uni-

Fig. 24.

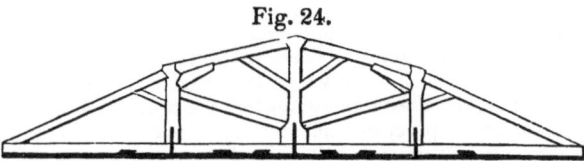

versity of Oxford, by Sir Christopher Wren. The span between the walls is 75 feet. This is accounted a very ingenious, and is a singular performance. The middle part of it is almost unchangeable in its form; but from this circumstance it does not distribute the horizontal thrust with the same regularity as the usual construction. The horizontal thrust on the tie-beam is about twice the weight of the roof, and is withstood by an iron strap below the beam, which stretches the whole width of the building in the form of a rope, making part of the ornament of the ceiling.

Cases in which the thrust cannot be discharged from the walls by the tie-beam. In all the roofs which we have considered hitherto, the thrust is discharged entirely from the walls by the tie-beam. But this cannot always be done. We frequently want great elevation within, and arched ceilings. In such cases, it is a much more difficult matter to keep the walls free of all pressure outwards, and there are few buildings where it is completely done. Yet this is the greatest fault of a roof. We shall just point out the methods which may be most successfully adopted.

We have said that a tie-beam just performs the office of a string. We have said the same of the king-post. Now suppose two rafters AB, BC (fig. 25), moveable about the point B, and resting on the top of the walls. If the line BD be suspended from B, and the two lines DA, DC be fastened to the feet of the rafters, and if these lines be incapable of extension, it is plain that all thrust is removed from the walls as effectually as by a common tie-beam; and by shortening BD to Bd, we gain a greater inside height, and more room for an arched ceiling. Now if we substitute a king-post BD (fig. 26), and two stretchers or hammer-beams DA, DC for the other strings, and connect them firmly by means of iron straps, we obtain our purpose.

Fig. 25.

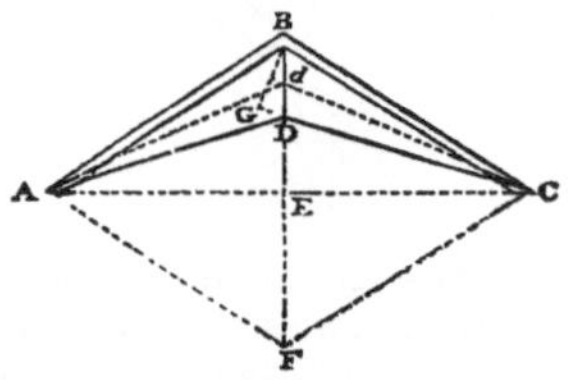

Fig. 26.

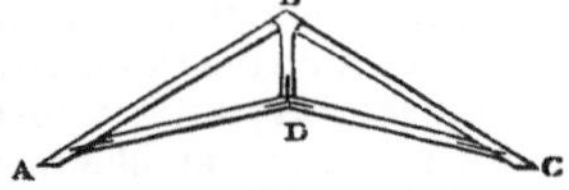

Let us compare this roof with a tie-beam roof in point of strain and strength. Recur to fig. 25, and complete the parallelogram ABCF, and draw the diagonals AC, BF, crossing in E. Draw BG perpendicular to CD. We have

[1] This church was burnt down after the present article was written.

seen that the weight of the roof, which we may call W, is to the horizontal thrust at C as BF to EC; and if we express this thrust by T, we have $T = \dfrac{W \times EC}{BF}$. We may at present consider BC as a lever moveable round the joint B, and pulled at C in the direction EC by the horizontal thrust, and held back by the string pulling in the direction CD. Suppose that the forces in the directions EC and CD are in equilibrio, and let us find the force S by which the string CD is strained. These forces must, by the property of the lever, be inversely as the perpendiculars drawn from the centre of motion on the lines of their direction.

Therefore $BG : BE = T : S$, and $S = T \times \dfrac{BE}{BG} = W \times \dfrac{BE \cdot EC}{BF \cdot BG}$.

Therefore the strain upon each of the ties DA and DC is always greater than the horizontal thrust or the strain on a simple tie-beam. This would be no great inconvenience, because the smallest dimensions that we could give to these ties, so as to procure sufficient fixtures to the adjoining pieces, are always sufficient to withstand this strain. But although the same may be said of the iron straps which make the ultimate connections, there is always some hazard of imperfect work, cracks, or flaws, which are not perceived. We can judge with tolerable certainty of the soundness of a piece of timber, but cannot say so much of a piece of iron. Moreover, there is a prodigious strain excited on the king-post when BG is very short in comparison of BE, namely, the force compounded of the two strains S and S on the ties DA and DC.

But there is another defect from which the straight tie-beam is entirely free. All roofs settle a little. When this roof settles, and the points B and D descend, the legs BA, BC must spread further out, and thus a pressure outwards is excited on the walls. It is seldom therefore that this kind of roof can be executed in this simple form, and other contrivances are necessary for counteracting this supervening action on the walls. Fig. 27 is one of the best which

Fig. 27.

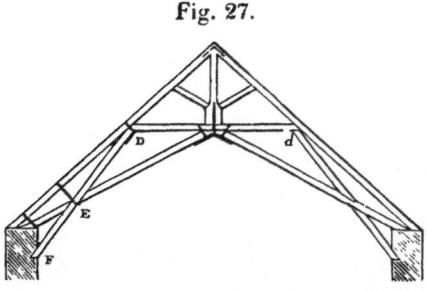

we have seen, and is executed with great success in the circus or equestrian theatre (now, 1809, a concert-room) in Edinburgh, the width being sixty feet. The pieces EF and ED help to take off some of the weight, and by their greater uprightness they exert a smaller thrust on the walls. The beam Dd is also a sort of truss-beam, having something of the same effect. Mr Price has given another very judicious one of this kind (British Carpenter, plate IK, fig. C), from which the tie-beam may be taken away, and there will remain very little thrust on the walls. Those which he has given in the following plate, K, are, in our opinion, very faulty. The whole strain in these last roofs tends to break the rafters and ties transversely, and the fixtures of the ties are also not well calculated to resist the strain to which the pieces are exposed. We hardly think that these roofs could be executed.

It is scarcely necessary to remind the reader, that in all that we have delivered on this subject, we have attended only to the construction of the principal rafters or trusses.

In small buildings all the rafters are of one kind; but in great buildings the whole weight of the covering is made to rest on a few principal rafters, which are connected by beams placed horizontally, and either mortised into them or scarfed on them. These are called *purlins*. Small rafters are laid from purlin to purlin; and on these the laths for tiles, or the skirting-boards for slates, are nailed. Thus the covering does not immediately rest on the principal frames. This allows some more liberty in their construction, because the garrets can be so divided that the principal rafters shall be in the partitions, and the rest left unencumbered. This construction is so far analogous to that of floors which are constructed with girders, binding, and bridging joists.

It may appear presuming in us to question the propriety of this practice. There are situations in which it is unavoidable, as in the roofs of churches, which can be allowed to rest on some pillars. In other situations, where partition-walls intervene at a distance not too great for a stout purlin, no principal rafters are necessary, and the whole may be roofed with short rafters of very slender scantling. But in a great uniform roof, which has no intermediate supports, it requires at least some reasons for preferring this method of carcass-roofing to the simple method of making all the rafters alike. The method of carcass-roofing requires the selection of the greatest logs of timber, which are seldom of equal strength and soundness with thinner rafters. In these the outside planks can be taken off, and the best part alone worked up. It also exposes to all the defects of workmanship in the mortising of purlins, and the weakening of the rafters by this very mortising; and it brings an additional load of purlins and short rafters. A roof thus constructed may surely be compared with a floor of similar construction. Here there is not a shadow of doubt, that if the girders were sawed into planks, and these planks laid as joists sufficiently near for carrying the flooring boards, they will have the same strength as before, except so much as is taken out of the timber by the saw. This will not amount to one-tenth part of the timber in the binding, bridging, and ceiling joists, which are an additional load, and all the mortises and other joinings are so many diminutions of the strength of the girders; and as no part of a carpenter's work requires more skill and accuracy of execution, we are exposed to many chances of imperfection. But, not to rest on these considerations, however reasonable they may appear, we shall relate an experiment made by one on whose judgment and exactness we can depend.

Two models of floors were made, eighteen inches square, of the finest uniform deal, which had been long seasoned. The one consisted of simple joists, and the other was framed with girders, binding, bridging, and ceiling joists. The plain joists of the one contained the same quantity of timber with the girders alone of the other, and both were made by a most accurate workman. They were placed in wooden trunks eighteen inches square within, and rested on a strong projection on the inside. Small shot was gradually poured in upon the floors, so as to spread uniformly over them. The plain joisted floor broke down with 487 pounds, and the carcass floor with 327. The first broke without giving any warning, and the other gave a violent crack when 294 pounds had been poured in. A trial had been made before, and the loads were 341 and 482; but the models having been made by a less accurate hand, it was not thought a fair specimen of the strength which might be given to a carcass floor.

The only argument of weight which we can recollect in favour of the compound construction of roofs is, that the plain method would prodigiously increase the quantity of work, would admit nothing but long timber, which would greatly add to the expense, and would make the garrets a mere thicket of planks. We admit this in its full force:

but we continue to be of the opinion that plain roofs are greatly superior in point of strength, and therefore should be adopted in cases where the main difficulty is to insure this necessary circumstance.

Of the roofs put on round buildings.

It would appear very neglectful to omit an account of the roofs put on round buildings, such as domes, cupolas, and the like. They appear to be the most difficult tasks in the carpenter's art. But the difficulty lies entirely in the mode of framing, or what the French call the *trait de char-penterie*. The view which we are taking of the subject, as a part of mechanical science, has little connection with this. It is plain, that whatever form of a truss is excellent in a square building, must be equally so as one of the frames of a round one ; and the only difficulty is how to manage their mutual intersections at the top. Some of them must be discontinued before they reach that length, and common sense will teach us to cut them short alternately, and always leave as many, that they may stand equally thick as at their first springing from the base of the dome. Thus the length of the purlins, which reach from truss to truss, will never be too great.

The truth is, that a round building which gathers in at top, like a glass-house, a potter's kiln, or a spire steeple, instead of being the most difficult to erect with stability, is of all others the easiest. Nothing can show this more forcibly than daily practice, where they are run up without centres and without scaffoldings ; and it requires gross blunders indeed in the choice of their outline to put them in much danger of falling from a want of equilibrium. In like manner, a dome of carpentry can hardly fall, give it what shape or what construction you will. It *cannot* fall, unless some part of it flies out at the bottom. An iron hoop round it, or straps at the joinings of the trusses and purlins, which make an equivalent to a hoop, will effectually secure it. And as beauty requires that a dome shall spring almost perpendicularly from the wall, it is evident that there is hardly any thrust to force out the walls. The only part where this is to be guarded against is where the tangent is inclined about forty or fifty degrees to the horizon. Here it will be proper to make a course of firm horizontal joinings.

We doubt not but that domes of carpentry will now be raised of great extent. The Halle du Bled at Paris, of two hundred feet in diameter, was the invention of an intelligent carpenter, the Sieur Moulineau. He was not by any means a man of science, but had much more mechanical knowledge than artisans usually have, and was convinced that a very thin shell of timber might not only be so shaped as to be nearly in equilibrio, but that, if hooped or firmly connected horizontally, it would have all the stiffness that was necessary ; and he presented his project to the magistracy of Paris. The grandeur of it pleased them, but they doubted of its possibility. Being a great public work, they prevailed on the Academy of Sciences to consider it. The members who were competent judges were instantly struck with the justness of M. Moulineau's principles, and astonished that a thing so plain had not been long familiar to every house-carpenter. It quickly became an universal topic of conversation, dispute, and cabal, in the polite circles of Paris. But the academy having given a very favourable report of their opinion, the project was immediately carried into execution, and soon completed ; and now stands as one of the great exhibitions of Paris.

The construction of this dome is the simplest thing that can be imagined. The circular ribs which compose it consist of planks nine feet long, thirteen inches broad, and three inches thick ; and each rib consists of three of these planks bolted together in such a manner that two points meet. A rib is begun, for instance, with a plank of three feet long standing between one of six feet and another of nine ; and this is continued to the head of it. No machinery was necessary for carrying up such small pieces, and the

whole went up like a piece of bricklayer's work. At various distances these ribs were connected horizontally by purlins and iron straps, which made so many hoops to the whole. When the work had reached such a height that the distance of the ribs was two thirds of the original distance, every third rib was discontinued, and the space was left open and glazed. When carried so much higher that the distance of the ribs is one third of the original distance, every second rib, now consisting of two ribs very near each other, is in like manner discontinued, and the void is glazed. A little above this the heads of the ribs are framed into a circular ring of timber, which forms a wide opening in the middle ; over which is a glazed canopy or umbrella, with an opening between it and the dome for allowing the heated air to get out. All who have seen this dome say that it is the most beautiful and magnificent object they have ever beheld.

The only difficulty which occurs in the construction of wooden domes is when they are unequally loaded, by carrying a heavy lanthorn or cupola in the middle. In such a case, if the dome were a mere shell, it would be crushed in at the top, or the action of the wind on the lanthorn might tear it out of its place. Such a dome must therefore consist of trussed frames. Mr Price has given a very good one in his plate OP, though much stronger in the trusses than there was any occasion for. This causes a great loss of room, and throws the lights of the lanthorn too far up. It is evidently copied from Sir Christopher Wren's dome of St Paul's Church in London ; a model of propriety in its particular situation, but by no means a general model of a wooden dome. It rests on the brick cone within it ; and Sir Christopher has very ingeniously made use of it for stiffening this cone, as any intelligent person will perceive by attending to its construction.

Fig. 28 presents a dome executed in the Register Office in Edinburgh by James and Robert Adam, and is very

Fig. 28.

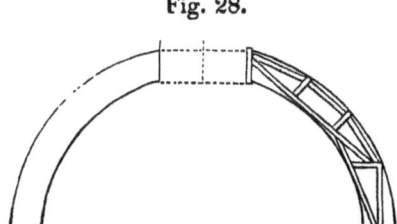

agreeable to mechanical principles. The span is fifty feet clear, and the thickness is only four and a half feet. (J. R.)

Causes of Failure in Roofs.

It has been shown in the preceding treatise that the simplest form of a roof is that given in fig. 12. Let us now inquire the method in which such a roof would fail, as deduced from the former treatise, and given there in scientific language. Let the dotted lines (fig. 29) show the original line of roof. If an undue weight be put on this, it has been shown the point B will descend, and A and C will spread or open to the right and left, just as pressing on the top of a pair of compasses makes them open. The rafters must then either slip off the top of the walls, or, if properly secured to them, which we ought to suppose, must push the walls over ; or if they be very strong, the rafters must bend or sag in the middle at D and E (fig. 30). Now, to prevent the walls being thrust over, an easy remedy, as our author shows, is fig. 14, to tie them together with either a piece of string or a rod of timber, or, as Mr Robert Stephenson (art. IRON BRIDGES, fig. 13) shows, by a chain. But whether the walls be kept upright by this tie or by their own size and strength, still the same bending at D and E will take place if the timbers be not strong enough to bear the weight. Now, in small roofs we have a ready

Fig. 29.

remedy; we put a collar beam DE (fig. 31) between them, which has a double effect,—it not only keeps these points

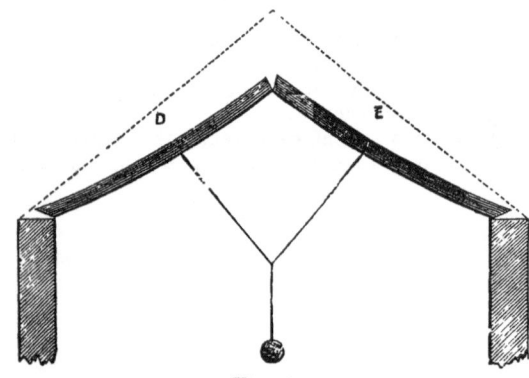

Fig. 30.

from coming towards each other, which they must do if the rafters bend, but it also assists very much to prevent the rafters AB, BC, going out, as is shown in fig. 29. If, however, the roof be too large and the timbers too weak, or, which is the same thing, the load be too heavy, the roof, though it cannot bend (fig. 31) between BD, BE, will yield be-

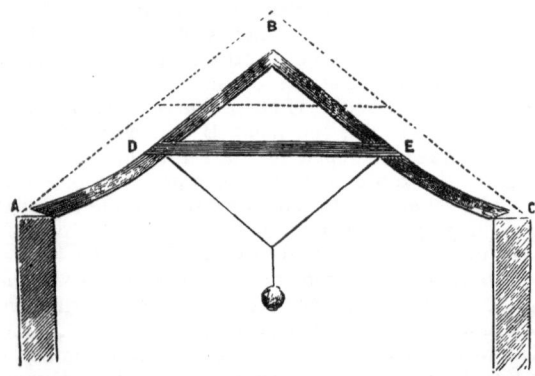

Fig. 31.

tween DA, EC; and either sag, as we have before said, or push out the walls if they are too weak. This is the great

cause of failure in the early mediæval roofs (see figs. 35, 38), where they are all weak between A and B. However trusses may be braced together at the apex of the triangle, it is clear nothing except excessive thickness of timber will prevent one of these results. We will now go back to fig. 14, and for the present lay aside the consideration of a collar-beam. We are here liable to fall into this contingency,—the beam AC must have considerable weight if strong enough to act as an efficient tie. Now wood is always weakest in horizontal position; it is therefore liable to sink or sag in the middle, and the effect of this will be to bring the points AC (fig. 32) closer together, and to

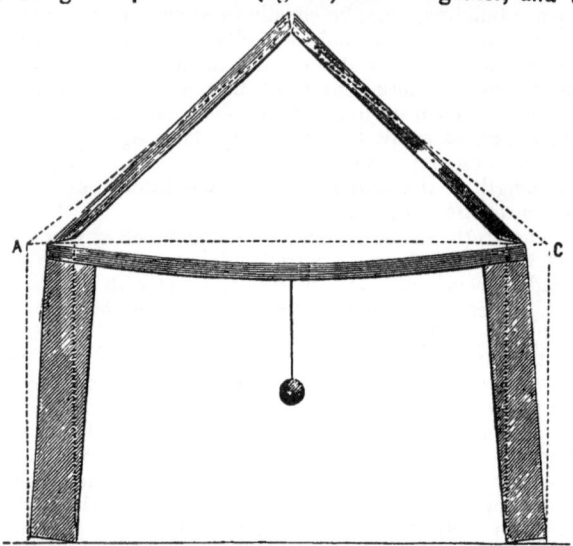

Fig. 32.

pull the walls in. It would interfere with our space to prop this beam up from below; but some ingenious carpenter at some time has thought of hanging it up to the ridge by another string or tie, BD (fig. 15), usually called a king-post, though it does not act as a post to prop up, as is shown above, but as a string or tie, to hang up the tie-beam and keep it straight. We will now suppose the rafters AB, BC (fig. 15) too weak for their weight. We must bear in mind, in all questions of carpentry, there are two difficulties to contend with. First, timber itself is limited in size; and next, if it be made too strong for its length, its own weight will cause it to bend. So we are between two difficulties. But we have several remedies: either we may resort to a collar-beam in addition to a tie-beam, as in fig. 31; or, what is better, we may employ two struts, DE, DF (fig. 17). An inspection of this diagram will show why, although a chain or rod would answer every purpose at BD, we prefer a post of wood. The struts FD, ED are supports and not ties, and require a good butment, which is best got by framing them into the king-post, as shown at D (fig. 17), and also in the various diagrams of the article CARPENTRY. Still referring to the same diagram (fig. 17), we will suppose the span still further increased; then the tie-beam may be too weak, and may sag between AD and DC. This must be remedied by again suspending the weak parts, which may be done by two rods or posts BE, CF (fig. 18), which are generally called queen-posts, as BD (figs. 15 and 17) is called a king. This system divides the tie into three parts instead of two, or if a king-post also be used, into four parts, each of which is suspended. Fig. 18, however, shows what is generally called a queen-post roof, and is framed with a collar BC, in the points of which the purlins are usually seated, and the common rafters run up, as AG, DG, unless, as in the figure, it is intended to make the roof flat between B and C. In the same manner, an inspection of the diagram following will show (as well as those in CAR-PENTRY), and *infra* fig. 51, how the principle of using queen-posts and struts may be multiplied almost to any extent.

Roofs sometimes fail in consequence of the trusses being placed too far apart; the purlins are then unable to sustain the weight, and the surface undulates between truss and truss, bringing down the ridge, and producing the most unpleasant and sometimes pernicious effects. The mediæval architects braced the purlins upon the principles as shown in fig. 40. A better plan, however, has been devised at the Lambeth Baths, which will be hereafter described.

Roofs also frequently fail from the weakness of walls, or the want of extra thickness under the principals; but this is rather matter for consideration under the article STONE MASONRY. They, however, often fail from a simpler cause. The plate or template is generally of timber, and bedded into the wet brickwork. It swells with the moisture, and afterwards shrinks, which causes it to lose its hold in the walls; so that in case of settlement it is easily pulled out; and the walls, which should be kept upright by the tension of the tie-beams, are unsupported, and settle outwards. The best remedy is to make the templates of stone, and pin down the tie-beams to them by strong iron pins, going some depth into the walls. There is another benefit about this system, that air may be allowed to come sideways to the rafter feet, or ends of the tie-beams, and so prevent their becoming rotten.

Mediæval Roofs.

Those of the Italian basilicas, erected before the tenth century, are framed much according to the present methods: of the same pitch, and covered with tiles like those which have been in use from the Roman period down to the pre-

no doubt it was hoisted to its place. It is of one huge single block, all chiselled out of the solid.

Of Saxon roofs we have no remains; but from the illuminations in the MSS. of the Ælfric Pentateuch, and to Cædmon, now in the British Museum, we should suppose they formed an angle of 45°. They seem to be covered with a sort of tiles rounded at the end, or they probably may have been of shingles. A very curious instance, taken from the Harleian MS. (fig. 34), shows not only a gabled roof,

Fig. 34.

but also a sort of dome. The subject is Lot entertaining the two angels.

Few Norman roofs now remain, if any, and those which are supposed to be so are evidently much altered. Judging from the string courses up the gables, and the water-

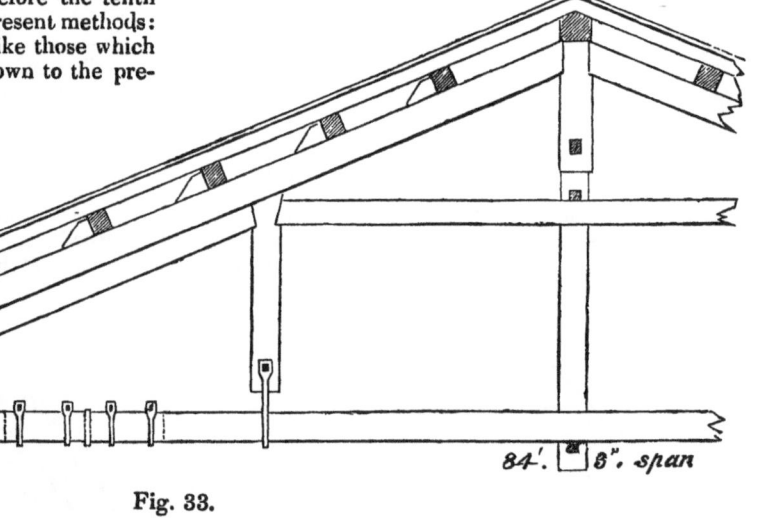

84' 8". span

Fig. 33.

sent day. Fig. 33 shows the roof of San Paolo Fuori le Mura, lately consumed by fire at Rome. It is upwards of 84 feet in span, and consists of a king-post and two queens, without struts, but with a collar. The principal rafter is doubled from the head of the queen to the plate, which adds immensely to its strength. The kings and queens are not framed into the tie-beam, but the latter is hung up to them by iron straps. This perhaps is the earliest instance known where iron has formed one of the chief features in the construction of a roof, and this is said to have stood upwards of 1400 years. A curious instance of a roof is found in the tomb of Theodosius at Ravenna, supposed to have been erected shortly after the year 526; this is composed of a circular dome of white marble 36 feet in diameter, surrounded by a number of ears or lugs, by which

tables, they seem to have been of about the same angle as those named above; but in the next style, the early English, as has before been stated, the roofs suddenly sprung up to equilateral, and often even to whole pitch, or an angle of 64°. The singularity is, that many of them are covered with lead, and groined beneath; so that if bare economy or dry utility were the only object of the Gothic architects, as has been stated by many, the roof might have been flat, and an immense expense both in timber and lead have been saved. It only serves to show that, like all other architects, they were not always guided by considerations of dry utility alone, but had high, bold, artistic, and æsthetical feeling besides. It is not improbable that, as the pointed arch and slender shaft are evidently of Saracenic origin, and their use no doubt brought to Europe by the Crusaders,

2 I

so the aspiring roofs of the refined orientals suggested those of the early English period.

Year by year, as styles changed, the roofs became of less pitch, till in the latter styles, as has been stated, many, if covered with lead, became almost flat. A very curious illustration is found in the tower of St Regulus' church at St Andrews, on which are the marks of the lines of three

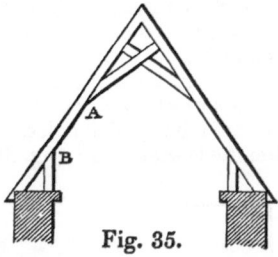

Fig. 35.

roofs which have covered the building at three different periods. The lower is probably the line of the original Norman roof, the upper that of the early English period, and the middle that of the decorated. Much of course depended at all periods on the covering, tiles, for instance,

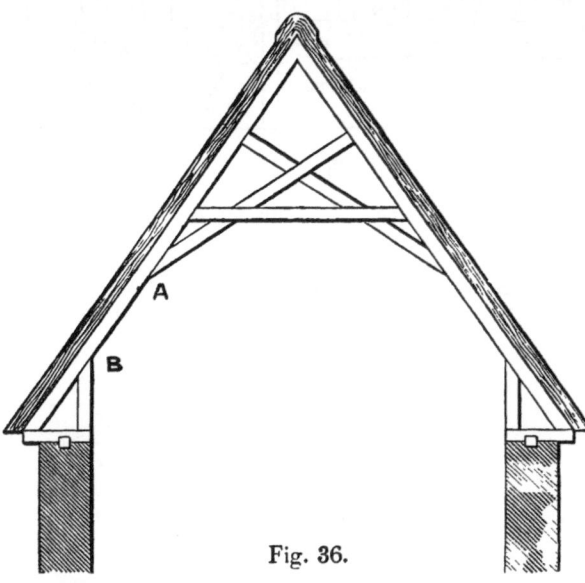

Fig. 36.

necessitating a sharper pitch than slates. No doubt the Norman roofs, like those of the Roman basilicas, from which they were derived, had level tie-beams. It is difficult to

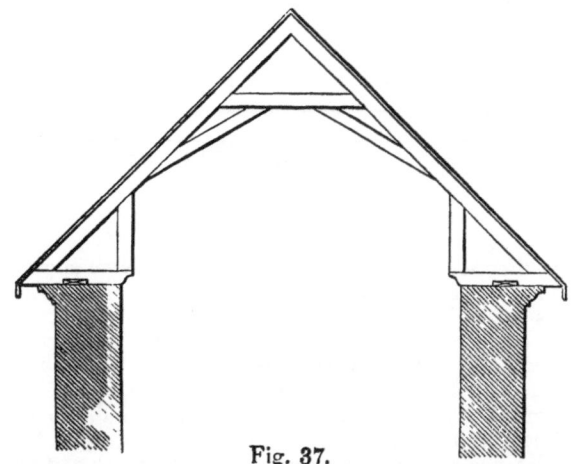

Fig. 37.

conceive how else they could have stood; but in the succeeding styles, when internal height became an object, these were often omitted, and as the thrust of the roof was enormously increased thereby, it was necessary to build large projecting buttresses to keep up the walls, instead of

using those of the Norman period, which are in fact mere shallow pilasters. The truth is, the buttresses, which were originally intended simply to stiffen the walls, were afterwards enlarged to that extent that they became struts to it (A, A, fig. 42), especially the flying buttresses, which continued the line of the principal rafters down to the ground, making the earth, as it were, the tie-beam. To such a degree was this system carried, that in many continental churches, and in some of our own, of which Henry VII.'s chapel is a known instance, there is literally no support derived from the walls, the windows filling the whole space between buttress and buttress, which last, from their vast mass and projection, sustain both roof and groining.

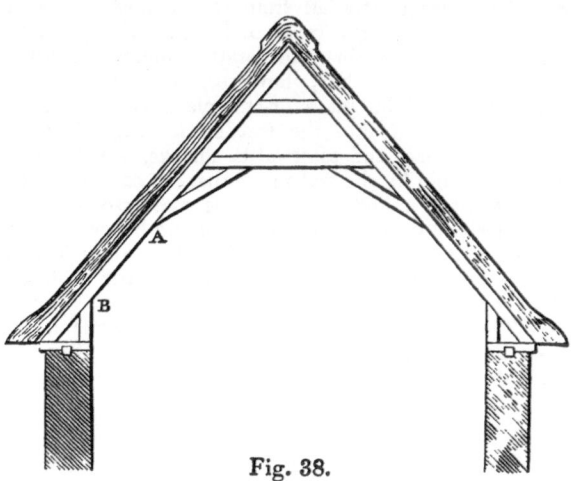

Fig. 38.

Mediæval roofs may be considered,—1st, As those in which every pair of common rafters is framed together, and forms of itself a separate truss, or, as it was called in those days, a "couple;" 2d, As those with common rafters and trusses framed with collar-beams; 3d, As those with hammer-beams; 4th, As those with tie-beams.

But before going into this subject, we must warn our readers who are accustomed to roofs framed of fir, that the mediæval timbers are almost invariably of oak or some hard wood; so the strength or the scantling of the timber must not be judged by our modern notions. That

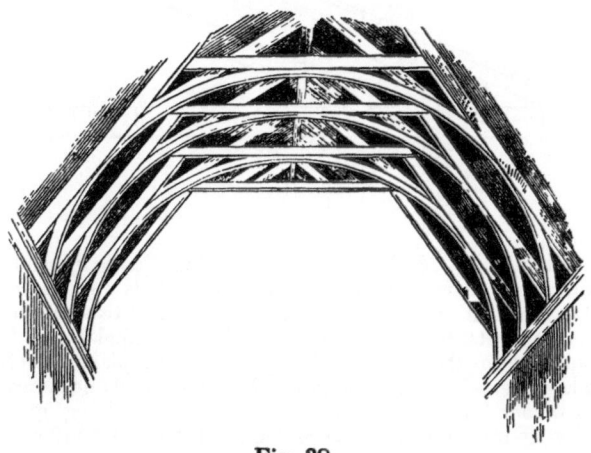

Fig. 39.

of oak to fir is assumed by Professor Robison to be as three to two. An inch of oak may be safely subjected to a strain of three tons for every square inch, while fir will bear but two.

The most common form of the first is a simple St An

drew's cross, as at fig. 35 ; a cross and collar, as fig. 36 ; a collar and struts, as fig. 37 ; or two collars and struts, as fig. 38. All these are taken from examples of the thirteenth and fourteenth centuries. As has been before explained, these roofs are liable to spread, as there is nothing to tie the plates together, and they are all weak at the point be-

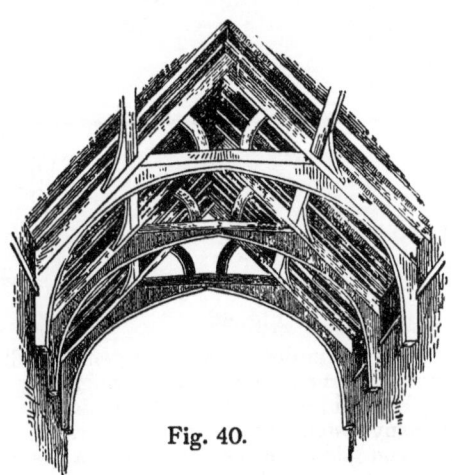

Fig. 40.

tween the end of the brace A, and the top of the ashler-piece B. To obviate this, various contrivances were used ;[1] one of the most elegant is the addition of curved braces, which not only strengthened each rafter, but tended to brace the whole together. Fig. 39 is a common example in England, particularly in Somersetshire.

As those roofs with simple couples could not be erected of any great span, they were commonly framed with trusses, purlins, and common rafters ; and then the defect of these open roofs became more felt, as each truss had not only to bear its own load, but that

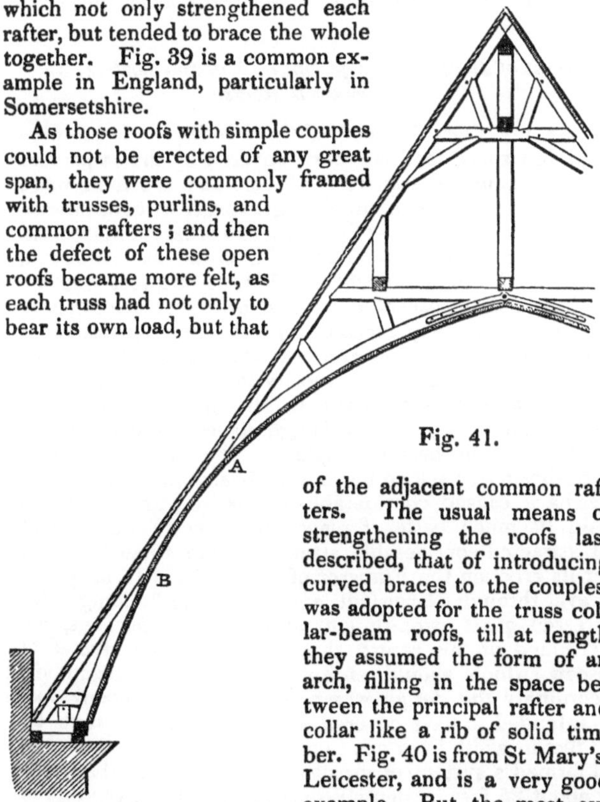

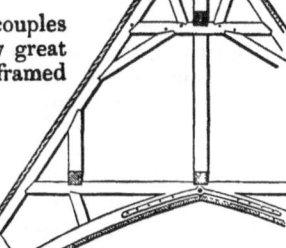

Fig. 41.

of the adjacent common rafters. The usual means of strengthening the roofs last described, that of introducing curved braces to the couples, was adopted for the truss collar-beam roofs, till at length they assumed the form of an arch, filling in the space between the principal rafter and collar like a rib of solid timber. Fig. 40 is from St Mary's, Leicester, and is a very good example. But the most extraordinary mediæval roof, on the principle of a collar-beam and curved braces, is that over the Salle des Pas Perdus at Rouen, which was built in 1493, and is of inconceivable lightness and boldness (see fig. 41). It is 54 feet 5 inches English in span, and covers a hall 155 feet long, the trusses being not quite 4 feet apart. It is close-boarded inside, so

as to show as one long, pointed vault below. The thrust, which is immense, is resisted by walls 6 feet thick, with huge buttresses. The upper part of the construction is very good ; but how the part between AB can sustain the weight of the upper part without bending has puzzled everybody. It can, in effect, be only due to the excellence of the workmanship and strength of the materials.

On looking back to fig. 40, it will be seen not only that the curved brace was intended to strengthen the rafters, but also to relieve the thrust by conveying it lower down the wall, and distributing it over a much larger portion of its surface, as well as bringing it more in a line with the

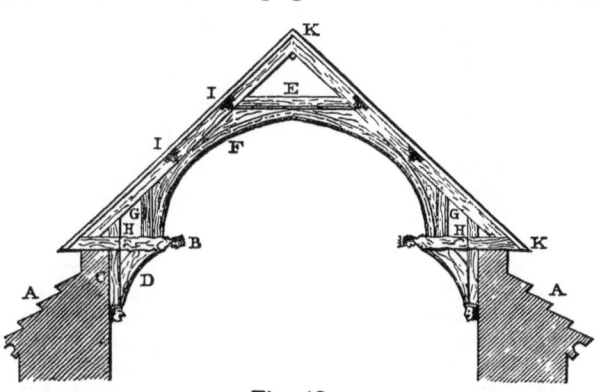

Fig. 42.

buttress AA (fig. 42). Fig. 43 shows the beautiful roof over the church at Wymondham in Norfolk. Here the

Fig. 43.

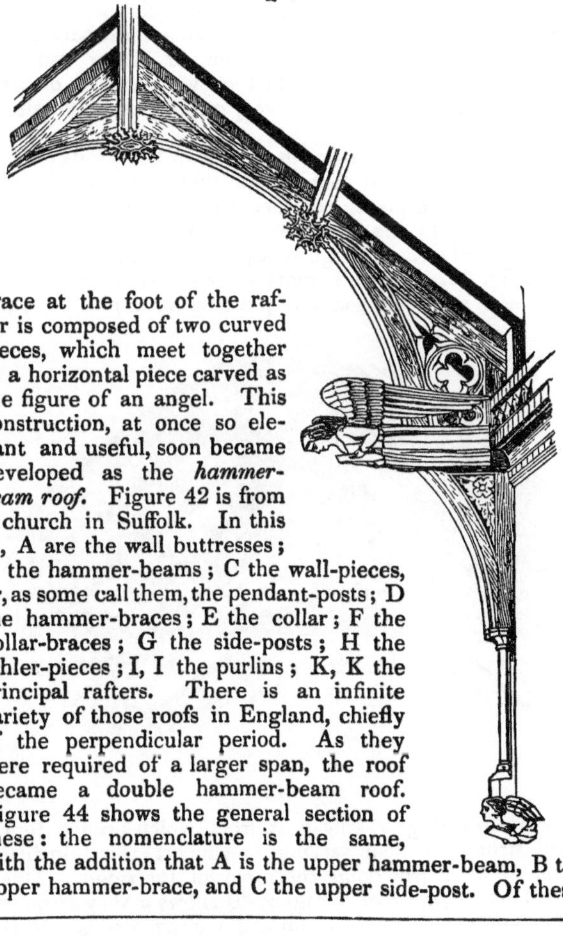

brace at the foot of the rafter is composed of two curved pieces, which meet together in a horizontal piece carved as the figure of an angel. This construction, at once so elegant and useful, soon became developed as the *hammer-beam roof*. Figure 42 is from a church in Suffolk. In this A, A are the wall buttresses ; B the hammer-beams ; C the wall-pieces, or, as some call them, the pendant-posts ; D the hammer-braces ; E the collar ; F the collar-braces ; G the side-posts ; H the ashler-pieces ; I, I the purlins ; K, K the principal rafters. There is an infinite variety of those roofs in England, chiefly of the perpendicular period. As they were required of a larger span, the roof became a double hammer-beam roof. Figure 44 shows the general section of these : the nomenclature is the same, with the addition that A is the upper hammer-beam, B the upper hammer-brace, and C the upper side-post. Of these,

---

[1] Most Gothic curved braces are cut out of boughs of large trees bent by nature.

the most celebrated are those over Westminster Hall, Hampton Court, the palace at Eltham, and very many of our col-

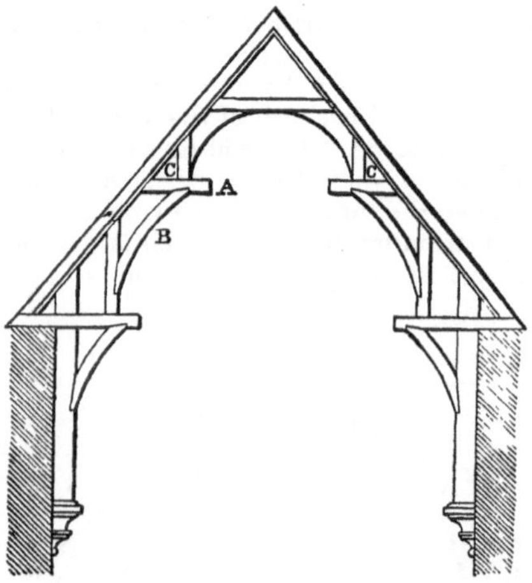

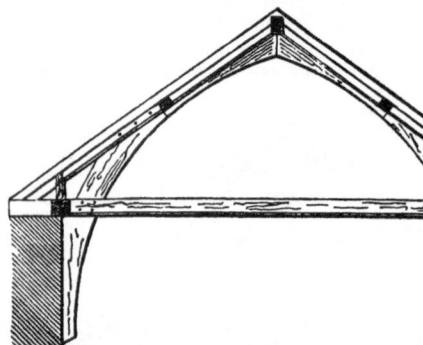

Fig. 44.

lege halls. It will be seen by the section, the use of the hammer-beam framing was to deepen, as it were, the principal rafter, and thereby prevent its bending between the collar and the plate; besides which it got better hold of the wall; but there was nothing to tie the walls together, nor to keep the truss from spreading; the consequence is, in spite of the buttresses, many of these roofs have opened, and the walls have gone over. A very curious roof, and one which is much stronger in point of construction, is that over the Parliament House at Edinburgh, which was begun in 1632. In this roof the pieces which act as hammer-beams are not horizontal but incline, or rather radiate towards a centre; they are filled in with arched pieces bearing pendants, and have a very original and pleasing effect.

The use of *level tie-beams* in roofs is of two kinds: one where the rafters are in couples, and they seem merely introduced to tie the plates together, and probably were inserted afterwards; the other where they form parts of the trusses themselves. In large roofs, we have already an example in fig. 35. In smaller roofs of the fourteenth century, fig. 45 is a very common and pleasing example. From

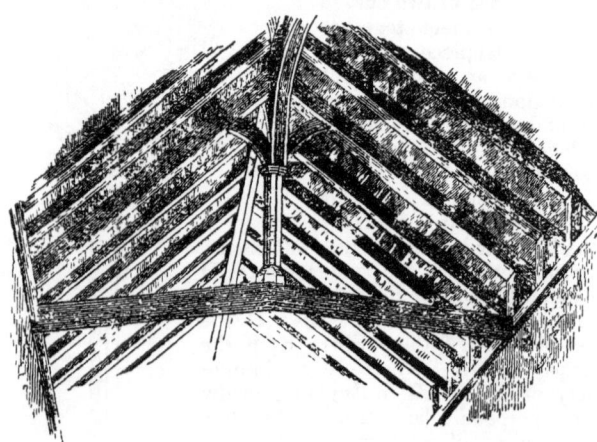

Fig. 45.

the tie-beam springs a king-post, from which branch four curved struts, one pair serving to support the principal rafters, the other pair doing the same for the ridge. This is called a tree-post roof. Figure 46 shows a very curious combination of a level tie-beam and curved struts. It is from the church of St Mary the Virgin at Pulham in Nor-

folk. But it was reserved for the perpendicular period to design the most elegant roofs with level tie-beams. These

Fig. 46.

are of infinite variety in design, and are generally filled in with tracery, and ornamented with carving. Figure 47 shows a very beautiful example, that of St Martin's at Leicester. In many instances these roofs are richly adorned with painting and gilding of the most brilliant description.

Another sort of mediæval roof is yet to be noticed, and that is where, instead of timber principals, arches of stone are thrown across from wall to wall, and carry the purlins and common rafters: among these may be named the great hall at Mayfield; but they are of very rare occurrence.

Account of Roofs of Large Span (à grand portées) and some modern Roofs.

For many years the roofs of the basilicas at Rome were the largest spans that had been covered by the carpenter's art; but about three-quarters of a century ago there was a great desire to roof over much larger spaces without internal supports, for the purpose of military and other riding-schools. The vast roof over the *Salle d'Exercise* at Darmstadt, erected by M. Schubknecht in 1771, is 228 feet long and 154 feet in the clear of the walls, or 2 feet more in span than that roof lately erected over the railway station at Lime Street, Liverpool, and was the largest roof in the world till that we are about shortly to describe was erected over the New Street station at Birmingham. It appears to have stood very well, although its construction is certainly not on the best principles. The thickening out of the tie-beam by packing beam after beam, one on the other, must have caused considerable shrinkage; and the struts would be much better if placed the reverse way. This roof attracted considerable attention; and about ten years after its erection the Emperor of Russia, Paul I., happened to travel through Darmstadt, and visited the building. He expressed great astonishment at its vast proportions, and determined on his return that one should be erected at Moscow which should entirely eclipse it in magnitude. Accordingly the design was prepared. This gigantic roof was intended to have covered a hall 852 feet in length, by 308 feet in width from out to out. The walls were double, and formed a system of arcaded galleries round the building about 25 feet in depth; so that the span of the roof in the clear was reduced to about 230 feet. The main support of this roof was the curved rib of three thicknesses of timber, notched on to each other *en cremaillière*. Krafft says it was executed, and was used in 1790 for the exercise of the Cossack cavalry and infantry. But M. de Bétancourt, of whom we shall speak presently, affirms that it never was finished. The dotted lines show a method proposed by Rondelet and Krafft to strengthen this roof and make it effective. The chief defect, however, seems to be, that the principal rafters are too weak, and receive no direct support from the cross struts. The troubles in Russia seem to have caused the matter to drop, till the year 1817, when the Emperor Alexander, being at Moscow, resolved on carrying out a roof that should rival the one at Darmstadt. A great number of designs were prepared, none exceeding spans of from 110 to 115 feet. They were referred to General de Bétancourt, before named, who was then chief director of public roads. After some study, this officer prepared the design of it. It is for a hall 501 feet long and 150 in width, which was executed in the short

outside, to prevent their shifting apart, and are of vast strength, and particularly adapted to rooms where the ceiling is intended to represent a large arch or barrel vault, or

Fig. 47.

space of five months. On striking the scaffold, which had supported the roof during its erection, the tie-beams went down about two inches. This depression increased in three months to eight inches, when the tie-beam of the twenty-fourth truss gave way, in consequence of the existence of a large knot. The roof was shored up and strengthened, and afterwards stood perfectly well. In justice to M. Bétancourt, it should be stated that such was the hurry in which the work was done that the greater part of the timbers were cut down and floated on the river only a few days before they were framed, which was done by 400 carpenters, or rather woodmen, who hewed the wood with axes—they being ignorant of the use of any other tool. The writer of this article has designed a roof for the first-class swimming bath in the Westminster Road. This is on entirely a new principle, being, in fact, the adaptation of the trellis or lattice principle to a roof. The trusses are about 18 feet apart, or nearly double the usual distance. The purlins, however, are prevented from bending by a series of light longitudinal trusses, likewise on the lattice principle. The roof is, in truth, trussed fore and aft, so that no part can move. It is of extraordinary lightness and cheapness, the trusses being so few and the timbers so slender. Its deflection, after receiving the load of slates, skylights, &c., was scarcely perceptible.

(a.) *Roofs Trussed with Curved Timbers.*—These are (a) with timbers side by side, flatwise, the ends breaking joint, or, as the system is called by French writers, " en bois plat." It is the invention of Philibert de Lorme, a French architect, who published it in 1561, in a work called *Traité sur la Manière de Bien Bàtir, et à Petits Frais.* In this system the rafters are in effect curved ribs, of several thicknesses of timber, nailed side by side, care being taken that the ends do not come in the same place; in other words, that they break joint. The rib then resembles a beam cut out of a crooked limb of a tree, and owes its strength simply to cohesion of the particles. These beams are often used in pairs joined together by cross pieces, which are keyed on the

to domes intended to be covered with lead or other metal. One of the most ingenious adaptations of this principle is at the Hotel Legion d'Honneur at Paris, where it forms a dome, with an inner half-dome or cove which carries a gallery. The roof over the great hall of the Pantheon in Oxford Street is a long vault on this principle, the middle thickness or flitch of each rib being of teak wood. That lately erected over the Surrey Music Hall is also a long vaulted ceiling; but here the main ribs are formed of four thicknesses of 1¼-inch deal, 1 foot 3 inches wide at bottom and 1 foot at top, in the middle of which is a rib of rolled ⅜ths boiler-plate, 6 inches wide, the whole of which are bolted together. The main ribs are 17 feet apart, between which are two intermediate ribs of slighter construction.

(b.) *Roofs constructed of Timber bent on the flat " en bois plat."*—This is the reverse of the former system; the boards being bent over a centre, and thickness added upon thickness till the beam is thought strong enough, when the whole is bolted together, care being taken, as in the former system, that the ends of the boards break joint. The difficulty is, to prevent the natural tendency of the wood to spring back to the straight, as well as to counteract the weight of the roof covering, which would cripple the curve either at the haunch or crown, as the pressure might be exercised. In this respect something analogous to the laws of arches will apply to explain the result.

The roof over the great riding-house at Libourne was designed by the celebrated Colonel Emy, and executed in 1826. Every rib is composed of five thicknesses of deal, each nearly 2 inches thick, about 6 inches wide and 40 feet long. The rib is semicircular, the springing about 24 feet from the ground, and the span about 70 feet. The ribs are not only bolted together, but clipped with a sort of stirrup-iron and bolts. From these ribs a number of struts radiate to and support the principal rafters, purlins, &c., and at the same time prevent the rib from being crippled at either the haunch or crown. The worst of this roof is, there is nothing to prevent their spreading at the foot. The consequence is, the wall was not only of unusual thickness (nearly 5 feet), but large buttresses were added to counteract the thrust. The roofs over the Great Northern Railway station at London are of about the same span, and are of similar construction, except that the ribs are of 1¼ deal in 16 thicknesses, and the spandrils are filled in with ornamental cast-iron work in form of circles, guilloches, &c., which have a very pretty effect. Like the Libourne

ROOF.

roof, the thrust was compensated by massive brick-work.

To prevent this last defect, as well as to stiffen the rib at the springing, M. Émy designed the roof for the cavalry school at Saumur. This is intended for a span of upwards of 130 feet. Each truss is composed of two sets of ribs, similar to those before described, kept apart at the foot by a series of trellis, and joining together as one rib about half-way up the curve. Instead of one thick wall, it was intended to carry the ends of the trusses partly on an outer wall, which has strong double pilasters, and partly on one large square pilaster, intended to stand under the inner part of the end of the truss, as shown in the figure given in his work; so that, though the space is available for spectators, and little room is lost, the truss itself has a level bearing 18 feet in width from whence to spring.

But of all roofs ever projected, either in wood or iron, the most gigantic, the most original, and the boldest, is one for another riding-house, also by Colonel Émy. This has a span of 328 feet, or at least half as much again as the largest existing roof in the world,—that at New Street, Birmingham. It is intended to be composed of two ribs similar to those before described, with another intermediate rib carried up about two-thirds of the span, and braced, as shown in the figure. The building was to have been surrounded by an ordinary wall, at right angles to which, under the foot of every truss, was a return wall 50 feet long and 4 feet thick, perforated below with arches to form passage-ways, and to serve for spectators, as in the former instance.

Iron Roofs.

Like most important discoveries, the use of iron in roofs has grown up from the smallest beginnings to the largest and most extraordinary results. From the simple substitution of an iron rod for a king or queen post in a wooden truss, we have attained the art of covering spaces so vast as to throw all other modes of construction into insignificance, so light as to appear the work of fairies rather than of human beings, and yet so strong as to bear their weight with ease, and to resist unshaken and unhurt the action of the roughest wind. The theory of these roofs, however, is just the same as those of timber, allowing only for the difference of weight and power of resistance of the materials. The tie keeps the walls together, the king and queen rods prevent the tie from sagging: all these are in a state of tension; while the struts prevent the rafters from bending, and are in a state of compression. Fig. 48

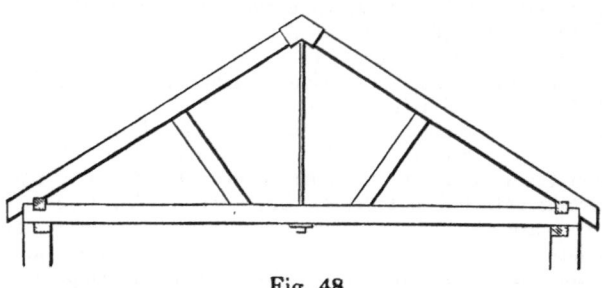

Fig. 48.

shows the earliest use of iron as the king-post of roofs suited to those of from about 20 to 30 feet span. In Plate LIV, CARPENTRY, figs 38, 40, is shown the method of employing iron as king or queen posts of still larger dimensions. Fig. 49 shows a very convenient method of using iron rods as king-post and tie-beam in a collar-beam roof where height or head-room is of consequence; but this was shortly after superseded by the form fig. 50, the adjustment of the screws, &c., at A being more convenient than at

fig. 49. With large spans, however, the wooden tie-beam continued in use for a long time. One of the best in-

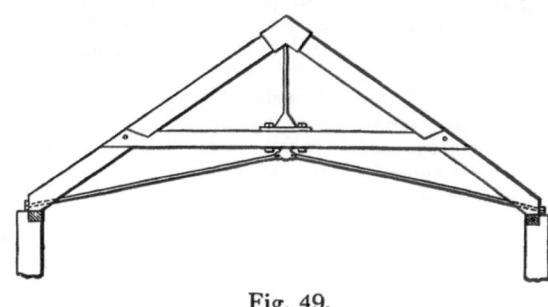

Fig. 49.

stances for the time it was executed may be found in the roof over the passenger-station at London Bridge of the

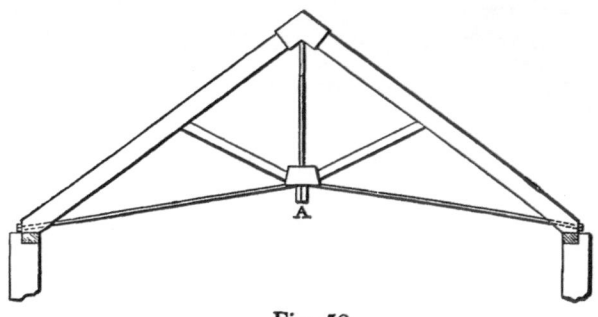

Fig. 50.

Croydon Railway. This roof, which is about 54 feet in span, is constructed with an iron king-post and ten sets of iron queens, with wooden struts. (See fig. 51.) It is, how-

Fig. 51.

Croydon Railway Roof.

ever, somewhat twisted on the face by the winding of the timber. The great fire at the Houses of Parliament, and subsequently at the Royal Exchange, drew attention strongly to the importance of fire-proof roofs; and the first of these structures of any size or importance was designed by Sir Charles Barry for the new palace at Westminster. Fig. 52 shows the section of that over the committee-rooms in the front next to the river. This is entirely of wrought iron, excepting the shoes, by which the ends of the various pieces are connected. The iron is flat bar, simply cut to lengths, and punched to receive the bolts that pass through them and the shoes. They vary in width from 2 to $3\frac{1}{2}$ inches, and in thickness from $\frac{3}{8}$ths to $\frac{5}{8}$ths. These light principals are not quite 3 feet apart, and are covered with a species of galvanized iron tile reaching from one to another, hung on a sort of connecting-rod without purlins. It is stated that the sulphurous acid in the smoke of London is already causing serious oxidation, and several methods have already been tried to prevent premature decay. This is to be regretted, as the material is very picturesque in character, as well as light, and not expensive. Some fault at the time was found with the construction of the shoes or connecting-plates, but they stand perfectly well, and there has been no failure. The suspension-rods at each side of the king are a peculiar feature; they not only keep up the tie, but bind the roof together at each intersection.

Fig. 53 shows the roof over the House of Lords. This is of much larger span, being 45 feet, while the former is 28 feet 3 inches. It is of exactly similar construction, except that the struts, wall-plates, purlins, and bearers are

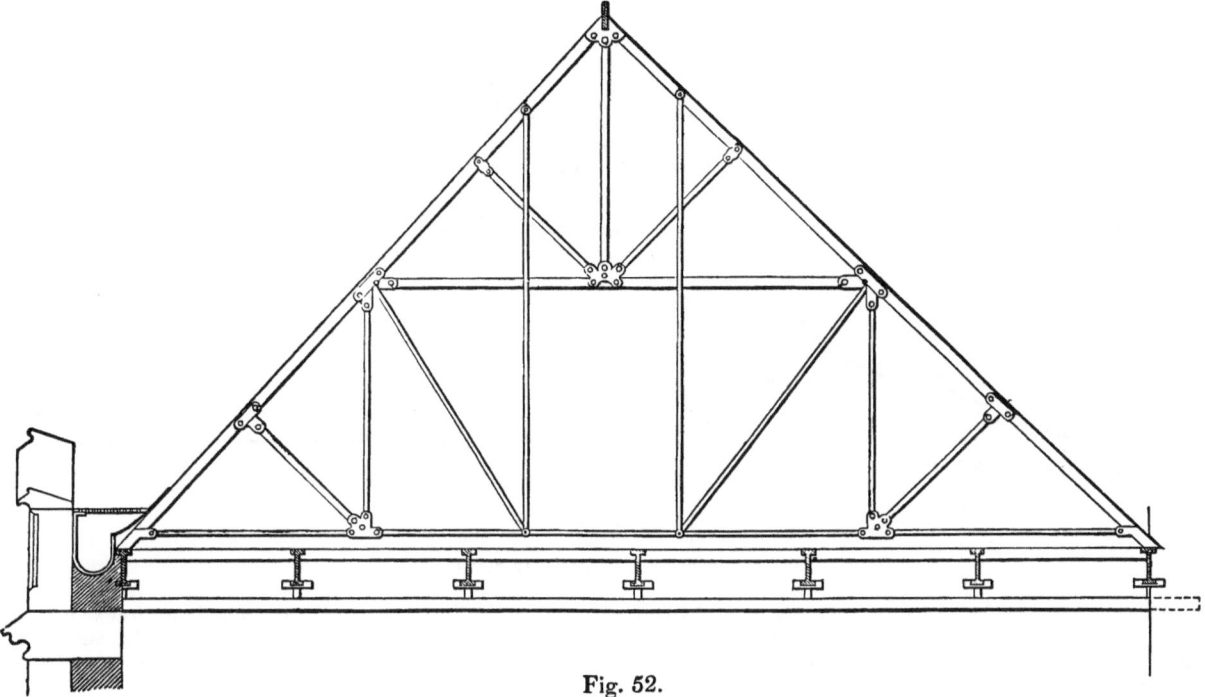

Fig. 52.

New Houses of Parliament—Roof over North and South Curtains.

of cast-iron, tne section of the struts being the form of a cross. The suspension-bars are double. The principals are 7 feet 6 inches apart—more than twice as much as the other roof, which of course necessitates the use of purlins.

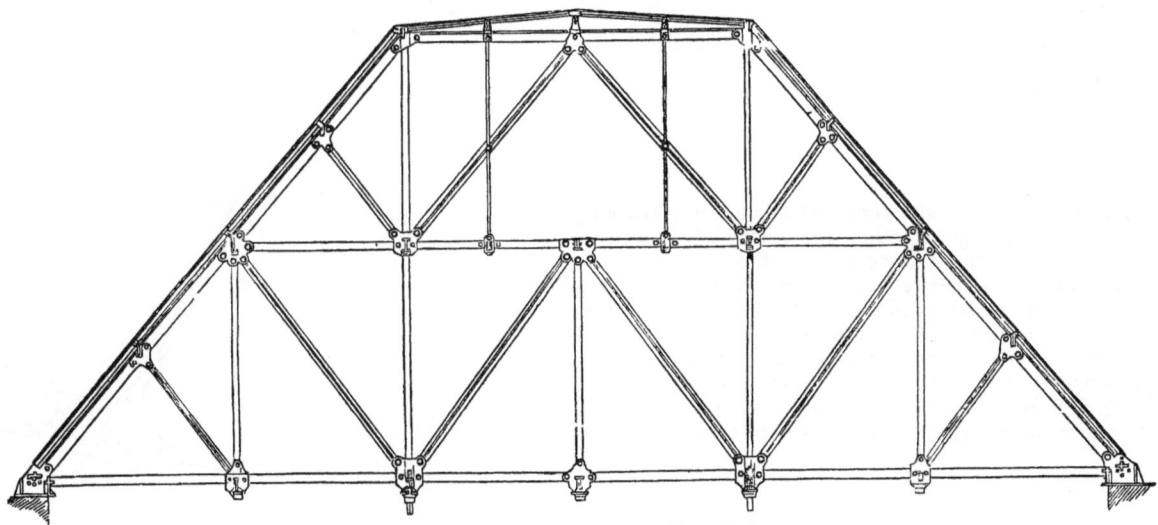

Fig. 53.

These carry two common rafters 2 feet 6 inches apart, this being the width of the iron plate or tile with which the roofs are covered.

Much animadversion has been expressed as to these roofs. The nearness of the principals has been criticised, as well as the small subdivision of detail; in fact, we shall shortly show roofs of three times the space with a less number of junctions, the principals of which are very much farther apart: so in this respect they are very much more economical. But it must be remembered these Gothic roofs are of high pitch, while the others are very flat; and therefore more struts and queens are necessary; and that the distance of the principals was regulated in great measure by the nature of the covering. If we also consider they were the first roofs of the kind, we think they may be regarded with great commendation. The progress of railways caused a great demand for light roofs of every span and length, and the facilities for obtaining rolled T and L

ROOF.

iron, and the machines invented for punching and rivetting, combined with the fact, that iron was not only the lightest material of its strength, but was at the same time incombustible, brought it day by day into more general use.

The awkwardness of fitting wooden struts to an iron tie-rod suggested the use of cast-iron for that purpose; and a very pretty light roof, first used on the Manchester and Birmingham line, was invented (fig. 54), which has since been

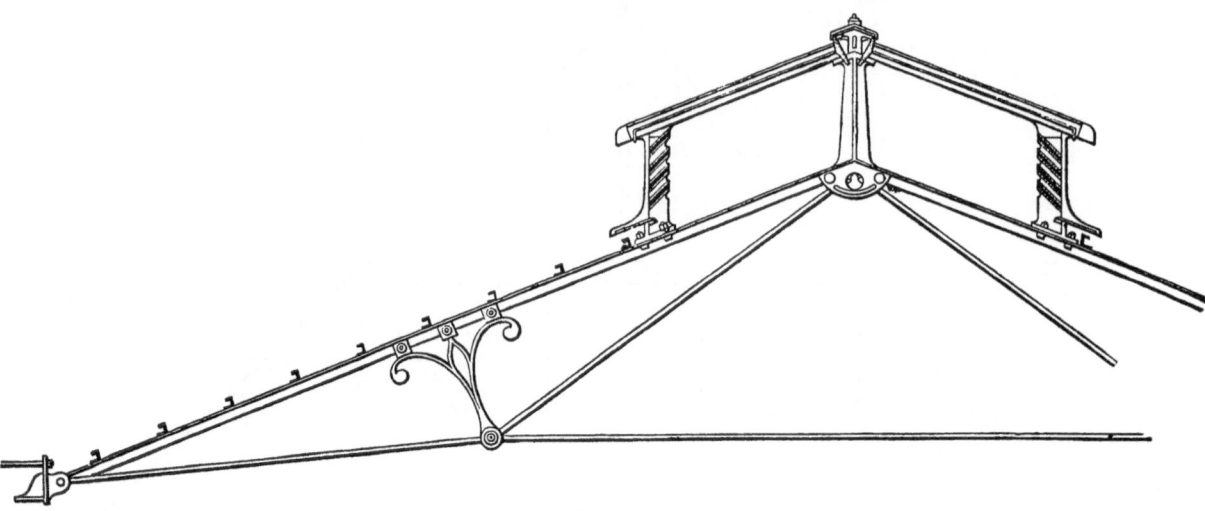

Fig. 54.

Engine-House Roof, Manchester Station.

of almost universal adoption for moderate spans. For small spans, a very simple and cheap contrivance was used (fig. 55). It is composed of ordinary wrought-iron tubing: A

the same principle as in fig. 54 is carried out, but to a much larger scale; fig. 57 shows the details of the strut AC, the principal rafter, and tie. This roof, however, is weak

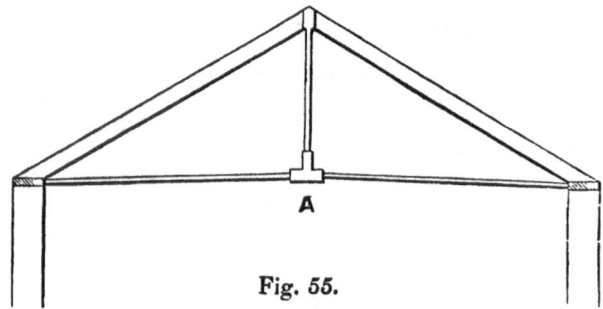

Fig. 55.

is a common tee-piece with right and left screws; the king and tie are thus brought together, the other ends first being put into the fire, and hammered into the form of straps. The considerations before named soon caused the abandonment of the use of wood as principal rafters altogether. Some important roofs were constructed with rafters of cast-iron; but this is a material which, if it gives way, breaks quite suddenly, and without any warning; and by degrees the rolled T iron superseded its use. A very large roof, 87 feet in span, was erected at Paris on the Quai Jemappes,

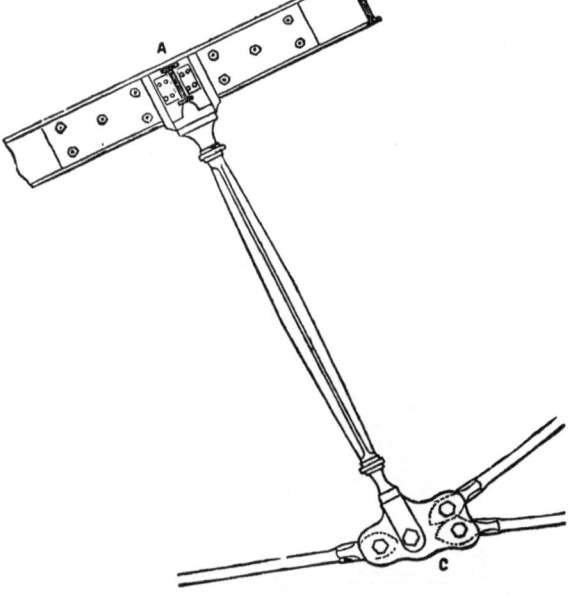

Fig. 57.

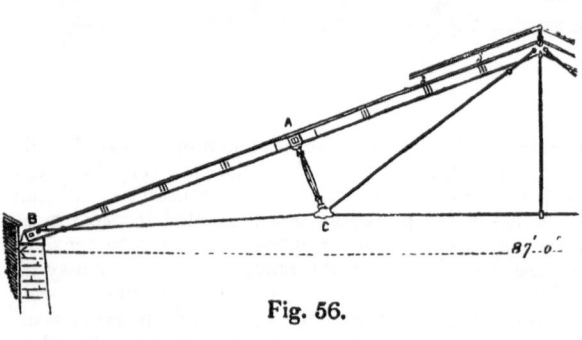

Fig. 56.

over the Providence magazine, as shown in fig. 56, where

between A and B, there being too long a bearing without a strut. It was, however, much improved at Paris at the terminus of the Rouen Railway, where one of a span of 88 feet 6 inches was erected, as shown in fig. 58; while fig. 59 shows the detail of the junction of the ties and struts. The introduction of more struts has much strengthened the principal rafter; but the triangle ABC is too large, and ought to have been subdivided by another strut from B to D, the point C being unfairly loaded, and there being a tendency to sink at the point D. A still larger roof, 97 feet 5 inches in span, was shortly after put up at the terminus of the Strasburg Railway at Paris (fig. 60). This is of an arched form, the principal rafter being a huge circular arc of wrought iron framed on the trellis principle. The top and bottom ribs are of rolled T iron, $3\frac{3}{4}$ inches deep, with a $3\frac{1}{2}$ flange, the web of the metal being $\frac{4}{4}$ths thick. The struts are of cast-iron. The rods which tie this vast structure together are only from $1\frac{3}{8}$ to $1\frac{5}{8}$ inch diameter. The walls which support it are above 40 feet high; and the effect is almost cobweb-like, so slight do the ties appear when viewed from beneath. There is perhaps a defect in making the centre part of the trussing two parallelograms, which could be easily prevented by the introduc-

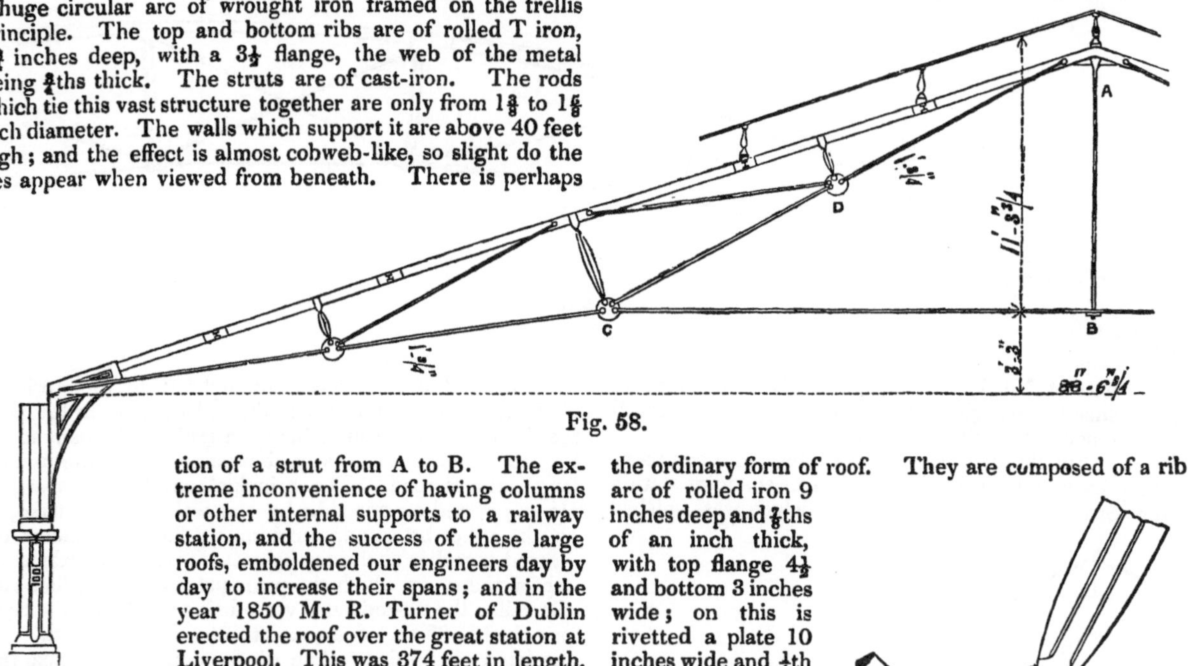

Fig. 58.

tion of a strut from A to B. The extreme inconvenience of having columns or other internal supports to a railway station, and the success of these large roofs, emboldened our engineers day by day to increase their spans; and in the year 1850 Mr R. Turner of Dublin erected the roof over the great station at Liverpool. This was 374 feet in length, of the enormous span of 153 feet, and of the principle shown in fig. 61. The height of the springing is 25 feet, and to the crown 55 feet. The principals are 21 feet 6 inches apart, and are more like a trussed girder than the ordinary form of roof. They are composed of a rib or arc of rolled iron 9 inches deep and $\frac{7}{8}$ths of an inch thick, with top flange $4\frac{1}{2}$ and bottom 3 inches wide; on this is rivetted a plate 10 inches wide and $\frac{1}{4}$th of an inch thick. From the springing to the haunches these ribs are strengthened by plates rivetted on both sides. The struts radiate as shown, and are constructed like the ribs, but are only 7 inches in depth. From strut to strut there are three sets of tie-rods between the two extreme radiating struts, and two between the others, varying from 2 inches to $1\frac{1}{2}$ inch in diameter. The diagonal braces are $1\frac{5}{8}$ inch. The ends of the principal are secured to a chair of cast-iron, resting partly on cast-iron columns, partly on the walls of the office, and partly on a box-girder Each purlin is composed of three pieces of T iron. The centre piece runs straight

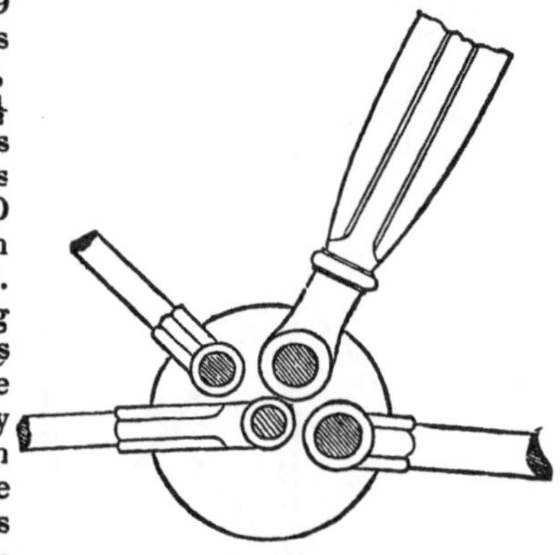

Fig. 59.

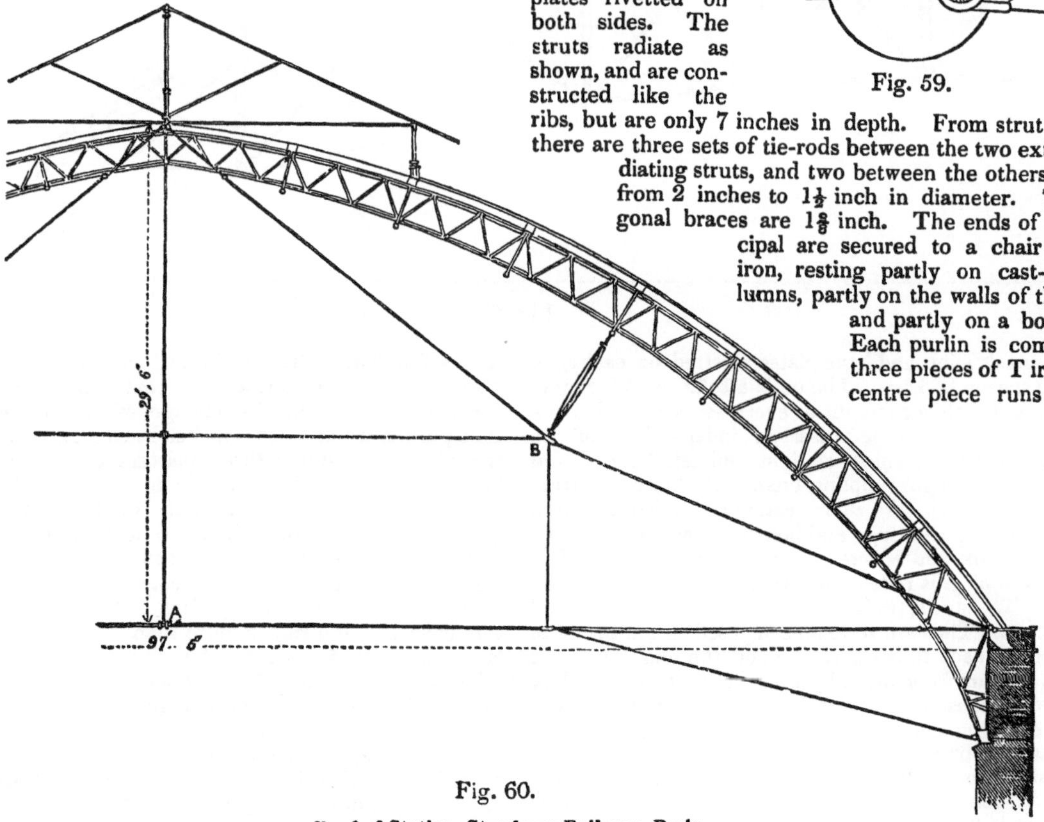

Fig. 60.

Roof of Station, Strasburg Railway, Paris.

from principal to principal, the other two branch off, so as to strut them in three points; besides this, they are crossed by diagonal braces; so that the whole roof is trussed fore and aft, and forms one solid mass of framing. The attachments are made by linking-plates, much like those in fig. 59. The roof is covered partly with corrugated galvanized

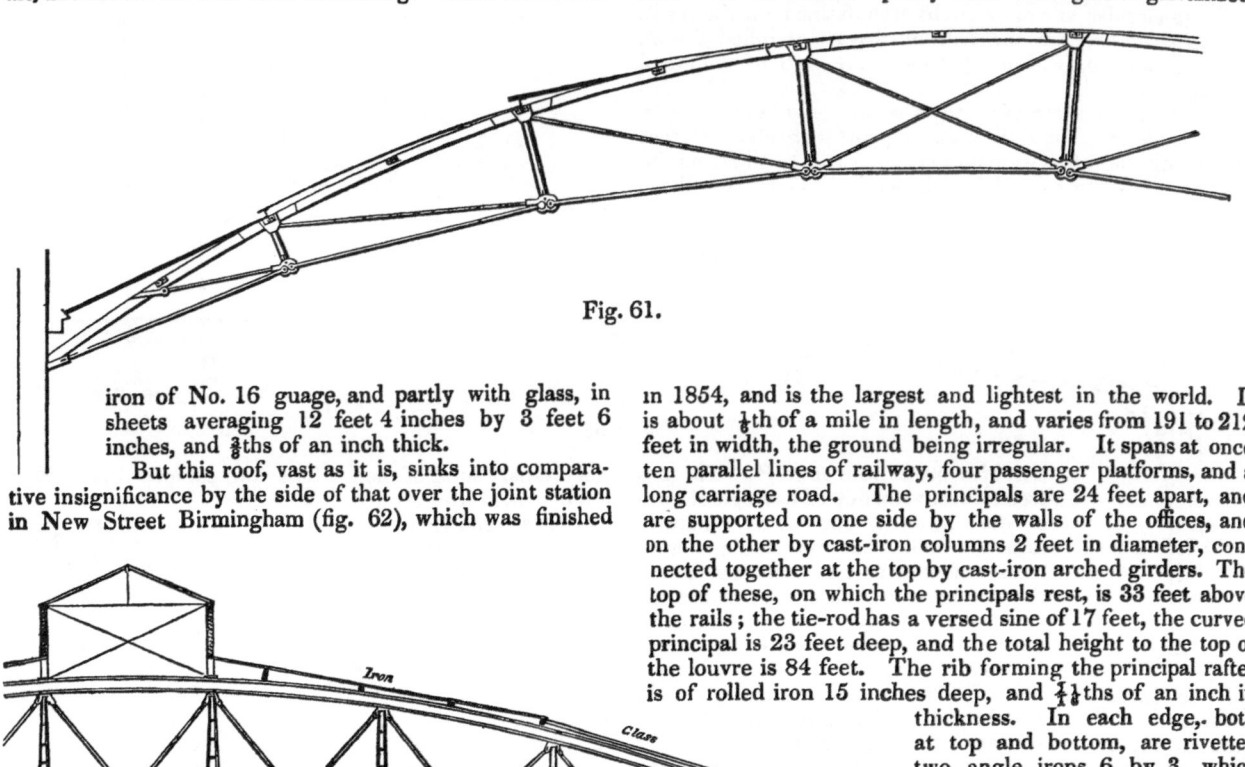

Fig. 61.

iron of No. 16 guage, and partly with glass, in sheets averaging 12 feet 4 inches by 3 feet 6 inches, and ⅜ths of an inch thick.

But this roof, vast as it is, sinks into comparative insignificance by the side of that over the joint station in New Street Birmingham (fig. 62), which was finished in 1854, and is the largest and lightest in the world. It is about ⅛th of a mile in length, and varies from 191 to 212 feet in width, the ground being irregular. It spans at once ten parallel lines of railway, four passenger platforms, and a long carriage road. The principals are 24 feet apart, and are supported on one side by the walls of the offices, and on the other by cast-iron columns 2 feet in diameter, connected together at the top by cast-iron arched girders. The top of these, on which the principals rest, is 33 feet above the rails; the tie-rod has a versed sine of 17 feet, the curved principal is 23 feet deep, and the total height to the top of the louvre is 84 feet. The rib forming the principal rafter is of rolled iron 15 inches deep, and $\frac{7}{16}$ths of an inch in thickness. In each edge, both at top and bottom, are rivetted two angle irons 6 by 3, which together form two flanges 12¾ inches wide. All junctions are

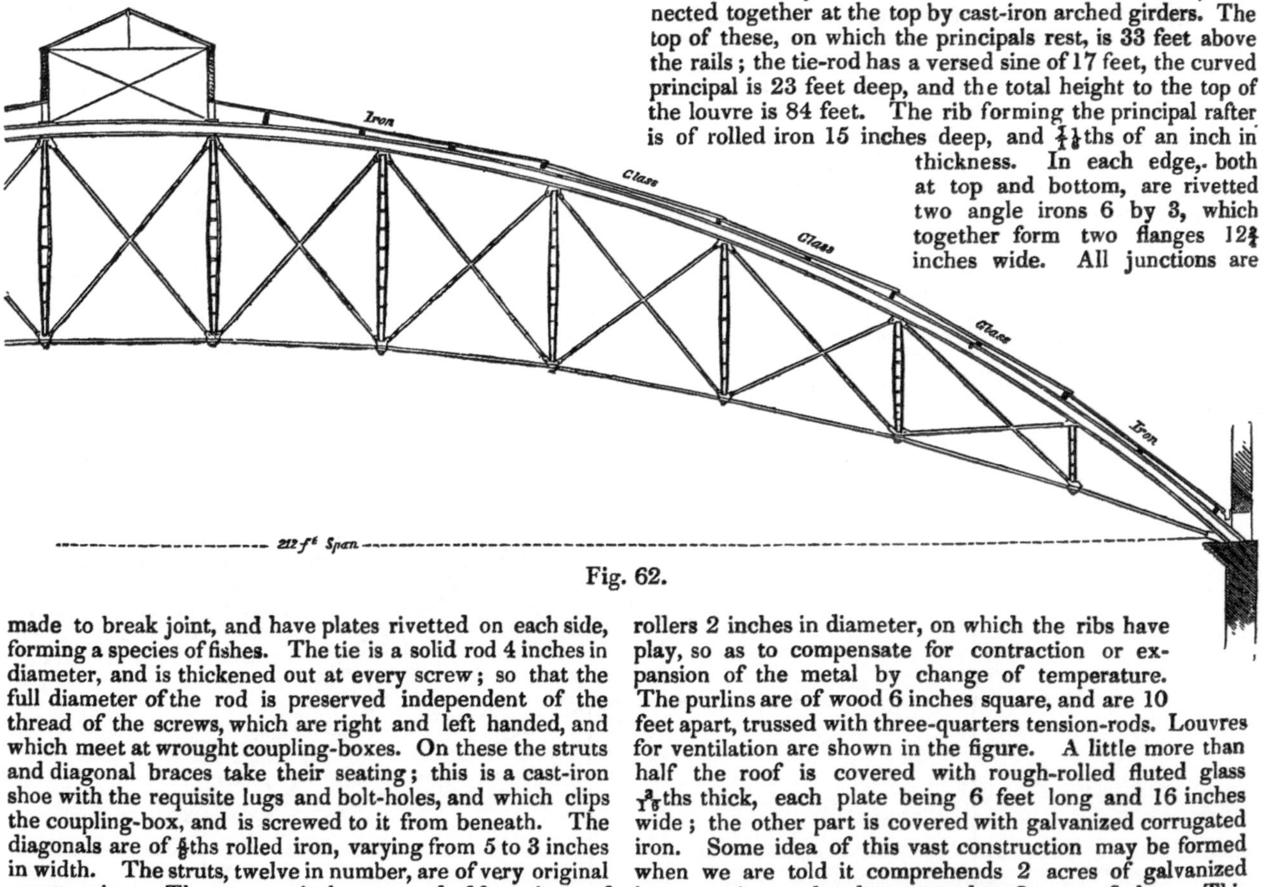

Fig. 62.

made to break joint, and have plates rivetted on each side, forming a species of fishes. The tie is a solid rod 4 inches in diameter, and is thickened out at every screw; so that the full diameter of the rod is preserved independent of the thread of the screws, which are right and left handed, and which meet at wrought coupling-boxes. On these the struts and diagonal braces take their seating; this is a cast-iron shoe with the requisite lugs and bolt-holes, and which clips the coupling-box, and is screwed to it from beneath. The diagonals are of ⅜ths rolled iron, varying from 5 to 3 inches in width. The struts, twelve in number, are of very original construction. They are vertical, composed of four pieces of angle iron set back to back, as if upon the four corners of a square, and are kept apart by iron crosses to which they are bolted. These crosses are larger in the middle of the strut than at the ends, so as to cause the angle irons to curve out each like a bow, and form a sort of open swelling strut, which is not only very strong, but has a very pleasing effect. At one end these vast ribs are secured to stones let into the wall; at the other, which is over the columns, are sets of plates, one attached to the foot of the ribs, and the other to the girders, between which are a series of rollers 2 inches in diameter, on which the ribs have play, so as to compensate for contraction or expansion of the metal by change of temperature. The purlins are of wood 6 inches square, and are 10 feet apart, trussed with three-quarters tension-rods. Louvres for ventilation are shown in the figure. A little more than half the roof is covered with rough-rolled fluted glass $\frac{3}{16}$ths thick, each plate being 6 feet long and 16 inches wide; the other part is covered with galvanized corrugated iron. Some idea of this vast construction may be formed when we are told it comprehends 2 acres of galvanized iron covering, and rather more than 2 acres of glass. This last weighed 115 tons, and the whole iron work 1050 tons. The cost was L.32,274, or about L.19 per square; but iron at that time was exceedingly low in price—more so, in fact, than had ever been known.

A very ingenious roof (fig. 63) has just been erected over the new Royal Italian Opera-House at Covent Garden. It was designed by Mr Edward Barry, and is 90 feet in span, and on the ridge-and-furrow principle. The spans being supported by a series of double-trellis girders 9 feet deep, and 19 feet 6 inches apart from centre to centre, between which are the painting-rooms, carpenters' shops, &c. They are entirely of wrought iron, the ends AB (fig. 63) being, in fact, box-girders for a length of 3 feet 9 inches. A double set of iron trellis is then constructed, as

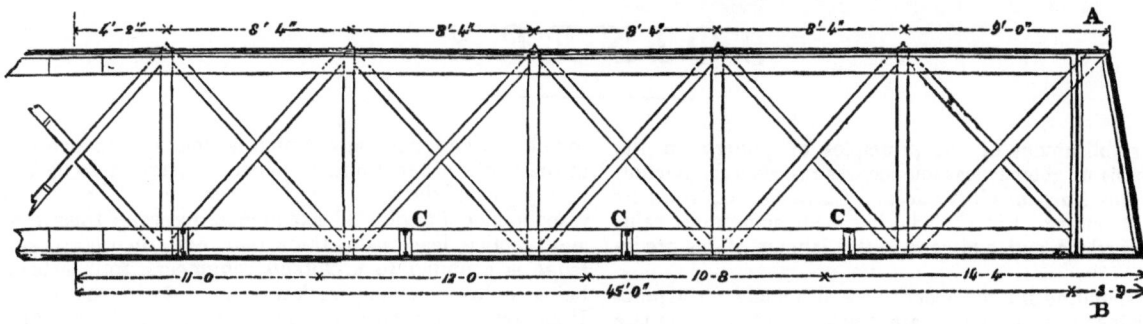

Fig. 63.

Trellis Girder and Roof, Royal Italian Opera.

in fig. 63, 9 feet deep, and 6 inches apart. On the top and bottom of each set, and on both sides (fig. 64, AB), a series of angle iron is rivetted, on the top and bottom of which, again, a plate of flat iron is also rivetted, forming

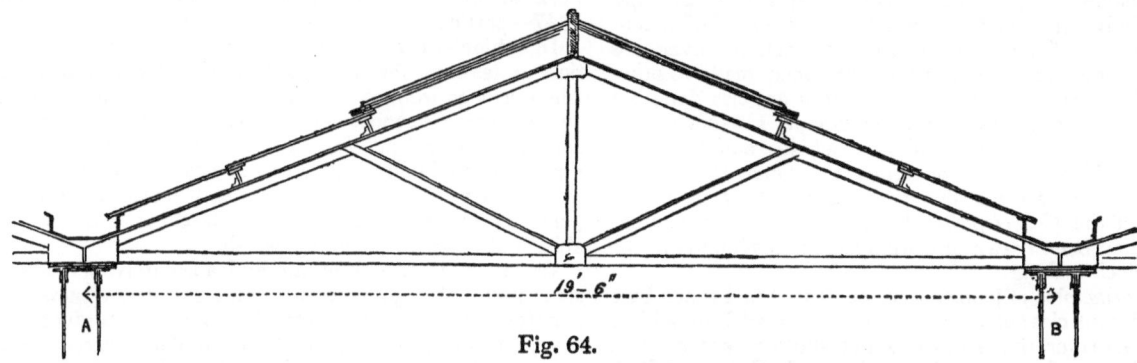

Fig. 64.

upper and lower compression and tension flanges 1 foot 6 inches wide. The whole may be regarded as a perforated hollow beam 90 feet long, 9 feet deep, and 6 inches in thickness, having, as has before been stated, top and bottom flanges 1 foot 6 inches wide. From each of these rises a light transverse roof 19 feet 6 inches in span, the gutter of which is on the top of the girder. The tie and king are of flat iron, and the struts, rafters, and purlins of T iron, all as shown in fig. 64. At the top are skylights which open for ventilation; the rest of the roof is covered with slabs of slate. The floor of the painting-rooms is supported by iron bearers (CC, fig. 63), on which rest common joists and the ordinary boarding. The strength is enormous, and the construction has these advantages: It is not only lighter than any known roof, but there is no thrust in the walls; on the contrary, it tends to tie them together, and there is no loss of room, as the interval between each girder forms a fine room 90 feet long, 19 feet wide, and 9 feet high. (A. A.)

JOINERY.

THE establishment of the principles of joinery on the sound basis of geometrical science was reserved for Nicholson. In his *Carpenter's Guide*, and *Carpenter and Joiner's Assistant*, published in 1792, he has made some most valuable corrections and additions to the labours of his predecessors.

Corresponding improvements were also made in the practice of joinery, for which we are much indebted to the late Mr James Wyatt. But the art is still far short of perfection. In fact, in some respects it seems to have retrograded. It is seldom we find large glued-up pannels will now stand well. Mouldings of great girth give at the mitres, doors wind, and skirtings shrink from the floors, in a way seldom seen in old houses. Our sashes, perhaps, are made better than the heavy barred windows of a century and a half ago. In no other respect, however, has joinery made the progress which has been made in other arts. The improved state of machinery has also done but little for its excellence, though the circular saw-bench, the planing machines, the moulding machines, and the mortising machines, have done much to reduce the cost of labour. This last machine was suggested by us in the edition of the *Encyclopædia Britannica* (1830), our attention having been drawn to it from the improvements in the art of block-making.

Progress of joinery in France. The principles of joinery were cultivated in France by a very different class of writers. The celebrated Blondel had given details for the construction of shutters, wainscoting, doors, hinges, fastenings, &c., in his work *Palais et Maisons de Campagne.* In the extensive work of Frezier, entitled *Coupé des Pierres et des Bois,* 3 vols. 4to, 1739, all the leading principles are given and explained with tedious minuteness, offering a striking contrast to the brevity of our English writers. The first elementary work on that part of geometrical science which contains the principles of joinery appeared in France in 1705, from the pen of the celebrated Gaspard Monge, who gave it the name of *Géometrie Déscriptive.* Much of what has been given as new in English works had been long known on the Continent; but there does not appear to have been much, if any, assistance derived from these foreign works by any writer prior to Nicholson. In fact, this writer has been the founder of all the subsequent works on the subject: Peter Nicholson's *Carpenter and Joiner's Assistant* has been published again and again, in various forms, with additions from time to time, by different hands. For revived mediæval and Elizabethan joinery, particularly as adapted to windows and staircases, Weale's *Carpentry,* 4to, 1849, will be found of great value; and most modern improvements are given in Laxton's *Examples of Building Construction,* now in course of publication.

The most celebrated French work which treats of joinery is Rondelet's *L'Art de Bâtir.* It is also the best foreign work on the subject that we have seen; but it is little adapted to the state of joinery in England. In practice the French joiners are very much inferior to our own. Their work is rough, slovenly, and often clumsy, and at the best is confined to external effect. The neatness, soundness, and accuracy, which is common to every part of the works of an English joiner, is scarcely to be found in any part of the works of a French one. The little correspondence, in point of excellence, between their theory and practice, leads us to think that their theoretical knowledge is confined to architects, engineers, &c., instead of being diffused among workmen, as it is in this country. Rondelet's work occupied fourteen years (from 1802 to 1816) in publication, since which *Nosban, Manuel de Menuiserie,* 4to, 1849; and *Thiollet et Roux, Nouveau Recueil de Menuiserie* for 1837, are the most celebrated works published in France. The latter is expressly to be commended. Much also may be learned from the famous work of Col. Emy—*Traité de la Charpente,* atlas fol., 1847—particularly with regard to framing.

In cabinet-work the French workmen are certainly superior, at least as far as regards external appearance; but when use, as well as ornament, is to be considered, our own countrymen must certainly carry away the palm. The appearance of French furniture is much indebted to a superior method of polishing, which is now generally known in this country.[1] For many purposes, however, copal varnish (such as coachmakers use) is preferable; it is more durable, and bears an excellent polish.

Geometrical knowledge necessary. Geometry is useful in all, and absolutely necessary in some parts of a joiner's business; but it is absurd to encounter difficulties in execution, and to sacrifice good taste, convenience, economy, and comfort, merely for the purpose of displaying a little skill in that science. It is, however, a common fault among such architects as are better acquainted with geometrical rules than with the production of visible beauties, to form designs for no other purpose than to create difficulties in the execution.

But, when geometrical science is properly directed, it gives the mind so clear a conception of the thing to be executed, that the most intricate piece of work may be conducted with all the accuracy it requires.

Practice of joinery. The practice of joinery is best learned by observing the methods of good workmen, and endeavouring to imitate them. But the sooner a workman begins to think for himself the better; he ought always to endeavour to improve on the processes of others, either so as to produce the same effect with less labour, or to produce better work.[2]

We intend, in this article, to give a plain and simple exposition of the most valuable principles of the art of joinery, which will, we hope, place many parts of the practice under a new point of view, and ultimately tend to improve them.

Cabinet-making. Cabinet-making, or that part of the art of working in wood which applies to furniture, has little affinity with joinery, though the same materials and tools be employed in both. Correctness and strict uniformity are not so essential in moveables as in the fixed parts of buildings; they are also more under the dominion of fashion, and therefore are not so confined by rules as the parts of buildings.

Cabinet-making offers considerable scope for taste in beautiful forms, and also in the choice and arrangement of

---

[1] The method of making and using the French polish is minutely described in Dr Thomson's *Annals of Philosophy,* vol. **xi.** pp. 119 and 371.

[2] Descriptions of the tools, with instructions for using them, may be found in Moxon's work before quoted, and in Nicholson's *Mechanical Exercises,* Taylor, London, 1812.

coloured woods. It requires considerable knowledge of perspective, and also that the artist should be able to sketch with freedom and precision.

If the cabinet-maker intend to follow the higher departments of his art, it will be necessary to study the different kinds of architecture, in order to make himself acquainted with their peculiarities, so as to impress his works with the same character as the rooms they are to furnish.

In as far as regards materials, and the principles of joining work, the cabinet-maker will find some useful information in the second and third sections of this article. Many curious works on furniture were published in the reigns of Louis XIV. and XV., also by Chambers and the Adams. Cruden's *Joiner's and Cabinet-Maker's Darling*, 8vo, 1770, is also a curious book. In ornamental composition he may derive much benefit from Tatham's *Etchings of Ancient Ornamental Architecture*, London, 1799; Percier and Fontaine's *Recueil des Décorations Intérieures comprenant tout ce qui a rapport à l'Ameublement*, Paris, 1812; and, for general information, the *Cabinet Dictionary* and the *Cabinet-Maker and Upholsterer's Drawing-Book*, of Sheraton, may be consulted. But the most important works that could be consulted are the various publications relative to the Great Exhibitions in London and Paris, 1851 and 1855, where some of the finest specimens in the world were exhibited. The most accessible to the English reader are those given in the *Art Journal* and *Illustrated London News*.

SECT. I.—ON MAKING WORKING DRAWINGS.

1. In this section we propose to lay before the reader the most important part of the principles of describing, on a plane surface, the lines necessary for determining bevels, forming moulds, or any other purpose required in the practice of joinery. The limits within which such an article as joinery must be confined, in a work like this, will not permit us to enter much into detail on the various points to be illustrated in this section; but we hope, by judicious selection, to place under one point of view the principles that are most useful to the joiner.

Projection of Bodies.

Nature of projection illustrated.

2. A clear idea of the nature of projection is so essential in making working drawings, that, in our endeavours to illustrate it, we cannot proceed upon principles too simple. In the first stage of such an inquiry, experiment furnishes at once the most clear and satisfactory evidence, particularly to those who are not familiar with mathematical subjects.

If some small pieces of wood, or pieces of wire, were joined together, so as to represent the form of a solid body, a cube for example, and if this figure were held between the sun and the surface of a plane board, then the shadow of the figure upon the board would be its projection upon that plane. From this simple experiment it will appear, that the projection of any line placed in the direction of the sun's rays will be a point: the projection of any line parallel to the plane will be of the same length as the line itself, and the projection of any line inclined to the plane will be always shorter than that line.

3. We have supposed the board to be placed at any angle with the direction of the rays of the sun; but, for our present purpose, it is sufficient to consider them to fall perpendicularly upon it; hence it is obvious, that to project a straight line upon a plane, a perpendicular to the plane should be let fall from each end of the line, and the line joining the points where the perpendiculars meet the plane will be the projection required.

When a projection is made upon a horizontal plane, it is usually called a *plan* of the body. When the projection is upon a vertical plane, it may be an *elevation* or a *section* of the body; it is a section when a portion is supposed to be cut off; and the plane of projection is usually parallel to the plane of the section.

4. Bodies may be divided into three classes, according to the kinds of surfaces by which they are bounded. The first class comprehends those which are bounded by plane surfaces, such as cubes, prisms, pyramids, and the like. The second class contains those which are bounded in part by plane surfaces, and the rest by curved surfaces, as cylinders, cones, &c. The third including those which are bounded by curved surfaces only, as spheres, spheroids, &c.

The projections of the first class of bodies will consist of straight lines; those of the second class, of curved as well as straight lines; and those of the third class, of curved lines only.

Projection of lines

5. Let ABCD and CDEF (fig. 1), be two plane surfaces, connected by a joint at CD, so that while the plane CDEF remains horizontal, the plane ABCD may be placed perpendicular to it, and thus represent a vertical plane. Then, if a line be so placed in space that ab is its projection on the vertical plane, and $a'b'$ its projection on the horizontal plane, its projection on any other vertical plane, HGEC, may be determined.

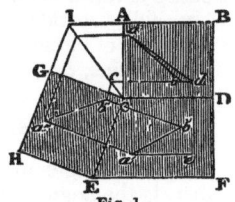

Fig. 1.

This is easily effected, for we have seen, that if a perpendicular be drawn to the plane from each end of the given line, they will give the positions of the ends of the line in the projection (art. 3.) Now, the same thing will be done, by drawing $a'a''$ and $b'b''$ perpendicular to EC, and setting off the points a'' and b'' at the same height above EC respectively, as a and b are above CD, then the line $a''b''$ is the projection required.

The heights may be transferred from one vertical plane to another when they are both supposed to be laid flat, by drawing the line IC, so as to bisect the angle ECD, and if cb be parallel to CD, meeting IC in c, then a line drawn parallel to EC, from the point c, will give the height of the point b'', and so may be found the height of any other point.

To determine the length of a projected line.

6. In the particular case we have drawn, none of the projections represent the real length of the given line. To obtain this length, draw $a'e$ parallel to CD, and with the radius ab' describe the arc $b'e$ cutting $a'e$ in e; draw de perpendicular to CD, cutting the line cb in d; join ad, and it is the length of the given line.

Projection of planes.

The real lengths of lines frequently are not given, therefore another general method of finding them will be useful, and which may be stated as follows:—the length of an inclined line projected upon a plane is equal to the hypothenuse of a right-angled triangle, of which one side is the projection upon the plane, and the other side is the difference between the perpendicular distances of the extremes of the line from the plane.

7. In fig. 2, $a'b'cd$ represents the horizontal projection, or plan, of a rectangular surface, and the elevation ab shows its inclination; and its projection against another vertical plane, making any angle ECD with the former, or plane of elevation, is shown by $a''b''c'd'$. GC being perpendicular to EC, and AC perpendicular to CD, the heights may be transferred by means of arcs of circles described from C as a centre. This is a better method than that by bisecting the angle given in fig. 1; but neither of them so good, in practice, as setting off the heights with the compasses, or with a lath. In our figures it is desira-

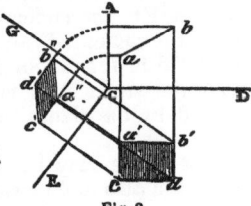

Fig. 2.

Joinery.

ble to show the connection of corresponding parts as much as possible; therefore, the reader will bear in mind that many of the operations we describe may be done with fewer lines when the operator is fully master of his subject.

8. It may be further noticed in this place, that when a point is to be determined in one line by the intersection of another, the lines should cross each other as nearly at right angles as possible; for, when the intersecting lines cross very obliquely, a point cannot be determined with any tolerable degree of accuracy.

Projection of curved surfaces.

9. A curved line can seldom be projected by any other means than by finding a number of points through which the curve may be traced. Rules for all these cases, and for the development of surfaces, will be found under the heads GEOMETRY, CONIC SECTIONS, PROJECTION, &c., &c., and would swell our work to too great a length were we to repeat them. They are also more applicable to the carpenter than the joiner. An exceedingly good practical treatise is furnished in Weale's *Carpentry*, 4to, 1849, vol. i., where full directions are given to find the lines for the development of circular niches, circular-headed sashes, domes, niches, groins, &c., &c.

To determine the Angle formed by two Inclined Planes.

To find the angle of planes inclined to one another.

10. The angle made by two planes which cut one another, is the angle contained by two straight lines drawn from any (the same) point in the line of their common section, at right angles to that line; the one in the one plane, and the other in the other. This angle is the same as that which the joiner takes with his bevel, the bevel being always applied so that its legs are square from the arris, or common section of the planes.

If two lines, AB and CD, be drawn upon a piece of pasteboard, at right angles to one another, crossing at the point E, and the pasteboard be cut half through, according to the line AB, so that it may turn upon that line as a joint; then, to whatever angle, CED (fig. 3), the parts may be turned, the lines EC and ED will be always in the same plane. Also, a line FD, drawn from any point D, in the line ED, to any point F, in the line EC, will be always in the same plane. From these self-evident properties of planes, it is easy to determine the angle formed by any two planes, when two projections, or one projection and the development of the surfaces, are given.

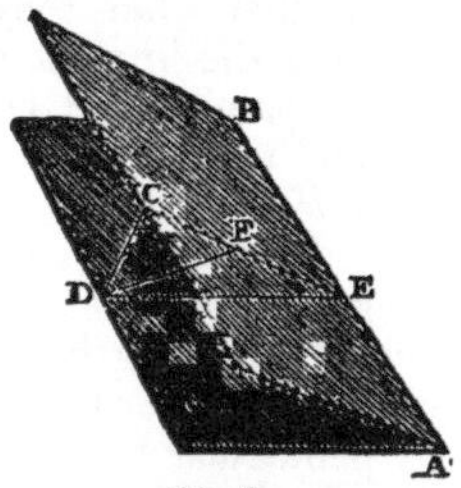

Fig. 3.

11. Let ABC (fig. 4), be the plan of part of a pyramid, and BD the elevation of the arris, or line formed by the common section of the planes in respect to the line EB; EB being the projection of that arris upon the plan.

Draw AC perpendicular to EB, cutting it in any point E, and from E draw EF perpendicular to DB. With the radius EF, and centre E, cross EB in *f*, and join A*f* and *f*C; then the angle A*f*C is the angle formed by the planes of the pyramid.

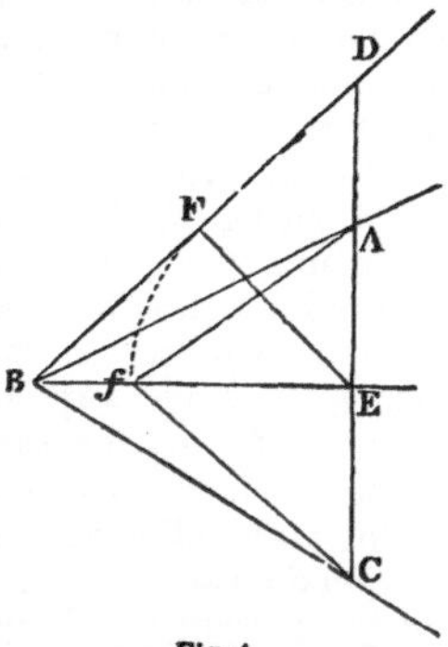

The angle may be constructed when the plan and elevation of any two lines drawn in the planes, so as to intersect in the arris, are given; but as these projections are not often given in drawings of joiners' work, we have inserted the preceding, though it be a less general method.

The backing, or angle for the back of hip-rafters in car-

Fig. 4.

pentry, and of hipped sky-lights, is found in this manner; ABC being, in that case, supposed to be the plan of an angle of the roof or sky-light, and DB the inclination of the hip-rafter.

12. To show how the angle formed by two planes may be found when the plan and development are given, let it be required to find the angle contained by the two faces of a square pyramid, fig. 5. Let ABC be the plan of a square pyramid, draw *a*C perpendicular to, and bisecting AB, and make *a*C equal to the slant height of the pyramid; then with the centre C, and radius AC, describe the arc A 1 2 3, and make A 1, 1 2, 2 3, respectively equal to AB. Join C 1, C 2, C 3, and the four triangles will show respectively the four surfaces of the pyramid.

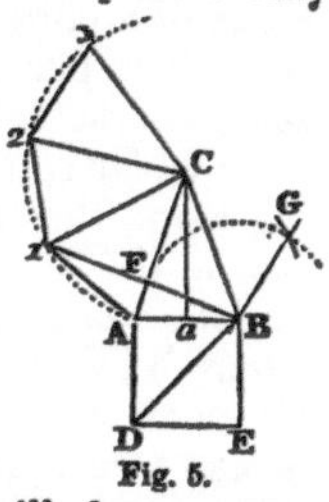

Fig. 5.

Next draw FB perpendicular to AC, and with the radius BF, and centre B, describe the arc FG. Then, with the radius DB, and centre F, cross the former arc in G, join BG, and FBG is the angle formed by the two inclined faces of the pyramid.

Mouldings

Used in joinery are generally composed of parts of circles, and differ somewhat from those used in stone. (See ARCHITECTURE.) Those which present the convex side to the eye are fig. 6, which is merely a rounded edge; fig. 7, if of small size, a bead; if large (fig. 8), a torus: fig. 9 shows the torus and bead together; if there is a deep sinking under a bead (fig. 10), it is called a quirked or cock bead; if there be two such sinkings, so as to show three quarters of a circle in the bead, it is called (fig. 11), a double quirked bead; two or more beads, side by side (fig. 12), are called reeds; the fourth part of a circle, or half a bead (fig. 13), is called an ovolo, or quarter round.

Fig. 6.

Fig. 7.

Fig. 8.    Fig. 9.

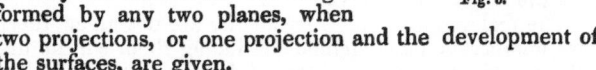

Fig. 10.    Fig. 11.    Fig. 12. 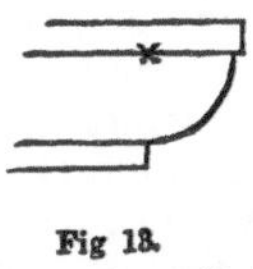   Fig 13.

13. A moulding composed of two convex parts is also called an ovolo, and is delineated thus,—Let the points A, B (fig. 14), represent the height of the moulding and extremities of the curves, from C as a centre draw any circle at pleasure, touching a perpendicular let fall from A, join AB and draw a line parallel to this, touching the circle at D, join DC and produce it towards E, then join DB and make the angle DBE, equal to BDE. From E as a centre draw the rest of the curve BD.

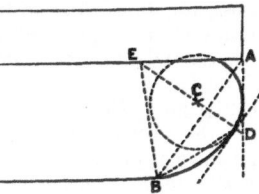

Fig. 14.

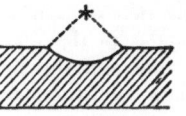

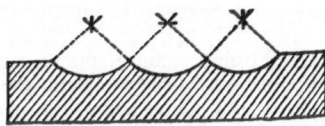

Fig. 15.   Fig. 16.

14. In concave mouldings, a simple curved grooving (fig.

13), is called a hollow, two or more such grooves (fig. 16), are called flutes; a hollow forming the fourth of a circle (fig. 17), is called a cavetto; if the curve die into a plane face, and the line be continued (fig. 18), it is a scape or listel, a deep hollow, generally used as a base moulding (fig. 19), is called a scotia, it is set out thus,—Let *ab* be the height of the required moulding, divide the same into three equal parts,

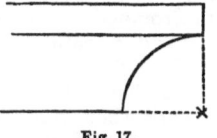

Fig. 17.

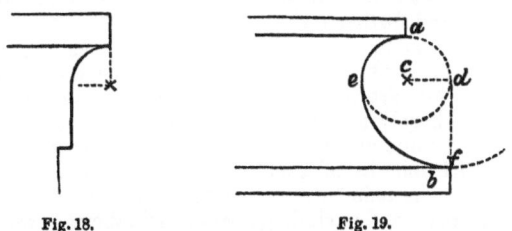

Fig. 18.   Fig. 19.

one of which will be at *c*, from which as a centre draw the circle *aed*; draw *ed* parallel through *c*, and from *d*, with the distance *de*, draw the quarter circle *ef*.

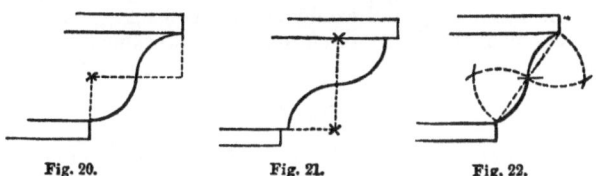

Fig. 20.   Fig. 21.   Fig. 22.

15. In mouldings which are partly convex and partly concave, there are two sorts, the cyma recta and cyma reversa, or ogee; these may be drawn in two ways, as they are required to be bolder or flatter (figs. 20, 21, 22, 23). Grecian mouldings are all drawn on similar principles, but are parts of conic sections instead of circles: they are often struck by hand.

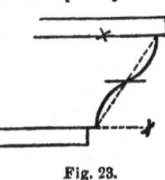

Fig. 23.

A plain square sinking on the edge of a board (fig. 24), for the purposes of framing, is called a *rebate*; if away from the edge (fig. 25), a *groove*; placed under a cap (fig. 26), or as a necking (fig. 27), it is called a *fillet*; three such

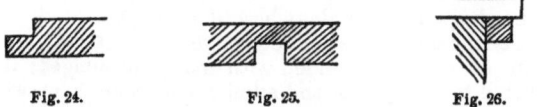

Fig. 24.   Fig. 25.   Fig. 26.

fillets under an ovolo, when composing part of the capital of a column (fig. 28), are called *annulets*.

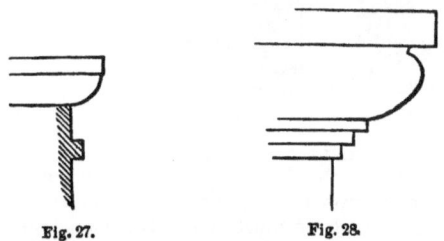

Fig. 27.   Fig. 28.

In all kinds of framing, the mouldings which rise above the styles are called bolextion mouldings. See *infra*, fig. 44.

Raking Mouldings.

16. When an inclined or raking moulding is intended to join with a level moulding, at either an exterior or an in-

terior angle, the form of the level moulding being given, it is necessary that the form of the inclined moulding should be determined, so that the corresponding parts of the surfaces of the two mouldings should meet in the same plane, this plane being the plane of the mitre. It may be otherwise expressed, by saying that the mouldings should mitre truly together.

If the angle be a right angle, the method of finding the form of the inclined moulding is very easy; and as it is not very difficult for any other angle, it may perhaps be best to give a general method, and to illustrate it by examples of common occurrence.

General Method of describing a Raking Moulding, when the Angle and the Rake, or inclination of the Moulding, is given.

17. Let ABC (fig. 29), be the plan of the angle of a build- ing, piece of framework, or any other body, which is to have a level moulding on the side AB; and this level moulding is to mitre with an inclined moulding on the side BC. Also, let CBD be the angle the inclined moulding makes with a level or horizontal line BC.

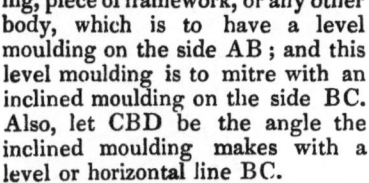

Fig. 29.

Produce AB to *b*, and draw C*b* perpendicular to AB; also make DC perpendicular to BC, and *d*C perpendicular to *b*C. Set off C*d* equal to CD, and join *bd*; then the inclined moulding must be drawn on lines parallel to *bd*.

Let 1, 2, 3, 4, &c., be any number of points in the given section of the level moulding; from each of these points, draw a line parallel to *bd*, and draw A 6′ perpendicular to *bd*. Set off the points 1′, 2′, 3′, 4′, &c., at the same distances, respectively from the line A 6′, as the corresponding points 1, 2, 3, 4, &c., are from the line AB, and through the points 1′, 2′, 3′, &c., draw the moulding. The moulding thus found will mitre with the given one; also, supposing the inclined moulding to be given, the level one may be found in like manner.

If the angle ABC be less than a right angle, the whole process remains the same; but when it is a right angle, BD coincides with *bd*; and the method of describing the moulding becomes the same as that usually given; as it does not then require the preparatory steps which are necessary when the angle is any other than a right angle.

18. It is in pediments, chiefly, that the method of form- ing raking mouldings is of use. Fig. 30 represents part of

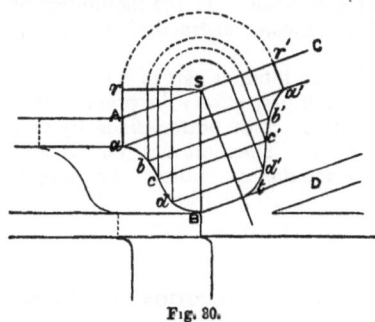

Fig. 30.

a pediment; AB is that part of the level moulding which mitres with the inclined moulding; all that part of the cornice below B being continued along the front, the lower members of the raking cornice stop upon it, and, therefore, do not require to be traced from the other.

In that part of the cornice marked AB, set off a sufficient number of points; and from each of these points draw a line parallel to the rake, or inclination of the pediment. Also, let a vertical line be drawn to each of the same points from the horizontal line *rs*. Make S*t* perpendicular to the inclination of the pediment, and with a slip of paper, or by means of arcs of circles, transfer the distances on *r*S to the line *r'*S, and from the points thus found, draw lines parallel to S*t* ; the intersection of these, with the inclined lines, will determine the form of the moulding, as is indicated by the letters.

When a pediment has a cornice with modillions, the caps of the modillions require to be traced by the same method.

19. It sometimes happens that an inclined base-mould-ing has to mitre with a level one at an angle ; and as the same thing occurs still more frequently with other mould-ing, such as cornices under the steps of stairs, &c., we shall give another example, which will serve still farther to illus-trate the method of proceeding in such cases.

In fig. 31 a raking base-moulding is shown, where the

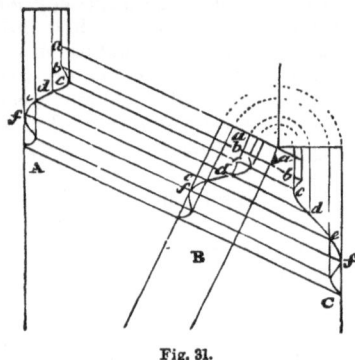

Fig. 31.

inclined moulding B is traced to mitre with the horizontal moulding C ; and the horizontal moulding A is traced to mitre with the inclined one B. The preceding examples being understood, the lines and letters in the figure will be sufficient to show the mouldings are traced. On the same principle the lines for the angle bars of shop sashes, may be readily found.

20. Mouldings being almost the only part of modern joiners' work which can, in strictness, be called ornamental, and consequently that in which the taste of the workman is most apparent, we shall offer a remark or two that may have their use. The form of moulding should be dis-tinct and varied, forming a bold outline of a succession of curved and flat surfaces, disposed so as to form distinct masses of light and shade. If the mouldings be of consi-derable length, a greater distinction of parts is necessary than in short ones.

Mouldings for the internal part of a building should not, however, have much projection ; the proper degree of shade may always be given, with better effect, by deep sinkings judiciously disposed. The light in a room is not sufficiently strong to relieve mouldings, without resorting to this me-thod ; and hence it is that quirked mouldings are so much esteemed.

SECT. II.—ON THE CONSTRUCTION OF JOINERS' WORK.

21. The goodness of joiners' work depends chiefly upon the care that has been bestowed in joining the materials. In carpentry, framing owes its strength to the form and position of its parts ; but in joinery, the strength of a frame depends upon the strength of the joinings. The import-ance, therefore, of fitting the joints together as accurately as possible, is obvious. It is very desirable that a joiner

should be a quick workman, but it is still more so that he should be a good one ; that he should join his materials with firmness and accuracy ; that he should make surfaces even and smooth, mouldings true and regular, and the parts intended to move so that they may be used with ease and freedom. It is also of the greatest importance that the work when thus put together, should be constructed of such sound and dry materials, and on such principles, that the whole should bear the various changes of temperature, and of moisture and dryness, so that the least possible shrinkage or swelling should take place. This last point will be treated further on.

Where dispatch is considered as the chief excellence of a workman, it is not probable that he will strive to improve himself in his art further than to produce the greatest quantity of barely tolerable work with the least quantity of labour. In some articles of short duration, dispatch in the manufacture may be of greater importance ; but in works that ought to remain firm for years, it certainly is bad eco-nomy to spare a few shillings' worth of labour at the risk of being annoyed with a piece of bad work as long as it will hold together.

We have seen, with no small degree of pleasure, the effect of encouraging good workmanship in the construc-tion of machinery, and would recommend that a like en-couragement should be given to superior workmen in other arts.

Joining Angles.

22. When the length of a joint at an angle is not consi-derable, it is sufficient to cut the joint, so that when the parts are joined, the plane of the joint shall bisect the angle. This kind of joint is shown for two different angles, by fig. 32, and is called a mitre.

When an angle of considerable length is to be joined, and the kind of work does not require the joining should be concealed, fig. 33 is often employed ; the small bead renders the appearance of the joint less objection-able, because any irregularities, from shrinkage, are not seen in the shade of the quirk of the head.

Fig. 32.

Fig. 33.

A bead upon an angle, where the nature of the thing does not determine it to be an arris, is attended with many advantages ; it is less liable to be injured, and admits of a secure joint, with-

Fig. 34.   Fig. 35.   Fig. 36.

out the appearance of one. Fig. 34 shows a joint of this description, which should always be used in passages.

Fig. 35 represents a very good joint for an exterior angle, whether it be a long or short one. Such a joint may be nailed both ways. But the joint represented by fig. 36 is superior to it ; the parts being drawn together by the form of the joint itself, they can be fitted with more accuracy, and joined with certainty. The angles of pilasters are often joined, as fig. 36.

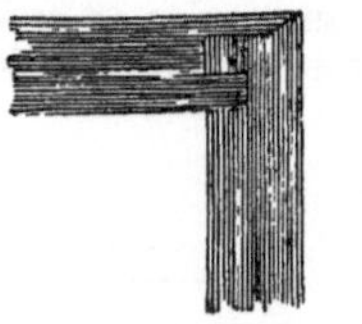

Interior angles are commonly joined, as

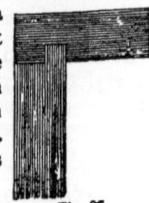

Fig. 37.

Joinery.

shown by fig. 37. If the upper or lower edge be visible, the joint is mitred, as in fig. 32, at the edge only, the other part of the joint being rebated as in fig. 35. In this manner are put together the skirting and dado at the interior angles of rooms, the backs, and back-linings of windows, the jambs of door-ways, and various other parts of joiners' work.

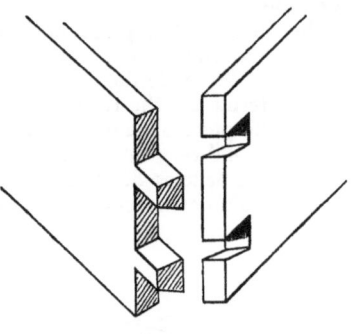

Fig. 38.

Fig. 38 is an excellent method of joining angles for drawers, frames for lead cisterns, boxes, &c., and is commonly called a dovetail; if a portion of the junction is cut off at an angle of 45° (fig.

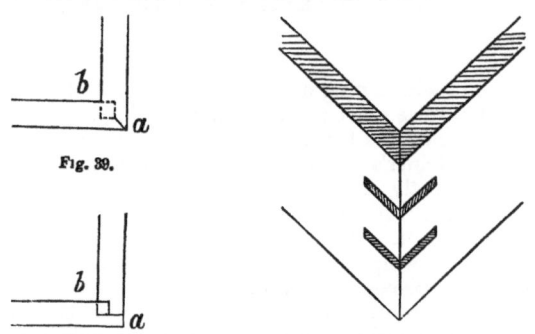

Fig. 39.

Fig. 40.   Fig. 41.

39), while the portion at *b* is dovetailed, it is called a *mitre dovetail*; while if the portion at *a* (fig. 40), passes the other portion at right angles, it is called a *lap-dovetail*.

A very good joint is shown at fig. 41, the angles being brought together at an angle of 45°, two or more saw curfs are cut with a dovetail saw, and thin pieces of wood glued in as shown; this is called a keyed mitre.

Framing.

The object of framing. 23. Frames in joinery are usually connected by mortise tenon joints, with grooves to receive pannels. Doors, window-shutters, &c., are framed in this manner. The object in framing is, to reduce the wood into narrow pieces, so that the work may not be sensibly affected by its shrinkage; and, at the same time, it enables us to vary the surface without much labour. Besides this, as the strains from the grain of the wood are in different directions, the work is prevented from winding on its face.

From this view of the subject, the joiner will readily perceive, that neither the parts of the frame nor the pannels should be wide. And as the frame should be composed of narrow pieces, it follows that the pannels should not be very long, otherwise the frame will want strength. The pannels of framing should not be more than 15 inches wide, and 4 feet long, and pannels so large as this should be avoided as much as possible.[1] The width of the framing is commonly about one-third of the width of the pannel.

It is of the utmost importance in framing that the tenons and mortises should be truly made. After a mortise has been made with the mortise chisel, it should be rendered perfectly even with a float; an instrument which differs from a single cut, or float file, only by having larger teeth. An inexperienced workman often makes his work fit too tight in one place, and too easy in another, hence the mor-

tise is split by driving the parts together, and the work is never firm; whereas if the tenon fill the mortise equally, without using any considerable force in driving the work together, it is found to be firm and sound. The thickness of tenons should be about one-fourth of that of the framing, and the width of a tenon should never exceed about five times its thickness, otherwise, in wedging, the tenon will become bent, and bulge out of the sides of the mortise. If the rail be wide, two mortises should be made, with a space of solid wood between; fig. 42 shows the tenons for a wide rail.

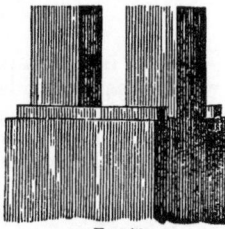

Fig. 42.

Joinery.

If the tenon occurs at the end of a piece of framing, it must be set back a little, so as to allow sufficient solid wood to form a sound mortise; this is called a haunching (see *e*, fig. 43).

In thick framing, the strength and firmness of the joint is much increased by putting a cross or feather tongue in on each side of the tenon; these tongues are about an inch in depth, and are easily put in with a plough proper for such purposes. The projected figure of the end of a rail (fig. 42), shows these tongues put in, in the style there are grooves ploughed to receive them.

Sometimes these projections are left in the solid wood itself, in which case they are called *stump tenons*.

Sometimes, in thick framing, a double tenon in the thickness is made; but we give the preference to a single one, when tongues are put in the shoulders, as we have described; because a strong tenon is better than two weak ones, and there is less difficulty in fitting one than two.

The pannels of framing should be made to fill the grooves, so as not to rattle, and yet to allow the pannels to shrink without splitting. When the mouldings are stuck on the framing, as is often the case in large stuff, it becomes necessary to find the lines to bring the angles together. In square framing, this is done simply by cutting *ab*, *cd* at a mitre; but if the framing be oblique angle, it is done by scribing; the angle at *ab* being determined by eye, *cd* is cut parallel thereto.

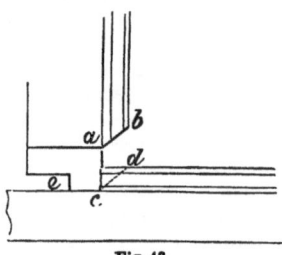

Fig. 43.

Where large projecting or bolexion mouldings are used the French have a very excellent way of framing (fig. 44), which it would be well to imitate in this country. Here C is the pannel round which the moulding B is framed and mitred, the whole is then framed into A, which is a section both of the styles and rails.

Fig. 44.

24. When a frame consists of curved pieces, they are often joined by means of pieces of hard wood called keys. Fig. 45 is the head of a Gothic window frame, joined with a key, with a plan of the joint below it. A cross tongue is put in on each side of the key, and the joint is tightened by means of the wedges *a, a*.

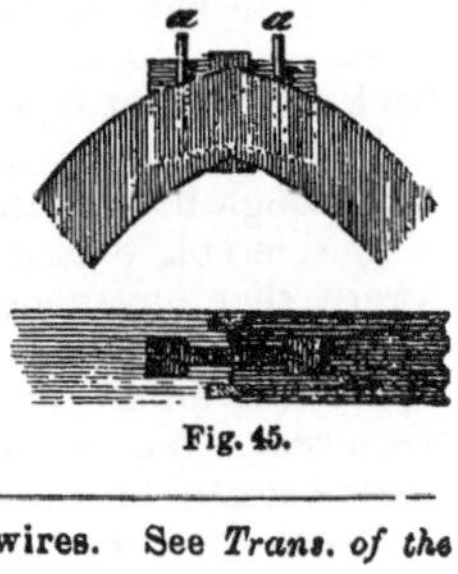

It is, however, a better method to join such pieces by means of a screw

Joining curved pieces

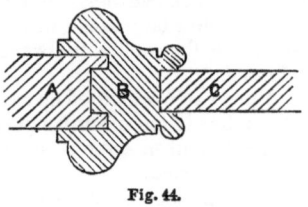

Fig. 45.

---

[1] Pannels of external doors and shutters may be rendered more secure by boring them, and inserting iron wires. See *Trans. of the Society of Arts*, vol. xxv., p. 106.

bolt instead of a key, the cross tongues being used which-ever method is adopted. Where the ends of the bolts cannot be allowed to project, they should be fixed as *bed-bolts*.

Joining with Glue.

25. It is seldom possible to procure boards sufficiently wide for pannels without a joint, on account of heart shakes, which open in drying. In cutting out pannels, for good work, shaken wood should be carefully avoided. That part near the pith is generally the most defective.

If the pannels be thick enough to admit of a cross or feather tongue in the joint, one should always be inserted, for then, if the joint should fail, the surfaces will be kept even, and it will prevent light passing through. A very good way also is to glue a piece of strong canvas on the back of the pannel when the work is not intended to be seen on both sides.

Sometimes plane surfaces of considerable width and length are introduced in joiners' work, as in dado, window backs, &c.; such surfaces are commonly formed of inch, or inch and quarter boards joined with glue, and a cross or feather tongue ploughed into each joint. When the boards are glued together, and have become dry, tapering pieces of wood, called keys, are grooved in, across the back, with a dovetail groove. These keys preserve the surface straight, and also allow it to shrink and expand with the changes of the weather.

26. It would be an endless task to describe all the methods that have been employed to glue up bodies of such varied forms as occur in joinery ; for every joiner forms methods of his own, and merely from his being most familiar with his own process, he will perform his work, according to it, in a better manner than by another, which, to an unprejudiced mind, has manifestly the advantage over it. The end and aim of the joiner, in all these operations, is to avoid the peculiar imperfections and disadvantages of his materials, and to do this with least expense of labour or material. The straightness of the fibres of wood renders it unfit for curved surfaces, at least when the curvature is considerable. Hence short pieces are glued together as nearly in the form desired as can be, and the apparent surface is covered with a thin veneer; or the work is glued up in pieces that are thin enough to bend to the required form. Sometimes a thin piece of wood is bent to the required form upon a cylinder or saddle, and blocks are jointed and glued upon the back; when the whole is completely dry it will preserve the form that had been given to it by the cylinder. The curve should be made a little "quicker" than the curve intended, as the stuff will always spring back a trifle on being released.

A piece of work glued up in thicknesses should be very well done; but it too often happens that the joints are visible, irregular, and in some places open; therefore other methods have been tried.

27. If a piece of wood be boiled in water for a certain time, then taken out and immediately bent into any particular form, and it be retained in that form till it be dry, a permanent change takes place in the mechanical relations of its parts; so that though, when relieved, it will spring back a little, yet it will not return to its natural form.

The same effect may be produced by steaming wood; but though both these methods have been long practised to a considerable extent in the art of ship-building, we are not aware that any general principles have been discovered, either by experiment or otherwise, that will enable us to apply it to an art like joinery, where so much precision is required. We are not aware that it has been tried; but, before it can be rendered extensively useful, the relation between the curvature to which it is bent, and that which it assumes, when relieved, should be determined, and also

the degree of curvature which may be given to a piece of a given thickness.

The time that a piece of wood should be boiled, or steamed, in order that it may be in the best state for bending, should be made the subject of experiments; and this being determined, the relation between the time and the bulk of the piece should be ascertained.

A novel and very simple and effective way of boiling sash-bars or thin articles has been done thus (fig. 46). Take a piece of common cast-iron pipe of sufficient diameter, stop

Fig. 46.

up one end with a plug of wood driven tight, fill the pipe with water, raise one end in a sloping position, leaning it on a pile of bricks, and kindle a fire as shown in the diagram.

For the joiner's purposes we imagine that the process might be greatly improved, by saturating the convex side of each piece with a strong solution of glue, immediately after bending it. By filling, in this manner, the extended pores, and allowing the glue to harden thoroughly before relieving the pieces, they would retain their shape better.

28. Large pieces of timber should never be used in joinery, because they cannot be procured sufficiently dry to prevent them splitting with the heat of a warm room. Therefore, the external part of columns, pilasters, and works of a like kind, should be formed of thin pieces of dry wood; and, if support be required, a post, or an iron pillar, may be placed within the exterior column. Thus, to form columns of wood, so that they shall not be liable to split, narrow pieces of wood are used, not exceeding five inches in width. These are jointed like the staves of a cask, and glued together, with short blocks glued along at each joint.

Fig. 47 is a plan of the lower end of a column glued up in staves; the bevel at A is used for forming the staves, that at B is used for adjusting them when they are glued together. A similar plan must be made for the upper end of the column, which will give the width of the upper end of the staves. The bevels taken from the plan, as at A and B, are not the true bevels; but they are those generally used, and are very nearly true,

Fig. 47.

when the columns are not much diminished. To find the true bevels, the principle we have given in art. 19 should be applied. The same method may be adopted for forming large pillars for tables, &c.

If a column have flutes, with fillets, the joints should be in the fillets, in order to make the column as strong as possible; also, if a column be intended to have a swell in the middle, proper thickness of wood should be allowed for it.

When columns or pillars are small, they may be made of dry wood; and to secure them against splitting a hole should be bored down the axis of each column.

Fixing Joiners' Work.

29. We have hitherto confined our remarks to that part of joinery which is performed at the bench; but by far the most important part remains to be considered. For, how-

Joinery.

ever well a piece of work may have been prepared, if it be not properly fixed, it cannot fulfil its intended purpose. As in the preceding part, we shall state the general principles that ought to be made the basis of practice, and illustrate those principles by particular examples.

If the part to be fixed consist of boards jointed together, but not framed, it should be fixed so that it may shrink or swell without splitting or winding. The nature of the work will generally determine how this may be effected. Let us suppose that a plain back of a window is to be fixed. Fig. 48 is a section showing B the back of the window, A the window-sill, D the floor, C the skirting, and E the wall of the house. The back is supposed to be prepared, as we have stated in art. 29, and that it is kept straight by a dovetailed key *a*. Now, let the back be firmly nailed to the window-sill A, and let a narrow piece *d*, with a groove, and cross tongue, in its upper edge, be fixed to bond timbers or plugs in the wall; the tongue being inserted also into a corresponding groove in the lower

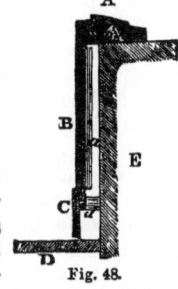

Fig. 48.

edge of the back of B. It is obvious, that the tongue being loose, the back B may contract or expand, as a pannel in a frame. The dado of a room should be fixed in the same

Fixing skirting for rooms.

manner. In the principal rooms of a house, the skirting C is usually grooved into the floor D, and fixed only to the narrow piece *d*, which is called a ground. By fixing in this manner, the skirting covers the joint, which would otherwise soon be open by the shrinking of the back, and from the skirting being grooved into the floor, but not fastened to it, there cannot be an open joint between the skirting and floor. When it is considered, that an open joint, in such a situation, must become a receptacle for dust, and a harbour for insects, the importance of adopting this method of fixing skirting will be apparent. As grooving a floor is attended with considerable labour, and as the boards will sometimes twist, it is more common now to nail a small fillet to the floor, against which the back of the skirting rests, and, of course, has every room for expansion.

In fixing any board above five or six inches wide, similar precautions are necessary; otherwise it is certain to split when the house becomes inhabited. We may, in general, either fix one edge, and groove the other, so as to leave it at liberty, or fix it in the middle, and leave both edges at liberty.

Fixing landings of stairs, tops of tables, &c.

Sometimes a wide board, or a piece consisting of several boards, may be fixed by means of buttons, screwed to the back, which turn into grooves in the framing, bearers, or joists, to which it is to be fixed. If any shrinking takes place the buttons slide in the grooves. In this manner the landing of stairs are fixed, and it is much the best mode of fixing the top of a table to its frame.

Forming architraves, &c.

30. The extension of the principle of ploughing and tongueing work together is one of the most important of the improvements that have been introduced by modern joiners. It is an easy, simple, and effectual method of combination, and one that provides against the greatest defect of timber work, its shrinkage. By means of this method, the bold mouldings of Gothic architecture can be executed with a comparatively small quantity of material; and even in the mouldings of modern architecture it saves much labour. For example, the moulded part of an architrave may be joined with the plain part, as shown by fig. 49. If this method be compared with the old method of glueing one piece upon another, its advantage will be more evident.

Fig. 49.

Fixing grounds.

31. The architraves, skirtings, and surbase mouldings, are fixed to pieces of wood called *grounds*; and as the

straightness and accuracy of these mouldings must depend upon the care that has been taken to fix the grounds truly; it will appear, that fixing grounds, which is a part often left to inferior workmen, in reality requires much skill and attention; besides, they are almost always the guide for the plasterer. Where the plasterer's work joins the grounds, they should have a small groove ploughed in the edge to form a key for the plaster. In old work the ground was generally hidden, but in modern work it is frequently shown, which is a saving of stuff: thus, instead of architraves being prepared as in fig. 49, they are made thus (fig. 50)—A is the rebated and beaded door-jambs, B the ground which is generally splayed at the back as a key to the plastering instead of being grooved. On this a thin piece of stuff is bradded to form the double-faced architrave, instead of sinking out of the solid, and on this the ogee, or ovolo moulding, is nailed. Again, with base mouldings A (fig. 51), is the ground fixed against the wall, on the top of which B is nailed as the upper moulding, and C forms the skirting and lower moulding.

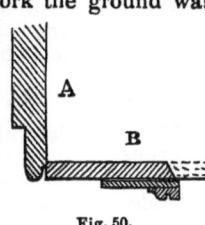

Fig. 50.

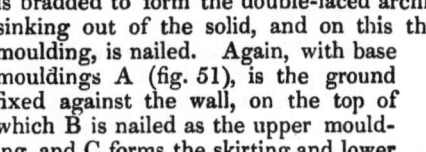

Laying floors.

Fig. 51.

32. In our remarks on construction, we must not omit to say a few words on laying floors, because it will give us an opportunity of pointing out a defect which might be easily remedied. The advice of Evelyn, to tack the boards down only the first year, and nail them down for good the next, is certainly the best, when it is convenient to adopt it; but, as this is very seldom the case, we must expect the joints to open more or less. Now, these joints always admit a considerable current of cold air, and also, in an upper room, unless there be a counter floor, the ceiling below may be spoiled by spilling a little water, or even by washing the floor. To avoid this, we would recommend a tongue to be ploughed into each joint, according to the old practice. When the boards are narrow, they might be laid without any appearance of nails, in the same way as a dowelled floor is laid, the tongue serving the same purpose as the dowels. In this case we would use cross or feather tongues for the joints. A new system of floors has lately been used in London, to which the name of "Victoria floors" has been given. A rough floor of boards, three quarters of an inch thick, is first laid, and the rest of the joiners' work fixed, and the plastering finished. When all is done, an inch or inch and quarter floor of plank, ripped down the middle, and consequently very little more than five inches wide, is laid; the rough boarding being first covered with a layer of shavings, or old newspapers, or other waste paper. The boards are then dowelled on one edge and nailed on the other, and a very sound floor is thus formed, which neither springs nor creaks. We should fear, however, that insects would harbour between the boards, and if frequently washed the damp would get in between the joints, and remain some time in the paper.

Folding floors censured.

There is a method sometimes used in laying floors, which workmen call folding; according to this method, two boards are laid, and nailed at such a distance apart, that the space is a little less than the aggregate width of the boards intended for it; these boards are then put to their places, and, on account of the narrowness of the space left for them, they rise like an arch between its abutments. The workmen force them down by jumping upon them. Accordingly, the boards are never soundly fixed to the joists, nor can the floor be laid with any kind of evenness or accuracy. We merely notice this method here, in order that it may be avoided, except in very common work.

As boards can seldom be got long enough to do without

Fig. 52.   Fig. 53.

joints, it is usual, except in very inferior work, to join the ends with a tongued joint, as shown in fig. 52, where B is the joist. The etched board is first laid, and nailed to the joist.

In oak floors, the ends are forked together sometimes, as shown at A (fig. 53), in order to render the joints less conspicuous.

The joints should be kept as distant from one another as possible.

Hinging.

Hinging. 33. It requires a considerable degree of care to hang a door, a shutter, or any other piece of work in the best manner. In the hinge, the pin should be perfectly straight, and truly cylindrical, and the parts accurately fitted together.

The hinges should be placed so that their axes may be in the same straight line, as any defect in this respect will produce a considerable strain upon the hinges every time the hanging part is moved, will prevent it from moving freely, and is injurious to the hinges.

In hanging doors, centres are often used instead of hinges; but, on account of the small quantity of friction in centres, a door moves too easily, so that a slight draft of air accelerates it so much in falling to, that it shakes the building, and is disagreeable. We have seen this in some degree remedied by placing a small spring to receive the shock of the door.

The greatest difficulty, in hanging doors, is to make them to clear a carpet, and be close at the bottom when shut. To do this, that part of the floor which is under the door, when shut, may be made to rise above a quarter of an inch above the general level of the floor, which, with placing the hinges so as to cause the door to rise as it opens, will be sufficient, unless the carpet should be a very thick one. Several mechanical contrivances have been used for either raising the door, or adding a part to spring close to the floor as the door shuts. The best method now in use, and the simplest, is the invention of the skew-butt hinge. The parts of this which bear on each other are made with a double bevel, so that, if more than half opened, the door falls against the wall by its own weight; if less than half open, it closes itself.

34. Various kinds of hinges are in use. Sometimes they are concealed, as in the kind of joints called rule joints; others project, and are intended to let a door fold back over projecting mouldings, as in pulpit doors. When hinges project, the weight of the door acts with an increased leverage upon them, and they soon get out of order, unless they be strong and well fixed.

Room doors. The door of a room should be hung so that, in opening the door, the interior of the room cannot be seen through the joint. This may be done by making the joint according to fig. 54. The bead should be continued round the door, and a common butt-hinge answers for it.

The proper bevel for joints of a door. The proper bevel for the edge of a door or sash may be

Fig. 54.

Fig. 55.

found by drawing a line from the centre of motion C (fig. 55) to *e*, the interior angle of the rebate; draw *ed* perpen-

dicular to C*e*, which gives the bevel required. In practice the bevel is usually made less, leaving an open space in the joint when the door is shut; this is done on account of the interior angle of the rebate often being filled with paint.

Stairs.

Stairs. 35. The construction of stairs is generally considered the highest department of the art of joinery, therefore we treat of it under a distinct head.

The principal object to be attended to in stairs is, that they afford a safe and easy communication between floors of different levels. The strength of a stair ought to be apparent as well as real, in order that those who ascend it may feel conscious of safety. In order to make the communication safe, it should be guarded by a railing of proper height and strength; in order that it may be easy, the rise and width, or tread, of the steps should be regular and justly proportioned to each other, with convenient landings; there should be no winding steps, and the top of the rail should be of a convenient height for the hand.

Proper proportion for stairs. The first person that attempted to fix the relation between the height and width of a step, upon correct principles, was, we believe, Blondel, in his *Cours d'Architecture.* His formula is applicable to very large buildings, but not to ordinary dwellings. Mr Ashpitel, who has investigated the subject at great length, gives the following rules for buildings of seven different classes :—

Tread breadth in inches.	Rise height in inches.	Tread breadth in inches.	Rise height in inches.
If 12	5½	If 10	6½
„ 11½	5¾	„ 9½	6¾
„ 11	6	„ 9	7
„ 10½	6¼		

These dimensions give angles of ascent varying from 24° to 37°. Of course the projection of the nosing is not reckoned.

Different kinds of stairs. 36. The forms of staircases are various. In towns, where space cannot be allowed for convenient forms, they are often made triangular, circular, or elliptical, with winding steps, or of a mixed form, with straight sides and circular ends. In large mansions, and in other situations, where convenience and beauty are the chief objects of attention, winding steps are never introduced when it is possible to avoid them. Good stairs, therefore, require less geometrical skill than those of an inferior character.

The best architectural effect is produced by rectangular staircases, with ornamented railing and newels. In Gothic structures scarcely any other kind can be adopted, with propriety, for a principal staircase. Modern architecture admits of greater latitude in this respect; the end of the staircase being sometimes circular, and the hand-rail continued, beginning either from a scroll or a newel.

Rectangular staircase. 37. When a rectangular staircase has a continued rail, it is necessary that it should be curved so as to change gradually from a level to an inclined direction. This curvature is called the *ramp* of the rail. The plan of a staircase of this kind is represented by ABCD (fig. 56); and fig. 58 shows a section of it, supposing it to be cut through at *ab*, on the plan.

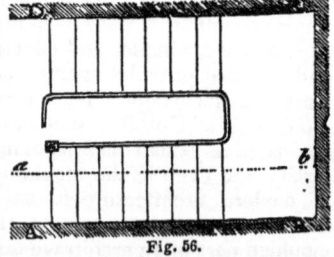

Fig. 56.

The hand-rail is supposed to begin with a newel at the bottom, and the form **To find the cap for newel.** of the cap of the newel ought to be determined, so that it will mitre with the hand-rail. Let H (fig. 57) be the section of the hand-rail, and *ab* the radius of the newel; then the form of the cap may be traced at C by the methods we have already described. (Arts. 17, 18, and 19).

The sections of hand-rails are of various shapes; some of the most common ones are too small; a hand-rail should never be less than would require a square, of which the side is 2¼ inches, to circumscribe it.

For the level landings of a staircase the height of the top of the hand-rail should never be more than about 40 inches, and in any part of the inclined rail the height of its upper side above the middle of the width of the step should be 40 inches, less the rise of one step, when measured in a vertical direction.

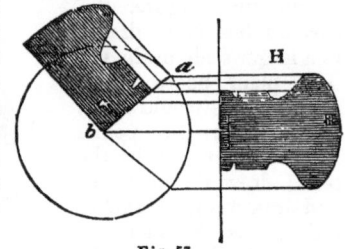

Fig. 57.

To describe the ramps, let *rs* (fig. 58) be a vertical line drawn through the middle of the width of the step; set *ru* equal to *rs*, and draw *ut* at right angles with the back of the rail, cutting the horizontal line *st* in *t*. From the point *t*, as a centre, describe the curve of the rail. When there is a contrary flexure, as in the case before us, the method of describing the lesser curve is the same.

38. The hand-rail of a stair generally begins with a scroll, and the first step

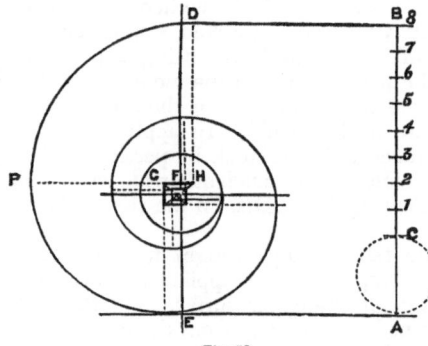

Fig. 58.

of the stair is generally finished with what is called a curtail or form, corresponding as much as possible to the scroll.

There are a great variety of geometrical spirals, which are described and investigated under their proper heads; but as they all finish on a point, and as all architectural scrolls and volutes finish on a circle or eye, the usual mathematical scrolls are inapplicable. The earliest spiral adapted to architecture was that of De Lorme. Since that several systems have been invented, particularly that of Goldmann, but the best is clearly that derived from the Ionic volute, and is drawn thus—

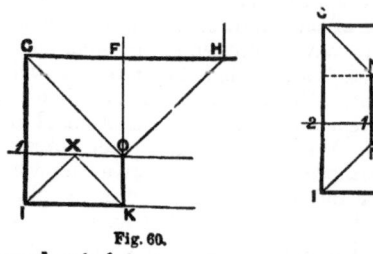

Fig. 59.

The height, eye, and number of revolutions of the im-

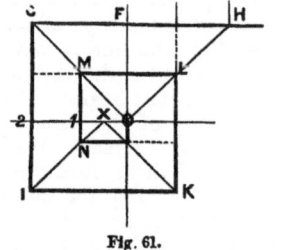

Fig. 60.   Fig. 61.

proved spiral being given to describe the curve, let AB

(fig. 59) be the total height, and AC the intended height of the eye, and let the spiral be required to make two revolutions. Divide BC into four times as many parts as there are revolutions required (4 × 2 = 8), because there are four quadrants in every revolution. Draw any line DE equal to the height of the spiral. Set down from D half the number of parts, and one other part (4 + 1 = 5), this is the top of the eye. Set down half AC at O, and describe the eye; then at O set up half a part to F, and make FG, FH = OF; then (fig. 60) draw OG, OH, GI, and from O draw a line parallel to GH, and divide the same into as many parts as there are to be revolutions. Fig. 60 is for one, fig. 61 for two revolutions. Divide the part O1 at X, and proceed to draw the quarter circles, as in the diagram; HD being the first opening of the compasses, HP the next, and H, G, I, K, L, M, and N being the centres. To describe the scroll let AB (fig. 62) be the width across, usually about 10 or 12 inches; let EB be the intended diameter of the eye; and let the scroll be required to make one revolution and a half, or six quadrants (these are shown at greater size by the side of fig. 63), then proceed as last directed,

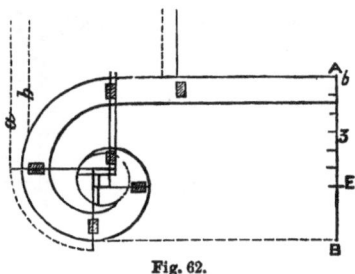

Fig. 62.

and complete the scroll, also dot in the lines of the nosings and risers.

For the curtail step transfer the lines of nosings *a*, and the lines of the risers *b*, to another place, as fig. 63, and set out the thickness of the veneer within the line of nosing, the part within this represents the solid block of the curtail. The places of the ballusters are shown in fig. 62.

It is obvious that in every geometrical staircase, the half of a cylinder placed upright in the well-hole

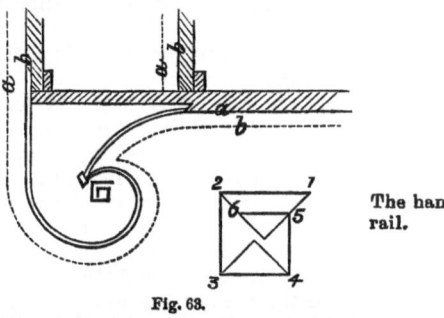

Fig. 63.

would touch the wreathed string in all parts, another a little less would touch all parts of the hand-rail. Let us suppose ACB (fig. 64), to be the plan of half a cylinder so set upright in the well-hole, and let us suppose A'E to be the height of the same. Divide the curved line ACB into any convenient number of parts, and set the same off by compasses on the straight line from C to A' and C to B'. Or, in case ACB is a semicircle, divide the line AB, draw the diameter CD, making *a*D equal to

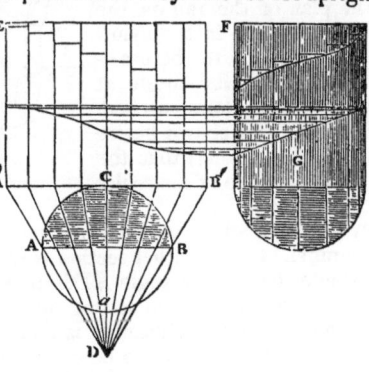

Fig. 64.

three-fourths of the radius, and draw DA, DB', and the rest of the lines through the points of division, as shown in the diagram. Then A'B' is the stretch out or length of the circumference ACB unrolled. But A'E is said to be the whole height. From E set down the respective heights of the winders, step by step, as shown. Now let G be the representation of the cylinder, with the different lines squared up and across, these will give a representation of the curve at which the winders must ascend, and which, of course, must regulate the hand-rail. The other faint lines show the edge of the covering, and is the same as finding a mould for a soffit.

39. Let us now turn to fig. 67. This represents the plan of a staircase, beginning with a scroll, and having steps winding round the circular part of the well-hole.

In the first place, let the end of the steps be developed according to the method we have just given (fig. 65 shows this development). Now the hand-rail ought to follow the inclination of a line drawn to touch the nosings of the steps,

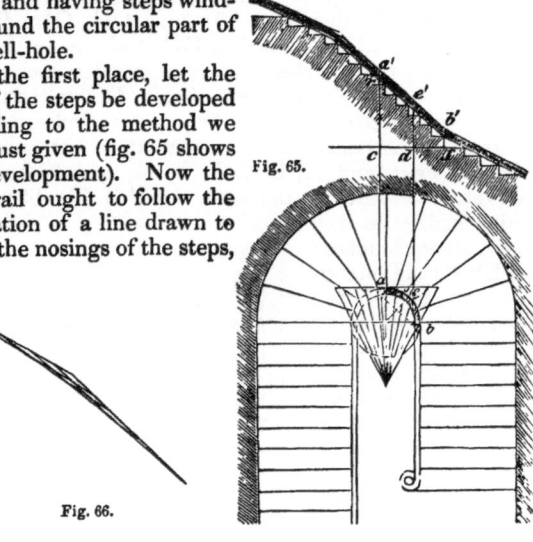

Fig. 65.

Fig. 66.

Fig. 67.

except where there is an abrupt transition from the rake of the winding to that of the other steps; at such places it must be curved,—the curve may be drawn by the help of intersecting lines, as in fig. 66, if the workman cannot trust to his eye.

Development of the circular part.

The part which is shaded in fig. 65 represents the hand-rail and ends of the steps when spread out, and the hand-rail is only drawn close to the steps for convenience, as it would require too much space to raise it to its proper position. This development of the rail is called the falling-mould.

The falling mould.

We will now refer to fig. 68, and will suppose the inner semicircle of ACB to be the plan of the well-hole; *e*A, *a*B, the width of the rail, then the outer shaded part ACB will be the plan of the rail on the level; ADEB is the cylinder referred to before—ADE being the angle at which the stairs ascend. Now we have shown before (CONIC SECTIONS) that the oblique section of a circular cylinder is an ellipse, if the cylinder be circular the lines may then be found by a trammel. Be it of what section it may the delineation of a cylinder cut at any angle ADE may be found by dividing it into equal parts, and setting up the ordinates *a*1, *b*2, &c., as shown. This delineation is a *plan on the oblique*, or the face-mould of the rail, to be cut "on the plumb."

Fig. 68.

The face mould.

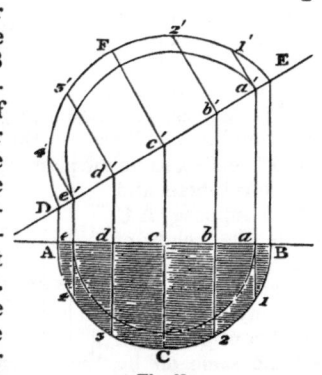

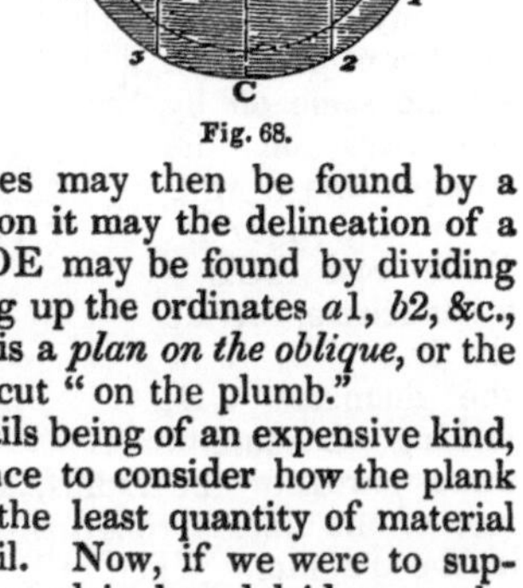

The wood used for hand-rails being of an expensive kind, it becomes of some importance to consider how the plank may be cut so as to require the least quantity of material for the curved part of the rail. Now, if we were to suppose the rail executed, and a plain board laid upon the upper side of it, the board would touch the rail at three points; and a plank laid in the same position as the board would be that out of which the rail could be cut with the least waste of material.

To find the face-moulds, by another method.

Let it be required to find the moulds for the part *ab* of the rail (fig. 67), and to avoid confusing the lines in our small figure, the part *ab* has been drawn to a larger scale in fig. 69. The plain board mentioned above would touch the rail at the points marked C and B in the plan; draw the line CB, and draw a line parallel to CB, so as to touch the curve at the point E. Then E is the other point on the plan; and a', e', and b', are the heights of these points in the development (fig. 65).

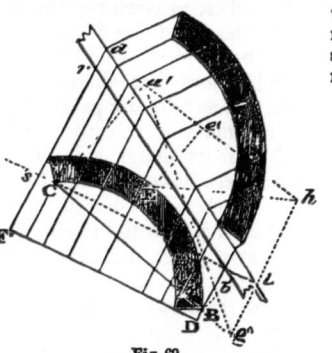

Fig. 69.

Erect perpendiculars to CB, from the points C, E, and B (fig. 69), and set off Ca', on fig. 69, equal to $a'c$ (fig. 65); Ee' equal to de', and Bb' equal to fb'. Through the points C and E, draw the dotted line C*h*; through $a'e'$ draw a line to meet CE in *h*; and through the points $a'b'$, draw a line to meet CB in *g*; then join *hg*, and make C*i* perpendicular to *hg*.

Now, if C*d* be equal to C*a*, and perpendicular to C*i*; and *di* be joined, it will be the angle which the plank makes with the horizontal plane, or plan. Therefore, draw FD parallel to C*i*, and find the section by the process before described. This section is the same thing as would be obtained by projecting vertical lines from each point in the hand-rail against the surface of a board, laid to touch it in three points. The inexperienced workman will be much assisted in applying the moulds if he acquires a clear notion of the position when executed.

To find the thickness of the plank.

To find the thickness of the plank, take the height to the under side of the rail *cr* in the development (fig. 65), and set it off from *s*, in the line C*i*, to *r*, in fig. 69; from the point *r* draw a line parallel to *di*, and the distance between those parallel lines will be the thickness of the plank.

To apply the moulds.

The mould (fig. 69), which is traced from the plan, is called the *face-mould*. It is applied to the upper surface of the plank, which being marked, a bevel should be set to the angle *id*C, and this bevel being applied to the edge will give the points to which the mould must be placed to mark out the under side. It is then to be sawn out, and wrought true to the mould. In applying the bevel, care should be taken to let its stock be parallel to the line *di*, if the plank should not be sufficiently wide for *di* to be its arris. In the method fig. 68, ADE, on the rise of the stair, is the bevel.

After the rail is truly wrought to the face-mould, the falling-mould (fig. 65), being applied to its convex side, will give the edge of the upper surface, and the surface itself will be formed by squaring from the convex side, holding the stock of the square always so that it would be vertical if the rail were in its proper situation. The lower surface is to be parallel to the upper one.

The sudden change of the width of the ends of the steps causes the soffit line to have a broken or irregular appearance; to avoid it, the steps are made to begin to wind before the curved part begins. Different methods of proportioning the ends of the steps are given by Nicholson, Roubo, Rondelet, and Krafft. We cannot in this place enter into a detail of these methods, nor can we give the varied systems of cutting the rail in the spring and in the plumb, about which so much has been written, but for the reader's information a list of the principal writers on staircases is subjoined:—

Joinery.

Price, in his *British Carpenter*, 4to, 1735; Langley, *Builders' Complete Assistant*, 8vo, 1738; Frezier, *Coupe des Pierres et des Bois*, 4to, 1739; Roubo, *L'Art de Menuisier*, folio, 1771; Skaife, *Key to Civil Architecture*, 8vo, 1774; Nicholson, *Carpenters' New Guide*, 4to, 1792; *Carpenters' and Joiners' Assistant*, 4to, 1792; *Architectural Dictionary*, 4to; *Transactions Society of Arts*, &c., for 1814; *Treatise on the Construction of Staircases and Hand-rails*, 4to, 1820; Rondelet, *Traité de l'Art de Bâtir*, tome iv. 4to, 1814; Krafft, *Traité sur l'Art de la Charpenter*, part ii., folio, 1820; Jeakes, *Orthogonal System of Handrailing*, 1849; Ashpitel on *Hand-rails and Staircases*, 4to, 1851; and Riddell, *Handrailing Simplified*, folio, Philadelphia, 1856.

SECT. III.—ON MATERIALS.

Importance of the subject.

40. There is no art in which it is required that the structure and properties of wood should be so thoroughly understood as in joinery. The practical joiner, who has made the nature of timber his study, has always a most decided advantage over those who have neglected this most important part of the art.

In the article ANATOMY, *Vegetable* (vol. iii., pp. 61 and 82), the structure of wood is described; in this place, therefore, we shall only show how the joiner may, in a great measure, avoid the warping caused by its irregular texture.

Boards cut in a particular direction will not retain their form.

41. It is well known that wood contracts less in proportion, in diameter, than it does in circumference; hence a whole tree always splits in drying. Mr Knight has shown that, in consequence of this irregular contraction, a board may be cut from a tree that can scarcely be made by any means to retain the same form and position when subjected to various degrees of heat and moisture. From the ash and the beech he cut some thin boards, in different directions relatively to their transverse septa, so that the septa crossed the middle of some of the boards at right angles, and lay nearly parallel with the surfaces of others. Both kinds were placed in a warm room, under perfectly similar circumstances. Those which had been formed by cutting across the transverse septa, as at A in fig. 70, soon changed their form very considerably, the one side becoming hollow, and the other round; and in drying, they contracted nearly 14 per cent. in width.

Difference in shrinkage.

The other kind, in which the septa were nearly parallel to the surfaces of the boards, as at B in fig. 70, retained, with very little variation, their primary form, and did not contract in drying more than three and a half per cent. in width.[1]

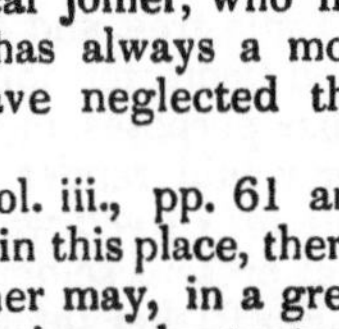

As Mr Knight had not tried resinous woods, two specimens were cut from a piece of Memel timber; and to render the result of our observation more clear, conceive fig. 70 to represent the section

Fig. 70.

of a tree, the annual rings being shown by circles. BD represents the manner in which one of our pieces was cut, and AC the other. The board AC contracted 3·75 per cent. in width, and became hollow on the side marked b. The board BD retained its original straightness, and contracted only 0·7 per cent. The difference in the quantity of contraction is still greater than in hard woods.

From these experiments, the advantages to be obtained merely by a proper attention in cutting out boards for pannels, &c., will be obvious; and it will also be found that pannels cut so that the septa are nearly parallel to their faces, will appear of a finer and more even grain, and require less labour to make their surfaces even and smooth.

Curving in the direction of the width.

But as this system would necessitate the rejection of all

but the heart of the tree for superior work, a method has Joinery. lately been pursued which it is said was first used by the billiard-table makers. Let AC (fig. 71) represent the piece above referred to by the same letters. It will become hollow on the side marked b, no doubt because the rings of the wood when cut across are relieved from tension, and endeavour to expand themselves. To counteract this it is customary, in all good work, to rip the plank down the centre, and then to "turn the stuff inside out" as it is popularly called. This is done by reversing the wood end for end, so as to bring the heart against heart, and the outside against outside (without which the glue joints are sometimes liable to fly); and also so as to reverse the circular parts of the grain, as is shown in fig. 72.

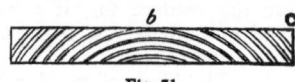
Fig. 71.

Fig. 72.

In wood that has the larger transverse septa, as the oak, for example, boards cut as BD will be figured, while those cut as AC will be plain.

Cause of pieces curving in the direction of their length.

42. There is another kind of contraction in wood whilst drying, which causes it to become curved in the direction of its length. In the long styles of framing we have often observed it; indeed, on this account, it is difficult to prevent the style of a door, hung with centres, from curving, so as to rub against the jamb. A very satisfactory reason for this kind of curving has been given by Mr Knight,[1] which also points out the manner of cutting out wood, so as to be less subject to this defect, which it is most desirable to avoid. The interior layers of wood, being older, are more compact and solid than the exterior layers of the same tree; consequently, in drying, the latter contract more in length than the former. This irregularity of contraction causes the wood to curve in the direction of its length, and it may be avoided by cutting the wood so that the parts of each piece shall be as nearly of the same age as possible. But as this would also necessitate the rejection of a great deal of stuff, a simpler method is found, which is always to turn the heart of the wood outwards. Thus, in framing a door, the heart should always go against the jambs, and the sap side to the pannels.

Changes produced by the weather.

43. Besides the contraction which takes place in drying, wood undergoes a considerable change in bulk with the variations of the atmosphere. In straight-grained woods the change in length is nearly insensible;[2] hence they are sometimes employed for pendulum rods; but the lateral dimensions vary so much, that a wide piece of wood will serve as a rude hygrometer.[3] The extent of variation decreases in a few seasons, but it is of some importance to the joiner to be aware, that even in very old wood, when the surface is removed, the extent of variation is nearly the same as in new wood.

It appears, from Rondelet's experiments,[4] that in wood of a mean degree of dryness, the extent of contraction and expansion, produced by the usual changes in the state of the atmosphere, was—

In fir wood, from $\frac{1}{360}$ to $\frac{1}{75}$ part of its width;

And, in oak, from $\frac{1}{412}$ to $\frac{1}{84}$ part of its width.

Consequently, the mean extent of variation in fir is $\frac{1}{124}$, and in oak, $\frac{1}{140}$; and, at this mean rate, in a fir board about $12\frac{1}{2}$ inches wide, the difference in width would be

---

[1] *Philosophical Transactions*, part ii. for 1817; or *Philosophical Magazine*, vol. l., p. 437.
[2] Mr Ramsden and General Roy made some experiments on the expansion in length. See *Account of the Trig. Survey*, vol. i., pp. 46 and 49.
[3] See *Philosophical Transactions*, Lowthorpe's Abridg., vol. ii., p. 37.
[4] *Traité Théoretique et Pratique de l'Art de Bâtir*, article MENUISERIE, tome iv., p. 425, 1814.

Joinery. $\frac{1}{10}$th of an inch. This will show the importance of attending to the maxims of construction we have already laid before the reader; for, if a board of that width should be fixed at both edges, it must unavoidably split from one end to the other.

Kinds of wood.

44. The kinds of wood commonly employed in joinery are,—the oak, the different species of pine, mahogany, and sometimes lime-tree and poplar.

Oak.

Of the oak, there are two species common in this island; that which Linnæus has named *Quercus robur* is the most valuable for joiners' work; it is of a finer grain, less tough, and not so subject to twist as the other kind. Oak is also imported from the Baltic ports, from Germany, and from America. These foreign kinds being free from knots, of a straighter grain, and less difficult to work, they are used in preference to our home species. Foreign oak is also much used for cabinet-work; and lately, the fine curled oak that is got from excrescences produced by pollard, and other old trees, has been used with success in furniture. When well managed, it is very beautiful, and makes a pleasing variety. It is relieved by inlaid borders of black or white wood, but these should be sparingly used. Borders of inlaid brass, with small black lines, give a rich effect to the darker coloured kinds.

Fir.

The greater part of joiners' work is executed in fir, imported from the north of Europe. Yellow fir is used for outside work, as doors, sashes, and for floors where there is likely to be much wear. Some very good red pine deals have been imported from Canada. Inside work is almost always framed of white fir. Some very good pannels when not too wide, and excellent mouldings, are made of American pine. White fir is often used for internal work, and yellow pine is much used for mouldings.

The forest of Braemar, in Aberdeenshire, furnishes yellow fir of an excellent quality, little inferior to the best foreign kinds.

Larch.

For the general purpose of joinery, the wood of the larch tree seems to be the best; this useful tree thrives well on our native hills. We have seen some fine specimens of this wood from Blair-Athol. It makes excellent steps for stairs, floors, framing, and most other articles.

Mahogany

Mahogany, in joinery, is only used where painted work is improper, as for the hand-rails of stairs, or for the doors and windows of principal rooms. For doors it is not now so often used as it was formerly; its colour is found to be too gloomy to be employed in large masses. In cabinet-work it is almost the only kind used for ornamental work.

Lime-tree

Lime-tree, and the different species of poplar, make very good floors for inferior rooms, and may often be used for other purposes, in places where the carriage of foreign timber would render it more expensive. Lime-tree is valuable for carved work, and does not worm-eat; but carving is at present seldom used in joinery.

For farther information on wood, in addition to the works referred to, the reader may consult Evelyn's *Silva*, Dr Hunter's edition; Duhamel, *Du Transport, de la Conservation, et de la Force des Bois*, Paris, 1767; Barlow's *Essay on the Strength and Stress of Timber*, 1817; Tredgold's *Elementary Principles of Carpentry*, sect. x., 1820; and the article DRY-ROT. (T. T—D.) (A. A.)

STRENGTH OF MATERIALS.

Importance of the subject.

IN *Mechanics*, is a subject of so much importance, that in a nation so eminent as this for invention and ingenuity in every species of manufacture, and in particular so distinguished for its improvements in machinery of every kind, it is somewhat singular that no writer has treated it in the detail which its importance and difficulty demand. The man of science who visits our great manufactories is delighted with the ingenuity which he observes in every part, the innumerable inventions which come even from individual artisans, and the determined purpose of improvement and refinement which he sees in every workshop. Every cotton-mill appears an academy of mechanical science; and mechanical invention is spreading from these fountains over the whole kingdom. But the philosopher is mortified to see this ardent spirit cramped by ignorance of principle, and many of those original and brilliant thoughts obscured and clogged with needless and even hurtful additions, and a complication of machinery which checks improvement even by its appearance of ingenuity. There is nothing in which this want of scientific education, this ignorance of principle, is so frequently observed, as in the injudicious proportion of the parts of machines and other mechanical structures; proportions and forms of parts in which the strength and position are nowise regulated by the strains to which they are exposed, and where repeated failures have been the only lessons.

Strength of materials arises from cohesion.

The strength of materials arises immediately or ultimately from the cohesion of the parts of bodies. Our examination of this property of tangible matter has as yet been very partial and imperfect, and by no means enables us to apply mathematical calculations with precision and success. The various modifications of cohesion, in its different appearances of perfect softness, plasticity, ductility, elasticity, hardness, have a mighty influence on the strength of bodies, but are hardly susceptible of measurement. Their texture, whether uniform like glass and ductile metals, crystallized or granulated like other metals and freestone, or fibrous like timber, is a circumstance no less important; yet even here, although we derive some advantage from remarking to which of these forms of aggregation a substance belongs, the aid is but small. All we can do in this want of general principles is, to make experiments on every class of bodies.

Experiments to ascertain it.

Accordingly philosophers have endeavoured to instruct the public in this particular. The Royal Society of London, at its very first institution, made many experiments at their meetings, as may be seen in the first registers of the society; and since then a vast multitude of experiments have been made by public bodies and private individuals. The best of these, perhaps, up to the date of the present edition, are those of Mr Barlow.

Rendered useful by generalization.

But to make use of any experiments, there must be employed some general principle by which we can generalize their results. They will otherwise be only narrations of detached facts. We must have some notion of that intermedium, by the intervention of which an external force applied to one part of a lever, joist, or pillar, occasions a strain on a distant part. This can be nothing but the cohesion between the parts. It is this connecting force which is brought into action, or, as we more shortly express it, excited. This action is modified in every part by the laws of mechanics.

Strength defined.

It is this action which we call the *strength* of that part, and its effect is the strain on the adjoining parts; and thus it is the same force, differently viewed, that constitutes both the strain and the strength. When we consider it in the light of a resistance to fracture, we call it *strength*.

We call every thing a *force* which we observe ever to be accompanied by a change of motion ; or, more strictly speaking, we infer the presence and agency of a force wherever we observe the state of things in respect of motion different from what we know to be the result of the action of all the forces which we know to act on the body. Thus when we observe a rope prevent a body from falling, we infer a moving force inherent in the rope, with as much confidence as when we observe it drag the body along the ground. The *immediate action* of this force is undoubtedly exerted between the immediately adjoining parts of the rope. The immediate effect is the keeping the particles of the rope together. They ought to separate by any external force drawing the ends of the rope contrariwise ; and we ascribe their not doing so to a mechanical force really opposing this external force. When desired to give it a name, we name it from what we conceive to be its effect, and therefore its characteristic, and we call it *cohesion*. This is merely a name for the fact ; but it is the same thing in all our denominations. We know nothing of the causes but in the effects ; and our name for the cause is in fact the name of the effect, which is *cohesion*. We mean nothing else by gravitation or magnetism. What do we mean when we say that Newton understood thoroughly the nature of gravitation, of the force of gravitation ; or that Franklin understood the nature of the electric force ? Nothing but this : Newton considered with patient sagacity the general facts of gravitation, and has described and classed them with the utmost precision. In like manner, we shall understand the nature of cohesion when we have discovered with equal generality the laws of cohesion, or general facts which are observed in the appearances, and when we have described and classed them with equal accuracy.

Let us therefore attend to the more simple and obvious phenomena of cohesion, and mark with care every circumstance of resemblance by which they may be classed. Let us receive these as the laws of cohesion, characteristic of its supposed cause, the force of cohesion. We cannot pretend to enter on this vast research. The modifications are innumerable ; and it would require the penetration of more than Newton to detect the circumstance of similarity amidst millions of discriminating circumstances. Yet this is the only way of discovering which are the primary facts characteristic of the force, and which are the modifications. The study is immense, but it is by no means desperate ; and we entertain great hopes that it will ere long be successfully prosecuted ; but, in our particular predicament, we must content ourselves with selecting such general laws as seem to give us the most immediate information of the circumstances that must be attended to by the mechanician in his constructions, that he may unite strength with simplicity, economy, and energy.

1. Then, it is a matter of fact that all bodies are in a certain degree perfectly elastic ; that is, when their form or bulk is changed by certain moderate compressions or distractions, it requires the continuance of the changing force to continue the body in this new state ; and when the force is removed, the body recovers its original form. We limit the assertion to *certain moderate* changes. For instance, take a lead wire of one fifteenth of an inch in diameter and ten feet long ; fix one end firmly to the ceiling, and let the wire hang perpendicular ; affix to the lower end an index like the hand of a watch ; on some stand immediately below let there be a circle divided into degrees, with its centre corresponding to the lower point of the wire ; now turn this index twice round, and thus twist the wire. When the index is let go, it will turn backwards again, by the wire untwisting itself, and make almost four revolutions before it stops ; after which it twists and untwists many times, the index going backwards and forwards round the circle, diminishing, however, its arch of twist

each time, till at last it settles precisely in its original position. This may be repeated for ever. Now, in this motion, every part of the wire partakes equally of the twist. The particles are stretched, require force to keep them in their state of extension, and recover completely their relative positions. These are all the characters of what the mechanician calls *perfect* elasticity. This is a quality quite familiar in many cases, as in glass, tempered steel, &c., but was thought incompetent to lead, which is generally considered as having little or no elasticity. But we make the assertion in the most general terms, with the limitation to moderate derangement of form. We have made the same experiment on a thread of pipe-clay, made by forcing soft clay through the small hole of a syringe by means of a screw, and we found it more elastic than the lead wire ; for a thread of one twentieth of an inch diameter and seven feet long allowed the index to make two turns, and yet completely recovered its first position.

2. But if we turn the index of the lead wire four times round, and let it go again, it untwists again in the same manner, but it makes little more than four turns back again ; and after many oscillations it finally stops in a position almost two revolutions removed from its original position. It has now acquired a new arrangement of parts, and this new arrangement is permanent like the former ; and, what is of particular moment, it is perfectly elastic. This change is familiarly known by the denomination of a *set*. The wire is said to have *taken a set*. When we attend minutely to the procedure of nature in this phenomenon, we find that the particles have, as it were, slid on each other, still cohering, and have taken a new position, in which their connecting forces are in equilibrio ; and in this change of relative situation, it appears that the connecting forces which maintained the particles in their first situation were not in equilibrio in some position intermediate between that of the first and that of the last form. The force required for changing this first form augmented with the change, but only to a certain degree ; and during this process the connecting forces always tended to the recovery of this first form. But after the change of mutual position has passed a certain magnitude, the union has been partly destroyed, and the particles have been brought into new situations ; such, that the forces which now connect each with its neighbour tend, not to the recovery of the first arrangement, but to push them farther from it, into a new situation, to which they now verge, and require force to prevent them from acquiring. The wire is now in fact again perfectly elastic ; that is, the forces which now connect the particles with their neighbours, augment to a certain degree as the derangement from this new position augments. This is not reasoning from any theory. It is narrating facts, on which a theory is to be founded. What we have been just now saying, is evidently a description of that sensible form of tangible matter which we call *ductility*. It has every gradation of variety, from the softness of butter to the firmness of gold. All these bodies have some elasticity ; but we say they are not perfectly elastic, because they do not completely recover their original form when it has been greatly damaged. The whole gradation may be most distinctly observed in a piece of glass or hard sealing-wax. In the ordinary form glass is perhaps the most completely elastic body that we know, and may be bent till just ready to snap, and yet completely recovers its first form, and takes no set whatever ; but when heated to such a degree as just to be visible in the dark, it loses its brittleness, and becomes so tough that it cannot be broken by any blow ; but it is no longer elastic, it takes any set, and keeps it. When more heated, it becomes as plastic as clay ; but in this state is remarkably distinguished from clay by a quality which we may call *viscidity*, which is something like elasticity, of which clay and other bodies purely plastic exhibit no appearance. This

Strength of Materials.

is the joint operation of strong adhesion and softness. When a rod of perfectly soft glass is suddenly stretched a little, it does not at once take the shape which it acquires after some little time. It is owing to this that, in taking the impression of a seal, if we take off the seal while the wax is yet very hot, the sharpness of the impression is immediately destroyed. Each part drawing its neighbour, and each part yielding, the prominent parts are pulled down and blunted, and the sharp hollows are pulled upwards and also blunted. The seal must be kept on till the wax has become not only stiff, but hard.

Observed in all homogeneous plastic bodies.

This viscidity is to be observed in all plastic bodies which are homogeneous. It is not observed in clay, because clay is not homogeneous, but consists of hard particles of argillaceous earth sticking together by their attraction for water. Something like it might be made of finely powdered glass and a clammy fluid such as turpentine. Viscidity has all degrees of softness, till it degenerates to ropy fluidity like that of olive oil. Perhaps something of it may be found even in the most perfect fluid with which we are acquainted, as we observed in the experiments for ascertaining specific gravity.

When ductility and elasticity are combined in different proportions, an immense variety of sensible modes of aggregation may be produced. Some degree of both are probably to be observed in all bodies of complex constitution; that is, which consist of particles made up of many different kinds of atoms. Such a constitution of a body must afford many situations permanent, but easily deranged.

Particles acted on by attractions and repulsions.

In all these changes of disposition which take place among the particles of a ductile body, the particles are at such distance that they still cohere. The body may be stretched a little; and on removing the extending force, the body shrinks into its first form. It also resists moderate compressions; and when the compressing force is removed, the body again swells out. Now the corpuscular *fact* here is, that the particles are acted on by attractions and repulsions, which balance each other when no external force is acting on the body, and which augment as the particles are made, by any external cause, to recede from this situation of mutual inactivity; for since force is requisite to produce either the dilatation or the compression, and to maintain it, we are obliged, by the constitution of our minds, to infer that it is opposed by a force accompanying or inherent in every particle of dilatable or compressible matter; and as this necessity of employing force to produce a change indicates the agency of these corpuscular forces, and marks their kind, according as the tendencies of the particles appear to be toward each other in dilatation, or from each other in compression; so it also measures the degrees of their intensity. Should it require three times the force to produce a double compression, we must reckon the mutual repulsions triple when the compression is doubled;

The great problem in corpuscular mechanism.

and so in other instances. We see from all this that the phenomena of cohesion indicate some relation between the centres of the particles. To discover this relation is the great problem in corpuscular mechanism, as it was in the Newtonian investigation of the force of gravitation. Could we discover this law of action between the corpuscles with the same certainty and distinctness, we might with equal confidence say what will be the result of any position which we give to the particles of bodies; but this is beyond our hopes. The law of gravitation is so simple, that the discovery or detection of it amid the variety of celestial phenomena required but one step; and in its own nature its possible combinations still do not greatly exceed the powers of human research. One is almost disposed to say that the Supreme Being has exhibited it to our reasoning powers as sufficient to employ with success our utmost efforts, but not so abstruse as to discourage us from the noble attempt. It seems to be otherwise with respect to cohesion. Mathematics informs us, that if it deviates sensibly from the

law of gravitation, the simplest combinations will make the joint action of several particles an almost impenetrable mystery. We must therefore content ourselves, for a long time to come, with a careful observation of the simplest cases that we can propose, and with the discovery of secondary laws of action, in which many particles combine their influence. In pursuance of this plan, we observe,

Strength of Materials.

Particles kept in their places by a balance of forces.

3. That whatever is the situation of the particles of a body with respect to each other, when in a quiescent state, they are kept in these situations by the balance of opposite forces. This cannot be refused, nor can we form to ourselves any other notion of the state of the particles of a body. Whether we suppose the ultimate particles to be of certain magnitudes and shapes, touching each other in single points of cohesion; or whether, with Boscovich, we consider them as at a distance from each other, and acting on each other by attractions and repulsions, we must acknowledge, in the first place, that the centres of the particles (by whose mutual distances we must estimate the distance of the particles) may and do vary their distances from each other. What else can we say when we observe a body increase in length, in breadth, and thickness, by heating it, or when we see it diminish in all these dimensions by an external compression? A particle, therefore, situated in the midst of many others, and remaining in that situation, must be conceived as maintained in it by the mutual balancing of all the forces which connect it with its neighbours. It is like a ball kept in its place by the opposite action of two springs. This illustration merits a more particular application. Suppose a number of balls ranged on the table in the angles of equilateral triangles, and that each ball is connected with the six which lie around it by means of an elastic wire curled like a cork-screw; suppose such another stratum of balls above this, and parallel to it, and so placed that each ball of the upper stratum is perpendicularly over the centre of the equilateral triangle below, and let these be connected with the balls of the under stratum by similar spiral wires. Let there be a third and a fourth, and any number of such strata, all connected in the same manner. It is plain that this may extend to any size, and fill any space. Now let this assemblage of balls be firmly contemplated by the imagination, and be supposed to shrink continually in all its dimensions, till the balls, and their distances from each other, and the connecting wires, all vanish from the sight as discrete individual objects. All this is very conceivable. It will now appear like a solid body, having length, breadth, and thickness; it may be compressed, and will again resume its dimensions; it may be stretched, and will again shrink; it will move away when struck; in short, it will not differ in its sensible appearance from a solid elastic body. Now when this body is in a state of compression, for instance, it is evident that any one of the balls is at rest, in consequence of the mutual balancing of the actions of all the spiral wires which connect it with those around it. It will greatly conduce to the full understanding of all that follows to recur to this illustration. The analogy or resemblance between the effects of this constitution of things and the effects of the corpuscular forces is very great; and wherever it obtains, we may safely draw conclusions from what we know would be the condition of a body of common tangible matter. We shall just give one instructive example, and then have done with this hypothetical body. We can suppose it of a long shape, resting on one point; we can suppose two weights A, B, suspended at the extremities, and the whole in equilibrio. We commonly express this state of things by saying that A and B are in equilibrio. This is very inaccurate. A is in fact in equilibrio with the united action of all the springs which connect the ball to which it is applied with the adjoining balls. These springs are brought into action, and each is in equilibrio with the joint action of

Illustration of this proposition.

By example.

Strength of Materials.

all the rest. Thus through the whole extent of the hypothetical body the springs are brought into action in a way and in a degree which mathematics can easily investigate. We need not do this: it is enough for our purpose that our imagination readily discovers that some springs are stretched, others are compressed, and that a pressure is excited on the middle point of support, and the support exerts a reaction which precisely balances it; and the other weight is, in like manner, in immediate equilibrio with the equivalent of the actions of all the springs which connect the last ball with its neighbours. Now take the analogical or resembling case, an oblong piece of solid matter, resting on a fulcrum, and loaded with two weights in equilibrio; for the actions of the connecting springs substitute the corpuscular forces; and the result will resemble that of the hypothesis.

Now, as there is something that is at least analogous to a change of distance of the particles, and a concomitant change of the intensity of the connecting forces, we may express this in the same way that we are accustomed to do in similar cases. Let A and B (fig. 1) represent the centres of two particles of a coherent elastic body in their quiescent inactive state, and let us consider only the mechanical condition of B. The body may be stretched.

Fig. 1.

In this case the distance AB of the particles may become AC. In this state there is something which makes it necessary to employ a force to keep the particles at this distance. C has a tendency towards A, or we may say that A attracts C. We may represent the magnitude of this tendency of C *towards* A, or this attraction of A, by a line C*c* perpendicular to AC. Again, the body may be compressed, and the distance AB may become AD. Something obliges us to employ force to continue this compression, and D tends *from* A, or A appears to *repel* D. The intensity of this tendency or repulsion may be represented by another perpendicular D*d*; and, to represent the different directions of these tendencies, or the different nature of these actions, we may set D*d* on the opposite side of AB. It is in this manner that Boscovich has represented the actions of corpuscular forces in his celebrated Theory of Natural Philosophy. Newton had said, that as the great movements of the solar system were regulated by forces operating at a distance, and varying with the distance, so he strongly suspected (*valde suspicor*) that all the phenomena of cohesion, with all its modifications in the different sensible forms of aggregation, and in the phenomena of chemistry and physiology, resulted from the similar agency of forces varying with the distance of the particles. The learned Jesuit pursued this thought; and has shown, that if we suppose an ultimate atom of matter endowed with powers of attraction and repulsion, varying, both in kind and degree, with the distance, and if this force be the same in every atom, it may be regulated by such a relation to the distance from the neighbouring atom, that a collection of such may have all the sensible appearance of bodies in their different forms of solids, liquids, and vapours, elastic or unelastic, and endowed with all the properties which we perceive, by whose immediate operation the phenomena of motion by impulse, and all the phenomena of chemistry, and of animal and vegetable economy, may be produced. He shows, that notwithstanding a perfect sameness, and even a great simplicity, in this atomical constitution, there will result from this union all that unspeakable variety of form and property which diversifies and embellishes the face of

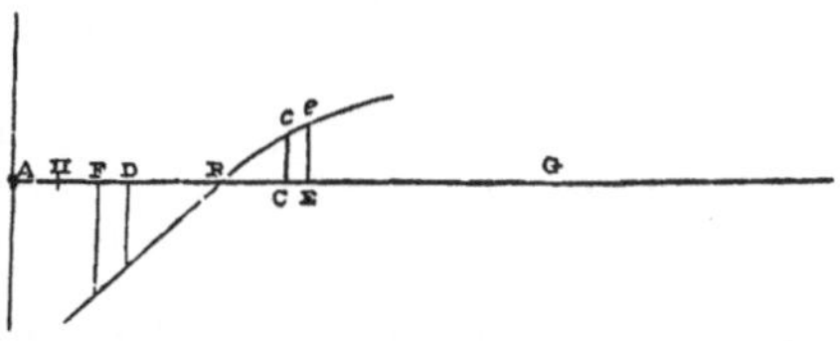

How Boscovich represents the action of corpuscular forces.

nature. We shall take another opportunity of giving such an account of this celebrated work as it deserves. We mention it only by the by, as far as a general notion of it will be of some service on the present occasion. For this purpose, we just observe that Boscovich conceives a particle of any individual species of matter to consist of an unknown number of particles of simpler constitution; each of which particles, in their turn, is compounded of particles still more simply constituted, and so on through an unknown number of orders, till we arrive at the simplest possible constitution of a particle of tangible matter, susceptible of length, breadth, and thickness, and necessarily consisting of four atoms of matter. And he shows that the more complex we suppose the constitution of a particle, the more must the sensible qualities of the aggregate resemble the observed qualities of tangible bodies. In particular, he shows how a particle may be so constituted, that although it act on one other particle of the same kind through a considerable interval, the interposition of a third particle of the same kind may render it totally, or almost totally, inactive; and therefore an assemblage of such particles would form such a fluid as air. All these curious inferences are made with uncontrovertible evidence; and the greatest encouragement is thus given to the mathematical philosopher to hope, that by cautious and patient proceeding in this way, we may gradually approach to a knowledge of the laws of cohesion, that will not shun a comparison even with the Principia of Newton. No step can be made in this investigation, but by observing with care, and generalizing with judgment, the phenomena, which are abundantly numerous, and much more at our command than those of the great and sensible motions of bodies. Following this plan, we observe,

4. It is a matter of fact, that every body has some degree of compressibility and dilatability; and when the changes of dimension are so moderate that the body completely recovers its original dimensions on the cessation of the changing force, the extensions or compressions are sensibly proportional to the extending or compressing forces; and therefore *the connecting forces are proportional to the distances of the particles from their quiescent, neutral, or inactive positions.* This seems to have been first viewed as a law of nature by the penetrating eye of Dr Robert Hooke, one of the most eminent philosophers of the last century. He published a cipher, which he said contained the theory of springiness and of the motions of bodies by the action of springs. It was this, *c e i i i n o s s s t t u u.* When explained in his dissertation, published some years after, it was *ut tensio sic vis.* This is precisely the proposition now asserted as a general fact, a law of nature. This dissertation is full of curious observations of facts in support of his assertion. In his application to the motion of bodies, he gives his noble discovery of the balance-spring of a watch, which is founded on this law. The spring, as it is more and more coiled up, or unwound, by the motion of the balance, acts on it with a force proportional to the distance of the balance from its quiescent position. The balance, therefore, is acted on by an accelerating force, which varies in the same manner as the force of gravity acting on a pendulum swinging in a cycloid. Its vibrations therefore must be performed in equal time, whether they are wide or narrow. In the same dissertation Hooke mentions all the facts which John Bernoulli afterwards adduced in support of Leibnitz's whimsical doctrine of the force of bodies in motion, as the doctrine of the *vires vivæ;* a doctrine which Hooke might justly have claimed as his own, had he not seen its futility.

Experiments made since the time of Hooke show that this law is strictly true in the extent to which we have limited it, viz. in all the changes of form which will be completely undone by the elasticity of the body. It is nearly true to a much greater extent. James Bernoulli, in his dis

Strength of Materials.

Every body compressible and dilatable.

Law of nature discovered by Dr Hooke,

and confirmed by the experiments of others.

Strength of
Materials.

sertation on the elastic curve, relates some experiments of his own, wnich seem to deviate considerably from it ; but on close examination they do not. The finest experiments are those of Coulomb, published in some late volumes of the Memoirs of the Academy of Paris. He suspended balls by wires, and observed their motions of oscillation, which he found accurately corresponding with this law.

This we shall find to be a very important fact in the doctrine of the strength of bodies, and we desire the reader to make it familiar to his mind. If we apply to this our manner of expressing these forces by perpendicular ordinates Cc, Dd (fig. 1), we must take other situations E, F, of the particle B, and draw Ee, Ff; and we must have Dd : Ff = BD : BF, or Cc : Ee = BC : BE. In such a supposition FdBce must be a straight line. But we shall have abundant evidence by and by that this cannot be strictly true, and that the line Bce, which limits the ordinates expressing the attractive forces, becomes concave towards the line ABE, and that the part Bdf is convex towards it. All that can be safely concluded from the experiments hitherto made is, that *to a certain extent* the forces, both attractive and repulsive, are *sensibly* proportional to the dilatations and compressions. For,

5. It is universally observed, that when the dilatations have proceeded a certain length, a less addition of force is sufficient to increase the dilatation in the same degree. This is always observed when the body has been so far stretched that it takes a set, and does not completely recover its form. The like may be generally observed in compressions. Most persons will recollect, that in violently stretching an elastic cord, it becomes suddenly weaker, or more easily stretched. But these phenomena do not positively prove a diminution of the corpuscular force acting on one particle : it more probably arises from the disunion of some particles whose action contributed to the whole or sensible effect. And in compressions we may suppose something of the same kind ; for when we compress a body in one direction, it commonly bulges out in another ; and in cases of very violent action some particles may be disunited, whose transverse action had formerly balanced *part* of the compressing force. For the reader will see on reflection, that since the compression in one direction causes the body to bulge out in the transverse direction, and since this bulging out is in opposition to the transverse forces of attraction, it must employ some part of the compressing force. And the common appearances are in perfect uniformity with this conception of things. When we press a bit of dryish clay, it swells out and cracks transversely. When a pillar of wood is overloaded, it swells out, and small crevices appear in the direction of the fibres. After this it will not bear half of the load. This the carpenters call *crippling;* and a knowledge of the circumstances which modify it is of great importance, and enables us to understand some very paradoxical appearances, as will be shown by and by.

This partial disuniting of particles formerly cohering, is, we imagine, the chief reason why the totality of the forces which really oppose an external strain does not increase in the proportion of the extensions and compressions. But sufficient evidence will also be given that the forces which would connect one particle with one other particle do not augment in the accurate proportion of the change of distance ; that in extensions they increase more slowly, and in compressions more rapidly.

But there is another cause of this deviation perhaps equally effectual with the former. Most bodies manifest some degree of ductility. Now what is this ? The fact is, that the parts have taken a new arrangement, in which they again cohere. Therefore, in the passage to this new arrangement, the sensible forces, which are the joint result of many corpuscular forces, begin to respect this new arrange-

ment instead of the former. This must change the simple law of corpuscular force, characteristic of the particular species of matter under examination. It does not require much reflection to convince us that the possible arrangements which the particles of a body may acquire, without appearing to change their nature, must be more numerous according as the particles are of a more complex constitution ; and it is reasonable to suppose that the constitution even of the most simple kind of matter that we are acquainted with is exceedingly complex. Our microscopes show us animals so minute, that a heap of them must appear to the naked eye an uniform mass with a grain finer than that of the finest marble or razor hone ; and yet each of these has not only limbs, but bones, muscular fibres, blood-vessels, fibres, and a blood consisting in all probability of globules organized and complex like our own. The imagination is here lost in wonder ; and nothing is left us but to adore inconceivable art and wisdom, and to exult in the thought that we are the only spectators of this beautiful scene who can derive pleasure from the view. But let us proceed to observe,

6. That the forces which connect the particles of tangible bodies change by a change of distance, not only in degree, but also in kind. The particle B (fig. 1) is attracted by A when in the situation C or E. It is repelled by it when at D or F. It is not affected by it when in the situation B. The reader is requested carefully to remark, that this is not an inference founded on the authority of our mathematical figure. The figure is an expression (to assist the imagination) of facts in nature. It requires no force to keep the particles of a body in their quiescent situations : but if they be separated by stretching the body, they endeavour (pardon the figurative expression) to come together again. If they be brought nearer by compression, they endeavour to recede. This endeavour is manifested by the necessity of employing force to maintain the extension or condensation ; and we represent this by the different position of our lines. But this is not all : the particle B, which is repelled by A when in the situation F or D, is neutral when at B, and is attracted when at C or E, may be placed at such a distance AG from A greater than AB that it shall be again repelled, or at such a distance AH that it shall be again attracted ; and these alterations may be repeated again and again. This is curious and important, and requires something more than a bare assertion for its proof.

In the article OPTICS we mentioned the most curious and valuable observations of Sir Isaac Newton, by which it appears that light is thus alternately attracted and repelled by bodies. The rings of colour which appear between the object-glasses of long telescopes showed, that in the small interval of $\frac{1}{1000}$th of an inch, there are at least an hundred such changes observable, and that it is highly probable that these alternations extend to a much greater distance. At one of these distances the light actually converges towards the solid matter of the glass, which we express shortly by saying that it is attracted by it, and that at the next distance it declines from the glass, or is repelled by it. The same thing is more simply inferred from the phenomena of light passing by the edges of knives and other opaque bodies. We refer the reader to the experiments themselves, the detail being too long for this place ; and we request him to consider them minutely and attentively, and to form distinct notions of the inferences drawn from them. And we desire it to be remarked, that although Newton, in his discussion, always considers light as a set of corpuscles moving in free space, and obeying the actions of external forces like any other matter, the particular conclusion in which we are just now interested does not at all depend on this notion of the nature of light. Should we, with Descartes or Huygens, suppose light to be the undulation of an elastic medium, the

Strength of
Materials.

conclusion will be the same. The undulations at certain distances are disturbed by forces directed towards the body, and at a greater distance the disturbing forces tend *from* the body.

The same alternations of attraction and repulsion observable in the particles of other bodies, as glass.

But the same alternations of attraction and repulsion may be observed between the particles of common matter. If we take a piece of very flat and well-polished glass, such as is made for the horizon-glasses of a good Hadley's quadrant, and if we wrap round it a fibre of silk as it comes from the cocoon, taking care that the fibre shall nowhere cross another, and then press this pretty hard on such another piece of glass, it will lift it up and keep it suspended. The particles therefore of the one do most certainly attract those of the other, and this at a distance equal to the thickness of the silk fibre. This is nearly the limit; and it sometimes requires a considerable pressure to produce the effect. The pressure is effectual only by compressing the silk fibre, and thus diminishing the distance between the glass plates. This adhesion cannot be attributed to the pressure of the atmosphere, because there is nothing to hinder the air from insinuating itself between the plates, since they are separated by the silk. Besides, the experiment succeeds equally well under the receiver of an air-pump. This most valuable experiment was first made by Huygens, who reported it to the Royal Society. It is narrated in the Philosophical Transactions, No. 86.

Here, then, is an attraction acting, like gravity, at a distance. But take away the silk fibre, and try to make the glasses touch each other, and we shall find a very great force necessary. By Newton's experiments it appears, that unless the prismatic colours begin to appear between the glasses, they are at least $\frac{1}{890}$th of an inch asunder or more. Now we know that a very considerable force is necessary for producing these colours, and that the more we press the glasses together the more rings of colours appear. It also appears from Newton's measures, that the difference of distance between the glasses where each of these colours appears is about the 89,000th part of an inch. We know further, that when we have produced the last appearance of a greasy or pearly colour, and then augment the pressure, making it about 1000 pounds on the square inch, all colours vanish, and the two pieces of glass seem to make one transparent undistinguishable mass. They appear now to have no air between them, or to be in mathematical contact. But another fact shows this conclusion to be premature. The same circles of colours appear in the top of a soap-bubble; and as it grows thinner at top, there appears an unreflecting spot in the middle. We have the greatest probability therefore that the perfect transparency in the middle of the two glasses does not arise from their being in contact, but because the thickness of air between them is too small in that place for the reflection of light. Nay, Newton expressly found no reflection where the thickness was $\frac{2}{3}$ths or more of the $\frac{1}{89000}$th part of an inch.

All this while the glasses are strongly repelling each other, for great pressure is necessary for continuing the appearance of those colours, and they vanish in succession as the pressure is diminished. This vanishing of the colours is a proof that the glasses are moving off from each other, or repelling each other. But we can put an end to this repulsion by very strong pressure, and at the same time sliding the glasses on each other. We do not pretend to account for this effect of the sliding motion; but the fact is, that by so doing, the glasses will cohere with very great force, so that we shall break them by any attempt to pull them asunder. It commonly happens (at least it did so with us), that in this sliding compression of two smooth flat plates of glass, they scratch and mutually destroy each other's surface. It is also worth remarking, that different kinds of glass exhibit different properties in this respect.

Flint glass will attract even though a silk fibre lies double between them, and they much more readily cohere by this sliding pressure.

Strength of
Materials.

Here, then, are two distances at which the plates of glass attract each other; namely, when the silk fibre is interposed, and when they are forced together with this sliding motion. And in any intermediate situation they repel each other. We see the same thing in other solid bodies. Two pieces of lead, made perfectly clean, may be made to cohere by grinding them together in the same manner. It is in this way that pretty ornaments of silver are united to iron. The piece is scraped clean, and a small bit of silver like a fish scale is laid on. The die which is to strike it into a flower or other ornament is then set on it, and we give it a smart blow, which forces the metals into contact as firm as if they were soldered together. It sometimes happens that the die adheres to the coin so that they cannot be separated: and it is found that this frequently happens when the engraving is such that the raised figure is not completely surrounded with a smooth flat ground. The probable cause of this is curious. When the coin has a flat surface all around, this is produced by the most prominent part of the die. This applies to the metal, and completely confines the air which filled the hollow of the die. As the pressure goes on, the metal is squeezed up into the hollow of the die; but there is still air compressed between them, which cannot escape by any passage. It is therefore prodigiously condensed, and exerts an elasticity proportioned to the condensation. This serves to separate the die from the metal when the stroke is over. The hollow part of the die has not touched the metal all the while, and we may say that the impression was made by air. If this air escape by any engraving reaching through the border, they cohere inseparably.

Lead and iron.

Probable cause why the die adheres to the coin.

We have admitted that the glass plates are in contact when they adhere thus firmly. But we are not certain of this: for if we take these cohering glasses, and touch them with water, it quickly insinuates itself between them. Yet they still cohere, but can now be pretty easily separated.

It is owing to this repulsion, exerted through its proper sphere, that certain powders swim on the surface of water, and are wetted with great difficulty. Certain insects can run about on the surface of water. They have brushy feet, which occupy a considerable surface, and if their steps be viewed with a magnifying glass, the surface of the water is seen depressed all around, resembling the footsteps of a man walking on feather beds. This is owing to a repulsion between the brush and the water. A common fly cannot walk in this manner on water. Its feet are wetted, because they attract the water instead of repelling it. A steel needle, slightly greased, will lie on the surface of water, make an impression as a great bar would make on a feather bed; and its weight is less than that of the displaced water. A dew-drop lies on the leaves of plants without touching them mathematically, as is plain from the extreme brilliancy of the reflection at the posterior surface; nay, it may be sometimes observed that the drops of rain lie on the surface of water, and roll about on it like balls on a table. Yet all these substances can be wetted; that is, water can be applied to them at such distances that they attract it.

Repulsion the cause of some bodies swimming in a fluid specifically lighter than themselves

What we lately remarked of water insinuating itself between the glass plates without altogether destroying their cohesion, shows that this cohesion is not the same that obtains between the particles of one of the plates; that is, the two plates are not in the state of one continued mass. It is highly probable, therefore, that between these two states there is an intermediate state of repulsion, nay, perhaps, many such, alternated with attractive states.

A piece of ice is elastic, for it rebounds and rings. Its particles, therefore, when compressed, resile; and when

Strength of Materials.

stretched, contract again. The particles are therefore in the state represented by B in figure 1, acted on by repulsive forces if brought nearer, and by attractive forces if drawn further asunder. Ice expands, like all other bodies, by heat. It absorbs a vast quantity of fire, which, by combining its attractions and repulsions with those of the particles of ice, changes completely the law of action, and the ice becomes water. In this new state the particles are again in limits between attractive and repulsive forces; for water has been shown, by the experiments of Canton and Zimmerman, to be elastic or compressible. It again expands by heat. It again absorbs a prodigious quantity of heat, and becomes elastic vapour; its particles repelling each other at all distances yet observed. The distance between the particles of one plate of glass and those of another which lies on it, and is carried by it, is a distance of repulsion; for the force which supports the upper piece is acting in opposition to its weight. This distance is less than that at which it would *suspend* it below it with a silk fibre interposed; for no prismatic colours appear between them when the silk fibre is interposed. But the distance at which glass attracts water is much less than this, for no colours appear when glass is wetted with water. This distance is less, and not greater, than the other; for when the glasses have water interposed between them instead of air, it is found, that when any particular colour appears, the thickness of the plate of water is to that of the plate of air which would produce the same colour, nearly as three to four. Now, if a piece of glass be wetted, and exhibit no colour, and another piece of glass be simply laid on it, no colour will appear; but if they are strongly pressed, the colours appear in the same manner as if the glasses had air between. Also, when glass is simply wetted, and the film of water is allowed to evaporate, when it is thus reduced to a proper thinness the colours show themselves in great beauty.

Particles of matter connected by forces acting at a distance.

These are a few of many thousand facts, by which it is unquestionably proved that the particles of tangible matter are connected by forces acting at a distance, varying with the distance, and alternately attractive and repulsive. If we represent these forces, as we have already done in fig. 1, by the ordinates Cc, Dd, Ee, Ff, &c. of a curve, it is evident that this curve must cross the axis at all those distances where the forces change from attractive to repulsive, and the curve must have branches alternately above and below the axis.

All these alternations of attraction and repulsion take place at small and insensible distances. At all sensible distances the particles are influenced by the attraction of gravitation; and therefore this part of the curve must be a hyperbola whose equation is $y = \dfrac{a^3}{x^2}$. What is the form of the curve corresponding to the smallest distance of the particles? that is, what is the mutual action between the particles just before their coming into absolute contact? Analogy should lead us to suppose it to be repulsion; for solidity is the last and simplest form of bodies with which we are acquainted. Fluids are more compounded, containing fire as an essential ingredient. We should conclude that this ultimate repulsion is insuperable, for the hardest bodies are the most elastic. We are fully entitled to say that this repelling force exceeds all that we have ever yet applied to overcome it; nay, there are good reasons for saying that this ultimate repulsion, by which the particles are kept from mathematical contact, is really insuperable in its own nature, and that it is impossible to produce mathematical contact.

Mathematical contact impossible.

We shall just mention one of these, which we consider as unanswerable. Suppose two atoms, or ultimate particles of matter, A and B. Let A be at rest, and B move up to it with the velocity 2; and let us suppose that it comes into

mathematical contact, and impels it (according to the common acceptation of the word). Both move with the velocity 1. This is granted by all to be the final result of the collision. Now the instant of time in which this communication happens is no part either of the duration of the solitary motion of A, nor of the joint motion of A and B: it is the separation or boundary between them. It is at once the end of the first and the beginning of the second, belonging equally to both. A was moving with the velocity 2. The distinguishing circumstance, therefore, of its mechanical state is, that it has a determination (however incomprehensible) by which it would move for ever with the velocity 2, if nothing changed it. This it has during the whole of its solitary motion, and therefore in the last instant of this motion. In like manner, during the whole of the joint motion, and therefore in the first instant of this motion, the atom A has a determination by which it would move for ever with the velocity 1. In one and the same instant, therefore, the atom A has two incompatible determinations. Whatever notion we can form of this state, which we call velocity, as a distinction of condition, the same impossibility of conception, or the same absurdity, occurs. Nor can it be avoided in any other way than by saying, that this change of A's motion is brought about by insensible gradations; that is, that A and B influence each other precisely as they would do if a slender spring were interposed. The reader is desired to look at what we have said in the article PHYSICS.

The two magnets there spoken of are good representatives of two atoms endowed with mutual powers of repulsion; and the communication of motion is accomplished in both cases in precisely the same manner.

If, therefore, we shall ever be so fortunate as to discover the law of variation of that force which connects one *atom* of matter with another atom, and which is therefore characteristic of matter, and the ultimate source of all its sensible qualities, the curve whose ordinates represent the kind and the intensity of this atomical force will be something like that sketched in fig. 2. The first branch *a n* B will have

Fig. 2.

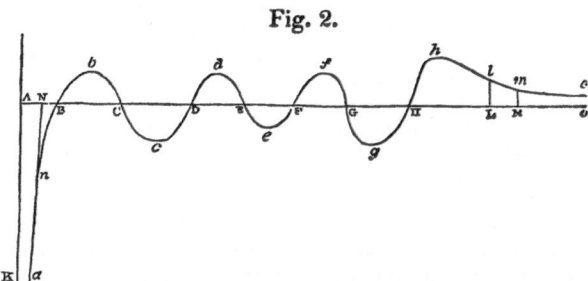

AK (perpendicular to the axis AH) for its assymptote, and the last branch *lmo* will be to all sense a hyperbola, having AO for its assymptote; and the ordinates *l*L, *m*M, &c. will be proportional to $\dfrac{1}{AL^2}$, $\dfrac{1}{AM^2}$, &c. expressing the universal gravitation of matter. It will have many branches B*b*C, D*d*E, F*f*G, &c. expressing attractions, and alternate repulsive branches C*c*D, E*e*F, G*g*H, &c. All these will be contained within a distance AH, which does not exceed a very minute fraction of an inch.

The simplest particle which can be a constituent of a body having length, breadth, and thickness, must consist of four such atoms, all of which combine their influence on each atom of another such particle. It is evident that the curve which expresses the force that connects two such particles must be totally different from this original curve, this hylarchic principle. Supposing the last known, our mathematical knowledge is quite able to discover the first; but when we proceed to compose a body of particles, each of

The simplest extended particle consists of four atoms.

which consists of four such particles, we may venture to say that the compound force which connects them is almost beyond our search, and that the discovery of the primary force from an *accurate* knowledge of the corpuscular forces of *this* particular matter is absolutely out of our power.

All that we can learn is, the possibility, nay, the certainty, of an innumerable variety of external sensible forms and qualities, by which different kinds of matter will be distinguished, arising from the number, the order of composition, and the arrangement of the subordinate particles of which a particle of this or that kind of matter is composed. All these varieties will take place at those small and insensible distances which are between A and H, and may produce all that variety which we observe in the tangible or mechanical forms of bodies, such as elasticity, ductility, hardness, softness, fluidity, vapour, and all those unseen motions or actions which we observe in fusion and congelation, evaporation and condensation, solution and precipitation, crystallization, vegetable and animal assimilation and secretion, &c.; while all bodies must be, in a certain degree, elastic, all must gravitate, and all must be incompenetrable.

This general and satisfactory resemblance between the appearance of tangible matter and the legitimate consequence of this general hypothetical property of an atom of matter, affords a considerable probability that such is the origin of all the phenomena. We earnestly recommend to our readers a careful perusal of Boscovich's celebrated treatise. A careful perusal is necessary for seeing its value; and nothing will be got by a hasty inspection. The reader will be particularly pleased with the facility and evidence with which the ingenious author has deduced all the ordinary principles of mechanics, and with the explanation which he has given of fluidity, and his deduction from thence of the laws of hydrostatics. No part of the treatise is more valuable than the doctrine of the propagation of pressure through solid bodies. This, however, is but just touched on in the course of the investigation of the principles of mechanics. We shall borrow as much as will suffice for our present inquiry into the strength of materials; and we trust that our readers are not displeased with this general sketch of the doctrine (if it may be so called) of the cohesion of bodies. It is curious and important in itself, and is the foundation of all the knowledge which we can acquire of the present article. We are sorry to say that it is as yet a new subject of study; but it is a very promising one, and we by no means despair of seeing the whole of chemistry brought by its means within the pale of mechanical science. The great and distinguishing agent in chemistry is heat, or fire the cause of heat; and one of its most singular effects is the conversion of bodies into elastic vapour. We have the clearest evidence that this is brought about by mechanical forces; for it can be opposed or prevented by external pressure, a very familiar mechanical force. We may perhaps find another mechanical force which will prevent fusion.

Having now made our readers familiar with the mode of action in which cohesion operates in giving strength to solid bodies, we proceed to consider the strains to which this strength is opposed.

A piece of solid matter is exposed to four kinds of strains, different in the manner of their operation.

1. It may be torn asunder, as in the case of ropes, stretchers, king-posts, tie-beams, &c.

2. It may be crushed, as in the case of pillars, posts, and truss-beams.

3. It may be broken across, as happens to a joist or lever of any kind.

4. It may be wrenched or twisted, as in the case of the axle of a wheel the nail of a press, &c.

1.—IT MAY BE PULLED ASUNDER.

This is the simplest of all strains, and the others are indeed modifications of it. To this the force of cohesion is *directly* opposed, with very little modification of its action by any particular circumstances.

When a long cylindrical or prismatic body, such as a rod of wood or metal, or a rope, is drawn by one end, it must be resisted at the other, in order to bring its cohesion into action. When it is fastened at one end, we cannot conceive it any other way than as equally stretched in all its parts; for all our observations and experiments on natural bodies concur in showing us that the forces which connect their particles in any way whatever are equal and opposite. This is called the *third law of motion;* and we admit its universality, while we affirm that it is purely experimental. Yet we have met with dissertations by persons of eminent knowledge, where propositions are maintained inconsistent with this. During the dispute about the communication of motion, some of the ablest writers have said, that a spring compressed or stretched at the two ends was gradually less and less compressed or stretched from the extremities towards the middle: but the same writers acknowledged the universal equality of action and re-action, which is quite incompatible with this state of the spring. No such inequality of compression or dilatation has ever been observed; and a little reflection will show it to be impossible, in consistency with the equality of action and re-action.

Since all parts are thus equally stretched, it follows that the strain in any transverse section is the same, as also in every point of that section. If therefore the body be supposed of a homogeneous texture, the cohesion of the parts is equable; and since every part is equally stretched, the particles are drawn to equal distances from their quiescent positions, and the forces which are thus excited, and now exerted in opposition to the straining force, are equal. This external force may be increased by degrees, which will gradually separate the parts of the body more and more from each other, and the connecting forces increase with this increase of distance, till at last the cohesion of some particles is overcome. This must be immediately followed by a rupture, because the remaining forces are now weaker than before.

It is the united force of cohesion, immediately before the disunion of the first particles, that we call the *strength* of the section. It may also be properly called its *absolute strength*, being exerted in the simplest form, and not modified by any relation to other circumstances.

If the external force have not produced any permanent change on the body, and it therefore recovers its former dimensions when the force is withdrawn, it is plain that this strain may be repeated as often as we please, and the body which withstands it once will always withstand it. It is evident that this should be attended to in all constructions, and that in all our investigations on this subject this should be kept strictly in view. When we treat a piece of soft clay in this manner, and with this precaution, the force employed must be very small. If we exceed this, we produce a permanent change. The rod of clay is not indeed torn asunder, but it has become somewhat more slender; the number of particles in a cross section is now smaller; and therefore, although it will again, in this new form, suffer or allow an endless repetition of a *certain* strain without any farther permanent change, this strain is smaller than the former.

Something of the same kind happens in all bodies which receive a *set* by the strain to which they are exposed. All ductile bodies are of this kind. But there are many bodies which are not ductile. Such bodies break completely whenever they are stretched beyond the limit of their perfect elasticity. Bodies of a fibrous structure exhibit very great

Strength of Materials.

Great varieties in cohesion, but

varieties in their cohesion. In some the fibres have no lateral cohesion, as in the case of a rope. The only way in which all the fibres can be made to unite their strength, is to twist them together. This causes them to bind each other so fast, that any one of them will break before it can be drawn out of the bundle. In other fibrous bodies, such as timber, the fibres are held together by some cement or gluten. This is seldom as strong as the fibre. Accordingly timber is much easier pulled asunder in a direction transverse to the fibres. There is, however, every possible variety in this particular.

In stretching and breaking fibrous bodies, the visible extension is frequently very considerable. This is not solely the increasing of the distance of the particles of the cohering fibre; the greatest part chiefly arises from drawing the crooked fibre straight. In this, too, there is great diversity; and it is accompanied with important differences in their power of withstanding a strain. In some woods, such as fir, the fibres on which the strength most depends are very straight. Such woods are commonly very elastic, do not take a set, and break abruptly when overstrained: others, such as oak and birch, have their resisting fibres very undulating and crooked, and stretch very sensibly by a strain. They are very liable to take a set, and they do not break so suddenly, but give warning by *complaining*, as the carpenters call it; that is, by giving visible signs of a derangement of texture. Hard bodies of an uniform glassy structure, or granulated like stones, are elastic through the whole extent of their cohesion, and take no set, but break at once when overloaded.

Notwithstanding the immense variety which nature exhibits in the structure and cohesion of bodies, there are certain general facts of which we may now avail ourselves with advantage. In particular,

the absolute cohesion or strength proportional to the area of the section perpendicular to the extending force

The absolute cohesion is proportional to the area of the section. This must be the case where the texture is perfectly uniform, as we have reason to think it is in glass and the ductile metals. The cohesion of each particle being alike, the whole cohesion must be proportional to their number, that is, to the area of the section. The same must be admitted with respect to bodies of a granulated texture, where the granulation is regular and uniform. The same must be admitted of fibrous bodies, if we suppose their fibres equally strong, equally dense, and similarly disposed through the whole section; and this we must either suppose, or must state the diversity, and measure the cohesion accordingly.

We may therefore assert, as a general proposition on this subject, that the absolute strength in any part of a body, by which it resists being pulled asunder, or the force which must be employed to tear it asunder *in that part*, is proportional to the area of the section perpendicular to the extending force.

Therefore all cylindrical or prismatical rods are equally strong in every part, and will break alike in any part; and bodies which have unequal sections will always break in the slenderest part. The length of the cylinder or prism has no effect on the strength. Also the absolute strengths of bodies which have similar sections are proportional to the squares of their diameters or homologous sides of the section.

The weight of the body itself may be employed to strain it and to break it. It is evident, that a rope may be so long as to break by its own weight. When the rope is hanging perpendicularly, although it is equally strong in every part, it will break towards the upper end, because the strain on any part is the weight of all that is below it. Its *relative*

Relative strength.

strength in any part, or power of withstanding the strain

which is actually laid on it, is inversely as the quantity below that part.

When the rope is stretched horizontally, as in towing a ship, the strain arising from its weight often bears a very sensible proportion to its whole strength.

These are the chief general rules which can be safely deduced from our clearest notions of the cohesion of bodies. In order to make any practical use of them, it is proper to have some measures of the cohesion of such bodies as are commonly employed in our mechanics, and other structures where they are exposed to this kind of strain. These must be deduced solely from experiment; therefore they must be considered as no more than general values, or as the averages of many particular trials. The irregularities are very great, because none of the substances are constant in their texture and firmness. Metals differ by a thousand circumstances unknown to us, according to their purity, to the heat with which they were melted, to the moulds in which they were cast, and the treatment they have afterwards received, by forging, wire-drawing, tempering, &c.

Strength of Materials.

The cohesion of metals depends on various circumstances.

It is a very curious and inexplicable fact, that by forging a metal, or by frequently drawing it through a smooth hole in a steel plate, its cohesion is greatly increased. This operation undoubtedly deranges the natural situation of the particles. They are squeezed closer together in one direction, but it is not in the direction in which they resist the fracture. In this direction they are rather separated to a greater distance. The general density, however, is augmented in all of them except lead, which grows rather rarer by wire-drawing; but its cohesion may be more than tripled by this operation. Gold, silver, and brass, have their cohesion nearly tripled; copper and iron have it more than doubled. In this operation they also grow much harder. It is proper to heat them to redness after drawing a little. This is called *nealing* or *annealing*. It softens the metal again, and renders it susceptible of another drawing without the risk of cracking in the operation.

We do not pretend to give any explanation of this remarkable and very important fact, which has something resembling it in woods and other fibrous bodies, as will be mentioned afterwards.

The varieties in the cohesion of stones and other minerals, and of vegetable and animal substances, are hardly susceptible of any description or classification.

We shall take for the measure of cohesion the number of pounds avoirdupois which are just sufficient to tear asunder a rod or bundle of one inch square. From this it will be easy to compute the strength corresponding to any other dimension.

1st, Metals.

		lbs.
Gold, cast		{ 20,000
		{ 24,000
Silver, cast		{ 40,000
		{ 43,000
Copper, cast,	Japan	19,000
	Barbary	22,000
	Hungary	31,000
	Anglesea	34,000
	Sweden	37,000
Iron, cast		{ 42,000
		{ 59,000
Iron, bar	Ordinary	68,000
	Stirian	75,000
	Best Swedish and Russian	84,000
	Horse-nails	71,000[1]

---

[1] This was an experiment by Muschenbroeck, to examine the vulgar notion that iron forged from old horse-nails was stronger than others, and shows its falsity.

Steel, bar	Soft 120,000
	Razor temper.................150,000
Tin, cast.	Malacca...........................3,100
	Banca..............................3,600
	Block...............................3,800
	English block....................5,200
	———— grain.......................6,500

Lead, cast..860
Regulus of antimony............................1,000
Zinc. 2,600
Bismuth...2,900

Tenacity of metals increased by mixtures

It is very remarkable, that almost all the mixtures of metals are more tenacious than the metals themselves. The change of tenacity depends much on the proportion of the ingredients, and the proportion which produces the most tenacious mixture is different in the different metals. We have selected the following from the experiments of Muschenbroeck. The proportion of ingredients here selected is that which produces the greatest strength.

Two parts of gold with one of silver....................28,000
Five parts of gold with one of copper.................50,000
Five parts of silver with one of copper...............48,500
Four parts of silver with one of tin.....................41,000
Six parts of copper with one of tin.....................41,000
Five parts of Japan copper with one of Banca tin...57,000
Six parts of Chili copper with one of Malacca tin...60,000
Six parts of Swedish copper with one of Malacca tin 64,000
Brass consists of copper and zinc in an unknown proportion ; its strength is 51,000
Three parts of block-tin with one part of lead........10,200
Eight parts of block-tin with one part of zinc.........10,000
Four parts of Malacca tin with one part of regulus
of antimony..................... 12,000
Eight parts of lead with one of zinc................... 4,500
Four parts of tin with one of lead and one of zinc...13,000

These numbers are of considerable use in the arts. The mixtures of copper and tin are particularly interesting in the fabric of great guns. We see that, by mixing copper, whose greatest strength does not exceed 37,000, with tin, which does not exceed 6000, we produce a metal whose tenacity is almost double, at the same time that it is harder and more easily wrought. It is, however, more fusible, which is a great inconvenience. We also see that a very small addition of zinc almost doubles the tenacity of tin, and increases the tenacity of lead five times ; and a small addition of lead doubles the tenacity of tin. These are economical mixtures. This is very valuable information to the plumbers, for augmenting the strength of water-pipes.

By having recourse to these tables, the engineer can proportion the thickness of his pipes, of whatever metal, to the pressures to which they are exposed.

2d, Woods.

We may premise to this part of the table the following general observations.

Tenacity or strength of wood.

1. The wood immediately surrounding the pith or heart of the tree is the weakest, and its inferiority is so much more remarkable as the tree is older. In this assertion, however, we speak with some hesitation. Muschenbroeck's *detail* of experiments is decidedly in the affirmative. M. Buffon, on the other hand, says that his experience has taught him that the heart of a sound tree is the strongest; but he gives no instances. From many observations of our own on very *large* oaks and firs, we are certain that the heart is much weaker than the exterior parts.

2. The wood next the bark, commonly called the *white* or *blea*, is also weaker than the rest; and the wood gradually increases in strength as we recede from the centre to the blea.

3. The wood is stronger in the middle of the trunk than at the springing of the branches or at the root; and the wood of the branches is weaker than that of the trunk.

4. The wood of the north side of all trees which grow in our European climates is the weakest, and that of the southeast side is the strongest; and the difference is most remarkable in hedge-row trees, and such as grow singly. The heart of a tree is never in its centre, but always nearer to the north side, and the annual coats of wood are thinner on that side. In conformity with this, it is a general opinion of carpenters that timber is stronger whose annual plates are thicker. The trachea or air-vessels are weaker than the simple ligneous fibres. The air-vessels are the same in diameter and number of rows in trees of the same species, and they make the visible separation between the annual plates. Therefore, when these are thicker, they contain a greater proportion of the simple ligneous fibres.

5. All woods are more tenacious while green, and lose very considerably by drying after the trees are felled.

The only author who has put it in our power to judge of the propriety of his experiments is Muschenbroeck. He has described his method of trial minutely, and it seems unexceptionable. The woods were all formed into slips fit for his apparatus, and part of the slip was cut away to a parallelopiped of $\frac{1}{5}$th of an inch square, and therefore $\frac{1}{25}$th of a square inch in section. The absolute strengths of a square inch were as follow :

Absolute strength of different kinds of wood,

	lib.		lib.
Locust tree..........20,100		Pomegranate..........9750
Juleb.................18,500		Lemon.................9250
Beech, oak.........17,300		Tamarind.............8750
Orange...............15,500		Fir....................8330
Alder.................13,900		Walnut................8130
Elm...................13,200		Pitch-pine 7650
Mulberry... 12,500		Quince................6750
Willow...............12,500		Cypress...............6000
Ash...................12,000		Poplar................5500
Plum.................11,800		Cedar.................4880
Elder.................10,000		

Muschenbroeck has given a very minute detail of the experiments on the ash and the walnut, stating the weights which were required to tear asunder slips taken from the four sides of the tree, and on each side, in a regular progression from the centre to the circumference. The number of this table corresponding to these two timbers may therefore be considered as the average of more than fifty trials made of each; and he says that all the others were made with the same care. We cannot therefore see any reason for not confiding in the results; yet they are considerably higher than those given by some other writers. Mr Pitot, on the authority of his own experiments, and of those of Mr Parent, avers that sixty pounds will just tear asunder a square line of sound oak, and that it will bear fifty with safety. This gives 8640 for the utmost strength of a square inch, which is much inferior to Muschenbroeck's valuation.

We may add to these,

and of other substances.

Ivory..16,270
Bone...5,250
Horn.. 8,750
Whalebone................................... 7,500
Tooth of sea-calf........................ 4,075

No substance to be strained in architecture above one half its strength.

The reader will surely observe, that these numbers express something more than the utmost cohesion; for the weights are such as will very quickly, that is, in a minute or two, tear the rods asunder. It may be said in general, that two thirds of these weights will sensibly impair the strength after a considerable while, and that one half is the utmost that can remain suspended at them without risk for ever; and it is upon this last allotment that the engineer should reckon in his constructions. There is, however, considerable difference in this respect. Woods of a very straight

Strength of fibre, such as fir, will be less impaired by any load which is
Materials not sufficient to break them immediately.

According to Mr Emerson, the load which may be safely
suspended to an inch square is as follows :

Iron	76,400
Brass	35,600
Hempen rope	19,600
Ivory	15,700
Oak, box, yew, plum-tree	7,850
Elm, ash, beech,	6,070
Walnut, plum	5,360
Red fir, holly, elder, plane, crab	5,000
Cherry, hazel	4,760
Alder, asp, birch, willow	4,290
Lead	430
Freestone	914

He gives us a practical rule, that a cylinder whose diameter is 1 inch, loaded to one fourth of its absolute strength, will carry as follows :

Iron	135	
Good rope	22	cwt.
Oak	14	
Fir	9	

The rank which the different woods hold in this list of Mr Emerson's is very different from what we find in Muschenbroeck's. But precise measures must not be expected in this matter. It is wonderful, that in a matter of such unquestionable importance the public has not enabled some persons of judgment to make proper trials. They are beyond the abilities of private persons.

II.—BODIES MAY BE CRUSHED.

It is of importance to know what will crush bodies.

It is of equal, perhaps greater, importance to know the strain which may be laid on solid bodies without danger of crushing them. Pillars and posts of all kinds are exposed to this strain in its simplest form; and there are cases where the strain is enormous, viz. where it arises from the oblique position of the parts, as in the struts, braces, and trusses, which occur very frequently in our great works. It is therefore most desirable to have some general knowledge of the principle which determines the strength of bodies, in opposition to this kind of strain. But, unfortunately, we are much more at a loss in this than in the last case. The mechanism of nature is, in the present case, much more complicated. It must be in some circuitous way that compression can have any tendency to tear asunder the parts of a solid body, and it is very difficult to trace the steps.

If we suppose the particles insuperably hard and in contact, and disposed in lines which are in the direction of the external pressures, it does not appear how any pressure can disunite the particles; but this is a gratuitous supposition. There are infinite odds against this precise arrangement of the lines of particles; and the compressibility of all kinds of matter in some degree shows that the particles are in a situation equivalent to distance. This being the case, and the particles, with their intervals, or what is equivalent to intervals, being in situations that are oblique with respect to the pressures, it must follow, that by squeezing them together in one direction, they are made to bulge out or separate in other directions. This may proceed so far that some may be thus pushed laterally beyond their limits of cohesion. The moment that this happens the resistance to compression is diminished, and the body will now be crushed together. We may form some notion of this by supposing a number of spherules, like small shot, sticking together by means of a cement. Compressing this in some particular direction causes the spherules to act among each other like so many wedges, each tending to penetrate through between the three which lie below it : and this is the simplest, and perhaps the only distinct, notion we can have of the matter.

We have reason to think that the constitution of very homogeneous bodies, such as glass, is not very different from this. Strength of Materials.

If this be the constitution of bodies, it appears probable that the strength, or the resistance which they are capable of making to an attempt to crush them to pieces, is proportional to the area of the section whose plane is perpendicular to the external force; for each particle being similarly and equally acted on and resisted, the whole resistance must be as their number, that is, as the extent of the section. Their strength or power of resistance to such a force,

Accordingly this principle is assumed by the few writers who have considered the subject; but we confess that it appears to us very doubtful. Suppose a number of brittle or friable balls lying on a table uniformly arranged, but not cohering nor in contact, and that a board is laid over them and loaded with a weight; we have no hesitation in saying that the weight necessary to crush the whole collection is proportional to their number or to the area of the section. But when they are in contact, and still more if they cohere, we imagine that the case is materially altered. Any individual ball is crushed only in consequence of its being bulged outwards in the direction perpendicular to the pressure employed. If this could be prevented by a hoop put round the ball like an equator, we cannot see how any force can crush it. Any thing therefore which makes this bulging outwards more difficult, makes a greater force necessary. Now this effect will be produced by the mere contact of the balls before the pressure is applied; for the central ball cannot swell outward laterally without pushing away the balls on all sides of it. This is prevented by the friction on the table and upper board, which is at least equal to one third of the pressure. Thus any interior ball becomes stronger by the mere vicinity of the others; and if we further suppose them to cohere laterally, we think that its strength will be still more increased.

The analogy between these balls and the cohering particles of a friable body is very perfect. We should therefore expect that the strength by which it resists being crushed will increase in a greater ratio than that of the section, or the square of the diameter of similar sections; and that a square inch of any matter will bear a greater weight in proportion as it makes a part of a greater section. Accordingly this appears in many experiments, as will afterwards be noticed. Muschenbroeck, Euler, and some others, have supposed the strength of columns to be as the biquadrates of their diameters. Euler deduced this from formulæ which occurred to him in the course of his algebraic analysis; and he boldly adopts it as a principle, without looking for its foundation in the physical assumptions which he had made in the beginning of his investigation. But some of his original assumptions were as paradoxical, or at least as gratuitous, as these results; and those, in particular, from which this proportion of the strength of columns was deduced, were almost foreign to the case; and therefore the inference was of no value. Yet it was received as a principle by Muschenbroeck and by the academicians of St Petersburg. We make these very few observations, because the subject is of great practical importance; and it is a great obstacle to improvements when deference to a great name, joined to incapacity or indolence, causes authors to adopt his careless reveries as principles from which they are afterwards to draw important consequences. It must be acknowledged that we have not as yet established, on solid mechanical principles, the relation between the dimensions and the strength of a pillar. Experience plainly contradicts the general opinion, that the strength is proportional to the area of the section; but it is still more inconsistent with the opinion, that it is in the quadruplicate ratio of the diameters of similar sections. It would seem that the ratio depends much on the internal structure of the body; and experiment seems the only method for ascertaining its general laws.

Strength of
Materials.

to be as-
certained
only by ex-
periment.

If we suppose the body to be of a fibrous texture, having the fibres situated in the direction of the pressure, and slightly adhering to each other by some kind of cement, such a body will fail only by the bending of the fibres, by which they will break the cement and be detached from each other. Something like this may be supposed in wooden pillars. In such cases, too, it would appear that the resistance must be as the number of equally resisting fibres, and as their mutual support, jointly; and, therefore, as some function of the area of the section. The same thing must happen if the fibres be naturally crooked or undulated, as is observed in many woods, provided we suppose some similarity in their form. Similarity of some kind must always be supposed, otherwise we need never aim at any general inferences.

In all cases therefore we can hardly refuse admitting that the strength in opposition to compression is proportional to a function of the area of the section.

As the whole length of a cylinder or prism is equally pressed, it does not appear that the strength of a pillar is at all affected by its length. If indeed it be supposed to bend under the pressure, the case is greatly changed, because it is then exposed to a transverse strain; and this increases with the length of the pillar. But this will be considered with due attention under the next class of strains.

Few experiments have been made on this species of strength and strain. Mr Pitot says that his experiments and those of Mr Parent show that the force necessary for crushing a body is nearly equal to that which will tear it asunder. He says that it requires something more than sixty pounds on every square line to crush a piece of sound oak. But the rule is by no means general: glass, for instance, will carry a hundred times as much as oak in this way, that is, resting on it; but will not *suspend* four or five times as much. Oak will suspend a great deal more than fir; but fir, as a pillar, will carry twice as much. Woods of a soft texture, although consisting of very tenacious fibres, are more easily crushed by their load. This softness of texture is chiefly owing to their fibres not being straight but undulated, and there being considerable vacuities between them, so that they are easily bent laterally and crushed. When a post is overstrained by its load, it is observed to swell sensibly in diameter. Increasing the load causes longitudinal cracks or shivers to appear, and it presently after gives way. This is called *crippling*.

In all cases where the fibres lie oblique to the strain, the strength is greatly diminished, because the parts can then be made to slide on each other when the cohesion of the cementing matter is overcome.

Muschenbroeck has given some experiments on this subject; but they are cases of long pillars, and therefore do not belong to this place. They will be considered afterwards.

The only experiments of which we have seen any detail (and it is useless to insert mere assertions) are those of Mr Gauthey, in the fourth volume of Rozier's *Journal de Physique*. This engineer exposed to great pressures small rectangular parallelopipeds, cut from a great variety of stones, and noted the weights which crushed them. The following table exhibits the medium results of many trials on two very uniform kinds of freestone, one of them among the hardest and the other among the softest used in building.

Column first expresses the length AB of the section, in French lines or 12ths of an inch; column second expresses the breadth BC; column third is the area of the section, in square lines; column fourth is the number of ounces required to crush the piece; column fifth is the weight which was then borne by each square line of the section; and column sixth is the round numbers to which Mr Gauthey imagines that those in column fifth approximate.

Strength of
Materials.

Experiments for
this purpose made
on free-
stone

Hard Stone.

	AB.	BC.	AB×BC.	Weight.	Force.	
1	8	8	64	736	11·5	12
2	8	12	96	2625	27·3	24
3	8	16	128	4496	35·1	36

Soft Stone.

	AB.	BC.	AB×BC.	Weight.	Force.	
4	9	16	144	560	3·9	4·
5	9	18	162	848	5·3	4·5
6	18	18	324	2928	9·	9·
7	18	24	432	5296	12·2	12·

Little can be deduced from these experiments: the first and third, compared with the fifth and sixth, should furnish similar results; for the first and fifth are respectively half of the third and sixth; but the third is three times stronger (that is, a line of the third) than the first, whereas the sixth is only twice as strong as the fifth.

It is evident, however, that the strength increases much faster than the area of the section, and that a square line can carry more and more weight, according as it makes a part of a larger and larger section. In this series of experiments on the soft stone, the individual strength of a square line seems to increase nearly in the proportion of the section of which it makes a part.

Mr Gauthey deduces, from the whole of his numerous experiments, that a pillar of hard stone of Givry, whose section is a square foot, will bear with perfect safety 664,000 pounds, and that its extreme strength is 871,000; and the smallest strength observed in any of his experiments was 460,000. The soft bed of Givry stone had for its smallest strength 187,000, for its greatest 311,000, and for its safe load 249,000. Good brick will carry with safety 320,000; chalk will carry only 9000. The boldest piece of architecture in this respect which he has seen is a pillar in the church of All-Saints at Angers. It is twenty-four feet long and eleven inches square, and is loaded with 60,000, which is not one seventh of what is necessary for crushing it.

We may observe here by the way, that Mr Gauthey's measure of the suspending strength of stone is vastly small in proportion to its power of supporting a load laid above it. He finds that a prism of the hard bed of Givry, of a foot section, is torn asunder by 4600 pounds; and if it be firmly fixed horizontally in a wall, it will be broken by a weight of 56,000 suspended a foot from the wall. If it rest on two props at a foot distance, it will be broken by 206,000 laid on its middle. These experiments agree so ill with each other, that little use can be made of them. The subject is of great importance, and well deserves the attention of the patriotic philosopher.

A set of good experiments would be very valuable, because it is against this kind of strain that we must guard by judicious construction in the most delicate and difficult problems which come through the hands of the civil and military engineer. The construction of stone arches, and the construction of great wooden bridges, and particularly the construction of the frames of carpentry called *centres* in the erection of stone bridges, are the most difficult jobs that occur. In the centres on which the arches of the bridge of Orleans were built, some of the pieces of oak were carrying upwards of two tons on every square inch of their scantling. All who saw it said that it was not able to carry the fourth part of the intended load. But the engineer understood the principles of his art, and ran the risk, and the result completely justified his confidence; for the centre did not complain in any part, only it was found too supple; so that it went out of shape while the haunches only of the arch were laid on it. The engineer corrected this by loading it at the crown, and thus kept it completely in shape during the progress of the work.

In the Memoirs of the Academy of St Petersburg for

Strength of
Materials.

1778, there is a dissertation by Euler on this subject, but particularly limited to the strain on columns, in which the bending is taken into the account. Mr Fuss has treated the same subject with relation to carpentry in a subsequent volume. But there is little in these papers besides a dry mathematical disquisition, proceeding on assumptions which (to speak favourably) are extremely gratuitous. The most important consequence of the compression is wholly overlooked, as we shall presently see. Our knowledge of the mechanism of cohesion is as yet far too imperfect to entitle us to a confident application of mathematics. Experiments should be multiplied.

How they are to be made useful.

The only way in which we can hope to make these experiments useful, is to pay a careful attention to the *manner* in which the fracture is produced. By discovering the general resemblances in this particular, we advance a step in our power of introducing mathematical measurement. Thus, when a cubical piece of chalk is slowly crushed between the chaps of a vice, we see it uniformly split in a surface oblique to the pressure, and the two parts then slide along the surface of fracture. This should lead us to examine mathematically what relation there is between this surface of fracture and the necessary force; then we should endeavour to determine experimentally the position of this surface. Having discovered some general law or resemblance in this circumstance, we should try what mathematical hypothesis will agree with this. Having found one, we may then apply our simplest notions of cohesion, and compare the result of our computations with experiment.

III.—A BODY MAY BE BROKEN ACROSS.

It is of importance to know what strain will break a body transversely

The most usual, and the greatest strain, to which materials are exposed, is that which tends to break them transversely. It is seldom, however, that this is done in a manner perfectly simple; for when a beam projects horizontally from a wall, and a weight is suspended from its extremity, the beam is commonly broken near the wall, and the intermediate part has performed the functions of a lever. It sometimes, though rarely, happens that the pin in the joint of a pair of pincers or scissors is cut through by the strain; and this is almost the only case of a simple transverse fracture. Being so rare, we may content ourselves with saying, that in this case the strength of the piece is proportional to the area of the section.

Experiments made to ascertain it.

Experiments were made for discovering the resistances made by bodies to this kind of strain in the following manner. Two iron bars were disposed horizontally at an inch distance; a third hung perpendicularly between them, being supported by a pin made of the substance to be examined. This pin was made of a prismatic form, so as to fit exactly the holes in the three bars, which were made very exact, and of the same size and shape. A scale was suspended at the lower end of the perpendicular bar, and loaded till it tore out that part of the pin which filled the middle hole. This weight was evidently the measure of the lateral cohesion of two sections. The side-bars were made to grasp the middle bar pretty strongly between them, that there might be no distance interposed between the opposite pressures. This would have combined the energy of a lever with the purely transverse pressure. For the same reason it was necessary that the internal parts of the holes should be no smaller than the edges. Great irregularities occurred in our first experiments from this cause, because the pins were somewhat tighter within than at the edges; but when this was corrected they were extremely regular. We employed three sets of holes, viz. a circle, a square (which was occasionally made a rectangle whose length was twice its breadth), and an equilateral triangle. We found in all our experiments the strength exactly proportional to the area of the section, and quite independent of its figure or position, and

Strength of
Materials.

we found it considerably above the direct cohesion; that is, it took considerably more than twice the force to tear out this middle piece that it did to tear the pin asunder by a direct pull. A piece of fine freestone required 205 pounds to pull it directly asunder, and 575 to break it in this way. The difference was very constant in any one substance, but varied from four thirds to six thirds in different kinds of matter, being smallest in bodies of a fibrous texture. But indeed we could not make the trial on any bodies of considerable cohesion, because they required such forces as our apparatus could not support. Chalk, clay baked in the sun, baked sugar, brick, and freestone, were the strongest that we could examine.

But the more common case, where the energy of a lever intervenes, demands a minute examination.

Let ABCD (fig. 3) be the longitudinal section of a beam inserted into a wall at the end AD, and supporting a weight at the free end BC: it is required to find what tendency the weight will have to break the beam over at the section EF.

Fig. 3.

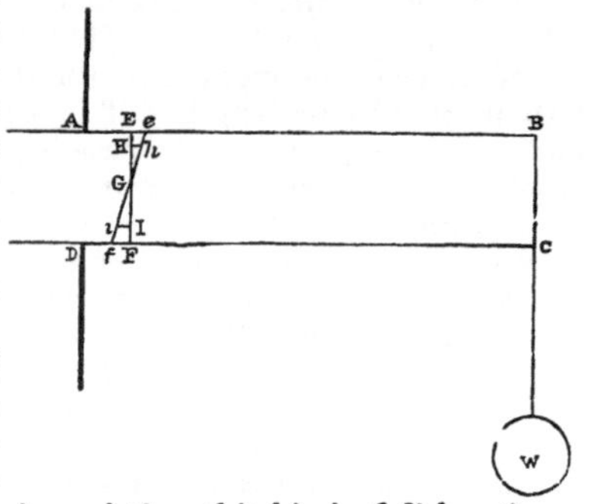

The weight at C will, in the first place, cause a tendency of the part EBCF to slide down on the surface EF, but the strength of the beam in resisting this kind of dislocation is much greater than its power of resisting common fracture: it is therefore unnecessary to examine this case.

The weight at C will cause the beam to bend; that is, it will distend the upper and compress the under part of the beam, and, acting at the extremity of the lever FC, its power of causing such compression and distention will be W × FC. Since the weight at C acts in a vertical direction, it cannot tend either to lengthen or to shorten the beam, and thus the repulsion of the compressed part must be exactly equal to the attraction of the distended parts.

Let G be the neutral point, and draw through it the line *f* G *e*, contiguous to FGE; draw also H*h* and I*i* parallel to CD. The lines I*i*, in the under side of the cross section, will represent the degrees of compression, and also (since within the limits of security the repulsion is proportional to the degree of compression) the force of repulsion, while the lines H*h* on the upper side will serve to represent the attractions. The sum of all the lines on the upper side, that is, the wedge EG*e*, will thus represent the entire amount of attraction, while the wedge FG*f* will represent the total amount of repulsion. These two wedges, then, must be equal to each other; and this equality determines the position of the axis of flexure represented by the point G. In the case of a rectangular beam, G must clearly be in the middle of the line EF; but when the cross section of the beam is irregular, the position of the axis of flexure is not quite so easily found.

Let EF be the cross section of an irregular beam, and OH the axis of flexure, or the place where the beam is neither compressed nor distended. Then the wedges generated by turning the section EF upon OH as an axis must be equal to each other. This is always the case when OH passes through the centre of gravity of the cross section, and thus it follows that those points which are neither compressed nor distended are always ranged in a straight line drawn through the centre of gravity of the cross section.

Fig. 4.

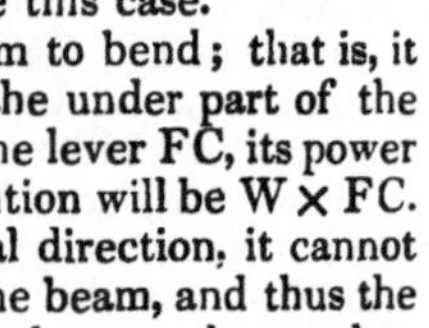

The position of the fulcrum of the lever being now known, we can proceed to ascertain the effects of the various forces.

Returning, for the sake of simplicity, to the rectangular beam; the sum of the repulsions iI will be represented by the triangle GfF: but the rectangle under GF and Ff would measure the entire strength of the under half of the beam, and therefore the force actually exerted when the beam is about to be broken across, is just half of the absolute strength of the beam; one quarter being exhibited as attraction, another quarter as repulsion. These forces act at different distances from the fulcrum G, and it is well known that the influence of a number of weights in turning a lever round is the same as if all these weights were to act at their common centre of gravity; so that to find the entire action in this case, we have to suppose one fourth of the whole strength of the beam to act at the distance $\frac{2}{3}$ of GF or $\frac{1}{3}$ of EF from the fulcrum G, and another quarter of the strength at $\frac{2}{3}$ of GE. Now, if s be the strength of one square inch of the beam, and if D, B, and L be its depth, breadth, and length, measured in inches, DBs will be the absolute strength of the whole beam; and therefore the tendencies of the above forces to straighten the beam will be

$$\tfrac{1}{4} \, DBs \times \tfrac{1}{3} \, D + \tfrac{1}{4} \, DBs \times \tfrac{1}{3} \, D \, ;$$

that is, $\frac{1}{6} \, D^2 \cdot B \cdot s$. And again, the tendency of the weight W to bend the beam is WL; so that

$$\tfrac{1}{6} \, D^2 \cdot B \cdot s = WL,$$

$$\text{or } 6L : D :: DBs : W \, ;$$

that is, *as six times the length of the beam is to its depth, so is the absolute strength of the beam to the weight which it can carry at the free end.*

Throughout this investigation we have supposed that the force needed to extend or compress a fibre is exactly proportional to the quantity of extension or compression. This hypothesis, though not perhaps strictly true, and though it certainly errs when we approach to dislocation or fracture, is yet confirmed by all experiments when the extensions have been kept within the limits of safety: the results of this hypothesis therefore are what must guide us in forming any structure.

It may now be interesting to inquire into the strength of a beam when bent in different directions.

EF being, as before, the cross section of an irregular beam, let that beam be bent in the direction OG, the axis of flexure being OH perpendicular to OG; and let us consider the action of a small portion ds of the surface situated at I. Having drawn the perpendiculars IK and IL, it is clear that the compression of the fibre I must be proportional to KI; and therefore the force exerted by it may be denoted by $C \cdot IK \cdot ds$, C being some constant depending on the nature of the material and the degree of flexure. This force, conceived to act at the extremity of the lever KI, will tend to bend the beam round the axis OH; but again acting at the end of the lever LI, it will tend to bend the beam on OG as an axis. Putting Σ to denote the integral for the entire surface, $C \cdot \Sigma \cdot IK^2 ds$ will be the entire tendency to rectify the form of the beam, and $C\Sigma IK \cdot IL \cdot ds$ will be the entire tendency to take a flexure in a plane at right angles to that which it actually has. A beam therefore will not bend in the direction in which the pressure is applied to it unless $\Sigma \cdot IK \cdot ILds = 0$. This is a circumstance overlooked in all treatises on flexure; but it is one that must be carefully attended to in practice. It may easily be illustrated thus: Take a thin slip of wood, such perhaps as is used for Venetian blinds, and fix it in a vice so that while its length is horizontal its flat sides may be inclined at a considerable angle. Attach now a weight to the free end, and it will be found that that end does not descend vertically, but that it moves obliquely, the flexure not happening in that direction in which the force is applied. The force necessary to bend the beam in the plane of OG is not a force in that

direction, but is the resultant of two forces, $\Sigma \, IK^2 \cdot ds$ in the direction OG, and $\Sigma \cdot IK \cdot IL \cdot ds$ in the direction OH. For the present we shall call $\Sigma IK^2 ds$ the *stiffness in the direction* OG, and $\Sigma \cdot IL^2 ds$ of course the stiffness in the direction OH. The sum of these two stiffnesses is manifestly $\Sigma \cdot OI^2 ds$, which is a constant quantity, depending not at all upon the directions of OG, OH, but only on the form of the cross section: hence follows the remarkable law, that the sum of the stiffnesses of a beam in two directions perpendicular to each other is constant; and that therefore, whatever may be the form of the beam, its directions of greatest and least stiffness are always perpendicular to each other.

For the purpose of discovering in what directions the greatest and least stiffness lie, let us refer all the points in the cross section to the axes OX and OY, putting the angle $XOG = \varphi$. We have then $IK = x \cos. \varphi + y \sin. \varphi$, $IL = -x \sin. \varphi + y \cos. \varphi$; and thus the tendency of the beam to redress itself in the direction GO is $\Sigma \cdot (x \cos. \varphi + y \sin. \varphi)^2 \, ds$, while the deflecting tendency in the direction HO is $\Sigma \cdot (x \cos. \varphi + y \sin. \varphi)(-x \sin. \varphi + y \cos. \varphi) \, ds$. Regarding φ as the variable quantity, and differentiating the former for the purpose of discovering its maximum, we obtain $\Sigma \cdot (x \cos. \varphi + y \sin. \varphi)(-x \sin. \varphi + y \cos. \varphi) \, ds = 0$; now it will be observed that this expression is just that for the deflecting tendency, and hence this law,

That when a beam is bent in the direction of greatest or of least stiffness, the pressure to be applied is exactly in the direction of the bending.

The value of φ may easily be found from the above equation; the result is

$$\tan. 2 \, \varphi = \frac{2 \, \Sigma \cdot xy \, ds}{\Sigma x^2 ds - \Sigma y^2 ds}.$$

The lines OX and OY will coincide with the directions of greatest and least stiffness when $\varphi = 0$, or when $\tan. 2 \varphi = 0$, that is, when $\Sigma \cdot xy \, ds = 0$.

If, then, the directions of greatest and least stiffness be taken for the axes of x and of y, we shall have $\Sigma \cdot xy \, ds = 0$, $\Sigma \, x^2 ds =$ greatest stiffness $=$ A, $\Sigma \, y^2 ds =$ least stiffness $=$ B. These being once known, the stiffness in any other direction, as well as the deflecting tendency, can readily be obtained. Putting P and Q respectively for these quantities, we have

$$P = A \cos. \varphi^2 + B \sin. \varphi^2,$$
$$Q = \tfrac{1}{2} \, (B - A) \sin. 2 \, \varphi.$$

The deflecting tendency is thus greatest when $\varphi = 45°$, that is, when the actual direction of flexure is equally inclined to the directions of greatest and least stiffness. In this case

$$P^1 = \tfrac{1}{2} \, (A + B), \quad Q = \tfrac{1}{2} \, (B - A).$$

Galileo, who was the first to investigate the law of transverse strain, conceived the lower edge of the beam to be the fulcrum, and each fibre to be exerting its whole strength; Professor Robison, in the former editions of this work, corrected the supposition in the case of rectangular beams: the above investigation extends it to beams of all forms.

We must now remark, that this correction of the Galilean hypothesis of equal forces was suggested by the bending which is observed in all bodies which are strained transversely. Because they are bent, the fibres on the convex side have been extended. We cannot say in what proportion this obtains in the different fibres. Our most distinct notions of the internal equilibrium between the particles render it highly probable that their extension is proportional to their distance from that fibre which retains its former dimensions. But by whatever law this is regulated, we see plainly that the actions of the stretched fibres must follow the proportions of some function of this distance, and that therefore the relative strength of a beam is in all cases susceptible of mathematical determination.

Bernoulli's
problem of
the elastic
curve.

We also see an intimate connection between the strain and the curvature. This suggested to the celebrated James Bernoulli the problem of the *elastic curve, i. e.* the curve into which an extensible rigid body will be bent by a transverse strain. His solution in the *Acta Eruditorum* 1694 and 1695, is a very beautiful specimen of mathematical discussion, and we recommend it to the perusal of the curious reader. He will find it very perspicuously treated in the first volume of his works, published after his death, where the wide steps which he had taken in his investigation are explained so as to be easily comprehended. His nephew, Daniel Bernoulli, has given an elegant abridgement in the Petersburg Memoirs for 1729. The problem is too intricate to be fully discussed in a work like ours, but it is also too intimately connected with our present subject to be entirely omitted. We must content ourselves with showing the leading mechanical properties of this curve, from which the mathematician may deduce all its geometrical properties.

Its leading
mechanical
property
described.

When a bar of uniform depth and breadth, and of a given length, is bent into an arch of a circle, the extension of the outer fibres is proportional to the curvature; for, because the curves formed by the inner and outer sides of the beam are similar, the circumferences are as the radii, and the radius of the inner circle is to the difference of the radii as the length of the inner circumference is to the difference of the circumferences. The difference of the radii is the depth of the beam, the difference of the circumferences is the extension of the outer fibres, and the inner circumference is supposed to be the primitive length of the beam. Now the second and third quantities of the above analogy, viz. the depth and length of the beam, are constant quantities, as is also their product. Therefore the product of the inner radius and the extension of the outer fibre is also a constant quantity, and the whole extension of the outer fibre is inversely as the radius of curvature, or is directly as the curvature of the beam.

The mathematical reader will readily see, that into whatever curve the elastic bar is bent, the whole extension of the outer fibre is equal to the length of a similar curve having the same proportion to the thickness of the beam that the length of the beam has to the radius of curvature.

Now let ADCB (fig. 5) be such a rod of uniform breadth and thickness, firmly fixed in a vertical position, and bent into a curve AEFB by a weight W suspended at B, and of such magnitude that the extremity B has its tangent perpendicular to the action of the weight, or parallel to the horizon. Suppose, too, that the extensions are proportional to the extending forces. From any two points E and F draw the horizontal ordinates EG, FH. It is evident that the exterior fibres of the sections E*e* and F*f* are stretched by forces which are in the proportion of EG to FH (these being the long arms of the levers, and the equal thicknesses E*e*, F*f* being the short arms). Therefore (by the hypothesis) their extensions are in the same proportion. But because the extensions are proportional to some similar functions of the distance from the axes of fracture E and F, the extension of any fibre in the section E*e* is to the contemporaneous extension of the similarly situated fibre in the section F*f*, as the extension of the exterior fibre in the section E*e* is to the extension of the exterior fibre in the section F*f*: therefore the whole extension of E*e* is to the whole extension of F*f* as EG to FH, and EG is to FH as the curvature in E to the curvature in F.

Here let it be remarked, that this proportionality of the curvature to the extension of the fibres is not limited to the

Fig. 5.

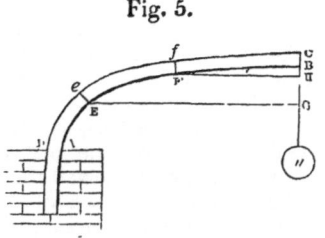

hypothesis of the proportionality of the extensions to the extending forces: it follows from the extension in the different sections being as some similar function of the distance from the axis of fracture; an assumption which cannot be refused.

This, then, is the fundamental property of the elastic curve, from which its equation, or relation between the abscissa and ordinate, may be deduced in the usual forms, and all its other geometrical properties. These are foreign to our purpose; and we shall notice only such properties as have an immediate relation to the strain and strength of the different parts of a flexible body, and which in particular serve to explain some difficulties in the valuable experiments of Buffon on the Strength of Beams.

It is not a
circle.

We observe, in the first place, that the elastic curve cannot be a circle, but is gradually more incurvated as it recedes from the point of application B of the straining forces. At B it has no curvature; and if the bar were extended beyond B there would be no curvature there. In like manner, when a beam is supported at the ends and loaded in the middle, the curvature is greatest in the middle; but at the props, or beyond them, if the beam extend farther, there is no curvature. Therefore, when a beam projecting twenty feet from a wall is bent to a certain curvature at the wall by a weight suspended at the end, and a beam of the same size projecting twenty feet is bent to the very same curvature at the wall by a greater weight at ten feet distance, the figure and the mechanical state of the beam in the vicinity of the wall is different in these two cases, though the curvature at the very wall is the same in both. In the first case every part of the beam is incurvated; in the second, all beyond the ten feet is without curvature. In the first experiment the curvature at the distance of five feet from the wall is three fourths of the curvature at the wall; in the second, the curvature at the same place is but one half of that at the wall. This must weaken the long beam in this whole interval of five feet, because the greater curvature is the result of a greater extension of the fibres.

Every beam
has a certain deter-
minate cur-
vature

In the next place we may remark, that there is a certain determinate curvature for every beam, which cannot be exceeded without breaking it; for there is a certain separation of two adjoining particles that puts an end to their cohesion. A fibre can therefore be extended only a certain proportion of its length. The ultimate extension of the outer fibres must bear a certain determinate proportion to its length, and this proportion is the same with that of the thickness (or what we have hitherto called the depth) to the radius of ultimate curvature, which is therefore determinate.

And when
of uniform
breadth and
depth, is
most incur-
vated where the
strain is
greatest.

A beam of uniform breadth and depth is therefore most incurvated where the strain is greatest, and will break in the most incurvated part. But by changing its form, so as to make the strength of its different sections in the ratio of the strain, it is evident that the curvature may be the same throughout, or may be made to vary according to any law. This is a remark worthy of the attention of the watchmaker. The most delicate problem in practical mechanics is so to taper the balance-spring of a watch that its wide and narrow vibrations may be isochronous. Hooke's principle *ut tensio sic vis* is not sufficient when we take the *inertia* and motion of the spring itself into the account. The figure into which it bends and unbends has also an influence. Our readers will take notice that the artist aims at an accuracy which will not admit an error of $\frac{1}{86400}$th, and that Harrison and Arnold have actually attained it in several instances. The taper of a spring is at present a nostrum in the hands of each artist, and he is careful not to impart its secret.

Again, since the depth of the beam is thus proportional to the radius of ultimate curvature, this ultimate or breaking curvature is inversely as the depth. It may be expressed by $\frac{1}{d}$.

To what
the curva-
ture is pro-
portional.

When a weight is hung on the end of a prismatic beam, the curvature is nearly as the weight and the length directly, and as the breadth and the cube of the depth inversely; for the strength is $= f\dfrac{bd^2}{6l}$. Let us suppose that this produces the ultimate curvature $\dfrac{1}{d}$. Now let the beam be loaded with a smaller weight w, and let the curvature produced be C; we have this analogy, $f\dfrac{bd^2}{6l} : w = \dfrac{1}{d} : C$, and $C = \dfrac{6lw}{fbd^3}$. It is evident that this is also true of a beam supported at the ends and loaded between the props; and we see how to determine the curvature in its different parts, whether arising from the load, or from its own weight, or from both.

Deflection.

When a beam is thus loaded at the end or middle, the loaded point is pulled down, and the space through which it is drawn may be called the *deflection*. This may be considered as the subtense of the angle of contact, or as the versed sine of the arch into which the beam is bent, and is therefore as the curvature when the length of the arches is given (the flexure being moderate), and as the square of the length of the arch when the curvature is given. The deflection therefore is as the curvature and as the square of the length of the arch jointly; that is, as $\dfrac{6lw}{fbd^3} \times l^2$, or as $\dfrac{6l^3w}{fbd^3}$. The deflection from the primitive shape is therefore as the bending weight and the cube of the length directly, and as the breadth and cube of the depth inversely.

In beams just ready to break, the curvature is as the depth inversely, and the deflection is as the square of the length divided by the depth; for the ultimate curvature at the breaking part is the same whatever is the length; and in this case the deflection is as the square of the length.

The theorems resulting from this subject afford the finest methods of examining the laws of corpuscular action

We have been the more particular in our consideration of this subject, because the resulting theorems afford us the finest methods of examining the laws of corpuscular action, that is, for discovering the variation of the force of cohesion by a change of distance. It is true it is not the atomical law, or *hylarchic principle* as it may justly be called, which is thus made accessible, but the specific law of the particles of the substance or kind of matter under examination. But even this is a very great point; and coincidences in this respect among the different kinds of matter are of great moment. We may thus learn the nature of the corpuscular action of different substances, and perhaps approach to a discovery of the *mechanism* of chemical affinities. For that chemical actions are insensible cases of local motion is undeniable, and local motion is the province of mechanical discussion; nay, we see that these hidden changes are produced by mechanical forces in many important cases, for we see them promoted or prevented by means purely mechanical. The conversion of bodies into elastic vapour by heat can at all times be prevented by a *sufficient* external pressure. A strong solution of Glauber's salts will congeal in an instant by agitation, giving out its latent heat; and it will remain fluid for ever, and retain its latent heat in a close vessel which it completely fills. Even water will by such treatment freeze in an instant by agitation, or remain fluid for ever by confinement. We know that heat is produced or extricated by friction, that certain compounds of gold or silver with saline matters explode with irresistible violence by the smallest pressure or agitation. Such facts should rouse the mathematical philosopher, and excite him to follow out the conjectures of the illustrious Newton, encouraged by the ingenious attempts of Boscovich; and the proper beginning of this study is to at-

tend to the laws of attraction and repulsion exerted by the particles of cohering bodies, discoverable by experiments made on their actual extensions and compressions. The experiments of simple extensions and compressions are quite insufficient, because the total stretching of a wire is so small a quantity, that the mistake of the 1000th part of an inch occasions an irregularity which deranges any progression so as to make it useless. But by the bending of bodies a distention of $\frac{1}{100}$th of an inch may be easily magnified in the deflection of the spring ten thousand times. We know that the investigation is intricate and difficult, but not beyond the reach of our present mathematical attainments; and it will give very fine opportunities of employing all the address of analysis. In the 17th century and the beginning of the 18th this was a sufficient excitement to the first geniuses of Europe. The cycloid, the catenaria, the elastic curve, the velaria, the caustics, were reckoned an abundant recompense for much study; and James Bernoulli requested, as an honourable monument, that the logarithmic spiral might be inscribed on his tombstone. The reward for the study to which we now presume to incite the mathematicians is the almost unlimited extension of natural science, important in every particular branch. To go no further than our present subject, a great deal of important practical knowledge respecting the strength of bodies is derived from the single observation, that in the moderate extensions which happen before the parts are overstrained, the forces are nearly in the proportion of the extensions or separations of the particles. To return to our subject.

Bernoulli calls in question this law,

James Bernoulli, in his second dissertation on the elastic curve, calls in question this law, and accommodates his investigation to any hypothesis concerning the relation of the forces and extensions. He relates some experiments of lute-strings where the relation was considerably different. Strings of three feet long,

stretched by	2, 4, 6, 8, 10 pounds,	
were lengthened	9, 17, 23, 27, 30 lines.	

But this is a most exceptionable form of the experiment. The strings were twisted, and the mechanism of the extensions is here exceedingly complicated, combined with compressions and with transverse twists, &c. We made experiments on fine slips of the gum caoutchouc, and on the juice of the berries of the white bryony, of which a single grain will draw to a thread of two feet long, and again return into a perfectly round sphere. We measured the diameter of the thread by a microscope with a micrometer, and thus could tell in every state of extension the proportional number of particles in the sections. We found, that through the whole range in which the distance of the particles was changed in the proportion of thirteen to one, the extensions did not *sensibly* deviate from the proportion of the forces. The same thing was observed in the caoutchouc as long as it perfectly recovered its first dimensions. And it is on the authority of these experiments that we presume to announce this as a law of nature.

which was first assumed by Dr Hooke.

Dr Robert Hooke was undoubtedly the first who attended to this subject, and assumed this as a law of nature. Mariotte indeed was the first who expressly used it for determining the strength of beams: this he did about the year 1679, correcting the simple theory of Galileo. Leibnitz, indeed, in his dissertation in the *Acta Eruditorum* 1684, *De Resistentia Solidorum*, introduces this consideration, and wishes to be considered as the discoverer; and he is always acknowledged as such by the Bernoullis, and others who adhered to his peculiar doctrines. But Mariotte had published the doctrine in the most express terms long before; and Bulfinger, in the *Comment. Petropol.* 1729, completely vindicates his claim. But Hooke was unquestionably the discoverer of this law. It made the foundation of his theory of springs, announced to the Royal Society about the year 1661, and read in 1666. On this occasion he mentions many things on the strength of bodies as quite familiar to his thoughts, which are immediate deductions from this principle; and among these *all* the facts which John Bernoulli so vauntingly adduces in support of Leibnitz's finical dogmas about the force of bodies in motion; a doctrine which Hooke might have claimed as his own, had he not perceived its frivolous inanity.

Though corrected by Mariotte, it does not properly explain the mechanism of transverse strain.

But even with this first correction of Mariotte, the mechanism of transverse strain is not fully nor justly explained. The force acting in the direction BW (fig 3), and bending the body ABCD, not only stretches the fibres on the side opposite to the axis of fracture, but compresses the side CD, which becomes concave by the strain. Indeed it cannot do the one without doing the other; for, in order to stretch the fibres at D, there must be some fulcrum, some support, on which the virtual lever BAD may press, that it may tear asunder the stretched fibres. This fulcrum must sustain both the pressure arising from the cohesion of the distended fibres, and also the action of the external force, which immediately tends to cause the prominent part of the beam to slide along the section EF.

as is fully verified by experiment

This is fully verified by experiment. If we attempt to break a long slip of cork, or any such very compressible body, we always observe it to bulge out on the concave side before it cracks on the other side. If it is a body of fibrous or foliated texture, it seldom fails splintering off on the concave side; and in many cases this splintering is very deep, even reaching half way through the piece. In hard and granulated bodies, such as a piece of freestone, chalk, dry clay, sugar, and the like, we generally see a considerable splinter or shiver fly off from the hollow side. If the fracture be slowly made by a force at B gradually augmented, the formation of the splinter is very distinctly seen. It forms a triangular piece, which generally breaks in the middle.

Consequences result ing from the state of the case.

Let us see what consequences result from this state of the case respecting the strength of bodies. Let DΔKC (fig. 6) represent a vertical section of a prism of compres-

Fig. 6.

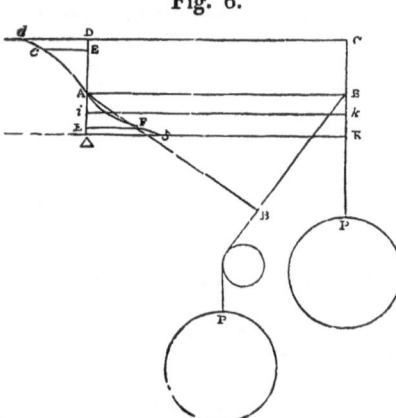

sible materials, such as a piece of timber. Suppose it loaded with a weight P hung at its extremity. Suppose it of such a constitution that all the fibres in AD are in a state of dilatation, while those in AΔ are in a state of compression. In the instant of fracture the particles at D and E are withheld by forces Dd, Ee, and the particles at Δ and F repel, resist, or support, with forces Δδ, Fε.

Some line, such as *de*Aεδ, will limit all these ordinates, which represent the forces actually exerted in the instant of fracture. If the forces be as the extensions and compressions, as we have great reason to believe, *de*A and Aεδ will be two straight lines. They will form *one* straight line dAδ, if the forces which resist a certain dilatation are equal to the forces which resist an equal compression. But this is

quite accidental, and is not strict y true in any body. In most bodies which have any considerable firmness, the compressions made by any external force are not so great as the dilatations which the same force would produce; that is, the repulsions which are excited by any supposed degree of compression are greater than the attractions excited by the same degree of dilatation. Hence it will generally follow, that the angle dAD is less than the angle δAΔ, and the ordinates Dd, Ee, &c. are less than the corresponding ordinates Δδ, Eε, &c.

An important consequence of the compressibility of body fully proved.

But whatever be the nature of the line dAδ, we are certain of this, that the whole area ADd is equal to the whole area AΔδ; for as the force at B is gradually increased, and the parts between A and D are more extended, and greater cohesive forces are excited, there is always such a degree of repulsive forces excited in the particles between A and Δ that the one set precisely balances the other. The force at B, acting perpendicularly to AB, has no tendency to push the whole piece closer on the part next the wall, or to pull it away. The sum of the attractive and repulsive forces actually excited must therefore be equal. These sums are represented by the two triangular areas, which are therefore equal.

The greater we suppose the repulsive forces corresponding to any degree of compression, in comparison with the attractive forces corresponding to the same degree of extension, the smaller will AΔ be in comparison of AD. In a piece of cork or sponge, AΔ may chance to be equal to AD, or even to exceed it; but in a piece of marble, AΔ will perhaps be very small in comparison of AD.

Now it is evident that the repulsive forces excited between A and Δ have no share in preventing the fracture. They rather contribute to it, by furnishing a fulcrum to the lever by whose energy the cohesion of the particles in AD is overcome. Hence we see an important consequence of the compressibility of the body. Its power of resisting this transverse strain is diminished by it, and so much the more diminished as the stuff is more compressible.

This is fully verified by some very curious experiments made by Duhamel. He took sixteen bars of willow two feet long and half an inch square, and supporting them by props under the ends, he broke them by weights hung on the middle. He broke four of them by weights of 40, 41, 47, and 52 pounds: the mean is 45. He then cut four of them one third through on the upper side, and filled up the cut with a thin piece of harder wood stuck in pretty tight. These were broken by 48, 54, 50, and 52 pounds; the mean of which is 51. He cut other four half through, and they were broken by 47, 49, 50, 46; the mean of which is 48. The remaining four were cut two thirds, and their mean strength was 42.

Another set of his experiments is still more remarkable. Six battens of willow thirty-six inches long and one and a half square were broken by 525 pounds at a medium.

Six bars were cut one third through, and the cut filled with a wedge of hard wood stuck in with a little force: these broke with 551.

Six bars were cut half through, and the cut was filled in the same manner: they broke with 542.

Six bars were cut three fourths through: these broke with 530.

A batten cut three fourths through, and loaded till nearly broken, was unloaded, and the wedge taken out of the cut. A thicker wedge was put in tight, so as to make the batten straight again by filling up the space left by the compression of the wood: this batten broke with 577 pounds.

From this it is plain that more than two thirds of the thickness (perhaps nearly three fourths) contributed nothing to the strength.

The point A is the centre of fracture in this case; and in order to estimate the strength of the piece, we may suppose that the crooked lever virtually concerned in the strain is DAB. We must find the point I, which is the centre of effort of all the attractive forces, or that point where the full cohesion of AD must be applied, so as to have a momentum equal to the accumulated momenta of all the variable forces. We must in like manner find the centre of effort i of the repulsive or supporting forces exerted by the fibres lying between A and Δ.

It is plain, and the remark is important, that this last centre of effort is the real fulcrum of the lever, although A is the point where there is neither extension nor contraction; for the lever is supported in the same manner as if the repulsions of the whole line AΔ were exerted at that point. Therefore let S represent the surface of fracture from A to D, and f represent the absolute cohesion of a fibre at D in the instant of fracture. We shall have $f S \times \overline{I + i} = pl$, or $l : I + i = f S : p$; that is, the length AB is to the distance between the two centres of effort I and i, as the absolute cohesion of the section between A and D is to the relative strength of the section.

It would be perhaps more accurate to make AI and Ai equal to the distances of A from the horizontal lines passing through the centres of gravity of the triangles of dAD and δAΔ. It is only in this construction that the points I and i are the centres of real effort of the accumulated attractions and repulsions. But I and i, determined as we have done, are the points where the full equal actions may be all applied, so as to produce the same momenta. The final results are the same in both cases. The attentive and duly informed reader will see that Mr Bulfinger, in a very elaborate dissertation on the strength of beams, in the *Comment. Petropolitan.* 1729, has committed several mistakes in his estimation of the actions of the fibres. We mention this because his reasonings are quoted and appealed to as authorities by Muschenbroeck and other authors of note. The subject has been considered by many authors on the continent. We recommend to the reader's perusal the very minute discussions in the Memoirs of the Academy of Paris for 1702 by Varignon, the Memoirs for 1708 by Parent, and particularly that of Coulomb in the *Mém. par les Sçavans Etrangers*, tom. vii.

It is evident from what has been said above, that if S and s represent the surfaces of the sections above and below A, and if G and g are the distances of their centres of gravity from A, and O and o the distances of their centres of oscillation, and D and d their whole depths, the momentum of cohesion will be $\dfrac{f S \cdot G \cdot O}{D} = \dfrac{f s \cdot g \cdot o}{d} = pl.$

If, as is most likely, the forces are proportional to the extensions and compressions, the distances AI and Ai, which are respectively $= \dfrac{G \cdot O}{D}$ and $\dfrac{g \cdot o}{d}$, are respectively $= \frac{1}{3} DA$ and $\frac{1}{3} \Delta A$, and when taken together are $= \frac{1}{3} D\Delta$. If, moreover, the extensions are equal to the compressions in the instant of fracture, and the body is a rectangular prism like a common joist or beam, then DA and ΔA are also equal; and therefore the momentum of cohesion is $f b \times \frac{1}{2} d \times \frac{1}{3} d = \dfrac{f b d^{2}}{6} = f b d \times \frac{1}{6} d = pl.$ Hence we obtain this analogy: " six times the length is to the depth as the absolute cohesion of the section is to its relative strength."

Thus we see that the compressibility of bodies has a very great influence on their power of withstanding a transverse strain. We see that in this most favourable supposition of equal dilatations and compressions, the strength is reduced to one half of the value of what it would have been had the body been incompressible. This is by no means obvious; for it does not readily appear how compressibi-

lity, which does not diminish the cohesion of a single fibre, should impair the strength of the whole. The reason, however, is sufficiently convincing when pointed out. In the instant of fracture, a smaller portion of the section is actually exerting cohesive forces, while a part of it is only serving as a fulcrum to the lever by whose means the strain on the section is produced. We see, too, that this diminution of strength does not so much depend on the sensible compressibility, as on its proportion to the dilatability by equal forces. When this proportion is small, AΔ is small in comparison of AD, and a greater portion of the whole fibre is exerting attractive forces. The experiments already mentioned, of Duhamel de Monceau, on battens of willow, show that its compressibility is nearly equal to its dilatability. But the case is not very different in tempered steel. The famous Harrison, in the delicate experiments which he made while occupied in making his longitude watch, discovered that a rod of tempered steel was nearly as much diminished in its length as it was augmented by the same external force. But it is not by any means certain that this is the proportion of dilatation and compression which obtains in the very instant of fracture. We rather imagine that it is not. The forces are nearly as the dilatations till very near breaking; but we think that they diminish when the body is just going to break. But it seems certain that the forces which resist compression increase faster than the compressions, even before fracture. We know incontestably that the ultimate resistances to compression are insuperable by any force which we can employ. The repulsive forces, therefore, in their whole extent, increase faster than the compressions, and are expressed by an assymptotic branch of the Boscovician curve formerly explained. It is therefore probable, especially in the more simple substances, that they increase faster, even in such compressions as frequently obtain in the breaking of hard bodies. We are disposed to think that this is always the case in such bodies as do not fly off in splinters on the concave side; but this must be understood with the exception of the permanent changes which may be made by compression when the bodies are crippled by it. This always increases the compression itself, and causes the neutral point to shift still more towards D. The effect of this is sometimes very great and fatal.

Experiment alone can help us to discover the proportion between the dilatability and compressibility of bodies. The strain now under consideration seems the best calculated for this research. Thus if we find that a piece of wood an inch square requires 12,000 pounds to tear it asunder by a direct pull, and that 200 pounds will break it transversely by acting 10 inches from the section of fracture, we must conclude that the neutral point A is in the middle of the depth, and that the attractive and repulsive forces are equal. Any notions that we can form of the constitution of such fibrous bodies as timber, make us imagine that the *sensible* compressions, including what arises from the bending up of the compressed fibres, is much greater than the real corpuscular extensions. One may get a general conviction of this unexpected proposition by reflecting on what must happen during the fracture. An undulated fibre can only be drawn straight, and then the corpuscular extension begins; but it may be bent up by compression to any degree, the corpuscular compression being little affected all the while. This observation is very important; and though the forces of corpuscular repulsion may be almost insuperable by any compression that we can employ, a *sensible* compression may be produced by forces not enormous, sufficient to cripple the beam. Of this we shall see very important instances afterwards.

It deserves to be noticed, that although the relative strength of a prismatic solid is extremely different in the three hypotheses now considered, yet the proportional

Strength of
Materials.

strengths of different pieces follow the same ratio, namely, the direct ratio of the breadth, the direct ratio of the square of the depth, and the inverse ratio of the length. In the first hypothesis (of equal forces) the strength of a rectangular beam was $\frac{fbd^2}{2l}$; in the second (of attractive forces proportioned to the extensions) it was $\frac{fbd^2}{3l}$; and in the third (equal attractions and repulsions proportional to the extensions and compressions) it was $\frac{fbd^2}{6l}$, or more generally $\frac{fbd^2}{ml}$, where m expresses the unknown proportion between the attractions and repulsions corresponding to an equal extension and compression.

Hence we derive a piece of useful information, which is confirmed by unexceptionable experience, that the strength of a piece depends chiefly on its depth, that is, on that dimension which is in the direction of the strain. A bar of timber of one inch in breadth and two inches in depth is four times as strong as a bar only one inch deep, and it is twice as strong as a bar two inches broad and one deep; that is, a joist or lever is always strongest when laid on its edge.

There is therefore a choice in the manner in which the cohesion is opposed to the strain. The general aim must be to put the centre of effort I as far from the fulcrum or the neutral point A as possible, so as to give the greatest energy or momentum to the cohesion. Thus if a triangular bar projecting from a wall is loaded with a weight at its extremity, it will bear thrice as much when one of the sides is uppermost as when it is undermost.

Hence it follows that the strongest joist that can be cut out of a round tree is not the one which has the greatest quantity of timber in it, but such that the product of its breadth by the square of its depth shall be the greatest possible. Let ABCD (fig. 7) be the section of this joist inscribed in the circle, AB being the breadth and AD the depth. Since it is a rectangular section, the diagonal BD is a diameter of the circle, and BAD is a right-angled triangle. Let BD be called a, and BA be called x; then $AD = \sqrt{a^2 - x^2}$. Now we must have $AB \cdot AD^2$, or $x(a^2 - x^2)$, or $a^2 x - x^3$, a minimum; its differential $(a^2 - 3x^2)dx$ must be 0, or $a^2 = 3x^2$, or $x^2 = \frac{a^2}{3}$. If therefore we make $DE = \frac{1}{3}$ DB, and draw EC perpendicular to BD, it will cut the circumference in the point C, which determines the depth BC and the breadth CD.

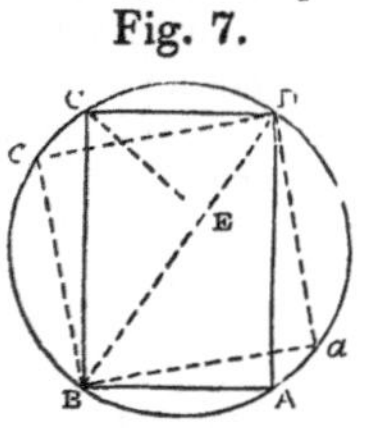

Because $BD : BC = CD : CE$, we have the area of the section $BC \cdot CD = BD \cdot CE$. Therefore the different sections having the same diagonal BD, are proportional to their heights CE. Therefore the section BCDA is less than the section $Bc\,Da$, whose four sides are equal. The joist so shaped, therefore, is stronger, lighter, and cheaper.

The strength of ABCD is to that of $a\,B\,c\,D$ as 10,000 to 9186, and the weight and expense as 10,000 to 10,607; so that ABCD is preferable to $a\,B\,c\,D$, in the proportion of 10,607 to 9186, or nearly 115 to 100.

From the same principles it follows that a hollow tube is stronger than a solid rod containing the same quantity of matter. Let fig. 8 represent the section of a cylindric tube, of which AF and BE are the exterior and interior diameters, and C the centre. Draw BD perpendicular to BC, and join DC. Then, because $BD^2 = CD^2 - CB^2$ BD is the radius of a circle containing the same quantity of matter with the ring. If we estimate the strength by

Fig. 7.

Fig. 8.

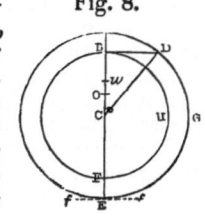

The proportional strengths of different pieces follow the same ratio

The strength of a piece depends chiefly on its depth;

and therefore a choice in the manner in which the cohesion is opposed to the strain.

The strongest joist has not the greatest quantity of timber.

A hollow tube stronger than a solid rod containing the same quantity of matter.

the first hypothesis, it is evident that the strength of the tube will be to that of the solid cylinder, whose radius is BD, as $BD^2 \times AC$ to $BD^2 \times BD$; that is, as AC to BD; for BD^2 expresses the cohesion of the ring of the circle, and AC and BD are equal to distances of the centres of effort (the same with the centres of gravity) of the ring and circle from the axis of the fracture.

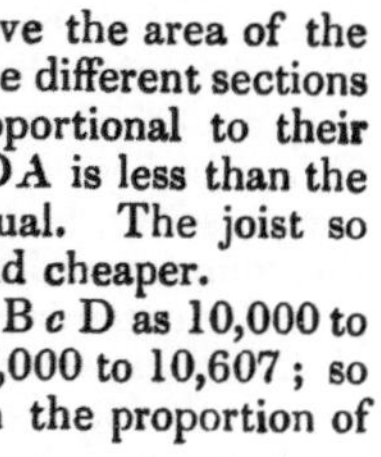

The proportion of these strengths will be different in the other hypothesis, and is not easily expressed by a general formula; but in both it is still more in favour of the ring or hollow tube.

The following very simple solution will be readily understood by the intelligent reader. Let O be the centre of oscillation of the exterior circle, o the centre of oscillation of the inner circle, and w the centre of oscillation of the ring included between them. Let M be the quantity of surface of the exterior circle, m that of the inner circle, and μ that of the ring.

We have $Fw = \dfrac{M \cdot FO - m \cdot Fo}{\mu} = \dfrac{5\,FC^2 + EC^2}{4\,FC}$, and the strength of the ring $= \dfrac{f\mu \times Fw}{2}$, and the strength of the same quantity of matter in the form of a solid cylinder is $f\mu \times \frac{5}{8}BD$; so that the strength of the ring is to that of the solid rod of equal weight as $F\,w$ to $\frac{5}{8}BD$, or nearly as FC to BD. This will easily appear by recollecting that FO is $= \dfrac{\text{sum of } p \cdot r^2}{m \cdot FC}$ (see Rotation), and that the momentum of cohesion is $\dfrac{fm \cdot FC \cdot Ca}{2\,FC} = \dfrac{fm \cdot Fo}{2}$ for the inner circle, &c.

Emerson has given a very inaccurate approximation to this value in his Mechanics, 4to.

This property of hollow tubes is accompanied also with greater stiffness; and the superiority in strength and stiffness is so much the greater as the surrounding shell is thinner in proportion to its diameter.

Here we see the admirable wisdom of the Author of nature in forming the bones of animal limbs hollow. The bones of the arms and legs have to perform the office of levers, and are thus opposed to very great transverse strains. By this form they become incomparably stronger and stiffer, and give more room for the insertion of muscles, while they are lighter and therefore more agile; and the same wisdom has made use of this hollow for other valuable purposes of the animal economy. In like manner the quills in the wings of birds acquire by their thinness the very great strength which is necessary, while they are so light as to give sufficient buoyancy to the animal in the rare medium in which it must live and fly about. The stalks of many plants, such as all the grasses, and many reeds, are in like manner hollow, and thus possess an extraordinary strength. Our best engineers now begin to imitate nature by making many parts of their machines hollow, such as their axles of cast iron, &c.; and the ingenious Mr Ramsden made the axes and framings of his great astronomical instruments in the same manner.

In the supposition of homogeneous texture, it is plain that the fracture happens as soon as the particles at D are separated beyond their utmost limit of cohesion. This is a determined quantity, and the piece bends till this degree of extension is produced in the uttermost fibre. It follows, that the smaller we suppose the distance between A and D, the greater will be the curvature which the beam will acquire before it breaks. Greater depth therefore makes a beam not only stronger, but also stiffer. But if the parallel fibres can slide on each other, both the strength and the stiffness will be diminished. Therefore, if, instead of one beam $D\triangle KC$ (fig. 6), we suppose two, DABC and

Strength of Materials.

and more stiff.

Hence the wisdom of God in forming the bones, &c. hollow.

Strength of
Materials.

How a strong compound beam may be formed.

A△KB, not cohering, each of them will bend, and the extension of the fibres AB of the under beam will not hinder the compression of the adjoining fibres AB of the upper beam. The two together therefore will not be more than twice as strong as one of them (supposing DA = A△), instead of being four times as strong; and they will bend as much as either of them alone would bend by half the load. This may be prevented, if it were possible to unite the two beams all along the seam AB, so that the one shall not slide on the other. This may be done in small works by gluing them together with a cement as strong as the natural lateral cohesion of the fibres. If this cannot be done (as it cannot in large works), the sliding is prevented by *joggling* the beams together, that is, by cutting down several rectangular notches in the upper side of the lower beam, and making similar notches in the under side of the upper beam, and filling up the square spaces with pieces of very hard wood firmly driven in, as represented in fig. 9. Some employ iron bolts by way of joggles. But when the joggle is much harder than the wood into which it is driven, it is very apt to work loose, by widening the hole into

Fig. 9.

which it is lodged. The same thing is sometimes done by scarphing the one upon the other, as represented in fig. 10; but this wastes more timber, and is not so strong, because the mutual hooks which this method form on each beam are very apt to tear each other up. By one or other of these

Fig. 10.

methods, or something similar, may a compound beam be formed, of any depth, which will be almost as stiff and strong as an entire piece.

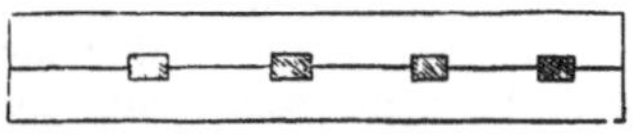

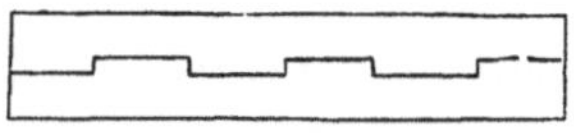

How strength may be combined with pliableness.

On the other hand, we may combine strength with pliableness, by composing our beam of several thin planks laid on each other, till they make a proper depth, and leaving them at full liberty to slide on each other. It is in this manner that coach-springs are formed, as is represented in fig. 11. In this assemblage there must be no joggles nor bolts of any kind put through the planks or plates, for this would hinder their mutual sliding. They must be kept together by straps which surround them, or by something equivalent.

Fig. 11.

Maxims of construction.

The preceding observations show the propriety of some maxims of construction, which the artists have derived from long experience.

Thus, if a mortise is to be cut out of a piece which is exposed to a cross strain, it should be cut out from that side which becomes concave by the strain.

If a piece is to be strengthened by the addition of another, the added piece must be joined to the side which grows convex by the strain.

Before we proceed any farther, it will be convenient to recall the reader's attention to the analogy between the strain on a beam projecting from a wall and loaded at the extremity, and a beam supported at both ends and loaded in some intermediate point. It is sufficient on this occasion to read attentively what is delivered in the article ROOF. We learn there that the strain on the middle point C (fig. 16 of the present article) of a rectangular beam AB, supported on props at A and B, is the same as if the part CA projected from a wall, and were loaded with the half of the weight W suspended at A. The momentum of the strain

is therefore $\frac{1}{2}$ W × $\frac{1}{2}$ AB = W × $\frac{1}{4}$ AB = $p \frac{1}{4} l$, or $\frac{pl}{4}$. The

momentum of cohesion must be equal to this in every hypothesis.

Strength of
Materials.

Having now considered in sufficient detail the circumstances which affect the strength of any section of a solid body that is strained transversely, it is necessary to take notice of some of the chief modifications of the strain itself. We shall consider only those that occur most frequently in our constructions.

The strain depends on the external force, and also on the lever by which it acts.

The strain depends on the external force,

It is evidently of importance, that since the strain is exerted in any section by means of the cohesion of the parts intervening between the section under consideration and the point of application of the external force, the body must be able in all these intervening parts to propagate or excite the strain in the remote section. In every part it must be able to resist the strain excited in that part. It should therefore be equally strong; and it is useless to have any part stronger, because the piece will nevertheless break where it is not stronger throughout; and it is useless to make it stronger (relatively to its strain) in any part, for it will nevertheless equally fail in the part that is too weak.

Suppose, then, in the first place, that the strain arises from a weight suspended at one extremity, while the other end is firmly fixed in a wall. Supposing also the cross sections to be all rectangular, there are several ways of shaping the beam so that it shall be equally strong throughout. Thus it may be equally deep in every part, the upper and under surfaces being horizontal planes. The condition will be fulfilled by making all the horizontal sections triangles, as in fig. 12. The two sides are vertical planes, meeting in an edge at the extremity L. For the equation expressing the balance of strain and strength is $pl = fbd^2$. Therefore, since d^2 is the same throughout, and also p, we must have $fb = l$, and b (the breadth AD of any section ABCD) must be proportional to l (or AL), which it evidently is.

Fig. 12.

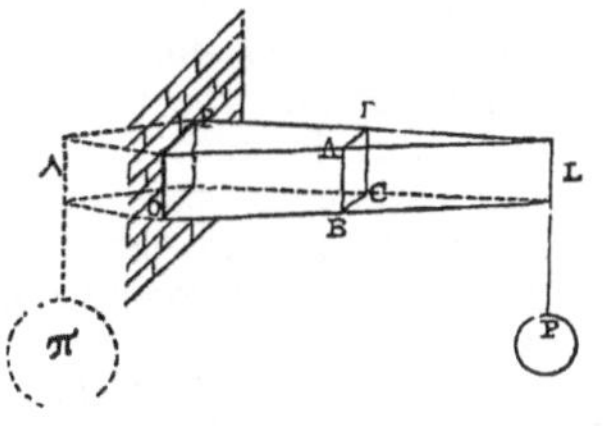

Or, if the beam be of uniform breadth, we must have d^2 everywhere proportional to l. This will be obtained by making the depths the ordinates of a common parabola, of which L is the vertex and the length is the axis. The upper or under side may be a straight line, as in fig. 13, or the middle line may be straight, and then both upper and under surfaces will be curved. It is almost indifferent what is the shape of the upper and under surfaces, provided the distances between them in every part be at the ordinates of a common parabola.

Fig. 13.

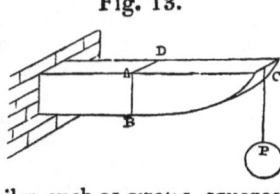

Or, if the sections are all similar, such as circles, squares, or any other similar polygons, we must have d^2 or b^3 proportional to l, and the depths or breadths must be as the ordinates of a cubical parabola.

and on the form of the levers by which it acts.

It is evident that these are also the proper forms for a lever moveable round a fulcrum, and acted on by a force at the extremity. The force comes in the place of the weight suspended in the cases already considered; and as such levers always are connected with another arm, we readily see that both arms should be fashioned in the same manner. Thus in fig. 12 the piece of timber may be supposed a kind of steelyard, moveable round a horizontal axis in the front of the wall, and having the two weights P and ϖ in equilibrio. The strain occasioned by each at the section in which the axis OP is placed must be the same, and each arm OL and Oλ must be equally strong in all its parts. The longitudinal sections of each arm must be a triangle, a common parabola, or a cubic parabola, according to the conditions previously given.

And, moreover, all these forms are equally strong; for any one of them is equally strong in all its parts, and they are all supposed to have the same section at the front of the wall or at the fulcrum. They are not, however, equally stiff. The first, represented in fig. 12, will bend least upon the whole, and the one formed by the cubic parabola will bend most. But their curvature at the very fulcrum will be the same in all.

It is also plain, that if the lever is of the second or third kind, that is, having the fulcrum at one extremity, it must still be of the same shape; for in abstract mechanics it is indifferent which of the three points is considered as the axis of motion. In every lever the two forces at the extremities act in one direction, and the force in the middle acts in the opposite direction, and the great strain is always at that point. Therefore a lever such as fig. 12, moveable round an axis passing horizontally through λ, and acting against an obstacle at OP, is equally able in all its parts to resist the strains excited in those parts.

The same principles and the same construction will apply to beams, such as joists, supported at the ends L and λ (fig. 12), and loaded at some intermediate part OP. This will appear evident by merely inverting the directions of the forces at these three points, or by recurring to the article Roof.

The external straining force may be distributed over the beam.
To make a beam strong which projects from a wall.

Hitherto we have supposed the external straining force as acting only in one point of the beam. But it may be uniformly distributed all over the beam. To make a beam in such circumstances equally strong in all its parts, the shape must be considerably different from the former.

Thus suppose the beam to project from a wall. If it be of equal breadth throughout, its sides being vertical planes parallel to each other and to the length, the vertical section in the direction of its length must be a triangle instead of a common parabola; for the weight uniformly distributed over the part lying beyond any section, is as the length beyond that section: and since it may all be conceived as collected at its centre of gravity, which is the middle of that length, the lever by which this load acts or strains the section is also proportioned to the same length. The strain on the section (or momentum of the load) is as the square of that length. The section must have strength in the same proportion. Its strength being as the breadth and the square of the depth, and the breadth being constant, the square of the depth of any section must be as the square of its distance from the end, and the depth must be as that distance; and therefore the longitudinal vertical section must be a triangle.

But if all the transverse sections are circles, squares, or any other similar figures, the strength of every section, or the cube of the diameter, must be as the square of the lengths beyond that section, or the square of its distance from the end; and the sides of the beam must be a semicubical parabola.

If the upper and under surfaces are horizontal planes, it is evident that the breadth must be as the square of the distance from the end, and the horizontal sections may be formed by arches of the common parabola, having the length for their tangent at the vertex.

By recurring to the analogy so often quoted between a projecting beam and a joist, we may determine the proper form of joists which are uniformly loaded through their whole length.

The strain upon a beam supported at both ends.

This is a frequent and important case, being the office of joists, rafters, &c.; and there are some circumstances which must be particularly noticed, because they are not so obvious, and have been misunderstood. When a beam AB

(fig. 14) is supported at the ends, and a weight is laid on any point P, a strain is excited in every part of the beam. The load on P causes the beam to press on A and B, and the props re-act with forces equal and opposite to these pressures. The load at P is to the pressures at A and B as AB to PB and PA, and the pressure at A is to that at B as BP to PA; the beam therefore is in the same state, with respect to strain in every part of it, as if it were resting on a prop at P, and were loaded at the ends with weights equal to the two pressures on the props: and observe, these pressures are such as will balance each other, being inversely as their distances from P. Let P represent the weight or load at P. The pressure on the prop P must be $P \times \frac{PA}{AB}$. This is therefore the re-action of the prop B, and is the weight which we may suppose suspended at B, when we conceive the beam resting on a prop at P, and carrying the balancing weights at A and B.

Fig. 14.

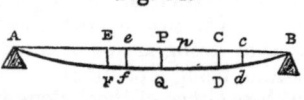

The strain occasioned at any other point C, by the load P at P, is the same with the strain at C, by the weight $P \times \frac{PA}{AB}$ hanging at B, when the beam rests on P, in the manner now supposed; and it is the same if the beam, instead of being balanced on a prop at P, had its part AP fixed in a wall. This is evident. Now we have shown at length that the strain at C, by the weight $P \times \frac{PA}{AB}$ hanging at B, is $P \times \frac{PA}{AB} \times BC$. We desire it to be particularly remarked, that the pressure at A has no influence on the strain at C, arising from the action of any load between A and C; for it is indifferent how the part AP of the projecting beam PB is supported. The weight at A just performs the same office with the wall in which we suppose the beam to be fixed. We are thus particular, because we have seen even persons not unaccustomed to discussions of this kind puzzled in their conceptions of this strain.

Now let the load P be laid on some point p between C and B. The same reasoning shows us that the point is, with respect to strain, in the same state as if the beam were fixed in a wall, embracing the part pB, and a weight $= P \times \frac{pB}{AB}$ were hung on at A, and the strain at C is $P \times \frac{pB}{AB} \times AC$.

A general proposition.

In general, therefore, the strain on any point C, arising from a load P laid on another point P, is proportional to the rectangle of the distances of P and C from the ends nearest to each. It is $P \times \frac{PA \times CB}{AB}$, or $P \times \frac{pB \times CA}{AB}$, according as the load lies between C and A or between C and B.

Cor. 1. The strains which a load on any point P occasions on the points C, c, lying on the same side of P, are as the distances of these points from the end B. In like manner the strains on E and e are as EA and eA.

Cor. 2. The strain which a load occasions in the part on which it rests is as the rectangle of the parts on each side. Thus the strain occasioned at C by a load is to that at D by the same load as AC × CB to AD × DB. It is therefore greatest in the middle.

The strain arising from a load distributed along the beam.

Let us now consider the strain on any point C arising from a load uniformly distributed along the beam. Let AP be represented by x, and Pp by dx, and the whole weight on the beam by a. Then

The weight on Pp is.................. $= a\dfrac{dx}{AB}.$

Pressure on B by the weight on Pp $= a\dfrac{dx}{AB} \times \dfrac{AP}{AB},$

Or..................... $= a\dfrac{xdx}{AB^2}.$

Pressure on B by whole wt. on AC $= a\dfrac{\frac{1}{2}AC^2}{AB^2} = a\dfrac{AC^2}{2AB^2}.$

Strain at C by the weight on AC $= a\dfrac{AC^2 \times BC}{2AB^2}.$

Strain at C by the weight on BC $= a\dfrac{BC^2 \times AC}{2AB^2}.$

Do. by whole weight on AB $= a\dfrac{AC^2 \times BC + BC^2 \times AC}{2AB^2}$

$= a\dfrac{AC \times BC \times \overline{AC + CB}}{2AB^2} = a\dfrac{AC \times BC}{2AB}.$

Thus we see that the strain is proportional to the rectangle of the parts, in the same manner as if the load a had been laid directly on the point C, and is indeed equal to one half of the strain which would be produced at C by the load a laid on there.

Mistakes on this subject committed by authors of reputation. It was necessary to be thus particular, because we see in some elementary treatises on mechanics, published by authors of reputation, mistakes which are very plausible, and mislead the learner. It is there said that the pressure at B from a weight uniformly diffused along AB, is the same as if it were collected at its centre of gravity, which would be the middle of AB; and then the strain at C is said to be this pressure at B multiplied by BC. But surely it is not difficult to see the difference of these strains. It is plain that the pressure of gravity downwards on any point between the end A and the point C has no tendency to diminish the strain at C, arising from the upward re-action of the prop B; whereas the pressure of gravity between C and B is almost in direct opposition to it, and must diminish it. We may however avoid the fluxionary calculus with safety by the consideration of the centre of gravity, by supposing the weights of AC and BC to be collected at their respective centres of gravity; and the result of this computation will be the same as above: and we may use either method, although the weight be not uniformly distributed, provided only that we know in what manner it is distributed.

This investigation is evidently of importance in the practice of the engineer and architect, informing them what support is necessary in the different parts of their constructions. We considered some cases of this kind in the article ROOF.

To form a joist which may have the same relative strength in all its parts. It is now easy to form a joist so that it shall have the same relative strength in all its parts.

I. To make it equally able in all its parts to carry a given weight laid on any point C taken at random, or uniformly diffused over the whole length, the strength of the section at the point C must be as AC × CB. Therefore,

1. If the sides be parallel vertical planes, the square of the depth (which is the only variable dimension), or CD^2, must be as AC × CB, and the depths must be ordinates of an ellipse.

2. If the transverse sections be similar, we must make CD^3 as AC × CB.

3. If the upper and under surfaces be parallel, the breadth must be as AC × CB.

II. If the beam be necessarily loaded at some given point C, and we would have the beam equally able in all its parts to resist the strain arising from the weight at C, we must make the strength of every transverse section between C and either end as its distance from that end. Therefore,

1. If the sides be parallel vertical planes, we must make CD^2 : EF^2 = AC : AE.

2. If the sections be similar, then CD^3 : EF^3 = AC : AE.

3. If the upper and under surfaces be parallel, then breadth at C : breadth at E = AC : AE.

The strain and strength of square or circular plates of different extent but of equal thickness may be determined from the same principles. The same principles enable us to determine the strain and strength of square or circular plates of different extent but equal thickness. This may be comprehended in this general proposition.

Similar plates of equal thickness supported all round will carry the same absolute weight, uniformly distributed, or resting on similar points, whatever be their extent.

Suppose two similar oblong plates of equal thickness, and let their lengths and breadths be L, l, and B, b. Let their strength or momentum of cohesion be C, c, and the strains from the weights W, w, be S, s.

Suppose the plates supported at the ends only, and resisting fracture transversely. The strains, being as the weights and lengths, are as WL and wl, but their cohesions are as the breadths; and since they are of equal relative strength, we have WL : wl = B : b, and WLb = wlB, and L : l = wB : Wb; but since they are of similar shapes, L : l = B : b, and therefore w = W.

The same reasoning holds again when they are also supported along the sides, and therefore holds when they are supported all round (in which case the strength is doubled).

And if the plates be of any other figure, such as circles or ellipses, we need only conceive similar rectangles inscribed in them. These are supported all around by the continuity of the plates, and therefore will sustain equal weights; and the same may be said of the segments which lie without them, because the strengths of any similar segments are equal, their lengths being as their breadths.

Therefore the thickness of the bottoms of vessels holding heavy liquors or grains should be as their diameters and as the square root of their depths jointly.

Also the weight which a square plate will bear is to that which a bar of the same matter and thickness will bear as twice the length of the bar to its breadth.

The strain of a beam arising from its own weight. There is yet another modification of the strain which tends to break a body transversely, which is of very frequent occurrence, and in some cases must be very carefully attended to, viz. the strain arising from its own weight.

When a beam projects from a wall, every section is strained by the weight of all that projects beyond it. This may be considered as all collected at its centre of gravity. Therefore the strain on any section is in the joint ratio of the weight of what projects beyond it, and the distance of its centre of gravity from the section.

General principle respecting it. The determination of this strain, and of the strength necessary for withstanding it, must be more complicated than the former, because the form of the piece which results from this adjustment of strain and strength influences the strain. The general principle must evidently be, that the strength or momentum of cohesion of every section must be as the product of the weight beyond it, multiplied by the distance of its centre of gravity. For example:

Suppose the beam DLA (fig. 15) to project from the wall, and that its sides are parallel vertical planes, so that the depth is the only variable dimension. Let LB = x and Bb = y. The element BbcC is = $y dx$. Let G be the centre of gravity of the part lying without Bb, and g be its distance from the extremity L. Then $x - g$ is the arm of the lever by which the strain is excited in the section Bb.

Fig. 15.

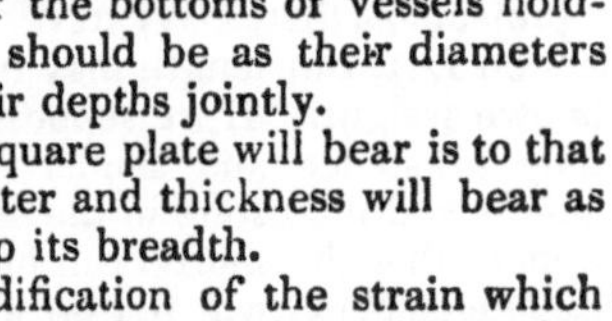

Strength of Materials.

Let Bb or y be as some power m of LB; that is, let $y = x^m$. Then the contents of LBb is $\frac{x^{m+1}}{m+1}$. The momentum of gravity round a horizontal axis at L is $yxdx = x^{m+1}dx$, and the whole momentum round the axis is $\frac{x^{m+2}}{m+2}$. The distance of the centre of gravity from L is had by dividing this momentum by the whole weight, which is $\frac{x^{m+1}}{m+1}$. The quotient or g is $\frac{x \times \overline{m+1}}{m+2}$, and the distance of the centre of gravity from the section Bb is $x - \frac{x \times \overline{m+1}}{m+2} = \frac{x \times \overline{m+2} - x \times \overline{m+1}}{m+2} = \frac{x}{m+2}$. Therefore the strain on the section Bb is had by multiplying $\frac{x^{m+1}}{m+1}$ by $\frac{x}{m+2}$. The product is $\frac{x^{m+2}}{m+2 \times \overline{m+1}}$. This must be as the square of the depth, or as y^2. But y is as x^m, and y^2 as x^{2m}. Therefore we have $m+2 = 2m$, and $m = 2$; that is, the depth must be as the square of the distance from the extremity, and the curve LbA is a parabola touching the horizontal line in L.

A conoid equally able in every section to bear its own weight.

It is easy to see that a conoid formed by the rotation of this figure round DL will also be equally able in every section to bear its own weight.

We need not prosecute this farther. When the figure of the piece is given, there is no difficulty in finding the strain; and the circumstance of equal strength to resist this strain is chiefly a matter of curiosity.

The more a beam projects, the less able it is to bear its own weight.

It is evident, from what has been already said, that a projecting beam becomes less able to bear its own weight as it projects farther. Whatever may be the strength of the section DA, the length may be such that it will break by its own weight. If we suppose two beams A and B of the same substance and similar shapes, that is, having their lengths and diameters in the same proportion; and further suppose that the shorter can just bear its own weight; then the longer beam will not be able to do the same; for the strengths of the sections are as the cubes of the diameters, while the strains are as the biquadrates of the diameters; because the weights are as the cubes, and the levers by which these weights act in producing the strain are as the lengths or as the diameters.

Small bodies more able to withstand the strain produced by the weight of the machine than great bodies.

These considerations show us, that in all cases where strain is affected by the weight of the parts of the machine or structure of any kind, the smaller bodies are more able to withstand it than the greater; and there seem to be bounds set by nature to the size of machines constructed of any given materials. Even when the weight of the parts of the machine is not taken into the account, we cannot enlarge them in the same proportion in all their parts. Thus a steam-engine cannot be doubled in all its parts, so as to be still efficient. The pressure on the piston is quadrupled. If the lift of the pump be also doubled in height while it is doubled in diameter, the load will be increased eight times, and will therefore exceed the power. The depth of lift, therefore, must remain unchanged; and in this case the machine will be of the same relative strength as before, independent of its own weight. For the beam being doubled in all its dimensions, its momentum of cohesion is eight times greater, which is again a balance for a quadruple load acting by a double lever. But if we now consider the increase of the weight of the machine itself, which must be supported, and which must be put in motion by the intervention of its cohesion, we see that the large machine is weaker and less efficient than the small one.

There is a similar limit set by nature to the size of plants and animals formed of the same matter. The cohesion of an herb could not support it if it were increased to the size of a tree, nor could an oak support itself if forty or fifty times bigger; nor could an animal of the make of a long-legged spider be increased to the size of a man; the articulations of its legs could not support it.

Even small animals are remarkable for strength and agility.

Hence may be understood the prodigious superiority of the small animals both in strength and agility. A man by falling twice his own height may break his firmest bones. A mouse may fall twenty times its height without risk; and even the tender mite or wood-louse may fall unhurt from the top of a steeple. But their greatest superiority is in respect of nimbleness and agility. A flea can leap above 500 times its own length, while the strength of the human muscles could not raise the trunk from the ground on limbs of the same construction.

The angular motions of small animals (in which consists their nimbleness or agility) must be greater than those of large animals, supposing the force of the muscular fibre to be the same in both. For supposing them similar, the number of equal fibres will be as the square of their linear dimensions; and the levers by which they act are as their linear dimensions. The energy therefore of the moving force is as the cube of these dimensions. But the momentum of inertia, or $\int p \cdot r^2$, is as the fourth power; therefore the angular velocity of the greater animals is smaller. The number of strokes which a fly makes with its wings in a second is astonishingly great; yet, being voluntary, they are the effects of its agility.

We have hitherto confined our attention to the simplest form in which this transverse strain can be produced. This was quite sufficient for showing us the mechanism of nature by which the strain is resisted; and a very slight attention is sufficient for enabling us to reduce to this every other way in which the strain can be produced. We shall not take up the reader's time with the application of the same principles to other cases of this strain, but refer him to what has been said in the article Roof. In that article we have shown the analogy between the strain on the section of a beam projecting from a wall and loaded at the extremity, and the strain on the same section of a beam simply resting on supports at the ends, and loaded at some intermediate point or points. The strain on the middle C of a beam AB (fig. 16) so supported, arising from a weight laid on there, is the same with the strain which half that weight hanging at B would produce on the same section C, if the other end of the beam were fixed in

Fig. 16.

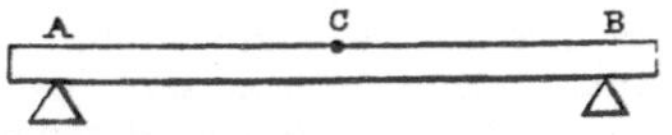

a wall. If therefore 1000 pounds hung on the end of a beam projecting ten feet from a wall will just break it at the wall, it will require 4000 pounds on its middle to break the same beam resting on two props ten feet asunder. We have also shown in that article the additional strength which will be given to this beam by extending both ends beyond the props, and there framing it firmly into other pillars or supports. We can hardly add any thing to what has been said in that article, except a few observations on the effects of the obliquity of the external force. We have hitherto supposed it to act in the direction BP (fig. 6) perpendicular to the length of the beam. Suppose it to act in the direction BP′, oblique to BA. In the article Roof we supposed the strain to be the same as if the force p acted at the distance AB′, but still perpendicular to AB: so it is. But the strength of the section AΔ is not the same in both cases; for by the obliquity of the action the piece DCKΔ is pressed to the other. We are not sufficiently acquainted with the corpuscular forces to say precisely what will be the effect of the pressure arising from this obliquity; but we can clearly see in general, that the point A, which in the instant of fracture

Effects of obliquity of the external force.

Strength of Materials. is neither stretched nor compressed, must now be farther up, or nearer to D; and therefore the number of particles which are exerting cohesive forces is smaller, and therefore the strength is diminished. Therefore, when we endeavour to proportion the strength of a beam to the strain arising from an external force acting obliquely, we make too liberal allowance by increasing this external force in the ratio of AB to AB'. We acknowledge our inability to assign the proper correction. But this circumstance is of very great influence. In many machines, and many framings of carpentry, this oblique action of the straining force is unavoidable; and the most enormous strains to which materials are exposed are generally of this kind. In the frames set up for carrying the ringstones of arches, it is hardly possible to avoid them; for although the judicious engineer dispose his beams so as to sustain only pressures in the direction of their lengths, tending either to crush them or to tear them asunder, it frequently happens that, by the settling of the work, the pieces come to check and bear on each other transversely, tending to break each other across. This we have remarked upon in the article ROOF, with respect to a truss by Mr Price (see ROOF, p. 244). Now when a cross strain is thus combined with an enormous pressure in the direction of the length of the beam, it is in the utmost danger of snapping suddenly across. This is one great cause of the carrying away of masts. They are compressed in the direction of their length by the united force of the shrouds, and in this state the transverse action of the wind soon completes the fracture.

The strain on columns. When considering the compressing strains to which materials are exposed, we deferred the discussion of the strain on columns, observing that it was not, in the cases which usually occur, a simple compression, but was combined with a transverse strain, arising from the bending of the column. When the column ACB (fig. 17), resting on the ground at B, and loaded at top with a weight A, acting in the vertical direction AB, is bent into a curve ACB, so that the tangent at C is perpendicular to the horizon, its condition somewhat resembles that of a beam firmly fixed between B and C, and strongly pulled by the end A, so as to bend it between C and A. Although we cannot conceive how a force acting on a straight column AB in the direction AB can bend it, we may suppose that the force acted first in the horizontal direction Ab till it was bent to this degree, and that the rope was then gradually removed from the direction Ab to the direction AB, increasing the force as much as is necessary for preserving the same quantity of flexure.

Fig. 17.

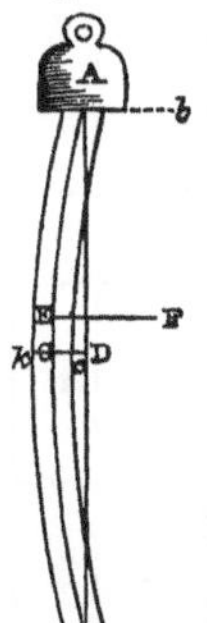

Observations on Euler's theory of the strength of columns. The first author, we believe, who considered this important subject with scrupulous attention was the celebrated Euler, who published in the Berlin Memoirs for 1757 his Theory of the Strength of Columns. The general proposition established by this theory is, that the strength of prismatical columns is in the direct quadruplicate ratio of their diameters, and the inverse duplicate ratio of their lengths. He prosecuted this subject in the Petersburg Commentaries for 1778, confirming his former theory. We do not find that any other author has bestowed much attention on it, all seeming to acquiesce in the determinations of Euler, and to consider the subject as of very great difficulty, requiring the application of the most refined mathematics. Muschenbroeck has compared the theory with experiment; but the comparison has been very unsatisfactory, the difference from the theory being so enormous as to afford no argument for its justness. But the experiments do not contradict it, for they are so anomalous as to afford no conclusion or general rule whatever.

To say the truth, the theory can be considered in no other light than as a specimen of ingenious and very artful *Strength of Materials.* algebraic analysis. Euler was unquestionably the first analyst in Europe for resource and address. He knew this, and enjoyed his superiority, and without scruple admitted any physical assumptions which gave him an opportunity of displaying his skill. The inconsistency of his assumptions with the known laws of mechanism gave him no concern; and when his algebraic processes led him to any conclusion which would make his readers stare, being contrary to all our usual notions, he frankly owned the paradox, but went on in his analysis, saying, *Sed analysi magis fidendum.* Mr Robins has given some very risible instances of this confidence in his analysis, or rather of his confidence in the indolent submission of his readers. Nay, so fond was he of this kind of amusement, that after having published an untenable Theory of Light and Colours, he published several Memoirs, explaining the aberration of the heavenly bodies, deducing some very wonderful consequences, fully confirmed by experience, from the Newtonian principles, which were opposite and totally inconsistent with his own theory, merely because the Newtonian theory gave him *occasionem analyseos promovendæ.* We are thus severe in our observations, because his Theory of the Strength of Columns is one of the strongest instances of this wanton kind of proceeding, and because his followers in the Academy of St Petersburg, such as Fuss, Lexill and others, adopt his conclusions, and merely echo his words. Since the death of Daniel Bernoulli, no member of that academy has controverted any thing advanced by their *Professor sublimis Geometriæ*, to whom they had been indebted for their places and for all their knowledge, having been (most of them) his amanuenses, employed by this wonderful man during his blindness, to make his computations and carry on his algebraic investigations. We are not a little surprised to see Mr Emerson, a considerable mathematician, and a man of very independent spirit, hastily adopting the same theory, of which we doubt not but our readers will easily see the falsity.

Euler considers the column ACB as in a condition precisely similar to that of an elastic rod bent into the curve by a cord AB connecting its extremities. In this he is not mistaken. But he then draws CD perpendicular to AB, and considers the strain on the section C as equal to the momentum or mechanical energy of the weight A, acting in the direction DE, upon the lever kcD, moveable round the fulcrum c, and tending to tear asunder the particles which cohere along the section $cC k$. This is the same principle (as Euler admits) employed by James Bernoulli in his investigation of the elastic curve ACB. Euler considers the strain on the section ck as the same with what it would sustain if the same power acted in the horizontal direction EF on a point E, as far removed from C as the point D is. We reasoned in the same manner (as has been observed) in the article ROOF, where the obliquity of action was inconsiderable. But in the present case this substitution leads to the greatest mistakes, and has rendered the whole of this theory false and useless. It would be just if the column were of materials which are incompressible. But it is evident, by what has been said above, that by the compression of the parts the real fulcrum of the lever shifts away from the point c, so much the more as the compression is greater. In the great compressions of loaded columns, and the almost unmeasurable compressions of the truss-beams in the centres of bridges, and other cases of chief importance, the fulcrum is shifted far over towards k, so that very few fibres resist the fracture by their cohesion, and these few have a very feeble energy or momentum, on account of the short arm of the lever by which they act. This is a most important consideration in carpentry, yet makes no element of Euler's theory. The consequence of this is, that a very small degree of curvature is sufficient

Strength of Materials. to cause the column or strut to snap in an instant, as is well known to every experienced carpenter. The experiment by Muschenbroeck, which Euler makes use of in order to obtain a measure of strength in a particular instance, from which he might deduce all others by his theorem, is an incontestable proof of this. The force which broke the column is not the twentieth part of what is necessary for breaking it by acting at E in the direction EF. Euler takes no notice of this immense discrepancy, because it must have caused him to abandon the speculation with which he was then amusing himself.

This theory false and useless. The limits of this work do not afford room to enter minutely upon the refutation of this theory; but we can easily show its uselessness, by its total inconsistency with common observation. It results legitimately from this theory, that if CD have no magnitude, the weight A can have no momentum, and the column cannot be broken. True, it cannot be broken in this way, snapped by a transverse fracture, if it do not bend; but we know very well that it can be crushed or crippled, and we see this frequently happen. This circumstance or event does not enter into Euler's investigation, and therefore the theory is at least imperfect and useless. Had this crippling been introduced in the form of a physical assumption, every topic of reasoning employed in the process must have been laid aside, as the intelligent reader will easily see. But the theory is not only imperfect, but false. The ordinary reader will be convinced of this by another legitimate consequence of it. Fig. 18 is the same with fig. 106 of Emerson's Mechanics, where this subject is treated on Euler's principles, and represents a crooked piece of matter resting on the ground at F, and loaded at A with a weight acting in the vertical direction AF. It results from Euler's theory that the strains at *b*, B, D, E, &c. are as *bc*, BC, DI, EK, &c. Therefore the strains at G and H are nothing; and this is asserted by Emerson and Euler as a serious truth; and the piece may be thinned *ad infinitum* in these two places, or even cut through, without any diminution of its strength. The absurdity of this assertion strikes at first hearing. Euler asserts the same thing with respect to a point of contrary flexure. Farther discussion is, we apprehend, needless.

Fig. 18.

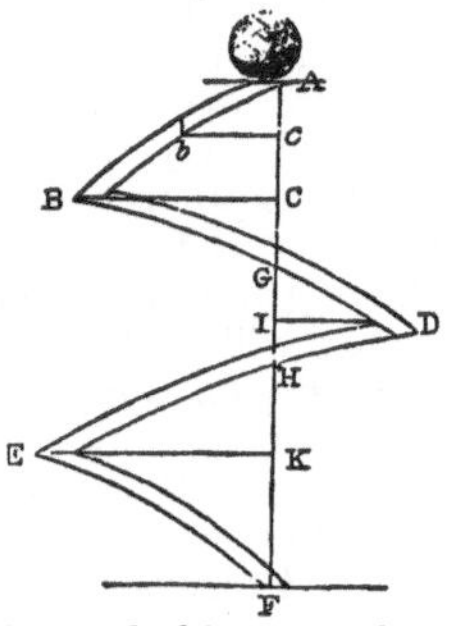

Yet Euler's dissertations deserve a perusal. This theory must therefore be given up. Yet these dissertations of Euler in the Petersburg Commentaries deserve a perusal, both as very ingenious specimens of analysis, and because they contain maxims of practice which are important. Although they give an erroneous measure of the comparative strength of columns, they show the immense importance of preventing all bendings, and point out with accuracy where the tendencies to bend are greatest, and how this may be prevented by very small forces, and what a prodigious accession of force this gives the column. There is a valuable paper in the same volume by Fuss on the Strains on framed Carpentry, which may also be read with advantage.

A new theory cannot be substituted in place of Euler's till many experiments be made. It will now be asked, what shall be substituted in place of this erroneous theory? what is the true proportion of the strength of columns? We acknowledge our inability to give a satisfactory answer. This can only be obtained by a previous knowledge of the proportion between the extensions and compressions produced by equal forces, by the knowledge of the absolute compressions producible by a given force, and by a knowledge of the degree of that derangement of parts which is termed crippling. These circumstances are but imperfectly known to us, and there lies before us a wide field of experimental inquiry. Fortunately

the force requisite for crippling a beam is prodigious, and a very small lateral support is sufficient to prevent that bending which puts the beam in imminent danger. A judicious engineer will always employ transverse bridles, as they are called, to stay the middle of long beams which are employed as pillars, struts, or truss-beams, and are exposed, by their position, to enormous pressures in the direction of their lengths. Such stays may be observed, disposed with great judgment and economy, in the centres employed by Mr Perronet in the erection of his great stone arches. He was obliged to correct this omission made by his ingenious predecessor in the beautiful centres of the bridge of Orleans, which we have no hesitation in affirming to be the finest piece of carpentry in the world.

It only remains on this head to compare these theoretical deductions with experiment.

Experiments on the transverse strength of bodies are easily made, and accordingly are very numerous, especially those made on timber, which is the case most common and most interesting. But in this great number of experiments there are very few from which we can draw much practical information. The experiments have in general been made on such small scantlings, that the unavoidable natural inequalities bear too great a proportion to the strength of the whole piece. Accordingly, when we compare the experiments of different authors, we find them differ enormously, and even the experiments by the same author are very anomalous. The completest series that we have yet seen is that detailed by Belidor in his *Science des Ingenieurs*. They are contained in the following table. The pieces were sound, even-grained oak. The column *b* contains the breadths of the pieces in inches; the column *d* contains their depths; the column *l* contains their lengths; column *p* contains the weights (in pounds) which broke them when hung on their middles; and *m* is the column of averages or mediums.

Strength of Materials.

Table of experiments made by Belidor.

No.	b	d	l	p	m	
1	1	1	18	400 / 415 / 405	406	The ends lying loose.
2	1	1	18	600 / 600 / 624	608	The ends firmly fixed.
3	2	1	18	810 / 795 / 812	805	Loose.
4	1	2	18	1570 / 1580 / 1590	1580	Loose.
5	1	1	36	185 / 195 / 180	187	Loose.
6	1	1	36	285 / 280 / 285	283	Fixed.
7	2	2	36	1550 / 1620 / 1585	1585	Loose.
8	$2\frac{1}{3}$	$2\frac{1}{3}$	36	1665 / 1675 / 1640	1660	Loose.

By comparing Experiments 1st and 3d, the strength appears proportional to the breadth.

Experiments 3d and 4th show the strength proportional to the square of the depth.

Experiments 1st and 5th show the strength nearly in the inverse proportion of the lengths, but with a sensible deficiency in the longer pieces.

Experiments 5th and 7th show the strengths proportional to the breadths and the square of the depth.

Experiments 1st and 7th show the same thing, compounded with the inverse proportion of the length: here the deficiency relative to the length is not so remarkable.

Experiments 1st and 2d, and experiments 5th and 6th, show the increase of strength, by fastening the ends, to be in the proportion of two to three. The theory gives the proportion of two to four. But a difference in the manner of fixing may produce this deviation from the theory, which only supposed them to be held down at places beyond the props, as when a joist is held in the walls, and also rests on two pillars between the walls.

The chief source of irregularity in such experiments is the fibrous, or rather plated texture of timber. It consists of annual additions, whose cohesion with each other is vastly weaker than that of their own fibres. Let fig. 19 represent the section of a tree, and ABCD, abcd the section of two battens that are to be cut out of it for experiment, and let AD and ad be the depths, and DC, dc the breadths. The batten ABCD will be the stronger, for the same reason that an assemblage of planks set edgewise will form a stronger joist than planks laid above each other like the plates of a coach-spring. M. Buffon found by many trials that the strength of ABCD was to that of abcd (in oak) nearly as eight to seven. The authors of the different experiments were not careful that their battens had their plates all disposed similarly with respect to the strain. But even with this precaution they would not have afforded sure grounds of computation for large works; for great beams occupy much, if not the whole, of the section of the tree; and from this it has happened that their strength is less than in proportion to that of a small lath or batten. In short, we can trust no experiments but such as have been made on large beams. These must be very rare, for they are most expensive and laborious, and exceed the abilities of most of those who are disposed to study this subject.

But we are not wholly without such authority. M. Buffon and M. Duhamel, two of the first philosophers and mechanicians of the age, were directed by government to make experiments on this subject, and were supplied with ample funds and apparatus. The relation of their experiments is to be found in the Memoirs of the French Academy for 1740, 1741, 1742, 1768; as also in Duhamel's valuable performances *Sur l'Exploitation des Arbres, et sur la Conservation et le Transport de Bois*. We earnestly recommend these dissertations to the perusal of our readers, as containing much useful information relative to the strength of timber, and the best methods of employing it. We shall here give an abstract of M. Buffon's experiments.

He relates a great number which, during two years, he had prosecuted on small battens. He found that the odds of a single layer, or part of a layer, more or less, or even a different disposition of them, had such influence that he was obliged to abandon this method, and to have recourse to the largest beams that he was able to break. The following table exhibits one series of experiments on bars of sound

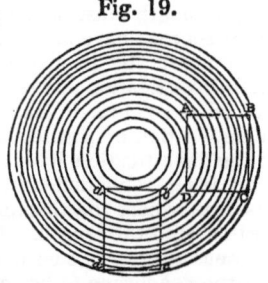

Fig. 19.

oak, clear of knots, and four inches square. This is a specimen of all the rest.

Column 1st is the length of the bar in clear feet between the supports.

Column 2d is the weight of the bar (the second day after it was felled) in pounds. Two bars were tried of each length. Each of the first three pairs consisted of two cuts of the same tree. The one next the root was always found the heaviest, stiffest, and strongest. Indeed M. Buffon says that this was invariably true, that the heaviest was always the strongest; and he recommends it as a certain (or sure) rule for the choice of timber. He finds that this is always the case when the timber has grown vigorously, forming very thick annual layers. But he also observes that this is only during the advances of the tree to maturity; for the strength of the different circles approaches gradually to equality during the tree's healthy growth, and then it decays in these parts in a contrary order. Our tool-makers assert the same thing with respect to beech: yet a contrary opinion is very prevalent; and wood with a fine, that is, a small grain, is frequently preferred. Perhaps no person has ever made the trial with such minuteness as M. Buffon, and we think that much deference is due to his opinion.

Column 3d is the number of pounds necessary for breaking the tree in the course of a few minutes.

Column 4th is the number of inches which it bent down before breaking.

Column 5th is the time at which it broke.

1	2	3	4	5
7	60 56	5350 5275	3·5 4·5	29 22
8	68 63	4600 4500	3·75 4·7	15 13
9	77 71	4100 3950	4·85 5·5	14 12
10	84 82	3625 3600	5·83 6·5	15 15
12	100 98	3050 2925	7· 8·	... ...

The experiments on other sizes were made in the same way. A pair at least of each length and size was taken. The mean results are contained in the following table. The beams were all square, and their sizes in inches are placed at the head of the columns, and their lengths in feet are in the first column.

	4	5	6	7	8	A
7	5312	11525	18950	32200	47649	11525
8	4550	9787	15525	26050	39750	10085
9	4025	8308	13150	22350	32800	8964
10	3612	7125	11250	19475	27750	8068
12	2987	6075	9100	16175	23450	6723
14	...	5300	7475	13225	19775	5763
16	...	4350	6362	11000	16375	5042
18	...	3700	5562	9245	13200	4482
20	...	3225	4950	8375	11487	4034
22	...	2975	...	...	...	3667
24	...	2162	...	...	...	3362
28	...	1775	...	...	...	2881

M. Buffon had found, by numerous trials, that oak-timber lost much of its strength in the course of drying or season-

2 P

ing; and therefore, in order to secure uniformity, his trees were all felled in the same season of the year, were squared the day after, and tried the third day. Trying them in this green state gave him an opportunity of observing a very curious and unaccountable phenomenon. When the weights were laid briskly on, nearly sufficient to break the ,og, a very sensible smoke was observed to issue from the two ends with a sharp hissing noise. This continued all the while the tree was bending and cracking. This shows that the log is affected or strained through its whole length. Indeed this must be inferred from its bending through its whole length. It also shows us the great effects of the compression. It is a pity M. Buffon did not take notice whether this smoke issued from the upper or compressed half of the section only, or whether it came from the whole.

We must now make some observations on these experiments, in order to compare them with the theory which we have endeavoured to establish.

M. Buffon considers the experiments with the five-inch bars as the standard of comparison, having both extended these to greater lengths, and having tried more pieces of each length.

Our theory determines the relative strength of bars of the same section to be inversely as their lengths. But, if we except the five experiments in the first column, we find a very great deviation from this rule. Thus the five-inch bar of twenty-eight feet long should have half the strength of that of fourteen feet, or 2650; whereas it is but 1775. The bar of fourteen feet should have half the strength of that of seven feet, or 5762; whereas it is but 5300. In like manner, the fourth of 11,525 is 2881; but the real strength of the twenty-eight feet bar is 1775. We have added a column A, which exhibits the strength which each of the five-inch bars ought to have by the theory. This deviation is most distinctly seen in fig. 20, where BK is the scale

Fig. 20.

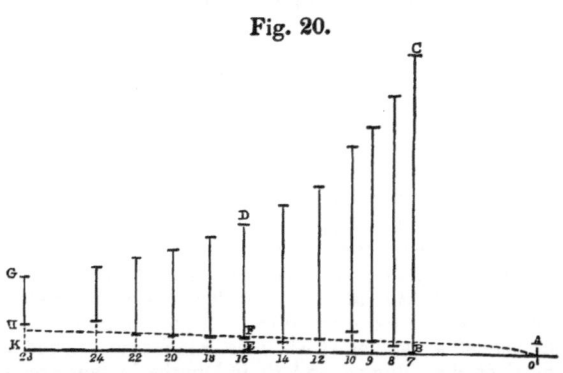

of lengths, B being at the point seven of the scale, and K at twenty-eight. The ordinate CB is $= 11,525$, and the other ordinates DE, GK, &c. are respectively $= \dfrac{7CB}{length}$. The lines DF, GH, &c. are made $= 4350, 1775,$ &c., expressing the strengths given by experiment. The ten-feet bar and the twenty-four feet bar are remarkably anomalous. But all are deficient, and the defect has an evident progression from the first to the last. The same thing may be shown of the other columns, and even of the first, though it is very small in that column. It may also be observed in the experiments of Belidor, and in all that we have seen. We cannot doubt therefore of its being a law of nature, depending on the true principles of cohesion and the laws of mechanics.

But it is very puzzling, and we cannot pretend to give a satisfactory explanation of the difficulty. The only effect which we can conceive the length of a beam to have, is to increase the strain at the section of fracture, by employing the intervening beam as a lever. But we do not distinctly

see what change this can produce in the mode of action of
the fibres in this section, so as either to change their cohesion or the place of its centre of effort: yet something of this kind must happen.

We see indeed some circumstances which must contribute to make a smaller weight sufficient, in M. Buffon's experiments, to break a long beam, than in the exact inverse proportion of its length.

In the first place, the weight of the beam itself augments the strain as much as if half of it were added in the form of a weight. M. Buffon has given the weights of every beam on which he made experiments, which is very nearly seventy-four pounds per cubic foot. But they are much too small to account for the deviation from the theory. The half weights of the five-inch beams of seven, fourteen, and twenty-eight feet length, are only forty-five, ninety-two, and 182 pounds; which makes the real strains in the experiments 11,560, 5390, and 1956; which are far from having the proportions of four, two, and one.

Buffon says that healthy trees are universally strongest at the root end; therefore, when we use a longer beam, its middle point, where it is broken in the experiment, is in a weaker part of the tree. But the trials of the four-inch beams show that the difference from this cause is almost insensible.

The length must have some mechanical influence which the theory we have adopted has not yet explained. It may not however be inadequate to the task. The very ingenious investigation of the elastic curve by James Bernoulli and other celebrated mathematicians is perhaps as refined an application of mathematical analysis as we know. Yet in this investigation it was necessary, in order to avoid almost insuperable difficulties, to take the simplest possible case, viz. where the thickness is exceedingly small in comparison with the length. If the thickness be considerable, the quantities neglected in the calculus are too great to permit the conclusion to be accurate, or very nearly so. Without being able to define the form into which an elastic body of considerable thickness will be bent, we can say with confidence, that in an extreme case, where the compression in the concave side is very great, the curvature differs considerably from the Bernoullian curve. But as our investigation is incomplete and very long, we do not offer it to the reader. The following more familiar considerations will, Probable
that the
relative
strength of
beams de-
creases
faster than
in the in-
verse ratio
of their
length. we apprehend, render it highly probable that the relative strength of beams decreases faster than in the inverse ratio of their length. The curious observation by M. Buffon, of the vapour which issued with the hissing noise from the ends of a beam of green oak, while it was breaking by the load on its middle, shows that the whole length of the piece was affected: indeed it must be, since it is bent throughout. We have shown above, that a certain definite curvature of a beam of a given form is always accompanied by rupture. Now suppose the beam A of ten feet long, and the beam B of twenty feet long, bent to the same degree, at the place of their fixture in the wall; the weight which hangs on A is nearly double of that which must hang on B. The form of any portion, suppose five feet, of these two beams, immediately adjoining to the wall, is considerably different. At the distance of five feet the curvature of A is half of its curvature at the wall. The curvature of B in the corresponding point is three fourths of the same curvature at the wall. Through the whole of the intermediate five feet, therefore, the curvature of B is greater than that of A. This must make it weaker throughout. It must occasion the fibres to slide more on each other (that it may acquire *this* greater curvature), and thus affect their lateral union; and therefore those which are stronger will not assist their weaker neighbours. To this we must add, that in the shorter beams the force with which the fibres are pressed laterally on each other is double. This must impede the

Strength of Materials. mutual sliding of the fibres which we mentioned a little ago; nay, this lateral compression may change the law of longitudinal cohesion (as will readily appear to the reader who is acquainted with Boscovich's doctrines), and increase the strength of the very surface of fracture, in the same way (however inexplicable) as it does in metals when they are hammered or drawn into wire.

The reader must judge how far these remarks are worthy of his attention. The engineer will carefully keep in mind the important fact, that a beam of quadruple length, instead of having one fourth of the strength, has only about one sixth; and the philosopher should endeavour to discover the cause of this diminution, that he may give the artist a more accurate rule of computation.

We cannot discover the precise relation between the curvature and the momentum of cohesion. Our ignorance of the law by which the cohesion of the particles changes by a change of distance, hinders us from discovering the precise relation between the curvature and the momentum of cohesion; and all we can do is to multiply experiments, upon which we may establish some *empirical* rules for calculating the strength of solids. Those from which we must reason at present are too few and too anomalous to be the foundation of such an empirical formula. We may however observe, that M. Buffon's experiments gave us considerable assistance in this particular; for if to each of the numbers of the column for the five-inch beams, corrected by adding half the weight of the beam, we add the constant number 1245, we shall have a set of numbers which are very nearly reciprocals of the lengths. Let 1245 be called *c*, and let the weight which is known by experiment to be necessary for breaking the five-inch beam of the length *a* be called P. We shall have $\frac{\overline{P+c}\times a}{l}-c=p$.

Thus the weight necessary for breaking the seven-feet bar is 11,560. This added to 1245, and the sum multiplied by 7, gives $\overline{P+c}\times a=89,635$. Let *l* be 18; then $\frac{89,635}{18}-1245$

$=3725=p$, which differs not more than $\frac{1}{40}$th from what experiment gives us. This rule holds equally well in all the other lengths except the 10 and 24 feet beams, which are very anomalous. Such a formula is abundantly exact for practice, and will answer through a much greater variety of length, though it cannot be admitted as a true one; because, in a certain very great length, the strength will be nothing. For other sizes the constant number must change in the proportion of d^3, or perhaps of *p*.

Relation between the strength and the square of the depth of the section. The next comparison which we have to make with the theory is the relation between the strength and the square of the depth of the section. This is made by comparing with each other the numbers in any horizontal line of the table. In making this comparison we find the numbers of the five-inch bars uniformly greater than the rest. We imagine that there is something peculiar to these bars; they are in general heavier than in the proportion of their section, but not so much so as to account for all their superiority. We imagine that this set of experiments, intended as a standard for the rest, has been made at one time, and that the season has had a considerable influence. The fact however is, that if this column be kept out, or uniformly diminished about one sixteenth in their strength, the different sizes will deviate very little from the ratio of the square of the depth, as determined by theory. There is however a small deficiency in the bigger beams.

We have been thus anxious in the examination of these experiments, because they are the only ones which have been related in sufficient detail, and made on a proper scale for giving us data from which we can deduce confidential maxims for practice. They are so troublesome and expensive that we have little hopes of seeing their number greatly increased; yet surely our navy board would do an unspeakable service to the public by appropriating a fund for such experiments under the management of some man of science.

Proportion between the absolute cohesion and the relative strength There remains another comparison which is of chief importance, namely, the proportion between the *absolute cohesion* and the *relative strength*. It may be guessed, from the very nature of the thing, that this must be very uncertain. Experiments on the absolute strength must be confined to very small pieces, by reason of the very great forces which are required for tearing them asunder. The values therefore deduced from them must be subject to great inequalities. Unfortunately we possess no detail of any experiments; all that we have to depend on are two passages of Muschenbroeck's *Essais de Physique*; in one of which he says, that a piece of sound oak $\frac{27}{100}$ths of an inch square is torn asunder by 1150 pounds; and in the other, that an oak plank twelve inches broad and one thick will just suspend 189,163 pounds. These give for the cohesion of an inch square 15,755 and 15,763 pounds. Bouguer, in his *Traité du Navire*, says that it is very well known that a rod of sound oak one fourth of an inch square will be torn asunder by 1000 pounds. This gives 16,000 for the cohesion of a square inch. We shall take this as a round number, easily used in our computations. Let us compare this with M. Buffon's trials of beams four inches square.

The absolute cohesion of this section is $16,000\times16=256,000$. Did every fibre exert its whole force in the instant of fracture, the momentum of cohesion would be the same as if it had all acted at the centre of gravity of the section at two inches from the axis of fracture, and is therefore 512,000. The four-inch beam, seven feet long, was broken by 5312 pounds hung on its middle. The half of this, or 2656 pounds, would have broken it, if suspended at its extremity, projecting $3\frac{1}{2}$ feet, or 42 inches, from a wall. The momentum of this strain is therefore $2656\times42=111,552$. Now this is in equilibrio with the actual momentum of cohesion, which is therefore 111,552 instead of 512,000. The strength is therefore diminished in the proportion of 512,000 to 111,552, or very nearly of 4·59 to 1.

As we are quite uncertain as to the place of the centre of effort, it is needless to consider the full cohesion as acting at the centre of gravity, and producing the momentum 512,000; and we may convert the whole into a simple multiplier *m* of the length, and say, *as* m *times the length is to the depth, so is the absolute cohesion of the section to the relative strength.* Therefore let the absolute cohesion of a square inch be called *f*, the breadth *b*, the depth *d*, and the length *l* (all in inches), the relative strength, or the external force, *p*, which balances it, is $\frac{fbd^2}{9\cdot18l}$, or, in round numbers, $\frac{fbd^2}{9l}$; for $m=2\times4\cdot59$.

This great diminution of strength cannot be wholly accounted for by the inequality of the cohesive forces exerted in the instant of fracture; for in this case we know that the centre of effort is at one third of the height in a rectangular section (because the forces really exerted are as the extensions of the fibres). The relative strength would be $\frac{fbd^2}{3l}$, and *p* would have been 8127 instead of 2656.

We must ascribe this diminution (which is three times greater than that produced by the inequality of the cohesive forces) to the compression of the under part of the beam; and we must endeavour to explain in what manner this compression produces an effect which seems so little explicable by such means.

As we have repeatedly observed, it is a matter of nearly universal experience that the forces *actually* exerted by the particles of bodies, when stretched or compressed, are very nearly in the proportion of the distances to which the particles are drawn from their natural positions. Now, although we are certain that, in enormous compressions, the forces increase faster than in this proportion, this makes no sensible change in the present question, because the body is broken before the compressions have gone so far; nay, we imagine that the compressed parts are crippled in most cases even before the extended parts are torn asunder Muschenbroeck asserts this with great confidence with respect to oak, on the authority of his own experiments. He says, that although oak will suspend half as much again as fir, it will not support, as a pillar, two thirds of the load which fir will support in that form.

We imagine therefore that the mechanism in the *present* case is nearly as follows:

Let the beam DCKΔ (fig. 21) be loaded at its extremity with the weight P, acting in the direction KP perpendicular to DC. Let DΔ be the section of fracture. Let DA be about one third of DΔ. A will be the particle or fibre which is neither extended nor compressed. Make $\Delta\delta$: Dd = DA : AΔ. The triangles DAd, ΔAδ, will represent the accumulated attracting and repelling forces. Make AI and Ai = $\frac{1}{3}$ DA and $\frac{1}{3}$ ΔA. The point I will be that to which the full cohesion Dd or f of the particles in AD must be applied, so as to produce the same momentum which the variable forces at I, D, &c. really produce at their several points of application. In like manner, i is the circle of similar effort of the repulsive forces excited by the compression between A and Δ, and it is the real fulcrum of a bended lever IiK, by which the whole effect is produced. The effect is the same as if the full cohesion of the stretched fibres in AD were accumulated in I, and the full repulsion of all the compressed fibres in AΔ were accumulated in i. The forces which are balanced in the operation are the weight P, acting by the arm ki, and the full cohesion of AD acting by the arm Ii. The forces exerted by the compressed fibres between A and Δ only serve to give support to the lever, that it may exert its strain.

We imagine that this does not differ much from the real procedure of nature. The position of the point A may be different from what we have deduced from Buffon's experiments, compared with Muschenbroeck's value of the absolute cohesion of a square inch. If this last should be only 12,000, DA must be greater than we have made it, in the proportion of 12,000 to 16,000. For Ii must still be made = $\frac{1}{3}$ AΔ, supposing the forces to be proportional to the extensions and compressions. There can be no doubt that a part only of the cohesion of DΔ operates in resisting the fracture in all substances which have any compressibility; and it is confirmed by the experiments of M. Duhamel on willow, and the inferences are by no means confined to that species of timber. We say, therefore, that when the beam is broken, the cohesion of AD alone is exerted, and that each fibre exerts a force proportional to its extension; and the accumulated momentum is the same as if the full cohesion of AD were acting by the lever Ii = $\frac{1}{3}$d of DΔ.

It may be said, that if only one third of the cohesion of oak be exerted, it may be cut two thirds through without weakening it. But this cannot be, because the cohesion of the whole is employed in preventing the lateral slide so often mentioned. We have no experiments to determine

Fig. 21.

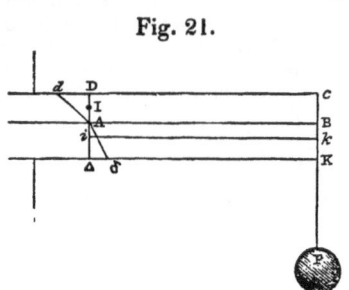

that it *may not* be cut through one third without loss of its strength.

This must not be considered as a subject of mere speculative curiosity. It is intimately connected with all the practical uses which we can make of this knowledge; for it is almost the only way that we can learn the compressibility of timber. Experiments on the direct cohesion are indeed difficult, and exceedingly expensive if we attempt them in large pieces. But experiments on compression are almost impracticable. The most instructive experiments would be, first to establish, by a great number of trials, the transverse force of a moderate batten; and then to make a great number of trials of the diminution of its strength, by cutting it through on the concave side. This would very nearly give us the proportion of the cohesion which really operates in resisting fractures. Thus if it be found that one half of the beam may be cut on the under side without diminution of its strength (taking care to drive in a slice of harder wood), we may conclude that the point A is at the middle, or somewhat above it.

Much lies before the curious mechanician, and we are as yet very far from a scientific knowledge of the strength of timber.

In the mean time, we may derive from these experiments of Buffon a very useful practical rule, without relying on any value of the absolute cohesion of oak. We see that the strength is nearly as the breadth, as the square of the depth, and as the inverse of the length. It is most convenient to measure the breadth and depth of the beam in inches, and its length in feet. Since, then, a beam four inches square and seven feet between the supports is broken by 5312 pounds, we must conclude that a batten one inch square and one foot between the supports will be broken by 581 pounds. Then the strength of any other beam of oak, or the weight which will just break it when hung on its middle, is 581 $\frac{bd^2}{l}$. *(margin: A useful practical rule may be deduced from M. Buffon's experiments.)*

But we have seen that there is a very considerable deviation from the inverse proportion of the lengths, and we must endeavour to accommodate our rule to this deviation. We found, that by adding 1245 to each of the ordinates or numbers in the column of the five-inch bars, we had a set of numbers very nearly reciprocal of the lengths; and if we make a similar addition to the other columns in the proportion of the cubes of the sixes, we have nearly the same result. The greatest error (except in the case of experiments which are very irregular) does not exceed $\frac{1}{13}$th of the whole. Therefore, for a radical number, add to the 5312 the number 640, which is to 1245 very nearly as 4^3 to 5^3. This gives 5952. The 64th of this is 93, which corresponds to a bar of one inch square and seven feet long. Therefore 93 × 7 will be the reciprocal corresponding to a bar of one foot. This is 651. Take from this the present empirical correction, which is $\frac{b\ 40}{b\ 4}$, or 10, and there remains 641 for the strength of the bar. This gives us for a general rule $p = 651 \frac{bd^2}{l} - 10\ bd^2$.

Example. Required the weight necessary to break an oak beam eight inches square and twenty feet between the props, $p = 651 \times \frac{8 \times 8^2}{20}\ 10 \times 8 \times 8^2$. This is 11,545, whereas the experiment gives 11,487. The error is very small indeed. The rule is most deficient in comparison with the five-inch bars, which, we have already said, appear stronger than the rest.

The following process is easily remembered by such as are not algebraists.

Multiply the breadth in inches twice by the depth, and

call this product *f*. Multiply *f* by 651, and divide by the length in feet. From the quotient take 10 times *f*. The remainder is the number of pounds which will break the beam.

We are not sufficiently sensible of our principles to be confident that the correction 10 *f* should be in the proportion of the section, although we think it most probable. It is quite empirical, founded on Buffon's experiments. Therefore the safe way of using this rule is to suppose the beam square, by increasing or diminishing its breadth till equal to the depth. Then find the strength by this rule, and diminish or increase it for the change which has been made in its breadth. Thus, there can be no doubt that the strength of the beam given as an example is double of that of a beam of the same depth and half the breadth.

The reader cannot but observe that all this calculation relates to the very greatest weight which a beam will bear for a very few minutes. M. Buffon uniformly found that two thirds of this weight sensibly impaired its strength, and frequently broke at the end of two or three months. One half of this weight brought the beam to a certain bend, which did not increase after the first minute or two, and may be borne by the beam for any length of time. But the beam contracted a bend, of which it did not recover any considerable portion. One third seemed to have no permanent effect on the beam; but it recovered its rectilineal shape completely, even after having been loaded several months, provided that the timber was seasoned when first loaded; that is to say, one third of the weight which would quickly break a seasoned beam, or one fourth of what would break one just felled, may lie on it for ever without giving the beam a set.

We have no detail of experiments on the strength of other kinds of timber: only M. Buffon says, that fir has about $\frac{6}{10}$ths of the strength of oak; Mr Parent makes it $\frac{10}{12}$ths; Emerson, $\frac{2}{3}$ds, &c.

We have been thus minute in our examination of the mechanism of this transverse strain, because it is the greatest to which the parts of our machines are exposed. We wish to impress on the minds of artists the necessity of avoiding this as much as possible. They are improving in this respect, as may be seen by comparing the centres on which stone arches of great span are now turned with those of former times. They were formerly a load of mere joists resting on a multitude of posts, which obstructed the navigation, and were frequently losing their shape by some of the posts sinking into the ground. Now they are more generally trusses, where the beams abut on each other, and are relieved from transverse strains. But many performances of eminent artists are still very injudiciously exposed to cross strains. We may instance one which is considered as a fine work, viz. the bridge at Walton on Thames. Here every beam of the great arch is a joist, and it hangs together by framing. The finest piece of carpentry that we have seen is the centre employed in turning the arches of the bridge at Orleans, described by Perronet. In the whole there is not one cross strain. The beam, too, of Hornblower's steam-engine is very scientifically constructed.

Strain produced by twisting. IV. The last species of strain which we are to examine is that produced by twisting. This takes place in all axles which connect the working parts of machines.

The resistance must be proportional to the number of particles. Although we cannot pretend to have a very distinct conception of that modification of the cohesion of a body by which it resists this kind of strain, we can have no doubt that, when all the particles act alike, the resistance must be proportional to the number. Therefore if we suppose the two parts ABCD, ABFE (fig. 22), of the body EFCD to be of insuperable strength, but cohering more weakly in the common surface

Fig. 22.

AB, and that one part ABCD is pushed laterally in the direction AB, there can be no doubt that it will yield only there, and that the resistance will be proportional to the surface.

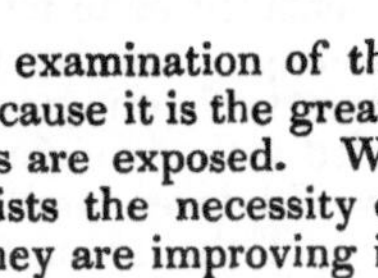

In like manner, we can conceive a thin cylindrical tube, of which KAH (fig. 23) is the section, as cohering more weakly in that section than anywhere else. Suppose it to be grasped in both hands, and the two parts twisted round the axis in opposite directions, as we would twist the joints of a flute; it is plain

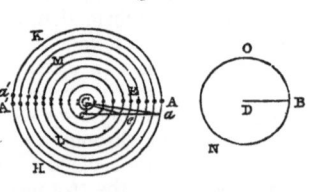

Fig. 23.

that it will first fail in this section, which is the circumference of a circle, and the particles of the two parts which are contiguous to this circumference will be drawn from each other laterally. The total resistance will be as the number of equally resisting particles, that is, as the circumference (for the tube being supposed very thin, there can be no sensible difference between the dilatation of the external and internal particles). We can now suppose another tube within this, and a third within the second, and so on till we reach the centre. If the particles of each ring exerted the same force (by suffering the same dilatation in the direction of the circumference), the resistance of each ring of the section would be as its circumference and its breadth (supposed indefinitely small), and the whole resistance would be as the surface; and this would represent the resistance of a solid cylinder. But when a cylinder is twisted in this manner by an external force applied to its circumference, the external parts will suffer a greater circular extension than the internal; and it appears that this extension (like the extension of a beam strained transversely) will be proportional to the distance of the particles from the axis. We cannot say that this is demonstrable, but we can assign no proportion that is more probable. This being the case, the forces simultaneously exerted by each particle will be as its distance from the axis. Therefore the whole force exerted by each ring will be as the square of its radius, and the accumulated force actually exerted will be as the cube of the radius; that is, the accumulated force exerted by the whole cylinder, whose radius is CA, is to the accumulated force exerted *at the same time* by the part whose radius is CE, as CA^3 to CE^3.

The whole cohesion now exerted is just two thirds of what it would be if all the particles were exerting the same attractive forces which are just now exerted by the particles in the external circumference. This is plain to any person in the least familiar with the fluxionary calculus. But such as are not may easily see it in this way.

Let the rectangle A C*ca* be set upright on the surface of the circle along the line CA, and revolve round the axis C*c*. It will generate a cylinder whose height is C*c* or A*a*, and having the circle KAH for its base. If the diagonal C*a* be supposed also to revolve, it is plain that the triangle *c*C*a* will generate a cone of the same height, and having for its base the circle described by the revolution of *ca*, and the point C for its apex. The cylindrical surface generated by A*a* will express the whole cohesion exerted by the circumference AHK, and the cylindrical surface generated by E*e* will represent the cohesion exerted by the circumference ELM, and the solid generated by the triangle CA*a* will represent the cohesion exerted by the whole circle AHK, and the cylinder generated by the rectangle A C*ca* will represent the cohesion exerted by the same surface if each particle had suffered the extension A*a*.

Now it is plain, in the first place, that the solid generated by the triangle *e*EC is to that generated by *a*AC as EC^3 to

Strength of Materials.

AC^3. In the next place, the solid generated by aAC is two thirds of the cylinder, because the cone generated by cCa is one third of it.

We may now suppose the cylinder twisted till the particles in the external circumference lose their cohesion. There can be no doubt that it will now be wrenched asunder, all the inner circles yielding in succession. Thus we obtain one useful information, viz. that a body of homogeneous texture resists a *simple twist* with two thirds of the force with which it resists an attempt to force one part laterally from the other, or with one-third part of the force which will cut it asunder by a square-edged tool; for to drive a square-edged tool through a piece of lead, for instance, is the same as forcing a piece of the lead as thick as the tool laterally away from the two pieces on each side of the tool. Experiments of this kind do not seem difficult, and they would give us very useful information.

With what force a body of a homogeneous texture resists a simple twist.

When two cylinders AHK and BNO are wrenched asunder, we must conclude that the external particles of each are just put beyond their limits of cohesion, are equally extended, and are exerting equal forces. Hence it follows, that in the instant of fracture the sum-total of the forces actually exerted are as the squares of the diameters.

The forces exerted in breaking two cylinders are as the squares of the diameters.

For drawing the diagonal Ce, it is plain that Ee = Aa expresses the distension of the circumference ELM, and that the solid generated by the triangle CEe expresses the cohesion exerted by the surface of the circle ELM, when the particles in the circumference suffer the extension Ee equal to Aa. Now the solids generated by CAa and CEe being respectively two thirds of the corresponding cylinders, are as the squares of the diameters.

Having thus ascertained the real strength of the section, and its relation to its absolute lateral strength, let us examine its strength relative to the external force employed to break it. This examination is very simple in the case under consideration. The straining force must act by some lever, and the cohesion must oppose it by acting on some other lever. The centre of the section may be the neutral point, whose position is not disturbed.

Relative strength of the section to the external force employed to break it.

Let F be the force exerted laterally by an exterior particle. Let a be the radius of the cylinder, and x the indeterminate distance of any circumference, and dx the indefinitely small interval between the concentric arches; that is, let dx be the breadth of a ring and x its radius. The forces being as the extensions, and the extensions as the distances from the axis, the cohesion actually exerted at any part of any ring will be $f\frac{xdx}{a}$. The force exerted by the whole ring (being as the circumference or as the radius) will be $f\frac{x^2dx}{a}$. The momentum of cohesion of a ring, being as the force multiplied by its lever, will be $f\frac{x^3dx}{a}$. The accumulated momentum will be the sum or fluent of $f\frac{x^3dx}{a}$; that is, when $x = a$, it will be $\frac{1}{4}f\frac{a^4}{a} = \frac{1}{4}fa^3$.

Hence we learn that the strength of an axle, by which it resists being wrenched asunder by a force acting at a given distance from the axis, is as the cube of its diameter. But, further, $\frac{1}{4}fa^3$ is $= fa^2 \times \frac{1}{4}a$. Now fa^2 represents the full lateral cohesion of the section. The momentum therefore is the same as if the full lateral cohesion were accumulated at a point distant from the axis by one fourth of the radius, or one eighth of the diameter of the cylinder.

The resistance of the axle is as the cube of its diameter.

Therefore let F be the number of pounds which measure the lateral cohesion of a circular inch, d the diameter of the cylinder in inches, and l the length of the lever by which the straining force p is supposed to act; we shall have $F \times \frac{1}{8}d^3 = pl$, and $F\frac{d^3}{8l} = p$.

We see in general that the strength of an axle, by which it resists being wrenched asunder by twisting, is as the cube of its diameter.

We see also that the internal parts are not acting so powerfully as the external. If a hole be bored out of the axle of half its diameter, the strength is diminished only one eighth, while the quantity of matter is diminished one fourth. Therefore hollow axles are stronger than solid ones containing the same quantity of matter. Thus let the diameter be 5, and that of the hollow 4; then the diameter of another solid cylinder having the same quantity of matter with the tube is 3. The strength of the solid cylinder of the diameter 5 may be expressed by 5^3, or 125. Of this the internal part (of the diameter 4) exerts 64; therefore the strength of the tube is $125 - 64 = 61$. But the strength of the solid axle of the same quantity of matter and diameter 3 is 3^3, or 27, which is not half of that of the tube.

Hollow axles more proper than solid ones

Engineers, therefore, have of late introduced this improvement in their machines, and the axles of cast iron are all made hollow when their size will admit of it. They have the additional advantage of being much stiffer, and of affording much better fixture for the flanches which are used for connecting them with the wheels or levers by which they are turned and strained. The superiority of strength of hollow tubes over solid cylinders is much greater in this kind of strain than in the former or transverse. In this last case the strength of this tube would be to that of the solid cylinder of equal weight as 61 to 32 and a half nearly.

and now generally used.

The apparatus which we mentioned on a former occasion for trying the lateral strength of a square inch of solid matter, enabled us to try this theory of twist with all desirable accuracy. The bar which hung down from the pin in the former trials was now placed in a horizontal position, and loaded with a weight at the extremity. Thus it acted as a powerful lever, and enabled us to wrench asunder specimens of the strongest materials. We found the results perfectly conformable to the theory, in as far as it determined the proportional strength of different sizes and forms; but we found the ratio of the resistance to twisting to the simple lateral resistance considerably different, and it was some time before we discovered the cause.

The ratio of resistance to twisting to the simple lateral resistence appears different.

We had here taken the simplest view that is possible of the action of cohesion in resisting a twist. It is frequently exerted in a very different way. When, for instance, an iron axle is joined to a wooden one by being driven into one end of it, the extensions of the different circles of particles are in a very different proportion. A little consideration will show that the particles in immediate contact with the iron axle are in a state of violent extension; so are the particles of the exterior surface of the wooden part, and the intermediate parts are less strained. It is almost impossible to assign the exact proportion of the cohesive forces exerted in the different parts. Numberless cases can be pointed out where parts of the axle are in a state of compression, and where it is still more difficult to determine the state of the other particles. We must content ourselves with the deductions made from this simple case, which is fortunately the most common. In the experiments just now mentioned, the centre of the circle is by no means the neutral point, and it is very difficult to ascertain its place; but when this consideration occurred to us, we easily freed the experiments from this uncertainty, by extending the lever to both sides, and by means of a pulley applied equal force to each arm, acting in opposite directions. Thus the centre became the neutral point, and the resistance to twist was found to be two thirds of the simple lateral strength.

But when the experiment was altered, it was exactly the same.

We beg leave to mention here, that our success in these

Strength of Materials.

Experiments on chalk, clay, and wax, satisfactory; but those on timber irregular.

experiments encouraged us to extend them much farther. We hoped by these means to discover the absolute cohesion of many substances, which would have required an enormous apparatus and a most unmanageable force to tear them asunder directly. But we could reason with confidence from the resistance to twist (which we could easily measure), provided that we could ascertain the proportion of the direct and the lateral strengths. Our experiments on chalk, finely prepared clay, and white bee's wax (of one melting and one temperature), were very consistent and satisfactory. But we have hitherto found great irregularities in this proportion in bodies of a fibrous texture like timber. These are the most important cases, and we still hope to be able to accomplish our project, and to give the public some valuable information. This being our sole object, it was our duty to mention the method which promises success, and thus excite others to the task; and it will be no mortification to us to be deprived of the honour of being the first who thus adds to the stock of experimental knowledge.

When the matter of the axle is of the most simple texture, such as that of metals, we do not conceive that the length of the axle has any influence on the fracture. It is otherwise if it be of a fibrous texture like timber; the fibres are bent before breaking, being twisted into spirals like a corkscrew. The length of the axle has somewhat of the influence of a lever in this case, and it is more easily wrenched asunder if long. Accordingly we have found it so; but we have not been able to reduce this influence to calculation.

Many useful deductions might be made from these premises respecting the manner of disposing and combining the strength of materials in our structures. The best form of joints, mortises, tenons, scarphs, the rules for joggling, tabling, faying, fishing, &c., practised in the delicate art of mast-making, are all founded on this doctrine; but the discussion of these would be equivalent to writing a complete treatise of carpentry.

The most recent experiments on the strength and elasticity of material give the results entered in the following tables :—

I.—*Distensions of Rods for a Strain of one Pound per Square Inch, computed from the results given in Tredgold, edition of 1840, as deduced from observations on transverse strain.*

British timber	Oak	$\frac{1}{55000}$
	Larch	$\frac{1}{48000}$
	Scotch fir	$\frac{1}{50000}$
	Ash	$\frac{1}{130000}$
Foreign timber	Memel fir	$\frac{1}{74000}$
	Norway fir	$\frac{1}{105000}$
	American pine	$\frac{1}{50000}$
	White spruce	$\frac{1}{57000}$
	Riga oak	$\frac{1}{100000}$
	American oak	$\frac{1}{55000}$
English malleable iron		$\frac{1}{21700000}$

II.—*Cohesive Strengths of Bars per Square Inch.*

	lb.		lb.
Oak	from 14,000	to	19,000
Beech	„ 11,000	„	22,000
Ash	„ 12,000	„	17,000
Elm	„ 13,000	„	14,000
Mahogany	„ 8,000	„	21,000
Teak	„ 8,000	„	15,000
Pine (Norway)	„ 7,000	„	14,000
Larch	„ 9,000	„	10,000
Iron wire	„ 94,000	„	113,000
Swedish iron	„ 53,000	„	78,000
English iron	„ 55,000	„	66,000
Cast-iron	„ 16,000	„	33,000

(J. R.)

Practical Remarks.

We now give a few remarks on the foregoing subjects, which may be of use to practical men who may not be acquainted with the higher branches of mathematics. It has been seen that the foregoing treatise has explained the limit of the strength of materials, and that they fail in three ways:—1st, By tension, or being pulled asunder; 2d, By compression, or being crushed; 3d, By transverse strain, or being broken across, either by heavy weights or other similar causes.

Strength of Materials.

Failure of material.

Tension.

With the first cause, the practical constructor has but little to do beyond consulting Mr Emerson's tables. They will afford a near approximation (quality being taken into the account) for weights to be suspended by iron-rods and other materials; but, since his time, iron-chains have been brought into much greater use, and iron wire-rope invented. It is impossible, in our space, to give anything like the information that has been published on this subject; our readers, however, will find full tables of the breaking weight, from 1 to 54 tons; the size in inches; the weight; and relative cost per fathom of wire-rope, hempen-rope, and chain, in Weale's *Engineers' Pocket-Book.*

For the tensile strength of iron and steel, we must refer to a very valuable paper in the *Transactions of the Institution of Engineers in Scotland,* by Mr Robert Napier, from experiments conducted by Mr David Kirkaldy, and the valuable tables subjoined thereto, just published. It would be impossible to give them in our limited space, as they extend over many columns of closely printed matter. It appears, however, that the breaking weight of bar-steel, per square inch of its original area, varied from 132,000 lb. to 62,000 lb., while that of iron has varied from 96,000 lb. to 44,000 lb. The like breaking weight of steel-plate varied from 96,000 lb. to 72,000 lb.; and that of iron from 56,000 lb. to 41,000 lb., and this is less than is generally given to iron; but it establishes the fact, that when used as bar and plate iron, for rough reckoning may be considered to have only about two-thirds the strength of steel.

Compression.

Practically it is seldom necessary to consider the weight any bodies would bear without crushing, except the wooden story posts, or iron columns that support the fronts of our houses, or the floors of our warehouses. The power of brick or stone to resist crushing being generally so very much more than those materials can have to carry, it does not come into the consideration of the builder, or engineer, except for very large works, where high professional assistance ought always to be sought. It will be seen, by an inspection of fig. 17, that posts generally fail first by crippling or buckling; and it is easy to perceive how any horizontal strut placed opposite to F, to prevent the timber bending in that direction, must add important strength to the post A B; and though it has been shown that Euler and his followers have pushed deductions drawn from such theories to the verge of absurdity, still, to the practical man, this is matter of grave consideration. If in a shop-front, for instance, we can get a strong transom between the story posts 3 feet from the pavement, those posts are much stronger than if they rose from floor to bressummer without any cross strut, or stay. Formulæ have been given again and again on the subject; but as they necessitate involution and evolution to cubic, and sometimes biquadratic, powers, they should not be used except by skilled mathematicians. There are, however, some very valuable tables given in Barlow's *Tredgold,* Nos. 16, 17, 18, and 19, for the scantlings of story posts to carry the fronts of ordinary houses, from two to five stories high, placed at various distances, from 4 to 8 feet apart, ranging from 8 to 16 feet in height; and some excellent information in Nicholson's *Carpentry,* chap. ii., 212, &c., on story posts generally.

Story posts.

Their superiority, in many points of view, has led to the extensive use of iron columns. The best engineers, for some years, have been occupied in making series of experiments on their bearing and breaking weights; but none on so large a scale as those of Mr Eaton Hodgkinson. The

Iron columns.

results of which were communicated to the Royal Society first in 1840, and again in 1857; and which, by the kindness of the author, are now before us; with an Appendix up to the present time. We regret we have not space for his valuable tables.

Mr Hodgkinson's experiments.
The results, however, may be stated that he found that, in cylindrical pillars of cast-iron, the top and bottom of which were turned perfectly true and flat, the breaking weight varied as the 3·55th power of the diameter, and inversely as the 1·7th power of the length, so that, instead of the theory of Euler (commented on before in page 760), which comes out $\frac{d^4}{l^2}$, it was found to be $\frac{d^{3\cdot55}}{l^{\cdot7}}$, d being the diameter, and l the length. In hollow pillars, as $\frac{d^{3\cdot55} - d_1^{3\cdot55}}{l^{1\cdot7}}$, d and d_1 representing the outer and inner diameters, and this for cases where the lengths varied from 30 times the dimensions of the diameters to 120 times. But as it continually happens that practical men wish to employ cast-iron columns, we must refer them to a series of eleven tables, given in Weale's *Engineers' Pocket-Book;* and subsequently in his *Contractors' Price-Book,* which give the lengths in feet, the load in hundredweights, and the extent of floor solid cast-iron columns will support from 2 inches to 8 inches in diameter, and hollow columns from 3 inches to 18 inches of various thicknesses of metal.

Cross strains.
In the third branch of our subject, transverse strains, we have first to regard those of beams of wood, and it has been shown that in breaking a beam of wood by any weight pulling downwards in the centre, that the upper half of the beam is in a state of compression, and the lower in a state

Tension and compression.
of tension. The practical man will understand this better if he will take a small lead tube, or a piece of a boy's rattan cane, and bend it downwards in the middle. He will find the top of the cane or tube will pucker up, the particles being crushed together, while the lower half will be torn asunder. In fact, we push together the fibres of the upper half and draw apart those of the lower half. From this simple fact all theories as to girders, whether of wood, or of cast or wrought iron are based. Several things must be borne in mind before we proceed; first, that a beam will bear twice the weight equally distributed over its surface than if col-

Distributed weight.
lected in the middle; that a beam of double width will

Proportion of depth to width.
only carry a double load, but a beam of double depth will carry four times the load of a single beam. This may be understood by an inspection of fig. 3. If another beam of the same size were placed alongside of it, the two beams would do double the work of one; but if the beam ABCD were of double depth, the two triangles EG*e*, FG*f* would also be of double depth, and the power of the leverage of *each* would be doubled.

If supported on both ends.
If a beam of any given length and width be fixed in a wall at one end only, it will break with half the weight a

At one end.
beam of half the length would bear, fixed at one end only. In other words, a fourth of its strength is taken away by removing the support from one end. If the ends are well

If ends secured.
wedged into a wall, instead of lying loose on the supports, the strength is much increased, as the ends which lie on a wall must "cock up;" if the middle bends down while they are loose, but should they be kept down tight by the superincumbent weight of the wall, they will counteract to a great degree the tendency to bending, and of course breaking. The securing the ends of beams into walls is said to increase their power to resist breaking weight from $\frac{1}{4}$ to $\frac{1}{3}$.

To calculate the strength of a beam.
The formulæ given by Tredgold are too abstruse for general purposes. There is, however, one given by Nicholson which is simple, and not far from the truth. A number of experiments were made on pieces of various woods, each one inch square and a foot long, and the weights which broke them recorded. Then as this weight c : is to the length of any given beam in feet l : : so is the weight the

beam will have to bear (in pounds) W : to the breadth b multiplied into the square of the depth d of the intended beam; or as $c : l : : W : b \times d^2$. Any of these three being given, the fourth is easily found. The breaking weight of Memel fir he gives as 330, that of oak he gives as 810, but this last seems too much. Suppose there is a warehouse 16 feet wide, the girders of which are 10 feet apart, and each superficial foot is to carry 3 cwt. or 336 lb. Then as each girder supports $16 \times 10 = 160$ feet superficial, and as each foot is to carry 336 lb., the total weight to be carried is 53,760 lb. distributed over the whole, or half this 26,880 lb. in the centre. Then as $330 : 16 : : 26,880 : 1303$, or the breadth multiplied into the square of the depth. But this is breaking weight, and no timber ought to be used of less strength than four times this. Then $1303 \times 4 = 5212$, the least amount we ought to reckon upon. Now we have our choice either to assume a breadth or a depth. Suppose we are confined to 17 inches for the latter. Then $\frac{5212}{17 \times 17} = 18$ inches very nearly.

If we assume 15 inches as our breadth, then $\frac{5212}{15} = 347$, the square root of which is nearly 19 inches; so that we may have a girder 18 inches wide and 17 inches deep, or one 15 inches wide and 19 inches deep, as we please.

Scantlings of timber.
As has been said before, formulæ for calculating these have been given, but too abstruse for general practice. Valuable tables, however, will be found both in *Gwilt, Encyclopædia of Architecture,* in *Tredgold,* and several other books, both for floors and roofs. For floors, in *Gwilt,* Art. 2015 to 2022; in *Tredgold,* Tables Nos. 1 to 4: for roofs, in *Gwilt,* Art. 2036 to 2040; in *Tredgold,* Tables Nos. 5 to 15. The former author seems to give stronger scantlings in proportion than the second, but of course considerable discretion must be used, as we must take into account the weight of the covering and flatness of pitch. It must be remembered, also, that common joists are much strengthened by cross strutting, and all girders above 25 feet in length should be trussed. (See BUILDING, pp. 157-158.)

Inclined beams.
The strength of an inclined beam as a rafter has been shown to be as the cosine of the angle it forms with the horizon. Therefore the strength of the rafter AB, Art. ROOF, fig. 3, is the same as that of a beam of similar scantling in a horizontal position of the length between the perpendiculars of its two ends, as AF.

Advantages and disadvantages of timber beams.
The advantages of timber beams are, they do not shrink in length, and they show signs of fracture before there is danger, and yield gradually if overloaded. The disadvantages—they shrink a great deal in thickness, occupy much space, become rotten by time or damp, and, worst of all, are combustible.

Iron beams.
These circumstances have led to the use of cast-iron as horizontal supports. The history of its introduction and use have been given in our foregoing articles, and at greater length in Mr Fairbairn's book, *The Application of Cast and Wrought Iron to Building,* just published. It will be therefore useless to enumerate the various stages of the invention. Suffice it to say, that from a simple flat plate of iron, which was the original form, and which in fact was a simple imitation of a beam of wood, a top and bottom flanch have by degrees been added, which have proved a great addition of strength, as well as a saving of metal. We have before explained in this page that the upper part of a beam, when it is intended to be bent, is in a state of compression, and the lower half in a state of tension. We must now add, that between these there is a point affected by neither of these forces, and which is called the neutral axis. If a parallel beam of the thickness of EG be widened as at AB, the power to resist compression becomes greater, and the same thing happens with regard to tension if

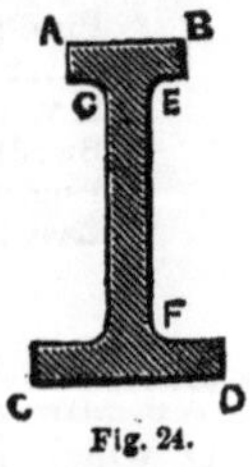

Fig. 24.

Use of flanges.

Strength of the thickness of the flitch EG be increased, as at CD. What
Materials. is called the "web" EF is principally of use to keep the
flanches apart, and increase the resistance by increased leve-

The web. rage. The works of Messrs Hodgkinson and Fairbairn
amply show examples of all these, which we should gladly give
in extenso did our space permit; and though these are
valuable for reference in special cases, the practical constructor needs some tolerably simple rule to give him an idea
what size an iron girder ought to be to support a given weight.

Rules to A number of formulæ have been published, the most elabo-
calculate. rate, and perhaps the nearest the truth, is that of Mr H.
Grissell, the eminent founder, in his evidence before the
House of Lords, in which he takes into consideration the
dimension of the top flange as well as that of the bottom.
But by far the simplest is that of Mr Hodgkinson, adopted
by Mr Fairbairn, viz.:—

$$W = \frac{cad}{l}$$

where W = breaking weight in the middle in tons.
        a = the section of the bottom rib or flange.
        d = the depth of the beam.
        l = the length or distance between the supports (all
            these in inches).
        c = a constant quantity derived from the result of
            various experiments, which varies according
to the quality of iron, and other causes, from 25 to 27·5,
and is generally taken at 26. An example is given of a
bridge girder at Manchester 26 feet or 312 inches in the
clear of the supports. The bottom flange CD is 16 inches
wide and 3 inches deep. The whole depth A to C 27½
inches. Then $l = 312$ $a = 48$ $d = 27·5$ and $c = 26$, as above.

Then $\dfrac{26 \times 48 \times 27·5}{312} = 110$ tons, which is the breaking

weight in the *middle* of the beam. No notice is here taken
of the upper flange, but the omission is justified by the fact
that the constant c has been taken from frequent experiments on girders having top and bottom flanges of proper
proportions, and that therefore it comes into the account
one way if not in another. In very particular instances,
however, we think it ought to be calculated, as, it is said,
cast-iron has eight times greater power to resist compression than tension.

Safe load. Much difference exists among engineers as to the proportion the permanent safe-load has to breaking weight;
some saying it should be three times as much, others
six, and some, where there is great momentum or impact,
ten times as much. It was confessed, however, in the
former cases the formulæ were fuller than ordinary. It
is more common lately to take six times the load as the
size on which to reckon.

Proof. But as there are so many hidden defects in cast-iron,
and when a failure takes place, as it gives no warning,
but all goes to ruin at once, and as there have been such
fearful accidents with this material, every girder should be
Over-proof. proved by placing actual weights on them. But this precaution may be carried to excess, for instances have been
known where a sound girder has been strained in the proof,
and has afterwards broken with a very small weight. A
good authority has laid it down as a rule, that a girder
should be designed to carry six times the load it is intended
at any time to bear, and should be proved to three times.
By this rule the girder just described should be proved in
the middle to 55 tons, and never have more than 18¼ tons
in the centre, or 37 tons if the weight be equally distributed. It is well also to sound the girder throughout with
a hammer, as they do railway axles.

Compound These are partly of cast trussed with wrought-iron,
girders. but are seldom used except in very large structures. We
fear, after all that has been said in their favour, that, as
temperature changes, the particles of the two materials are
often in a different state of tension, and therefore they are Strength of
very unsafe. We have, however, no space here to enter Materials.
into the argument.

The sudden accidents arising from the failure of cast- Wrought-
iron, and their fearful results, have suggested the use iron gird-
of wrought-iron; and as its manufacture is now so much ers.
cheaper from the great improvements in rolling T and L
iron, and in punching and riveting machines, the wrought-
iron girder is now a very important consideration to the
designer of a building. We shall endeavour to give some
short formulæ to guide the practical man, though, as has
been said before, in all important works reference should
be had to the higher branches of science.

There are two sorts of wrought-iron beams now Wrought
employed, generally called either plate-girders or box- girder.
girders. Fig. 25 shows the former, and fig. 26 the latter.
Neither have been in
use long enough, nor
have there been such a
multiplicity of experi-
ments on them, to afford
such certain formulæ as
have been published for
cast-iron beams. But
we give what we gather
to be a near approxi-
mation of the theories

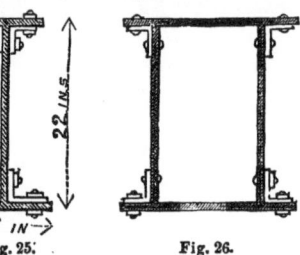

Fig. 25.   Fig. 26.

just published as far as they go, and venture, under correction of such authorities, to suggest they are incomplete as regards the upper flange, just as in the former
theories. Nay, more so, inasmuch as the upper flange
requires to be the *strongest* in wrought-iron, that material being more easily compressed than rent by tension
(which is just the contrary with cast-iron, where we have
shown the lower flange ought to be of largest area). The
constant c is given as 75. Taking the former formula,

$W = \dfrac{cad}{l}$ in a case where the length between the supports

is 30 feet (or 360 inches), the depth 22 inches of $\dfrac{5}{16}$ metal,

and the bottom flanges 3 × 3 each, of 1½ inch metal (see

fig. 25), we have $\dfrac{75 \times 6 \times 22}{360} = 27\frac{1}{2}$ tons as breaking weight.

From Mr Fairbairn's careful experiments, however, we
should gather that 75 is a fair constant where the top flange
has double the strength of the lower flange.

The formulæ collected for these is quite on another prin- Box gird-
ciple from those of the plate-girder. Instead of regarding ers.
the bottom flange alone, writers base their theories in
the sectional area of the metal of top, bottom, and sides,

and given as a formula $W = \dfrac{Ad\,C}{l}$. Here A is the area

of the entire cross-section (fig. 26). Now, as the girders vary
very much in form, some being square in section, some flat,
and some oblong, it follows that the constant C must vary
according to each section; and not only so, but with the
relative strength of the top and bottom flanges. Experiments have shown where the top has been only half the
strength of the bottom flange, the constant was as low as
under 10, but turning the very same girder bottom upwards,
so that the top flange was then double the strength of the
lower, a constant of nearly 18, or nearly double, was got
out of the same section. Mr Fairbairn thinks that 21·5 is
a fair constant to be reckoned for rectangular tubes, when
the upper flange is double the strength of that below. Of
course, these girders, like all others, ought to be proved;
but, as has been before explained, there is always a greater
security about them than those of cast-iron.

The formulæ for these are the same as for plate-girders, Trellis
but the constants vary enormously. Experiments show a girders.

Strength of Materials. difference from 44 to 17. It seems, however. to be the opinion of many, that a well constructed trellis girder may be calculated to bear about half what a close plate-girder of the same dimension would carry; but as these are little used in building, we shall not go further into the investigation.

See Barlow, *Experiments on Timber*, 1835; do.. 1837. Booth, *A popular treatise on the Strength of Materials used in Building*, 1836. Camus de Meziéres, *Traité de la Force de Bois*, 1782. Fairbairn, *Useful Information, &c.*, and several other very valuable works up to 1859. Hodgkinson, the like, also up to 1859. Lea, *Strength of Timber* (Tables of), 1850. Morin, *Resistance, &c.*, 1853. Penn, *Tables, &c.*, 1825. Rennie—Many works, especially the experiments on Strength of Materials in the *Office-Book for Architects, &c.* Tate *On Strength of Materials*, 1851. Tredgold, *Elementary Instructions*, 1836, and many other works. Turnbull, *Strength and Stress of Timber*, 1833. Ware, *Dynamics*, 1851. Wedeke and Romberg, *Die baumateriel, &c.* We can also refer to *the Report of the Royal Commissioners on Iron, &c., &c.*, 1849; and generally to the works of Rondelet, Gauthey. Nicholson, Barlow, Tredgold, Hodgkinson, Fairbairn, Gregory, Emerson, Eytelwein; almost all of which have before been cited in our pages. There are also many excellent articles interspersed in the *Transactions* of the Institution of Civil Engineers; Royal Institute of British Architects; the various French *Annales*; the Franklin Institute of America, &c.; and in the Architect and Civil Engineers' *Pocket-book*, the *Office-Book*, and various works of this description. (A. A.)

DESCRIPTIONS AND EXPLANATIONS OF THE PLATES.

Descrip-
tions of
Plates.

Plate I. This plate exhibits the varieties of columns and columnar composition which the ancient architecture of various countries presents, and is intended to elucidate their presumed derivation from the single pillar of the earliest records; together with specimens of ancient modes of structure.

Fig. 1 presents an example of the single pillar or stone of memorial, the Maenhir or Monolithon; fig. 2 of the demi-Dolmen or Bilithon; fig. 3 of the Trilithon, an example afforded by Stonehenge; and fig. 4 exhibits the immediately succeeding arrangement of pillars, with a continuous entablature. (*See* page 6.)

Fig. 5 shows the flank of the portico of the temple at Amada in Nubia, consisting of square piers or pillars as in fig. 4, and a cylindrical column, which is evidently formed of a similar pillar by working off its angles, the abacus and plinth remaining of the same size and form of which the pillars are; as also at Beni-Hassan, page 34.

Fig. 6, pillars with a plain entablature as in fig. 4, from the Rhamesseion at Thebes. The statues placed before the pillars most probably gave rise to the use of such figures to support an entablature, which these have the appearance of doing when seen in front. (*See* page 34.)

Fig. 7, an early Egyptian columnar composition, from Thebes also. In this, as in the example at Amada, the square abacus shows the form and size of the original pillar out of which the singular bulbous column has been sculptured.

Fig 8, piers of one of the cavern temples of Ellora. These likewise exhibit the tendency to the cylindrical form, and may be assumed as an example of the style of architectural columnar composition at the time they were executed. (*See* page 38.)

Fig. 9, ancient Hindoo columnar piers, in the Mokundra Pass, from Colonel Tod's second volume of the *Annals of Rajast'han*. The similarity in character which exists between these and the piers at Ellora in the preceding example, tends to strengthen the remark accompanying them, and affords proof of their contemporaneousness.

Fig. 10, Doric columns and their architrave from the ruins at Corinth, being the earliest known example of their style.

Fig. 11, ancient Persian columns from Persepolis, in front and in profile, the latter showing the mode in which they were probably made to receive an entablature, though it is stated that the capitals are wrought on the backs in such a manner as to render it improbable that they were ever intended to have anything placed on them.

Fig. 12, columns in front of the rock sculptures at Mundore, in Marwar, from Colonel Tod's first volume.

Fig. 13, from the ruins of Bheems Chlori, also in the Mokundra Pass, from Colonel Tod's second volume. These present another variety of Hindoo columnar composition of early date, though later, it is probable, than the example, fig. 9, *supra*. Figs. 14 and 16 exhibit the modes of structure described in the text at page 147; and fig, 15 is a view of the entrance to the great pyramid at Memphis, from Denon, and shows the mode of its structure.

Plate II. An example of the Egyptian style, sufficiently explained at page 31, *et seq.*

Plate III. The view of the Parthenon in its present state is from an original drawing made on the spot in the year 1821, by Mr W. W. Jenkins. It consequently exhibits the appearance of the splendid ruin before the disasters of the last revolution befell it, as the restored view, under the same aspect, does of the structure in its original state. This is introduced as a frontispiece to the subject, as being an acknowledged master-work of architecture, as well as to enable the reader the better to understand the details of the style of which it is an example, and the composition of that class of structures of which it may be reckoned the principal.

Plate IV. A Greek Doric octastyle, peripteral, and hypæthral temple, with the details of the Parthenon. The plan (fig. 3) is that of the Parthenon (*vide* Plate III.) slightly modified, the better to include the class to which it belongs. In the Parthenon, the opisthodomus has six columns, as in the pronaos, and not four *in antis*, as here laid down: this, however exhibits the ordinary mode of arrangement. The internal columns are arranged in this plan as they are generally found in other similar structures; and the pedestal for the statue of the divinity is placed in its most probable position.

Fig. 1 shows part of the flank of the temple and the internal composition of the hypæthral cella with its upper range of columns or attic, of the inner chamber or treasury, and of the opisthodomos and posticum: much of this, however, is necessarily taken at a venture, because of the imperfect state of the remains of the Grecian edifices.

Fig. 2 exhibits an elevation of the opisthodomus behind the outer range of the portico, not according to the Parthenon, but *in antis*.

Fig. 3 is the plan. In front, on the left-hand side, is the entrance porticus; behind this is the pronaos; within the pronaos is the hypæthral naos or cella, the middle space between the columns being open; the spaces between the columns and the walls on either side are covered; doors (these are not generally laid down to the Parthenon, but are assumed as probable) lead to the inner chamber, said to be the treasury,—this is by some called the opisthodomus, into which it opens, and the opisthodomus stands in the same relation to the posticum that the pronaos does to the porticus.

Fig. 4 is the external order of the Parthenon; fig. 5, the profile of its corona to a larger scale, to show its detail; fig. 6, a half-capital of the same, enlarged also, with its annulets larger still.

Fig. 9 is the order of the pronaos; fig. 8, the profile of its corona enlarged; fig. 7, its capital enlarged, with the annulets still larger.

Fig. 10, the antæ cap enlarged; and fig. 12 a half-plan of a column of the Parthenon, showing the contour of its flutes. (*Vide* page 45, *et seq.*)

Plate V. A Greek Doric hexastyle, peripteral, and cleithral temple, with the details of the temple of Theseus at Athens.

Fig. 1, front elevation of the temple.

Fig. 2, section behind the outer range of the portico, showing the elevation of the pronaos.

Fig. 3, plan of the temple. The arrangement of the

porticus here (to the left) is pseudo-dipteral; a space equal to two intercolumniations and the intervening column being left between the external range and the front of the pronaos,—the projection of the posticum is irregular.

Fig. 4, the external order of the temple of Theseus, with a half-plan of the column; fig. 5, the profile of the corona enlarged; fig. 6, half the capital enlarged; fig. 7, half the capital of the order of the pronaos enlarged also; fig. 9, the antæ, with profiles of the outer and inner entablatures of the pronaos,—this shows also the arrangement of the ceilings.

Fig. 10, enlarged profile of the antæ cap.

Fig. 11, inverted plan of part of the ceilings of the porticus and pronaos, showing the arrangement of the coffers, lacunæ, or cassoons.

Fig. 12, inverted plan of the planceer of the cornice, showing the form and arrangement of the mutules of the external entablature.

Fig. 13 is a plan of the triglyphs of the same on an external angle.

Figs. 8 and 14 are enlarged plans of the flutings of the columns, to show their contours. (*Vide* p. 45, *et seq.*)

Plate VI. A Greek Ionic hexa-prostyle apteral temple, with details of the temple of Erechtheus at Athens.

Fig. 1, elevation of the portico.

Fig. 2, rear elevation of the temple, showing an attached tetrastyle *in antis*, with windows as they exist in that of the temple of Erechtheus.

Fig. 3, flank elevation. The dotted projection to the right of the posticum indicates the amphiprostylar arrangement, which is shown on the plan fig. 4 also, and in the same manner.

Fig. 5, the order of the temple of Erechtheus, except the two lowest steps of the stylobate, which may be easily supplied, to a larger scale, with indications of the carved mouldings, &c.

Figs. 6, 7, and 8 are enlarged profiles of those parts of the entablature which are immediately behind and above them.

Fig. 10, the antæ of the same example, showing the ornament which enriches its necking, and runs along the flank of the edifice; fig. 11, profile of the antæ cap enlarged.

Fig. 12, flank elevation of the capital; all the vertical beads in this are carved. Fig. 13, transverse section of the capital.

Fig. 14, half the longitudinal section of the capital.

Fig. 15, an inverted plan of the capital, showing the arrangement of the flutings.

Fig. 16, an inverted plan of one of the angular capitals. (*Vide* p. 47, *et seq.*)

Fig. 9, The Ionic volute, enlarged to show the mode of striking it, and the contour of its face.

Plate VII. Fig. 1, the elevation, fig. 2 the plan, and fig. 3 the details, of the order of the Choragic Monument of Lysicrates at Athens. (*Vide* p. 48.)

Fig. 4 presents the elevation, and fig. 5 the plan, of the Caryatic prostyle, which is attached to the flanks of the temple of Erechtheus at Athens.

Fig. 6 shows the details of the hands and feet of the figure, and of the entablature and stereobate of the same. (*Vide* p. 49.)

Plate VIII. contains Greek and Roman mouldings, with their usual enrichments, all drawn from ancient examples, and detached profiles of them all, together with two examples of Greek and one of Roman ornament. The specimen of Greek ornament on the left hand of the centre is from the neck of the antæ cap of the tetrastyle portico on the flank of the temple of Erechtheus, gene-

rally known as that of Minerva Polias; and the other half of the same is the enrichment of the neck of the antæ of the temple of Erechtheus itself, as shown in Plate X. figs. 3 and 10. The Roman specimen of ornament is that of the frieze of the temple of Antoninus and Faustina in Rome. (*Vide* Plate IX. Ex. 3, fig. 1, and page 71.)

Plate IX. Four Roman examples of the Corinthian order. Ex. 1 is that of the temple of Jupiter Stator in Rome; Ex. 2 is that of the temple of Vesta at Tivoli (*vide* Plate XI. fig. 9); Ex. 3 is that of the temple of Antoninus and Faustina (*vide* Plate IX. fig. 1); and Ex. 4 is the example of the portico of the Pantheon in Rome (*vide* Plate XI. figs. 2, 3, 4, and 5). To every example fig. 1 shows the details enlarged, the shafts being cut away; and fig. 2 the elevation of the column and entablature. In every case, also, the distance from the inner surface of the column, fig. 2, to the vertical line dividing the examples is one half the intercolumniation at which that example is composed. (*Vide* pp. 57, 70, *et seq.*)

Plate X. Examples of the Roman orders. Ex. 1 is the Corinthian of the temple of Mars Ultor; Ex. 2 the Composite of the Arch of Titus (*vide* Plate XI. fig. 11); Ex. 3 the Ionic of the temple of Fortuna Virilis (*vide* Plate XI. fig. 12); and Ex. 4 the Doric of the Theatre of Marcellus, completed from that of the Colosseum. All of these are in Rome. Fig. 1, as in Plate IX., shows the entablatures, capitals, and bases, &c., on an enlarged scale; and fig. 2 the complete elevation of each order, except their stylobates, some of which are not ascertained, and those which are may be obtained from the structures they are referred to in Plate XI. (*Vide* pp. 57, 70, *et seq.*)

Plate XI. Elevations, plans, and sections of sundry Roman edifices, all drawn to the same scale.

Fig. 1 is a longitudinal elevation of the Colosseum. (*Vide* p. 53.)

Fig. 2 is the front elevation, fig. 3 the flank elevation, fig. 4 a section, and fig. 5 the plan, of the Pantheon. The dotted lines before the recess opposite the entrance, fig. 5, show the places the outstanding columns originally occupied. (*Vide* pp. 54 and 58.)

Fig. 6 is the front elevation, fig. 7 the plan, and fig. 8 the flank elevation, of the temple of Antoninus and Faustina; of this the front steps and stylobate are restorations. (*Vide* p. 58, and Pl. IX. Ex. 3.)

Fig. 9 is the plan and elevation of the temple of Vesta at Tivoli; of this the antefixæ and roof are restorations. (*Vide* Pl. IX. Ex. 2.)

Fig. 10 is the plan and elevation of the triumphal arch of Septimius Severus. (*Vide* p. 61.)

Fig. 11 is a plan and elevation of the Arch of Titus. (*Vide ut sup.* and Pl. X. Ex. 2.)

Fig. 12 is a plan and elevation of the temple of Fortuna Virilis. (*Vide* p. 59, and Pl. X. Ex. 3.)

Plate XII. Plans, sections, elevations, &c., of Roman mansions from Pompeii.

Fig. 1 is a plan of one of the most extensive and most regular of the domestic structures of Pompeii, with its immediate vicinage; it is known as the house of Pansa. The following nomenclature is generally that of Sir William Gell :—1, The entrance or recessed porch; 2, the vestibule; 3, the cavædium or atrium; 4, the compluvium or well for receiving the rain from the roof covering this part of the house (*vide* fig. 2); 5, penaria, or perhaps cubicula; 6, alæ or wings; 7, tablinum or parlour; 8, pinacotheca, or perhaps the library; 9, a passage from the first to the second atrium without passing through the tablinum; 10, cubiculum or bed-chamber; 11, persistylium or oicus—the house; 12, impluvium (*vide sup. in*

4, *et* fig. 2); 13, exhedræ or alæ—in these the siesta was taken—they were also used for conversation; 14, cellæ familiaricæ; 15, triclinium—here couches and seats were placed, and company received; 16, lararium or receptacle for the family gods; 17, cubiculum; 18, hall to the gynæceum or women's apartment; 19, the gynæceum—this is believed by some to be a distinct house, and not a part of that of Pansa; 20, porticus or pegula; 21, hortus or garden; 22, a passage from the oicus to the pegula and garden, to avoid the necessity of passing through the triclinium; 23, kitchen; 24, storeroom or larder; 25, an open court, communicating with the street by a doorway. This comprehends the whole of the apartments, &c., appropriated to domestic use—the residence; the other portions of the edifice are distinct from it. 26 is another smaller house; 27, a passage leading to the house of Pansa from the street on the right-hand side; all the places marked 28 are shops open to the street, as shown in the elevation, fig. 3; the rooms marked 29 are storerooms to the shops into which they open; 30 is a bakehouse, in which the mills, &c., are indicated as they exist; 31 is the oven; in the angle of the two adjoining streets on the left-hand (33) is the shop of a seller of wine and hot drinks; 33 is a fountain. The walls indicated on the other sides of the streets surrounding the houses, &c., of Pansa, are the fronts of shops and of some private houses, &c.

Fig. 2. is a section through the house of Pansa from the street to the garden, showing the manner in which it is probable the roofs, &c., were arranged.

Fig. 3 is the probable elevation of the entrance front of this mansion, though the sketch (fig. 4) of part of the same in its present state shows how slight the evidence for it is.

Fig. 5 is an outline of the side of a room, with the ornaments, &c., with which it is decorated. This is an average specimen: many were much plainer, and some were more enriched.

Fig. 6 is the plan of an ordinary sized house in one of the private streets of Pompeii: the uses of the various parts may be generally gathered from those of the similar portions of the house of Pansa. The word SALVE, printed across the threshold, is there wrought in mosaic.

Fig. 7 presents the presumed arrangement of the roofs, &c., of this house in section.

Fig. 8 is the elevation of it towards the street. This cannot really have been better than it appears here, and such must have been the ordinary average appearance of the street fronts of Pompeian houses. (*Vide* p. 55, *et seq.*)

Plate XIII. Fig. 1, an example to show how the term order is applied, and what parts of it the various technical terms are applied to, or are intended to indicate.

Figs. 2, 3, 4, 5, and 6, are the orders of the Italo-Vitruvian school as arranged by Palladio; fig. 2, the Tuscan; fig. 3, the Doric; fig. 4, the Ionic; fig. 5, the Corinthian; and fig. 6, the Composite. (*Vide* p. 61.)

Plate XIV. Varieties of Italian composition from existing structures in Italy and elsewhere, in the Italian style.

Figs. 1, 2, 3, 4, and 5, are windows of various form and arrangement.

Figs. 6, 7, 8, and 9, are doors of various composition, with plans to show their arrangement and ichnographic projections, &c.

Figs. 10, 11, 12, and 13, are arches and arcades, rusticated and with columns, &c. The plans show their forms and ichnographic projections. (*Vide* p. 62, *et seq.*)

Plate XV. Front elevations alone of the fronts of St Paul's in London and St Peter's in Rome. These two struc-

tures exhibit many of the peculiarities of the ecclesiastical architecture of the Italian school. In this plate their comparative magnitude has not been attended to; they are drawn to different scales to bring them more nearly of the same size, so as to render the contrast more effective. (*Vide* pp. 27, 67, *et seq.*)

Plate XVI. Flank elevations of St Peter's and St Paul's, drawn to the same scale, to show their comparative magnitude, and to enable the reader to judge of their respective merits, as well as to elucidate observations which will be found in the text *passim*. (*Vide* pp. 27, 67, &c.)

Plate XVII. Elevations of three esteemed Italian mansions. The merit of this (the principal) elevation of the Farnese Palace is divided between Antonio Sangallo and M. A. Buonarotti. The Villa Giula, near Rome, is esteemed one of the best works of Giacomo Barozzi da Vignolia; and the villa Capra, near Vicenza, by Palladio, is, by the admirers of his style, considered the most perfect of his works. (*Vide* p. 62, &c.)

Plate XVIII. A series of arches in the Norman and Pointed styles, from various structures in England. It exhibits the advance of the circular arch from the plainness exhibited in figs. 1 and 2, to the richer and more complicate arrangements of those examples which follow, until the ingrafting and gradual advance of the pointed arch. This first appears in fig. 10. Fig. 12 shows the substitution of the latter for the circular of fig. 9 in a similar composition. Fig. 13 exhibits the pointed arch on Gothic pillars or columns; and fig. 14 the perfected pointed arch with the clustered shafts which become identified with the Pointed style. (*Vide* "Glossary.")

Plate XIX. The elevation of the south transept of Beverley Minster. This affords a perfect and beautiful example of external composition of the first period of Pointed architecture. The presence of the circular arch embracing the pointed arches of the doorway, and composing with others, shows how gradual the advance of the new style was; the upper part of the front showing also how completely it was already systematised when the circular arch was not yet quite discarded. The plan of this front shows the various ichnographic projections, and the arrangement of the clustered shafts of the doors and windows. Fig. 2 is a niche in front of, and fig. 3 a pinnacle to, one of the buttresses of the nave of the same edifice: these are of the second period. Figs. 4, 5, 6, 7, and 8 are windows from various edifices, showing the gradual advance from the plain lancet arch of the Beverley Minster transept to the arch the most elaborately enriched with tracery. Fig. 4 is but a modification of the composition of the doorways of fig. 1, as that of figs. 9 and 12, Plate XVIII.; and the advance from that may be almost termed natural.

Plate XX. Fig. 1 is a sectional compartment of the nave of Lincoln Cathedral; it exhibits the mode of internal composition peculiar to the style of the first period; tending, however, to the Transition, it will be observed, in many particulars, and as a comparison of it with the adjacent example, of the next period, will more clearly show.

Fig. 2 is a similar sectional compartment of the choir of Lincoln Cathedral, exemplifying the internal composition of the second period of the Pointed style; the plans of the shafts to both examples show their forms and arrangement. The subject of the last three plates are drawn entirely, by his kind permission, from Mr Britton's *Chronological History of Ecclesiastical Architecture.*

Plate XXI., the front of York Minster, exemplifies the external composition of the second period, as that of Beverley Minster transept (Plate XIX. fig. 1) does that of the first period ; and the difference will be rendered very clear by comparing them. The upper parts of the towers of the front of York Minster, however, it must be remembered (*vide* pp. 66, 67), are of the third period, and so is the central tower which appears in the distance between them.

The front of Pisa Cathedral is here introduced in contrast with that of York Minster, to show the striking difference which exists between the real Gothic architecture of Italy and the Pointed style which superseded it so completely, in this country particularly, and to elucidate our observations to that effect at pages 19, 23. The cupola which appears behind and in the distance is surrounded at the base by pointed arches and pinnacles, all of which are evidently of much later date than the Gothic front.

Plate XXII. Fig. 1 is an elevation of Westminster Hall. It exemplifies the style of external composition of the third period. It was selected because of the variety of matter it contains elucidatory of the period to which it belongs particularly, and of the Pointed style generally. The door, windows, and canopied tabernacles on the second story of the towers, are peculiar ; the lower tabernacles are more general, and the pinnacles, crockets, corbels, tablets, &c.,[1] may also be taken in exemplification of such things in the style generally.

Fig. 2 is a plan of the front, showing the ribs of the groined entrance, the lower projections of the tabernacles, &c.

Fig. 3 is one of the flying buttresses of the flank of the edifice.

Fig. 4 is a spandrel of the entrance porch enlarged.

Fig. 5, crockets of the gable running from the towers to the crowning turret, enlarged.

Fig. 6, part of the head of one of the upper windows of the towers, enlarged.

Fig. 7, a foliated heraldic panel from under the pedestals of the lower tabernacles or niches of the front, enlarged.

Fig. 8, canopies and pinnacles, &c., of the lower tabernacles, enlarged ; the buttresses on which they rest are also shown at large in intercepted lengths.

Fig. 9, an enriched foliage pendent of the foregoing example, marked *a*, at a still larger scale.

Fig. 10, one of the pedestals for the reception of statues within the niches or tabernacles, enlarged.

Fig. 11, part of one of the canopies, &c., of the tower tabernacles, enlarged.

Fig. 12, one of the foliated pendents, marked *b*, of the foregoing ; and fig. 13 the corbel, marked *c*, of the same, on a still larger scale.

Plate XXIII.—*Arcades.*—Fig. 1, from the refectory in Westminster Abbey, built by Edward the Confessor ; figs. 2 and 3, from Devizes and from Canterbury, specimens of interlaced Norman ; fig. 4, from Wells Cathedral ; fig. 5, the chapter house, Lichfield ; fig. 6, from Amiens ; fig. 7, Notre Dame, Paris ; fig. 8, arcade in front of the Duomo Lucca. *Bases.*—Fig. 9, Norman from Rochester ; fig. 10, early English, Lincoln ; fig. 11, Decorated, Beverley ; fig. 12, do. Tintern ; fig. 13, Cresset, Normandy. *Bell Cots.*—Fig. 14, Northampton ; fig. 15, Wiltshire ; fig. 16, Somersetshire.

Plate XXIV.—*Bell Cots.*—Fig. 1, Froissy, France ; fig. 2, Peakirk, Lincolnshire ; fig. 3, on the Rhine near Cologne. *Bell Towers.*—Fig. 4, Brescia ; fig. 5, Torre della Signoria, Siena ; fig. 6, Lausanne ; fig. 7, Sainte Chapelle, Paris ; fig. 8, Evreux ; fig. 9, Freiburg. *Crest Ridge.*—Fig. 10, Sainte Chapelle ; figs. 11 and 12, Rouen ;

fig. 13, Dogal Palace, Venice ; figs. 14 and 15, French examples of stancheon heads.

Plate XXV.—*Bosses.*—Fig. 1, St Sepulchre's, Cambridge, Norman example ; fig. 2, Winchester ; fig. 3, Sainte Chapelle, Paris ; fig. 4, Notre Dame ; do., fig. 5, York ; 6, Windsor ; fig. 7, Sicilian ; fig. 8, Mars d' Agen. *Buttresses.*—Fig. 9, Beauvais ; fig. 10, Brasted, Kent ; figs. 11 and 12, Perpendicular examples ; fig. 13, Hotel de la Tremouille, Paris.

Plate XXVI.—*Capitals.*—Fig. 1, from the Saxon crypt at Repton, probably the oldest example extant in England ; fig. 2, Caen ; fig. 3, Folkestone ; fig. 4, Buildwas ; fig. 5, Canterbury ; fig. 6, the White Tower, London ; fig. 7, Alton ; fig. 8, Durham ; figs. 9 and 11, Sutton, Kent ; figs. 10 and 12, Buildwas ; fig. 13, Mantes ; fig. 14, Blois ; —these are all late Norman or Transition.

Plate XXVII.—*Capitals.*—Fig. 1, Salisbury ; figs. 2 and 3, Stone, Kent ; fig. 4, Warrington ; fig. 5, Freiburg ; fig. 6, Sainte Chapelle, Paris ; figs. 7, 8, and 9, Westminster ; figs. 10 and 12, Beverley ; fig. 11, Stone ;—these are all early English, Transition, and Decorated.

Plate XXVIII.—*Chamfer Stops.*—Figs. 1 to 5, various examples of various periods and richness. *Corbel.*—Fig. 6, Buildwas ; fig. 7, Stone ; fig. 8, Winchester ; fig. 9, Melrose. *Crocket.*—Fig. 10, Amiens ; fig. 11, Strasbourg ; fig. 12, from the Palazzo della Signoria, Siena ; fig. 13, Milan ; fig. 14, Salisbury ; fig. 15, Westminster ; fig. 16, Canterbury ; fig. 17, St Albans ; fig. 18, Litcham. *Crosses.*—Fig. 19, Freshton ; fig. 20, Dodington. *Cusps.*—Fig. 21, Higham Ferrers ; fig. 22, Raunds ; fig. 23, Lincoln.

Plate XXIX.—*Diapers.*—Figs. 1, 2, and 5, Westminster ; figs. 3 and 7, Auxerre ; fig. 4, Canterbury ; fig. 6, Beverley. *Drip Stone Terminations.*—Fig. 8, Shoreham ; fig. 9, Debenham ; figs. 10, 11, and 12, Perpendicular examples.

Plate XXX.—Fig. 1, San Pietro Màrtire, Verona ; fig. 2, lately opened in the cloisters, Westminster ; fig. 3, Higham Ferrers ; fig. 4, Assisi ; fig. 5, York ; fig. 6, Tournay ; fig. 7, Crick ; fig. 8, the Duomo, Florence ; fig. 9, Magdalen College, Oxford.

Plate XXXI.—*Finials.*—Fig. 1, Amiens ; fig. 2, Lincoln ; fig. 3, Westminster ; fig. 4, Hotel Cluny ; fig. 5, Winchester ; fig. 6, the French lily, probably the origin of the Gothic finial ; fig. 7, York ; fig. 8, Bourges. *Gargoyle.*—Fig. 9, Ruislip, and the like at Petherton ; fig. 10, Glastonbury ; figs. 11 and 12, Amiens ; fig. 13, York ; fig. 14, Troyes.

Plate XXXII.—*Mouldings.*—Figs. 1 and 2, Saxon ; figs. 3 and 4, Norman ; figs. 5 and 6, Transition, the latter from Shoreham ; fig. 7, Grimsby ; fig. 8, Rasen ; fig. 9, Sempringham ; figs. 10, 11, 12, 13, 14, early English examples ; figs. 15 to 19, Transition and Decorated ;—these styles overlap each other, and care must be taken to observe whether the order of moulding is based on the principle of recessed squares, or on that of bevelled faces, (see Glossary, *Moulding*)—the rest of the examples are Perpendicular and Tudor, for the peculiar hollow of which see fig. 24, which is from St Alban's ; fig. 25, St Sepulchre's ; and fig. 26, from Grandchester ; a more elegant is fig. 28, from Skirlaw. Keel mouldings are given, figs. 11 and 18, the latter from a Lincoln example ; a simple notched moulding at fig. 17, a Suffolk example.

Plate XXXIII.—*Mouldings Ornamented.*—Figs. 1 and 2, Canterbury ; fig. 3, Deeping ; fig. 4, Southwell ; fig. 5, Hadiscoe ; fig. 6, Leicester (this much resembles some North Italian examples) ;—these a. e all Norman or Transition ; fig. 7, Soissons ; fig. 8, a star ornament, probably the origin of the dog tooth ; figs. 9 and 12, pure dog tooth,

[1] See all these articles in the *Glossary*.

the latter from Lincoln; figs, 10 and 13, the boll flower, the latter from Rouen; fig. 14, a richer example from St Cross, which leads to the supposition that the dog tooth violet was the type of this ornament; fig. 11, a curious moulding from Oadby, Lincolnshire;—these are early English or Transition; fig. 14, a fine example from Amiens; figs. 16 and 17, Rouen; fig. 18, La Sainte Chapelle; fig. 19, Ameins; fig. 20, an example of the Tudor flower, a common cresting over tombs, "vinettes in casements;" fig. 21, from Bourges; fig. 22, from Westminster.

Plate XXXIV.—*Capitals, Sections of.*—Figs. 1 and 2, Norman; fig. 3, Anglo-Norman, from Folkeston. The rest exhibit the gradual transition to Perpendicular, as explained in the Glossary.—Fig. 4 is from the early work at Lincoln; figs. 5, 6, 7, and 8, early English, the latter from Skelton; fig. 9, Waltham; figs. 10 to 14, Transition and Decorated, chiefly from Lincoln. *Bases.*—Fig. 1, common in early Italian work, and no doubt derived from the "Atticurges" classic base; figs. 2 and 3, Norman, the latter from Peterboro'; figs. 4 and 5, Transition; fig. 6, Decorated, showing the "lip moulding." *Rib Moulds.*—*a*, St Mary, Hunts; *b*, Roberts-bridge. *Mullion.*—A, a Decorated example, showing the "double · order" of moulding. *Strings.*—Figs. 1 to 4, Norman and Transition; figs. 5 to 11, early English and Decorated; fig. 5 is the pure roll moulding; fig. 6, a very fine example from Middleton Chaney, Oxford; figs. 12 to 17 are Perpendicular examples. The rest are foreign examples of mouldings.—Fig. 1, door jamb from Verona; fig. 2, string from Notre Dame, Paris; fig. 3, St Stefano, Verona; fig. 4, 12th century example from d'Ebrueil; figs. 5 and 6, examples showing the theories of M. Viollet-le-Duc as to working out mouldings; fig. 8, a base from Rheims; fig. 9, the like, from Monreale; fig. 10, Dijon; figs. 12 and 13, Laon; fig. 16, a 14th century example. *Strings.*—Figs. 14 to 18, various examples, which should be compared with the English given above.

Plate XXXV.—*Towers and Spires.*—Fig. 1, Sompting; fig. 2, Earl's-Barton;—these are supposed to be Saxon (see Glossary under that head); fig. 3, St Alban's central tower, early Norman, the work of Abbot Paul; fig. 4, the early English tower and broach spire at Raunds, Northamptonshire; fig. 5, Ellington, Huntingdonshire; fig. 6, Witherby, Lincolnshire, with crocketed spire.

Plate XXXVI.—*Windows*, early English examples.—Fig. 1, Oxford, the Cathedral; fig. 2, Stanton, St John's, one of the early cusped examples, afterwards so numerous; fig. 3, Wenham, an early example of the branched mullion; fig. 4, Warrington, a triplet under a single hood mould. The rest of the examples are from the choir of Chester Cathedral, which was built by degrees, extending over a period of about three centuries, and is perhaps the most unique and instructive example in England. Fig. 5 is an example of branching mullions with three plain circles in the heads; fig. 6, the like, but the circles are cusped; fig. 7, the like, but richer; fig. 8, positively Decorated tracery; fig. 9, the like, inclined to flowing; fig. 10, positively flowing Decorated tracery; fig. 11, partly Perpendicular; figs. 12 and 13, positively so,—in the latter example A is the ordinary "day" or light, B the "batement light," and C the "anglet," which see in Glossary.

Plate XXXVII.—*Windows*, continued.—Fig. 1, the house of Jacques Cœur; fig. 2, Duomo at Lucca; fig. 3, Le Mans; fig. 4, circular window in transept at Chichester; fig. 5, Amiens; fig. 6, a very peculiar circular window from San Antonio, at Padua; fig. 7, Santa Maria dell' Orto, at Venice; fig. 8, circular from Beverley; fig. 9, clerestory window at Westminster Abbey, a cusped circle in a spherical triangle;—the foreign examples should be compared with the English (see the articles in the Glossary). *Window Transom.*—Fig. 10, from Amiens; fig. 11, Dorchester, Oxon, a very peculiar example, as the upper mullion is over the middle of the lower light; fig. 12, from Evesham, a very elegant example.

Plate XXXVIII.—*Groining and Vaulting.*—Fig. 1, Norman, from Bow Church, London; fig. 2, plan of the same; fig. 3, example of quadripartite groining showing the different ways of filling up the spandrels,—*a, b, c,* being the foreign method,—*b, c, d*, the English; fig. 4, skeleton isometrical perspective, with plan below of quadripartite groining with tiercerons and lierne ribs; fig. 5, the like, of hexpartite groining; fig. 6, an elevation of the same groining taken from the chapel of St Blaize at Westminster; fig. 7, the plan of the same, showing the filling in of the spandrels, and the peculiar use of a harder stone on the ridge rib as shown by the darker tints; fig. 8, plan and elevation of *fan tracery* from Ely Cathedral; fig. 9, skeleton isometrical perspective, showing the lines of the intrados, or the inner joints, or lines visible from below of fan tracery vaulting; fig. 10, the like of the extrados or joints as visible from above; fig. 11, plan of the rich groining of St George's Chapel, Windsor, with numerous tiercerons and lierne ribs—a style which was, without doubt, the origin of fan groining; fig. 12, plan of pure rich fan groining from the Dean's Chapel at Canterbury.

Plate XXXIX.—*Groining*, continued.—Fig. 1, perspective view of tierceron and lierne groining at Canterbury; fig. 2, the like, from groining at Winchester. *Parapet.*—Fig. 3, from Le Mans Cathedral. *Flying Buttress.*—Fig. 4, from Amiens; fig. 5, earlier and simpler from Clermont;—these are, in fact, double flying buttresses forming two springs, of which it is believed there is no example in England. *Gable Boards or Verge Boards.*—Fig. 6, from Caen; fig. 7, the Moat House at Ightham; fig. 8, the triforium at Lichfield (see the article in the Glossary); fig. 9, a peculiar moulding in use chiefly at Venice, the example is from San Marco; figs. 10, 11, and 13, examples of parapets in Italy, partly brick and partly stone,—fig. 10 is from near Monza, fig. 11 is from Milan, and fig. 13 from Rome; fig. 12 is similarly constructed, common about Verona, &c.

Plate XL.—*Modern French Architecture — Villas.*—Fig. 1, a villa in the environs of Paris; fig. 2, the like, at Neuilly; fig. 3, large house in the Avenue de l'Imperatrice at Paris; fig. 4, a beautiful summer pavilion lately erected near St Cloud; fig. 5, lucarne lately executed in zinc in Paris.

Plate XLI.—*Modern French Architecture — Street Fronts.*—Fig. 1, Boulevard de Sebastopol, the mezzanine forms a gallery round and above the lower shop; fig. 2, Rue de Turin; fig. 3, Rue de Rivoli.

Plate XLII.—*Modern French Architecture—Details.*—Fig. 1, window head, Rue de Conservatorie, the decoration is incised; fig. 3, the like, Rue de Sebastopol. The rest, various decorations.—Fig. 2, Boulevard de Sebastopol; fig. 5, Rue St Placide; fig. 6, chimney top, Boulevard Monçeau; fig. 7, Boulevard de Vincennes; fig. 8, Rue de Turin; figs. 9 and 11, Avenue de l'Imperatrice; fig. 10, Rue de Valois.

Plates XLIII. to XLVII.—Plans to illustrate the ordinary routine in Building.

Plates XLVIII. and XLIX.—Illustrations of the section Arch.

Plates L. to LV.—Plans and sections to illustrate the text in Carpentry.

PRINTED BY NEILL AND COMPANY, EDINBURGH.

ARCHITECTURE.

Plate 1.

Fig. 1.   Fig. 2.   Fig. 3.   Fig. 4.   Fig. 5.

Fig. 6.   Fig. 7.

Fig. 8.   Fig. 9.   Fig. 10.

Fig. 11.   Fig. 12.   Fig. 13.

Fig. 14.   Fig. 15.   Fig. 16.

Published by A. & C. Black, Edinburgh.

ARCHITECTURE.

Plan, Section and Elevations of the Temple of Apollinopolis Magna in Upper Egypt.

Plate

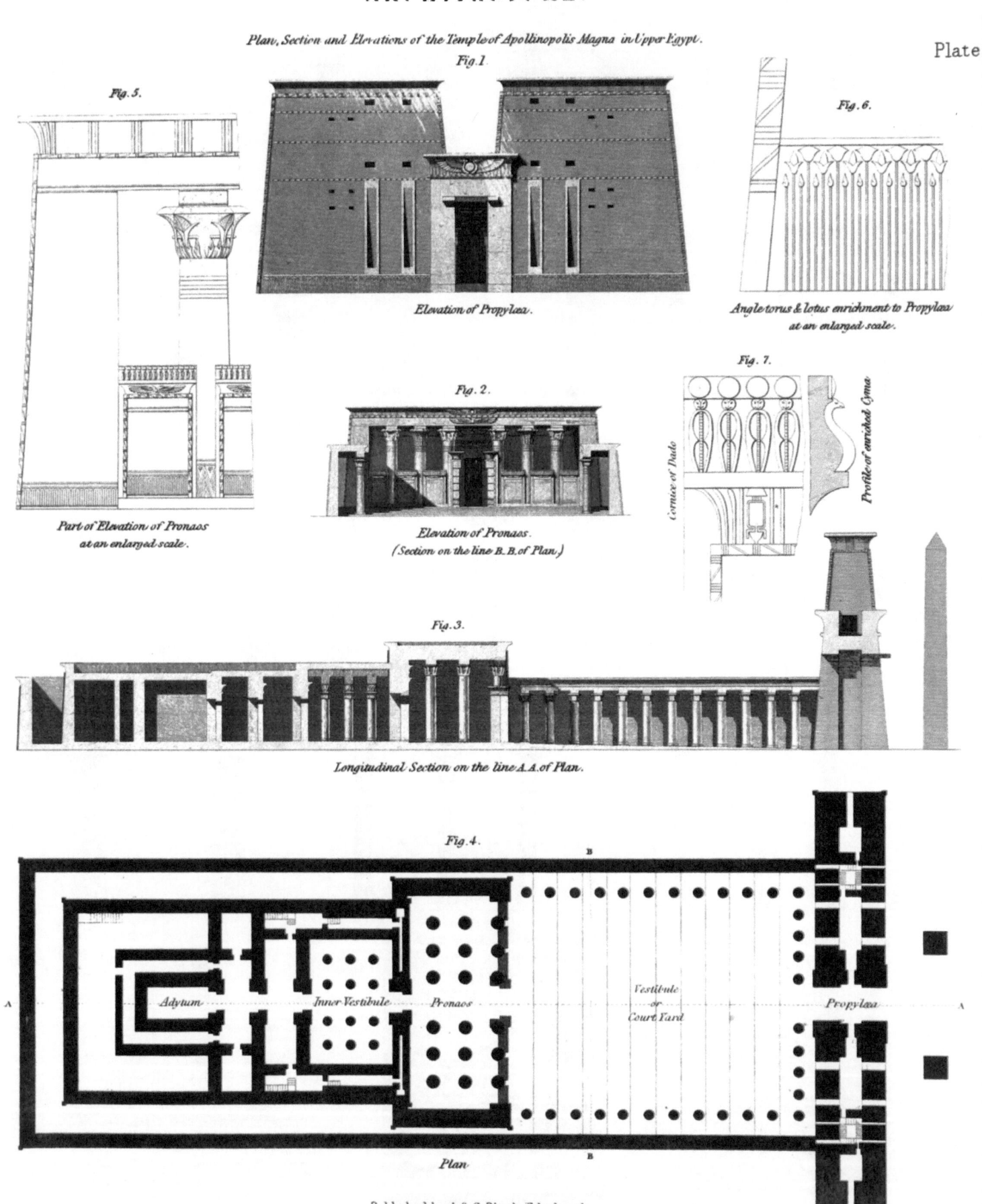

Fig.1.

Fig.5.

Fig.6.

Elevation of Propylæa.

Angle torus & lotus enrichment to Propylæa
at an enlarged scale.

Fig.7.

Part of Elevation of Pronaos
at an enlarged scale.

Fig.2.

Elevation of Pronaos.
(Section on the line B.B.of Plan.)

Cornice of Podo

Profile of enriched Cyma

Fig.3.

Longitudinal Section on the line A.A.of Plan.

Fig.4.

B

Adytum   Inner Vestibule   Pronaos   Vestibule
or
Court Yard   Propylæa

A   A

Plan

B

Published by A.& C Black. Edinburgh

ARCHITECTURE.

The Temple of Minerva Parthenon at Athens
(North-west view.)

Plate 3

In its present state.

W.W. Jenkins Esq.ᵗ Arch.ᵗ Del.ᵗ

Restored

Eng.ᵈ by W.H. ...son Edin.ᵗ

Published by A. & C. Black, Edinburgh.

ARCHITECTURE.

Flank & Sectional Elevation of a Greek Doric Peripteral & Hypæthral Temple.
(Section on the dotted lines of Plan below)
Fig. 1.

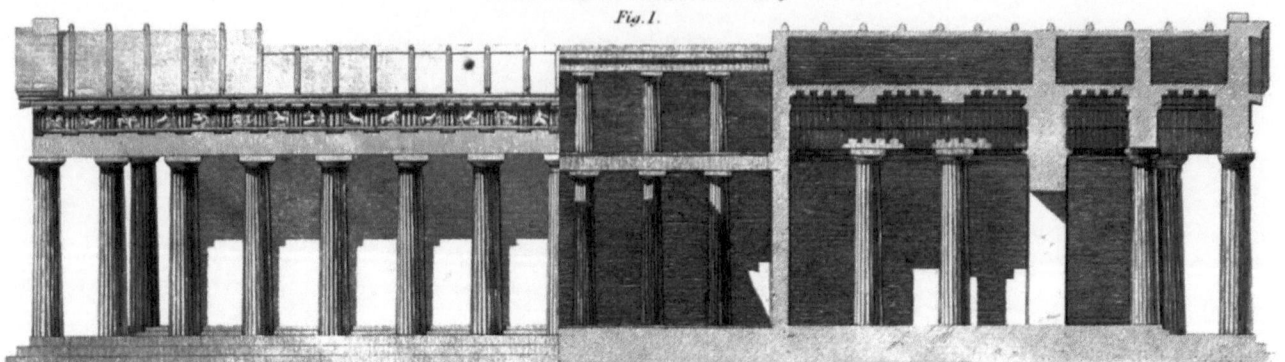

Sectional Elevation of the Pronaos of a Greek Doric Octastyle Temple.

Fig. 5. — *Fig. 2.* — *Fig. 8.*

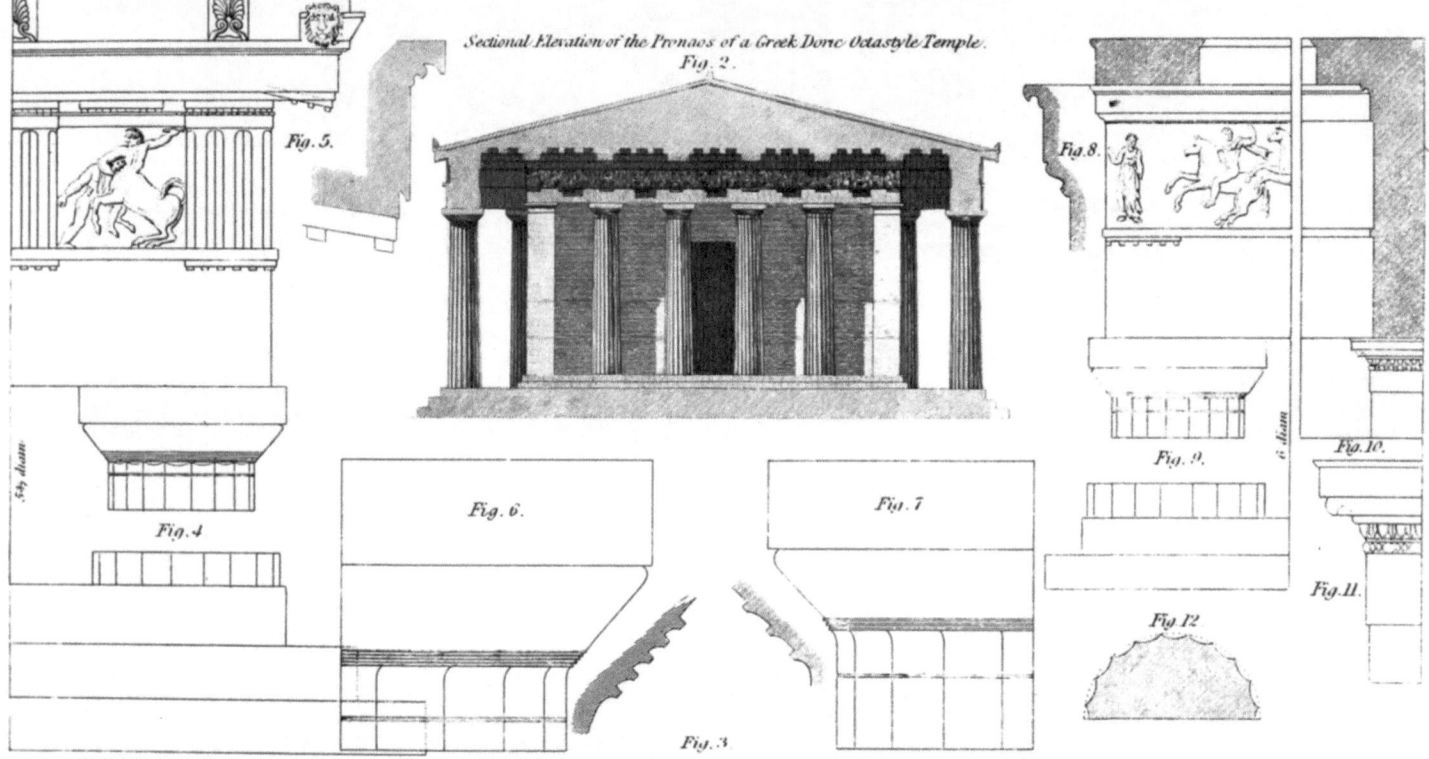

Fig. 4. — *Fig. 6.* — *Fig. 7.* — *Fig. 9.* — *Fig. 10.* — *Fig. 11.* — *Fig. 12.* — *Fig. 3.*

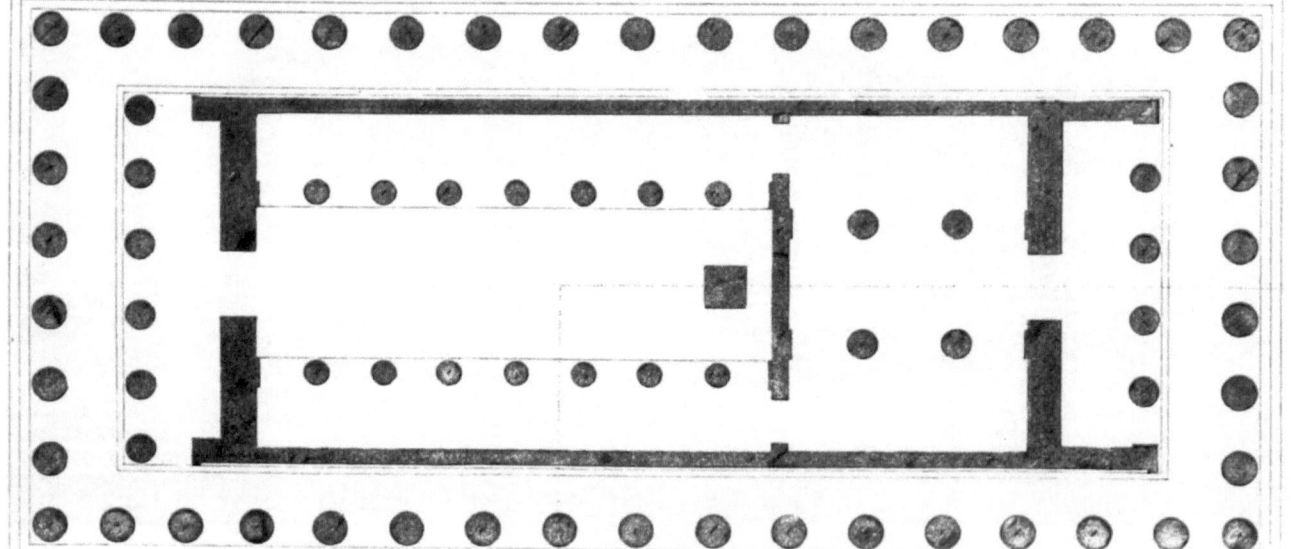

Plan of a Greek Octastyle Peripteral & Hypæthral Temple.

W.H. del.ᵗ

Eng.ᵈ by G. Aikman Edin.ʳ

Published by A. & C. Black, Edinburgh.

Plate 4

ARCHITECTURE.

Plate 5

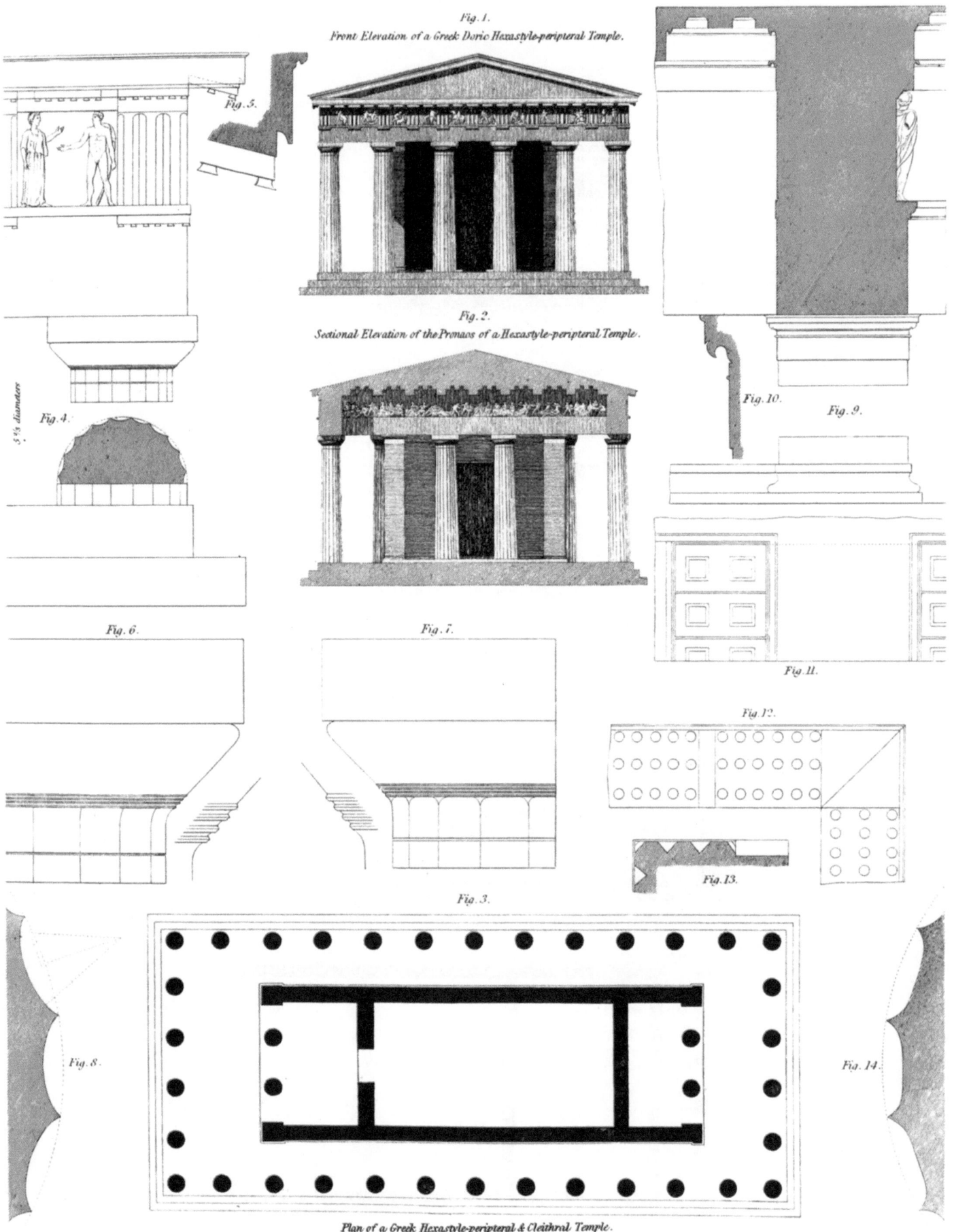

Fig. 1.
Front Elevation of a Greek Doric Hexastyle-peripteral Temple.

Fig. 2.
Sectional Elevation of the Pronaos of a Hexastyle-peripteral Temple.

Fig. 5.

Fig. 4.

Fig. 6.

Fig. 7.

Fig. 10.

Fig. 9.

Fig. 11.

Fig. 12.

Fig. 13.

Fig. 3.

Fig. 8.

Fig. 14.

Plan of a Greek Hexastyle-peripteral & Cleithral Temple.

W.H. delt.

Eng'd by G. Aikman Edin'r

Published by A.& C.Black, Edinburgh.

ARCHITECTURE.

Plate 6.

Fig. 6.

Fig. 7.

Fig. 8.

Fig. 1.
Elevation of a Greek Ionic Hexa-prostyle

Fig. 9.

Fig. 2.
Elevation of a Greek Ionic attached Tetrastyle in Antis

Fig. 11.

Fig. 5.

9 ½ diam

Fig. 10.

Fig. 3.
Flank of a Greek Ionic Prostyle Temple

Fig. 12.

Fig. 15.

Fig. 13.

Fig. 4.

Fig. 16.

Fig 14

Plan of a Greek Hexa-prostyle Temple

W. H. del.

Eng⁴ by G. Aikman, Edin.

Published by A. & C. Black, Edinburgh.

Plate 7.

ARCHITECTURE.

Choragic Monument Greek Corinthian.

Fig. 1.

Plan of Choragic Monument.

Fig. 2

Fig. 3

Fig. 4.
Portico of Caryatides.

Fig. 5

Plan of Portico.

Fig. 6.

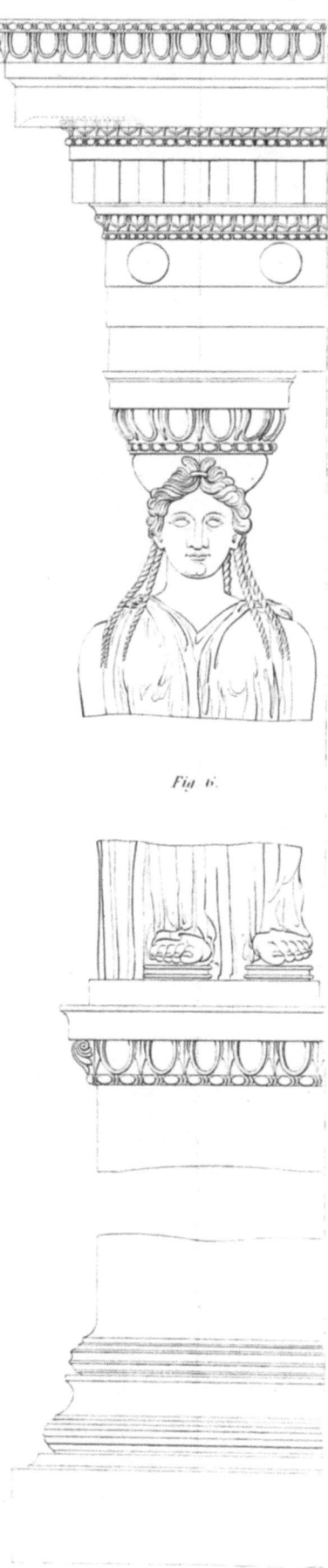

Published by A.& C.Black, Edinburgh.

ARCHITECTURE.

Plate 8.

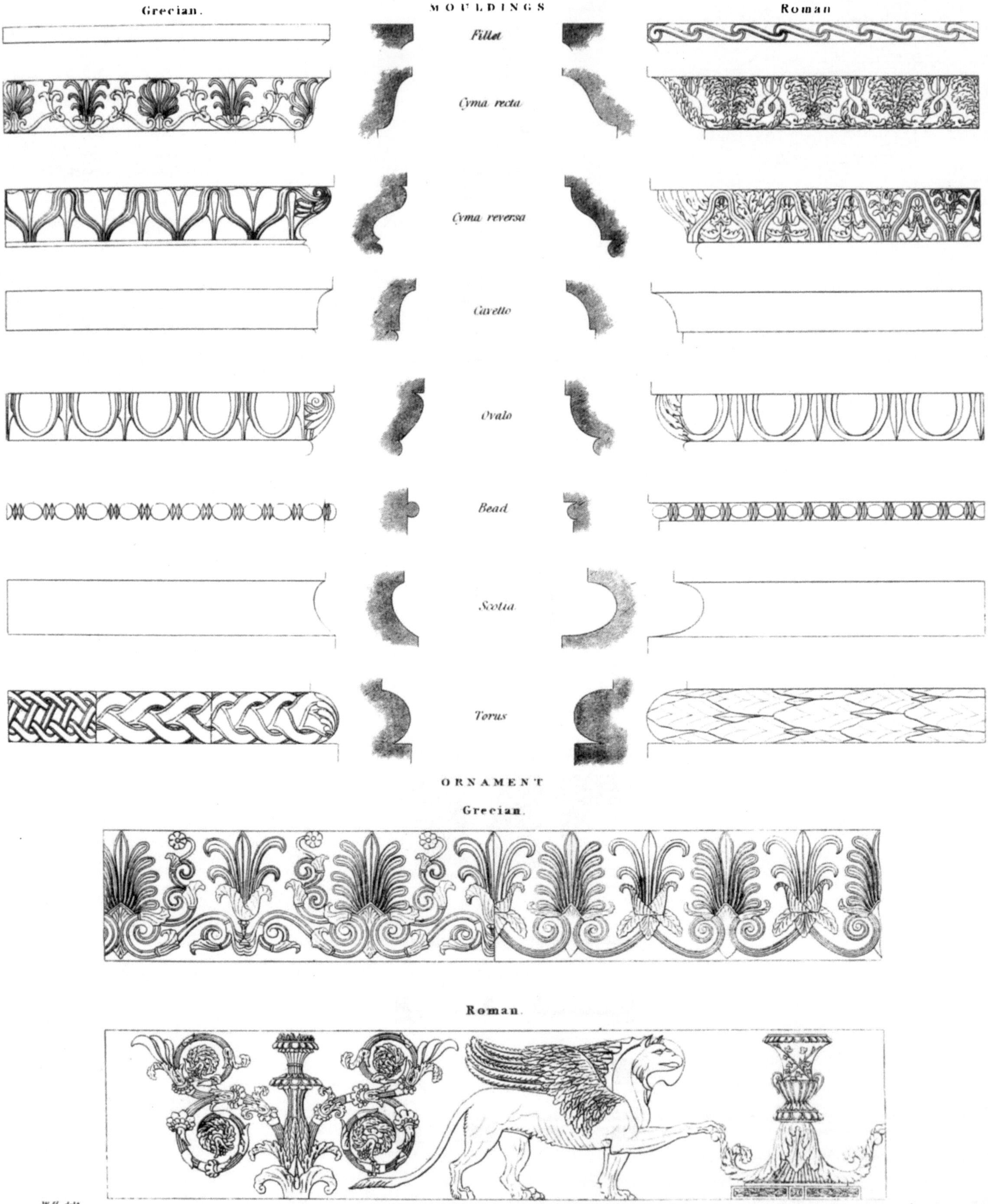
Grecian.
MOULDINGS
Roman
Fillet
Cyma recta
Cyma reversa
Cavetto
Ovalo
Bead
Scotia
Torus
ORNAMENT
Grecian.
Roman.

W.H. del^t

Eng^d by G. Aikman Edin^r

Published by A.& C. Black, Edinburgh.

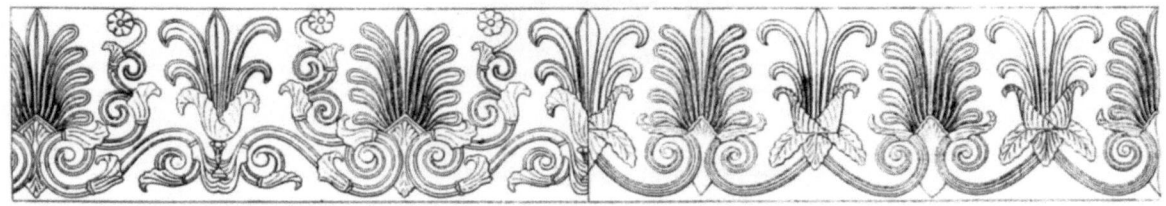

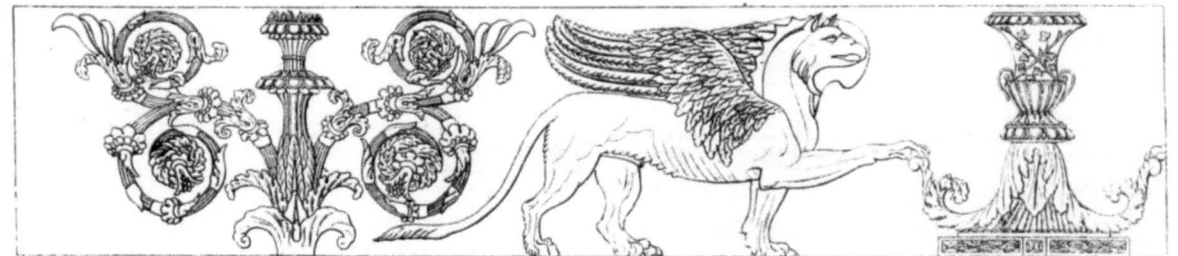

ARCHITECTURE.

Examples of the Roman Corinthian.

Plate 9.

Ex. 1.

Fig. 1.   Fig. 2.   Fig. 2.

Ex. 2.

Fig. 1.

Ex. 3.

Fig. 1.   Fig. 2.   Fig. 2.

Ex. 4.

Fig. 1.

Drawn by G. Maddox Esqr. Archt.   Published by A.& C. Black, Edinburgh.   Engd. by G. Aikman.

Ex. 1.

Fig. 1.

Fig. 2.

Fig. 2.

Ex. 2.

Fig. 1.

Ex. 3.

Fig. 1.

Fig. 2.

Fig. 2.

Ex. 4.

Fig. 1.

W.H. del.

Examples of the Roman Orders
Published by A.& C. Black, Edinburgh.

ARCHITECTURE.

Plate II.

Fig. 1.

Fig. 2.

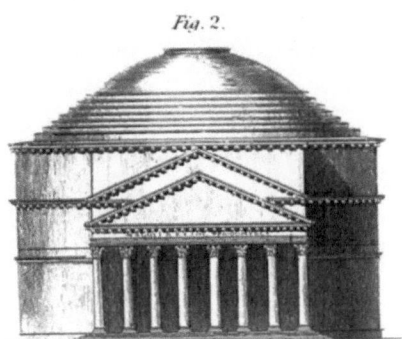

Fig. 3.

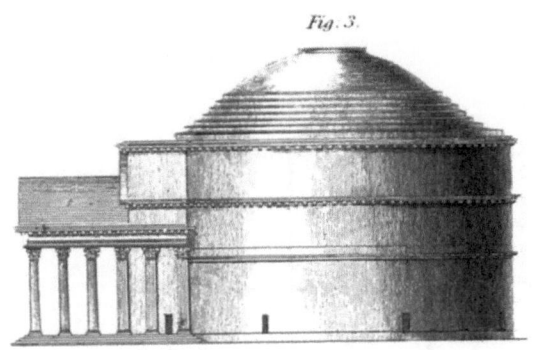

Fig. 4.

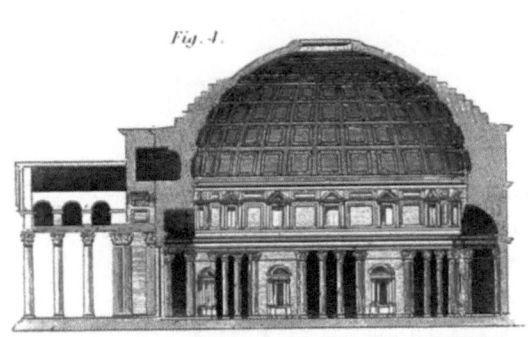

Fig. 5.

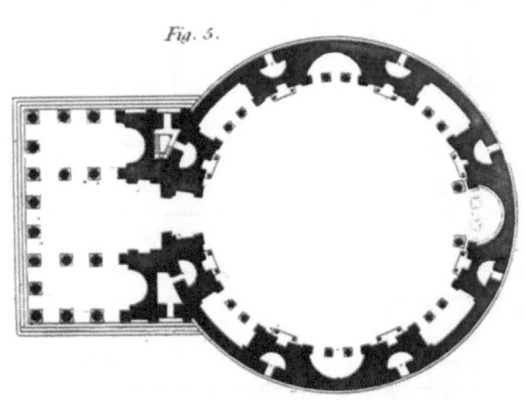

Fig. 6.

Fig. 7.

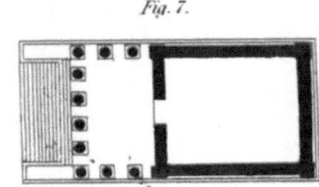

Fig. 8

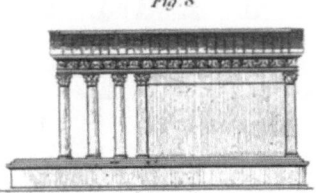

Fig. 9.

Fig. 10.

Fig. 11.

Fig. 12.

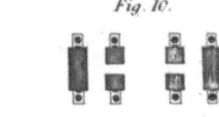

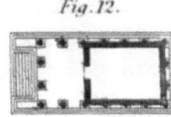

W.H. del.

Published by A.& C. Black, Edinburgh.

Engd by G. Aikman, Edin.

ARCHITECTURE.

[Plate 1

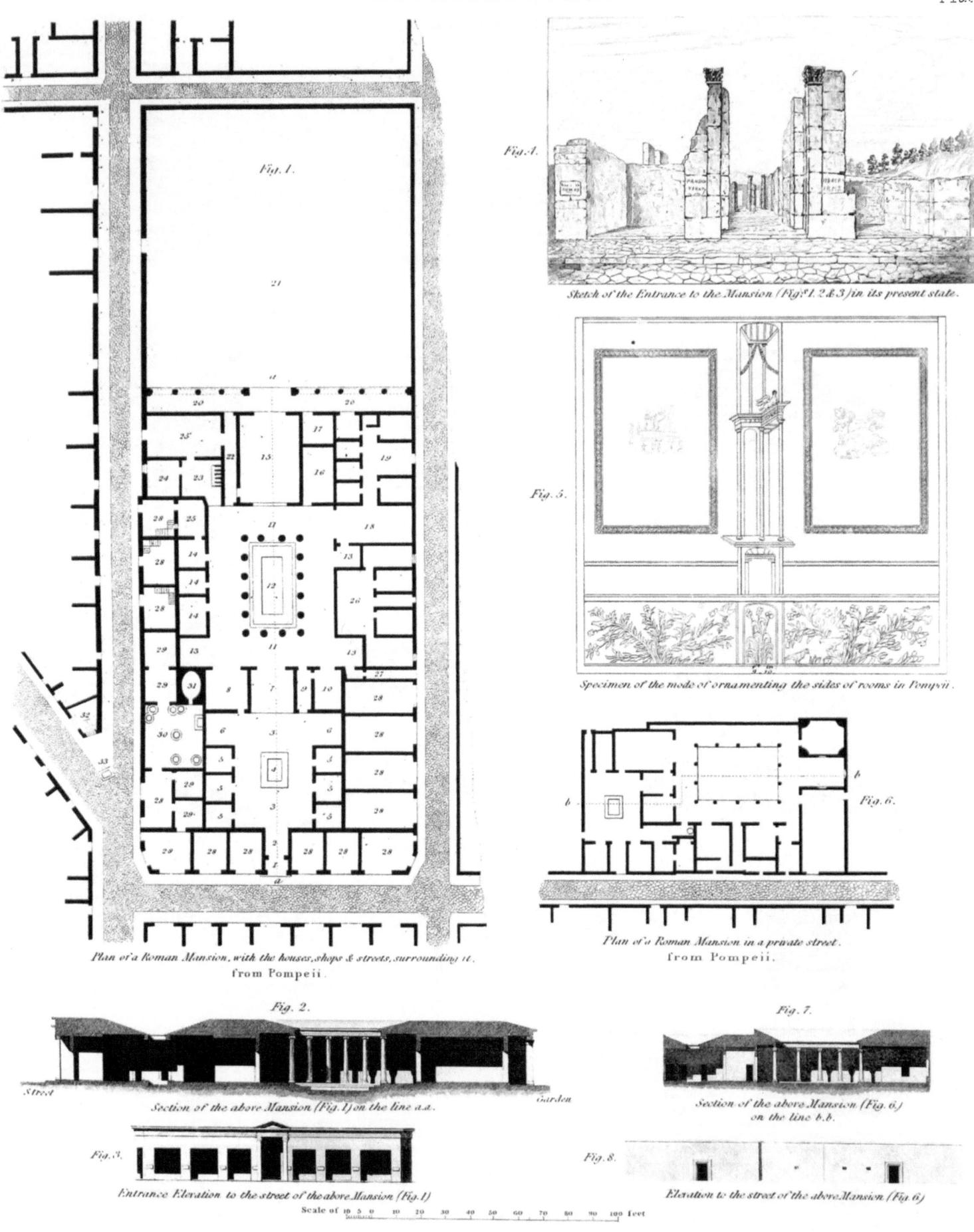

Fig. 1.

Fig. 4.

Sketch of the Entrance to the Mansion (Figs. 1, 2 & 3) in its present state.

Fig. 5.

Specimen of the mode of ornamenting the sides of rooms in Pompeii.

Fig. 6.

Plan of a Roman Mansion, with the houses, shops & streets, surrounding it.
from Pompeii.

Plan of a Roman Mansion in a private street.
from Pompeii.

Fig. 2.

Street

Section of the above Mansion (Fig. 1) on the line a.a.

Garden

Fig. 7.

Section of the above Mansion (Fig. 6)
on the line b.b.

Fig. 3.

Entrance Elevation to the street of the above Mansion (Fig. 1)

Fig. 8.

Elevation to the street of the above Mansion (Fig. 6)

Scale of 10 5 0 10 20 30 40 50 60 70 80 90 100 feet

W. H. delt.

Eng.d by G. Aikman, Edin.r

Published by A. & C. Black, Edinburgh.

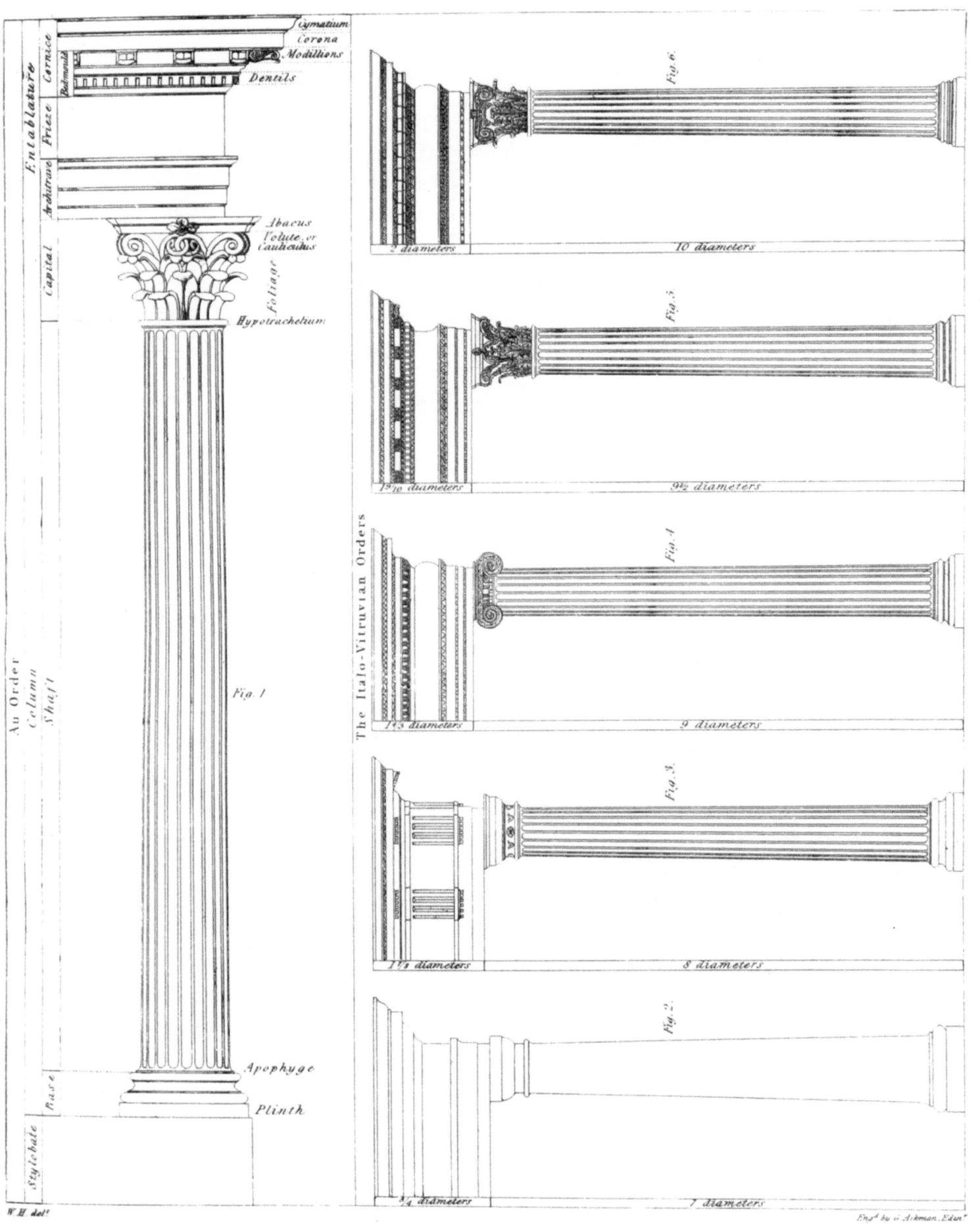

The Italo-Vitruvian Orders

W.H. del.

Eng. by G. Aikman, Edin.

Published by A.& C.Black, Edinburgh.

ARCHITECTURE.

Plate 14.

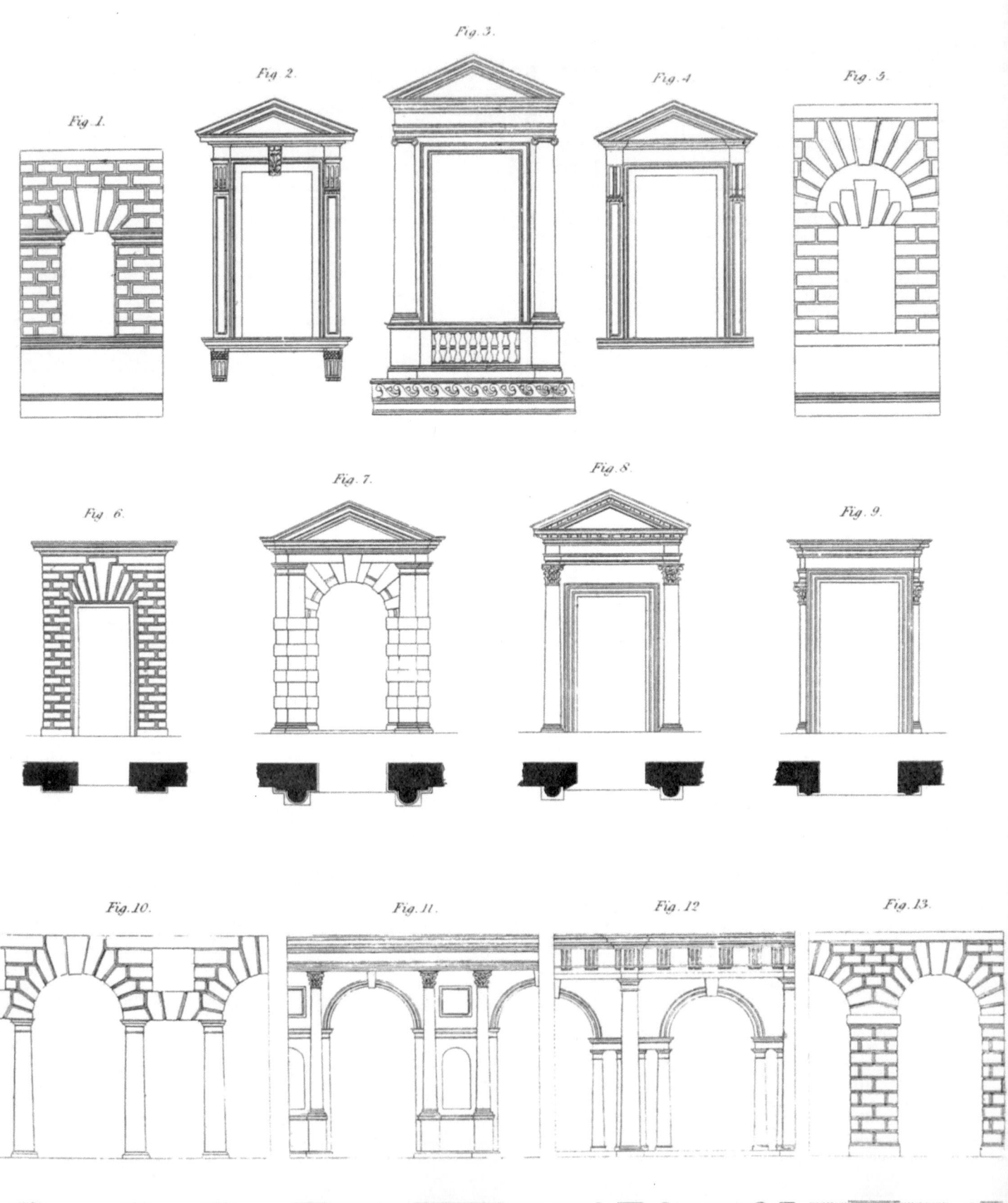

Fig. 1.   Fig. 2.   Fig. 3.   Fig. 4.   Fig. 5.

Fig. 6.   Fig. 7.   Fig. 8.   Fig. 9.

Fig. 10.   Fig. 11.   Fig. 12.   Fig. 13.

Published by A. & C. Black, Edinburgh.

ARCHITECTURE.

Plate 15

ST PAULS.

west front elevation

SCALE
10 5 0   10   20   30   40   50   60   70   80   90   100 Feet.

ST PETERS.

east front elevation

Drawn by J. P. Wise Esqr Archt

Engd by G. Aikman, Edinr

Published by A.& C. Black, Edinburgh.

ARCHITECTURE.

Plate 16

ST PAUL'S

South Flank Elevation .

ST PETER'S

North Flank Elevation .

SCALE

0   50   100   200   400 feet

Published by A.& C. Black Edinburgh

ARCHITECTURE.

Plate 17.

Fig. 3.

Scale of 10 5 0 10 20 30 40 50 60 feet.

VILLA CAPRA
near Vicenza.

Fig. 2.

Scale of 10 5 0 10 20 30 40 50 60 feet.

VILLA GIULIA
near Rome.

Fig. 1.

Scale of 10 5 0 10 20 30 40 50 60 70 80 90 100 feet.

FARNESE PALACE — ROME
Front elevation.

Drawn by J. P. Ware, Esq. Arch.t

Eng.d by S. Johnson, Edin.r

Published by A. & C. Black, Edinburgh.

ARCHITECTURE.

Plate 18.

Fig. 1.

Arch from the Nave of St Alban's Abbey Church.

Fig. 2.

Arches from a Chapel in the White Tower, London.

Fig. 3.

Arches from the Conventual Church, Ely.

Fig. 4.

Arches from Waltham Abbey Church, Herts.

Fig. 5.

Window from Steyning Church, Sussex.

Fig. 6.

Window from Steyning Church, Sussex.

Fig. 7.

West

Doorway from Iffley Church, Oxfordshire.

Fig. 8.

South

Doorway from Iffley Church, Oxfordshire.

Fig. 9.

Window from Pythagoras's School, Cambridge.

Fig. 10.

Windows from Barfreston Church, Kent.

Fig. 11.

Arches from Romsey Church, Hampshire.

Fig. 12.

Window from Chichester Cathedral.

Fig. 13.

Arches from the Nave of Shoreham Church, Sussex.

Fig. 14.

Arches from the Nave of Salisbury Cathedral.

Published by A. & C. Black, Edinburgh.

ARCHITECTURE.

Plate 19

Fig. 2.

Fig. 3.

Fig. 1.

Fig. 8.

Fig. 4.

Fig. 5.

Fig. 6.

Fig. 7.

Published by A. & C. Black, Edinburgh.

ARCHITECTURE.

Plate 20

Sectional compartment of the Nave of Lincoln Cathedral.

Sectional compartment of the Choir of Lincoln Cathedral.

Fig. 1.

Fig. 2.

Eng.ᵈ by G. Aikman, Edin.ʳ

Published by A.& C. Black, Edinburgh.

ARCHITECTURE.

Plate 21.

PISA CATHEDRAL.
(West Front)

YORK CATHEDRAL.
(West Front)

SCALA

Published by A & C Black Edinburgh

ARCHITECTURE.

Plate 22

Fig. 3.

Fig. 1.

Fig. 4

Fig. 5.

Fig. 6.

Scale of 10   5   0   10   20   30   40 feet

Fig. 9.

Fig. 2.

Fig. 12.

Fig. 8.

Fig. 11.

Fig. 7.

Fig. 10.

Fig. 13.

Eng.d by G. Aikman Sc.

Published by A.& C.Black, Edinburgh.

Plate 23.

C.S. Edmeston Del

M&R Bros Lith^{rs} Castle St London, E.C.

1. TO 8. ARCADE, 9. TO 13. BASE, 14. TO 16. BELL COT.

Plate 24.

S. Edmeston Del.

Nell Bros Ltd Castle of London E.C

I TO 3 BELL COT, 4 TO 9 BELL TOWER, 12,13,16, 17, CREST RIDGE
14, 15, STANCHION HEAD.

Plate 25.

J S Edmeston Del

1 TO 6. BOSS, 7 & 8. BRACKET, 9 TO 13. BUTTRESS.

Plate 26

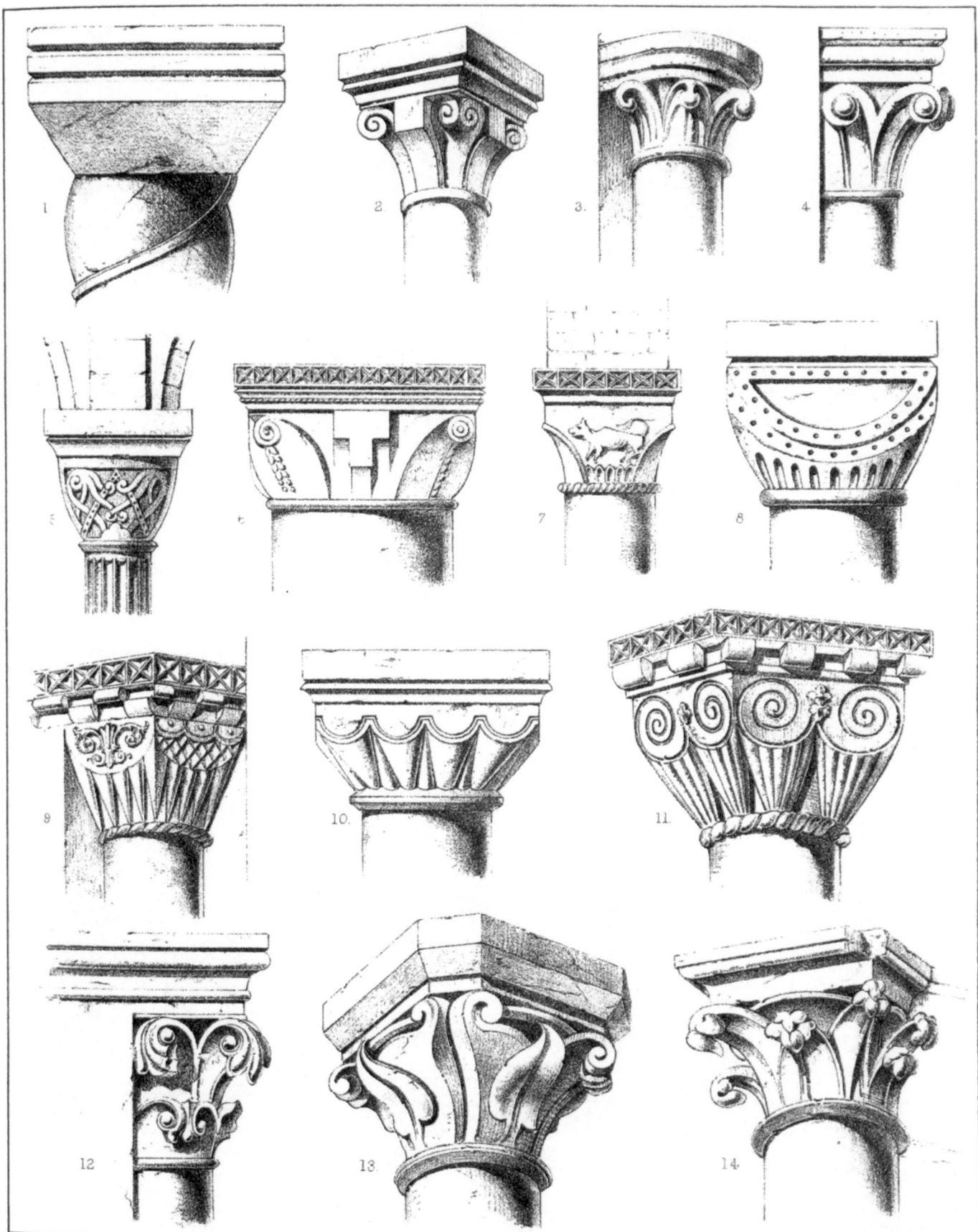

CAPITAL

Plate 27.

CAPITAL

Plate 28.

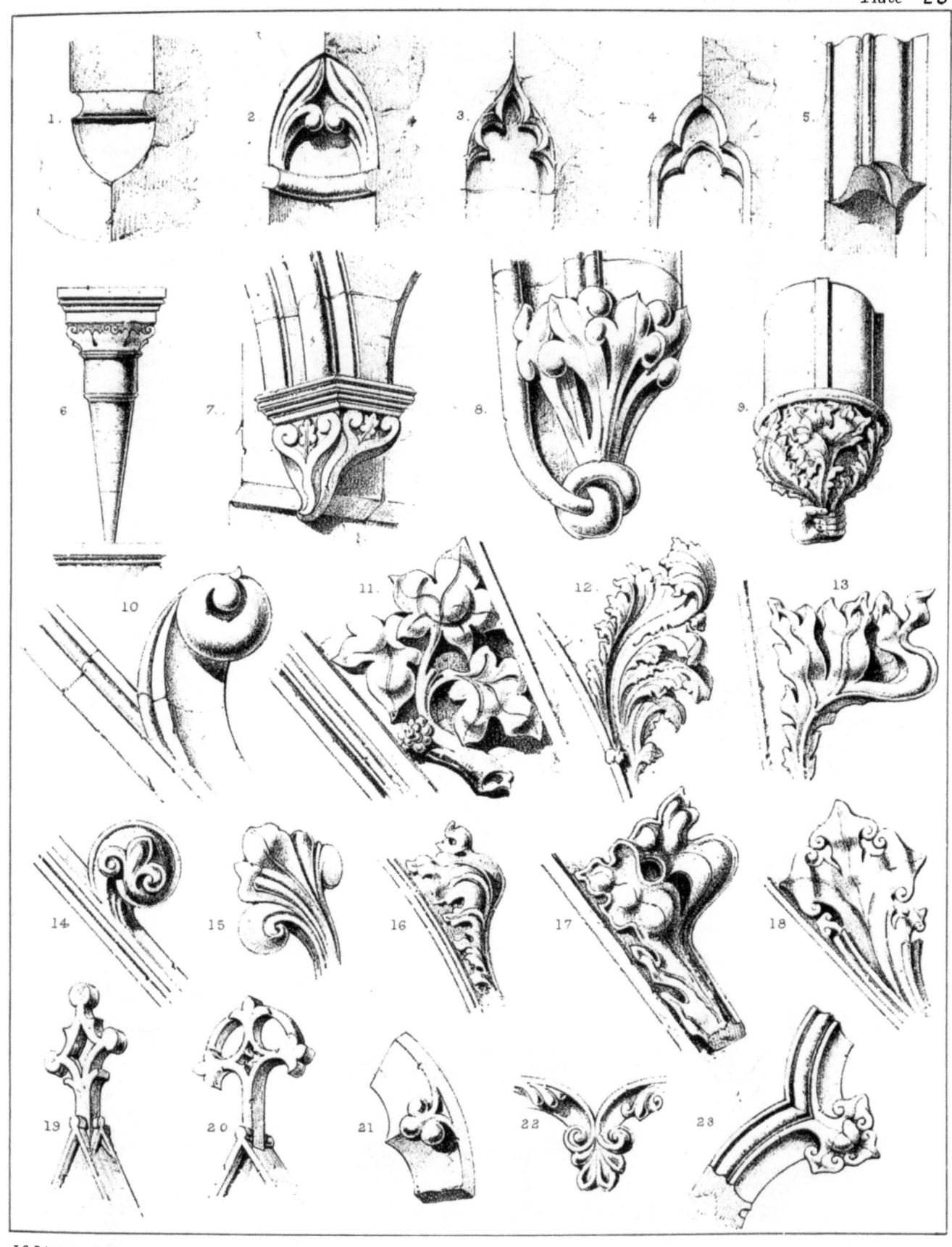

1 TO 5 CHAMFER, 6 TO 9 CORBEL, 10 TO 18 CROCKET,

19.20 CROSS, 21 TO 23 CUSP.

J.S.Edmeston Del

Plate 29.

I TO 7. DIAPER, 8 TO 12. DRIP STONE TERMINATION.

Plate 30.

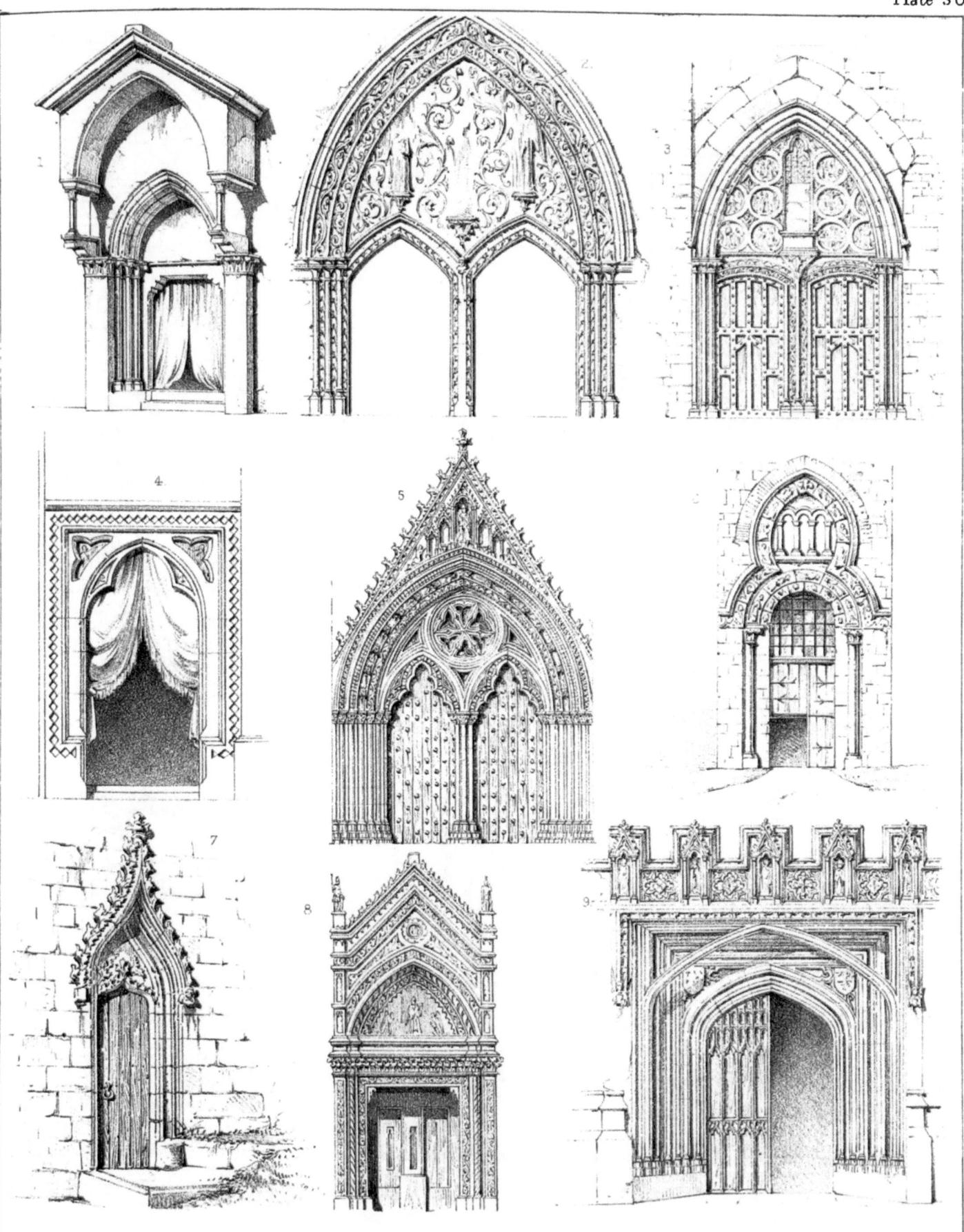

J S Edmeston Del.

DOORWAY.

Plate 31.

J S Edmeston De.

Kell Bros Lith Castle St London E.C

8 FINIAL, 9 TO 14 CARGOYLE

Plate 32.

MOULDING.

Kell Bros Lith

Plate 33.

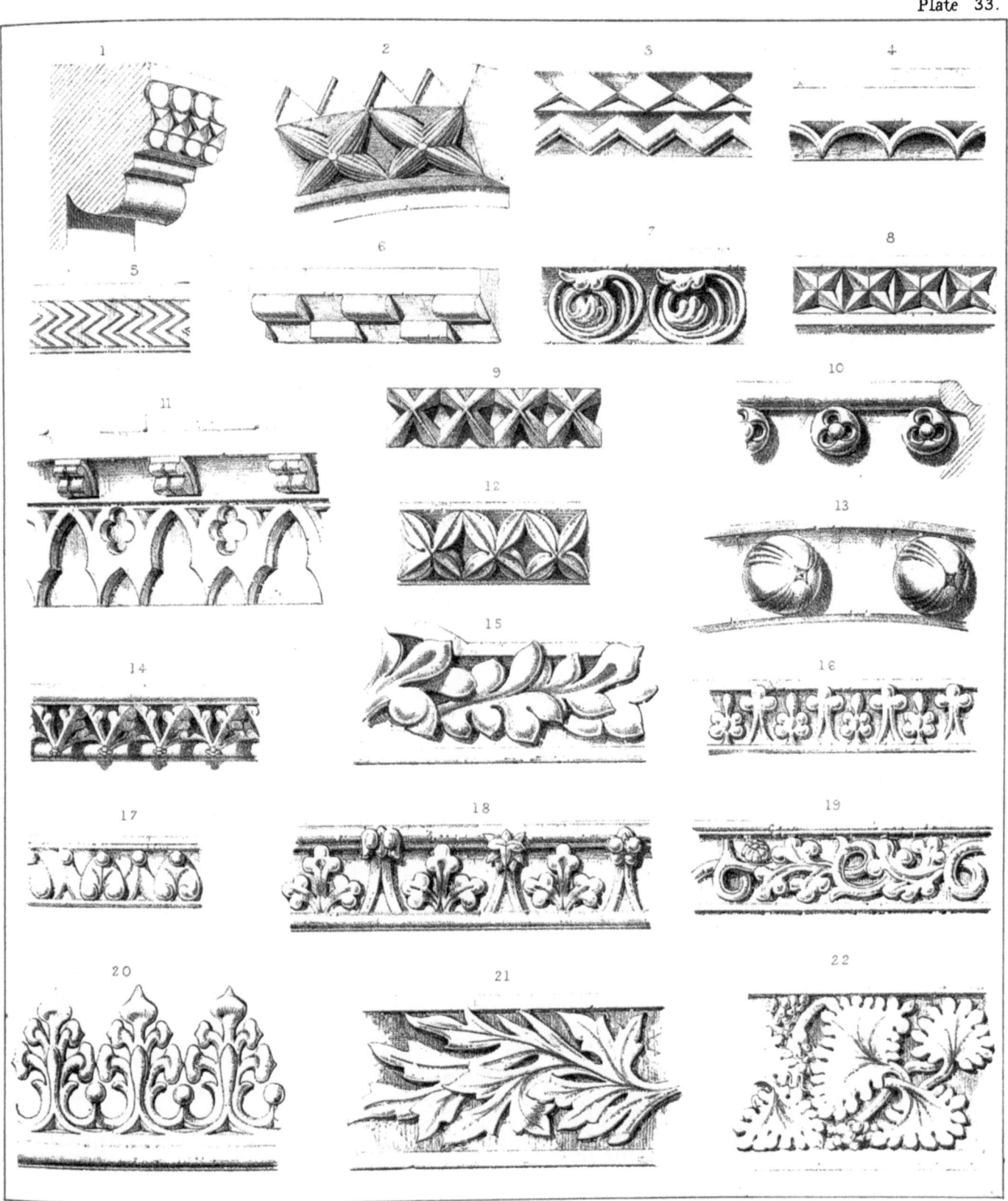

S. Edmiston Del.

Kell Bros Lith^{rs} Castle St London E C

MOULDING, ENRICHED.

Plate 34.

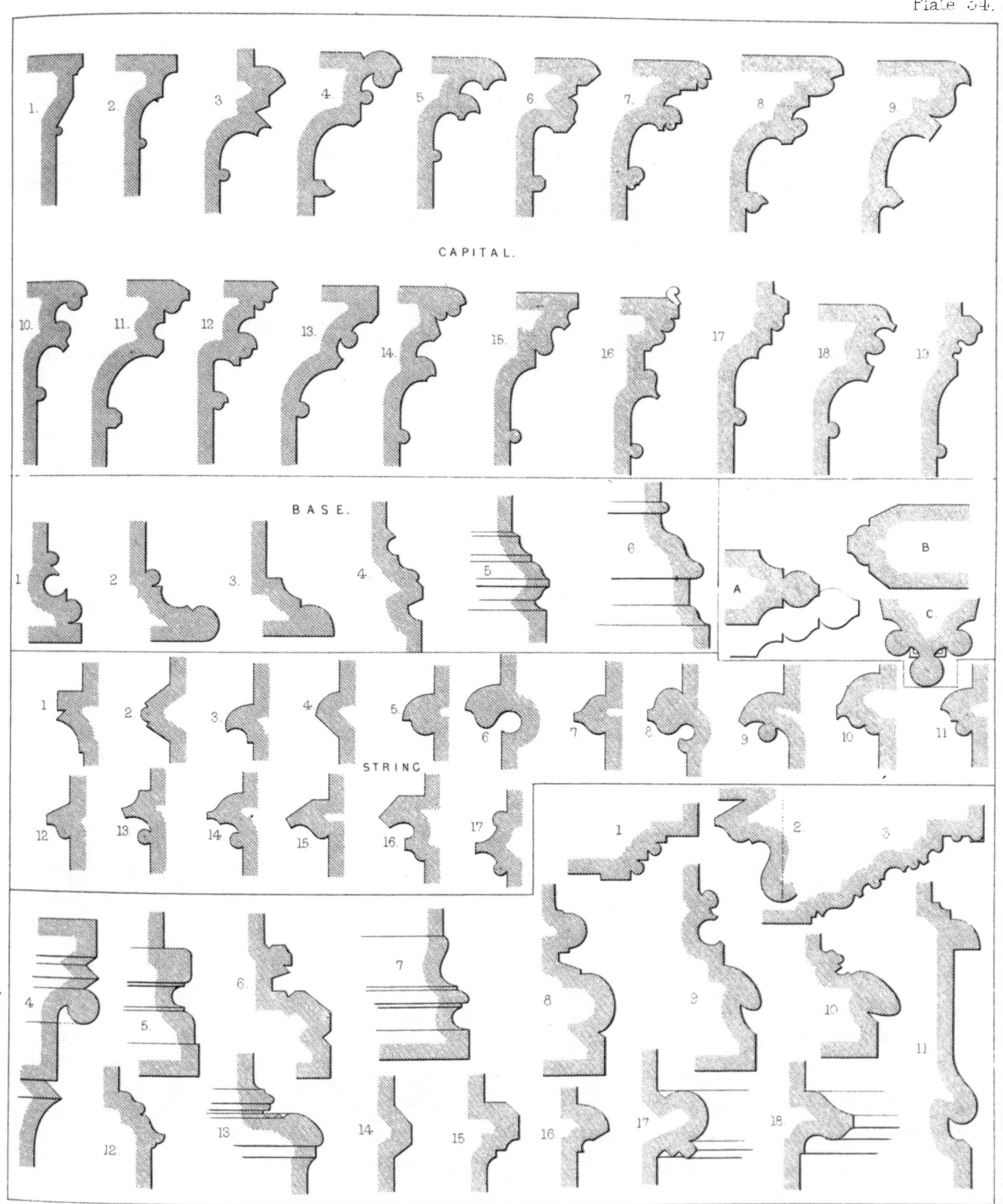

Kell.Bros.Lith.rs

MOULDING.

1 to 19 CAPITAL.   1 to 6. BASE.   1 to 17. STRING.   1 to 18. VARIOUS FOREIGN MOULDINGS.   A B. MULLION C. GROIN RIB.

Plate 35.

J.S.Edmeston Del.

Kell Bros Liths Castle St London E.C.

TOWER & SPIRE.

Plate 36.

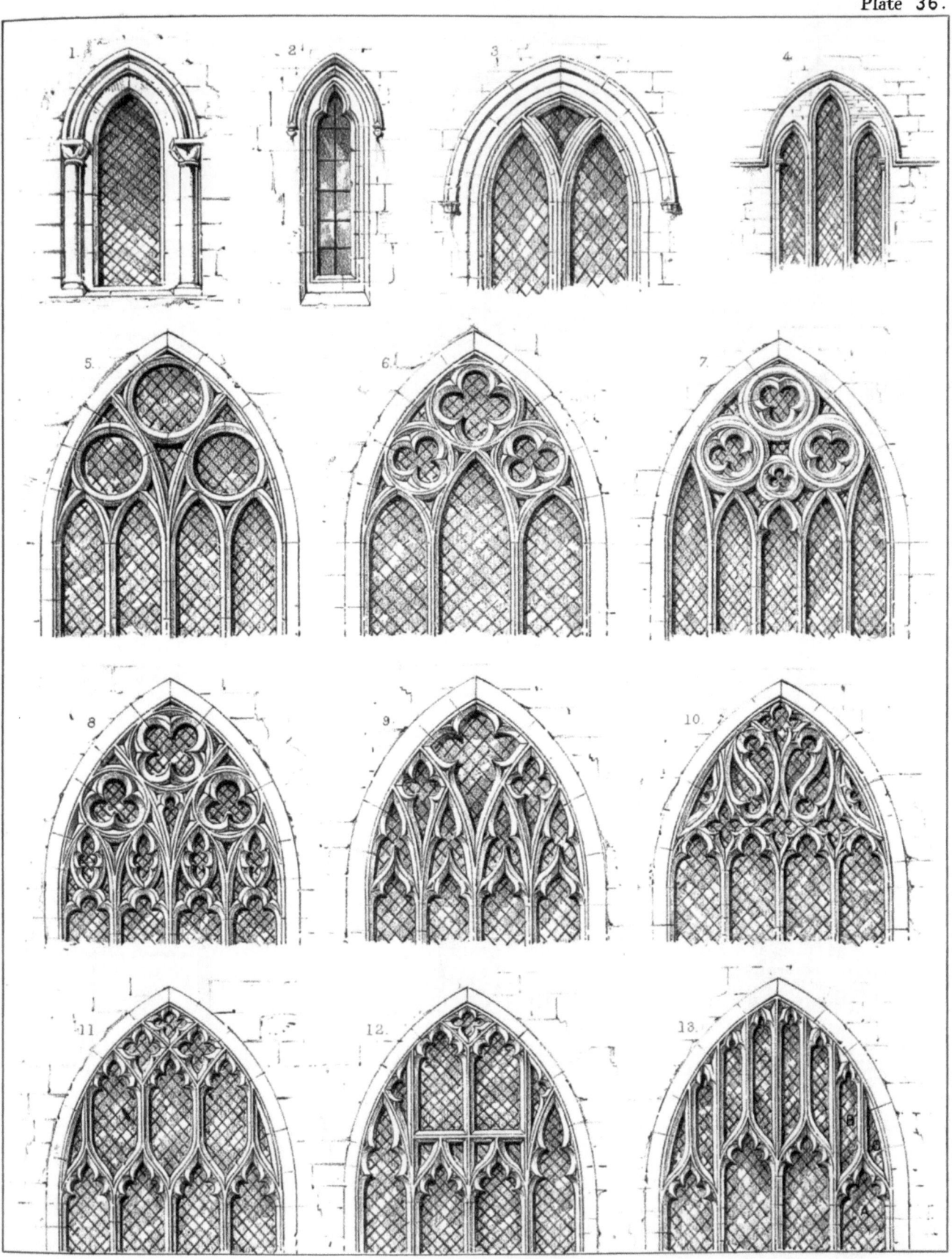

J S Edmeston Del.

Bell Bros Lith'd Castle St London, E.C.

WINDOW.

Plate 37.

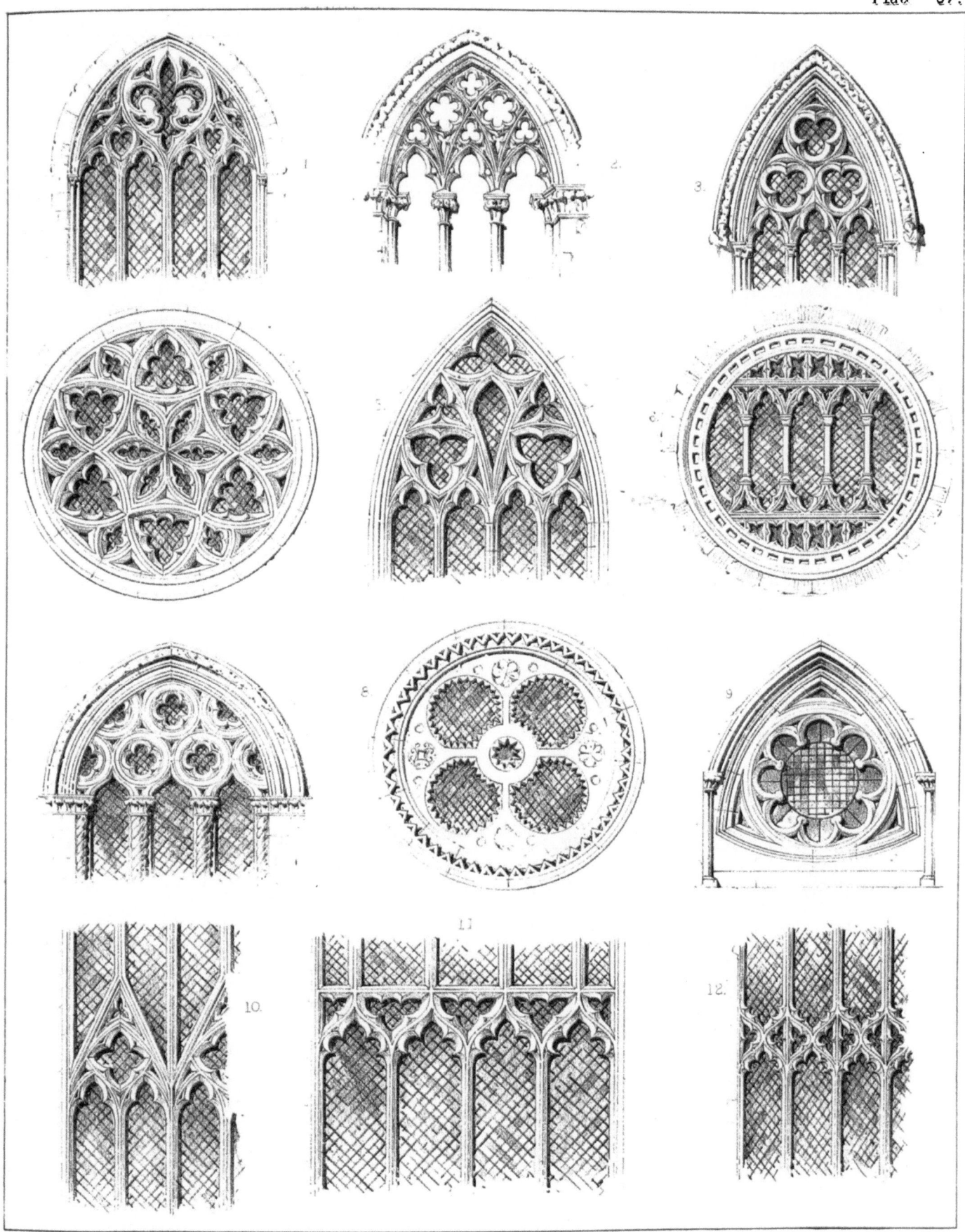

J S Edmeston Del

Kell Bros Lith.rs Castle St London E.C

I TO 9 MULLION 10 TO 12 TRANSOM

Plate 38.

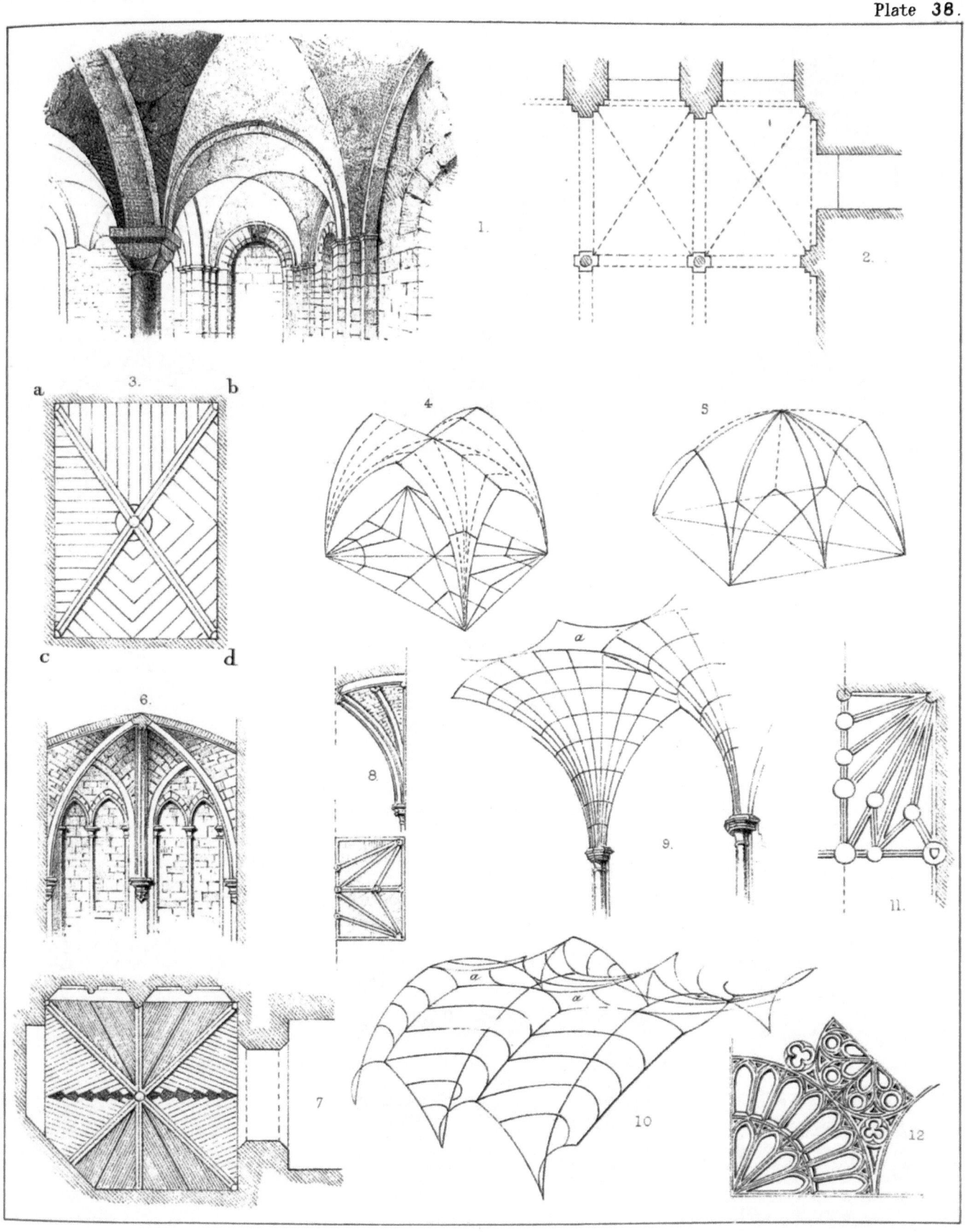

J S Widdineson Del.

Kell Bros Lith Castle St London E C

VAULTING CROIN'D.

Plate **39**.

I. 2. VAULTING GROIN'D, 3. PARAPET. 4. 5. FLYING BUTTRESS,
8 TRIFORIUM, 9. TO 13 NORTH ITALIAN.

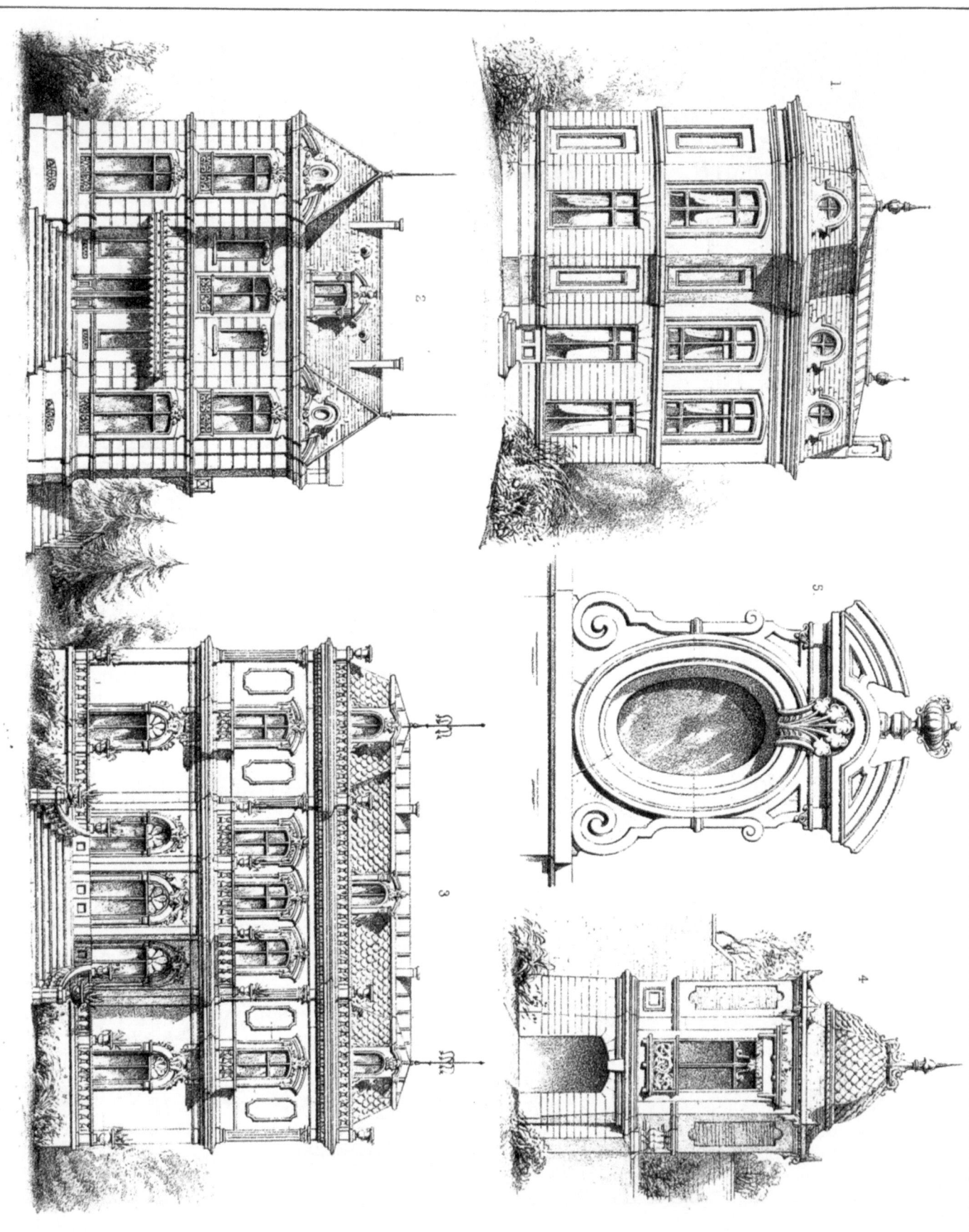

MODERN FRENCH ARCHITECTURE.

Plate 40.

Plate 41.

1

2

3

J.S. Edmeston Del

Bro⁵ Lith⁵ Castle S⁵ London, E.C.

MODERN FRENCH STREET ARCHITECTURE.

Plate 42.

J S Edmeston Del

Kell Bros Lith?? Castle St London E.C

MODERN FRENCH ARCHITECTURE

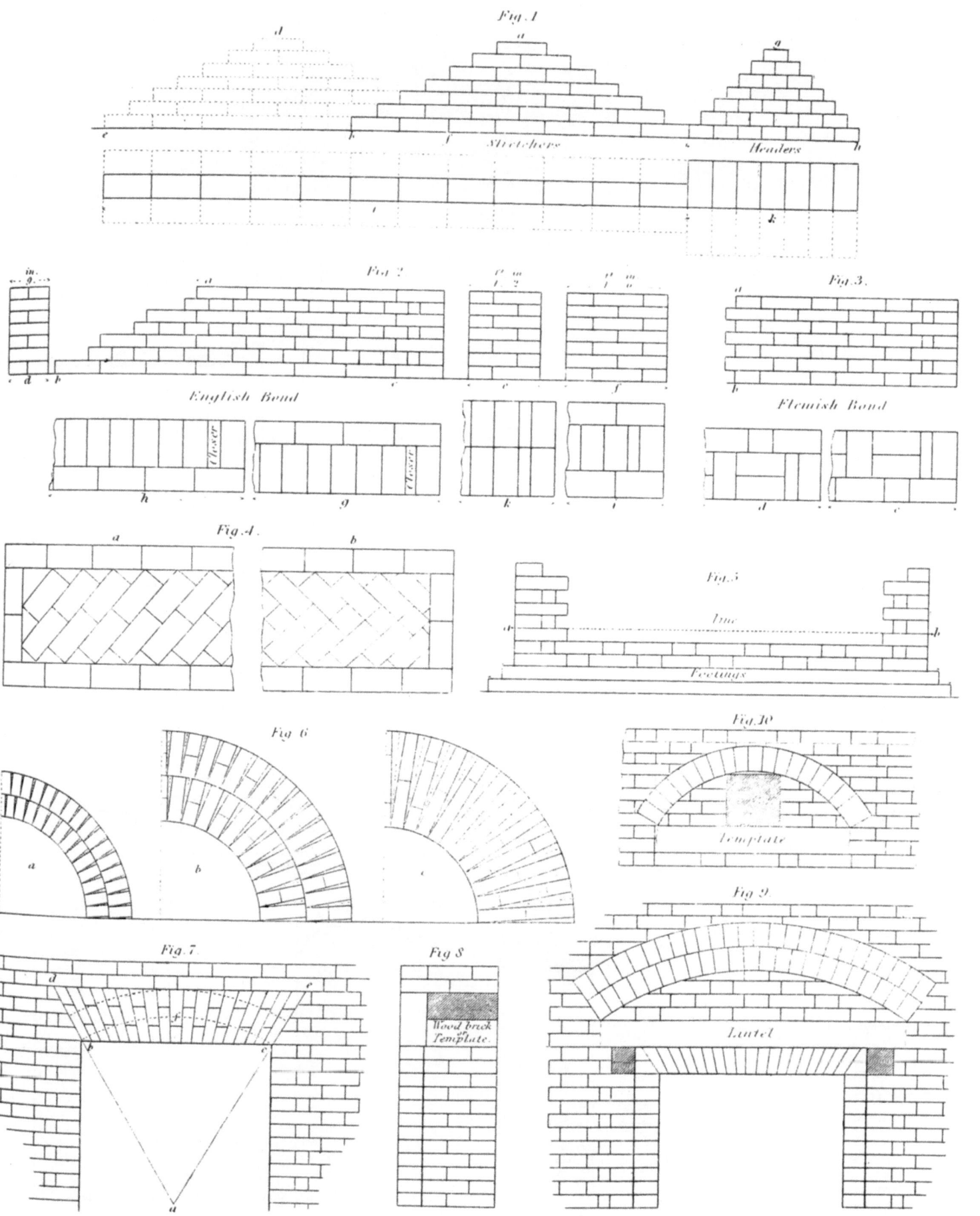

Fig. 1

Stretchers

Headers

Fig. 2

English Bond

Fig. 3

Flemish Bond

Closer

Closer

Fig. 4

Fig. 5

line

Footings

Fig. 6

Fig. 10

Template

Fig. 9

Fig. 7

Fig. 8

Wood brick Template.

Lintel

BUILDING.

Fig. 14.

Plate 44

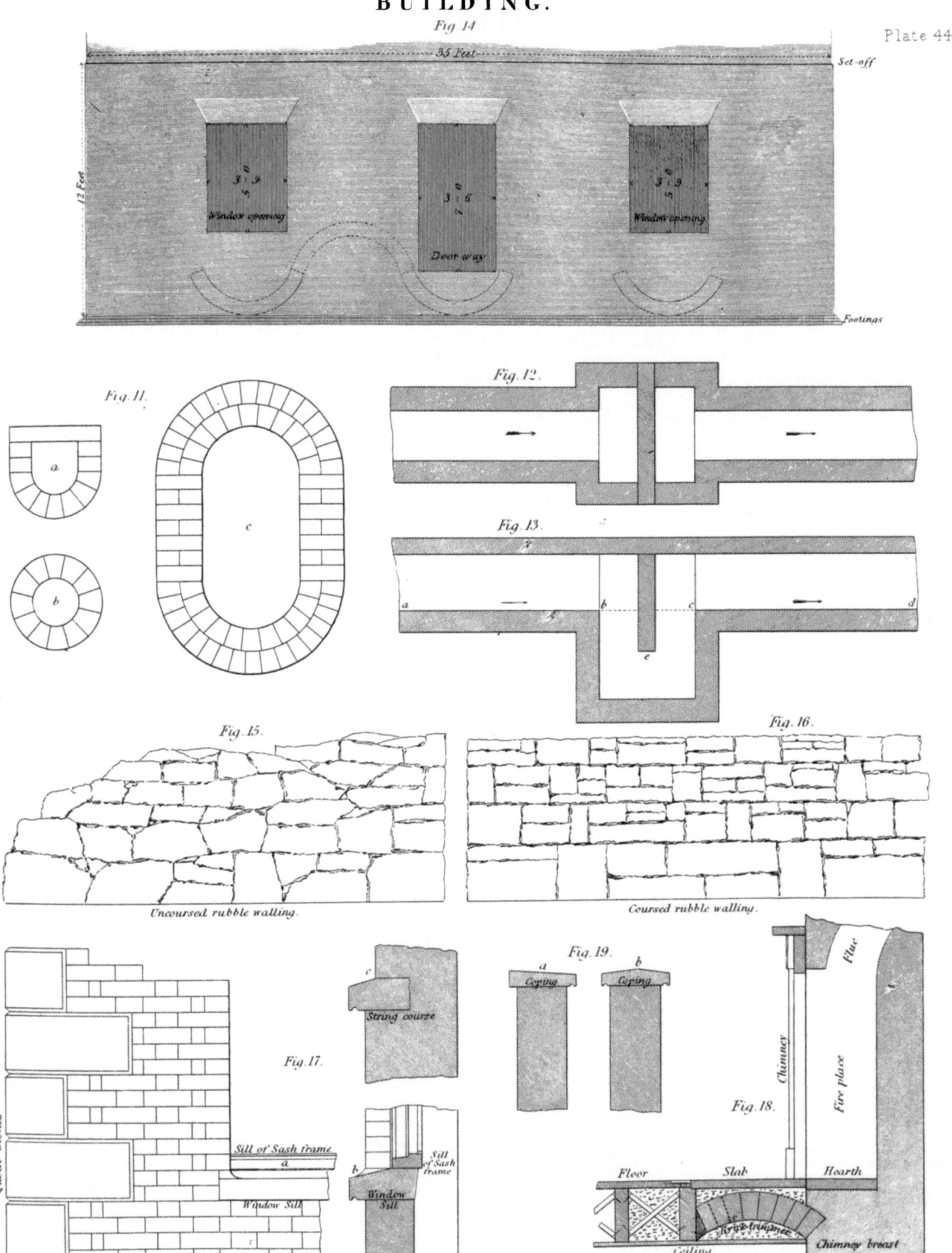

Published by A.&C Black, Edinburgh

Eng.d by G. Aikman, Edin.t

Plate 45

BUILDING.

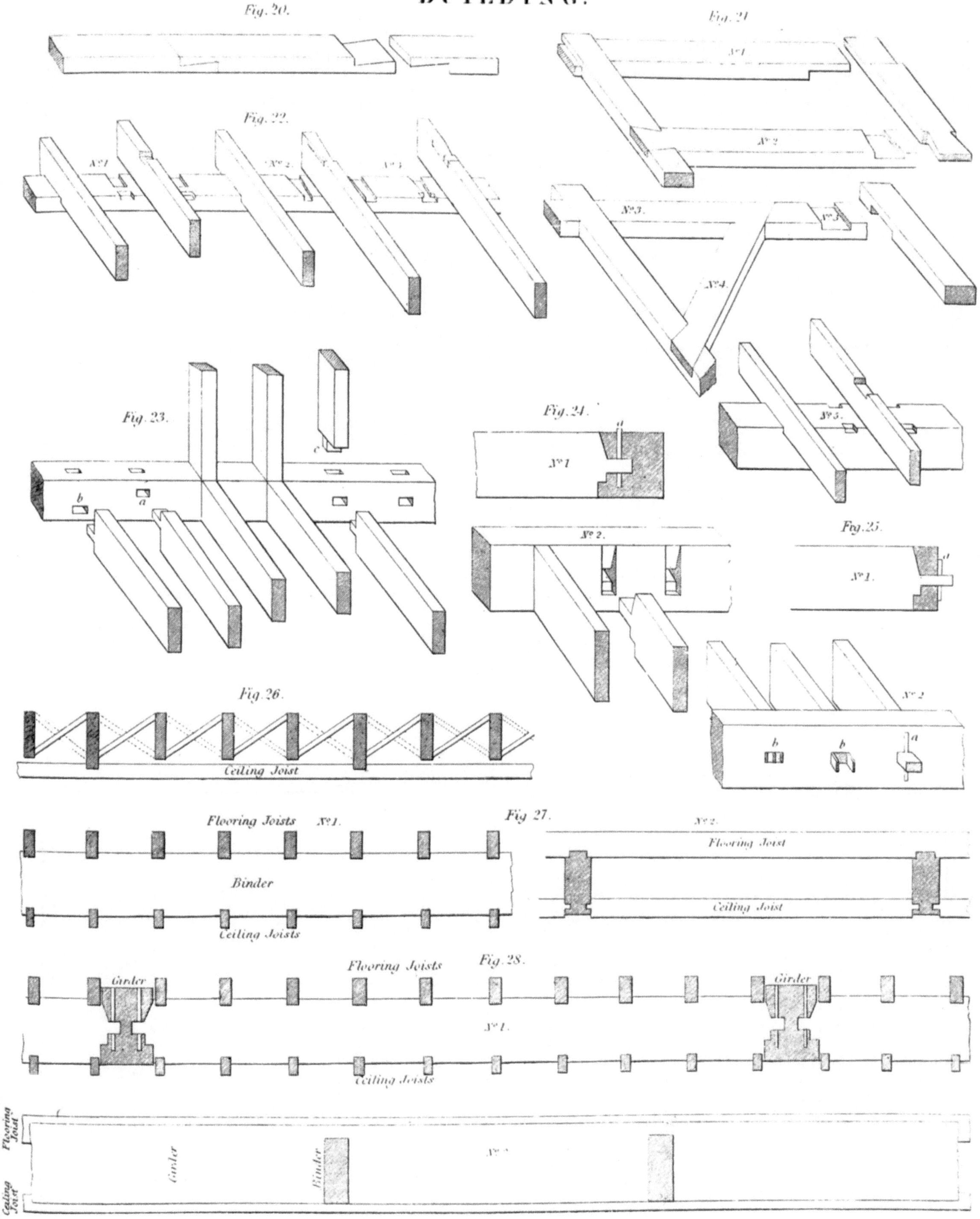

Fig. 20.

Fig. 21.

Fig. 22.

N.° 1. *N.° 2.* *N.° 3.*

N.° 1.

N.° 2.

N.° 3. *N.° 4.*

N.° 4.

Fig. 23.

c

b *a*

N.° 5.

Fig. 24.

a

N.° 1.

Fig. 25.

a

N.° 1.

N.° 2.

Fig. 26.

Ceiling Joist

b *b* *a*

N.° 2.

Flooring Joists *N.° 1.*

Fig 27.

N.° 2.

Flooring Joist

Binder

Ceiling Joist

Ceiling Joists

Girder

Flooring Joists

Fig. 28.

Girder

N.° 1.

Ceiling Joists

Flooring Joist

Ceiling Joist

Girder

Binder

N.°

W. H. del.t

Eng.d by G. Aikman & Co.

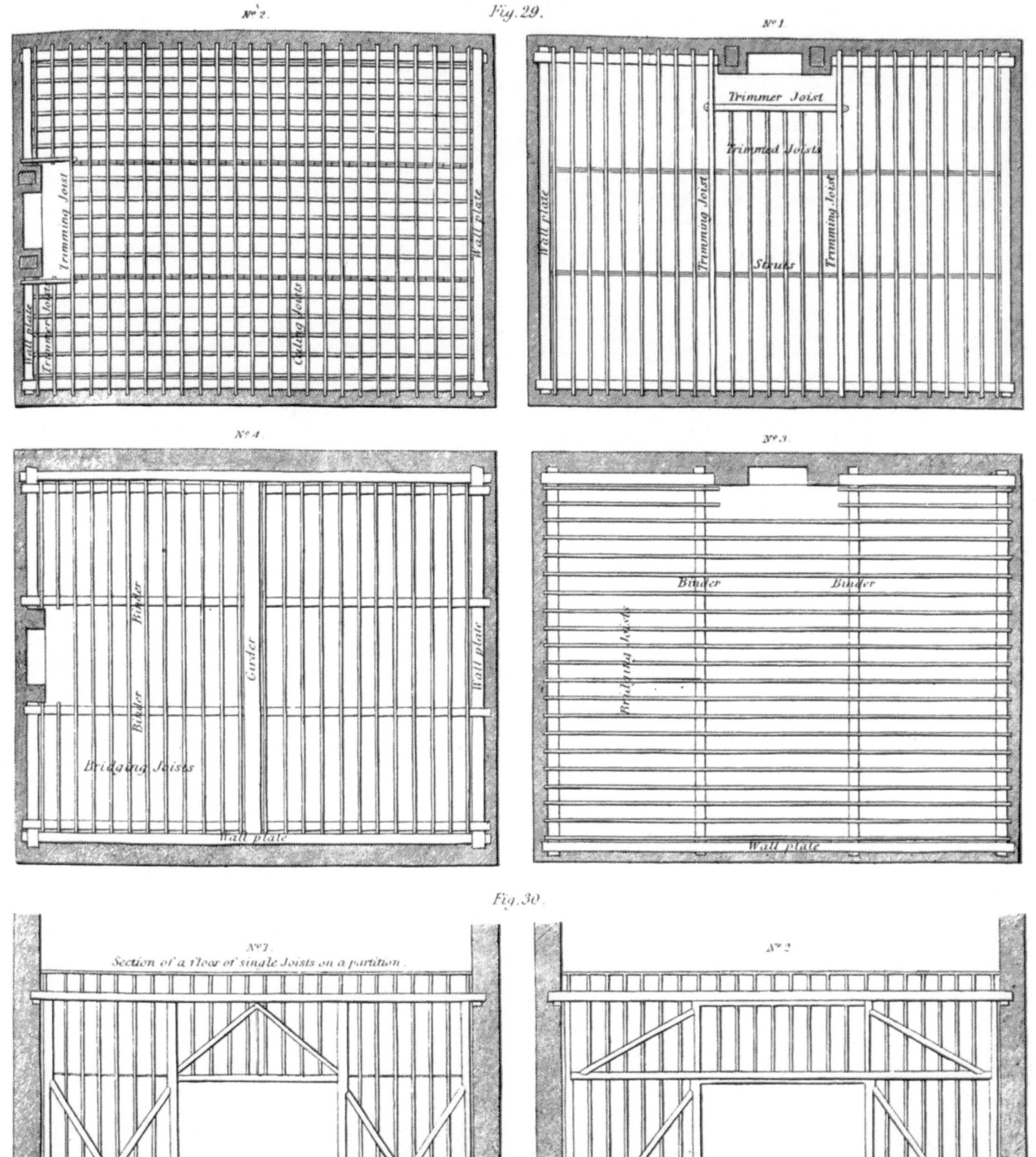

Fig.29.

Nº 2.

Nº 1.

Nº 4.

Nº 3.

Fig.30.

Nº 1.

Section of a floor of single Joists on a partition.

Elevation of a trussed partition.

Nº 2.

Elevation of a trussed partition.

W.H. del.

Engᵈ by G. Aikman, Edinᵗ

BUILDING.

Plate 47

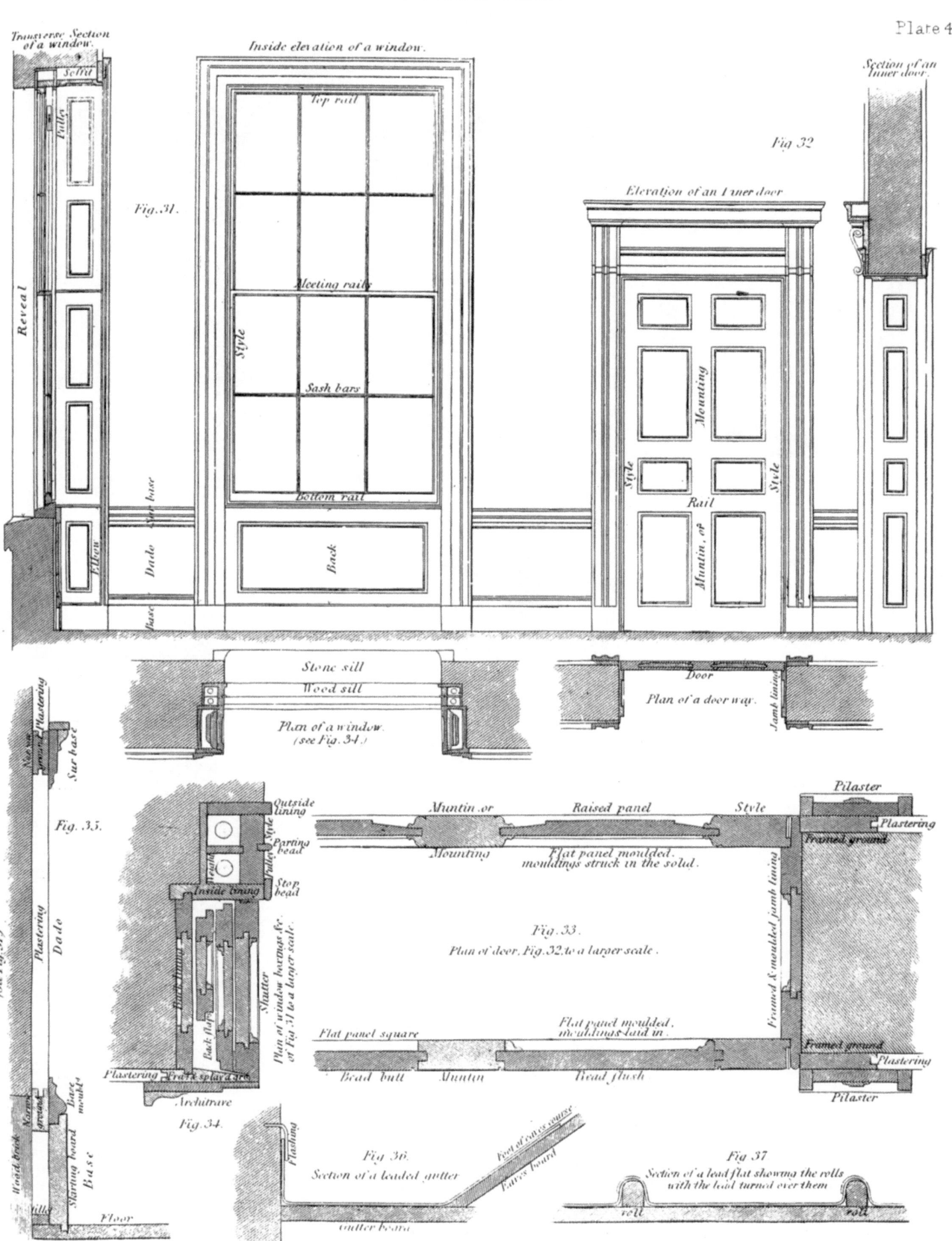

Transverse Section of a window.

Soffit

Pulley

Reveal

Elbow

Dado

Sur base

Base

Fig. 31.

Inside elevation of a window.

Top rail

Style

Meeting rails

Sash bars

Style

Bottom rail

Back

Stone sill

Wood sill

Plan of a window.
(see Fig. 34.)

Section of an Inner door.

Fig. 32

Elevation of an Inner door

Mounting

Style

Rail

Muntin, or

Style

Plastering

Door

Plan of a door way.

Jamb lining

Fig. 35.

Plastering

Sur base

Plastering

Dado

Wood brick

Plastering ground

Skirting board

Base mould

Base

Floor

Sill

(See Fig. 31.)

Norm ground

Outside lining

Parting bead

Stop bead

Inside lining

Back lining

Back flap

Shutter

Plan of window box bars &c.
of Fig 31 to a larger scale.

Architrave

Plastering Frate splayed &c

Fig. 34.

Muntin or

Mounting

Raised panel

Flat panel moulded.
mouldings struck in the solid.

Style

Pilaster

Plastering

Framed ground

Fig. 33.
Plan of door, Fig. 32. to a larger scale.

Flat panel moulded.
mouldings laid in

Flat panel square

Bead butt

Muntin

Bead flush

Framed & moulded jamb lining

Framed ground

Plastering

Pilaster

Flashing

Fig. 36.
Section of a leaded gutter

gutter board

Foot of cave course

Eaves board

Fig 37
Section of a lead flat showing the rolls
with the lead turned over them

roll

roll

W.H del.

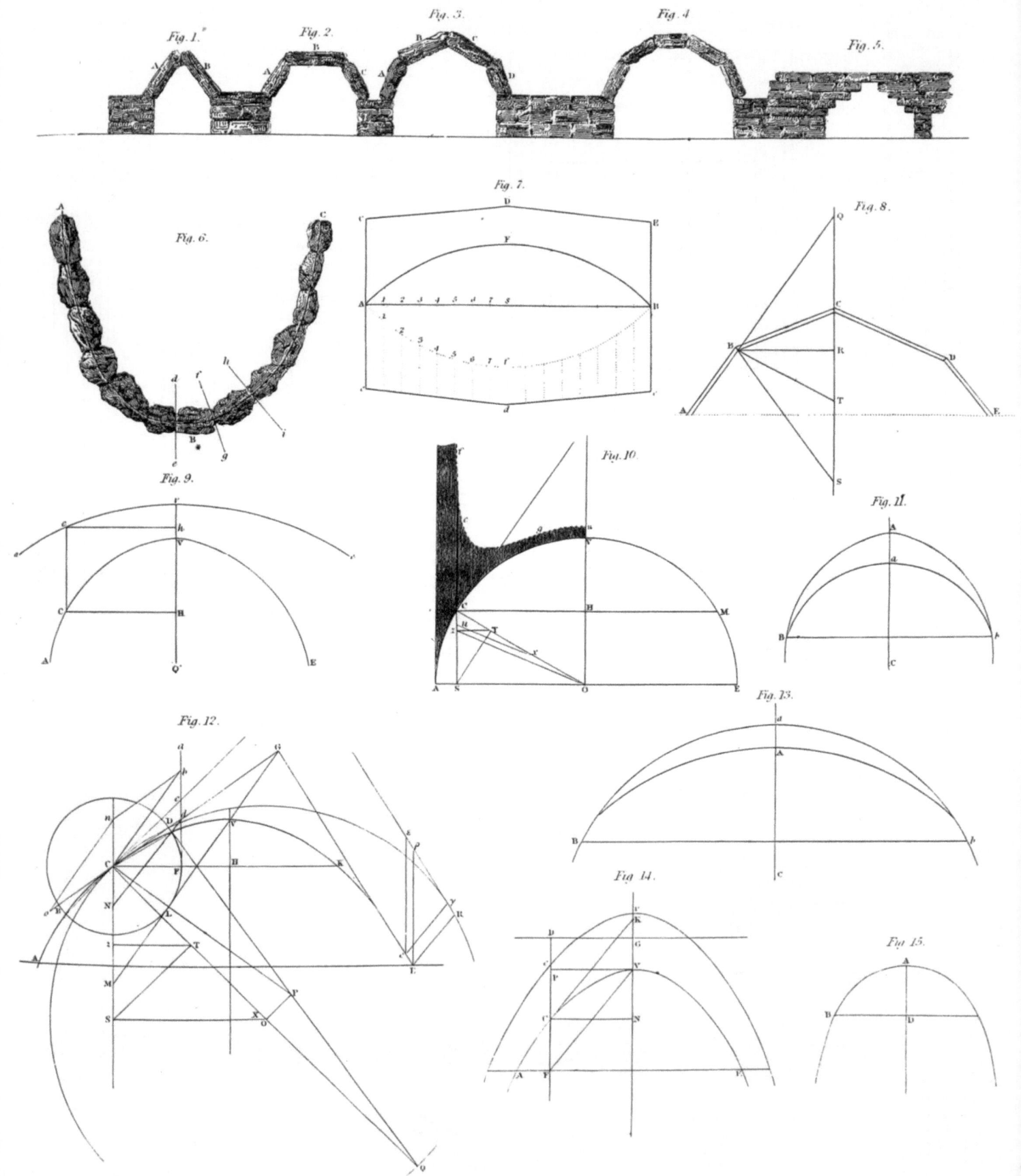

Fig. 1. Fig. 2. Fig. 3. Fig. 4. Fig. 5.

Fig. 6. Fig. 7. Fig. 8.

Fig. 9. Fig. 10. Fig. 11.

Fig. 12. Fig. 13.

Fig. 14. Fig. 15.

Published by A. & C. Black, Edinburgh.

ARCH.

Plate 49.

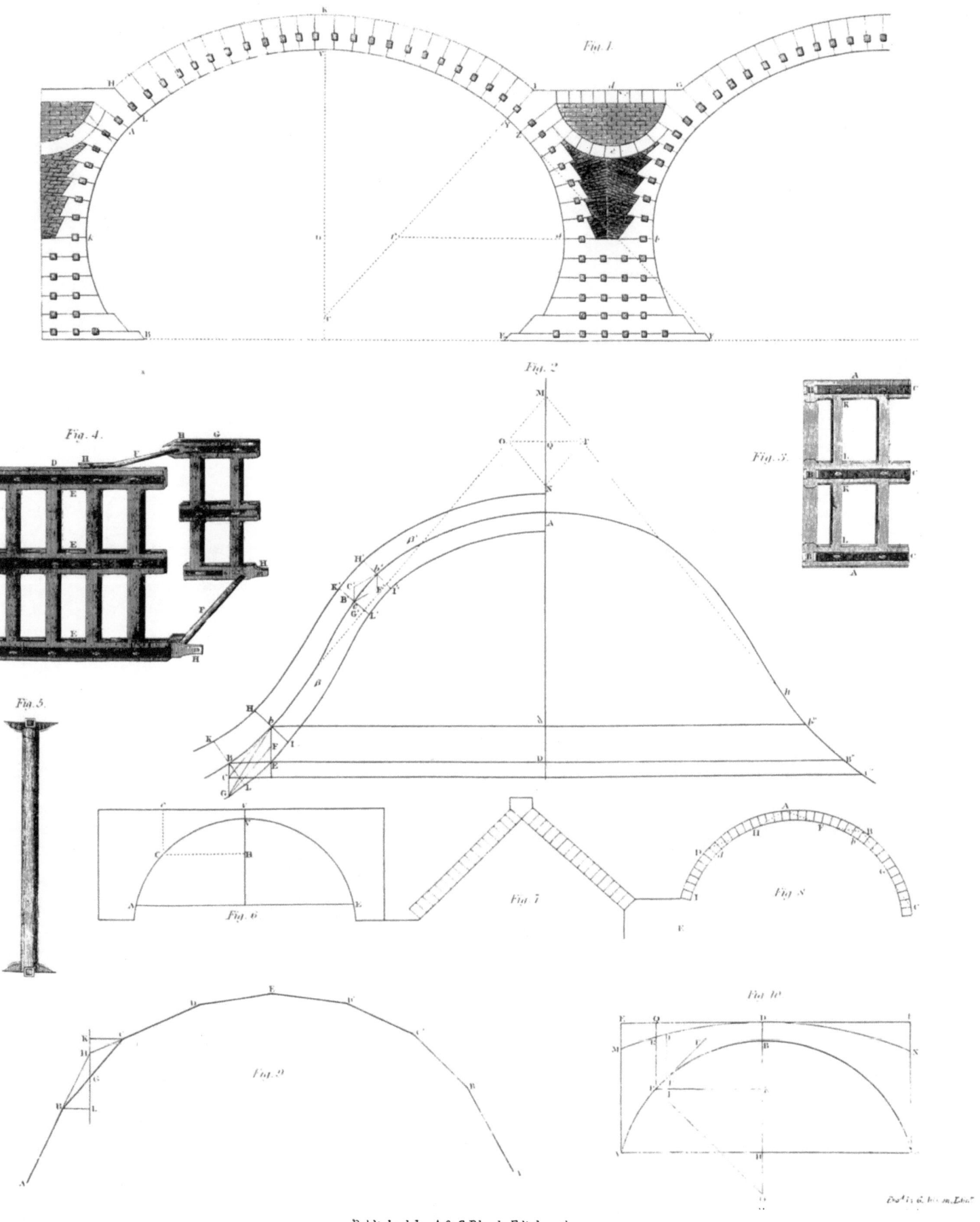

Fig. 1.

Fig. 2.

Fig. 3.

Fig. 4.

Fig. 5.

Fig. 6.

Fig. 7.

Fig. 8.

Fig. 9.

Fig. 10.

Published by A & C. Black, Edinburgh.

CARPENTRY.

Plate 50

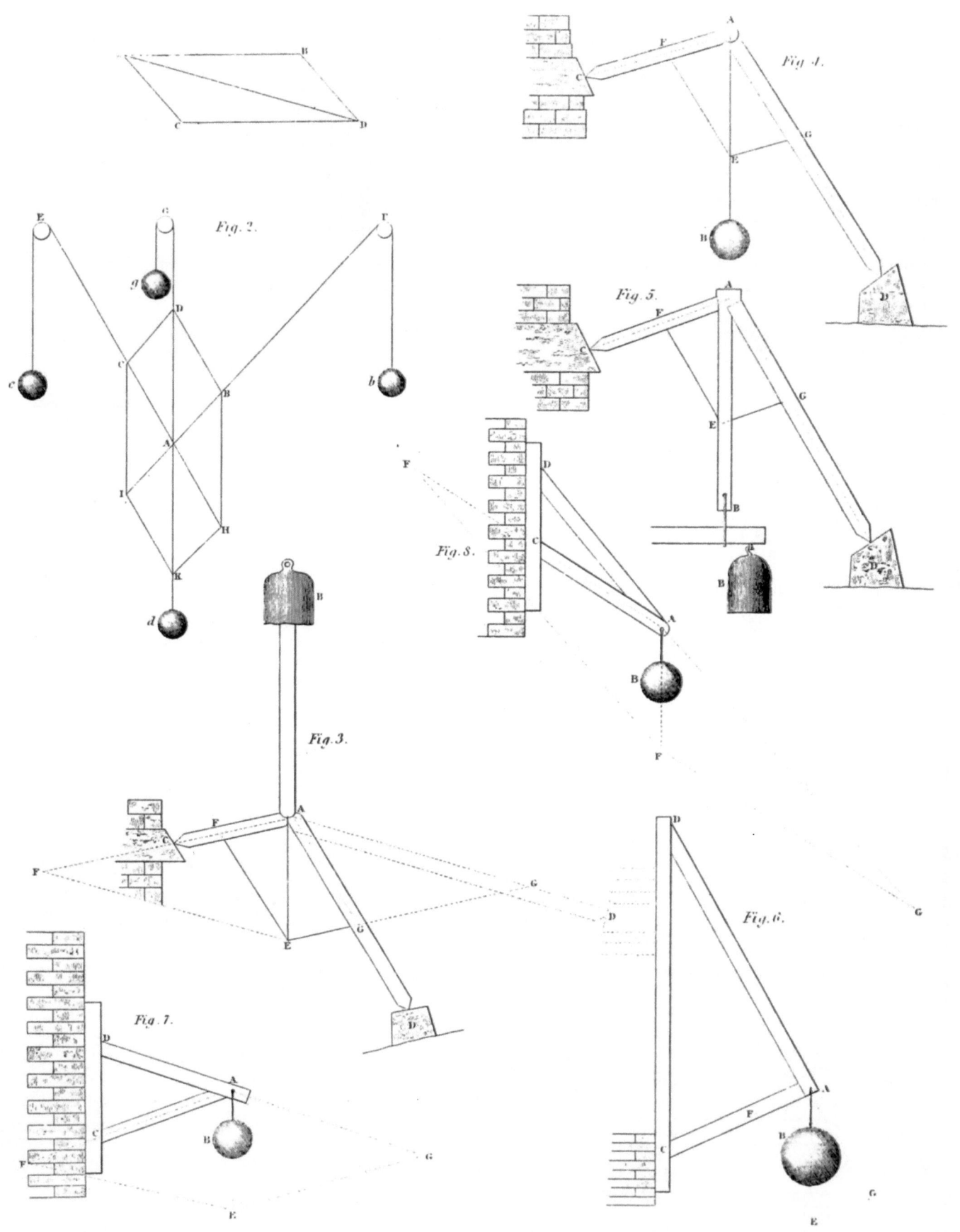

Fig. 1.

Fig. 2.

Fig. 3.

Fig. 4.

Fig. 5.

Fig. 6.

Fig. 7.

Fig. 8.

Published by A. & C. Black, Edinburgh.

Eng⁴ by G. Aikman.

CARPENTRY.

Plate 51

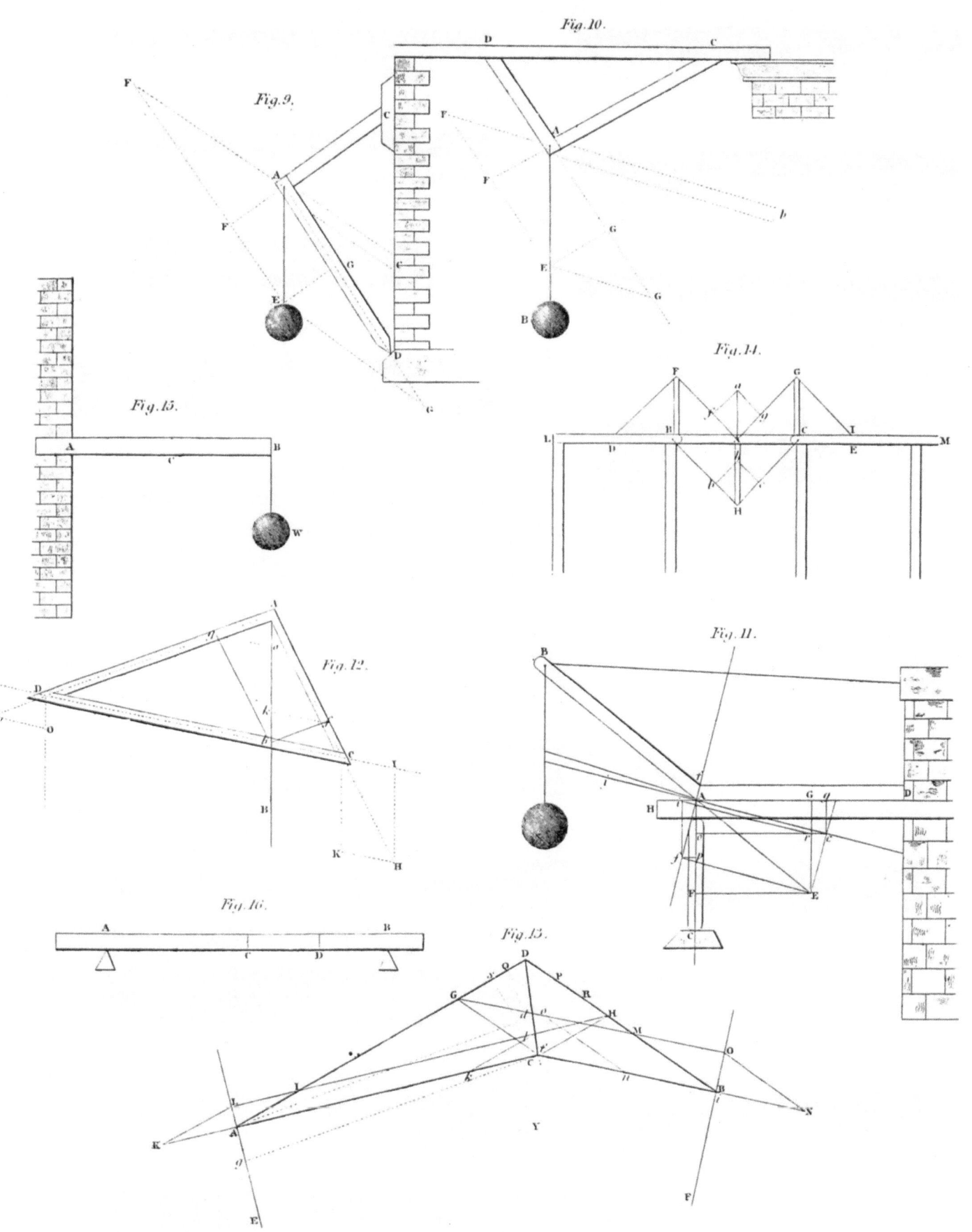

Published by A.& C.Black, Edinburgh.

Eng.ᵈ by G.Aikman.Edin.ʳ

CARPENTRY.

Plate 52

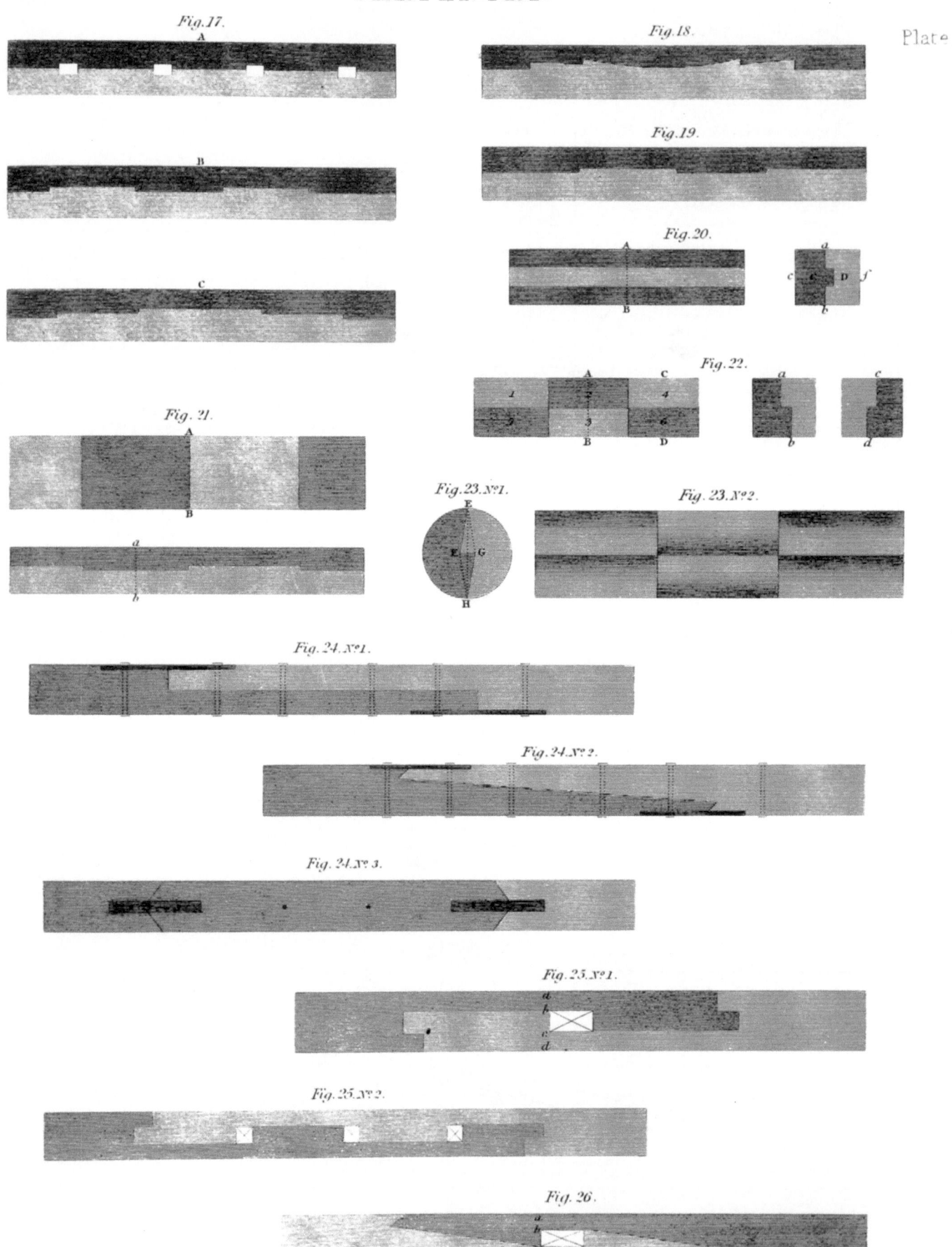

Fig. 17.

Fig. 18.

Fig. 19.

Fig. 20.

Fig. 21.

Fig. 22.

Fig. 23. Nº 1.

Fig. 23. Nº 2.

Fig. 24. Nº 1.

Fig. 24. Nº 2.

Fig. 24. Nº 3.

Fig. 25. Nº 1.

Fig. 25. Nº 2.

Fig. 26.

Published by A & C Black, Edinburgh

Engd by G. Aikman, Edinr.

CARPENTRY.

Plate 53

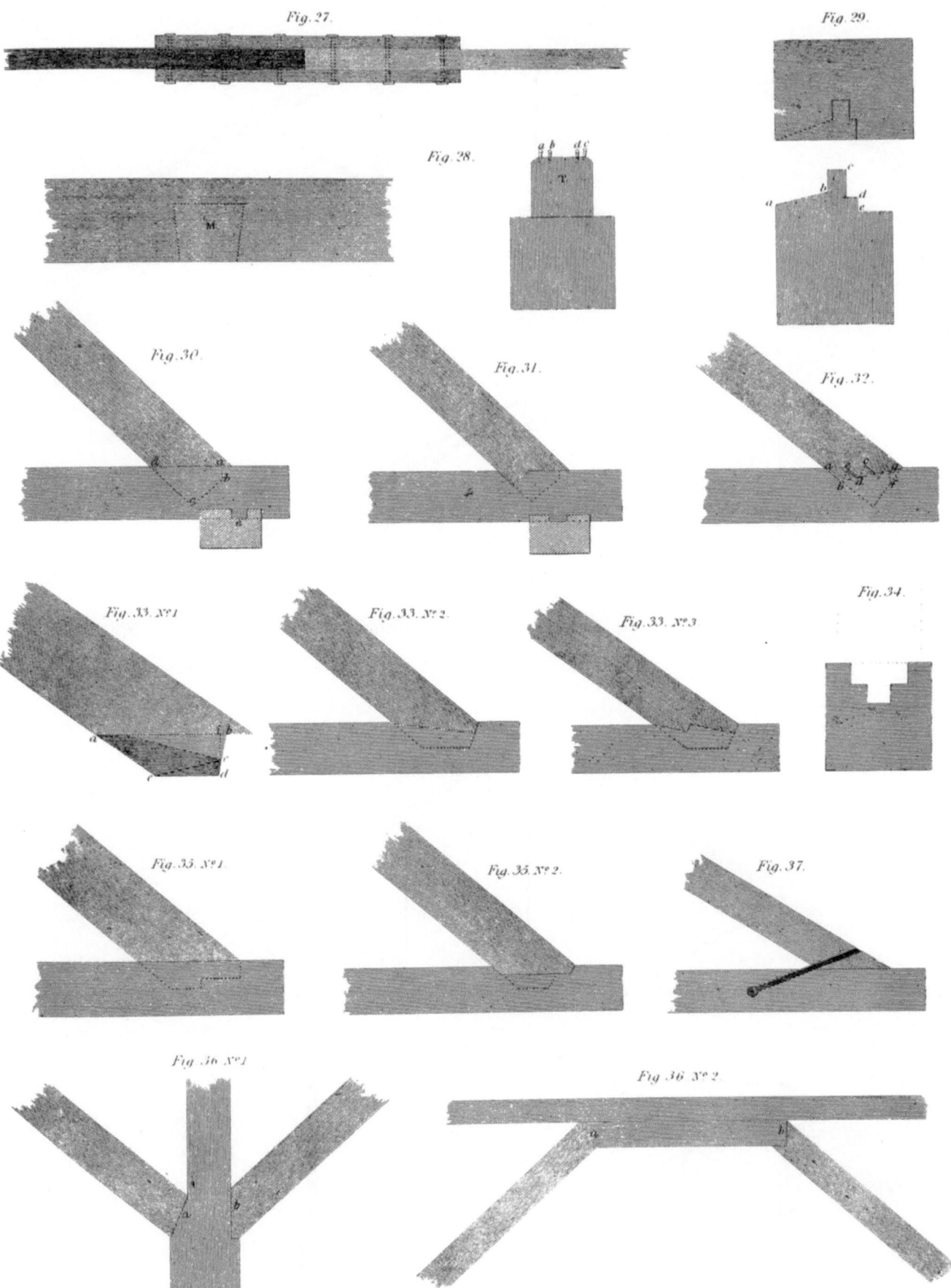

Fig. 27.

Fig. 29.

Fig. 28.

Fig. 30.

Fig. 31.

Fig. 32.

Fig. 33. Nº 1.

Fig. 33. Nº 2.

Fig. 33. Nº 3.

Fig. 34.

Fig. 35. Nº 1.

Fig. 35. Nº 2.

Fig. 37.

Fig. 36. Nº 1.

Fig. 36. Nº 2.

Published by A. & C. Black, Edinburgh.

Engᵈ by G. Aikman, Edinᵗ

CARPENTRY.

Plate 54

Fig. 38.

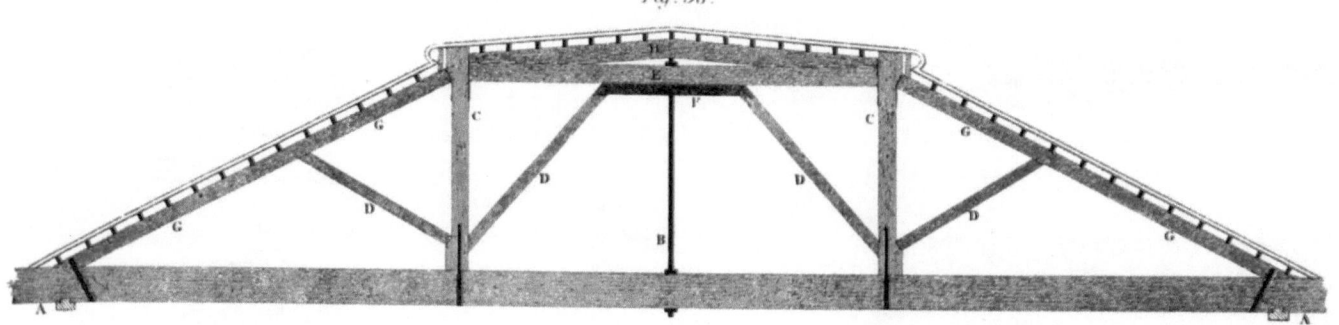

Fig. 39.

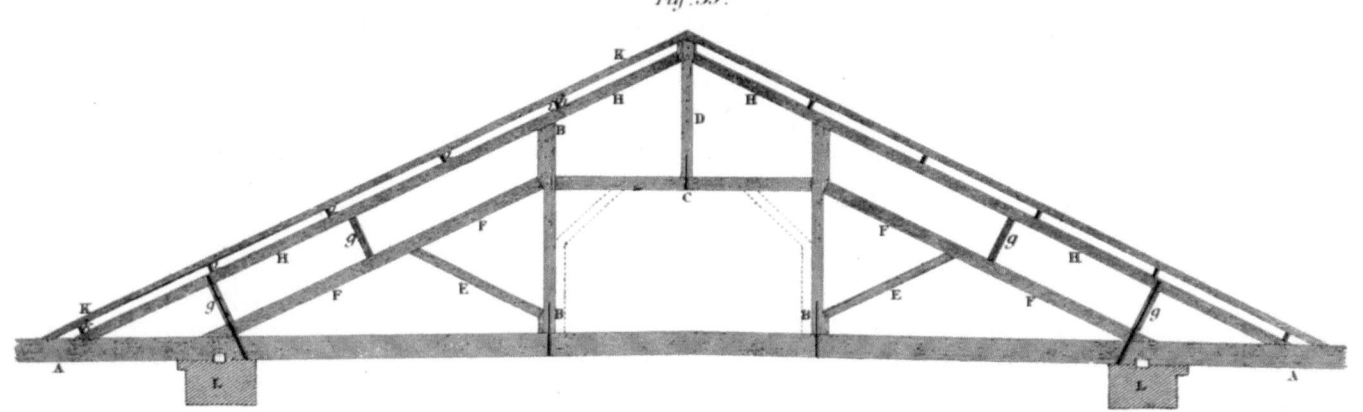

Fig. 40.

Fig. 41.

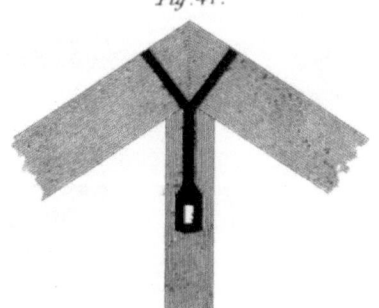

Fig. 42.

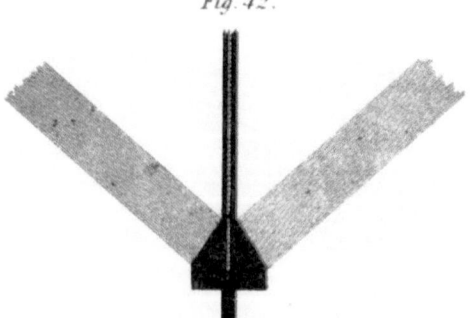

Published by A & C Black, Edinburgh

Eng.d by G. Aikman Edin.r

CARPENTRY.

Plate 55.

Fig. 43.

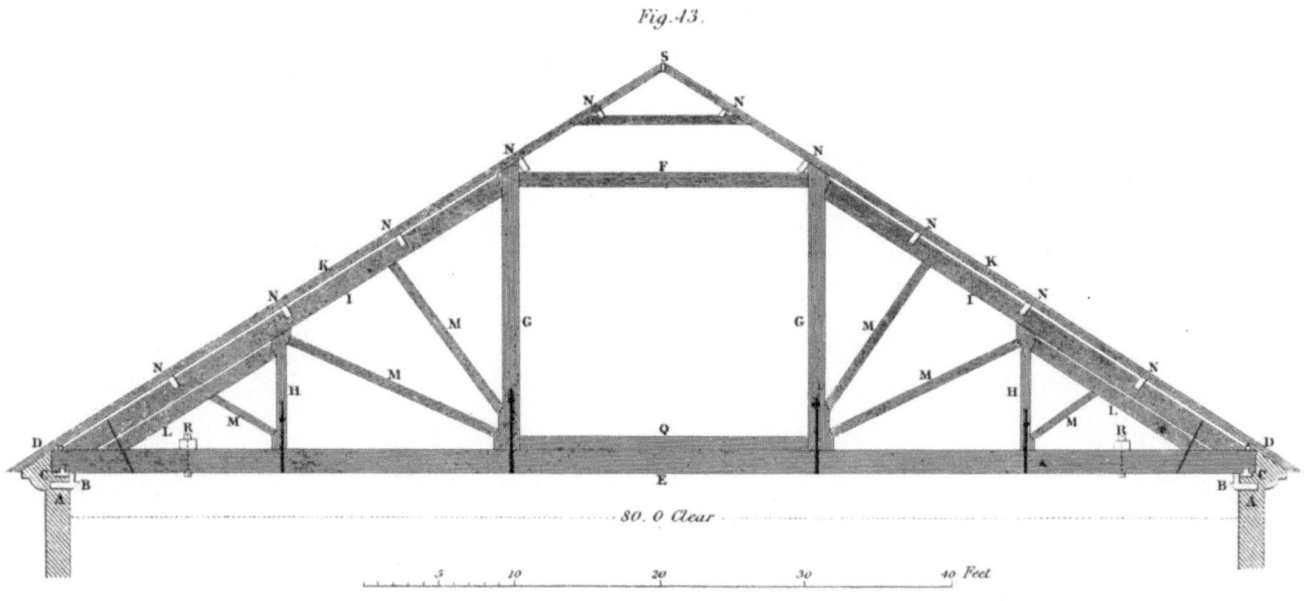

80. 0 Clear

5   10   20   30   40 Feet

Fig. 44.

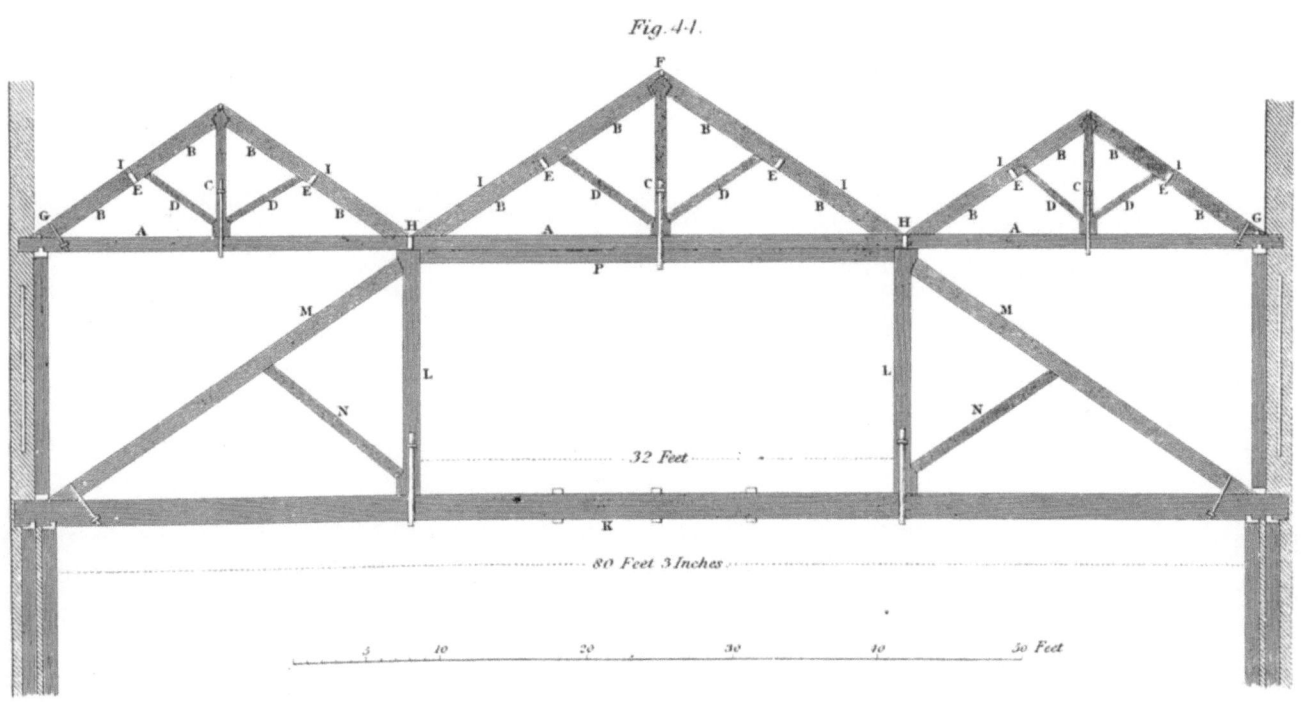

32 Feet

80 Feet 3 Inches

5   10   20   30   40   50 Feet

Fig. 45.

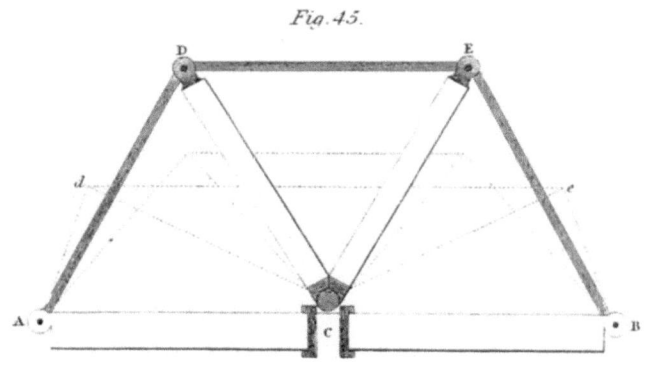

Published by A. & C. Black, Edinburgh.

Eng⁴ by G. Aikman, Edin⁰